Computational Approaches
for Aerospace Design

Computational Approaches for Aerospace Design
The Pursuit of Excellence

Andy J. Keane
Prasanth B. Nair

University of Southampton

John Wiley & Sons, Ltd

Copyright © 2005 John Wiley & Sons Ltd, The Atrium, Southern Gate, Chichester,
West Sussex PO19 8SQ, England

Telephone (+44) 1243 779777

Email (for orders and customer service enquiries): cs-books@wiley.co.uk
Visit our Home Page on www.wiley.com

This publication is designed to provide accurate and authoritative information in regard to the subject matter
covered. It is sold on the understanding that the Publisher is not engaged in rendering professional services. If
professional advice or other expert assistance is required, the services of a competent professional should be
sought.

Other Wiley Editorial Offices

John Wiley & Sons Inc., 111 River Street, Hoboken, NJ 07030, USA

Jossey-Bass, 989 Market Street, San Francisco, CA 94103-1741, USA

Wiley-VCH Verlag GmbH, Boschstr. 12, D-69469 Weinheim, Germany

John Wiley & Sons Australia Ltd, 42 McDougall Street, Milton, Queensland 4064, Australia

John Wiley & Sons (Asia) Pte Ltd, 2 Clementi Loop #02-01, Jin Xing Distripark, Singapore 129809

John Wiley & Sons Canada Ltd, 22 Worcester Road, Etobicoke, Ontario, Canada M9W 1L1

Wiley also publishes its books in a variety of electronic formats. Some content that appears
in print may not be available in electronic books.

British Library Cataloguing in Publication Data

A catalogue record for this book is available from the British Library

ISBN-13 978-0-470-85540-9 (HB)
ISBN-10 0-470-85540-1 (HB)

Typeset in 10/12pt Times by Laserwords Private Limited, Chennai, India
Printed and bound in Great Britain by Antony Rowe Ltd, Chippenham, Wiltshire
This book is printed on acid-free paper responsibly manufactured from sustainable forestry
in which at least two trees are planted for each one used for paper production.

This book is dedicated to our wives, whose patience and encouragement has been vital through the three years we have been writing. Without their support we would not have finished it.

Contents

IV Case Studies   387

Foreword

Following a brief review of the history of the aerospace sector over its first hundred years, Keane and Nair illustrate how the design process has evolved and demonstrate the dominant role that computation now plays in aerospace design, having advanced exponentially over the last 50 years. They focus on design synthesis as the important ingredient for producing new and better designs and contrast this with analysis, pointing out that, although much effort has gone into making computational analysis ever more accurate, it is in synthesis that design actually takes place. Having worked in the aerospace sector all my life, it is good to see the balance between the analysis of existing designs and the creation of new and better designs discussed in this way – all engineering is ultimately concerned with producing new designs and bringing them to the market. A deep understanding of these interactions is the fundamental hallmark of all good engineers.

Having provided an outline of modern computationally based design, the authors then cover in greater depth a number of more advanced topics appropriate to those working at the leading edge of their fields. These techniques lie outside the classical syllabus undertaken by most engineering graduates, focusing instead on the search for improved design performance by computational means. This includes sensitivity analysis, approximation concepts and their management, uncertainty and the computational architectures needed to harness these various tools. The presentation is, however, not overly mathematical and should appeal to those working with parametric CAD tools, optimization systems and modern analysis codes.

Lastly, and perhaps most importantly, the authors illustrate all these concepts by a series of extended case studies drawn from satellite, aircraft and aero-engine design where they have worked in partnership with EADS Astrium, BAE Systems and Rolls-Royce, all leading international organizations working in these fields. While the studies inevitably have an academic bent, the authors have often carried out the work in partnership with the companies concerned, using the firms' design codes and applying their tools and techniques to industrially sourced problems. The book is therefore a welcome addition to the published work on design and helps to bridge the gap between academic research and industrial practice in a very helpful way.

Phil Ruffles, Derby, 2005

Preface

This book is concerned with engineering design. The distinguishing feature of engineering, as opposed to other forms of design is that decisions are based on analysis. So, a precursor to carrying out such design is the ability to analyze potential configurations. Over the last 50 years, this ability to carry out analysis in aerospace design has increased dramatically, mostly owing to the advent of modern computing systems, but also because of rapid advances in numerical methods for the solution of the partial differential equations that govern aerospace systems. These developments have led to new ideas on how engineering design should be carried out. Such ideas have focused on the design process as much as on the product being designed itself. To see this, one need only compare the approaches of the pioneers of flight with those commonly found in modern aerospace companies.

The great aerospace engineers of the past had relatively little ability to make use of analysis, although tables of the forces acting on simple wings in airflows existed from the late 1800s. Unfortunately, as the Wright brothers noted, such tables were often misleading and they found it easier to begin from scratch. It was really only from the Second World War that the use of design calculations began to dominate aerospace design. Thus, when Mitchell decided on the elliptical shape of the Spitfire wings immediately prior to the Second World War, this shape was dictated by analytical knowledge of the performance of airflows. Even so, he was forced to use the slide rule to exploit this knowledge. Today, aerospace designers have massive computing power at their disposal and must consider how to make best use of this power. The science and practice of optimization is becoming central to this process. This book is therefore much concerned with understanding the place of optimization and related tools in design.

Engineering design, of course, requires much more than just analysis and optimization. Even so, if one reflects on just these two ideas, it is clear that they arise because the designer must understand each given design intimately, as well as be able to meet tough performance targets. Moreover, there are likely to be many constraints that must also be satisfied. Engineering design is inevitably concerned with specifying goals and meeting constraints. Often, the goals and constraints are difficult to define numerically and sometimes are even interchangeable. For example, it may be desirable to reduce the weight in an aero-engine component. Such minimum weight, however, cannot be achieved at the expense of all other characteristics: stress levels will need to be maintained within working limits. Alternatively, a low-stress design may be the goal within a given weight budget. Even a basic problem can become quite complex when framed in an engineering context. The case studies that are examined in Part IV of this book illustrate that quite simple problems will often have many constraints and several, possibly conflicting, goals.

To add to these complexities, it is also clear that most engineering design is carried out by teams of designers working in a whole range of different disciplines. It is quite normal to consider structural strength, aerodynamic performance, manufacturing and cost issues all at the same time. Each of these disciplines is likely to have its own experts and analytical tools that must be brought to bear on the design to achieve a balanced product. In this book, we set out to describe the tools that modern designers need to have at their disposal and how these tools interact with each other. We give sufficient detail and references for those new to the field to be able to grasp the core ideas but we do not claim to offer an exhaustive survey of the topics covered. Moreover, we deliberately focus on those aspects that can be broadly considered as mechanics – design of software systems lies outside the scope of this book, although such design is increasingly important in building integrated systems.

The problems we describe and most of the examples we cite are taken from the field of aerospace engineering because we are most familiar with that world – we believe, however, that the ideas presented are relevant to all fields of engineering.

Further material, corrections and internet links related to the work described in this book may be found at http://www.aerospacedesign.org .

Andy J. Keane and Prasanth B. Nair
January 2005, Southampton

Acknowledgments

We would like to acknowledge the efforts of all those students, researchers and colleagues who have helped with the overall production of this book. We list in the introduction to Part IV those who have worked on the various case studies described there. In addition, we would like to thank the following individuals for their efforts in proofreading the manuscript: Vijaya Bhaskar, Phani Chinchapatnam, Alex Forrester, Apurva Kumar, Joaquim Martins, Sachin Sachdeva, Daisuke Sasaki and Andreas Sóbester. We would also particularly like to thank our secretary Ros Mizen for all her efforts in supporting us and making our work happen smoothly.

Part I

Preliminaries

1

Introduction

We begin this treatise on computational methods in aerospace design by first examining the process of design itself. After all, from the designer's point of view, computational methods are only of value if they aid in the design process. Computation for its own sake may be of interest to researchers and analysts – a designer will always wish to remember that it is his or her business to specify a product that is to be made. Unless a firm eye is kept on this goal, it is far too easy to undertake interesting work that ultimately does not bring one any closer to a successful piece of engineering – and it is by their products that engineers should wish to be remembered. This introductory chapter aims to set out the place of computation in design, both as it is now practiced and also as it has developed over the last 100 years.

1.1 Objectives

The fundamental objective of this book is to set out the main computational methods available to aerospace designers. The intention is to capture the state of the art in this field some 100 years after the first powered flight of the Wright brothers in 1903. Here, the primary focus is on tools that aid the engineer in making design decisions as opposed to those that aid engineers analyze new designs (although such analysis, of course, plays a vital role in supporting design choice). It is therefore not intended to be a guide to computational methods in fluid dynamics or structural mechanics – many excellent texts already exist covering these topics. It is also not a text on optimization, although it does discuss optimization and related methods in some detail. Rather, it aims to set optimization and associated decision support aids in context when viewed from the needs of design and the capabilities of modern distributed computing systems. Lastly, it should be noted that the focus throughout the book is on the design of flight hardware rather than on any embedded computers or software systems, i.e., on those aspects that can be modeled using partial differential field equations, such as structures and fluid dynamics. This is not to downplay the importance of control systems but simply to indicate the perspective taken in this work. Many of the techniques described can be applied to detailed models of control systems and embedded software.

Computational Approaches for Aerospace Design: The Pursuit of Excellence.   A. J. Keane and P. B. Nair
© 2005 John Wiley & Sons, Ltd

The book is laid out in four major parts: the first aims to provide a broad introduction to the field without recourse to much mathematics so that it is easily readable. It does, however, contain a few simple examples and also references to the primary literature, which may be used as the starting point for further study. Parts II and III then delve more deeply into some of the key technologies that aid design decision making and the computational frameworks that support such processes, respectively. Although this necessitates a more mathematical approach, the material presented should not be beyond any graduate engineer. Finally, part IV describes a number of case studies familiar to the authors that illustrate the techniques currently available, again in a highly readable form. Some of these are rather idealized studies and some are based on work carried out in collaboration with aerospace companies.

The intention is that the book will serve a number of roles: hopefully, practicing design engineers will find it useful as a compact summary of their trade and a reminder of the full tool set that is available to them. Those with just a passing interest in design should find parts I and IV of interest even if they skip the central core of the book at first reading. Finally, and perhaps most importantly, it is intended that the central portion of the book should set out the material that specialists in this field should be familiar with, which although not exhaustive, is sufficiently detailed to form a starting point for more advanced study via the references provided. Inevitably, its usefulness will decline with the passage of years but, hopefully, it will serve to indicate where this field has reached after 100 years of powered, heavier-than-air flight.

1.2   Road Map – What is Covered and What is Not

Part I introduces aerospace design, multidisciplinary analysis and optimization in a simple and straightforward fashion, especially when describing design. It is relatively self contained and should thus be read by all who use this book. This first chapter on design can easily be read at a single pass and will allow the reader to understand the viewpoint taken in writing the book. It is not a full history of aerospace design, a thorough guide to the theory of design or a complete description of the practices of a typical aerospace company, although it touches on all these aspects. The chapter on design-oriented analysis rehearses many of the issues faced by designers and should be familiar to most people working in the aerospace sector. It does, however, illustrate to those new to the area just how systematic an approach is needed in producing modern aerospace products. The chapter dealing with optimization, while being necessarily less detailed than a dedicated text on that subject, is intended to give a relatively complete description of the current state of search tools, illustrating their use on simple problems. It should suffice to give the reader an understanding of the various methods commonly in use, their underpinnings, qualities and limitations.

Part II deals with sensitivity studies and approximation concepts. This is the most mathematical part of the text and describes a series of methods that allow designers to take straightforward search and analysis tools and exploit them in a relatively sophisticated way. It thus represents the intellectual heart of the book and the area in which much of the authors' research has been carried out. On first reading, this material might be considered somewhat advanced but every attempt has been made to avoid overly complex theoretical approaches.

Part III examines frameworks that make use of the tools described in earlier sections of the book. This focuses on the systems needed to deliver flexible tools to design engineers. It deals with systems for accelerating optimization algorithms for expensive problems, how to manage uncertainty in the design process and how to conduct optimization when multiple disciplines are involved. It thus tends to take a slightly more computer science–based view of design than is common in many texts in this area. Increasingly, it is not sensible, or even desirable, to decouple the tasks faced by designers and the computing systems used to support their work – often, the best tools will be those that map well onto the available infrastructure.

Part IV contains a number of case studies that can be read in isolation, although to gain the best insight into why they have been tackled in the particular ways used, the reader should consult the earlier parts of the text. They seek to span both academic test cases that may be reproduced in a university setting and also problems taken from the authors' contacts with the U.K. aerospace sector. The latter ones are therefore hopefully relatively realistic in what they cover and the approaches taken. Nonetheless, even the most complex case described here remains much simpler than those routinely dealt with by a large modern design office. To set out such full-scale processes lies outside the scope of this book. Those wishing to learn such things will need to seek employment in one of the leading aerospace companies.

1.3 An Historical Perspective on Aerospace Design

At the time of writing, modern aerospace design is almost exactly 100 years old. It begins with the desire of mankind to achieve powered, heavier-than-air flight. Early aerospace designs were characterized not only by a great deal of trial and error but also by a not inconsiderable amount of analysis. The Wright brothers fully understood that control of their vehicle would be a fundamental requirement before it would be able to fly successfully. They also knew that designing an efficient propeller would be critical. In short, they knew they had to understand their designs and how they would be likely to perform before they built them. So, one may say that analysis and understanding have had a key place in aerospace design from the outset.

It is also clear that the early pioneers had clear goals they wished to meet. At first, these were simply measures of distances covered. As the technologies used in aerospace applications have developed, the goals have become vastly more sophisticated. The advent of the First World War and the potential impact of flight on the fighting of war led to rapid improvements in the capabilities of aircraft. What these early designers lacked, however, were modern computational facilities to predict the behavior of their designs. Thus, they were forced to experiment and use simple back-of-the-envelope calculations. Even so, some of these experiments were quite sophisticated. For example, the Wright brothers developed an early wind tunnel in which to test their wings.

At the same time, the fundamental understanding of fluid flow and stress analysis were advancing in the world's universities. By the 1930s, designers realized that if they could only exploit knowledge of fluid flow and stress analysis systematically they would be able to significantly improve the quality of their designs. The wings of R.J. Mitchell's Spitfire fighter aircraft were elliptical in shape because it was known that such shapes lead to minimum downwash in the flow regime. At the same time, the development of early jet engines was critically hampered by the inability to design successful axial flow compressors. Such compressors cannot be built without a detailed understanding of the boundary layer flow over the blades. Advances in aerospace design remain intimately involved with advances in science – and great design engineers have often had strongly scientific backgrounds.

A brief description of the work of six aerospace designers and researchers serves to provide an historical perspective to aerospace design and the place that analysis and computation play in it. Here, we start with the Wright brothers and mention just two of the many great designers of the interwar years (R.J. Mitchell and John K. Northrop) and two great researchers (Sir Frank Whittle and Theodore von Kármán).

The Wright brothers were intensely practical men but also keen to begin the use of scientific method in aircraft design. Mitchell was not a graduate, though he was clearly analytically very capable. He believed mainly in full-scale experiments where possible and his use of the rapidly accumulating science base was limited. Northrop was more inclined to draw on the expertise of research scientists in attempts to gain theoretical insights, calling on von Kármán, for example, when designing flying wings. Whittle began with a highly inventive mind, patented a jet engine design before going to university, and ultimately became a great research engineer. Von Kármán was highly gifted mathematically and though heavily involved with engineering all his life was unmistakably and foremost a great scientist. These men represent the spectrum of influential individuals seen in the early years of the aerospace sector.

By examining the activities of such people, it is possible to view the first 75 years of aerospace design as a series of stages: first came the early pioneers, working at a time when many were skeptical about the benefits of flight or even if it was possible at all. Next came the phase where outstanding individuals dominated aeronautical science and design during the interwar and early postwar years. This was followed in the 1950s and 1960s by a greater emphasis on design teams and the massive expansion of the great aerospace companies that we recognize today, such as Lockheed Martin and Rolls-Royce.

Following these stages, the advent of computer-aided engineering, design and manufacture in the mid-1970s transformed activities in aerospace with an increasing focus on faster design cycles, more accurate predictive capabilities and the rise of computer science. At the same time, the great drawing offices of the 1950s declined and fewer people were involved in the design of each individual component. These developments led to the techniques that are the subject of this book. It is now the case that designers must be masters of computational methods if they are to produce competitive products at affordable costs in reasonable timescales.

1.3.1 A Pair of Early Pioneers

The Wright Brothers

It is generally accepted that the Wright brothers were the first men to achieve controlled, powered, heavier-than-air flight. This took place on December 17, 1903 when Orville Wright flew their prototype aircraft at Kitty Hawk in North Carolina, watched by his brother Wilbur. The longest flight achieved on the day covered nearly half a mile. The brothers had designed and built their aircraft, including its engine and propellers, over the previous four years with the aid of a bicycle mechanic, Charlie Taylor; see Figure 1.1. Although these bare facts are very widely known, what is less commonly appreciated is the technical advances made by the brothers that enabled them to achieve this success. For example, they built some of the first wind tunnels to enable them to study the behavior of wing sections and to gather the data needed for design. They realized that propeller blades are just a form of wing that must be twisted to allow for the varying air speed seen by each section along the blade. Their first aircraft had remarkably efficient propellers. They clearly understood that by varying

Figure 1.1 "The first flight" photo, 10:35 a.m. December 17, 1903, showing the Wright 1903 Flyer just after liftoff from the launching dolly at Kill Devil Hill, Kitty Hawk, North Carolina. Orville Wright is at the controls, and Wilbur Wright watches from near the right wing tip. National Air and Space Museum, Smithsonian Institution (SI 2002-16646).

the angle of attack on a section the amount of lift generated could be controlled. They used this knowledge to gain control of the aircraft by "wing warping", so that by twisting the entire wing the effective center of lift could be moved from side to side noting "*That with similar conditions large surfaces may be controlled with not much greater difficulty than small ones, if the control is effected by manipulation of the surfaces themselves, rather than by a movement of the body of the operator*".

It is apparent that the Wright brothers took a rather thoughtful and analytical approach to aircraft design and that they treated their design as an integrated whole – a very modern approach, albeit one that made limited use of computation. This essentially scientific approach to design is obvious in the brother's writings: two extended quotes from work published slightly before their historic first powered flight (Wright 1901) and several years later (Wright and Wright 1908) make this very clear.

"*The difficulties which obstruct the pathway to success in flying-machine construction are of three general classes: (1) Those which relate to the construction of the sustaining wings; (2) those which relate to the generation and application of the power required to drive the machine through the air; (3) those relating to the balancing and steering of the machine after it is actually in flight. Of these difficulties two are already to a certain extent solved. Men already know how to construct wings or airplanes which, when driven through the air at sufficient speed, will not only sustain the weight of the wings themselves, but also that of the engine and of the engineer as well. Men also know how to build engines and screws of sufficient lightness and power to drive these planes at sustaining speed. As long ago as 1884 a machine weighing 8,000 pounds demonstrated its power both to lift itself from the ground and to maintain a speed of from 30 to 40 miles per hour, but failed of success owing to the inability to balance and steer it properly. This inability to balance and steer still confronts students of the flying problem, although nearly eight years have passed. When this one feature has been worked out, the age of flying machines will have arrived, for all other difficulties are of minor importance.*" and

"*To work intelligently, one needs to know the effects of a multitude of variations that could be incorporated in the surfaces of flying-machines. . . . The shape of the edge also makes a difference, so that thousands of combinations are possible in so simple a thing as a wing. . . . Two testing-machines were built, which we believed would avoid the errors to which the measurements of others had been subject. After making preliminary measurements on a great number of different-shaped surfaces, to secure a general understanding of the subject, we began systematic measurements of standard surfaces, so varied in design as to bring out the underlying causes of differences noted in their pressures.*

Our tables made the designing of the wings an easy matter; and as screw propellers are simply wings traveling in a spiral course, we anticipated no trouble from this source. . . . When once a clear understanding had been obtained, there was no difficulty in designing suitable propellers, . . . Our first propellers, built entirely from calculation, gave in useful work 66 per cent of the power expended. This was about one third more than had been secured by Maxim or Langley.

As soon as our condition is such that constant attention to business is not required, we expect to prepare for publication the results of our laboratory experiments, which alone made an early solution of the flying problem possible."

Although there is no doubt that the Wright brothers' efforts significantly influenced the early take-up of powered flight, it is rather less well understood that their strong commitment to a careful scientific approach in design had such a strong impact. They would no doubt have been firm believers in computational engineering.

1.3.2 A Pair of Great Designers

R.J. Mitchell

R.J. Mitchell was the chief designer engineer for the Supermarine Aircraft Company from 1918 to 1937. During that time, he was responsible for the design of 24 different types of aircraft. Of these, perhaps the most famous are the racing seaplanes that were used to win the Schneider Trophy outright for Britain in 1931, and the Spitfire fighter aircraft of World War II. It is interesting to note that Mitchell became chief designer at Supermarine at the age of 25. It is also interesting to realize that at that time, the design team at Supermarine consisted of just 12 people. It is also noteworthy that Mitchell had won prizes for mathematics while studying at night class as an apprentice. He was clearly not afraid of analysis and calculation, as his few early papers to the Royal Aeronautical Society make clear. Nonetheless, he was not interested in the use of things like the Navier–Stokes equations to understand the performance of his aircraft. Moreover, he also distrusted wind tunnel and model scale testing and preferred full-scale trials to assess any new design (Mitchell 1986). He may thus be seen as a natural successor in his ways of working to the Wright brothers; he might even be judged conservative by their standards.

Perhaps the most interesting thing about Mitchell's winning seaplane designs was their dramatic departure from earlier competitors in the Schneider Trophy series. The S4 seaplane had fully cantilevered wings, and used floats to carry it when on the water. Its successors, the S5, S6 and S6B, that finally won the trophy outright, used a number of advanced features including monocoque aluminum structures, highly advanced engine cooling systems with hollow wing skins to carry the cooling fluid and internal airflows in the wing cavities to gain extra heat transfer. Mitchell was clearly capable of coming up with bold designs. The S4 was, however, not without its problems and crashed before taking part in the races, for causes that are still not fully understood, but are most likely associated with wing flutter and a loss of control.

At the time that Mitchell was designing the Spitfire in the late 1930s, his design office was producing design concepts at the rate of around one per month. Early drawings of the Spitfire prototypes show that these did not use the famous elliptical wing planform of the final aircraft (as seen in Figure 1.2). This was introduced on the recommendation of the Canadian aerodynamicist Beverly Shenstone, newly appointed to the Supermarine Company. Nonetheless, it is clear that understanding of potential flow theory and downwash had led to the advantages of such a shape being appreciated at around this time. Science was beginning to intrude on designers.

John K. Northrop

At around the same time as Mitchell joined Supermarine, Jack Northrop was helping the Loughead brothers to design flying boats. He went from there to the Douglas Aircraft Company before setting up the Northrop Aircraft Corporation in 1929. Northrop Corporation became a subsidiary of Douglas in 1932 with Douglas holding 51% of the stock in the Corporation. Northrop himself was a major figure in the design of the Douglas DC-1 and the famous DC-3, where his ideas on advanced wing structures were put to good effect. He also helped design a number of other famous aircraft including the "Vega" monoplane,

Figure 1.2 One-half right front view of Supermarine Spitfire HF VII (s/n EN474) in flight. National Air and Space Museum, Smithsonian Institution (SI 93-8430).

Figure 1.3 NASM restored Northrop N-1M (Northrop Model 1 Mockup) Flying Wing. April 28, 1983 at Paul E. Garber Facility, NASM (outside Bldg. 10). Photograph by Dale Hrabak, National Air and Space Museum, Smithsonian Institution (SI 83-2944).

for the then Lockheed Aircraft Corporation, and the P-61 "Black Widow" night fighter of World War II. He made great contributions in the development of all-metal monocoque stressed skin systems for wings and fuselages in aircraft such as the Northrop Delta, adopting approaches that have subsequently become standard practice for all commercial aircraft. Perhaps more importantly, he was the main promoter of flying wing aircraft, believing from the outset that clean aerodynamic lines were of central importance in building efficient aircraft. These included the N-1M, the first flying wing aircraft built in America (1940), now in the Smithsonian National Air and Space Museum (NASM) (see Figure 1.3) and the XP-56 flying wing fighter, the first all-magnesium aircraft. His ideas for flying wings were much ahead of their time but have now been adopted in modern stealth aircraft designs. It remains to be seen if flying wing aircraft will be adopted for passenger traffic.

It is clear from Northrop's lecture to the Royal Aeronautical Society in 1947 that the fundamental drive to build flying wings was the reduction in drag that could be obtained by doing without a fuselage, tail plane and other excrescences. Interestingly, when starting work on the N-1M in 1949, Northrop enlisted the help of Theodore von Kármán from the California Institute of Technology and a great deal of effort was devoted to various aerodynamic devices to gain control stability, such as using split flaps that could be opened at either wing tip to give yaw control.

Almost all of the refinements to these flying wing designs were achieved from wind tunnel and full-scale testing. For example, the test pilot noted that the initial N-1M design could only be flown by holding it at very precise angles of attack. It is reported that this problem was solved by von Kármán making adjustments to the elevon trailing edges. The increased use of the outputs of university research labs was clearly changing aircraft design. As Northrop noted in his 1947 lecture (Northrop 1947), great care was still taken to allow for testing when designing experimental aircraft: "*Because of the many erratic answers and unpredictable flow patterns which seemed to be associated with the use of sweepback, it was decided to try to explore most of these variables full scale, and the N-1M provided for changes in planform, sweepback, dihedral, tip configuration, C.G. location, and control surface arrangement. Most of these adjustments were made on the ground between flights; some, such as C.G. location, were undertaken by the shift of ballast during flight.*"

1.3.3   A Pair of Great Researchers

Sir Frank Whittle

Frank Whittle was born in 1907 into a family that owned a small general engineering business. He joined the RAF in 1923 to train as a fitter but was selected for pilot training at the RAF College, Cranwell. While there, he wrote a thesis entitled "Future Developments in Aircraft Design" and filed his first patent on jet engine design at the age of 23. This patent was granted in 1932 and published in many countries including Germany, where research on jet engines began a year later. Despite contacting many aero-engine manufacturers at that time, none showed any interest in his ideas.

In 1932, Whittle undertook the RAF officer's engineering course at Henlow and obtained an aggregate of 98% in all subjects of the preliminary examination and distinctions in all subjects except drawing of the final examination. The RAF allowed Whittle to go to Cambridge to study Mechanical Sciences, where he graduated with first-class honors in just two years. During this time, he further developed his ideas on jet propulsion, gaining the backing of two retired RAF officers, an engineer and a merchant bank. With this support, the company Power Jets Ltd. was formed and work started on experimental engines in 1936. Their first engine, the WU (short for Whittle Unit) was run in April 1937 and the first aircraft powered by one of their jets, the W1 flew in May 1941, achieving a speed of 338 mph. This was the experimental Gloster E28/39, which before long was achieving speeds of 450 mph with the W2 engine; see Figure 1.4. Given the limited resources of Power Jets and the pressures of war, further developments were undertaken first with Rover and then Rolls-Royce, who produced the Welland and Derwent turbojets. The Welland was Rolls-Royce's version of the Power Jet W2 engine and was named after a river to indicate the flowing nature of the jet engine (the tradition of naming Rolls-Royce jet engines after rivers continues to this day with the Trent series of engines).

The first jet powered RAF fighter, the Gloster Meteor, entered service toward the end of the World War II, and with Derwent engines could fly at 493 mph, some 200 mph faster than the Spitfire. In 1945, a Meteor with a Derwent V engine set the world speed record at 606 mph. After the war, Power Jets was nationalized and merged with the U.K. government's research laboratories at Farnborough to form the U.K.'s National Gas Turbine Establishment. Most of the engineering staff resigned in protest at this point and Whittle retired from the RAF in poor health. He then worked as a technical adviser to BOAC is British Overseas Airways Corporation before he eventually left the United Kingdom for America, where he worked for Shell on drilling research, before finally becoming a research professor at the U.S. Naval Academy. In 1981, he summarized his knowledge of the field in a book entitled

Figure 1.4   Gloster E28/39 experimental aircraft powered by Whittle's W1 jet engine. National Air and Space Museum, Smithsonian Institution (SI 72-4720).

"Gas turbine aero-thermodynamics" (Whittle 1981). At the time of his death in 1996, he was working on schemes for large supersonic passenger aircraft (Fielden and Hawthorne 1988).

Theodore Von Kármán

Von Kármán was a mathematical prodigy who was steered into engineering by his father to broaden his interests. He graduated from Budapest with the highest honors before moving on to Gottingen to take his PhD, having spent some time there working with Prandtl on a new wind tunnel for airship research. It was while at Gottingen that he studied the stability of rows of vortices behind circular cylinders and the associated drag of what has ever since been known as a Kármán vortex street; see Figure 1.5. Von Kármán was always especially gifted at thinking through the physical concepts behind problems. This approach allowed him to construct analytical models from which calculations could be made. They were often based on a close familiarity with experimental results although he was not a good experimentalist himself.

In 1912, he left Gottingen to become Professor of Aerodynamics and Mechanics and Director of the Aeronautical Institute at Aachen. During his time at Aachen, he carried out research in fluid mechanics and aerodynamics. His reputation grew and he traveled widely and lectured in many countries including the United States. Von Kármán was a founder member of the International Congresses of Applied Mechanics and the International Union of Theoretical and Applied Mechanics (IUTAM). He was a consultant to aircraft makers in Germany, England and Japan. In 1926, he was invited to America to act as an adviser to the California Institute of Technology. In 1930, he accepted the offer of the Directorship of

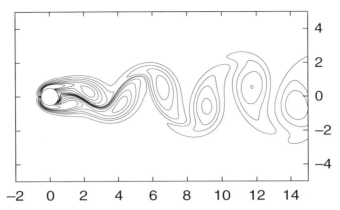

Figure 1.5 Flow visualization of a von Kármán vortex street behind a circular cylinder at a Reynolds number of 100. The contours are constant vorticity (negative for vortices above the axis, positive below). The values were generated numerically using a Discrete Vortex Method. Based on the work of Takeda et al. (1997), courtesy of O. Tutty.

the Guggenheim Aeronautical Laboratory at CalTech. Under his leadership, the laboratory at Pasadena became one of the foremost centers for research in aeronautical sciences. He was also involved in the earliest considerations of supersonic and hypersonic flight and the formation of the Jet Propulsion Laboratory.

During his life, von Kármán contributed widely and profoundly to the science and applied mathematics that support aerospace engineering and design. He was also extremely committed to education and international scientific exchange and was instrumental in founding the NATO aeronautical research activity, AGARD, the International Council of the Aeronautical Sciences, ICAS, the International Academy of Astronautics and the von Kármán Institute. In 1963, he received the first National Medal of Science from President Kennedy, awarded in recognition for his "incomparable contributions not only in the field of applied mechanics, aerodynamics and astronautics, and to education in general, but also to industrial, national, international, and human affairs in their broadest sense". His contributions included work on turbulence (von Kármán 1937), airfoil theory, high-speed aerodynamics (von Kármán 1955), stability and nonelastic buckling. Without input from men like von Kármán, aeronautical engineering could not have come so far as it has so quickly.

1.3.4 Two Great Aerospace Companies

The origins of the great aerospace companies that exist today can mostly be traced back to the efforts of a few key individuals, such as those mentioned above. Their growth to major international operations owes much to the large numbers of aircraft built during World War II and subsequently to the consolidation of the many firms that were operating in the industry at that time. Here, we very briefly outline the history of just two such corporations: Rolls-Royce and Lockheed. In both cases, the companies were formed at the very beginnings of the age of powered flight, while today they are very large international concerns. Their history has also been linked at critical stages.

The Lockheed Corporation

The history of the Lockheed Corporation begins with the two Loughead brothers, Allan and Malcolm, who started flying in 1913. They set up their own company, the Loughead Aircraft Manufacturing Company in Santa Barbara, California, in 1916, working with Jack Northrop and others. The company was soon renamed Lockheed to avoid mispronunciation. It folded just after the First World War, restarted in late 1926 and then became a division of Detroit Aircraft in 1929, having produced the record breaking "Vega" monoplane and its derivatives. These were used by such famous aviators as the Lindberghs, Amelia Earhart and the explorers Wilkins and Eielson. Detroit went bankrupt in 1931 in the aftermath of the great stock-market crash and an investment banker, Bob Gross and some coinvestors then bought Lockheed for $40,000.

The company experienced dramatic expansion through the war years reaching over 90,000 employees by the middle of World War II, during which they built nearly 20,000 aircraft. This then rapidly scaled down following the end of the war so that by 1946 there were less than 20,000 staff on the payroll. Three aircraft dominated this era in the company: the P38 Lightning, the Hudson and the Constellation transport – it was in a Constellation that Orville Wright made his last flight. During the 1950s and 1960s, the company remained heavily dependent on military sales with a rapid expansion in missile systems beginning in 1954, which grew to account for over 50% of company turnover within seven years. It was also during this time that the legendary "Skunk Works" came to prominence with the production of the U-2 and SR-71 spy planes. In the 1960s, the company became closely involved with Rolls-Royce as it selected the Rolls-Royce RB-211 engine for its new TriStar triple engine passenger jet. The maiden flight of this aircraft took place in November 1970, but before the end of 1971, the strain of developing its engines had driven Rolls-Royce into receivership. The president of Lockheed at the time, Dan Haughton, was a key player who then helped ensure the survival of Rolls-Royce (Boyne 1999).

Today, the company is called Lockheed Martin (having merged with the company originally started in 1912 by another great pioneer, Glenn Martin) and works in a diverse range of sectors including design, production and support of strategic and cargo aircraft; command, control, communication, computers and intelligence systems integration; design and production of fighter aircraft; supply of communications and military satellites; missile systems; naval electronics and surveillance systems; space launch vehicles; strategic missile systems; and systems integration and engineering, logistics and information services.

Rolls-Royce plc

Rolls-Royce was founded by Charles Rolls, who was educated in Mechanical Engineering at Cambridge University, and Henry Royce, who had served an apprenticeship with the Great Northern Railway. Royce had begun making cars in 1903 while Rolls sold cars in London. They met in the spring of 1904 and agreed that they would go into business together. Royce would make a range of cars that Rolls would sell. These would be called Rolls-Royce cars and the new company started trading in March 1906. From the outset, Rolls tried to persuade the company to manufacture aircraft or aircraft engines, but Royce was reluctant to do so. In fact, the company began the manufacture of aero engines only with the outbreak of the First World War in 1914. Five thousand engines were delivered during the war and these powered 54 separate aircraft types. Following the end of hostilities, Rolls-Royce aero engines were used extensively to power civil aircraft in Europe. Interestingly, Rolls was killed while piloting a Wright Flyer in 1910; this was the flyer which he had spent some time persuading the Wright bothers to sell to him. His was the first recorded fatality of a British pilot in the United Kingdom – but not before he had succeeded in being the first to fly both ways across the English channel.

During the interwar years, the Schneider Trophy races had a significant impact on the company. At this time, the Rolls-Royce Kestrel engine was developed through supercharging to become the "R" series engine, which finally won the trophy outright in 1931. This engine was ultimately developed into the Merlin engine that powered the Spitfire, Hurricane, Lancaster, and Mustang aircraft during World War II. By the end of the war, over 160,000 Merlin engines been manufactured, a good many of them in America under license. Its power increased from 1,030 hp to over 2,000 hp during the war years, mainly because of technical improvements in its supercharger design led by the scientist and engineer Sir Stanley Hooker.

In the latter stages of World War II Rolls-Royce was closely involved in the development of jet engines, so that at the end of the war the company had a world lead in gas turbine technology. In 1958, the Rolls-Royce Avon engine became the first gas turbine to power a civil aircraft carrying passengers across the Atlantic. During the 1960s, the development of new aircraft in America was extremely competitive. Boeing, Lockheed and Douglas were all bidding to sell new aircraft to American Airlines, Eastern Airlines, TWA and Pan Am. In turn, Pratt and Whitney, General Electric and Rolls-Royce were all bidding to sell new engine designs to the aircraft manufacturers.

Rolls-Royce finally won the contract to design a new three-shaft engine for Lockheed to power their TriStar airliner. The scale of advance promised in the RB 211 engine was enormous. It had a fan diameter almost twice that of the previous Rolls-Royce Conway engine and the bypass ratio increased from 0.7 to nearly 5, with significantly increased turbine inlet temperatures. At the same time, Rolls-Royce offered a number of commercial concessions to win the business. The strain of these technical advances and commercial guarantees ultimately lead to the company going into receivership in 1971. Even so, flight-ready engines were just four months late of the promised schedule and the three-shaft design developed laid the foundation for the current, very successful Rolls-Royce series of Trent engines. The fact that Rolls-Royce was not wound up by the U.K. government of the day owed much to the support from senior staff at Lockheed (Pugh 2001).

Today, Rolls-Royce plc manufactures aero engines over a wide spread of power ranges as well as industrial power systems, marine propulsion systems and submarine nuclear power plant.

1.3.5   Rationalization and Cooperation

During the postwar years, the very many aerospace companies that existed at the end of World War II were inevitably rationalized into fewer, larger corporations. This rationalization can be illustrated by considering the changes in the British aerospace companies over this period: in 1945, there were no fewer than 27 airframe manufacturers in the United Kingdom and eight aero-engine makers. Moreover, each company lobbied government for contracts to sustain itself into the future. By 1957, the U.K. government decided to rationalize this situation into three main airframe groups and two engine groups. The aircraft groups were: the British Aircraft Corporation (which incorporated English Electric, Vickers-Armstrong, Bristol and Hunting), the Hawker-Siddeley Group (which incorporated Avro,

Armstrong-Whitworth, Blackburn, Folland, Gloster, Hawker and De Havilland) and the Westland Aircraft Group (which incorporated Westland, Saunders Roe, Fairey and the Bristol Aircraft Company's helicopter division). The two engine groups were built around Rolls-Royce and Bristol Siddeley engines. Even so, this still left four independent companies: Short Brothers, Handley Page, Scottish Aviation and Auster Aircraft.

At the time of writing, in the United Kingdom, there is only one main engine group, Rolls-Royce plc (which now includes Allison in the United States.), and four aircraft manufacturers of size: BAE Systems (which has largely divested itself of civil aircraft manufacture) the U.K. arm of Airbus (which contains the majority of the U.K. civil aircraft manufacturing capability and is part of a European consortium), Westland (which focuses on helicopters) and Short Brothers (which is now the European unit of Bombardier Aerospace). Each of these companies is sufficiently specialized that they are no longer in such strongly direct competition with each other. There remain, of course, many other U.K. aerospace companies specializing in various sectors, whether at the component level or in complete systems, such as satellites. Nonetheless, it is clear that a fiercely competitive peacetime market has enforced significant rationalization. The situation in America has followed broadly similar lines so that now there are just two principal American jet-engine manufacturers (General Electric and Pratt and Whitney) and only one major builder of large civil aircraft (Boeing).

At the same time, new aerospace projects are now so complex and their costs so high that the current trend in design and manufacture is to form multicorporation international projects. Such collaborations are also driven by the inevitable desires of countries to purchase products designed and made, at least in part, by their own companies. So, for example, if one tries to sell aero engines to Japan or China, it is perhaps inevitable that Japanese and Chinese companies wish to be involved in their design and manufacture. An example of this is the V-2500 engine, which involved collaborators in the United Kingdom (Rolls-Royce plc), America (Pratt and Whitney), Germany (MTU), Italy (Fiat) and Japan (Ishikawajima Harima, Kawasaki and Mitsubishi). These engines are used to power the Airbus A320 aircraft. Similarly, the competition to design and build the Joint Strike Fighter for the U.S. military saw teams formed with member companies from around the world. The winning team was led by the Lockheed Martin Corporation and includes Rolls-Royce, among many others.

1.3.6 The Dawn of the Computational Era

It is clear from these brief sketches that the small companies and design teams that were brought into being by enthusiasts during the early years of flight have been transformed into major activities by the demands of war. Instead of a few key men who often knew each other, working with small teams of 10 or 20 designers, the aerospace industry is now dominated by corporations with thousands of employees and very large-scale design and manufacturing capabilities. Inevitably, such large corporations have substantial research budgets and were early users of computers.

When the great advances in computing took place in the 1960s, this technology was combined with the scientific understanding that had been achieved during the previous 50 years by people like von Kármán and others. The result was that rapid progress became possible in the practice of aerospace design. New ways of working became available, not because designers had been unaware of the possibilities of direct analysis before but because now they had a means to systematically exploit the fundamental knowledge discovered in the universities and research labs. Detailed computer modeling of the physics involved in engineering gave the design engineer a completely new approach for solving problems. Today, it is inconceivable to consider aerospace design without recourse to modern finite element codes and computational fluid dynamics (CFD) models, albeit still backed up by experiments at model and full scale.

At the same time, Computer-aided Design (CAD) tools were being developed that enabled draftsman to produce drawings electronically. During the 1970s and 1980s, these tools were little more than electronic drawing boards. However, as technology improved, designers have increasingly worked with more sophisticated geometric entities rather than simple lines and arcs. Today, modern CAD systems mainly work with solid models and may be used to produce extremely high quality images of almost photographic realism for any part of the design being produced. Such systems can be used to make layout and assembly decisions and help develop maintenance schedules long before any metal is actually cut. They are increasingly being used to aid costing assessments and to decide on manufacturing options.

Alongside developments in analysis and product representation tools, progress has also been made in the field of automated optimization. Although crude to begin with, attempts have been made from the earliest availability of computers to use search methods to improve, and ideally optimize, designs. Significant improvements to real designs have been made during this period in a number of fields (Vanderplaats 1984). Tools in this area are now beginning to catch up with the sophistication of those in other fields and it is the increasingly widespread availability of commercial Design Search and Optimization (DSO) toolkits that has been one of the reasons for writing this book. At the time of writing, DSO packages are beginning to be accepted as everyday tools rather than as specialist capabilities used only by consultants.

The design work undertaken in aerospace companies is now also characterized by very large distributed teams of engineers working over extended periods. Currently, Rolls-Royce aims to design a new engine from initial concept to approved product in around two years. This is in stark contrast to Mitchell's small group of staff designing a new aircraft in a few weeks and reflects the vastly increased complexity and much greater scientific sophistication of modern aerospace products. It also means, however, that explicitly managing the design process and the tools used to underpin design is a key concern in the aerospace sector. It is now necessary to carefully consider the impact of new computational packages on the way designers work before they are introduced. Such tools must serve as aids to designers and not as their replacements. The place of computational tools as design aids is the central theme of this book.

1.4 Traditional Manual Approaches to Design and Design Iteration, Design Teams

1.4.1 Design as a Decision-making Process

In writing this book, design has been considered from a largely computational standpoint, with a view to the impact that computational tools can have on design practice. Aerospace engineering design is one of the most complex activities carried out by mankind, and it is therefore necessary to take a somewhat idealized perspective. Even so, design should fundamentally always be seen as a decision-making process. Despite all the advances that have been made in computational analysis, it must be remembered that the fundamental hallmark of design is not analysis but synthesis – the choice of appropriate mechanism types, power sources, the setting of dimensions, the choice of features and subcomponents, the selection of materials – all these are acts of synthesis and it is the skillful making of decisions in these areas that is the hallmark of the good designer. Of course, designers use analysis all the time, but design is about decision making and analysis is, by contrast, an act of gaining understanding, and not of making decisions. Moreover, to be a good designer, the most often cited personal prerequisite is experience – this view is backed up by observational studies in engineering design offices (Ahmed and Hansen 2002).

So, even though design decision making is commonly preceded by a great deal of information gathering, and although the gathering of such information, often using computational models, may be a very skilled and time-consuming activity, it should be made clear that whatever the cost, this remains just a precursor to the decisions that lie at the heart of design. This decision-making process may be thought of as being built from four fundamental components:

1. *Taxonomy*: the identification of the fundamental elements that may be used, be they gears, pumps, airfoils, and so on;

2. *Morphology*: the identification of the steps in the design process;

3. *Creativity or synthesis*: the creation of new taxonomies, morphologies or (more rarely) fundamental elements (such as the linear induction motor) and

4. *Decision making*: the selection of the best taxonomy, morphology and design configurations, often based on the results of much analysis.

Perhaps the simplest and most traditional way of representing the process is as a spiral; see Figure 1.6. The idea behind this view is that design is also iterative in nature with every aspect being considered in turn, and in ever more detail as the design progresses. It begins with an initial concept that attempts to meet a (perceived) customer need specified in very few major requirements. For an aircraft, this might cover payload, range, speed and anticipated cost; for an engine, it might be thrust, weight, fuel efficiency and cost. This phase of the design process is often called concept design. It is then followed by preliminary design, detailed design and the generation and verification of manufacturing specifications before production commences. Once the product is operational, a continuing "in-service" design team takes over to correct any emerging problems and deal with any desired through-life enhancements. Finally, decommissioning and waste disposal/reuse must be considered. In Rolls-Royce, for example, this is called the Derwent process (Ruffles 2000), and is characterized by significant business decision gates at each stage.

Nowadays, even a traditional manual approach to design will probably make use of extensive computing facilities, but it will generally not draw on modern search and optimization strategies, knowledge-based systems, or Grid-based computing[1], to carry out design synthesis. It will, in all probability, make heavy use of extremely experienced design staff

---

[1] widely distributed computing systems networked together to form a single resource

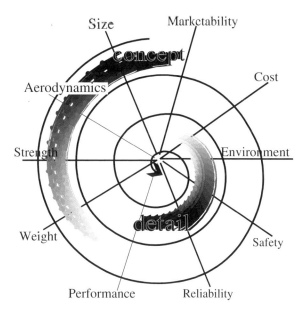

Figure 1.6 The design spiral.

and detailed experimental and analysis activities. Thus, the use of the term manual or traditional here implies not that computers are not used, but rather that changes to the design specification are made as the result of direct human intervention, with the computer being used very much to support this process, i.e., manual decision making.

1.4.2 Concept Design

When embarking on a concept design (sometimes termed preliminary concept design), the designers must first decide which aspects of the product must be considered from the outset, and which tools will be used to make a start. Often, these are based on considerations of previous designs: it is rare to design from a blank sheet of paper – even Whittle knew that attempts had been made to design and build jet engines before he filed his first patent; the Wright brothers were by no means the first to attempt powered, heavier-than-air flight.

Given a few previous designs, it is usually possible to construct some basic design rules that will allow a designer to make outline predictions on a "back-of-the-envelope" basis; see, for example, the many approximate formulae and design curves in the books by Torenbeek (1984) or Stinton (1983). If a great deal of data is available or the product very complex, such design rules may well be encoded into statistically based concept tools, such as that developed by Airbus (Cousin and Metcalfe 1990).

Stinton (1983) lists a number of aspects to be considered in aircraft design:

- Environment (altitude, temperature, distances, humidity, noise)

- Human need (passengers, load, cost, work, recreation)

- Airfield or airstrip (type and size)

- Cockpit, cabin air-conditioning and pressurization

- Fuselages or hull design

- Wing type and size

- Performance, power plant type and arrangements

- Flaps, slats and high lift devices.

- Handling qualities, control and stability

- Structure design

- Equipment and services.

Clearly, this list is not exhaustive, but it begins to set out some of the topics that might be considered for a new aircraft concept. If these are to be assessed using computational tools, then models of some kind must be available for every aspect. Even today, it is not possible to provide first principles analysis for all of the topics given in the list. In a traditional design process, the best that can be hoped for, particularly during the early concept stages, are empirical relationships.

Concept design is usually undertaken by teams of up to 30 individuals who between them cover all of the disciplines being considered. They will probably be dedicated to this kind of activity and may consider many widely different concepts over the course of a single year. If they work in a traditional way, using empirical relationships their needs for computational support will be modest; even their collection of data will tend to be rather small since it will usually be based on summary descriptions of just the previous products of the company – few companies have made more than 50 aircraft or engines in a form that will be applicable to newly evolving designs. It is quite possible to carry out the preliminary design of a light aircraft using a spreadsheet.

The designs produced during concept work can usually be characterized by relatively few quantities. A typical aircraft design may be summarized at this stage by less than 100 numbers. Such designs are commonly produced fairly rapidly, in timescales measured in days and weeks. In the work described by Keane (2003), a wing is characterized by just 11 variables, in that by Walsh et al. (2000) at NASA Langley a little under 30. The designs produced during concept design will normally be used to decide whether to proceed further in the design process. As Torenbeek (1984) notes, the "*aim is to obtain the information required in order to decide whether the concept will be technically feasible and possess satisfactory economic possibilities*". Thus, concept design may be seen as part of forming a business-case, as opposed to an engineering process. The designs considered following the decision to proceed may well differ radically from those produced by the concept team. The concept designers should ideally allow for this dislocation so that performance specifications can be delivered in practice by those charged with preliminary and detailed design.

The concept team must also allow for the fact that the design process may take years to complete and so they must make allowances for any likely improvements in technological capabilities that will occur. They must also make judgments concerning risks involved in any new technologies that are assumed will be available before production begins. Technology-acquisition activities, including any necessary applied research, will also be initiated toward the end of this stage of design. A close working relationship with established customers will also be important during concept design.

It is in the area of concept design where optimization has perhaps had its greatest effect so far. Even so, the decision-making process, where many competing designs are traded off against each other, is still normally carried out manually. This aspect of design is beginning to change with the advent of multiobjective optimization and the use of game theory and search methods in decision making (Parmee 2002; Sen and Yang 1998).

1.4.3   Preliminary Design

Once the (usually economic/business) decision to proceed further with the design is taken, the preliminary design stage can begin (this is sometimes termed full concept or development definition or design). At this point, it is traditional for the different aspects or components of the design to be considered by dedicated teams. For aircraft, this might consist of an aerodynamics division, a structures division, a control systems division, costing teams, and so on, or it might consist of a team considering the fuselage, another considering wing design, others looking at the tail plane, propulsion systems, and so on. For aero engines, it is common to break the design teams up into divisions based on the various components of the engine such as fans, compressors, turbines, combustors, and so on. Managing the interrelationships between such teams becomes a key part of managing the design process, especially if they are geographically widely dispersed, as is now often the case.

An alternative, and more modern, approach to managing design is via the use of Integrated Project Teams (or IPTs). Such teams are normally formed specifically for the product being designed and grow in size progressively throughout the preliminary design phase. If an IPT-based approach is used, it is then usually supported by specialist divisions that are charged with providing technical input across a range of project teams. These specialist divisions are responsible for the retention and development of core technologies and capabilities. They will also interface directly with any research activities and engage in technology-acquisition programs.

The IPT will be led by a chief engineer supported by a program manager, a chief of design and a chief service engineer. The program manager will be expected to keep the project to deadline, cost and technical specification in line with the strategic business plan. The service engineer will be responsible for ensuring that the design can be serviced in use and will provide increasing support to the customer as the product enters service. The chief of design is responsible for integrating together the teams looking at each subsystem in the product and making sure that the design comes together in a coherent fashion. Such an IPT will encompass all of the specializations needed to define the product, being drawn from specialist support divisions, other teams and perhaps even major subsystem suppliers. This then leads to a matrix structure of management with one direction being made up of the project teams and the other of specialist support divisions.

In whatever form the design teams take, the tools used by them will be much more sophisticated than during concept work. For example, the designers considering structures will, as a matter of routine, make use of quite detailed stress analysis, normally by means of Finite Element Analysis (FEA). Those considering the wings will pay close attention to predictions of the airflow. This may well involve significant experimental programs as well as the extensive use of computational methods (CFD).

A key observation about how computational tools are used in the traditional approach to design is that they tend to have no *direct* impact on the geometry of the design being produced. They are used instead for analysis and it is left to the designer to make decisions on how to change the design as a result of studying the resulting outputs. Of course, this means the designers must be highly experienced if they are to make effective use of the results obtained. It is often by no means obvious how to change a design to improve its performance or reduce its cost simply by studying the results of computational analysis. Trading performance improvement for cost saving is even more difficult to carry out in a rational fashion. Even so, it is common to find that design teams spend enormous amounts of time during the preliminary design stage preparing the input to CFD and FEA runs and studying the results.

At the present time, preparing the input for computational analysis is commonly very far from automated. It is quite normal for design teams to take two or three weeks to prepare the meshes needed for the CFD analysis of a complete aircraft configuration, to run the analysis, and to assimilate the results. This severely restricts the number of different configurations that can be considered during preliminary design. The result is that the design freezes early on during the design process, so that even when the design team knows that the design suffers from some shortcomings, it is often very difficult to go back and change it. Developments in advanced CAD systems are beginning to make some inroads into this problem. Using the ICAD system (ICAD 2002), for example, Airbus has been able to make significant changes to wing geometries and these then lead to the automated modification of many of the components in the wing design. This process does, however, require considerable changes to the traditional methods of working usually found in design offices.

1.4.4 Detailed Design

Once preliminary design is complete, detailed production or embodiment design takes over. This stage will also focus on design verification and formal approval or acceptance of the designs. The verification and acceptance process may well involve prototype manufacture and testing. Such manufacturing needs to fully reflect likely productions processes if potentially costly reengineering during production is to be avoided. Apart from concerns over product performance, issues such as reliability, safety and maintainability will be major concerns at this stage. The objective of detailed design is a completely specified product that meets both customer and business needs. At this stage in the process, the number of staff involved in the design team will increase greatly.

In large aerospace companies, detailed design is dominated by the CAD system in use. Moreover, the capabilities of that CAD system can significantly affect the way that design is carried forward. If the system in use is little more than a drafting tool, then drafting becomes the fundamental design process. Increasingly, however, more advanced CAD tools are becoming the norm in aerospace companies. These may allow parametric descriptions of features to be produced, which permit more rapid changes to be carried through. They may also allow for information other than simple geometry to be captured alongside the drawings. Such information can address manufacturing processes, costing information, supplier details, and so on. The geometric capabilities of the system may also influence the way that complex surfaces are described: it is very difficult to capture the subtleties of modern airfoils with simplistic spline systems for example. Furthermore, as the CAD systems used in the aerospace sector develop, companies occasionally need to change system to take advantage of the latest technology. This can have a profound effect on the design team and result in a severe dislocation to the design process. Such a decision to change tool is never made lightly but, nonetheless, occurs from time to time and may take many months to take effect. Even then, the need to support existing products may require legacy systems to be maintained for decades after they cease to be used in new design work.

In the traditional approach, detailed design revolves around the drafting process, albeit one that uses an electronic description of the product. It is, additionally, quite usual for the CAD system description and the analysis models used in preliminary design to be quite separate from each other. Even in detailed design, these two worlds often continue to run in parallel. So, when any fresh analysis is required, the component geometries must be exported from the CAD system into the analysis codes for study. This conversion process, which commonly makes use of standards such as STEP (STEP 1994) or IGES (IGES 1996), is often very far from straightforward. Moreover, even though such standards are continuously being updated, it is almost inevitable that they will never be capable of reflecting all of the complexity in the most modern CAD systems, since these are continually evolving themselves.

The effort required to convert full geometries into descriptions capable of being analyzed by CFD or FEA codes is often so great that such analyses are carried out less often than otherwise might be desirable. Current developments are increasing the ability of knowledge-based systems to control CAD engines so that any redrafting can be carried out automatically. Nonetheless, the analyst is often faced with the choice of using either an analysis discretization level that is far too small for preference, or of manually stripping out a great deal of fine detail from a CAD model to enable a coarser mesh to be used. The farther that the design is into the detailed design process, the greater this problem becomes. In many cases, it leads to a parallel analysis geometry being maintained alongside the CAD geometry.

To appreciate how restricted the scope for computational engineering becomes in detailed design, one must only consider the decision-making processes of a typical detailed designer. Such a designer is commonly considering a single component or feature in the design and is usually driven by a desire to reduce cost or weight or both, alongside a pressing workload of other features or components that must be considered in due course. The designer will have a considerable list of constraints or test cases that this particular aspect of the design must be engineered to deal with. Knock-on effects on other components being detailed by other designers will also need considering. In such circumstances, he or she must examine the CAD representation and decide where changes can be made. Moreover, consideration must also be given to the time delay that is inherent in any use of computational methods. If it is a simple matter to run a further FEA job, this will be done. However, if the decision to use analysis leads to a complex data transfer process, followed by a reduction in unnecessary detail and then a manual meshing task before a time consuming analysis can even begin, it is unlikely that designers will engage in this process unless absolutely necessary – the benefits to be gained are simply outweighed by the effort required. Instead, judgment based on experience and a certain degree of conservatism are more likely to be used.

These concerns and the increasing freezing of the design during detailed design mean that analysis tends to have a smaller and smaller ability to significantly change the design as it continues to be refined. Therefore, computational methods currently have a rather smaller impact on the detailed design process than might be imagined. It may be concluded that computational methods are always destined to have their greatest impact during the preliminary design phase. This need not be the case: if the transfer of data between CAD systems and analysis programs were routine and straightforward, such analysis would be more heavily used. If, in addition, it were possible to rapidly change the geometry being considered without a major redrafting effort, no doubt designers would be more prepared to undertake such modifications. Finally, it would clearly be useful if advice systems could be made available to designers that suggest possible alternatives that might not otherwise be considered. All these aspects are currently being investigated by those researching the design process.

1.4.5   In-service Design and Decommissioning

When a product reaches manufacture, the design process does not stop. Very few products prove to be completely free of design faults when first produced, and equally, it is common for operators to seek enhancements to designs throughout the life of a product. The lifetime of current aerospace systems can be extremely long, sometimes as much as 50 or even 60 years, and so this phase of design is commonly the longest, even if it is carried through with a relatively small team of engineers. Even when a product reaches the end of its life, it must be de-commissioned and any waste disposed of or recycled. This process itself requires the input of designers, who may need to design specialist facilities to deal with waste or with stripping down equipment.

With the rapid advances occurring in design, manufacturing and computational technology, a major issue at this stage of design is legacy systems, whether these are old Fortran codes used in analysis, superseded CAD systems or obsolete manufacturing plant. For example, it is no longer possible to source many of the transistor-based electronic components that were common in avionics systems in the 1970s. Thus, if systems fail, the design team may need to change the design to adopt more modern equipment and then study any likely knock-on effects. If the drawings for an airframe were produced in an early release of a CAD tool that is no longer supported, it may even be necessary to resort to printed drawings of the design. Enabling modern design tools to support the curation of designs over such long time frames is a problem currently causing much concern in aerospace companies. In consequence, when considering computational methods in design, one must always consider how any products produced with them will be supported over extended periods with evolving standards, and so on.

1.4.6   Human Aspects of Design Teams

Although human factors and team working aspects of design lie outside the scope of this book, one must continually remember that design is a human activity carried out by ordinary people who have career aspirations, are defensive over their acquired stores of knowledge and who must be motivated if they are to give off their best. When considering the various stages of the design process, one must therefore not lose sight of the design teams engaged to carry forward the work.

A great deal of effort has been expended on computational aids over the years that have not realized the benefits that were hoped for because they were often designed without paying due regard to the needs and desires of the design teams that were ultimately to use them. The introduction of IPTs has gone some way to address these concerns. Even so, it is worth noting that the introduction of project teams has been cited (Pugh 2001) as one of the difficulties that made Rolls-Royce's development of the RB-211 harder than it might have been: not that such teams were harmful, but rather that the change in working practices occurred at a time of great stress in the company and its ramifications took some time to settle down. Introducing changes to working practices inevitably leads to knock-on consequences that should be allowed for when they are introduced. It must be realized that there is a great temptation for new software systems to be designed by computer-science graduates without regard to the needs of engineering designers. Design is a very interactive and team-based activity that requires a good deal of discussion between those involved on the project, the experts supporting them, component suppliers and even the customer.

At the time of writing, the introduction of product data models (PDM) and product life-cycle models (PLM) is an issue affecting design teams. Such models are intended to benefit the companies using them over the medium to long term. However, they require input from designers, who will get little short-term gain from their introduction. Designers need a good deal of sensitive management input to get new PDM and PLM systems to work well and this is often not explicitly allowed for. This is typical of much modern technology; it is introduced with the best of intentions but often without regard to the impact it will have on those using the tools when they hit the design office.

1.5   Advances in Modeling Techniques: Computational Engineering

1.5.1   Partial Differential Equations (PDEs)

Perhaps the greatest impact that computers have so far had in the aerospace sector stems from their use in modeling the behavior of engineering systems. Commonly, such models are based on partial differential equations (PDEs). Although computers are used for many other functions in the sector, ranging from automating corporate payrolls through to the driving of numerically controlled machining facilities, it is the ability to predict performance before physical systems are built, by solving systems of PDEs, that has been one of the key factors in allowing designers to make such rapid advances since the 1960s.

Of course, calculation has played a role in aerospace design from the earliest times, but before the advent of computing, such calculations were inevitably limited in scope and had to resort to approximations, curve fits, scaling laws and the like. The ability to tackle the fundamental equations governing the behavior of solids, fluids, thermal systems, and so on, all require fast and large computers that can deal with the underlying PDEs that govern these problems. In almost all cases, this leads to the need to consider equations applying over the large and complex domains that represent the products being designed. Such approaches are now very sophisticated and in some areas sufficiently mature for airworthiness certification to be based solely on computational modeling. There are other areas, however, such as the prediction of unsteady fluid flows using the Navier–Stokes equations or nonlinear impact studies, where accurate calculations still cannot be made in all cases and progress is limited by computing power. In the following sections, some of the advances being made in computational engineering are briefly assessed to illustrate progress in analysis capability.

1.5.2   Hardware versus Software

It is, of course, commonly observed that computers are getting ever faster. What is less often remarked, however, is that the gains that have been made in the power and speed of software are of equal importance (Gallopoulos et al. 1992). Significant reductions in solution time have been gained for structural mechanics problems due to advances in preconditioned Krylov methods. Similar advances have been made in CFD using domain decomposition and Newton–Krylov methods. As Jameson (2001) observes, improvements in algorithm design alone resulted in a two order of magnitude improvement in computational speed for Euler-based fluid dynamics calculations between 1980 and 1990. Consider the fundamental matrix operations that lie at the heart of a linear finite element solver. Such solvers essentially have to calculate the solution to a large set of linear equations of the form $\mathbf{Ax} = \mathbf{B}$. Fifty years ago, such an operation would probably have been carried out by Gaussian elimination. Today, banded solvers take advantage of the structure commonly found in such problems to solve these equations in a fraction of the time. Such advances are found in all corners of computational engineering – the Lanczos solver for eigenproblems is another example, as is multigrid acceleration in CFD.

Nonetheless, developments in computer hardware have also had a profound effect on the practice of engineering design. The ability to handle large matrices in core memory; large, fast and cheap disk arrays; increased bandwidth between distributed design offices; parallel processing and, above all, the dramatic increases in processor speed have led to a major increase in the scale of problems that can be tackled by analysis and the regularity with which this can be done. For example, in around 1950, the Swedish Royal Institute took delivery of its first "Binary Electronic Sequence Calculator", a valve-based device, which allowed for up to 4,096 instructions or numbers (the user's choice) with 80 bits each, while today, gigabytes of memory are quite commonplace. By 1970, the Control data 6600 had a speed of about one megaflop while by 1990 the eight-processor Cray YMP achieved around one gigaflop – a three order of magnitude increase in speed in two decades. At the time of writing, gigaflop speeds are available on powerful personal computers.

Unfortunately, future increases in processor speed look to be approaching the physical limits of devices made in silicon – unless there are breakthroughs in device technology, this will probably mean that increased effort will be needed in making use of extremely large arrays of parallel machines.

Advanced parallel machines will need to be built from tens of thousands of cheap individual processors if desirable improvements in processing capacity are to be realized. Many aerospace companies already have computing networks with such large numbers of machines connected to them. Their use in harness together to solve engineering design problems is forcing even more effort to be directed toward advanced software methods. Such computing "grids" need a whole new class of software systems to enable their management and exploitation by engineers and these environments are actively being developed at the time of writing (Berman et al. 2003; Foster and Kesselman 1999; Foster et al. 2002, 2001).

Skills in distributed computing systems are becoming increasingly important to the prosecution of engineering analysis and add yet a further set of capabilities to those that design teams will need to encompass. This is in addition to a rather thorough understanding of the kinds of techniques being used in the tools that actually solve the PDEs of interest. A brief examination of two of the most important areas of computational modeling reveals just how complex the latest generation of solvers has now become.

1.5.3 Computational Solid Mechanics (CSM)

Computational Solid Mechanics is concerned with the analysis of structures subjected to loads. Such analyses depend on the geometry of the object being studied, the nature of the materials involved, the forces applied and the aims of the computational study. It will be clear that if the object under study has a complex shape with many features, this will necessarily lead to a large and complex computational model. Similarly, if it is made of materials with highly nonlinear properties, this will also lead to increased complexities. Sophisticated loading cases will further increase the difficulties of analysis as will any desires to carry out optimization or trade-off studies. When trying to design a high-quality aerospace component that has highly curved surfaces, is made of advanced, possibly composite materials, and is subject to the many loads arising in flight, it is clear that the problems to be dealt with can be formidable.

Nowadays, FEA is the fundamental computational technique used in Computational Solid Mechanics (CSM), although FEA can of course be used in other disciplines, and other methods are used in CSM from time to time. FEA may be used for static strength analysis in both linear and nonlinear forms, for structural dynamics, and for input into aeroelasticity and aeroservoelasticity studies. In all cases, the structure to be analyzed is represented by an array of elements that model the geometry being considered. Forces and boundary conditions are applied to the structure and solutions sought by the application of various matrix methods. The key idea behind FEA is to approximate the deflections within an element in terms of those at nodal points (the so-called Degrees of Freedom in the model) located at the corners, edges and sometimes within the interior of the elements. These approximations are known as shape functions and are commonly polynomial in form. The nodal displacements can then be linked to the applied forces using the fundamental equations governing the structure (such as those of linear elasticity). Then, when a complex geometry is built out of simpler elements, algebraic equations may be set up for the whole problem that relate all the (unknown) nodal displacements to the externally applied forces and any boundary conditions (Reddy 1993; Zienkiewicz 1977).

For some linear problems such as those involving simple beam structures, the shape functions in use may exactly solve the governing equations and then exact solutions can be obtained provided that loads and constraints are correctly modeled. Even when this is not possible, as is normally the case, if sufficiently small elements are used and the forces and constraints are modeled in fine detail, very good approximations can be obtained. The refinement of FEA models can also be achieved by adopting more complex shape functions within individual elements. Both approaches can of course be combined and it is also possible to use solvers that automatically refine the mesh or element complexity in areas of high curvature/stress gradient, and so on, to directly control error bounds in the solution.

The whole FEA process transforms complex differential equation problems in structural analysis into algebraic ones arising from forming mesh geometries. These may readily be tackled given sufficient computing power to solve the resulting matrix equations. For buckling and linear vibration analysis, the FEA leads to an eigenvalue analysis and these, although more difficult to deal with than simple linear problems, can now routinely be solved with millions of degrees of freedom (DoF). For nonlinear problems such as those involving contact, friction and large-scale deformation, FEA can still be applied, usually by adopting time stepping or iterative approaches (Plimpton et al. 1996).

A great deal of the fundamental speed advances that have taken place in structural applications of FEA are concerned with the rapid manipulation and solution of the matrix equations formed. This has allowed dramatically increased numbers of elements to be used routinely in design. At the same time, more complex element types have been introduced that use higher-order approximations for shape functions or which allow for geometrical complexities directly within the elements themselves. Such advances have further increased the scope and accuracy of structural studies.

The types of solvers involved can be broadly categorized by the specific equations being solved. The simplest of these are the equations of structural statics, where stiffness is linearly related to deflection. The fundamental advance that has taken place in this area of FEA is in the size of the matrices that can be dealt with in a routine fashion. For example, Waanders et al. (2001) show that the number of DoF tackled in electronic package design has increased by a factor of 4,000 in the last 15 to 20 years. In the 1980s, such studies might well have addressed several thousands of DoF. It is now commonplace to deal with systems with many millions of DoF. This has allowed problems to move from two dimensions into three and to have increasingly fine resolution so that stresses can be resolved in great detail.

At the next level of difficulty are studies that involve buckling or vibrations. In either case, an eigenvalue analysis is usually required. Although more difficult than linear analysis, similar increases in model size have taken place. In 1989, Keane et al. (1991) carried out a study of a marine vehicle where a full three-dimensional finite element calculation was used to predict the vibrational behavior of the vessel. To do this, second-order differential equations are solved either to find the natural frequencies and modes of the structure (which can then be used to predict forced responses) or to predict the forced responses at given frequencies directly. The model adopted consisted of approximately 4,700 elements and 34,000 DoF: it required the main computational facilities of the university involved for some three weeks to derive the required modal values. Today, the same calculation can be carried out on a workstation in an hour or two. Apart from the obvious increases in processor speed and the size of core memory available, a significant difference between these calculations is that when first reported the modes were derived using simultaneous vector iteration whereas now a Lanczos solver would be used. This difference alone speeds up the calculation of the first ten modes by an order of magnitude.

Another change in the area of CSM has been the significant increase in the complexity of nonlinear problems that can be studied. In particular, it is now routine to deal with problems involving contact and friction, large deformation and nonlinear materials. In all cases, this leads to nonlinear systems of equations that are usually solved in an iterative fashion (Thole et al. 1997). When designing a modern gas turbine, it is commonplace to examine the contact effects between components that may suffer from fretting fatigue, the impacts on fan blades and the use of fiber-reinforced materials with carefully aligned ply directions to make components. For example, the analysis of bird strike and blade loss in a running gas turbine is currently one of the largest single calculations carried out in the analysis of new engine designs. It must allow for impact between the elements of the engine, large-scale deformations of components, vibrations and element failures. At the time of writing, such calculations typically take several days to carry out using large, multiprocessor computers. Figure 1.7 shows one-time step in the solution to such a calculation.

FEA-based CSM studies are now also used as key building blocks in the area of aeroelasticity, where structural responses to fluid loading and the consequent changes to the flow field are studied. This enables problems like flutter to be studied computationally. Engineers are also beginning to tackle the even more complex problems that arise when actively controlled aerodynamic surfaces are modeled, leading to the field of aeroservoelasticity. In

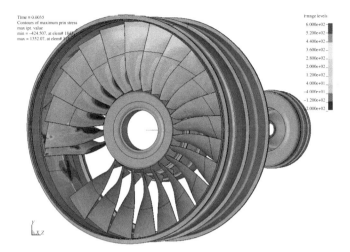

Figure 1.7   Results from a dyna3d calculation on Rolls-Royce engine during bird strike (courtesy Rolls-Royce plc).

both these areas, there has been a significant increase in the size of problems that may be solved on an everyday basis (Borglund and Kroo 2002; Done 1996; Marshall and Imregun 1996).

1.5.4   Computational Fluid Dynamics (CFD)

Computational Fluid Dynamics (CFD) is a fundamentally more difficult area for analysis than much of CSM, because the equations that govern fluid flow problems are so fundamentally nonlinear in nature. In aerospace applications, the fluid problems that must be addressed are usually described by the (Reynolds Averaged) Navier–Stokes equations: these allow for both the compressibility of the fluid and its viscosity. Compressible flows can lead to shocks where the flow jumps between supersonic and subsonic conditions, while viscosity can give rise to thin but important boundary layers and also to separation.

The full Navier–Stokes equations are far too difficult to deal with in most real aerospace design problems because to solve them in three dimensions it would be necessary to use computational grids with the order of $Re^{9/4}$ cells: the Reynolds numbers (Re) typical of aircraft operation are in the millions and so the meshes needed in design would be beyond the capabilities of currently foreseeable computing systems (i.e., 10^{16} cells per mesh). It is therefore normal to simplify the equations in a number of ways.

The most common approach to simplifying the Navier–Stokes equations is based on the so-called Reynolds Averaged (RANS) equations, which still allow for compressibility and viscosity, but use turbulence models to estimate the effect of fluctuating components. If the flow is considered to be inviscid but still compressible, the Euler equations result. Because such equations are slightly simpler in nature and also because the resulting flows do not exhibit boundary layers (and can thus use much coarser grids), the solution of the Euler equations is normally much simpler than those for the RANS equations (they often also converge much faster). This can sometimes lead to an order of magnitude saving in computational effort. Moreover, corrections can be made to such calculations in the form of so-called viscous-coupled models to approximate for the effects of viscosity.

If, in addition to viscosity, rotation in the flow is not an issue, it is possible to model the flow using a full nonlinear potential method and, if also incompressible, a linearized potential method. Potential calculations can offer a further order of magnitude saving in run times. At the time of writing, almost all the major aircraft in production have been designed using no more than viscous-coupled Euler-based flow simulations with limited use of RANS simulations. This is mainly because of mesh preparation difficulties rather than because of solver run times.

The nature of the RANS equations leads to the need for complex domain discretization schemes as well as complex modeling with large numbers of elements or cells. Even with Reynolds Averaging, such schemes cannot usually afford to adopt a uniform resolution mesh throughout the domain of interest if they are to have a high enough mesh density to deal with the features of interest. Therefore, meshes must usually be adapted to provide fine resolution in areas where boundary layers and shocks are likely to occur and a coarser grid elsewhere. This often leads to complex mesh structures on which the equations must be solved, and building such meshes can be a very time consuming business. It is therefore perhaps not surprising that the use of CFD in design is at a relatively less mature stage than structural FEA studies.

There are three main approaches to solving the RANS equations numerically: these are via finite differences, finite volumes or finite elements. Although much effort has been expended on all three approaches, nowadays it is almost universally the case that solution techniques are based on the finite volume approach. At the same time, there are three fundamentally different ways of constructing the meshes used for such calculations. These are based on regular Cartesian grids, regular body-fitted grids, or unstructured body-fitted grids. Of these, unstructured body-fitted grids are much the easiest to construct (usually being based on unstructured tetrahedral meshes). However, they are also usually the least efficient computationally. Conversely, regular body-fitted grids can be a very efficient and accurate way of describing the fluid domain. They are, however, the most difficult to construct, especially when dealing with complex geometric entities such as those commonly found in aerospace design. Since computing power is increasing more rapidly than improvements in automated meshing techniques, unstructured approaches are increasingly becoming the design tool of choice. At the same time, there is currently an expanding interest in so-called meshless methods (Belytschko et al. 1996). These work without using elements by adopting approximating functions that span over the whole, or large parts of, the domain and thus need only clouds of points to discretize the problem. Whether these prove to be an effective substitute for unstructured meshes remains to be seen, but such efforts are indicative of the ongoing developments in software methods.

To indicate the sort of complexity involved in using CFD, it is useful to consider likely mesh sizes and run times. Jameson (2001) notes that "*accurate three-dimensional inviscid calculations can be performed for a wing on a mesh, say with* $192 \times 32 \times 48 = 294,912 \, cells$". Such calculations can now be carried out in an hour or so on desktop-based workstations. To solve the same problem at high Reynolds number with a RANS solver would require of the order of eight million cells. Even then, Jameson considers it doubtful if a universally valid turbulence model could be devised. Therefore, even given suitable meshes, RANS methods must be carefully used to yield more accurate results than well-calibrated viscous-coupled Euler solvers. Nonetheless, the world's major aerospace companies are currently all moving to RANS solvers for design work. This is particularly true for aero-engine designers, because it is almost impossible to accurately predict the boundary layer–dominated flows over the fans of modern engines without using RANS methods.

Noise problems are also being tackled with various CFD codes, and these can be even more demanding to deal with than those involving body forces because the quantities being dealt with are so sensitive. When assessing the behavior of the flow over a wing, the lift forces generated are very significant. The drag forces are typically an order of magnitude smaller but are still quite large: most of the propulsive effort is, of course, used to overcome this drag. By contrast, the energy dissipated as noise is tiny and also often at high frequency. Constructing models to accurately resolve such terms is extremely challenging.

Despite all these difficulties, improvements in CFD solvers still continue to emerge. In particular, much effort continues to be invested in improvements in solver speeds by the development of more advanced software techniques. Most notable of these has been the rise of multigrid solvers. When solving CFD problems, iterative techniques are needed because of the nonlinear nature of the fundamental equations of fluid flow. The convergence of such iterative methods is fundamentally linked to the number of points in the meshes used to discretize the flow. Given that fine meshes are needed to resolve many of the features of interest, this is a fundamental problem in solver design. Multigrid methods tackle this problem by working at various levels of discretization as the solution proceeds, since convergence speeds on fine meshes can be improved by taking results from coarser meshes that, though less accurate, stabilize more quickly. The benefits achieved are particularly great in diffusion-dominated flows and thus yield most benefits in RANS codes, being typically more than one order of magnitude faster than nonmultigrid approaches. In a recent survey of opportunities in design computation by Alexandrov et al. (June 2002), comparison is made with so-called textbook multigrid efficiency solvers and it is noted that such solvers can be up to two orders of magnitude faster than those currently used in commercial CFD codes. The existence of such solvers, albeit currently only applicable to simplistic problems, holds out the prospect of further speed-ups in the ability of designers to use CFD.

Finally, before leaving CFD, mention should be made of so-called adjoint codes. These are codes that are set up to calculate the sensitivities of the solution to changes in input parameters. They have been studied in the domain of CFD because of the desire to have gradient information available for slope-based hill-climbing searches. A full discussion of such methods is left until Part II; at this stage, it is simply noted that such formulations can lead to methods for calculating all the gradients in a problem at much less cost and with greater accuracy than would be achieved by using standard finite differencing approaches (Reuther et al. 1999a).

1.5.5 Multilevel Approaches or 'Zoom' Analysis

Given the need for information on some aspect of a new design's performance, and before reaching for the tools of CSM or CFD, the designer should first decide on the most appropriate level of modeling – will curve fits to the performance of previous designs be sufficient or must a full Navier–Stokes solve be attempted? Can empirical stress formulae be adopted or must a full nonlinear FEA study be made? Often, the designer has no choice in such matters, either because there is insufficient data to feed into the more elaborate methods or because there is insufficient time to use them. Increasingly, however, automated systems are being brought on line that allow such options to be considered and even to allow multiple levels of fidelity to be used simultaneously. Even within a single CFD or CSM code, it may be possible to adopt various levels of mesh discretization. This leads to the concept of multilevel or 'zoom' analysis where the degree of fidelity is a variable open to the designer's manipulation – such a choice becomes part of the design process setup and needs managing along with all the other issues facing the designer.

Variable fidelity modeling also opens the possibility of data fusion, whereby information coming from various sources may be blended together to either improve the reliability of analysis predictions or to improve the speed with which they can be made (or both) (Keane 2003). For example, it may be possible to use a complex RANS CFD code to calibrate a much simpler panel code so that initial design decisions can be made using the simpler code before switching back to the full code, rather than using the expensive calculation at every step. Moreover, rapid increases in the size of on-line data storage are making the extensive reuse of previously computed results sets more and more feasible.

1.5.6 Complexity

Given the wide set of computational methods used in engineering and the range of problems being tackled, it is useful to try and classify calculations by their complexity. Venkataraman and Haftka (2002) use three axes to make this classification. These are model complexity, analysis complexity and optimization complexity. The first two of these can readily be described given the preceding sections. Model complexity deals with the degree of realism with which the problem being studied is described. So, for example, in finite element studies, the number of DoF and the choice of one-dimensional, two-dimensional or three-dimensional elements all affect the model complexity. For a given level of model complexity, a designer may also choose the degree of analytical complexity used in the study. At its simplest in structural mechanics, this will be a linear elastic analysis. Following this would come buckling and vibration analysis, both of which lead to eigenvalue problems. Beyond this lie large deformation calculations, friction and contact studies and stochastic or reliability based approaches. Similarly, in fluid mechanics, the simplest approaches may be one or two dimensional and usually adopt linear potential or perhaps full potential codes. Next come Euler methods, Reynolds Averaged Navier–Stokes solvers and large eddy simulations, all usually applied to three-dimensional meshes, often with very fine levels of modeling in the fluid boundary layers or near shock waves. These variations provide a zoom capability and, as has already been noted, variations in model and analysis complexity often lie outside the designer's choice but sometimes can be exploited to speed up the design process – much depends on the design support aids available.

Less obvious is the optimization axis introduced by Venkataraman and Haftka (2002). This essentially deals with the design process being invoked, i.e., the subject of this book. At its simplest, a designer may wish to carry out a single analysis to predict the performance of a system or structure. At the next level of complexity, it may be necessary to study a small number of competitive designs to assess which best meets current requirements. Beyond this, there may be a desire to carry out a local optimization over a number of variables against a single goal together with a few constraints. If wider-ranging searches are required, this may necessitate the use of global search methods. Also, the number of constraints may rise from two or three into the thousands. It is also often the case that multiple operating conditions must be considered to identify those that will dictate key decisions. Further, the designer may wish to carry out a multiobjective optimization study where competing goals must be balanced against each other and a so-called Pareto front constructed (Fonseca and Fleming 1995). Beyond this, the design team may be required to address various subject areas or disciplines simultaneously, leading to multidisciplinary optimization.

Venkataraman and Haftka (2002) observe that, as has already been noted, over time the ability of computing systems to support studies of increasing complexity in all three directions has grown enormously as computing power has increased. They suggest that the combined effect of increasing processor speed, parallel processing and more sophisticated solver techniques over the last 40 years has led to a four-million-fold increase in the degree of problem complexity that can now be tackled. This is in line with the figures quoted in the previous sections. Despite all this, the desires and requirements of engineers have kept pace with this growth, so that what is considered an adequate study continually changes to reflect available resources – often the definition of reasonableness is taken to mean a calculation that can be performed on local computing resources overnight. It is possible to conclude from this that it is unlikely ever to be possible to carry out studies that are of maximal sophistication in all the three aspects of complexity. Thus, in practical design work, the engineer will need to make sensible choices on the models and methods being used, rather than simply opting for the most complex approach currently available. He or she will then probably have to contend with criticism from specialists that the work carried out is deficient in terms of what could be achieved in at least one of the axes of complexity.

It should not be forgotten, however, that one of the consequences of all these developments is that problems need to be tackled in an appropriate fashion to exploit the gains that can be made. If finite element meshes are not set up correctly to exploit banded solvers, then much of the speed gains possible will not be realized. If the division of load across parallel computing facilities is not organized to minimize data flows between processors, again speed advantages will not be gained. If the designer's computing Grid is not well configured and managed, progress will be hampered. Thus, it is true to say that, although the hardware and software available to support computational engineering has improved radically, the changes that have taken place have also made increased demands on the knowledge base that designers must possess to take advantage of the latest tools. This has led to the emerging discipline of Computational Engineering, which brings together traditional engineering subjects and knowledge of the computational facilities needed to solve real problems. This new field spans topics as diverse as software integration, data handling, systems management, parallel computing, advanced solver techniques and visualization systems alongside the core engineering disciplines such as CSM and CFD.

1.6 Trade-offs in Aerospace System Design

1.6.1 Balanced Designs

Even assuming that the design team has mastery of all the relevant tools of computational engineering, good design still requires a balance between a number of competing aspects: to do this, the designers must trade-off desirable characteristics against each other. In aerospace applications, these commonly involve cost, weight and performance – aspects that mostly pull in opposite directions, since lightweight, high-performance designs are typically costly ones, while cheap heavy ones usually have poor performance. Issues such as manufacturability, supportability, environmental impact and ultimate disposal of products must also be considered. Moreover, since it is increasingly common for the company that designs and makes an aerospace product to be asked to retain ownership and merely lease it to the operator, such issues are steadily moving further up the designer's agenda–the simple pursuit of improved nominal performance at any cost is no longer commercially viable, even in military or space applications.

As has already been noted, the designer rarely has the luxury of being able to call on fully detailed analysis capabilities to study the options in all areas before taking important design decisions. Moreover, given the iterative nature of the design spiral, thought must also be given to when and how such comparisons and trade-offs are made. If trade-offs are made too early, opportunities may be missed that have not been revealed by simple models and rapid studies; if they are made too late, the effort required to radically change a design may be too great given the resources already expended and the remaining time available. Moreover, the designer must decide if individual disciplines can be considered independently, either serially or in parallel, or if an integrated approach must be taken using tightly coupled models. Some of the issues involved in trade-off studies can be illustrated by a few very simple examples.

1.6.2 Structural Strength versus Weight

In an early paper on design search, Nowacki (1980) considered the design of a tip-loaded encastre cantilever beam: the aim was to design a minimum weight beam carrying a load, F, of 5 kN at a fixed span, l, of 1.5 m; see Figure 1.8. The beam was taken to be rectangular in section, breadth b, height h, cross-sectional area A, and subject to the following design criteria:

1. a maximum tip deflection, $\delta = Fl^3/3EI_Y$, of 5 mm, where $I_Y = bh^3/12$;

2. a maximum allowable direct (bending) stress, $\sigma_B = 6Fl/bh^2$, equal to the yield stress of the material, σ_Y;

3. a maximum allowable shear stress, $\tau = 3F/2bh$, equal to one half the yield stress of the material;

4. a maximum height to breadth ratio, h/b, for the cross section of 10 and

5. the failure force for twist buckling, $F_{CRIT} = 4/l^2 \sqrt{GI_T EI_Z/(1 - v^2)}$, to be greater than the tip force multiplied by a safety factor, f of two, where I_T is $(hb^3 + bh^3)/12$ and I_Z is $hb^3/12$.

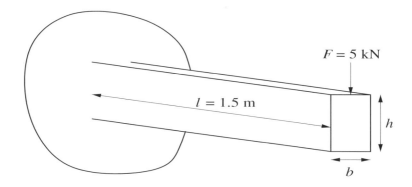

Figure 1.8 Encastre cantilever beam.

Table 1.1 Optimum design achieved by search using APPROX (Siddall 1982), requiring 32 iterations

Lower Limit	Value	Upper Limit	Item	Units
5	15.8	50	Breadth, b	mm
2	15.8	25	Height, h	mm
	114.1	200	Direct stress, σ_B	MPa
	3.0	100	Shear stress, τ	MPa
	5.0	5	Tip deflection, δ	mm
	10.0	10	Height-to-breadth ratio, h/b	–
10	132		Buckling force, F_{CRIT}	kN
	2,496		Cross-sectional area, A	mm^2

The material used was mild steel with a yield stress of 240 MPa, Young modulus, E, of 216.62 GPa, bulk modulus, G, of 86.65 GPa, and Poisson ratio, ν, of 0.27. Here, the variables defining the design are, of course, the cross section breadth and height and the aim is to minimize their product. Notice that it is not clear from the above specification which of the design limits will control the design, although obviously at least one will, if the beam is to have a nonzero cross-sectional area. Table 1.1 gives the details of the optimum design for this case, which may readily be located with a simple down-hill search in less than 30 steps.

The table shows that in this case two of the constraints are active at the optimum: that on tip deflection and that on breadth to depth ratio. Figure 1.9 illustrates the design space, showing the variation of cross-sectional area with the two variables (a simple hyperbolic surface) and the active constraints. Note that by convention here the constraints are shown in staircase fashion to indicate the density of data in the plot – the map is produced by sampling the problem on a rectangular grid of data points and only if all four corners of a sample area are feasible is the enclosed area considered feasible and contoured on the plot – this avoids errors in plotting/interpretation.

This beam problem can be reformulated, of course, by allowing any one of the constraints to become the objective and the previous objective a constraint with a suitable limit. For example, the goal may be set as minimizing tip deflection for a given maximum cross-sectional area. If the area is set as being no more than 2,500 mm^2, a slightly different optimum result arises; see Figure 1.10. Not surprisingly, the area limit is active in this plot, and now the bending stress constraint is also visible, since the deflection constraint has been removed, which was masking it before.

This simple example illustrates the obvious point that there is a trade-off that can be made between deflection and cross-sectional area: the designer must choose which is a constraint and which is a goal. If he or she is not certain about which is more important, then the best that can be done is to construct a Pareto front (Fonseca and Fleming 1995), whereby a set of solutions with low deflection and low cross-sectional area are produced; see Figure 1.11. Here, these have been found using an evolution strategy optimizer (Knowles and Corne 1999) that has been set the goal of producing a front with evenly spaced coverage in the design space between the limits of 5 mm deflection and 3,000 mm^2 cross-sectional area.

The designs that make up the Pareto front are said to be "nondominated" in that, if one starts from any design point on the front, an improvement in one goal can only be achieved by degrading the other, that is, each point in the figure represents an optimal design for some weighting between deflection and area. The figure shows the way that deflection may

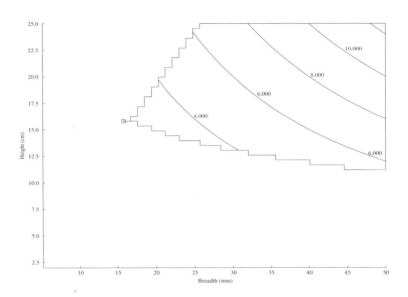

Figure 1.9 Contour map of beam cross-sectional area versus breadth and height, showing constraint boundaries and optimum design (square marker).

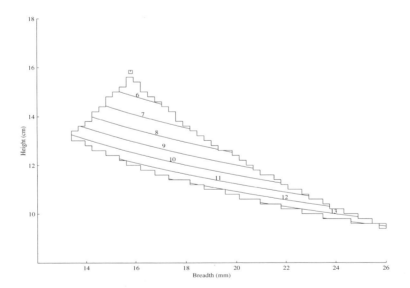

Figure 1.10 Contour map of beam tip deflection versus breadth and height, showing constraint boundaries and optimum design (square marker).

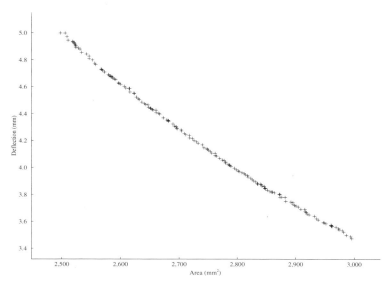

Figure 1.11 Pareto front showing the trade-off between minimum deflection and minimum cross-sectional area while meeting all other constraints.

be traded off against area while satisfying all the other design criteria. Notice that the previous two designs/figures lie at the high deflection end of this front and that it is clear that deflection may be significantly reduced if the cross section is allowed to become greater (as would be expected). What are not shown on the Pareto front are the design variables (breadth and height) for each design point – these must be extracted from the Pareto set when the design trade-off is chosen.

At the time of writing, few designers use this kind of formalism, probably because it is unfamiliar to many and also because constructing the Pareto front can be very time consuming in more realistic cases. Nonetheless, it is clear that trade-offs exist in almost all engineering problems and that producing balanced designs is a key aim of design. Importantly, achieving this balance in a robust fashion requires decisions on how to formulate and solve a problem as well as on the relative importance of the multiple goals being dealt with. We return to the subject of multiobjective optimization later on in this introduction and in later parts of the book.

1.6.3 Aerodynamics versus Structural Strength

A more realistic, multidiscipline trade-off can be seen when designing aircraft wing structures. The trade-off between aerodynamic performance and structural efficiency is one of the most fundamental compromises that must be made in aircraft design. It can readily be seen when considering the thickness-to-chord ratio of the wing. It is easy to show that for lightly loaded wings at low subsonic speeds thin, high aspect ratio designs give the best aerodynamic performance – long slender wings are the norm on high performance gliders. However, wings are of course cantilever beams that are subject to distributed loads (weight and pressure). As such, long slender wings tend to have poor structural efficiency as the section modulus tends to be low and the bending moments high. High aspect ratio wings can usually only be justified if aerodynamic efficiency dominates all other concerns (as it does in competition gliders).

When designing a wing for general transport purposes, other issues come to the fore. A major parameter governing such designs is the total aircraft take-off weight, and of course, the wing structural weight can be a significant part of this total. For this type of aircraft, adopting a higher thickness-to-chord ratio, which degrades the lift and drag performance, may actually result in a better design because the total weight is reduced. In some cases, this trade-off may still not favor short wings while in other cases the reverse may be the case, depending on the exact mission profile. Figure 1.12 shows the variation of wing lift to drag ratio and take-off weight for a typical modern wide-body jet airliner as the wing aspect ratio is changed (Figure 1.13). Here, the total wing area remains constant as does the planform shape and section details. Due allowance is made for the effects of the structural efficiency on the wing weight and the amount of fuel needed for a fixed payload/range combination. As can be seen from the figure, the lightest aircraft has an aspect ratio of around 7.8, at which point the lift-to-drag ratio is still rising quickly. Further increases in aspect ratio improve aerodynamics and reduce the fuel load, but only at the cost of ever increasing structural weight with consequent penalties in total take-off weight.

Most designers are more familiar with results presented in this way than via the mechanism of Pareto fronts. However, if one also allows the planform shape to change during such a study (and this is commonly done), the trade-off between aerodynamics and structures becomes much more complex. In addition, the designers may well have multiple aerodynamic and structural goals – a simple structure may be desired to reduce manufacturing or service costs; several operational points may need to be considered during flight, so that

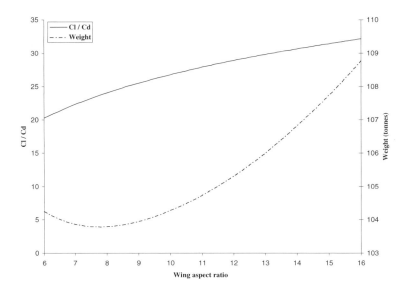

Figure 1.12 Variation of wing performance and aircraft weight with aspect ratio for a typical modern wide-body airliner.

Figure 1.13 Typical modern wide-body airliner wing geometry.

low drag at cruise speed is considered alongside high lift during takeoff, and so on. In such cases, the formalism of Pareto sets may become more helpful.

Another issue that must be considered is the accuracy of design calculations. Although the results given here come from a well-calibrated industrial strength concept design tool, the analysis does not involve any CFD calculations or use fully detailed structural models. Rather, it is based heavily on past practice and uses empirical expressions with calibrations coming from previous designs. One should therefore expect that the exact point at which take-off weight is a minimum would vary if more complex tools were employed – note that the variation in weight for aspect ratios between seven and nine is not very great. However, unless highly experienced, the designer cannot easily be sure as to how much in error this simple calculation might be. This leads naturally to considerations of robustness in the design process.

1.6.4 Structures versus Control

Another area where design trade-offs are very apparent is in the design of control systems for satellites. Many scientific satellite systems must address pointing accuracy and alignment issues in the face of flexible structures that are prone to distortion and vibrations. Problems arise because adverse vibrations can affect the performance of sensitive instruments such as multielement interferometric telescopes or synthetic aperture radars. Moreover, these telescopes or radars must be directed with great accuracy if they are to meet increasingly severe performance targets. In order to overcome such problems and to meet design targets, designers have resorted to the use of active direction and vibration control systems.

Of course, space-based control systems must necessarily have very low power consumption. Additionally, space structures must also be very light to minimize launch payloads. Unfortunately, lightweight structures tend to have poor structural rigidity and thus tend to be particularly prone to distortion and low frequency vibrational modes. To gain control of such structures can require significant control effort – again the best trade-offs are not obvious and a range of viable designs typically need examining.

This is a problem that has been studied quite intensively for at least 20 years (Balas 1982), and one of the key findings of work in this area is that not only do important trade-offs exist but also that the way they are tackled during the design process can significantly affect the outcome. The traditional approach to designing such systems has been to consider the structure and control system separately, that is, to design a lightweight structure with as much rigidity as possible within a given weight budget using finite element tools and then to use modal models of the structure to optimize the parameters of the control system. It turns out that a better approach is to try and optimize the structural and control system in an integrated fashion which enables the sizes of structural components to be fixed at the same time as the parameters of the control system.

Although it is perhaps obvious that following such a combined approach is bound to permit improved designs, because the integrated model is able to exploit the available design choices better, what is less obvious is the scale of improvements that can be obtained over the traditional approach. Major improvements are possible because the structural mass budget can be used more effectively giving mode shapes and frequencies that are better suited to the control system (Maghami et al. 1993). In that work, the first 14 flexible modes of the structure were used to characterize the low frequency structural dynamics, as predicted using traditional finite element methods with 3,216 DoF – a considerable model size at the time the work was undertaken. The integrated approach allowed an almost two-fold increase in controlled performance while reducing the satellite mass by 5%.

More recently, this kind of trade-off has been extended to radical geometric redesign and active control of much higher modes, again using a combined structural and control model, and once again significant improvements were obtained by adopting a fully integrated approach (Moshrefi-Torbati et al. 2003). It is clear that the interaction between satellite structural dynamics and their controls systems are so complex that any attempts to carry out design work where they are modeled separately and sequentially is bound to lead to suboptimal designs. This is a key point: a designer must be fully aware of any strong interactions between domains before deciding if they can be dealt with separately or alternately, if only a combined approach will suffice. Clearly, separate analysis is more convenient, but it may not be tenable in the face of strong interactions.

1.6.5   Robustness versus Nominal Performance

Finally, when carrying out analysis, consideration should be given to whether decisions are to be based solely on nominal performance or if instead allowance will be made for the likely variations that can occur because of inaccuracies in modeling, uncertainties in manufacturing tolerances and variations in operating conditions. In the previous examples, it has been assumed that the nominal performance of a design can be accurately predicted and decisions made on the basis of such predictions. In fact, real engineering systems almost never operate at nominal performance points. The manufacturing process is never wholly exact, so that finished products will always differ from the design specification to some degree. Operating conditions will never be identical to those used in calculations. Moreover, degradation in service will occur. A gas turbine blade will not be manufactured with complete precision; it will be operated in cross winds, rain and atmospheres with high dust content; it will suffer from leading edge erosion and fatigue. Engineers must obviously allow for such facts during design and produce products that are robust to such uncertainties. Figure 1.14 shows a turbine blade that has just finished its normal period of service life – the erosion to its leading edge and the build-up of debris show just how far from nominal conditions such components must be capable of operating in.

It is also commonly the case that the designs whose nominal performance has been heavily optimized, using inevitably approximate analysis models, can exhibit significantly degraded performance during operation, especially if the optimization process has been too narrowly focused on simple measures of nominal performance. It is therefore usually better to try and find a robust design point, even at the expense of poorer nominal performance.

Figure 1.14 A gas turbine blade at the end of its normal service life, illustrating erosion damage and debris buildup.

This is an area that has been studied from the earliest days of computer aided engineering (CAE) (Haugen 1968; Phadke 1989) and continues to be the subject of much active research.

The traditional way of dealing with uncertainty in design has been via the use of safety factors that allow margins to be built into products. These factors are ultimately an admission by the designer of a lack of complete understanding and are increasingly being subject to scrutiny. Better understanding leads to more efficient designs – nonetheless, it remains the case that some problems faced by the designer are probabilistic in nature, and so the production of robust designs is intimately involved with stochastic modeling. Such models are closely related to statistical approaches and so-called Design of Experiment (DoE) methods. The beginnings of this field stem from work in the 1920s and 1930s, when agricultural scientists developed statistical tools for planning crop trials that would yield maximum amounts of information from the limited numbers of plantings that could be carried out (Yates 1937). Such ideas have by now been applied to engineering experiments for many years, perhaps most famously by the Japanese engineer G. Taguchi (Taguchi and Wu 1980). More recently, these methods have been grouped together in the approach known as "design for six sigma" (Pyzdek 2000).

One way of dealing with uncertainty in design is to assess multiple operating points simultaneously when carrying out analysis – essentially averaging over likely operating conditions. Even this must be done with care, however – as many workers note (see for example, Huyse and Lewis (2001)), optimizers can sometimes simply then give designs that are sensitive at all the operating points used in the averaging process. One way of dealing with this problem is to adopt a Bayesian approach and allow the operating points to be treated as stochastic variables that change in a random fashion as the design proceeds. This approach can be no more expensive than using a fixed set of design points, although the design process must then tolerate repeated assessments carried out with different operating point sets yielding different results.

It is clear in all this that trade-offs lie at the center of design decision making and that the design team will therefore need tools that will help them carry out efficient and effective trade-off studies. This leads naturally to considerations of design automation, evolution and innovation.

1.7   Design Automation, Evolution and Innovation

To carry out high-quality trade-off studies, designers must synthesize and analyze alternative design configurations. To do this cost effectively and quickly requires tools that support automation, evolution and innovation. As has already been noted, the principal tool used in aerospace (and much other) design is the CAD workstation: drawings remain the fundamental language of engineering and are used to record and communicate design decisions. It is also noticeable that designers make use of relatively few tools other than CAD systems to help in this process of synthesis (in contrast to analysis via the various tools of CSM, CFD, etc.). The most common additional aid used by designers is probably the spreadsheet, which has replaced the calculator and before that the slide rule. The carefully compiled logbooks that were once to be seen on every engineer's desk are nowadays more likely to contain notes of meetings than detailed design calculations.

Although they may be the main tools of designers, it is by no means clear how much a typical CAD workstation, or access to a spreadsheet, helps support automation, evolution or innovation in design, or indeed how they support the winning of experience or its reuse. Nonetheless, the CAD tool has now almost completely replaced the drafting machine of earlier years – the photograph in Figure 1.15 seems to come from another age, but in fact represents typical design offices from the wartime period through to the 1960s. To understand the degree to which IT-based methods now support design, it is useful to very briefly rehearse the capabilities of current CAD systems and their development.

The earliest CAD systems began to be introduced in the 1970s. These systems worked in two dimensions, used specialist hardware and had little greater capability than a draftsman working with pencil and paper. Today, they are fully functioning three-dimensional solid modeling systems within photo-realistic visualization and sophisticated multiworker integration capabilities. Even though impressive, such systems still suffer from some limitations. For example, CAD systems are still not very good at allowing engineers to sketch, use free-form shapes or formulate rough ideas that will later be worked up into fully formed drawings or specifications – something that was readily possible using pencil and paper.

Another characteristic is that CAD systems typically work with screens that have viewing areas less than 600 mm across while an A0 drawing board is nearly 1,200 mm by 840 mm. This means that a considerably greater amount of information can be displayed at one time on an A0 drawing board than is now possible on a CAD screen.

Most CAD systems are also not very good at holding the collective experience of design teams – neither of course were A0 drawing boards, although the plan chests of old-fashioned design offices were much used resources. CAD systems are usually not well set up to capture the changing ideas considered by designers: rather, they tend to record the final outcome of design without the rationale behind it.

Figure 1.15 Engineering drafting section of the Glenn L. Martin Company captioned as the third floor, one of two floors devoted to engineering, with 627 people. Photo dated 12-7-1937. Lockheed Martin Corporation via National Air and Space Museum, Smithsonian Institution (SI 88-6627).

1.7.1 Innovation

The ability to sketch, to hold many ideas in front of the designer at once and to record and recall corporate wisdom can all be crucial to supporting innovative design. These activities are all highly knowledge dependent and it has been widely accepted for some time that to be useful in supporting innovation, systems must be able to record, organize, reuse and curate knowledge (Rouse 1986a,b). Design researchers have, for a number of years, envisaged systems that would assist engineers in being creative, by bringing appropriate information to the attention of designers when and where they needed it (Cody et al. 1995). These have at times seemed more like exercises in science fiction than realizable systems but efforts in computer-aided knowledge management are now beginning to bear fruit in many fields (Milton et al. 1999).

Using these evolving technologies, researchers are starting to produce tools that can directly support design innovation (Ahmed et al. 1999; Bracewell 2002) although, as yet, relatively few such ideas are in industrial use. Further progress in providing on-line knowledge services will no doubt increase the take-up of such systems and allow relatively inexperienced designers access to corporate stores of wisdom.

Another requirement for innovation in aerospace CAD systems is the ability to easily represent the products being designed. Often, these involve the highly complex curved surfaces that are needed to guide and exploit aerodynamic flows. Progress in this area has been quite significant: the earliest CAD systems could only represent simple cubic curves in 2-D; the ability to loft these into 3-D surfaces then followed. These early systems often had difficulties in intersecting such surfaces either with other cubic surfaces or even with simpler geometric objects such as cylinders. Modern systems can now normally represent objects using b-splines, Coons patches, nonuniform rational b-splines (NURBS) and Bezier curves, although manipulation of such objects is often not simple and problems can still occur. If a NURBS surface is built in one CAD system, intersected with a cubic spline–based surface and the result then exported via an exchange standard to another package, few experienced users of current CAD systems would be surprised if errors arose.

1.7.2   Evolution

In design, evolution is the natural counterpart to innovation. If a design problem has to be solved, the engineer has to choose between adapting an existing solution (evolution) or starting afresh (innovation). Usually, problems are best solved by a judicious mix of the two. While innovation is difficult to support with IT-based systems, it is perhaps in the area of evolution that CAD systems have made the greatest impact. It is of course a fundamental capability of all CAD systems to be able to erase elements from an existing model and to add new ones. This naturally encourages the reuse and adaptation of existing drawings to suit new problems, the collection and reuse of libraries of features and parts and the sharing of these collections among members of design teams.

The reuse and evolution of geometric models is something that fits quite naturally into formal optimization strategies. When carrying out design optimization, a search engine will be given a list of features in the design that may be changed and it then attempts to vary these systematically so as to improve the design. If a CAD system is coupled to an optimizer, it may then be possible to exploit its ability to evolve designs while at the same time seeking to optimize them. To do this, the CAD system needs to have a parametric modeling capability: such capabilities allow geometric descriptions to be manipulated by varying just a few key parameters. Most current-generation CAD systems have some capabilities in this area.

Of course, parametric models are more difficult to construct in the first place than simple drawings. It is also not obvious in many cases how best to parameterize a complex three-dimensional shape like a turbine blade or wing – but the representational issues raised can be critical to successful design improvement. If, however, the parametric model is planned with some care in the first place, it may be possible to use the CAD system to evolve many of the details and features of the product as the key parameters are altered (Song et al. 2002a). This can mean that instead of using the CAD system as a drawing tool it instead becomes a programming environment, where the engineer sets out the interrelationships between the various components and features of the design using some high-level symbolic language such as Lisp. Such changes can have important consequences for the skill sets needed by the design team.

It should be noted, however, that the ability to readily evolve and update existing designs can stifle innovation: a completely blank screen or drawing board can be a frightening prospect for all but the boldest designers. It is therefore important that the ready ability to adapt existing material is not used to drive out new ideas or ways of working. It is often too simple to say "we always do it that way in this company" without having to provide any more solid justification for decisions. Established design practices may not always retain the validity they had when first set out, probably in rather different circumstances.

1.7.3 Automation

Automation stems mainly from the desire to reduce the high costs associated with professional manpower and, at the same time, to reduce design cycle times. There are, as yet, relatively few ways of doing this with many design tasks, which are often, by their very nature, highly knowledge intensive. At the time of writing, significant effort is being made by many aerospace companies to build parametric CAD descriptions of their products so as to automate routine design tasks. These descriptions are beginning to contain more than just geometries and their interrelationships: rather they are starting to encompass design knowledge and practice. In this way, it is possible to embed significant sets of rules and guides within a parametric description so that these are automatically invoked whenever needed by changes in key dimensions (Rondeau et al. 1996). Such rule sets are currently no substitute for the decision-making capabilities of a trained design engineer – rather they are best at removing repetitive and mundane tasks in a prescriptive manner.

Knowledge engineering like this can certainly speed up the making of changes to product descriptions. If carried too far it may, however, lead to a significant part of a company's corporate expertise being embedded in systems that are not very visible to the engineers using them. This can result in a de-skilling of the design work force, which would then find itself unable to react when problems arise which the automated systems cannot cope with.

Automation has been tried in many fields of human endeavor, and it has often been found that attempts to replace humans are counterproductive. Automated systems should rather seek to work with the design team and appropriate divisions of activity found between human and machine, bearing in mind the capabilities of both. The updating of minor features in a design consequent on leading dimension changes can often be sensibly left to the computer, provided a clear set of rules on how to do this can be formulated. When this is not the case, the computer must act as an aid and not insert changes blindly (few users of word processors are happy to make global edits to documents, preferring instead to review changes one at a time, for example).

Developments in automation go hand in hand with the use of optimization systems in evolving better designs. In fact, it is often the case that one of the greatest benefits to be gained when setting up a formal optimization process is that the basic design sequence must first be automated. Even if optimization is then not used very often, or even at all, the time savings gained from automation can still be very significant (Rondeau and Soumil 1999). Of course, used well, optimization tools provide extremely powerful methods for suggesting design improvements and highlighting the effects of the trade-offs being made – they are thus central to the many integrated design environments currently being developed.

1.8 Design Search and Optimization (DSO)

Design search and optimization is the term used to describe the use of formal optimization methods in design. The use of optimization methods is steadily increasing in design, and here, we very briefly set out some of the more important aspects concerning these tools. We leave a detailed discussion of such methods to Chapter 3.

1.8.1 Beginnings

Optimization is a very old topic in mathematics, which goes back hundreds of years. For example, it has been known since Newton's time that calculus may be used to find the maximum or minimum values of simple functions. Even in the aerospace world, it has been a topic of interest almost since the beginnings of flight: a very early study on maximizing the height of rocket trajectories was described by Goddard (1919) just after the First World War. In that publication, the author calculates the required rocket nozzle velocities to maximize the altitude obtained while carrying a fixed payload. The behavior of the rocket was approximated over discrete intervals and the optimal nozzle velocity found by a trial and error search. In many ways, the process described is very similar to that found in many modern aerospace design offices, albeit that nowadays calculations are not made by hand or with tables of logarithms. With the advent of computing, this approach was reinvigorated; see, for example, the early work of Schmidt on structural synthesis (Schmit 1960). Progress in DSO has continued steadily since that time, and by now, a formidable range of optimization methods is available to designers – so much so that most current design-related optimization research focuses on how such methods should be deployed, rather than on developing new search methods *per se*.

At its simplest, optimization concerns the finding of the inputs to functions that cause those functions to reach extreme values, and of course, early uses of optimization theory were inevitably limited by the computational models available. Even so, given a little patience, approaches such as Newton's method (Avriel 1976) can be applied to find the turning values of simple functions with just pencil and paper. It is no surprise, however, that the real beginnings of applied optimization start with arrival of the first computers in the late 1950s and early 1960s: the first issue of the Journal of Optimization Theory and Applications was produced in 1967, for example.

Of course, designers are often not interested in finding optimal designs in the strictest sense: rather, they wish to improve their designs using the available resources within bounds set by the desire to produce balanced, robust, safe and well-engineered products that can be made and sold profitably. It is never possible to include all the factors of concern explicitly in computational models and so any results produced by numerical optimization algorithms must be tempered by judgments that allow for those aspects not encompassed by the models used. Moreover, there will often be multiple ways of solving design issues that all have benefits and drawbacks and no single clear wining design will emerge: instead, human judgment will be needed. Thus, it is more accurate to say that designers use optimization methods in the search for improved designs – hence the term "Design Search and Optimization". Nonetheless, modern optimization methods now provide an essential tool for designers seeking to produce state-of-the-art aerospace products.

1.8.2 A Taxonomy of Optimization

Before briefly reviewing the field of optimization, it is useful to set out a taxonomy that may be used to classify the problems to be tackled and the techniques that have been developed to solve them. In its most general form, optimization may be defined as the search for a set of inputs $\mathbf{x}$ (not necessarily numerical) that minimize (or maximize) the outputs of a function $f(\mathbf{x})$ subject to inequality constraints $g(\mathbf{x}) \geq 0$ and equality constraints $h(\mathbf{x}) = 0$. The functions may be represented by simple expressions, complex computer simulations, analog devices or even large-scale experimental facilities.

The first distinction that can be made in optimization is between problems in which some or all of the inputs are nonnumeric and those in which they are not: in the former case, the process requires optimal selection rather than optimization *per se*. Such cases occur quite commonly in design: for example, the selection of a compressor type as being either axial or centrifugal, the selection of a material between aluminum alloy and fiber-reinforced plastic for use in wing construction. Optimal selection is the subject of an entirely distinct set of approaches.

If all the inputs are numeric, the next division that occurs in problems is between those with continuous variable inputs and those that take discrete values (including the integers). Examples of discrete variables are things such as commercially available plate thicknesses, standardized bolt diameters, or the number of blades on a compressor disk. Continuous variables would include wing-span and fan-blade length.

Having dealt with problem inputs, the outputs can also be categorized. First, is there a single goal to be maximized or minimized (the so-called objective function) or are there multiple objectives? If there are multiple goals, these will lead to what is termed a *Pareto front*: a set of solutions that are all equally valid until some weight or preference is expressed between the goals. For example, if it is desired to reduce weight and cost, these aims may pull the design in opposite directions – until it is known which is more important or some weighting between them is chosen, it will not be possible to decide on the best design.

Next, one can categorize by the presence or absence of constraints. Such constraints may simply apply to the upper and lower values the inputs may take (when they are commonly known as bounds) or they may involve extremely complex relationships with the inputs (as in the limits typically found on stress levels in components). If equality constraints are present, then ideally these should be eliminated by using the constraint functions to reduce the number of free variables – in design work, this is often not possible and sometimes all that can be done is to substitute a pair of opposed inequality constraints to try and hold outputs close to the desired equality.

The final problem dependent type of categorization concerns the type of functional relationship between the inputs and outputs. These can be linear, nonlinear or discontinuous. They can be stationary, time dependent or stochastic in nature. Clearly, discontinuous functions make optimization most difficult, while linear relationships may lead to some very efficient solutions. Unfortunately, in most design work, which is riven through with decision taking, discontinuous relationships are the norm. Moreover, most aerospace products are subject to variability in manufacture and many must also deal with changing loads. Nonetheless, in most current optimization work, the functions being dealt with are usually taken to be stationary and deterministic (so as to simplify the problems being studied), although this is beginning to change as computing facilities become more powerful.

We can therefore categorize the beam design problem discussed in Section 1.6.2 as having continuous inputs, one or two outputs, various constraints and stationary nonlinear (but not discontinuous) relationships linking these quantities.

Search methods themselves may also be classified in a number of ways. The first division that can be made is between methods that deal with optimal selection, those that solve linear problems and the rest. Optimal selection routines commonly stem from the operational research (OR) community and typically are set up to deal with idealized traveling salesman or knapsack problems. These can often be readily applied to design work if sufficiently fast analysis methods are used to model the problem and generally work by using some form of exchange algorithm or list-sorting process; see for example Martello and Toth (1990).

Linear problems are nowadays almost universally solved using linear programming and the so-called simplex or revised simplex methods, which can efficiently deal with thousands of variables; see, for example, Siddall (1982). Optimal selection methods and linear programming, while valuable, essentially lie outside the scope of this book, as they do not find much practical application in aerospace design, which is dominated by large scale CFD and CSM models; they are not discussed further here.

The remaining methods that deal with nonlinear numeric problems form much the largest collection of approaches and are considered in some detail in later chapters. At the most basic level, such searches may be divided between those that need information on the local gradient of the function being searched and those that do not. Searches that will work without any gradient information may be termed zeroth order while those needing the first derivatives are first order, and so on.

Among these methods, a further distinction may be made between approaches that can deal with constraints, those that cannot and those that just need feasible starting points. Methods that cannot deal with constraints directly can be augmented by the addition of penalty functions to the objectives being optimized that aim to force designs toward feasibility as the search progresses by modifying the goal functions.

Methods may also be categorized by whether they are deterministic in nature or have some stochastic element. Deterministic searches will always yield the same answers if started from the same initial conditions on a given problem; stochastic methods make no such guarantees. While it might well be thought that results that vary from run to run ought to be avoided, it turns out that stochastic search methods are often very robust in nature: a straightforward random walk over the inputs is clearly not repeatable if truly random sequences are used – nonetheless, such a very simple search is the only rational approach to take if absolutely no information is available on the functions being dealt with. A random search is completely unbiased and therefore cannot be misled by features in the problem. Although it is almost always the case in design that some prior information is available on the functions being dealt with, the pure random search can be surprisingly powerful and it also forms a benchmark against which other methods can be measured: if a search cannot improve on a random walk, few would argue that it was a very appropriate method!

A distinction may further be made between searches that work with one design at a time and those that seek to manipulate populations or sets of designs. Population-based search has gained much ground with the advent of cluster and parallel-based computing architectures since the evaluation of groups of designs may be readily parallelized on such systems. Perhaps, the most well known of such methods are those based on the so-called evolutionary methods and those that use groups of calculations to construct approximations to the real objectives and constraints, the so-called Design of Experiment and Response Surface methods.

1.8.3   A Brief History of Optimization Methods

The development of nonlinear numerical optimization methods is a very large field, but may be grouped into three main themes: the classical gradient-based methods and hill-climbers of the 1960s; the evolutionary approaches that began to appear in the late 1970s and the adaptation of Design of Experiment and Response Surface methods to computer simulations in the 1980s and 1990s.

Classical hill-climbers all seek to move uphill (or downhill) from some given starting point in the direction of improved designs. Sometimes, they use simple heuristics to do this (as in the Hooke and Jeeves search (Hooke and Jeeves 1960)) and sometimes they use the local gradient to establish the best direction to move in (as in quadratic programming (Boggs and Tolle 1989)). Where needed, such gradients are commonly obtained by finite differencing but can sometimes be obtained directly (e.g., by using an adjoint code). Few new heuristic approaches seem to have found favor in the literature since the early work of the 1960s. Instead, most effort since that time has been focused on improving the speed of convergence of quadratic methods and getting them to deal efficiently with constraints. Perhaps the best known of these is the feasible sequential quadratic programming (FSQP) method developed by Panier and Tits (1993).

The so-called evolutionary methods are linked by the common thread of a stochastic approach to optimization. That is, the sequence of designs tested while carrying out the search has some random element – the main benefit is that such methods can avoid becoming stuck in local basins of attraction and can thus explore more globally. The downside is that they are often slower to converge to any optima they do identify. The most well known methods in this class are the genetic algorithms (GAs), which seek to mimic the processes of natural selection that have proved so successful in adapting living organisms to their environments (Goldberg 1989). A variety of alternative stochastic global search methods have also been developed over the same period and these methods are now much interwoven in their development. For example, a series of developments made in Germany at around the same time led to the so-called evolutionary algorithms; see, for example, Schwefel (1995) or the work on simulated annealing (Kirkpatrick et al. 1983). Some of these techniques work with populations of designs and are thus efficiently handled by clusters of computers: they are thus increasingly popular because of developments in modern computing grid architectures where hundreds of processors may be used simultaneously.

Design of Experiment and Response Surface methods are not really optimizers *per se*: rather, they are techniques that allow complex and computationally expensive optimization problems to be transformed into simpler tasks that can be tackled by the methods outlined in the previous two paragraphs. Essentially, these are curve fitting techniques that allow the designer to replace calls to an expensive analysis code by calls to a curve fit that aims to mimic the real code. Such methods all work in two phases: first, data is gathered on the nature of the function being represented by making a judiciously selected series of calls to the full code, usually in parallel – the placing of these calls in the design space is often best achieved using formal DoE methods, which aim to cover the search space in some statistically acceptable fashion (Montgomery 1997). Then, when the resulting data is available, the second phase consists of constructing a curve fit through or near the data – such curve fits are often termed "Response Surfaces" (Myers and Montgomery 1995). The choice of response surface models (RSMs) that may be used is quite wide and will depend on the nature of the problem being tackled and the quantities of data available. Sometimes, the initial RSM will not be sufficiently accurate in all the areas of interest and so an iterative updating scheme may then be used where fresh calls to the full analysis code are used to provide additional information – various updating schemes have been proposed (Jones et al. 1998). The use of DoE and RSM methods in optimization is relatively recent but, even so, has yielded some impressive results (Buck and Wynn 1993; Sacks et al. 1989).

As probably goes without saying, very many hybrid approaches that try and combine methods from these classes have been experimented with. So much so that current practice is often to use a collection of methods in some complicated workflow to try and achieve results tailored to the problem in hand. Most commercial DSO toolkits support such an approach, often providing powerful graphical workflow editors to set up and control such hybrid searches.

1.8.4   The Place of Optimization in Design – Commercial Tools

At the time of writing, optimization methods from each of the three classes mentioned in the previous section are now sufficiently mature that they are available in commercial optimization packages and are in routine use by designers – academic research is currently more focused on how to deploy optimizers most effectively, given the desire to minimize the use of time-consuming and complex analysis codes, or to search very large design spaces. Armed with a set of well-engineered search methods that encompasses these classes, and a suitably automated product representation and analysis process, it is possible to make real progress in DSO if suitable strategies are employed (Hale and Mavris 1999). The aim of current research is to enable rapid and more wide-ranging trade-off studies to be made between competing aspects in the design, hopefully leading to better and more-balanced designs, and also to exploit emerging features of analysis methods such as the sensitivities coming from CFD adjoint codes. The capabilities of commercial tools in this regard are also advancing rapidly.

An example of one such suite of tools is ModelCenter®[2] (Malone and Papay 2000; Phoenix Integration Inc 2000), marketed by Phoenix Integration, Inc. With this suite, a designer's existing codes are wrapped and placed in a catalog that all design team members can access. Then, at run time, a second tool is used by an individual engineer to build an application that links appropriate codes together from the catalog to deliver a workflow. This workflow can then be executed via a third program across a distributed heterogeneous collection of computing resources. Once set up, the internal optimization capabilities of the suite may be invoked to drive design change and trade-off studies. A key feature of such an approach is that after the design team's codes are wrapped and installed it is a simple matter to connect these together in a way suitable to the current needs of the team, at run time, without significant further effort. They are thus much more than just optimization tools – rather, they might better be described as design integration systems.

Another powerful DSO tool is that currently marketed by Engineous Software, Inc., and known as iSIGHT®[3]. This tool has a rather larger collection of inherent optimization tools and perhaps less focus on the systems integration process, although at the time of writing Engineous Software are known to be working on tools with much greater capabilities in this regard. Going in the other direction, the SpineWare®[4] system that was developed in part by the Dutch National Aerospace Laboratories and is currently marketed by NEC of Japan focuses on systems integration (Baalbergen and van der Ven 1998). This toolkit allows for collections of 'Commercial Of The Shelf' (COTS) and in-house analysis tools to be combined together by the design team. There are, of course, many other commercial DSO systems available and all are evolving very rapidly – the authors make no recommendation as to which is most suitable for any given task.

One key decision that any company engaged in engineering design has to make concerns the choice of tool that will dominate the design engineers' activities – is this to be a CAD package, with optimization and search being add-ons/plug-ins or, instead, is a commercial

---

[2] http://www.phoenix-int.com/
[3] http://www.engineous.com/products.htm
[4] http://www.spineware.com

DSO and systems integration tool to be the main tool in design with the manipulation of geometry via CAD being seen as the secondary function? Underlying both approaches, thought must also be given to the PDM/PLM systems, costing tools and all the analysis functionality that designers need. Resolving such issues requires corporations to have a clear view of their own strategies – and, in particular, whether design is to be as automated as possible with reduced scope for human input or, alternatively, if the human designer is to retain the key role, with all the tools acting as designer aids. Such decisions depend heavily on the complexity of the products being produced and also any regulatory framework that must be adhered to. While it may be possible to almost fully automate the design of a subsystem such as radiator or hydraulic motor, it seems unlikely that main systems such as engines, wings, and so on, are best treated in this way. Thus, leading contractors in the aerospace sector may well continue to prefer DSO systems to be controlled from within parametric CAD packages in a closely supervised fashion rather than to place geometric definition in the background with designers focusing mainly on automated search mechanisms.

1.9   The Take-up of Computational Methods

1.9.1   Technology Transfer

As has been made clear throughout this introduction, the software methods and decision support techniques that can be used in aerospace design are continuing to advance at a significant pace. However, by no means all of the advances made in academic research are taken up and used by industry. This lack of take-up occurs for a number of reasons. In part, it occurs because of issues concerned with technology transfer. It is now the case that the stock of knowledge acquired by newly qualified engineers, while at university, is outdated long before such engineers cease to be engaged on practical design activities. Technology transfer is naturally limited by the fact that busy senior designers must spend time being introduced to new technology and become convinced about clear benefits before they will begin to consider adopting it in practice. Even after it is decided to adopt a new way of working, there will still be issues arising from turning university, or research center, based technology into industrial-strength tools, and then of training engineers in their use.

Even where industrial designers are quite clear on the benefits that can be gained by the adoption of new design technology, take-up can still often be slow because of a lack of available resources within industry. Few companies can afford to have significant groups of engineers deployed in installing or understanding new ways of working, rather than on front line profit winning activities. To divert design teams from their efforts on the latest product into helping develop new design methods is always difficult. In practice, significant high-level management commitment is usually needed before new technology can be brought into a company of any size.

This potential disconnect between newly emerging methods and industrial take-up is particularly severe where whole new domains of technology must be introduced. Thus, it is not too difficult to introduce a novel element formulation into computational structural mechanics if all that is required is that the new formulation be inserted into existing FEA tools. When entire new ways of working are required, as with electronic PDM, or multiobjective, multilevel optimization tools, progress is much slower. Similar problems can arise with new IT infrastructure – who will own and manage new types of distributed computing clusters or database systems? Industry is naturally skeptical of any major changes to ways of working before it is completely clear that any disruption will be more than compensated for by gains in productivity or cost-effectiveness (Wiseall and Kelly 2002).

1.9.2 Academic Design Research

Another issue that bears on the gap between academic work and industrial practice is the position that design research takes in the academic world. A cursory study of the activities of the world's major university engineering departments reveals that the vast majority are engaged in significant efforts to improve the quality and sophistication of their computational modeling capabilities, that is, analysis. Relatively few are engaged in researching the issues that surround the design process or the take-up of new methods in that process, that is, synthesis.

Furthermore, because it is much less easy to define or describe in mathematical terms, the study of design is intrinsically harder to assess than that of, say, computational structural mechanics. If one postulates a new element type for use with a finite element code, it is a relatively simple matter to see whether the element works in practice and achieves the desired goals. By comparison, the assessment of a new design process often requires elements of social science as well as engineering to assess its benefits. Moreover, it is almost impossible to carry out controlled experiments on the efficacy of new design processes with real design teams in a realistic industrial setting: they rarely have the time to spare and academics can rarely provide their new ideas in a setting that is realistic enough to demonstrate unequivocal value. In some senses, design research has as much in common with the social sciences as engineering. This does not mean that design research is not valuable, however – better design processes are increasingly more important than improved analysis capability – just that it is harder to carry out.

1.9.3 Socio-technical Issues

There are also a number of socio-technical issues that arise when trying to transfer new ideas and ways of working into design teams. Aerospace companies often tend to be too technology focused and do not always pay enough attention to human and organizational issues. Wallace et al. (2001) note that

- the people using new systems tend to have too little say in their planning and set-up;

- new processes often fail to fit in well with existing ways of working and the users must then bridge any mismatches;

- engineering firms commonly view new systems as being about installing new technology, rather than setting up new ways of doing things that have a technological content; and

- there is often a failure to evaluate new systems against clear goals so that mistakes can be identified and past lessons learned.

There is also a strong supply community trying to persuade companies to try new things (i.e., to sell software, etc.) – these companies try to increase technology take-up but can be much less interested in the human and organizational issues (Clegg et al. 1997; Pepper et al. 2001; Wallace et al. 2001).

These factors often lead to new technologies failing to bring about the gains expected by those who invest in them. To be effective, new design systems must aid the designer: they must be simple to use, fit in with existing ways of working and not encroach too heavily on areas that designers, often rightly, see as their own fields of expertise. New methods must be strongly backed up by extended training and staff support, with key product champions put in place at the outset. If a new knowledge-based system is difficult to navigate, gives information in dense and unhelpful prose and is incomplete, one should not be surprised if take-up is not great – this has been seen many times with web search engines – only those that reliably find relevant information without swamping the user with masses of unconnected hits remain in use. The introduction of new design tools requires much care if they are to be successfully adopted.

2

Design-oriented Analysis

As has been made clear in the introduction, a great deal of the high fidelity performance prediction capability now open to aerospace design teams stems from the solution of PDEs based on geometric models. Although such models can be augmented by statistical analysis of past designs, experimental data, flight trials, and so on, the use of physics-based codes using detailed geometric descriptions lies at the heart of current capabilities and likely future progress in design. So, for example, in developing new airfoil sections, CFD codes have largely replaced approaches based on the development of methodical airfoil series in wind tunnels. Since balanced design requires that many different and often competing aspects of a system must be considered during the design process, engineers must now take a multidisciplinary perspective in their work. This leads to the subject of multidisciplinary systems analysis, which is introduced in this chapter. The aim here is to discuss some of the particular issues that arise when dealing with computational models coming from a variety of disciplines. Since so much of aerospace design and analysis is concerned with the geometry of the systems being studied, geometry modeling is taken as the starting point.

Before considering geometry modeling, however, it is of course important to note that the embedded computing systems used in many aerospace systems are of major importance in helping achieve design goals and that detailed models of such systems will also be built and refined during a multidisciplinary approach. Consideration of such models lies outside the main scope of this book, however, and so they will not be discussed in detail, although many of the techniques introduced here are applicable in computer and software engineering. Indeed, it is becoming increasingly understood that the best design practices attempt to consider both the flight hardware and the onboard computing and software systems as an integrated whole. This is particularly true of advanced flight control systems, which, necessarily, interact heavily with the structural and aerodynamic performance of the vehicle.

2.1 Geometry Modeling and Design Parameterization

It is clear that manipulation of the geometries being worked on is of fundamental importance in aerospace design – so much so that CAD packages are commonly seen as the main tool for much of the aerospace design community. The use of CAD in design may be viewed

Computational Approaches for Aerospace Design: The Pursuit of Excellence.   A. J. Keane and P. B. Nair
© 2005 John Wiley & Sons, Ltd

as a process whereby sets of geometric parameters are chosen and suitable values for them selected. In doing this, features are included and scaled according to current needs. This process can be an intensely manual one: its origins lie with the draftsman using a pencil and eraser. The pressure for automated approaches has, however, led to this process becoming highly structured in many parts of the aerospace community. In such approaches, as much as possible of the available and agreed design practice is codified and embedded in the design system. This is aimed at enabling rapid design changes to be considered and analyzed so that appropriate decisions can be made in a timely fashion. The latest-generation parametric CAD engines are set up with just this approach in mind.

Even groups solely dedicated to analysis are now accepting that there are distinct advantages to be had by the close coupling of commercial CAD packages and their state-of-the-art analysis codes. Moreover, the use of bespoke parameterization codes written in Fortran, C, C++, and so on, to set up geometries for analysis is increasingly becoming unacceptable. It is very wasteful of manpower to try and maintain separate programs to manipulate geometry for design and for analysis. Duplication of information also inevitably leads to errors and potential conflicts when multiple models disagree about the design. It is therefore important to understand the issues stemming from geometric modeling when considering computational approaches to design – so much so that NASA operates a geometry laboratory focusing on just this issue (GEOLAB: the NASA GEOmetry LABoratory[1]).

Much current geometry research is therefore aimed at producing design-oriented CAD systems that enable accurate models to be rapidly set up and parameterized and for these systems to then allow the resulting geometry to be passed to automated meshing systems and advanced solvers without further manual input. Such an approach views the CAD system as a kind of data repository that all solvers draw on when making performance calculations. Ideally, such a server would provide models with appropriate levels of detail given the stage in the design process and the needs of the analysis to be performed. This requires the ability to distinguish between different features and their relative importance to the various aspects of analysis and stages of design being worked on. It also needs a recognition that different disciplines require different geometric descriptions when analysis is carried out (the internal structure of a wing does not affect drag, for example; the precise curvature of the external cladding may have little impact on cost, and so on).

Some capabilities of this kind are now available in integrated design and analysis packages such as the I-DEAS NX Series®[2]. If all the capabilities needed by a corporation can be found in such a product, their integrated nature can bring real benefits. Integrating the disparate products of a range of corporations and software providers is, of course, much more difficult – it may nonetheless be necessary if specific analysis codes must be used alongside dedicated CAD tools chosen for reasons not directly associated with analysis. Currently, most of the world's major aerospace companies work in this way, using CAD tools such as Catia®[3], ProEngineer®[4], Unigraphics®[5], and so on, with separate major codes for analysis, such as Fluent®[6] or Ansys®[7]. The transfer of data around design systems built up from such components is still not completely straightforward. Moreover, designers are always

---

[1]http://geolab.larc.nasa.gov
[2]http://www.ugs.com/products/nx/ideas/
[3]http://plm.3ds.com
[4]http://www.ptc.com/
[5]http://www.ugs.com/products
[6]http://www.fluent.com
[7]http://www.ansys.com

looking to deal with ever more complex geometries, which leads to constantly increasing demands on capabilities in this field. Researchers are thus always investigating interesting new ways to manipulate geometries – these commonly appear as research codes long before being adopted in mainstream CAD packages. The availability of open-source commercial-grade CAD systems may help address a number of these issues – the OpenCascade project is attempting to provide just such a capability[8].

There is perhaps a greater heritage in the parameterization of shapes concerned with aerodynamic flows than of structures, largely because such shapes are often the most subtle and difficult to deal with – thus, much of what follows in this section stems from aerodynamic needs. It remains the case, however, that parameterization issues affect all uses of computational analysis in design, because design of this kind always involves geometric changes.

Many of the issues that occur when using multiple analysis codes first arose in the study of aeroelasticity, which fundamentally requires the results of CSM and CFD simulations to be brought together. Much of the early work in this field was characterized by the use of customized programs to help transfer results between the different meshes used with the structural and fluids codes (Dusto 1974). Recently, it has been proposed that CAD models can provide a useful bridge between analysis codes, again seeing the CAD tool as a geometry data service (Samareh and Bhatia 2000). Clearly, it would be even simpler if such essentially geometric activities as mesh production could be embedded within the CAD tool itself – most of the major CAD vendors now offer some capabilities of this kind.

It is also worth noting that to help with systems integration a range of new tools is beginning to emerge. These aim to deal with data transfer, workflow planning, job scheduling and execution; see, for example, the SpineWare®[9] product developed by the Dutch National Aerospace Lab and currently marketed by NEC (Baalbergen and van der Ven 1999). Such Problem-solving Environments (PSEs) (Gallopoulos et al. 1992) typically allow a graphical approach to tool selection and workflow construction and fit naturally with the way many design engineers wish to work. We discuss some of these issues further in Section 2.3. At the same time, most large companies are adopting various product-data and life-cycle management systems that enable them to store all the information relevant to designs in a consistent way. These can be seen as being rather bureaucratic in their approach but the use of both sorts of tools can be expected to increase substantially in the aerospace sector over the coming years.

2.1.1   The Role of Parameterization in Design

As has been made clear throughout this book, the basic process of design involves the making of decisions that change the product definition. The purpose of geometric parameterization is to aid in this process by providing increasingly powerful manipulation schemes for changing design definitions. At the lowest level, design changes can be achieved by altering the most primitive geometric entities being used to describe the design. This becomes increasingly time consuming as the design model grows in complexity and the effort required then stifles change. Consequently, for as long as systematic trade-off studies have been carried out, engineers have sought ever more powerful geometry manipulation tools – the famous

---

[8]http://www.opencascade.org
[9]http://www.spineware.com

NACA[10] series of airfoil sections are one of the earliest manifestations of this desire for efficient parameterizations when manipulating geometry. They were developed experimentally between the 1930s and 1950s and were heavily used in aircraft design up until the 1980s.

At the same time, computers have completely changed the way that engineering drawings are produced in aerospace design offices. Initially, CAD systems focused on replacing the drawing board and little more. They began with simple two-dimensional drawing systems, which made little use of the available calculating power of the computer. These were followed by three-dimensional surface modeling approaches – these were often difficult to use, required the designer to keep track of which surfaces combined to make which parts, could not unambiguously define voids, needed specialized hardware, and so on. These early systems rapidly moved on to working with solid models, since these allow a completely unambiguous description of the items being drawn. By concentrating on solid modeling, many useful capabilities become possible with little user interaction, including volume intersections, keeping track of voids and topology, reuse of standard parts, the automated production of engineering drawings and machine instruction sets for manufacturing purposes, right through to the use of rapid prototyping systems to produce physical models that can be handled, inspected and, in some cases, even used in service. Nowadays, such tools can easily be run on the desktop workstations of all engineers.

Throughout this evolution of CAD tools, their developers have attempted to adopt some of the higher-order functions and parameterization schemes produced by analysts for rapid geometry manipulation, bringing these ideas into the hands of increasing numbers of practicing designers. These have necessarily focused on the more generic capabilities, such as spline curves, rather than methods devoted to specific shapes such as airfoils. For example, Figure 2.1 shows variations in the sections through a turbine blade fir-tree root structure

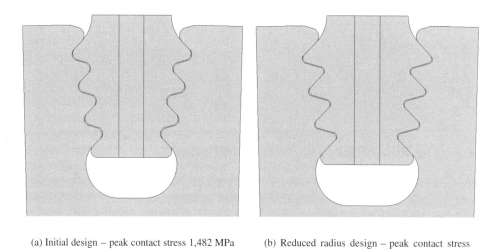

(a) Initial design – peak contact stress 1,482 MPa   (b) Reduced radius design – peak contact stress 1,107 MPa

Figure 2.1 Effect of varying tooth radius on parametric spline-based ICAD® model of a turbine blade fir-tree root structure.

---

[10]National Advisory Committee for Aeronautics, which became NASA in 1958

obtained using the spline based parametric capabilities of the ICAD®[11] system, see Song et al. (2002b).

Even so, there are still weaknesses in most commercial CAD tools: models that initially look to be in good order can suffer from low-level imperfections that can make subsequent activities difficult or impossible to carry out. Typical errors that can arise in solid models (Samareh 1999) include: free edges; bad loops (inconsistent face or surface normals); unacceptable vertex-edge gaps; unacceptable edge-face gaps; unacceptable loop closure gaps; minute edges; sliver faces and transition cracks. Such errors can be invisible to the designer – even so, they can cause mesh generation codes to fail or data exchange algorithms to give errors when transferring information from one system to another. Nonetheless, the latest general-purpose CAD tools are now the common standard of large aerospace companies, and these will continue to evolve as researchers produce improved parameterization schemes.

2.1.2   Discrete and Domain Element Parameterizations

As has already been noted, the most basic method for manipulating geometry is to control directly the lowest-level components in the model definition. Movement of the nodal data points that define individual lines, elements and low-level features of a model gives complete flexibility over the design. However, for all but the most basic models, this will mean that the designer is working with thousands or perhaps even millions of individual data items. In a multidisciplinary environment, it is also unlikely that each discipline will adopt the same discretization, which further complicates matters. Moreover, considerations of aerodynamic flows and structural stress concentrations mean that designs that lack smoothness will generally not be desirable. Thus, when manipulating the basic elements of a model, the designer will usually need to simultaneously maintain control of local curvature, smoothness, and so on. To do this manually is often extremely tedious. An alternative that might be considered is the use of automated optimization procedures to control data points, with the designer merely placing constraints on performance targets such as maximum stress or drag levels. Unfortunately, while this might be possible in principal, the task the optimizer then faces is usually far too difficult given current systems and computing resources. The search space to be assessed contains just too many possibilities, most of which are of no practical use.

The limitations in such an approach can be easily illustrated by considering a simple example. Imagine trying to design a low-drag two-dimensional airfoil section by specifying its shape using, say, 200 points around the circumference, specified in (x, y) Cartesian coordinates. Typically, the required lift would be fixed but not the angle of attack. If using a modern Euler solver (i.e., ignoring viscosity), estimation of the required angle of attack and resulting drag for a section might take only a few seconds. It might thus be concluded that a direct search using a modern stochastic search engine would be able to achieve good designs within a modest amount of time. In practice, this turns out not to be so. First, a starting point for the design must be specified – if a population-based algorithm is to be used, an initial set of designs would be needed. Commonly, random starting points might be considered – however, a random collection of 200 points defined in no particular order will usually specify loops that cross themselves and contain large concavities and abrupt changes in surface curvature. Even a modest set of point-based changes to a sensible airfoil shape rapidly lead to nonsensical geometries; see Figure 2.2. This will mean that for many

---

[11]http://www.ds-kti.com

Figure 2.2 Illustrating changes to an airfoil specified as a series of points.

configurations a normal CFD code will not even be able to converge to steady flow solutions around the section, let alone estimate lift or drag.

If a simple gradient descent search is used with such a model, staring from a reasonably well-defined initial section, it will need to calculate 400 finite differenced gradients at each step unless an adjoint code is available. Even given this kind of capability, such searches will only identify local minima – when dealing with real airfoil problems, it turns out that most of the existing sections currently in use lie at or near such local minima already, as they have usually been manually optimized over many years in wind tunnels and by flight trials. Gradient-based searches then tend not to give very worthwhile improvements as they get stuck in local optima, not being good at identifying step changes in designs. Note also that producing adjoint codes for multidiscipline problems is extremely complex and rarely tackled.

These considerations immediately led researchers to try and link together the points in a model in some way, both to reduce the number of variables being considered and also to enforce some kind of structure on the data sets. One way of doing this is via so-called domain elements; see Leiva and Watson (1998). Domain elements are superelements that contain and control collections of smaller items. Then, when the geometry of the domain element is changed, all of the smaller items contained within it are scaled and shifted to maintain their relative positions, keeping their topological relationships fixed; see Figure 2.3. This can be particularly effective in allowing for parameterized changes to mesh layouts, for example.

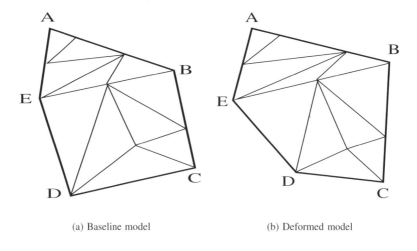

(a) Baseline model   (b) Deformed model

Figure 2.3 Domain element modeling.

It does not, however, make any guarantees regarding the smoothness of geometry from one domain element to the next. Such smoothness is particularly important in shapes that are exposed to aerodynamic flows such as airfoils or in highly loaded structural components where sharp changes give rise to stress concentrations.

2.1.3   NACA Airfoils

The need for smooth airfoil shape definitions led to some of the oldest geometry parameterizations used in the aerospace sector. These are the classic NACA 4-digit and 5-digit airfoil sections, as described by Abbot and Von Doenhoff (1959) and also by Ladson and Brooks Jr. (1975). These 2-d sections were introduced so that families of airfoil geometries could be easily studied. They were defined before the advent of computers and so are relatively simple to use, even with slide rules. Currently, these sections are rarely adopted for practical applications, although they are still used by academic and industrial aerospace teams to calibrate computational methods, and also to help in generating parametric descriptions of sections for use with CFD.

The NACA airfoils are composed of a camber line plus a thickness distribution. The thickness distribution is given by a single equation, while the camber is usually two joined quadratics. For the 4-digit airfoils, the first digit in the designation is the value of the maximum camber (as a percentage of the chord), the second digit is the position of the maximum camber from the leading edge in tenths of the chord, and the last two digits denote the maximum thickness of the airfoil (again as a percentage of the chord). So, for the NACA 3415 airfoil, the maximum camber is 3%, the position of the maximum camber is 0.4 times the chord from the leading edge and the thickness is 15% of the chord. The NACA 5-digit airfoils are set up in a similar manner, with the same thickness distribution but a changed camber line. In a 5-digit airfoil, 1.5 times the first digit is the design lift coefficient in tenths, the second and third digits are one-half the distance from the leading edge to the location of maximum camber as a percentage of the chord, and the fourth and fifth digits are again the thickness as a percentage of the chord. For example, a NACA 23015 airfoil has a design lift coefficient of 0.3, maximum camber at 0.15 times the chord from the leading edge, and is 15% thick. Additionally, the first three digits indicate the mean line used – in this case, the mean line designation is 230. The equations for the upper and lower surfaces of the 4-digit airfoils are given by

$$x(\text{upper}) = x - y_t \sin(\theta) \qquad y(\text{upper}) = y_c + y_t \cos(\theta) \qquad (2.1)$$

$$x(\text{lower}) = x + y_t \sin(\theta) \qquad y(\text{lower}) = y_c - y_t \cos(\theta) \qquad (2.2)$$

where $\tan(\theta) = dy_c/dx$, y_c is the camber line and y_t is the thickness distribution. The camber line and thickness are given by

$$y_c = f(1/(x_m^2))(2x_m x/c - (x/c)^2) \quad \text{for} \quad 0 \leq x/c \leq x_m, \qquad (2.3)$$

$$y_c = f(1/(1 - x_m)^2)(1 - 2x_m + 2x_m x/c - (x/c)^2) \quad \text{for} \quad x_m \leq x/c \leq 1 \qquad (2.4)$$

$$\text{and} \quad y_t = 5t(0.2969x^{0.5} - 0.1260x - 0.3516x^2 + 0.2843x^3 - 0.1015x^4) \qquad (2.5)$$

where t is thickness, x is the position along the x-axis, x_m is the position of maximum camber as a fraction of the chord, c and f is the maximum camber. As can be seen from this definition, the thickness distribution is particularly simple, while the camber line

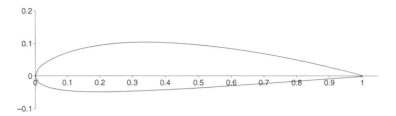

Figure 2.4 NACA 3415 airfoil section.

incorporates coefficients found to be effective by experimental means. Figure 2.4 illustrates the 3415 airfoil. The great strength of these NACA foil definitions is that they make use of the ideas of camber and thickness to describe the shape – ideas that are natural to engineers, while still allowing considerable flexibility in shape. Their primary limitation is the lack of local curvature control. Such curvature has been found to be vital in controlling shock drag in transonic airfoils, for example. To achieve such control requires more sophisticated parameterizations.

A number of other series of NACA airfoils have therefore been developed, mainly for use at higher Mach numbers; these are commonly referred to as the NACA 6-digit, 7-digit and 8-digit airfoils; see, for example, Ladson and Brooks Jr. (1974) or Graham (1949). These have increasingly complex definitions that are more difficult to use without dedicated computer codes.

2.1.4   Spline-based Approaches

The desire to link multiple points together, while at the same time controlling local curvature and smoothness, naturally leads to consideration of spline curves. At the most basic level, it is possible to use polynomials, typically quadratic (as in the NACA airfoils) or cubic, to specify curves that span some region of the model. These can then be linked together by enforcing slope continuity at the joints between curves (note that the NACA airfoils have smooth leading edges where the upper and lower surfaces meet). Such curves can either be defined in a simple (x, y) Cartesian fashion or parametrically along the curve. Polynomials are very simple to use and compact in form. They do, however, suffer from two drawbacks. First, the coefficients in the polynomial expressions bear little direct relationship to the geometry being produced and so are not readily understood by designers. Secondly, it is quite common in polynomial forms to end up with coefficients of opposing signs that tend to give round-off errors when the order of the polynomial gets high (both these features can be seen in the previously noted NACA airfoil equations).

Some of these limitations can be overcome by the mathematically equivalent Bezier representation where control points and Bernstein polynomials are used to define the individual curves. The control points form a polygon whose convex hull contains the curve – this leads to a very simple and direct geometric relationship between the controls and the geometry. Also, the Bernstein form leads to reduced round-off errors. Even so, if too high an order Bezier is used, round-off errors can become a problem, so a series of such curves can be stitched together to form a composite curve, which is usually termed a *B-spline*.

The main drawback of the normal B-spline is its inability to represent conic sections. This leads to the use of a special form, the nonuniform rational B-spline or NURBS curve,

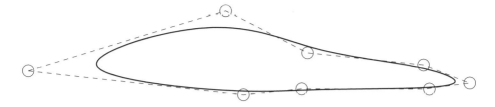

Figure 2.5 Airfoil defined by NURBS curves using control points.

which can represent most of the curve types needed in aerospace design without loss of accuracy (certain helicoidal forms cannot be dealt with directly, but this is rarely of great consequence). A NURBS curve is defined as

$$
\frac{\displaystyle\sum_{i=1}^{n} N_{i,p}(u) W_i P_i}{\displaystyle\sum_{i=1}^{n} N_{i,p}(u) W_i} \tag{2.6}
$$

where the P_i are the control points, the W_i are the weights and $N_{i,p}(u)$ is the ith B-spline basis function of degree p.

All of these forms are easy to use directly in two dimensions and for simple three-dimensional geometries; see Figure 2.5. When representing complex three-dimensional shapes they do, however, require a great deal of effort to set up and manipulate, particularly if working from first principles. NURBS are, nonetheless, the basic representation tool of most modern CAD systems, and it is from within CAD packages that they are perhaps most readily manipulated. If used to represent complex aerodynamic shapes such as modern transonic airfoils, they tend to need large numbers of control points. Then, when these are manipulated by optimization algorithms, they can lead to wavy or irregular shapes (Braibant and Fleury 1984). One way around this is to carry out any desired optimization in stages. To begin with, when the most radical changes are being introduced, a few control points are used and then extra controls are inserted between the existing points as the process proceeds and increasing levels of local refinement are required.

2.1.5 Partial Differential Equation and Other Analytical Approaches

Another way to represent surfaces efficiently is to describe them via solutions to partial differential equations (PDEs), which are solved subject to boundary conditions applied on a bounding curve. This approach was pioneered by Bloor and Wilson, initially to provide the means to blend objects into each other (Bloor and Wilson 1995). As originally proposed, lines are drawn on the existing bodies to be blended. These define the extent of the blending patch and then, by using the position of the boundary and the slopes on the existing bodies at this boundary, a set of boundary conditions can be derived that the PDE then has to meet when solved. To do this, only a very few parameters need specifying, which control the rates at which one surface blends into the next.

This approach has been extended to allow for complete wing and even aircraft models to be built wholly from PDE patches; see, for example, Figure 2.6. Such surface descriptions

Figure 2.6 PDE double delta aircraft geometry produced using the approach of Bloor and Wilson (1995) (From Bloor and Wilson; reprinted by permission of the American Institute of Aeronautics and Astronautics, Inc.).

can have very high levels of smoothness and complete accuracy at their intersections, and so on. Their main deficiency again stems from the lack of local control – in some senses, the resulting surfaces can be too smooth. Most modern airfoil sections include various bulges and curves that can appear counterintuitive but are, in fact, needed to control highly non-linear flow features such as boundary layer separation or the presence of shock waves. To gain such control of PDE-generated surfaces, it is usually necessary to convert them to piecewise collections of NURBS representations and then manipulate these local functions. The approach is thus perhaps best suited for use during preliminary design where over-all planform characteristics are being set, rather than at later stages where detailed section geometry is being manipulated. PDE surfaces are also better suited to aerodynamic work than for internal structural models, where spars, stringers, and so on, need modeling and controlling. It is, however, possible to remove part of the PDE surface and insert a patch formed from either some other spline representation or another PDE surface. Slope discon-tinuities are then accommodated by a character or singular line with appropriate boundary conditions.

There have been other attempts to use mathematical functions to help describe and parameterize smooth aerodynamic surfaces. Perhaps, the best known of these are the Hicks–Henne functions (Hicks and Henne 1978). These produce a localized bump that tapers away on either side to zero gradients and can thus be used to add or subtract cur-vature to an existing shape. They have the form $A[\sin(\pi x^{\ln 2/\ln x_p})]^t$ for x in the range 0 to 1. A is the height of the bump, x_p is the location of its peak and t controls the width (large values of t correspond to sharp bumps). Each function is thus controlled by three variables. They are very effective for modifying aerodynamic shapes, but again are of less use in structural models. They have mainly been used to describe airfoil sections and wings

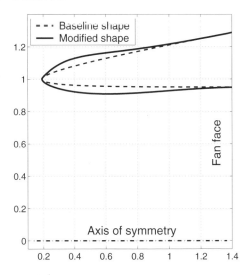

Figure 2.7 Cross section through an aero-engine inlet optimized using two Hicks–Henne functions (one on either side of the nose).

but have also been used to deal with aircraft engine nacelles, for example (Sóbester and Keane 2002); see Figure 2.7.

Various other analytical functions and geometries have been proposed in the literature for use in this way. These include combinations of straight lines, ellipses, cubics, and so on, to produce complex built up geometries; see, for example, the missile inlet geometry described by Zha et al. (1997). Such combinations tend, of course, to sacrifice the simplicity that is the main attraction of mathematical function modeling of geometry.

2.1.6 Basis Function Representation

As just noted, one of the fundamental problems with most mathematically derived geometrical models is that they do not incorporate the local shape information that has often been found to be useful over decades of aerospace research. One way around this problem, while still trying to preserve the extreme efficiency of compact mathematical forms, is to use existing geometry as the starting point for constructing functional representations. One way of doing this is to take a family of airfoil geometries and analyze them for the variations between family members and the underlying commonality in their forms. It is then usually possible to reconstruct the original airfoils from the functional data extracted. For example, in the work reported by Robinson and Keane (2001), a set of NASA transonic airfoils was analyzed to produce a series of orthogonal basis functions that, when appropriately combined, could be used to reconstruct the original airfoils. Because orthogonal functions were used in these descriptions, each basis function represents a different kind of geometric feature: the first described thickness-to-chord ratio, the second the camber, the third a measure of twist in the section, and so on; see Figure 2.8. This model is described in more detail in the case study of Chapter 11.

Another feature of such orthogonal sets of functions is that they converge very rapidly. For those shown in the figure, it is possible to construct quite realistic sections from just the

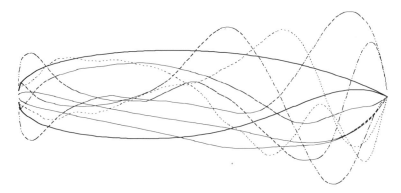

Figure 2.8 First five orthogonal basis functions for a NASA transonic airfoil section family (underlying airfoil shape plus variations).

first two functions in the series. This allows a very parsimonious description of the section and this proves particularly useful in reducing the number of variables needed to describe an entire wing (Keane and Petruzelli 2000). Moreover, it is also possible to gradually increase the sophistication of the section model by judiciously adding further terms from the orthogonal series at appropriate points in the design process.

Of course, such approaches are still best suited to aerodynamic shapes and are not very convenient for dealing with structures. They are also limited to cases where families of similar shapes can be used as a starting point. Moreover, they mitigate against any radical changes being produced in the resulting designs. They also tend to be limited to two-dimensional work.

2.1.7 Morphing

A more radical approach for modifying geometries is shape morphing. Morphing essentially implies that a fully formed and representative initial geometry is taken as a start point and this is then distorted in some way, while maintaining its original topological form. This necessarily requires that some aspect of the original design is chosen to which the transforming morph is applied, with the relationships between this feature and the overall geometry being linked in some way. When dealing with an airfoil, for example, the camber line might be morphed while the thickness distribution was fixed, so that camber line changes give a variety of airfoil shapes (in terms of the NACA foils, this would amount to using a whole range of camber polynomials). These control frameworks are, in some senses, similar in function to the knots and control points of traditional NURBS-based models. Setting up the control systems and defining their relationships to the geometry being changed represents a preprocessing stage in morphing-based methods just as it does in NURBS models. When dealing with more complex geometries in three dimensions, setting up this aspect of the morph becomes crucial.

One approach to shape morphing is the use of Soft Object Animation (SOA). This is a technology that was pioneered by computer graphics animation specialists (Watt and Watt 1992). It was originally used to allow animators to smoothly alter images between key frames, so that many fewer hand produced frames were necessary when making movies. SOA algorithms allow for the manipulation of geometry as though it were made of rubber that can be twisted, stretched, compressed, and so on, while preserving the original topology. SOA systems generally work with three-dimensional objects because the scenes that animators wish to consider are invariably populated by characters that make three-dimensional movements. To work efficiently, these techniques require that changes to relatively few key parameters control the evolution of the entire scene. This ability to make gross changes to geometry by the control of a small number of key parameters makes SOA techniques potentially very powerful for use in engineering design parameterization.

As already noted, shape-morphing methods start with the baseline geometry. To this is added the three-dimensional control framework that specifies how changes made to the control parameters will be mapped on to the underlying geometry of the model being manipulated. The degree to which any distortions then achieve the designer's aims is closely linked to how well this step is carried out. It turns out to be relatively easy to define controls for objects such as airfoils, which need to be smooth but contain few straight edges. When dealing with internal structures, such as wing spars, which need to maintain straightness, more care is required. As Samareh (2001a) shows, in practice, it is useful to define customized individual controls for each aspect of the object being considered, paying close attention to the desired outcomes and any additional constraints that may affect the objects under consideration.

In the work described by Samareh, he describes variants to the original free-form deformation (FFD) algorithms of Sederberg and Parry (1986) that allow the control parameters used to take physically meaningful forms that are more natural for aerospace designers – this helps ensure a compact and easy to use set of parameters. For example, it allows wing deformations to be described in terms of camber, thickness, twist and shear deformations.

Another key benefit of shape morphing is the ability to simultaneously control meshes that have been constructed for solution by CFD or CSM codes. Moreover, it is possible to define regions of such meshes that connect undisturbed parts to the geometry being changed in a controlled fashion. Thus, when the geometry is morphed, regions of the mesh can be adapted to match the changed geometry back to the remaining mesh. Recent work in this field by Samareh (2001a) has produced some very impressive results – Figure 2.9 shows the

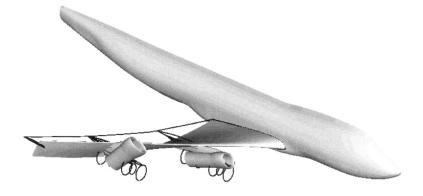

Figure 2.9 Result of a 45° twist applied on a transport aircraft wing tip (from J.A. Samareh, "Novel multidisciplinary shape parameterization approach", J. Aircraft, Vol. 38(6), pp. 1015–1024, 2001, courtesy of the author).

kind of geometry perturbations that can be made when applied to an entire aircraft model. These approaches are, as yet, still in their infancy in the engineering community – they seem likely, however, to take an increasingly important role.

2.1.8 Shape Grammars

Shape grammars offer another radical way of manipulating geometry, and one that is quite well suited to structural systems that contain beams, stringers, plates, and so on. They provide a means for constructing complex geometric entities from simpler elements by the repeated application of rules. They were introduced into the architectural literature to provide a formal means of shape generation (Stiny 1975) and work by matching elements within an existing shape and then substituting these with slightly more involved or reoriented or scaled elements. If the available rules in the grammar are then applied recursively many times, startling complex shapes may be constructed. The main tasks in using shape grammars for engineering lie in selecting the appropriate base elements, a useful set of rules and the order in which to apply them. However, given the right ingredients, shape grammars provide a powerful way of building complex shapes from first principles. They would appear to be particularly appropriate for dealing with structural geometry, since they are good at reusing and repeating regular elements – a characteristic often found in efficient structural design. They can also be combined with NURBS curves to generate airfoil shapes – Figure 2.10 shows a series of airfoils evolved in this way.

This approach has been examined by various workers, for example, Gero and coworkers (Gero et al. 1994) have applied genetic algorithm (GA) techniques to both the planning of rule sequences and the evolution of sets of rules to be used in trying to design structural shapes in this way. This involves a further encoding whereby the initial shapes, rules and application order have to be mapped into the genetic encoding used by the GA, typically as binary sequences. Then, the GA can be used to alter and amend the grammar as well as the application of its elements. It is clear that such approaches could be used to control the location, shape, number and sizes of structural elements in a wing box, for example. Moreover, it is to be expected that such approaches would be capable of rather radical design changes – stiffeners applied to stiffeners, for example. This again points

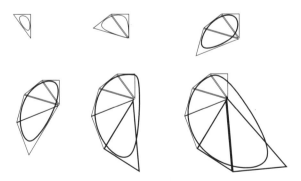

Figure 2.10 A series of airfoil shapes evolved using a shape grammar – the grammar consists of the base triangle and the operations of rotation and scaling; the vertices formed by the evolving complex are then use as the control points of a NURBS curve.

up one of the fundamental difficulties in design parameterization — if a system is capable of giving very radical, or even nonsensical, shapes, then it is likely that optimizers and other search systems will find difficulties in locating good designs that are much different from those already known. Conversely, too conservative an approach will restrict the design process and limit the gains that might be possible. One of the most important skills in design optimization is the selection and implementation of appropriately flexible parameterizations.

2.1.9   Mesh-based Evolutionary Encodings

A related series of methods for describing structural problems, particularly where the geometries are essentially two dimensional in nature, are the so-called Evolutionary Structural Optimization (ESO) methods (Xie and Steven 1997), artificial density approaches (Bendsoe 1989) and Solid isotropic microstructure with penalization (SIMP) approaches (Rozvany 2001). These schemes are used to evolve structural models from oversized, initially regular geometries by modifying or eliminating the elements used to discretize the domain during (finite element) analysis. In their simplest form, a rectangular block is broken down into thousands of rectangular elements and then elements are removed from the domain on the basis of repeated finite-element–based stress analysis, to reveal the final structure much in the same way that a sculptor reveals a statue from an initial block of marble; see Figure 2.11 (the dotted area shows the extent of the original block in this example).

The rules used for selecting elements for removal are usually based on eliminating unstressed material or material that does not impact greatly on rigidity. This removal can be done element by element or more gradually by steadily reducing the density or other properties of individual elements. The number of changes made per finite element (FE) calculation can also usually be adjusted by the designer, with more aggressive approaches

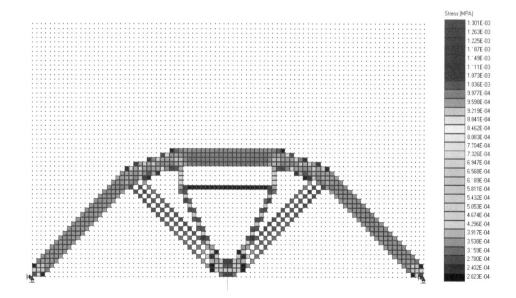

Figure 2.11   Optimal Michell arch produced by Evolve 97.

working faster but sometimes leading to less good designs. It is also possible to arrange schemes where material is reinstated so that the structure "grows" into areas where additional material is needed. This can avoid designs becoming stuck in local optima. When dealing with the linear elastic behavior of two-dimensional structures, these processes can be shown to yield fully optimal designs, with resolutions governed only by the accuracy of the elements used to discretize the domains.

Of course, this approach tightly couples the analysis process and the parameterization scheme. If the elements used are able to change in size, shape and orientation, as the optimization proceeds, extremely accurate shapes and topologies can be revealed. Conversely, if the basic elements used are fixed in size and shape, this limits what can be achieved and also impacts on the smoothness of the final design (although smoothing operators can be applied after the basic shape and topology are defined).

Schemes have been proposed that work in similar iterative ways to reveal structural topologies but by moving the shape boundaries; see, for example, Chen et al. (2002). These schemes can generate smooth geometry at the same time as revealing interesting shapes; see Figure 2.12, which shows a mounting link evolution, and Figure 2.13, which illustrates a Michell type cantilever design.

Processes of this kind can also be used to impact on the thickness of essentially laminar components, but this is more difficult to control. Their use in designing fully three-dimensional objects is still an area of active research. The approach efficiently combines the parameterization and solver meshing stages, but relies on relatively simple metrics for reshaping the design at each iteration. It is also computationally quite expensive in terms of the numbers of analyses needed (although, if these are just linear elastic solvers, this may not be a concern).

2.1.10 CAD Tools versus Dedicated Parameterization Methods

Ultimately, perhaps the best place to develop geometry parameterizations of all kinds is within the emerging feature-based CAD tools that are becoming the basic building blocks of most aerospace design offices. These CAD tools have capabilities that may be identified as coming from two distinct backgrounds. First, they are the direct descendants of the draftsman's large A0 drawing board with its mixture of pens, pencils, drafting instruments, French curves, battens, and so on. Secondly, they incorporate the parametric modeling capabilities that address the needs of computer analysts, and that have been set out in previous sections. The latest codes allow sophisticated design environments to be constructed that build on both these traditions. Their emergence stems from the limitations of traditional CAD tools in supporting parameterizations and of custom geometry manipulations tools in supporting information flows suitable for product definition and manufacture via CAD.

If the designer wishes to capture the internal structure of an aerospace vehicle as well as the (usually) smooth outer skin, by modeling items such as the ribs and stringers within a wing box, then a traditional CAD solid model can become enormously complex – much more so than is needed by the aerodynamicist. Even for structural analysis purposes, such models can be far too complex since the simplest way to produce an FE mesh for CSM given a solid model is to use an unstructured tetrahedral mesh – clearly, a very inefficient way of analyzing a plated stiffened structure where it is usually more sensible to use dedicated

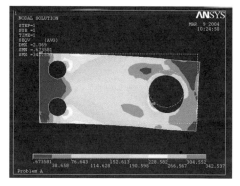

(a) Initial domain

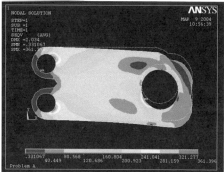

(b) After 25 iterations

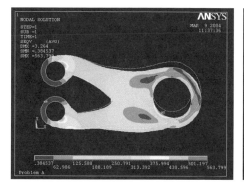

(c) After 200 iterations

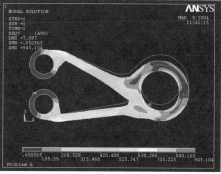

(d) After 350 iterations

Figure 2.12 Evolution history of mounting link design produced using nodal-based evolutionary structural optimization. The link is constrained by the two left-hand holes and loaded through the right-hand one.

element types. These difficulties may be seen as stemming from the fact that while CAD-based solid models may capture the product very well, they have little to say about design intent – structural engineers know that a plate and stiffener structure is efficient and that it is best dealt with in FEA by plate/shell elements combined with beam stiffeners – the basic CAD geometry has no way of capturing this information. While some CAD tools can automate minor feature elimination processes (such as the removal of fillets, holes, etc., to aid in meshing), it is almost impossible to automate the kind of gross changes needed to distinguish between differing disciplines. At best, such model abstraction requires an extensive set of rules to be defined, and these may not carry from project to project or even remain unchanged during a single project.

Typical analysis preprocessing tools, on the other hand, are commonly built on the engineer's intuition and knowledge, seeking to exploit this knowledge as much as possible. They are usually handcrafted for the problem under consideration, being focused directly

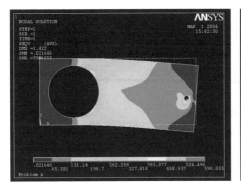

(a) Initial domain

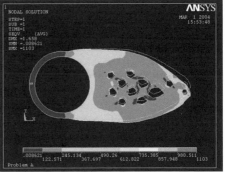

(b) After 50 iterations

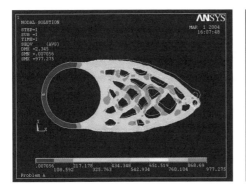

(c) After 100 iterations

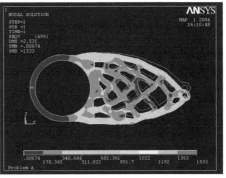

(d) After 150 iterations

Figure 2.13 Michell cantilever design produced using nodal-based evolutionary structural optimization. The cantilever is constrained by the large left-hand hole and loaded through the small right-hand one.

on providing for the needs of the solvers to be used and representing the minimum of geometry in the most parsimonious way possible. They are most often written in traditional programming languages and developed by staff used to programming complex mathematical ideas (in contrast to the drafting and geometry manipulation skills of the majority of CAD system users). This leads to simple and highly flexible models, but ones that can suffer from a restricted range of applicability (see, e.g., the geometry approaches based on the Hicks–Henne functions (Hicks and Henne 1978)). They are often tightly coupled to dedicated mesh generation tools and are then capable of yielding rapid and robust solver input. They usually suffer from a comparative lack of investment. However, since being specialized, the effort needed to document and develop them can rarely be justified in the way that is possible for general-purpose CAD systems – the installed user base is just too small.

Inevitably, the distinction in approaches between general-purpose CAD systems and dedicated geometry parameterization codes has led to somewhat different approaches to dealing with geometry – CAD systems aim at producing interactive tools with very high quality graphics and good GUI design; analysis preprocessors aim at speed of operation and the ability to rapidly alter input parameters while generating robust meshes with limited user interactions.

When trying to use either traditional CAD tools or customized shape parameterization tools/solver preprocessors, integrated design teams commonly face problems. Typically, the CAD geometry cannot be used directly in the analysis code because it has too many small details, does not provide input in the desired formats or cannot readily be adapted to model the range of designs needed to support trade-off studies. At the same time, the analysts' specialist preprocessors can be difficult to use and maintain, often being suitable for only a small range of problems. Moreover, after trade-off studies are completed, such codes do not always facilitate the transfer of the resulting geometries back into the CAD world.

Over time, both approaches have improved. There have been significant advances in the capabilities of CAD systems by way of spline systems, Bezier curves, Coons patches, and so on, so that current NURBS-based systems can model almost all the features of interest to engineers in a consistent way. Additionally, latest-generation CAD systems now incorporate feature-based parametric modeling systems that begin to allow the design team to specify design intent, at least at the detail level. Thus, holes and fillets may be specified and linked to overall dimensions in a way that takes account of the engineering needs. At the same time, dedicated geometry manipulation tools have improved. Also, practices in programming have evolved via the use of subroutine libraries, through object oriented methods, to the adoption of coding control systems and data transfer standards such as IGES and STEP. There is also ongoing work on producing unified CAD system interfaces to help deal with data transfer issues. A recent survey paper sets out some of the pros and cons of these various approaches (Samareh 2001b).

Despite all these improvements, the meshing of geometry prior to analysis using CSM, CFD or CEM (Computational Electromagnetics) solvers is still normally carried out by separate mesh generation tools. These tend to have their own scripting or programming capabilities, adding yet further complexity to the design environment. Moreover, different meshes and even mesh generators may be needed by different codes, even within a single discipline. Much progress is still needed before the parameterization, manipulation and transfer of product geometry and associated meshes becomes completely routine.

Currently, there are a number of CAD packages such as Catia®[12], ProEngineer®[13] or Unigraphics®[14], that offer the ability to combine many of the features needed in this area, although they often have a dominant leaning in one direction or another. Those such as Catia, whose origins lie in aerospace geometry modeling, now have very significant inherent parametric and feature-based modeling capabilities and may even be considered as programming environments in their own right – even so, they are still first and foremost CAD systems. Conversely, packages such as ICAD (ICAD 2002) are primarily software development environments that are designed to describe and manipulate geometry – in this case, the code editor and interpreter are usually the main tools used in

---

[12]http://plm.3ds.com
[13]http://www.ptc.com/
[14]http://www.ugs.com/products

driving the system (in ICAD, these are EMACS and Lisp, respectively). Both approaches offer a highly structured hierarchical way of considering the geometries being manipulated, allowing for the reuse of components, for multiple copies of features, for the specification of design intent (albeit only in a limited way), and so on. At the time of writing, Dassault Systemes, who develop Catia, have taken over the ICAD package, bringing both codes into the ownership of a single company – clearly, these worlds are continuing to merge.

Having said all this, it remains the case that custom-built geometry manipulation tools are still widely used by both researchers and those working on practical design tasks in industry. They are seen to be lightweight and flexible, do not incur licensing costs and do not require long training sessions to gain experience. For the moment, the best approach appears to be a judicious mix of the two approaches, with CAD-based work being used for those tasks that will be deeply embedded in commercial design systems and used alongside other drafting tasks, while custom-built codes tend to be the preserve of research and development divisions (Robinson and Keane 2001). Certainly, the authors are increasingly using CAD-based tools in support of their own research and software development.

2.2 Computational Mesh Generation

Throughout this book, most of the analysis discussed has involved the solution of PDEs over geometrical models of the systems being designed. Almost all the computational methods used to carry out such analyses work by employing discretization schemes to transform the relevant PDEs into sets of algebraic equations which can then be solved using matrix methods, combined with appropriate iterative schemes where necessary. Setting up the necessary discretization for the geometry under consideration is here termed *mesh generation* (or sometimes "meshing")[15]. It is currently one of the most demanding tasks carried out by analysts and can take many months of highly skilled manual intervention to carry out on a complex aerospace configuration. It is thus the subject of intense research effort that aims to reduce the required level of human input, increase the quality of automatically generated meshes and extend such capabilities to meshes with greater numbers of cells/elements over ever more complex geometries. There are a number of highly sophisticated commercial tools that designers can use in this area, such as Patran®[16], Gambit®[17] or GridGen®[18]. The capabilities of these tools are rapidly evolving.

2.2.1 The Function of Meshes

The basic function of computational meshes is to allow the PDEs that govern structural, fluid, electromagnetic behavior, and so on, to be transformed into sets of algebraic equations that can then be solved computationally. Normally, this is achieved using one of three basic approaches: finite elements, finite differences or finite volumes, although other schemes are

---

[15]N.B., to avoid confusion, the expression "grid generation" has been avoided here, as the term "grid" is reserved throughout this book for connected systems of computers, for example, a "computational grid" is a collection of machines, usually connected via TCP/IP protocols, which may be used to carry out the calculations required in design and analysis, as opposed to a "computational mesh", which is the discretization scheme adopted for solution of a set of PDEs.

[16]http://www.mscsoftware.com/products/

[17]http://www.fluent.com

[18]http://www.pointwise.com

possible (such as boundary element methods, spectral approaches, etc.). In all cases, the aim is to transform a set of PDEs that apply over a continuum into algebraic equations that apply at specific locations – the mesh points or nodes. The resulting algebraic equations are then solved at these points and the results can then be used to infer the behavior of the variables of interest at other locations.

Apart from selection of the basic equations being solved, the fundamental input to PDEs solvers is, of course, the description of the boundaries that define the domains of interest (and any conditions that apply at them). Depending on the precise equations being solved and the needs of the analyst, the meshes produced may represent the system and its boundaries in a variety of ways. They may represent only the geometry, and in a highly idealized fashion, or they may represent just the surface of the geometry, or the solid interior of the three-dimensional geometry itself or the fluid domain surrounding the system (and in some cases both of the latter). It is also possible to use meshes that do not follow the natural boundaries in the problem. Such meshes may be easier to set up than those that conform to the bodies being dealt with, but they then place additional difficulties on the solver as interpolation schemes are then required to allow for any areas where elements cross boundaries or each other. Figures 2.14 to 2.17 illustrate a number of these cases.

Figure 2.14 shows a highly idealized structural model built of beam and shell elements. This might be used to assess quickly the overall stiffness of an aircraft structure – such models are commonly built during the early stages of design when a full geometrical description of the aircraft is not yet complete, or when rapid turn around in design/analysis cycles is needed. Meshes like this are normally produced by hand and require significant understanding and intuition to construct if they are to yield useful results. They are not discussed further in this chapter – they can, however, be very effective during design exploration and

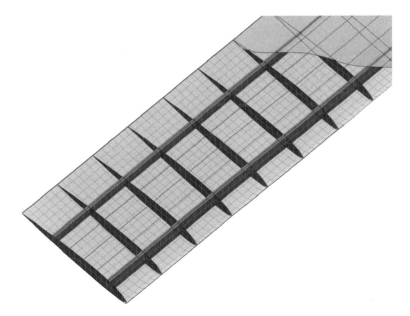

Figure 2.14   A simplified beam and shell element model of an aircraft wing structure.

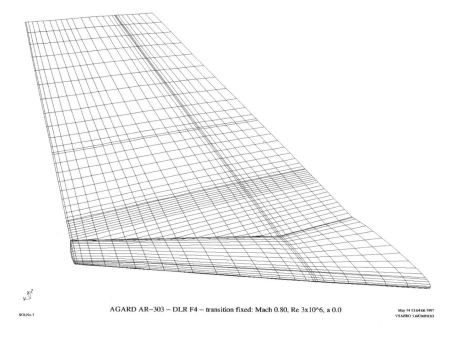

AGARD AR–303 – DLR F4 – transition fixed: Mach 0.80, Re 3x10^6, a 0.0

Figure 2.15 A panel solver mesh (N.B., both wing and wake are modeled here).

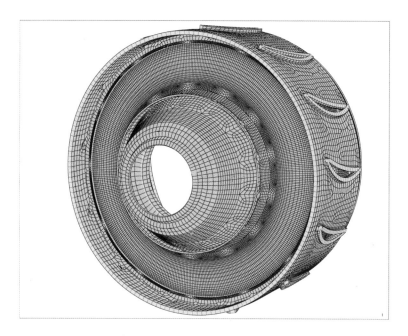

Figure 2.16 A full brick-based FEA model of a gas turbine structure. Reproduced by permission of Henrik Alberg.

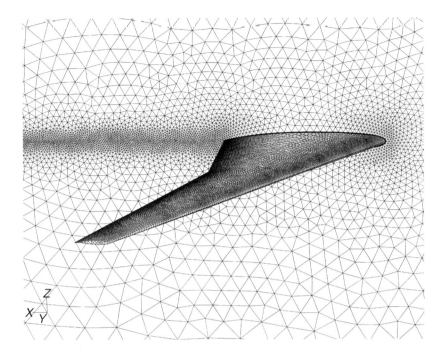

Figure 2.17 A mesh for a RANS solver.

should not be forgotten when carrying out initial studies. The analysis of such meshes is usually extremely rapid, and this can enable wide-ranging searches using optimizers, for example, provided the validity of the meshes is not stretched too far.

The next figure shows a typical surface mesh that might be used with a panel-based potential flow solver – such codes solve for the fluid flow around the geometry but do not need the mesh to discretize the flow-field itself. They are only suitable for potential flow models, which, of course, have limited accuracy in terms of quantities such as drag recovery. They are also not appropriate where viscous boundary layers must be considered directly by the solver (N.B., some panel methods use coupled boundary layer models to yield approximate solutions for viscous problems).

The third figure in the series shows a brick-built version of an aero-engine component, designed to study stress levels within the structure. Such meshes are very commonly used for accurate stressing of cast or forged items where through thickness stresses must be calculated. They are not efficient at representing plate/stiffener combinations where the thickness of elements is small compared to the other dimensions but can be rapidly constructed from CAD solid models using auto-meshing systems. When dealing with plated structures, which are common in many aerospace designs, the designer should first distinguish between parts of the structure that must be meshed in a fully three-dimensional way and those that can be idealized as beams or plates – this process then forms a preprocessing stage to full mesh generation. Again, this is often a highly skilled, manual activity.

The final mesh shown is for use with a Reynolds averaged Navier–Stokes (RANS) solver that will calculate lift and drag on the body being studied (only the surface paneling is shown in the figure). It includes allowance for the boundary layer and any shock waves that may form. Notice that to give accurate results the mesh must be able to resolve both the boundary layer and the detailed surface geometry of the configuration – normally, the desire to capture the boundary layer is the more demanding requirement. Such meshes are among the most complicated encountered in routine computational design work and can have tens or even hundreds of millions of cells. Even bigger meshes are occasionally encountered in computational electromagnetic studies, because of the very wide range of length scales involved, given the need to resolve individual cables in a full aircraft configuration with perhaps 10 points per wavelength or feature.

Clearly, when setting up a mesh, the analyst will want to mold the shape of the mesh and the density of elements to suit the problem in hand, always aiming for the simplest acceptable solution. This process draws on an understanding of the geometry, the physics involved and their interaction. The more complex the geometry being modeled, the more complex the mesh must be. Perhaps less obviously, the more rapidly the solutions to the PDE change across the domain, again the more complex the mesh must be if the resulting variations are to be accurately resolved. Such behavior commonly occurs where nonlinearities arise, such as around shock waves or boundary layers in CFD studies and contact points or cracks in CSM studies. Essentially, the analyst must temper the natural desire for a simple, regular, low-density mesh that is easy to produce, with the needs of the problem at hand, which may require highly accurate and expensive results to be of practical use in design.

2.2.2 Mesh Types and Cell/Element/Volume Geometries

In all mesh generation work, the aim is to create an assembly of mesh points (or nodes) and elements (or cells or volumes) that make up the mesh and (usually) define the boundaries of the geometry being studied. The exact use that the mesh will be put to depends ultimately on the type of solver being used (finite difference, finite volume, finite element, etc.). When working with different solver types, it is normal to refer to the components of the meshes in slightly different ways: as elements in FE work, cells or grids in finite difference (FD) solvers and volumes with finite volume (FV) codes – in all cases, the assembly of such components is referred to here as the mesh. In consequence of the variation in approaches, for some classes of solvers, focus is placed on elements defined by adjacent nodes, while in others, the emphasis is on the mesh points themselves. Whatever solver is in use, however, the mesh is specified by a combination of volume, cell or element shapes, types and topologies; volume, cell, edge and node numbers and nodal or mesh point coordinates.

Note also that all meshing approaches can be split into two very broad classes: structured and unstructured. Figure 2.15 shows a structured mesh, while that in Figure 2.17 is unstructured. In a structured approach, the mesh is made up of families of mesh lines where members of a single family do not cross themselves and may cross members of other sets only once. The position of a mesh point or node is thus fixed uniquely in the mesh by its indices and the resulting algebraic equations have considerable order that can be exploited during solution. By contrast, an unstructured mesh consists of arbitrarily shaped elements joined together in any fashion and nodes may have any number of neighbors (note that even in unstructured meshes, it is normal to use only one or two basic element types, such as triangles or quadrilaterals, and then to assemble them with widely varying vertex angles and with differing numbers of edges meeting at nodes, etc.). This leads ultimately to matrices of equations with no underlying structure – these can be more difficult to solve but the meshes are easier to manipulate and thus lend themselves to adaptive meshing schemes, mesh enrichment and interactive mesh generation tools, since new elements may readily be inserted or removed at any point with few ramifications for other parts of the model[19].

It is of course possible to combine the structured and unstructured approaches to produce so-called block structured or multiblock meshes, where blocks of structured (or indeed unstructured) elements are combined in an unstructured fashion, either by butting them up against each other or by overlaying different blocks in regions where they meet and then taking various forms of averages. It is also sometimes possible to restore some form of order to unstructured meshes by node renumbering schemes; a number of modern solvers incorporate such techniques.

Another distinction that can be made between mesh types is based on how they reflect the geometrical boundaries in the problem being tackled. Perhaps, the most intuitive approach is to ensure that some of the edges in the mesh align with boundaries in the problem – such meshes are said to be "body fitted". However, this is not the only approach that can be taken – an alternative is to adopt a mesh that is more convenient to set up but that overlaps the problem boundaries – a classic example is the use of a regular rectangular grid to mesh the fluid flow around a cylinder; see Figure 2.18. Clearly, some of the cells/elements in this mesh span both the fluid domain and the object around which it flows. Solvers can be constructed to work with such meshes but they are difficult to develop and often less accurate than those working with body-fitted schemes. They require schemes to interpolate between cells or the insertion of additional points at the intersections of the geometrical

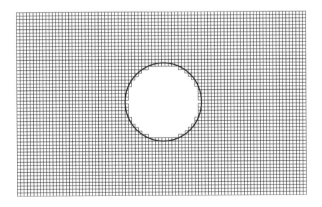

Figure 2.18 A rectangular mesh for studying flow around a cylinder.

---

[19]Adaptive meshes are ones that change to reflect emerging features in the solution such as shock waves; mesh enrichment means the insertion of extra elements in regions of a mesh to increase the accuracy of the solution in such areas.

boundaries and cell edges, and so on. Even so, such "Cartesian" solvers still find use in a variety of areas. For example, the commercial Euler CFD code MGaero, used in the case study of Chapter 12, works in this way and has been applied to complete aircraft geometries (Tidd et al. 1991). Cartesian methods also find use in specialist roles, for example, in the so-called hydrocode Euler solvers used in studying the ultrahigh velocity impacts that arise in weapon design (Cornish et al. 2001).

The basic shape of building block used in mesh construction fundamentally alters the topology of the meshes that can be created, of course. The most commonly used shapes are based on triangles or quadrilaterals. As the geometries and domains normally considered in design are three dimensional in nature, this leads mainly to tetrahedral, wedge and brick (sometimes termed hexahedral) shapes. Even so, some two-dimensional work is still necessary in design – airfoil section design, for example, is often carried out in two dimensions and many structural models use essentially one-dimensional and two-dimensional elements within three-dimensional models, to represent beams and plates. Here, discussion is, however, mainly restricted to three-dimensional meshes – the two-dimensional equivalents are usually obvious.

When a mesh is used by a PDE solver, equations are set up that map the field variables of the equations to the shapes of the mesh components – these mappings are referred to variously as weight, shape or interpolation functions, difference equations, and so on. For some types of analysis, a variety of different schemes may be available to make this mapping for each shape of element or volume, and these may also have midside or even midvolume mesh points or nodes or may be based on averages taken across the points or nodes. Such variations may be useful when trading cell/element/volume complexity against overall mesh complexity – simple meshes built using complex shapes or mappings may be more accurate or faster to solve than complex meshes made of simpler shapes. Whatever the shape or sophistication of the basic building blocks, for these to accurately model the underlying PDEs, there is normally a requirement that aspect ratios be kept near to unity and also that shapes should not be highly skewed or twisted. For example, ideally all triangles should be as close as possible to equilateral, all quadrilaterals close to square and all surfaces flat, and so on. Note, however, that for solutions with highly elongated features, such as shock waves, stretched elements may in fact be more appropriate if aligned using adaptive schemes.

The most basic three-dimensional shape that can be used is, of course, the tetrahedron and it turns out that the simplest automated meshing schemes that can be constructed for complex geometries are built using such elements, i.e., fully unstructured meshes. Defining high-quality formulations for tetrahedral meshes is thus vital to the efficient solution of many design calculations. Even so, it is often the case that more accurate or sometimes simpler models can be constructed with bricklike (hexahedral) elements and so it is also important that meshes can be constructed with these as well. Often, there is a need to use a mixture of types of building blocks and so transition elements such as wedges can also be useful. In practice, most commercial PDE solvers have built-in models for a variety of geometries and many have the capability to adopt user-defined types as well.

For FD-based methods (which is the oldest approach to discretization and which is found mostly in CFD work), the simplest schemes that can be constructed are built on rectangular Cartesian meshes with evenly spaced mesh points in all directions, leading to rectangular cells. More complex meshes can be dealt with via finite differences and, in principle, such schemes can be constructed for almost any form of mesh – in practice, however, they are only really used for structured meshes or blocks of structured meshes in more complex overall schemes.

Usually, a simple Cartesian mesh will not be directly useful because the boundaries in most problems of interest are curved and accuracy requirements will be simpler to satisfy if the mesh points are located along the boundaries. Moreover, in many cases, it is inefficient to use an even spacing of mesh points along boundaries or close to them; rather, it is better to have a greater density of mesh points in regions where the field variables are changing rapidly. In such cases, it is normal to transform the geometric mesh into a computational domain where uniform spacing can be arranged, modifying the differential equations and underlying boundary conditions along the way. Such transformations can sometimes be carried out by simple algebraic stretching operations but more often are achieved by relatively elaborate schemes that involve the solution of auxiliary sets of (often elliptic, sometimes parabolic or hyperbolic) differential equations. Figures 2.19 and 2.20 illustrate simple uniform and stretched meshes and a mesh transformed using elliptic equations, respectively. Note that alongside the geometrical transformations required for the stretched and transformed meshes, equivalent transformations are also required to the PDEs being solved.

The precise form of transformation to be used when constructing complex geometries is, itself, a sophisticated field of study. Often, Laplace's equation forms the underlying transformation in such schemes because of its inherent smoothing effects and also because it can, under certain circumstances, give rise to meshes with features that tend to align with the flow features themselves. The basic equation is, however, commonly augmented with source and dipole terms that can be used to modify the densities of mesh points along curved boundaries. The choice of position and strength of such terms is typically manipulated by the user directly or by automated mesh generation software (Weatherill et al. 2005). Sometimes, solution of the transforming equations is time consuming and parabolic or hyperbolic time marching schemes can be more efficient. Perhaps, not surprisingly, such schemes suffer

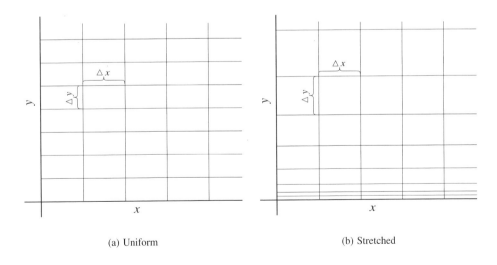

(a) Uniform   (b) Stretched

Figure 2.19 Simple finite difference meshes.

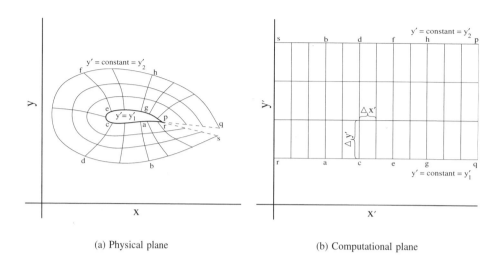

(a) Physical plane   (b) Computational plane

Figure 2.20 A finite difference mesh produced using an elliptic transformation.

from other problems: parabolic approaches find it difficult to deal with the most complex geometries while hyperbolic systems are not well suited for dealing with far-field boundary conditions.

In any case, once a regular mesh has been formed in the computational domain, either directly or by some form of stretching or transformation (either as a single mesh or assembly of multiple blocks of meshes), a series of simple FD equations can be set up. If an explicit solver is then used in a time marching fashion to solve the underlying PDEs, varying orders of accuracy can be adopted, depending on the number of mesh points involved in the formulation at each point. Alternatively, implicit formulations can be used that are much more stable with respect to time steps but which involve more complex coding and solution – they typically lead to series of simultaneous equations. In either case, a second-order central difference scheme is commonly adopted, where, for example, a simple partial differential would be written as $(\partial u/\partial x)_i = (u_{i+1} - u_{i-1})/2\Delta x + O(\Delta x)^2$ where at some time step u_i is the velocity at mesh point i and Δx is the spacing between mesh points in the x direction. Such schemes allow the (transformed) PDEs to be cast as algebraic relationships and solved in the usual ways.

To recapitulate, setting up a mesh for an FD solver using bricklike elements involves defining a topologically equivalent mesh (or series of meshes) that fits the boundaries, transforming these and the governing PDEs into a computational domain where the mesh is uniformly spaced in all three directions and the boundaries map to straight lines, solving the equations in this domain using a series of difference equations, either explicitly (and commonly with fine time steps to ensure stability) or implicitly (and thus operating on larger matrices but with fewer time steps) and then finally transforming the results back to the original geometric domain. In such approaches, the cell geometry cannot be considered in isolation from the boundary shapes and the transformations in use. Because they are difficult to apply to complex geometries, FD schemes are currently not very popular: it is perhaps where higher order accuracies are required that they are most competitive.

An alternative solution type, and one increasingly common in CFD studies, is the FV-based solver. Such solvers do not deal with the PDEs directly; rather, their equivalent integral form is used instead. These are set up to apply to a contiguous series of cells that cover the entire domain of interest. The nodal points used in the computations normally lie at the centroids of the cells and interpolation schemes are used to establish values at the cell faces, edges and vertices (N.B., it is possible to use dual mesh schemes that do not do this, but discussion of such approaches lies outside the scope of this brief introduction). The integrals involved are approximated via suitable quadrature formulae. Finite volume schemes have a number of advantages: they are physically easy to understand, they guarantee conservation of the various terms in the underlying equations (provided adjacent cells use equivalent formulations) and they can deal easily with arbitrarily complex mesh topologies, mesh refinement, enrichment, and so on. They are, however, difficult to program for approximations of higher than second order. Nonetheless, most commercial CFD codes now use FV approaches, although there is still some variation in the way these work.

Finite volume solvers require approximations for integrals taken over the volumes used and their surfaces, based on variable values at the cell centers (i.e., at the computational mesh points). Most commonly, this is done by simply taking the volume and multiplying it by the function value at the cell center (or by taking the surface area and multiplying it by the value at the surface centroid). These give second-order models but require interpolation schemes to approximate the surface centroid values. Higher-order schemes are possible but are rarely used in commercial codes.

The most basic FV interpolation scheme is the so-called upwind approach, which unconditionally satisfies boundedness criteria so that it never yields oscillatory behavior. This approach is only first-order accurate, however, and leads to significant numerical diffusion. The effect is rather like that of heavy damping in a mechanical system: while there is no overshoot, the response will not accurately capture sudden changes in the flow, such as those that occur through a shock wave. Figure 2.21 illustrates the difference between an oscillatory solution and one where there is significant numerical diffusion. Clearly, to maintain second-order accuracy overall, at least second-order interpolation schemes are necessary. Of these, the two most important are based on central differencing using linear interpolation, which gives a second-order result, and quadratic

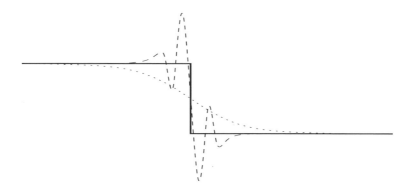

Figure 2.21 Diffusion and instability – solid line shows the exact properties through a shock, dashed line is for an oscillatory solution and dotted line one with significant diffusion.

upwind interpolation, which gives third-order accuracy (and is commonly referred to as the QUICK scheme). The differences between these two schemes are rarely large, however, as the overall accuracy is still second order if mean values at centroid positions are used in the integrations. Nonetheless, the combination of integration and interpolation does favor the use of brick (hexahedral) volumes in FV schemes as these shapes yield more accurate results for a given order of approximation. Many commercial packages treat wedges (and occasionally even tetrahedra) as reduced forms of hexahedra for convenience.

When using FV solvers for sets of coupled equations like the RANS equations, it is not necessary to use the same computational mesh points for all the variables in the problem (the colocated approach), although this may seem the most obvious choice. Of course, colocated meshes clearly have advantages: they minimize computational complexity and the number of coefficients that must be stored, and so on. They do, however, suffer from some problems that can lead to oscillations in the pressure field due to difficulties with pressure–velocity coupling. Such difficulties led to the development of so-called staggered meshes, where it is possible to avoid having to carry out some of the interpolations required by colocated meshes and where conservation of kinetic energy and strong coupling between velocities and pressure can be achieved, leading to improved accuracy and convergence properties. Even so, with the development of improved pressure–velocity coupling algorithms in the 1980s, colocated arrangements have tended to dominate recent developments, especially because they are much easier to set up for the unstructured meshes required to deal with more complex geometries.

In whatever form the interpolation and integration expressions are set up, the coupled equations found in most CFD work are almost always then solved using some form of iterative (often predictor-corrector) approach. Perhaps, the most commonly used is the "SIMPLER" algorithm of Patankar (1980). This starts with guessed values for the pressures and velocities, which are then used to find a set of improved pseudovelocities, which in turn can be used to solve the pressure equation. The resulting pressures are then used to solve the discretized momentum equations. Next, a pressure correction equation is set up and solved and used to correct the guessed values of velocities. The resulting values are then used to solve all other discretized transport equations before iterating by returning to recalculate the pseudovelocities, this loop being repeated until convergence is reached. Sometimes, underrelaxation is used when updating the pressures and velocities to prevent oscillations that can slow convergence. It is common to also adopt multigrid strategies when using such methods on RANS problems to enhance their convergence speeds.

For FE-based solvers (which are the almost universal choice in CSM work), a third approach is used. In this approach, weight and shape functions are employed. These are used to scale the underlying equations before integration. In FE models, the simplest three-dimensional element is a four-noded tetrahedron using linear weight and shape functions. If a simple CSM element is being considered, each node has displacements in the three coordinate directions and these contribute to displacements throughout the element, varying linearly away from the nodes, that is, linear shape functions are used so that if u, v and w are the displacements in the x, y and z-directions and x, y and z define the position in the element, then $u = \alpha_1 + \alpha_2 x + \alpha_3 y + \alpha_4 z$, $v = \alpha_5 + \alpha_6 x + \alpha_7 y + \alpha_8 z$, and $w = \alpha_9 + \alpha_{10} x + \alpha_{11} y + \alpha_{12} z$. The 12 unknown coefficients here are then solved by setting up three simultaneous equations at each of the four nodes in terms of the nodal displacements $\mathbf{a}_i = \{u_i, v_i, w_i\}^T$, $i = 1$ to 4, such as $u_1 = \alpha_1 + \alpha_2 x_1 + \alpha_3 y_1 + \alpha_4 z_1$, and so on.

This ultimately leads to solutions like $u - 1/6V\{(a_1 + b_1x + c_1y + d_1z)u_1 - (a_2 + b_2x + c_2y + d_2z)u_2 + (a_3 + b_3x + c_3y + d_3z)u_3 - (a_4 + b_4x + c_4y + d_4z)u_4\}$, where

$$6V = \det \begin{bmatrix} 1 & x_1 & y_1 & z_1 \\ 1 & x_2 & y_2 & z_2 \\ 1 & x_3 & y_3 & z_3 \\ 1 & x_4 & y_4 & z_4 \end{bmatrix}$$

and the symbol V is used as it represents the volume of the tetrahedron. When used to solve linear elasticity problems such linear shape functions ensure that there is continuity between the faces of adjacent elements. Four-noded tetrahedral elements can then be used directly to create eight-noded brick elements (using either five or six tetrahedra). Alternatively, such eight-noded elements can be constructed directly by using more complex shape functions. These two basic types of elements can be made more complex by allowing for nodes along each edge (to give 10-noded tetrahedra or 20-noded bricks). Such elements necessarily have more complex (usually quadratic) shape functions and these can give improved accuracy with fewer degrees of freedom. Higher-order elements with even more nodes are possible but, in practice, most users and most commercial FE codes stick to just these basic geometries. More complex formulations are generally encountered only when the equations being solved are more involved, such as when large deformations or contact occur or where combined structural and thermal effects are being modeled.

Whichever solution scheme is adopted, the discretization and order of the approximations used has a profound bearing on the subsequent computations. All numerical schemes must be studied for

- consistency (so that the discretized equations become exact as mesh spacing goes to zero),

- stability (the solution method does not amplify any errors as solution proceeds),

- convergence (the solution tends to the correct solution as the discretization spacing tends to zero – consistent and stable solutions are generally convergent),

- conservation (the scheme respects the conservation laws present in the original PDEs),

- boundedness (the scheme yields results that lie within the physical bounds for the problem, e.g., quantities like density remain strictly positive),

- realizability (features that cannot be treated directly by the models being used, such as turbulence in a RANS code, should be dealt with in such a way as to guarantee physically realistic results) and

- accuracy (modeling, discretization and convergence errors should all decrease with increasing mesh resolution).

A full discussion of these aspects lies outside the scope of this book, but nonetheless, practitioners of computational analysis should not rely on the commercial codes being used to guarantee that all these issues have been addressed, or even that any failures will be immediately obvious. It is perfectly possible to gain seemingly sensible results from commercial solvers that are in error because the user has overlooked or provoked a limitation that the code developer has not managed to trap and warn about. Instead, suitable training should be put in place and reference made to the many standard texts that discuss these issues, for example, Ferziger and Peric (1996); Zienkiewicz (1977).

2.2.3   Mesh Generation, Quality and Adaptation

Before a solution method can begin using any of the discretization schemes just described, the analyst must first prepare the computational mesh. This is still an extremely demanding and time-consuming activity for all but the most trivial of geometries and one that requires considerable skill to carry out well. In most design work, it usually means starting from a (typically three-dimensional) CAD model, importing this into the meshing tool and then generating the mesh. In passing, it is worth noting that the apparently simple task of just exporting geometry from the CAD model into a meshing tool can be a quite lengthy and difficult procedure, since such tools rarely share exactly the same internal methods for describing the shapes being considered. The transfer may involve a range of standards such as STEP or IGES or, increasingly, proprietary schemes for the direct interchange of information. Having loaded the geometry into the mesher, it may then need various cleanup processes to be applied before meshing can begin.

Whether or not the mesher is being used to produce a structured or unstructured mesh, two fundamental facts drive all meshing processes. First, the size of the elements used must be compatible with the features present in the geometry – it is almost inevitable that some accuracy will be lost from the geometric description by the meshing process, but obviously, this should be reduced as far as possible, especially near features that are thought to be of importance in the solution. Clearly, if a structural component has circular holes machined into it, these may act as stress raisers but their effects will only be accurately captured if the adjacent elements approximate the circularity well. Similarly, the flow over the nose of an airfoil is highly dependent on local curvature and so accurate capture of the leading edge shape by the computational mesh will be fundamental to gaining accurate solutions.

Secondly, the number of elements/cells used will fundamentally affect the accuracy of the solution. In general, and until round-off errors, and so on, begin to accumulate, the finer the mesh, the more accurate the solution. Thus, an experienced analyst will usually have some idea of the number of elements/cells that will be needed before meshing begins. With modern mesh generation software, mesh resolution can be controlled in a variety of fashions. Most commonly, the minimum and maximum edge lengths to be used on the boundary surface of the problem are specified. These may be directly controlled, linked to local curvature, or manipulated in a variety of ways, the precise details of which may be commercial secrets that give one tool its edge in the marketplace over another.

Since the mesh used to solve any PDE is clearly an artificial construct introduced by the analyst, its precise topology and geometry should not affect the results of the computation (see also the following section on meshless methods). Equally, the simpler the mesh is, the easier it will be to set up and, of course, the faster computations can be carried out. Perhaps not surprisingly, the desire for solutions independent of mesh details and that for simple meshes pull in opposite directions. It is therefore important to be able to assess mesh quality so that high density is used only where strictly necessary. It is also important to have some measures of likely errors at the end of any solution, and that such errors should not be high in areas of importance. Mesh quality measures are also fundamental to adaptive meshing schemes that seek to automatically tune mesh density during the solution procedure.

These requirements have lead to the development of various measures of quality that may be used when building or refining computational meshes. Such measures fall into two basic types: those that may be used before solution begins and those that need a solution of some form. Not surprisingly, the former are simpler to set up and quicker to apply than the latter but they are also usually less powerful. They are, however, directly available during initial mesh generation. The fundamental limitation of *a priori* measures is that they cannot account for the need to adapt a mesh to suit highly nonuniform features in the solution. This situation is made worse if the problem being dealt with contains solution features with extremely elongated dimensions such as shocks or cracks, since, as has already been remarked, in such cases, highly stretched, but carefully aligned, mesh geometries may then be most efficient/appropriate.

The fundamental checks that can be carried out using *a priori* measures begin with angles – such methods generally attempt to ensure that all cells/elements are as nearly equilateral/square as possible and that high degrees of twist or skewness are avoided. It is also normal for such algorithms to ensure correct connectivity, the avoidance of gaps and the accurate modeling of boundary shapes before solution begins. Almost all commercial mesh generation tools provide such capabilities, and their use is nowadays routine in competent analysis. Commercial tools also usually have the ability to record the essentially manual process of initial mesh generation so that this process may be tracked, replayed, edited and used in any auditing or automated meshing schemes of the sort so fundamental to computer-controlled optimization methods – the fully automated mesh generator that may be employed on unseen geometry without any manual intervention still being some way off.

An alternative to producing a fully detailed mesh at the outset of the solution process is to use adaptive mesh generation schemes. In these, both the topology and geometry of the mesh change as results become available during solution of the problem being studied. To do this, the solvers are almost invariably set up to work in an iterative way so that both the mesh and results converge to a final solution. In principle, such approaches should allow very simple initial meshes to be specified by the designer and then for these to be updated as the solution progresses, with increased mesh detail only being added where needed (and even allow extraneous detail to be removed where not needed). Since in many cases the precise location of interesting phenomena such as shock waves will not be known *a priori*, adaptive meshes may be the only way to proceed if very fine overall meshes are to be avoided.

To achieve an adaptive capability, however, the mesh generator and field solver must be closely coupled so that the results from each solution iteration can be made available to the mesh generation tool, which can then decide if and where mesh adaptation/refinement is required. Clearly, this significantly complicates the development and operation of such methods. At present, the costs and difficulties involved in using adaptive meshes are only really justifiable if no alternative is possible. This commonly arises either because a suitably fine mesh everywhere would be too slow or because the boundary conditions are themselves changing during the solution process. This can arise in a variety of fields. In structural mechanics, such problems occur in studies of impact, penetration, fragmentation, large deformation and crack growth. In fluid mechanics, they mostly arise from unsteady fluid flow, particularly where there are separated boundary layers, changing boundaries (such as during deployment of flaps and slats on a wing), large vortices or shock waves. In all fields, if the geometry being studied changes, then remeshing is clearly necessary. Even when used, adaptive meshing by itself is usually of little benefit unless the basic field and geometry being studied are adequately resolved in the initial iterations. If they are not, far too much effort is wasted adapting the mesh for them to be feasible in practice. They are thus not a complete panacea to the difficulties of meshing.

Adaptive meshes that allow for complex physical behavior with changing geometries thus represent the most complex work normally encountered in mesh generation. Even so, the techniques needed to adapt a mesh are similar to those needed to create one in the first place. They rely on automated mesh generation and measures of errors in the solution that are being computed and schemes for altering the mesh to reduce them. Their study and development is thus closely intertwined. The basic building blocks from which an adaptive mesh control scheme can be assembled may be summarized as

- geometry assessment tools (checking angles, aspect ratios and skewness);

- error estimation tools (usually edge or cell based);

- mesh point movement tools (to reposition mesh points);

- mesh refinement and coarsening tools (to add/remove mesh points and cells/elements);

- mesh reconnection or topology manipulation tools (which use existing mesh points but redefine their connectivities);

- methods of coupling these tools to the solver; and

- algorithms for sequencing the tools in a consistent and convergent fashion.

Error estimation tools seek to establish the error in the prediction of some quantity of interest and the true value. Because error estimation must be relatively fast compared to overall solution, it is usually not appropriate to use excessively cumbersome methods in this part of the combined mesh refinement/analysis process. Traditionally, errors have been computed using interpolation-based methods for each cell or element in a solution. Increasingly, focus is now placed on computing errors for each edge, since such schemes allow for elements to become elongated if this is warranted by the solution; see, for example, Habashi et al. (2000). In either case, however, the analyst must decide on which quantity in the solution errors will be computed for. In CFD, for example, it is common to track errors in local pressure or Mach number. It is also possible to focus on overall errors in the solution by setting up adjoint models that link changes in the mesh to errors in integral quantities such as drag or lift; see, for example, the work of Giles and coworkers (Giles and Pierce 2002).

Having established where changes are needed in a mesh, the most fundamental alteration that may be considered is to change the locations of mesh points. For structured meshes, this may be the only simple change that can be invoked if a complete remeshing is to be avoided. In all cases, movements in the interior of a domain are usually quite straightforward, while on the edges, care must be taken to ensure that the original geometry of the model is adhered to (and this may additionally involve recourse to the original CAD model from which the mesh was derived).

The addition of new mesh points or their removal (in areas where accuracy is higher than required and to save computational effort) is usually only possible in unstructured meshes, where it can enable a relatively low density mesh to be used to start calculations that give early solutions to initiate the mesh adaptation process. This fact alone is increasingly driving designers to use unstructured meshes despite the greater computational efficiency possible with body-fitted, structured systems.

The final point worth noting about mesh generation is the impact of geometry optimization systems on the meshing process. If a gradient descent optimizer is being used to reduce some quantity such as drag while maintaining lift, then there will be a need to assess a series of related geometries. Moreover, the changes between successive geometries may be very small and carried out to gain sensitivity information such as objective function or constraint gradients. In such circumstances, the best approach to mesh generation is probably to maintain a topologically identical mesh and to just stretch and deform the cells/elements – such an approach is commonly called *morphing*. This can be very easy to set up and highly robust until such times as the meshes generated have elements that are too highly skewed or stretched to satisfy error measures. Then, a remesh may be needed with consequent small-scale discontinuities in finite differenced gradients, and so on. It may well then make most sense to tie the remeshing processes intimately into the search engine so that any topology changes occur when the search algorithm is least sensitive to it.

This process becomes much more difficult when using stochastic optimizers, which make significant jumps around the design space, such as genetic algorithms, simulated annealing, and so on. Although such methods do not use gradients, they still require the meshing system to give robust results. In either case, when a mesh or remesh is required that uses a new topology, there is a significant chance that manual intervention may be required. In such circumstances, it may be best to prepare, by hand, a series of master meshes that span the likely designs of interest and then to morph all other meshes from the nearest member of such a set. Of course, if the changes required between geometries are too wide ranging and complex, this may not be possible *a priori* and a fully automated search may then simply not be possible unless the search engine itself is able to deal with mesh failure cases in a robust sense (and such difficulties may well dictate the type of searches that are adopted).

2.2.4 Meshless Approaches

Given that meshing is such a difficult and often intensely manual process, a number of researchers are now focusing on schemes for solving PDEs without meshes. These are usually referred to as meshless or mesh-free approaches. The attractions are obvious: the need to construct mesh topologies that have appropriate element/cell densities and shapes is completely removed. There are, of course, a further set of difficulties that must be mastered to make such schemes work.

The fundamental characteristic of meshless methods is the absence of explicit connectivity information between the nodes or points used to define the domain of study. Instead, some form of interpolation scheme is set up with unknown coefficients and substituted into the PDE to be solved. Perhaps, the most commonly used interpolators in meshless methods are currently radial basis functions (RBFs) see, for example, the papers of Wendland (1999) or Larsson and Fornberg (2003), although a number of other approaches have been proposed (e.g., moving least square averages, see Belytschko et al. (1996), Li and Liu (2002) and the cloud methods of Liszka et al. (1996) – these references contain extensive bibliographies outlining the origins of such methods). RBFs originated from work on curve fitting scattered data in high-dimensional spaces and take a common overall form with a number of different kernels. They are constructed from a series of functions ϕ based at the nodes or sample points, $\mathbf{x}_j$:

$$y(\mathbf{x}) = \sum_{j=1}^{n} b_j \phi(\mathbf{x} - \mathbf{x}_j). \tag{2.7}$$

Here, the coefficients b are chosen so that the model interpolates the desired function at the nodes, for example, n equations are set up solve for the unknown coefficients b. The functions ϕ can take many forms: when applied to solving PDEs, multiquadric, Gaussian and thin plate spline kernels seem to be the most commonly used approaches. For thin plate splines, for example, ϕ is defined as:

$$\phi(\mathbf{x} - \mathbf{x}_j) = ||\mathbf{x} - \mathbf{x}_j||^2 \ln(||\mathbf{x} - \mathbf{x}_j||). \tag{2.8}$$

This method can be numerically poorly conditioned and so it is often better to use a modified approach that combines a simple polynomial with the basis functions:

$$y(\mathbf{x}) = a_0 + \sum_{i=1}^{m} a_i x_i + \sum_{j=1}^{n} b_j \phi(\mathbf{x} - \mathbf{x}_j). \tag{2.9}$$

In this form, the model has $m + n + 1$ unknowns and so a set of $m + 1$ additional constraints are added:

$$\sum_{j=1}^{n} b_j = 0 \quad \text{and} \quad \sum_{j=1}^{n} b_j x_{ij} = 0 \quad \text{for} \quad x_{ij} = 1, \ldots, m. \tag{2.10}$$

Now the basis functions model the differences between the polynomial and the function values. The resulting model still strictly interpolates the function data, but is numerically better behaved. The fundamental downside of this model is that here each RBF has the ability to affect the solution everywhere, albeit only to a limited extent. This global property leads to dense matrices, and so such schemes become very inefficient and often ill-conditioned as the number of mesh points rises. A number of nonglobal, multilevel and smoothing schemes have been proposed in the literature to overcome these limitations (e.g., see Fasshauer (1999)) but, as yet, no universally accepted approach has emerged.

Whatever scheme is used, the coefficients b_j are computed using a collocation scheme that ensures that the governing equations are satisfied at a scattered set of points in the interior of the domain. Additional equations are set up at nodes placed at the boundary of the domain to ensure that the specified boundary conditions are also satisfied. The density of nodes is chosen, as in traditional mesh-based schemes, to be higher in areas where there are complex features, either in the geometry being studied or in the behavior of the solution. The key difference from ordinary mesh-based schemes is that the nodes can be specified in any order and with any spacing and are not linked to each other by the definition of elements or cells. This brings great flexibility, particularly in providing adaptive meshes. Impressive results have been reported in the field of crack propagation in CSM, for example (Belytschko et al. 1996). Nonetheless, meshless methods have as yet had only a fraction of the development effort of more traditional methods such as FEA and so far are numerically often not so efficient during the solution process. Even so, if the tedious process of mesh construction can be simplified in this way, a lack of absolute computational efficiency is perhaps a small price to pay. In practical design office terms, however, they still have some way to go.

2.3   Analysis and Design of Coupled Systems

Having considered the manipulation of geometric models and the associated meshes required for discipline analysis, attention is next turned to how the needs of different disciplines impact on an integrated design process. The development of such integrated processes is, of course, a precursor to any attempts at multidisciplinary design optimization (MDO). The first thing to grasp about the analysis of coupled systems is that their design goals are often in tension with each other – reduced weight may lead to higher stresses, for example. At the same time, while many of the parameters used in a design may be under the control of, and of interest to, just a single set of disciplinary specialists, others may be of interest to multiple disciplines. Such competing concerns must be addressed and accommodated by the design system if the whole design team is to find the resulting process acceptable, particularly if all disciplines are to be considered concurrently so as to minimize design cycle times.

Given these problems, the starting point for multidisciplinary analysis (MDA) is to distinguish between those parameters, constraints and goals that are of concern to just a single discipline and those that impact on more than one. Next, the relationships between coupled parameters must be established: are the inputs to the analysis in one discipline the outputs from another, with no feedback or influences arising from the follow-on calculations, or must some kind of iterative scheme be adopted so that converged results can be obtained? Lastly, are there any constraints that must invariably be met or can those from one discipline be compromised, perhaps against goals from another? This last point is perhaps the single most difficult issue to deal with in MDA – all disciplines naturally wish for their own goals and constraints to be met and can often have a good insight as to which of their own targets might best be compromised for doing so – it is much harder to balance the needs of competing disciplines, especially if the levels of uncertainty inherent in any calculations vary widely between disciplines – why should an almost certain gain in aerodynamic efficiency be sacrificed because it is believed that the resulting design *might* be harder or more costly to manufacture?

Having considered the relationships between the quantities used to represent the design, attention must be turned to any disparities in the sophistication, run times, accuracies and uncertainties of the analysis codes used to represent the disciplines under consideration. If some of the analyses needed have significantly longer run times than others, then it is probable that fewer such calculations will be possible than in those areas where faster models are available – this may influence the sequencing of such calculations. Equally, if an analysis has significant uncertainties, it may be necessary to carry out sensitivity studies to assess the impact of any variations in the results coming from such a discipline before it is used to impact on those areas where there is less doubt about calculated results.

Finally, it is necessary to consider how tightly coupled the analysis codes coming from different disciplines need to be, to operate meaningfully. It will be clear that when studying the aeroelasticity of flexible wing structures the interaction between structural and aerodynamic codes is very significant. Conversely, the same two areas of analysis are relatively weakly coupled when considering the flows over essentially rigid missile fuselages. Notice that the degree of coupling depends both on how directly the variables of one discipline affect another and also how many such connections there are, i.e., one should allow for the breadth as well as the strength of any coupling. In this section, we introduce some of the issues that arise in dealing with coupled systems and some of the ways that they can be addressed – we leave more detailed consideration of multidisciplinary design optimization architectures to Chapter 9.

2.3.1 Interactions between Geometry Definition, Meshing and Solvers – Parallel Computations

As has already been made clear several times, geometry manipulation tools are of fundamental importance in the types of design considered here. When dealing with coupled MDA, decisions have to be taken on where and how to set up such tools. It is one of the central tenets of this book that where possible geometry manipulation should be carried out using the latest-generation feature-based parametric CAD tools rather than bespoke codes that are heavily customized to the task at hand. Although adopting CAD tools brings the most powerful geometry engines available to bear on the problem, the approach is not without its difficulties, however – and it is in dealing with the analysis of coupled systems that these are most severe. Nonetheless, we still believe that the benefits of using CAD tools to integrate the analysis needs of multiple disciplines are amply sufficient to justify their use.

To understand these issues more fully, consider briefly one of the most complex aspects of airframe design – that of studying aeroservoelasticity using distributed computational facilities and high definition CFD and CSM codes. Such studies are now commonly undertaken for military aircraft, since these are often dynamically unstable in flight without their active control systems – such instabilities allowing the highest rates of maneuver under combat conditions (Karpel et al. 2003). An aeroservoelasticity analysis requires a CSM solver to be run to understand the stiffness of the aerodynamic surfaces, a CFD solver to be run to model the aerodynamic forces and a (usually time domain state-space) control model to understand the behavior of the system under the action of controller inputs and control surface deflections.

Clearly, the various analysis codes involved work with differing systems of equations, differing geometrical models and differing forms of discretization. The CFD calculations will need highly accurate geometrical modeling around the leading edges of aerodynamic surfaces if boundary layers and shocks are to be accurately resolved. The CSM modeling may well be hierarchical and employ implicit structural optimization. For example, as major structural or geometrical parameters are changed, detail structural stiffening may be adapted to provide a locally optimal configuration of beams and stringers to prevent local buckling and carry point loads arising from payload and control surface attachment considerations. Such local optimizations may well involve hundreds of constraints and variables.

Dealing with all the required geometrical data within a large-scale commercial CAD package in such circumstances often requires a considerable investment of staff effort by engineers intimately familiar with the internal programmatic workings of the CAD code. Moreover, there will be a need to couple the geometry to the mesh generation schemes needed by the CSM and CFD codes and these (different) meshes must then be mapped onto the available computing resources, most probably using parallel solution techniques built on tool sets such as MPI (the message passing interface system); see, for example, Tang et al. (2003). Figure 2.22 illustrates the typical data flows found when carrying out this process serially, with each discipline search following the next and returning to the beginning after each geometry optimization. Although such an approach is easy to understand and generally

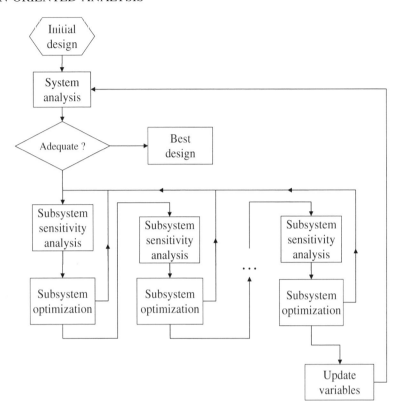

Figure 2.22 Sequential multidisciplinary optimization process workflow (conventional block diagram).

stable, it can be rather slow in practice. Figure 2.23 shows the same process but now using a so-called N-square diagram, which can be helpful in laying out complex data flows. In N-square diagrams, all data inflows to processes are shown as vertical lines and all outflows as horizontal lines. Such a diagram can be readily turned into a data-dependency matrix, of course – the blobs on the diagram showing where entries occur in the matrix. An analysis of such data dependencies is an important early step in designing computer-based design processes, as this can help highlight any bottlenecks or inconsistencies that might otherwise only become apparent late in the day.

If it is necessary to run large numbers of design configurations through such a system to establish the robustness of proposed designs, either to uncertainties in the analysis or the environmental or payload conditions, the run times that can be tolerated for each full aeroelastic calculation will then be limited – this may force the adoption of various approx-imations or metamodeling techniques to provide workable design systems. For example, a detailed CSM model may well reveal a wide variation in the modal behavior of the structure – many of the resulting modes may have only limited impact on the aerodynamic behavior. Therefore, such modes may typically be ignored to reduce the order of the result-ing structural dynamics model. Alternatively, it may be more efficient to replace much of the available detail in the CAD model with equivalent smeared elements before carrying out

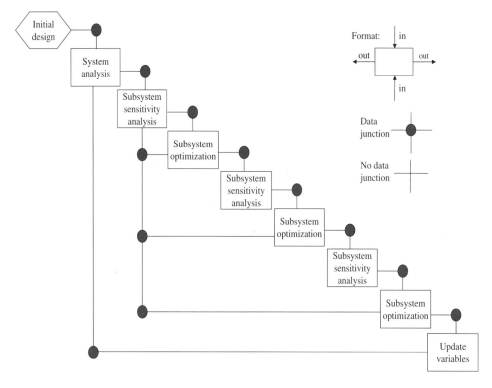

Figure 2.23   Sequential multidisciplinary optimization process workflow (N-square diagram).

the structural analysis. Further, the structural model may in reality be nonlinear (Patil et al. 2000) and linearization may be required. Similarly, it may well be impossible to always use the highest resolution RANS codes to study the aerodynamic behavior and less accurate but much faster panel codes may be used much of the time during design iterations – this may bring the added benefit of simpler meshing requirements. Moreover, the results of the CSM analysis must be linked to those of the CFD runs, despite the fact that rather different geometrical representations are used in the two forms of analysis. It may thus be seen that this entire process should be viewed as a whole when planning what codes to use on what machines and at what fidelity, even if full resolution CSM and CFD models are available in principle.

Whatever the fidelity of the codes used, it will be clear that in aeroservoelasticity there is a natural order in the analysis procedure – the effects of the aerodynamic forces on the structure can only be found after the structural analysis is complete. Similarly, a control model can only be built when the fluid-structure interaction behavior is captured in some way. Even so, the results of the control calculations will be felt in the structural analysis, since the forces required by the controller must be borne by the structure, and so on. A concurrent evolution of the design in structural, aerodynamic and control spheres is thus difficult to reconcile with the needs of aeroservoelasticity analyses. The serial approach of Figures 2.22 and 2.23 may then have to be accepted and, moreover, iterated a number of times, despite the natural desire for shortened design times (Karpel et al. 2003). In fact, it turns out to be rare for calculations in aerospace design to be purely hierarchical, so that the effects of downstream calculations have little or no effect on those that precede them – a jet engine in a nacelle mounted on the rear of the fuselage of a transport aircraft comes close, but such configurations are unusual.

An alternative view of the issues arising in MDA can be gained by examining the Bi-Level Integrated System Synthesis (BLISS) approach of Sobieszczanski-Sobieski and coworkers (Sobieszczanski-Sobieski et al. 1998, 2003, 2000). In the BLISS approach, design parameters are divided into key overall variables and local ones: the local variables are optimized by domain-specific procedures, often concurrently with other domains and these searches are then interleaved with overall optimization steps carried out on the overall variable set. While local searches are under way, the system-level quantities are frozen and vice versa – when the system-level variables change, local variables are kept fixed. The two search steps are coupled and information passes between them; see Figure 2.24. They are mapped onto the available processing power in a way that best matches the various module requirements, exploiting parallel processing where possible. In the original variants of BLISS, the coupling was effected via the use of system sensitivity derivatives of the system-level objectives with respect to the subsystem variables. Later variants attempt to work without such sensitivities since they can be difficult and expensive to compute. A further emerging thread in the BLISS system is the use of Design of Experiment and

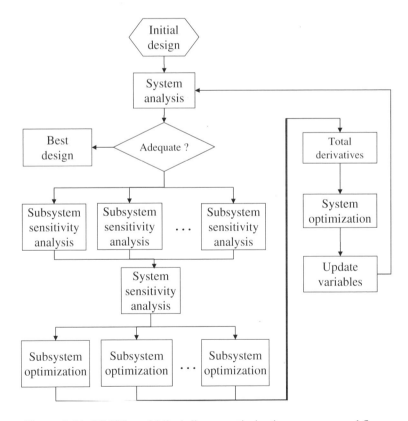

Figure 2.24 BLISS multidisciplinary optimization process workflow.

Response Surface Methods to represent subsystem activities by metamodels so as to allow off-line use of full fidelity codes and reduced search run times with the resulting metamodels. A similar approach using neural network metamodels has been proposed by Bloebaum and coworkers in their Data Fusion Analysis Convergence (DFAC) system (Hulme et al. 2000). This focuses on the use of metamodels combined with MPI parallel architectures. The trade-off between short run times and convergence stability tend to dominate such schemes. Moreover, as the complexity and sophistication of the process increases, the design team need to learn an additional set of skills related to the integration architecture and how best to apply it to the problem in hand – this can be a key element affecting the take-up of new processes – if they are not simple to understand, the learning threshold involved can prevent their adoption by designers.

In the attempt to even further decouple the domains in MDA, a number of workers have proposed using delegate or "blackboard" based design methods (Nair and Keane 2002a). In these systems, each domain is fully decoupled from all the others and works autonomously except that it sends a delegate to each other discipline to represent the variables it is working with and receives similar delegates from the other disciplines (i.e., each domain "chalks" its decisions on a common "blackboard" while continuing to refine its variables – other disciplines read any information they require from the "blackboard" whenever they need it); see Figure 2.25. This approach is modeled on the way human design teams often work in

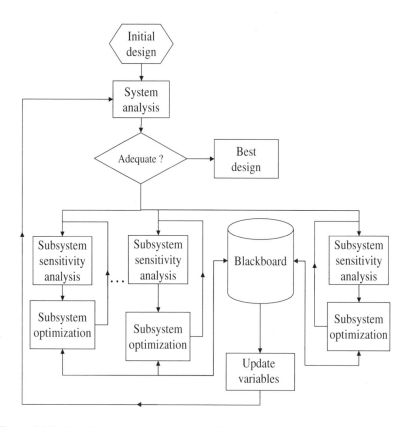

Figure 2.25 Blackboard-based multidisciplinary optimization process workflow.

practice – In companies with separate disciplinary design offices, it is quite normal for the aerodynamics design team to provide information on their favored design to the stress team while continuing to experiment with a variety of alternatives. At the same time, they might be using structural information that the stress office has supplied based on their favored layouts. Such schemes of course have multiple values in use for those variables that are shared across disciplines. Unfortunately, no formal convergence properties are guaranteed when working in this way and such systems can become stuck in loops where variables are tugged backward and forward by competing disciplines – compromise and negotiation between disciplines then needs to be invoked. Further research using various relaxation schemes and game theory approaches is under way to try and make this kind of computational design more robust and widely applicable. When it can be made to work, however, it allows highly decoupled, concurrent progress across a range of disciplines. Such an approach is, moreover, not far from the latest variants of the BLISS approach (Sobieszczanski-Sobieski et al. 2003) and so it may be that some kind of a consensus on how best to proceed in this field is beginning to emerge.

2.3.2 Simple Relaxation and Newton Techniques

In all the above schemes, there is a fundamental need to reconcile or converge the values of the variables affecting multiple disciplines. To illustrate in a more concrete fashion some of the issues arising from such coupled analysis codes, consider a simple example with just two coupled equations $f(x, y) = 0$ and $g(y, x) = 0$, each being a function of two variables x and y. $f(x, y)$ is primarily dominated by x but also depends on y and similarly $g(y, x)$ is mainly a function of y but also depends on x. We also assume that the equations are nonlinear. Here, we chose

$$f(x, y) = x^3 + 3 - y = 0 \quad \text{together with} \quad g(y, x) = y^2 - 2 + x = 0 \qquad (2.11)$$

and the goal is to find a solution that satisfies both equations by some kind of iterative scheme. In reality, the functions $f(x, y)$ and $g(y, x)$ would be sophisticated black-box CSM or CFD codes and the relationships being dealt with would be implicitly defined within these codes and vastly more complex than that arising from these simple equations. Moreover, x and y would both be vectors of perhaps hundreds of variables.

The simplest way to solve such a problem is via relaxation (or fixed point iteration) whereby we guess a solution, substitute one half of it into one problem and solve for the other, which we then use to update the opposite half of the solution, and so on. This is in the spirit of the BLISS scheme mentioned above, in that each analysis takes turns to move the solution forward. To do this, we set up an auxiliary or response process that links the variables in use, that is, a way to calculate the values of one subset of the variables that satisfy part of the overall problem, given values of the remaining variables. By way of example, assume we start with an initial estimate of $y = 0$. We then solve the equation for $f(x, y)$ to gain an estimate of x, that is, we set up the response equation $\hat{x}(y) = \sqrt[3]{y - 3}$. So, for our initial guess of $y = 0$, this is the cube root of -3, or -1.4422. We then substitute this estimate of x into the response equation for $g(y, x)$ to gain an improved estimate of y, that is, $\hat{y}(x) = \sqrt{2 - x}$, or here, the square root of 2 plus 1.442, or 1.8553. We can then iterate this process and after about seven loops we recover the solution of $(x, y) = (-1.0761, 1.7539)$. Such a process is extremely simple to set up, even using precompiled third party black-box analysis codes, but can be slow to converge and, if the coupling between the systems is too strong, can fail to converge at all. It is also inherently sequential in nature: first one code is analyzed and then the other – clearly this limits the scope for concurrent processing.

As a result of such limitations, a great deal of research has been carried out to find better ways to deal with coupled problems. These can be broken down into various categories. The most basic distinction is between problems where the relationships are linear or nonlinear – if they are linear, they are usually simple to deal with – unfortunately, most of the problems of interest in aerospace design are nonlinear. One important set of exceptions, however, is linear statics problems being solved by FEA means where, because of problem size, it is convenient to decompose the overall stiffness matrix into subdomains that can be partitioned over multiple computers to give parallel speed-ups. Such schemes lie outside the scope of this section and are perhaps best thought of as computationally expedient extensions to linear algebra methods. Here, focus lies on nonlinear problems where the coupling is between wholly different domains of interest, such as fluid/structure interactions. Even so, further subdivisions are possible.

Most commonly, the problems being tackled are not set down explicitly and the user does not have direct access to the computational methods being used to solve each problem domain. Sometimes, however, the codes being dealt with may be specifically set up to deal with a coupled problem and advantage can be taken of access to the inner workings of the solvers – many aeroelasticity codes work in this way. Another variant that may sometimes be found is where sensitivity information may be available directly, either via direct numerical differentiation or by adjoint codes – in such cases, the solution method can, if chosen carefully, take significant benefit from this information.

The simplest way to use sensitivity information is via Newton-based schemes, where the updates to the current estimates of the problem variables are found using gradients, that is, we solve the linear set of equations given by

$$\mathbf{DR}(\mathbf{u})\Delta\mathbf{u} = -\mathbf{R}(\mathbf{u}) \tag{2.12}$$

where $\mathbf{R}(\mathbf{u})$ are the residuals between the current estimates of the variables and the response variables, $\Delta\mathbf{u}$ are the updates to the variables and $\mathbf{D}$ is the Jacobian (Aluru and White 1999); see also Haftka and Gurdal (1992). For our two-variable problem, this leads to

$$\left[\begin{array}{c} \Delta x \\ \Delta y \end{array}\right] = \left[\begin{array}{cc} 1 & -\partial\hat{x}(y)/\partial y \\ -\partial\hat{y}(x)/\partial x & 1 \end{array}\right]^{-1} \left[\begin{array}{c} x - \hat{x}(y) \\ y - \hat{y}(x) \end{array}\right] \tag{2.13}$$

where $\partial\hat{x}(y)/\partial y$, and so on, are the sensitivities of the auxiliary or response processes to the variables, that is, here,

$$\partial\hat{x}(y)/\partial y = \partial/\partial y((y-3)^{1/3}) = \frac{1}{3(y-3)^{2/3}} \quad \text{and} \tag{2.14}$$

$$\partial\hat{y}(x)/\partial x = \partial/\partial x(\sqrt{2-x}) = \frac{1}{2\sqrt{2-x}}. \tag{2.15}$$

To use this method, the initial guesses for x and y are used to calculated the responses $\hat{x}(y)$ $= (y-3)^{1/3} = -3^{1/3}$ and $\hat{y}(x) = \sqrt{2-x} = \sqrt{2}$ together with their sensitivities $\partial\hat{x}(y)/\partial y$ $= -\frac{1}{3^{2/3}}$ and $\partial\hat{y}(x)/\partial x = -\frac{1}{2\sqrt{2}}$. Solution of the matrix equation involving the Jacobian then leads to the values for the updates Δx and Δy as being -1.1504 and 1.8210, which in turn leads to the new estimates of x and y of $0 + (-1.1504) = -1.1504$ and $0 + 1.8210 =$

1.8120. This process can then be iterated until convergence and it yields $(-1.0765, 1.7542)$ at step two and then $(-1.0761, 1.7539)$ at step three, that is, the same answers as before in three loops rather than seven (for more complex problems, the improvement in convergence can be even more striking). The equation solution steps can also be carried out in parallel (although the Jacobian step requires that the calculations be synchronized at that stage). Two disadvantages are, however, first that sensitivity information is required (and it is of course not normal to be able to calculate such sensitivities by direct differentiation) and secondly a matrix equation has to be solved, although unless very many variables are being used this is not usually a severe problem (there are many efficient schemes for solving such systems of equations, of course).

The calculation of the required sensitivities in Newton schemes is normally provided by forward finite differencing, and since this is a noisy process that requires good control of the FD steps, it often prevents the quadratic convergence that is theoretically possible with these methods as they approach the minimum (note how rapidly the previous example converged – if the problem equations used had not involved a cubic term, one iteration would have been sufficient to guarantee convergence when using exact expressions for the differentials). This is a topic of current research (see, for example, Kim et al. (2003)) and a number of methods have been proposed to try and overcome such problems by numerical means, such as via generalized Jacobi or Gauss–Seidel iterative schemes.

It remains the case, however, that unless one has direct access to the sensitivity information either via numerical differentiation or adjoint methods, there are always likely to be limitations in the convergence properties that can be achieved in practice with black-box analysis codes. It is also the case that if many problem areas are being dealt with simultaneously by finite differencing, the number of analysis calculations required rises substantially and the speed-up in convergence over simple relaxation schemes can then be lost.

It should also be noted that the final solution converged to (assuming there is one) will depend very strongly on the starting point for the iterative procedure, whatever its nature. If significantly different starting values are chosen, it may well be possible for a range of stable converged solutions to emerge. This may require the design team to impose additional constraints on the process to ensure that particular goals are met or perhaps that all the available solutions should be considered when making design choices – certainly, it is always wise to try several start points before assuming that any converged solution is the only one available. Such studies are in some senses equivalent to the convergence studies always recommended when carrying out CSM or CFD work to ensure that mesh dependency issues have been addressed – and it is notoriously true that such studies are often omitted in design offices when tight schedules are imposed on the design process.

2.3.3 Systems Integration, Workflow Management, Data Transfer and Compression

Given a number of analysis modules, some kind of design database (here, CAD tools are recommended), a strategy to ensure compatibility between the competing disciplines and suitable computing facilities, integration of the available tools to provide a complete design system requires that issues of data transfer and execution control be addressed. For small-scale problems using simple tools and few variables, such considerations are rarely of great concern – almost any architecture can be made to work satisfactorily. However, when dealing with large models with thousands of variables and multiple full-scale analysis codes, real difficulties can arise. The earliest systems designed to tackle such problems tended to be large, hard-coded monolithic systems, often running on the largest available mainframe computers in Fortran or C – some such codes still exist, but they are increasingly considered to be too cumbersome to operate or maintain. The latest approaches to these problems are being built following distributed object oriented approaches using graphical workflow composition tools, scripting languages and open standards communication protocols such as XML (the eXtensible Markup Language, that can be used for building web applications, etc.) and SOAP (the Simple Object Access Protocol, a simple XML-based protocol to let applications exchange information over the web).

The designs of modern frameworks seek to provide a range of capabilities that are increasingly considered necessary for corporations to remain competitive. The following list of desiderata is based on the work of Marvis and colleagues (Zentner et al. 2002), but also reflects the authors' own views:

1. design (as opposed to analysis) orientation allowing "what if" type approaches;

2. use of existing PDE-based modeling codes distributed across multiple and dispersed computing systems;

3. approximations via metamodels (using Kriging, RBFs, neural networks, etc.);

4. decomposition to allow concurrency and variability in the fidelity of the modules used;

5. search and optimization to allow automated design improvement at both overall system level and also at domain level;

6. good human-centric interfaces to support the activities of the (probably dispersed) design team.

It may be seen from this list that the overall architecture of the design system needs to be considered from the outset. Moreover, emphasis needs to be given to the ability to "plug and play" various modules and to distribute them across heterogeneous computing environments. By using appropriate environments, XML wrapping and SOAP protocols, such systems can be set up in just this way. Of course, the emerging Grid and web services standards are also impacting on the way such systems are constructed. What is common across modern approaches is the need to provide a backbone that is flexible and usable by noncomputer scientists and a high level of abstraction in the consideration of modules, data flows and on-line documentation. It is crucial that the core design team be able to assemble and modify the components used at any time in a speedy fashion and without recourse to specialist computer science staff.

Design system architectures are clearly impacted by the supporting computing environment and the capabilities/suitability of the analysis and CAD codes in use. Perhaps less obviously, they also heavily depend on the complexity of the workflows to be used and the way these will be set up and executed. Large design systems almost always need some kind of scripting language to allow high-level programming of the work to be undertaken. It is rarely the case that design teams wish to always run the same codes in the same sequences coupled in the same ways. It is much more common for such workflows to be tailored to the problem in hand, albeit being based on previous processes from earlier, similar studies.

Further, it is quite routine for workflows to need adapting in the face of results coming from the ongoing design process, emerging design needs and analysis outputs, and so on. Given that the designers involved are often not computer scientists and that they will wish to carry out this assembly process in a familiar and easy to use environment, it is clear that languages such as C++ and Java will not always be the tools of choice – the growth in the use of graphical programming tools and spreadsheets instead of more traditional programming languages emphasizes this way of working. However, care must be taken to provide systems that favor robust ways of working and do not store up debugging and maintenance problems for the future – many engineers will know how easy it is to make mistakes when building spreadsheet-based approaches and how complex these can become if used without appropriate discipline.

At the time of writing, a number of environments are becoming popular in this role, including Python[20], Tcl/Tk[21] and Matlab®[22]. There are also a number of tools now available that aim to aid the transfer of data between other codes and which allow mapping between differing levels of discretization and of geometry detail modeling, such as MpCCI[23] or Dakota[24]. The OpenCascade[25] open-source CAD system can also be useful in passing data between applications since, being open source, user can customize its capabilities to the problem in hand. These various tools can then be combined with graphical programming tools that allow visual inspection (and sometimes editing) of the workflows being proposed. Such workflows can be considered from a high-level perspective to see if they meet the team's requirements. Good design systems need to allow the finalized workflow to be rapidly and easily mapped onto the available computing languages and infrastructure, analysis modules and databases. Figure 2.26 illustrates the kind of graphical workflow map that designers find useful to work with. Notice that the ability to expand high-level modules in such flows by "zooming in" can be extremely effective in allowing detailed control while still providing a clear overall view of the main design process. Such views fit very naturally with the kinds of multistage optimization strategies that are needed when applying response surface based approaches over multiple disciplines, and so on. It should of course be noted, that even if the workflows to be used are specified in an appropriate high-level or graphical environment, this need not mean that the final processing be carried out without recourse to the most powerful computing languages – it is quite normal for workflow enactment engines to be built using things like C++ or Java, for example.

When executing workflows, it is also important to provide good monitoring tools to allow detailed investigations of any emerging features of interest. Some of the most creative moments in design can be stimulated by observing the way automated scripts are driving a design and then probing further at opportune points and posing "what if" type questions. Further, any database tools in use should support comparison and roll-back of design decisions, without the need to recompute all calculations. This can be particularly important when compute times become long since good database tools can then allow rapid "animations" of design changes, which again can be extremely stimulating and beneficial for the design team.

---

[20]http://www.python.org
[21]http://www.tcl.tk/man
[22]http://www.mathworks.com
[23]http://www.scai.fraunhofer.de/mpcci.0.html
[24]http://endo.sandia.gov/DAKOTA
[25]http://www.opencascade.org

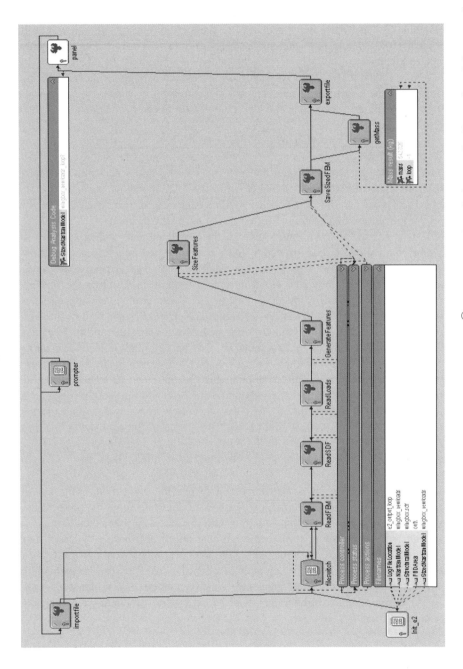

Figure 2.26　Typical graphical workflow design tool (ModelCenter® http://www.phoenix-int.com/). Reproduced by permission of Peter. S. Young.

Of course, the direct mapping of script-based workflows and the use of open communication standards with XML wrappers and the like can significantly add to the overheads of computational processes when compared to monolithic code structures. Thus, if tens of thousands of variables need to be passed between analysis modules running in parallel at high speeds to allow rapid concurrent processing, it may be necessary to take mitigating steps if performance is not to suffer, even given advances in network speeds and bandwidth. A number of approaches can be taken to deal with such issues. First, data compression and reduction methods can be employed. Alternatively, it may be possible to link modules at a metadata level with the metadata referring to common large data repositories that are optimized for speed of access rather than for open standards exchanges – although such an approach limits the flexibility of the resulting tools, for work within a single corporation, this may allow enough choice in the composition of the tool set to be used while not impinging on the system's performance or greatly restricting what can be achieved when rebuilding the available blocks in new ways. At the same time, it may be possible to design the workflow enactment engine to exploit its knowledge of the available computing hardware (and perhaps even other loads on it) so as to maximize throughput. In this way, much of the computer science needed to support more advanced ways of working can be "hidden under the hood" of the design system – ideally, they should appear almost simplistic to the user while providing much inbuilt sophistication – just as self-optimizing compilers now do in ways that engineers simply take for granted. The Geodise toolkit[26] written by the authors and colleagues and described in Chapter 16 sets out to adopt this way of working.

Alongside the need to be able to specify, reuse and adapt workflows, integrated MDA also requires that the differing discretization needs of the various analysis domains be met. As has already been pointed out, the meshing processes that are a precursor to almost all PDE solvers are carried out in a fashion that is highly dependent on the equations being solved and the geometrical representation of the product being analyzed. This will mean that if the results coming from one code are needed as inputs by another it is quite likely that some form of data interpolation/extrapolation scheme will be required. For example, in aeroelastic analysis, the deformation of the structure caused by fluid loading must be used to influence the shape analyzed using CFD and, similarly, the loadings on the structure used in CSM must be taken from the CFD solution (Giunta 2000). This will require essentially geometrical operations to map the relevant data from one application to the other. A number of approaches have been proposed for making such transfers (Smith et al. 1995) and schemes based on RBFs have much to offer in this context. There are also several commercial tools available to carry out such transfers. For example, the MpCCI tool[27] allows transfer between varying discretizations, automatic neighborhood computation, a variety of interpolation schemes, multiple coordinate systems, and so on.

Again, it is the view of the authors that such tools are best written within the underlying CAD package, even if this means providing tools that are not normally directly used in CAD systems to manipulate geometry. The fact that the CAD system in use has a good NURBS engine does not always mean that it is best to create a NURBS surface to map this kind of data between applications. Nonetheless, holding the data and mapping process in a single tool alongside the geometry itself offers many advantages in terms of data control and visualization, to say nothing of helping reduce the proliferation of tools on the designer's desktop.

---

[26]http://www.geodise.org/
[27]http://www.scai.fraunhofer.de/mpcci.0.html

3

Elements of Numerical Optimization

Having outlined the design processes common in much of the aerospace sector and highlighted some of the issues that arise when trying to bring together multiple disciplines in a single computational environment, including the subject of design search and optimization (DSO), attention is next focused on the subject of formal optimization methods and the theory underlying them. These increasingly lie at the heart of much computationally based design, not because designers necessarily believe that such methods can yield truly optimal designs, or even very good designs, but rather because they provide a systematic framework for considering some of the many decisions designers are charged with taking. Nonetheless, it is helpful when looking at computational approaches to design to have a thorough understanding of the various classes of optimization methods, where and how they work and, more importantly, where they are likely to fail.

As has already been noted in the introduction, optimization methods may be classified according to the types of problems they are designed to deal with. Problems can be classified according to the nature of the variables the designer is able to control (nonnumeric, discrete, real valued), the number and type of objectives (one, many, deterministic, probabilistic, static, dynamic), the presence of constraints (none, variable bounds, inequalities, equalities) and the types of functional relationships involved (linear, nonlinear, discontinuous). Moreover, optimizers can be subclassified according to their general form of approach. They can be gradient based, rule based (heuristic) or stochastic – often, they will be a subtle blend of all three. They can work with a single design point at a time or try to advance a group or "population" of designs. They can be aiming to meet a single objective or to find a set of designs that together satisfy multiple objectives. In all cases, however, optimization may be defined as the search for a set of inputs $\mathbf{x}$ that minimize (or maximize) the outputs of a function $f(\mathbf{x})$ subject to constraints $g_i(\mathbf{x}) \geq 0$ and $h_j(\mathbf{x}) = 0$.

In essence, this involves trying to identify those regions of the design space (sets of designer chosen variables) that give the best performance and then in accurately finding the minimum in any given region. If one thinks of trying to find the lowest point in a geographical landscape, then this amounts to identifying different river basins and then

Computational Approaches for Aerospace Design: The Pursuit of Excellence. A. J. Keane and P. B. Nair
© 2005 John Wiley & Sons, Ltd

tracking down to the river in the valley bottom and then on to its final destination – in optimization texts, it is common to refer to such regions as "basins of attraction" since, if one dropped a ball into such a region, the action of gravity would lead it to the same point wherever in the basin it started. The search for the best basins is here termed *global optimization*, while accurately locating the lowest point within a single basin is referred to as *local optimization*. Without wishing to stretch the analogy too far, constraints then act on the landscape rather like national borders that must not be crossed. It will be obvious that many searches will thus end where the "river" crosses the border and not at the lowest point in the basin. Of course, in design, we are often dealing with many more than two variables, but the analogy holds and we can still define the idea of a basin of attraction in this way. The aim of optimization is to try and accomplish such searches as quickly as possible in as robust a fashion as possible, while essentially blindfolded – if we had a map of the design space showing contours, we would not need to search at all, of course.

Notice that even when dealing with multiple design goals it is usually possible to specify a single objective in this way, either by weighting together the separate goals in some way, or by transforming the problem into the search for a Pareto optimal set where the goal is then formally the production of an appropriately spanning Pareto set (usually one that contains designs that represent a wide and evenly spread set of alternatives). It is also the case that suitable goals can often be specified by setting a target performance for some quantity and then using an optimizer to minimize any deviations from the target. This is often termed *inverse design* and has been applied with some success to airfoil optimization where target pressure distributions are used to drive the optimization process. Note that throughout this part of the book we will tackle the problem of minimization, since maximization simply involves a negation of the objective function.

3.1   Single Variable Optimizers – Line Search

We begin this more detailed examination of optimization methods by first considering the problem of finding an optimal solution to a problem where there is just one real-valued variable – so-called line search problems. Although rarely of direct interest in design, such problems illustrate many of the issues involved in search and serve as a useful starting point for introducing more complex ideas. Moreover, many more sophisticated search methods use line search techniques as part of their overall strategy. We will leave the issue of dealing with constraints, other than simple bounds, until we come to multivariable problems.

3.1.1   Unconstrained Optimization with a Single Real Variable

The first problem that arises when carrying out line search of an unbounded domain is to try and identify if a minimum exists at all. If we have a closed form mathematical expression for the required objective, this can of course be established by the methods of differential calculus – the first differential is set equal to zero to identify turning points and the sign of the second differential checked at these to see if it is positive there (consider $f\{x\} = x^2 - 4x - 1$; then, $df/dx = 2x - 4$ and $d^2f/dx^2 = 2$, so there is a minimum at $x = 2$). Clearly, in most practical engineering problems, the functional relationship between variables and goals is not known in closed form and so numerical schemes must be adopted. In such circumstances, the simplest approach to finding a minimum is to make a simple step in either positive or negative direction (the choice is arbitrary) and see if the function

reduces. If it does, one keeps on making steps in the same direction but each time of twice
the length until such time as the function starts to increase, a bound on the variable is
reached or the variable becomes effectively infinite (by virtue of running out of machine
precision). If the initial step results in the function increasing, one simply reverses direction
and keeps searching the other way. Given the doubling of step length at each step, one
can begin with quite small initial moves and yet still reach machine precision remarkably
quickly. Many alternative methods for changing the step size have been proposed and that
adopted in practice will depend on the presence or absence of bounds. If the problem is
bounded, then continuously doubling the step length is clumsy, since it is likely to lead
to unnecessarily coarse steps near the bounds. At the other extreme, using fixed steps may
mean the search is tediously slow. An alternative is to fit the last three points evaluated to a
parabola and then to take a step to the location predicted as the turning point (as indicated
by the parabola's differential and assuming that it has a minimum in the direction of travel).

As with all numerical approaches to optimization, one must next consider when this
kind of strategy will fail. Simply stepping along a function is only guaranteed to bracket
a minimum if the domain being searched contains a single turning point, as Figure 3.1
makes clear. If there are multiple minima, or even just one minimum, but also an adjacent
maximum (as in the figure), the strategy can fail. This brings us immediately to the idea
of local search versus global search. Local search can be defined as searching a domain
in which there is just a single minimum and no maxima (and its converse when seeking
a maximum) – anything else leads to global search. Thus, we may say that the strategy
of trying to bracket a minimum by taking steps of doubling lengths is a form of *local*
search since it seeks to locate a single turning point. Starting with a smaller initial step
size, or at a location closer to the minimum, will, of course, tend to bracket the optimum
more accurately. This will also better cope with situations where there are multiple turning
values, although there are still no guarantees that any minima will be found. It turns out
that, even with only one variable, no search method can guarantee to find a minimum in

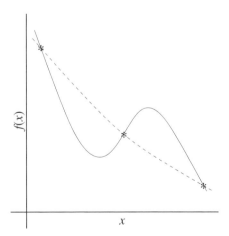

Figure 3.1 Attempting to bracket a function with two turning points (solid line) – the *'s
mark the sampled points and the function appears to reduce smoothly to the right (dashed
line).

the presence of multiple minima and maxima. The best that can be hoped for when dealing with *global* search problems is that a method is *likely* to find a good optimum value.

Assuming, however, that it is possible to bracket the desired minimum (by finding a triple x_1, x_2, x_3 where $f(x_2)$ is less than both $f(x_1)$ and $f(x_3)$), the next requirement in line search is that the bounds enclosing the turning point be squeezed toward the actual optimum as quickly as possible. This immediately raises issues of accuracy – if a highly accurate solution is required, it is likely that more steps will be needed, although this is not always so. Note also, that in the limit, all smooth, continuous functions behave quadratically at a minimum, and therefore the accuracy with which a minimum can generally be located is fixed by the square root of the machine precision in use, that is, to about 2×10^{-4} in single precision and 2×10^{-8} in double precision on 32-bit machines. Next, in keeping with our classification of optimizers as being gradient based, rule based or stochastic, we describe methods for homing in on the optimum of all three types.

The simplest scheme that can be adopted is to sample the bracketed interval using a series of random locations, keeping the best result found at each stage until as many trials as can be afforded have been made. This approach is easy to program and extremely robust, but is slow and inefficient – such characteristics are often true of stochastic schemes. It can be readily improved on by using successive iterations to tighten the bounds for random sampling on the assumption that there is a single local minimum in the initial bracket (although this might be argued as shifting the search into the category of heuristic methods). Random one-dimensional searches should ideally be avoided wherever possible – certainly, they should not be used if it is believed that the function being dealt with is smooth in the sense of having a continuous first derivative and it is possible to bracket the desired minimum. Sometimes, however, even in one dimension, the function being studied may have many optima and the aim of search is to locate the best of these, that is, global search. In such cases, it turns out that searches with random elements have an important role to play. We leave discussion of these methods until dealing with problems in multiple dimensions.

The next simplest class of methods, the rule based or heuristic methods, lead to a more structured form of sampling that directly seeks to move the bounds together as rapidly as possible. There are a number of schemes for doing this and all aim to replace one or other (or both) of the outer bounds with ones closer to the optimum and thus produce a new, tighter bracketing triplet. The two most important of them are the golden section search and the Fibonacci search. The fundamental difference between the two is that to carry out a Fibonacci search one must know the number of iterations to be performed at the outset, while the golden section search can be carried on until sufficient accuracy in the solution is achieved. If it can be used, the Fibonacci is the faster of the two but is also the less certain to yield an answer of given accuracy. This is another common distinction that arises in search methods – does one have to limit the search by the available time or can one keep on going until the answer is good enough. In much practical design work, searches will be limited by time, particularly if the functions being searched are computationally expensive to deal with. In such circumstances, the ability to use a method that will efficiently complete in a fixed number of steps can be a real advantage.

The golden section search places the next point to evaluate $(3 - \sqrt{5})/2 (\approx 0.38197)$ into the larger of the two intervals formed by the current triplet (measured from the central point of the triplet as a fraction of the width of the larger interval). Provided we start with a triplet where the internal point is 0.38197 times the triplet width from one end, this will mean that whichever endpoint we give up when forming the new triplet, the remaining three points will still have this ratio of spacing. Moreover, the width of the triplet is reduced by the same amount whichever endpoint is removed, and thus the search progresses at a uniform (linear) rate. To see this, consider points spaced out at $x_1 = 0.0$, $x_2 = 0.38197$ and $x_3 = 1.0$. With these spacings, x_4 must be placed at 0.61804 $(= 0.38197 \times (1.0 - 0.38197) + 0.38197)$ and the resulting triplet will then be either $x_1 = 0.0$, $x_2 = 0.38197$ and $x_4 = 0.61804$ or $x_2 = 0.38197$, $x_4 = 0.61804$ and $x_3 = 1.0$. In either case, the triplet is 0.61804 long and the central point is 0.23607 $(= 0.38197 \times 0.61804)$ from one end. Even if the initial triplet is not spaced out in this fashion, the search rapidly converges to this ratio, which has been known since ancient times as the "golden section" or "golden mean", which leads to the method's name.

The Fibonacci search proceeds in a similar fashion but operates with a series due to Leonardo of Pisa (who took the pseudonym Fibonacci). This series is defined as $f_0 = f_1 = 1$ and $f_k = f_{k-1} + f_{k-2}$ for $k \geq 2$ (i.e., 1, 1, 2, 3, 5, 8, 13, 21, ...). To apply the search, the bounds containing the minimum are specified along with the intended number of divisions, N. Then the current triplet width is reduced by moving both endpoints toward each other by the same amount, given by multiplying the starting triplet width by a reduction factor t_k at each step k, that is, the two endpoints are each moved inward by $t_k \times (x_2 - x_1)/2$. The reduction factor is controlled by the Fibonacci series such that $t_k = f_{N+1-k}/f_{N+1}$, which ensures that (except for the first division) when moving in each endpoint, one or other or both will lie at previously calculated interior points; see Figure 3.2 (note that when both endpoints lie at previously calculated locations this amounts to a null step and the search immediately proceeds to the next level). The process continues until $k = N$ and it may be shown that this approach is optimal in the sense that it minimizes the maximum uncertainty in the location of the minimum being sought. To achieve a given accuracy ε, N should

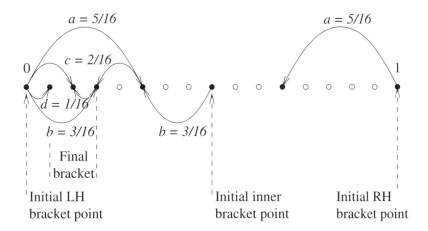

Figure 3.2 Steps in a Fibonacci line search of four moves, labeled a, b, c and d, with triplets of $(0/16, 8/16, 16/16)$, $(0/16, 5/16, 8/16)$, $(0/16, 3/16, 5/16)$ and $(1/16, 2/16, 3/16)$ – that is, there are four moves, so that $N = 4$ and the corresponding reductions factors are $t_1 = f_4/f_5 = 5/8$, $t_2 = (f_3/f_5)/(8/16) = (3/8)/(8/16) = 3/4$, $t_3 = (f_2/f_5)/(5/16) = (2/8)/(5/16) = 4/5$, $t_4 = (f_1/f_5)/(3/16) = (1/8)/(3/16) = 2/3$ – therefore, only five new points are sampled before the final bracket is established – those shown filled on the figure.

be chosen such that $1/f_{N+1} > \varepsilon/(x_2 - x_1) \geq 1/f_{N+2}$. As already noted for most functions (certainly those that are smooth in the vicinity of the optimum) the accuracy with which a turning point can be trapped is limited by the *square root* of the machine precision (i.e., it is limited to 2×10^{-4} in single precision) – attempting to converge a solution more closely than this, by any means, is unlikely to be useful.

An alternative to these two rule-based systems is to fit a simple quadratic polynomial to the bracketing triple at each step, and then, by taking its differential, calculate the location of its turning point and place the next point there. This is called *inverse parabolic interpolation* and the location of the next point x_4 is given by

$$x_4 = x_2 - \frac{(x_2 - x_1)^2(f(x_2) - f(x_3)) - (x_2 - x_3)^2(f(x_2) - f(x_1))}{2(x_2 - x_1)(f(x_2) - f(x_3)) - (x_2 - x_3)(f(x_2) - f(x_1))}. \tag{3.1}$$

This process can easily be mislead if the bracketing interval is wide, and so a number of schemes have been proposed that seek to combine this formula and the more robust process of the golden section search – that by Brent is perhaps the most popular (Brent 1973; Press et al. 1986). Brent's scheme alternates between a golden section search and parabolic convergence depending on the progress the algorithm is making, and is often used within multidimensional search methods.

The most sophisticated line searches are gradient-based methods, which work, of course, by explicitly using the gradients of the function when calculating the next step to take. Sometimes, the gradients can be found directly (for example, when dealing with closed form expressions, numerically differentiated simulation codes or adjoint codes) but, more commonly, they must be found by finite differencing. All such methods are sensitive to the accuracy of the available gradient information, but where this is of good quality, and particularly where the underlying function is approximately polynomial in form, they offer the fastest methods for carrying out line searches. Where they work, they are superlinear in performance. The degree to which they outperform the linear golden section and Fibonacci approaches depends on the order of the scheme adopted. The order that can be successfully adopted, in turn, depends on the quality of the available gradients and the degree to which the underlying function can be represented by a polynomial – again, we see the trade-off between robustness and speed.

The most aggressive gradient approach to take is to simply fit a polynomial through all the available data with appropriate gradients at the points where these are available. Then, differential calculus may be applied to locate any turning values and to distinguish between minima, maxima and points of inflexion. Assuming a minimum exists within the current triple, the next point evaluated is placed at the predicted location. This process can be iterated with the new points and gradients being added in turn. If all three gradients are known for the initial bracketing triple, then a fifth-order polynomial can be fitted and its turning values used. It will be immediately obvious that there are some serious flaws with such an approach. First, even if all the data is noise free and the quintic is a good model, the roots of its first differential must be obtained by a root search; secondly, high-order polynomials are notoriously sensitive to any noise and can become highly oscillatory and, thirdly, there is no guarantee that a minimum value will exist within the initial bracket when it is modeled by a quintic. Finally, as extra points are added, the order of the complete polynomial rapidly rises and all these issues become ever more problematic. It is therefore normal to take a more conservative approach.

Rather than fit the highest-order polynomial available, one can use a cubic instead. If the chosen polynomial is to be an interpolator, this requires that a subset of the available data be used from the bracketing triple. For a cubic model, one might use the three data values and the gradient of the central point of the triple to fix the four coefficients. It is then simple to solve for the roots of the gradient equation (which is of course second order) and jump to the minimum that must lie within the bracket. This will lead to a new triple and the process can be repeated, with the gradient being calculated at the central point each time (or to save time, at one or other end if the gradient is already known there). Note also, that if one always has gradients available, there is no need to seek a triple to bracket the minimum – provided the gradients of two adjacent points point toward each other, this suffices to bracket the turning point. Moreover, a cubic (or Hermitian) interpolant can be constructed from the two endpoints and their gradients and an iterative scheme constructed that will be very efficient if the function is smooth and polynomial-like.

An alternative gradient-based scheme is to use the Newton–Raphson approach for finding the zeros of a function to find the zeros of its derivative – this is commonly termed the *secant method*. It requires the gradient at the current point and an estimate of the second derivative. Again, if the minimum is bracketed by a pair of points where both derivatives are known, then the next location to chose can be based on a forward finite difference of the derivatives, that is, $x_3 = x_1 - (x_2 - x_1)\mathrm{d}f(x_1)/\mathrm{d}x/\{\mathrm{d}f(x_2)/\mathrm{d}x - \mathrm{d}f(x_1)/\mathrm{d}x\}$, which, since $\mathrm{d}f(x_2)/\mathrm{d}x$ is positive and $\mathrm{d}f(x_1)/\mathrm{d}x$ is negative, always yields a new location within the initial bracket. Of course, if the second derivative is also known at each point, this information can be used instead.

It is possible to use gradients in an even more conservative fashion, however. If gradients are available, then bracketing the minimum does not require that these be known very accurately; rather, one merely searches for a pair of adjacent points where the sign of the gradient goes from negative to positive. If one then carries out a series of interval bisections, keeping the pair of points at each time based only on the signs of the gradients, a more robust, if slower, search results. This can be particularly advantageous if the derivative information is noisy. Notice, however, that if the gradients are being found by finite differencing it hardly makes sense to evaluate four function values (i.e., two closely spaced pairs) in order to locate the next place to jump to – it is far better to use the golden section or Fibonacci methods. A slightly more adventurous approach based on triplets is to use a simple straight line (or secant) extrapolation of the gradient from that at the central point and the lower of the two edge points and see if this indicates a zero within the triplet. If the zero is within the triplet, this point is adopted, while if not, a simple bisection of the best pairing is used, again followed by iteration. This is the approach recommended by Press et al. (1992) for line searches with gradients.

3.1.2 Optimization with a Single Discrete Variable

Sometimes, when carrying out optimization problems, the free variables that the optimizer controls are discrete as opposed to being continuous. Such problems obviously arise when dealing with integer quantities. But they also arise when working with stock sizes of materials, fittings, screw sizes, and so on. In practice, since when dealing with such discrete variables no gradient information exists, all these problems are equivalent and optimizers that work with integers can solve all problems in this class by a simple scaling of the variables (i.e., we use the integers as indices in a look-up table of the discrete values being studied). The important point with such discrete problems is that the series of design variables be *ordered* in some way. Therefore, only the problem of optimization with integer variables is considered further here.

There are two fundamentally different approaches to the problem of integer optimization. The first is to work directly in integer arithmetic while the second is to assume that the design variables are actually continuous and simply to round to the nearest integer when evaluating the objective function (or constraints). Methods that work with integers can be constructed from those already described in a number of cases, albeit they may work rather less efficiently. If a global search in a multimodal space is being performed, some form of stochastic or exhaustive search over the integers being dealt with will be required – these are dealt with later. Conversely, if a local search is being conducted then, as before, the first goal is to find a bracketing triple that captures the optimum being sought. For example, it is easy to devise a one-dimensional bracketing routine equivalent to that already described for continuous variables: all that is required is that the first step taken be an integer one – when this is doubled, this will still of course result in an integer value and so such a search will naturally work with integers.

Having bracketed the desired minimum with a triplet, methods must then be devised to reduce the bracket to trap the optimum. Here, in some sense, the integer problem is easier than the real-valued one as no issues of accuracy and round-off arise. The simplest, wholly integer strategy that can be proposed is to bisect the larger gap of the bracket (allowing for the fact that this region will from time to time contain an even number of points and thus cannot be exactly bisected and either the point to the right or left of the center must be taken, usually alternately). Then, the bracket triple is reduced by rejecting one or other endpoints. This process is carried out until there are no more points left in the bracket at which time the value with lowest objective becomes the optimum value. This search can be refined by not simply bisecting the larger gap in the bracket at each step, but by instead using the point closest to the golden section search value. This is equivalent to using a true golden section search and rounding the new value to the nearest integer quantity at each iteration.

The Fibonacci search is rather better suited to integer problems than both of these methods as, with a little adaptation, it may be directly applied without rounding; see Figure 3.3. We first identify the pair of Fibonacci numbers that span the total integer gap between the upper and lower limits of the triplet to be searched (i.e., the width of the triplet) – in the figure, the initial bracket is 18-units wide and so the pair of Fibonacci numbers that span this range are 13 and 21. We then move the limit farthest from the initial inner point of the triple inward so that the remaining enclosed interval width to be searched is an exact Fibonacci number (i.e., here, we move the point at 18 inward to 13) – if this triplet does not bracket the minimum, we take the triplet formed from the original inner point, the new endpoint and the new endpoint's previous position and repeat the process of reducing to the nearest Fibonacci number until we do have a triplet that brackets the minimum and has a length equal to one of the Fibonacci numbers. Once the triplet is an exact Fibonacci number in width, we simply move both bounds inward in a sequence of Fibonacci numbers until we reach unity and hence identify the location of the minimum – here, 5, 3, 2 and 1 (we start with largest number that will fit twice into the interval remaining after making the first move).

Polynomial-based and gradient-based searches can only be applied to integer problems if the fiction of a continuous model is assumed as part of the search, i.e., that variable values between the integers *can* be accommodated by the search. As already noted, the

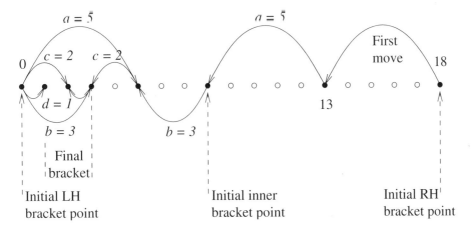

Figure 3.3 Steps in a Fibonacci integer search of five moves, an initial move to reduce the interval to a Fibonacci number followed by those labeled a, b, c and d, with triplets of (0, 8, 18), (0, 8, 13), (0, 5, 8), (0, 3, 5) and (1, 2, 3) – therefore, only six new points are sampled before the final bracket is established – those shown filled on the figure.

simplest way to do this is to proceed as for a continuous variable and then to round to the nearest integer value when calculating the objective and constraints. It is also sometimes possible to treat the whole problem as being continuous, including the calculation of the objective and any constraints and just to round the final design to the nearest whole number – such approaches can be effective when dealing with stock sizes of materials. For example, if the search is to find the thickness of a plate that minimizes a stress, this can be treated as a continuous problem. Then, the final design can be selected from the available stock sizes by rounding up. In fact, in many engineering design processes, the designer is often required to work with available materials and fittings, and so, even if most of this work is treated as continuous, it is quite common to then round to the nearest acceptable component or stock size. If rounding is used for each change of variable, however, care must be taken with gradient-based routines as they can be misled by the fictional gradients being used – this can often make them no faster than the Fibonacci approach or, worse, sometimes this means they become trapped in loops that fail to converge. Again a trade-off between speed and robustness must be made.

3.1.3 Optimization with a Single Nonnumeric Variable

Sometimes, searches must be carried out over variables that are not only discontinuous but not even numeric in nature. The most common such case arises in material selection. For example, when choosing between metals for a particular role, the designer will be concerned with weight, strength, ductility, fatigue resistance, weldability, cost, and so on. There may be a number of alloys that could be used, each with its own particular set of qualities, and the objective being considered may be calculated by the use of some complex nonlinear CSM code. The most obvious way to find the optimum material is to try each one in turn and select the best – such a blind search is bound to locate the optimum in the end. If a complete search cannot be afforded, then some subset must be examined and the search stopped after an appropriate number have been considered – very often, this is done manually, with the designer running through a range of likely materials until satisfied that a good choice has been identified. The difficulty here arises in ordering the choice of material to consider next. Clearly, each could be given an index number and the search performed as over integers for a finite number of steps, or even using one of the search methods outlined in the previous section. It is, of course, the lack of any natural ordering between index number and performance that makes such a search difficult, and even somewhat random. Unless the designer has a good scheme for ordering the choices to be made when indexing them, it is likely that an exhaustive search will be as efficient as any other method. Such observations apply to all problems where simple choices have to be made and where there is no natural mapping of the choices to a numerical sequence.

3.2   Multivariable Optimizers

When dealing with problems with more than one variable, it is, of course, a perfectly viable solution to simply apply methods designed for one variable to each dimension in turn, repeating the searches over the individual variables, until some form of convergence is reached. Such methods can be effective, particularly in problems with little interaction between the variables and a single optimum value, although they tend to suffer from convergence problems and find some forms of objective landscapes time consuming to deal with (narrow valleys that are not aligned with any one variable, for example). Unfortunately, most of the problems encountered in engineering design exhibit strong coupling between the variables and multiple optima and so more sophisticated approaches are called for. This has led to the development of a rich literature of methods aimed at optimization in multiple dimensions. Here, we will introduce a number of the more established methods and also a number of hybrid strategies for combining them.

Most of the classical search methods work in a sequential fashion by seeking to find a series of designs that improve on each other until they reach the nearest optimum design. This is not true of all methods, however – a pure random search, for example, does not care in which order points are sampled or if results are improved and is capable of searching a landscape with multiple optima. Here, we begin by examining those that work in an essentially sequential fashion, before moving on to those that can search entire landscapes (i.e., local and global methods, respectively). Note that in some of the classical optimization literature the term "global convergence" is used to indicate a search that will find its way to an optimal design wherever it starts from, as opposed to one that will search over multiple optima.

Local searches have to deal with two fundamental issues: first, they must establish the direction in which to move and, secondly, the step size to take. These two considerations are by no means as trivial to deal with as they might seem and very many different ways of trying to produce robust optimizers that converge efficiently have been produced. Almost all local search methods seek to move downhill, beginning with large steps and reducing these as the optimum is approached, so as to avoid overshoot and to ensure convergence. Global search methods must additionally provide a strategy that enables multiple optima to be found, usually by being able to jump out of the basin of attraction of one optimal design and into that of another, even at the expense of an initial worsening of the design.

We will illustrate these various issues by carrying out searches on a simple test function proposed some years ago by one of the authors – the "bump" problem. This multidimensional problem is usually defined as a maximization problem but can be set up for minimization as follows:

$$\textbf{Minimize}: \quad -\text{abs}\left(\sum_{i=1}^{n} \cos^4(x_i) - 2\prod_{i=1}^{n} \cos^2(x_i)\right) \Big/ \sqrt{\sum_{i=1}^{n} i x_i^2}$$

$$\textbf{Subject to}: \quad \prod_{i=1}^{n} x_i > 0.75 \quad \text{and} \quad \sum_{i=1}^{n} x_i < 15n/2 \tag{3.2}$$

$$0 \le x_i \le 10, \quad i = 1, 2, \ldots, n,$$

where the x_i are the variables (expressed in radians) and n is the number of dimensions. This function gives a highly bumpy surface where the true global optimum is usually defined by the product constraint. Figure 3.4 illustrates this problem in two dimensions. When dealing with optimizers designed to find the nearest optimal design, we start the search at (3.5, 2.5) on this landscape, and see if they can locate the nearest optimum, which is at (3.0871, 1.5173) where the function has a value of -0.26290. When applying global methods, we start at (5, 5) and try to find the locations of multiple optima. Finally, when using constrained methods, we try and locate the global constrained optimum at (1.5903, 0.4716) where the optimum design is defined by the intersection of the objective function surface and the

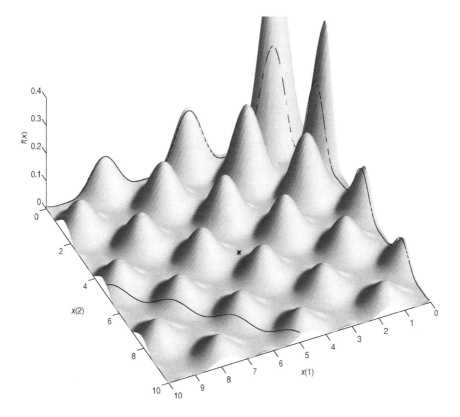

Figure 3.4 Constrained bump function in two dimensions (plotted in its original maximization form).

product constraint and the function value is 0.36485. All of the search methods discussed in this section are unconstrained in their basic operation, and so mechanisms have to be added to them to deal with any constraints in design problems being studied. As many of these are common to a variety of search methods, they are dealt with in a subsequent section.

Before examining the search methods in detail, we briefly digress, however, by discussing a key aspect of the way that searches are actually implemented in practice.

3.2.1   Population versus Single-point Methods

The single variable methods that have been outlined above are all serial in nature, in that new design points are derived and processed one at a time. This is natural when working with one variable and it is also the easiest way to understand many search tools but, increasingly, it is not the most efficient way to proceed in practice. Given that so many of the analysis tools used in design can be time consuming to run, there is always pressure to speed up any search method that is running large computational analyses. The most obvious and simple way of doing this is to consider multiple design points in parallel (we note also that many large analysis codes are themselves capable of parallel operation, and this fact too can often be exploited). The idea of searching multiple solutions in parallel leads to the idea of population-based search. Such searches can be very readily mapped onto clusters of computers with relatively low bandwidth requirements between them, and computing clusters suitable for this kind of working are increasingly common in design offices, formed either by linking together the machines of individual designers when they are not otherwise in use (so-called cycle harvesting) or in the form of dedicated compute clusters.

The populations of possible designs being examined by a search method can be stepped forward in many ways: in a gradient-based calculation, they may represent small steps in each variable direction and be used to establish local slopes by finite differencing; in a heuristic pattern-based search, multiple searches may be conducted in different areas of a fitness landscape at the same time; in an evolutionary search, the population may form a pool of material that is recombined to form the next generation of new designs. Sometimes, the use of populations fits very naturally into the search and no obvious bottlenecks or difficulties arise (evolutionary searches fall into this category, as do multiple start hill climbers, and design of experiment based methods), sometimes they can only be exploited for part of the time (as in the finite differencing of gradients before a pattern move takes place in a gradient-based search) and sometimes it is barely possible to accommodate simultaneous calculations at all (such as in simulated annealing). In fact, when choosing between alternatives, it turns out that the raw efficiency of a search engine may well be less important than its ability to be naturally mapped to cluster-based computing systems.

It goes without saying that to be of practical use in parallel calculations a population-enabled search must have a mechanism for deploying its activities over the available computing facilities in a workable fashion – and this often means that issues of job management form an intrinsic part of the implementation of such methods. In what follows, however, we will assume that, where needed, a population-based method can deploy its calculations in an appropriate fashion. Finally, it should be noted that what mostly matters in design is the elapsed time required to find a worthwhile improvement, including the time needed to define and set up the search, that needed to carry it out and that involved in analyzing results as they become available so as to be able to steer, terminate or restart the search. Methods that fit naturally with a designer's way of working and desire to be involved in the product improvement will obviously be more popular than those that do not.

3.2.2 Gradient-based Methods

We begin our study of multivariable search methods with those that exploit the objective function gradient since the fundamental difference between multiple variable and single variable searches is, of course, the need to establish a direction for the next move rather than just the length of the step to take, that is, one must know which way is down. Given a suitable direction of search, optimization in multiple dimensions is then, at least conceptually, reduced to a single variable problem: once the direction of move is known, the relative magnitudes of the input variables can be fixed with respect to each other. An obvious approach to optimization (and one due to Cauchy) is to use gradient information to identify the direction of steepest descent and go in that direction.[1] Then, a single move or entire line search is made in that direction. Following this, the direction of steepest descent is again established and the process repeated. If a line search is used, the new direction of descent will be at right angles to the previous one (the search will, after all, have reached a minimum in the desired direction) so that a series of repeated steepest descent line searches leads to a zigzag path; see Figure 3.5. If a line search is not used, issues then arise as to how long each step should be in a particular direction – if they are too long, then the best path may be overshot; conversely, if they are too small, the search will be very slow.

Although steepest descent optimization methods will clearly work, there are rather better schemes for exploiting gradients: all are based on the idea that, near a minimum, most functions can be approximated by a quadratic shape, that is, in two dimensions, the objective

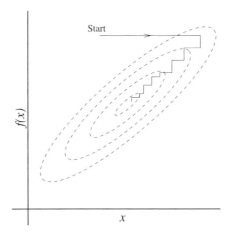

Figure 3.5 A steepest descent search in a narrow valley.

---

[1]We assume that suitable steps are taken to make the required gradients available – in fact, the calculation of such gradients by finite differencing schemes is by no means trivial and is also often prohibitively expensive if many variables are involved; conversely, if good quality gradients can be found from efficient adjoint or direct numerical differentiation methods, this will not be a problem.

function surface is an ellipsoid bowl and the objective function contours are ellipses. For a continuous smooth function, this is always true sufficiently close to any minimum (consider a Taylor series expansion taken about the optimum). Unfortunately, many of the optima being sought in engineering design are either defined by constraint boundaries or by discontinuities in their objective functions – this means that many of the gradient-based methods that give very good performance in theory do much less well in practice. Nonetheless, it is useful to review their underlying mechanisms as many effective search methods draw on an understanding of this basic behavior. We leave to Chapter 4 a detailed analysis of methods for computing the gradients of functions.

Newton's Method

The simplest way to exploit the structure of a quadratic form is via Newton's method. In this, we assume that $f(\mathbf{x}) \approx \frac{1}{2}\mathbf{x}^T\mathbf{A}\mathbf{x} + \mathbf{b}^T\mathbf{x} + c$ where now $\mathbf{x}$ is the vector of variables being searched over and $f(\mathbf{x})$ remains the objective function. $\mathbf{A}$, $\mathbf{b}$ and c are unknown coefficients in the quadratic approximation of the function, $\mathbf{A}$ being a positive definite square matrix known as the Hessian, $\mathbf{b}$ a vector and c a constant (it is the positive definiteness of $\mathbf{A}$ that ensures that the function being searched is quadratic and convex, and thus does contain a minimum – clearly we cannot expect such methods to work where there is no minimum in the function or where pathologically it is not quadratic, even in the limit as we approach the optimum). Notice that simple calculus tells us that the elements of the Hessian $\mathbf{A}$ are given by the second differentials of the objective function, that is, $A_{ij} = \partial^2 f/\partial x_i \partial x_j$ and that the gradient vector is given by $\mathbf{A}\mathbf{x} + \mathbf{b}$. At a minimum of $f(\mathbf{x})$, the gradient vector is of course zero.

Newton's search method is very similar to that for finding the zero of a function – here, we are seeking the zero of the gradient, of course. If our function is quadratic and we move from $\mathbf{x}_i$ to $\mathbf{x}_{i+1}$, then we can write

$$f(\mathbf{x}_{i+1}) \approx f(\mathbf{x}_i) + (\mathbf{x}_{i+1}-\mathbf{x}_i)^T\nabla f(\mathbf{x}_i) + 1/2(\mathbf{x}_{i+1} - \mathbf{x}_i)^T\mathbf{A}(\mathbf{x}_{i+1} - \mathbf{x}_i) \qquad (3.3)$$

$$\text{or} \quad \nabla f(\mathbf{x}_{i+1}) \approx \nabla f(\mathbf{x}_i) + \mathbf{A}(\mathbf{x}_{i+1} - \mathbf{x}_i). \qquad (3.4)$$

Newton's method simply involves setting $\nabla f(\mathbf{x}_{i+1})$ to zero so that the required step becomes $(\mathbf{x}_{i+1} - \mathbf{x}_i) \approx -\mathbf{A}^{-1}\nabla f(\mathbf{x}_i)$.

In this method, we assume that the Hessian is known everywhere (or can be computed) and so we can then simply move directly to the minimum of the approximating quadratic form in one step. If the function is actually quadratic, this will complete the search, and if not, we simply repeat the process with a new Hessian. This process converges very fast (quadratically) but relies on the availability of the Hessian. In most practical optimization work, the Hessian is not available directly and its computation is both expensive and prone to difficulties. Therefore, a range of methods have been developed that exploit gradient information and work without the Hessian, or with a series of approximations that converge toward it.

Conjugate Gradient Methods

Conjugate gradient searches work without the Hessian but try to improve on the behavior of simple steepest descent methods. The key problem with the steepest descent type approach is that the direction of steepest gradient rarely points toward the desired minimum, even if the function being searched is quadratic (in fact, only if the contours are circular does this direction always point to the minimum – an extremely unlikely circumstance in real problems). Conjugate gradient methods are based on the observation that a line through the minimum of quadratic function cuts all the contours of the objective function at the same angle. If we restrict our searches to run along such directions, we can, in theory, minimize an arbitrary quadratic form in the same number of line searches as there are dimensions in the problem, provided the line search is accurate. In practice, badly scaled problems can impact on this and, moreover, few real problems are actually quadratic in this way. Nonetheless, they usually provide a worthwhile improvement on simple steepest descent methods.

To design a conjugate gradient search, we must set out a means to decide on the conjugate directions and then simply line search down them. The conjugate gradient method starts in the current direction of steepest descent $\mathbf{v}_0 = -\nabla f(\mathbf{x}_0)$ and searches until the minimum is found in that direction. Thereafter, it moves in conjugate directions such that each subsequent direction vector $\mathbf{v}_i$ obeys the conjugacy condition $\mathbf{v}_i \mathbf{A} \mathbf{v}_j = 0$. These are found without direct knowledge of $\mathbf{A}$ by use of the recursion formula $\mathbf{v}_{i+1} = \lambda \mathbf{v}_i - \nabla f(\mathbf{x}_{i+1})$, where

$$\lambda = \nabla f(\mathbf{x}_{i+1})^{\mathrm{T}} \nabla f(\mathbf{x}_{i+1}) / \nabla f(\mathbf{x}_i)^{\mathrm{T}} \nabla f(\mathbf{x}_i) \quad \text{or} \tag{3.5}$$

$$\lambda = (\nabla f(\mathbf{x}_{i+1}) - \nabla f(\mathbf{x}_i))^{\mathrm{T}} \nabla f(\mathbf{x}_{i+1}) / \nabla f(\mathbf{x}_i)^{\mathrm{T}} \nabla f(\mathbf{x}_i). \tag{3.6}$$

The first of these two expressions for the factor λ is due to Fletcher and Reeves and is the original version of the method, while the later is due to Polak and Ribiere and helps the process deal with functions that are not precisely quadratic in nature. Notice that in this process there is no attempt to build up a model of the matrix $\mathbf{A}$ (requiring storage of the dimension of the problem size squared) – rather the idea is to be able to identify the next conjugate direction whenever a line minimization is complete using storage proportional to the dimension of the problem. The behavior of this kind of search is directly dependent on the quality of the gradient information being used and the accuracy of the line searches carried out in each of the conjugate directions.

Conjugate gradient searches are illustrated in Figures 3.6(a) and (b) for both forms of the recursion formula and with gradients either found by finite differencing or explicitly from the formulation of the problem itself. In all cases, Brent's line search is used along each conjugate direction, the code being taken from the work of Press et al. (1992). Rather coarse steps are used for the finite differencing schemes to illustrate the effect that inaccuracies in gradient information can have on the search process, but in fact, the consequences are here minor, amounting to around 5% more function calls and less than 0.1% error in the final solution (although this does help distinguish the traces in the figures). Table 3.1 details the results of these four searches. It is clear from the figures and the table that even though the function being dealt with here is very smooth and its contours quite close to elliptic, the recursion formula due to Polak and Ribiere leads to a much more efficient identification of the true conjugate directions. Further, the use of exact gradient information not only speeds up the process, but also leads to more accurate answers.

It should be noted in passing that the code implementations due to Press et al. have been criticized in some quarters as being rather inefficient and sometimes not completely reliable. There are faster implementations but these are usually more complex to code and not always as robust. To illustrate the speed differences, Table 3.1 also contains results from the version described by Gilbert and Nocedal (1992), which are available from the web-based

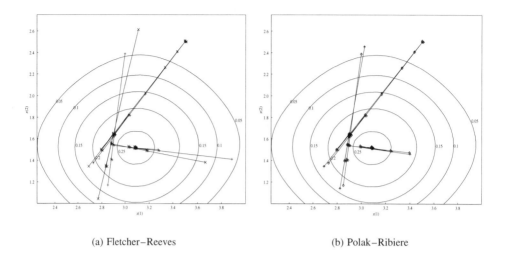

(a) Fletcher–Reeves   (b) Polak–Ribiere

Figure 3.6 Conjugate gradient searches.

Table 3.1   Results of conjugate gradient searches on local optimum in the bump function of (3.2)

Method	Gradient Calculation	Source of Code	Final Objective Function Value (% of true optimal objective)	Number of Function Calls (gradient calls)
Fletcher–Reeves	Direct	Press et al.	−0.26201 (99.7)	215 (84)
		Gilbert and Nocedal	−0.26289 (100)	99 (99)
	Finite difference	Press et al.	−0.26106 (99.3)	385 (0)
		Gilbert and Nocedal	−0.26267 (99.9)	282 (0)
Polak–Ribiere	Direct	Press et al.	−0.26290 (100)	52 (19)
		Gilbert and Nocedal	−0.26290 (100)	17 (17)
	Finite difference	Press et al.	−0.26289 (100)	113 (0)
		Gilbert and Nocedal	−0.26286 (100)	45 (0)

NEOS service.[2] These routines are substantially quicker (particularly when direct gradient information is available), but also, in our experience, slightly more sensitive to noise and rounding errors – the main differences stemming from the use of Mor'e and Thuente's line

---

[2]http://www-neos.mcs.anl.gov/neos/

search (Mor'e and Thuente 1994). The routine also seems not to like the function discontinuities introduced by the one pass external penalty functions often used when constrained
problems are being tackled (see section 3.3.4). In all cases, similar tolerances on the final
solution were specified. Interestingly, the Press et al. code does not invoke the gradient
calculation at every step whereas the Gilbert and Nocedal version does. Choice in these
areas is often a matter of personal preference and experience and we make no particular
recommendation – we would also note that different workers' implementations of the same
basic approaches often differ in this way.

Quasi-Newton or Variable Metric Methods

Quasi-Newton or variable metric methods also aim to find a local minimum with a minimal
sequence of line searches. They differ from conjugate gradient methods in that they slowly
build up a model of the Hessian matrix $\mathbf{A}$. The extra storage required for most problems
encountered in engineering design is of no great consequence (modern PCs can happily deal
with problems where the Hessian has dimensions in the hundreds with no great difficulty).
There is some evidence that they are faster than conjugate gradient methods but this is
problem specific and also not so overwhelming that it renders conjugate gradient methods
completely obsolete. Again, choice between such approaches often comes down to personal
preference and experience.

The basic idea of quasi-Newton methods is to construct a sequence of matrices that
slowly approximate the inverse of the Hessian ($\mathbf{A}^{-1}$) in the region of the local minimum
with ever greater accuracy. Ideally, this sequence converges to $\mathbf{A}^{-1}$ in as many line searches
as there are dimensions in the problem. Given an approximation to the inverse Hessian
$\mathbf{A}^{-1}$, it is possible to compute the next step in the sequence in the same fashion as a
Newton search, although since we know it is only an approximation, line searches are used
rather than simple moves directly to the presumed optimal location. A key factor in the
practical implementation of this process is that the approximations to the inverse Hessian
are limited to being positive definite throughout, so that the search always moves in a
downhill direction – recall that far from the local minimum, the function being searched
may not behave quadratically and so the actual inverse Hessian at such a point may in fact
point uphill. Even so, it is still possible to make an uphill step by overshooting the minimum
in the current direction. Therefore, when implementing the algorithm, the step sizes must
be limited to avoid overshoot. The Broyden–Fletcher–Goldfarb–Shanno (BFGS) updating
strategy achieves this (Press et al. 1992). Figure 3.7 and Table 3.2 illustrate the process
on the bump problem, c.f., Figures 3.6(a) and (b) and Table 3.1. The results obtained are
essentially identical to those yielded by the Polak–Ribiere conjugate gradient search.

Quasi-Newton routines can suffer from problems if the variables being used are poorly
scaled, as the inverse Hessian approximations can then become singular. By working with
the Cholesky decomposition of the Hessian rather than updates to its inverse, it is possible
to make sure that round-off errors do not lead to such problems and it is also then simple
to ensure that the Hessian is always positive definite – a version of this approach is due to
Spellucci (1996). An alternative is simply to restart the basic quasi-Newton method once it
has finished to see if it can make any further improvement – although untidy, such a simple
strategy is very easy to implement, of course. The results from using Spellucci's code are
also given in Table 3.2. This works much faster when direct gradients are available but
slightly slower when they are not (although it then returns a slightly more accurate answer).

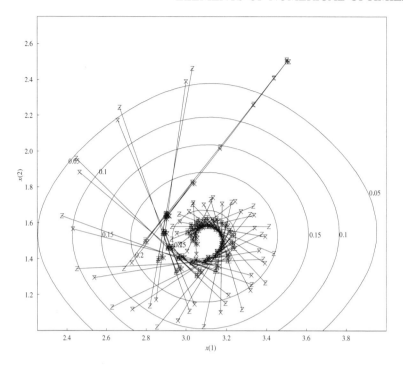

Figure 3.7 Broyden–Fletcher–Goldfarb–Shanno quasi-Newton searches.

Table 3.2   Results of quasi-Newton searches on local optimum in the bump function of (3.2)

Method	Gradient Calculation	Source of Code	Final Objective Function Value (% of true optimal objective)	Number of Function Calls (gradient calls)
Broyden–	Direct	Press et al.	−0.26290 (100)	54 (20)
Fletcher–		Spellucci	−0.26290 (100)	13 (12)
Goldfarb–	Finite	Press et al.	−0.26288 (100)	124 (0)
Shanno–	difference	Spellucci	−0.26290 (100)	157 (0)

This is because it uses a rather complex and accurate six-point method to compute the gradients numerically when they are not available analytically.

The impact of the large finite difference steps we have used with the Press et al. code (and thus the less accurate gradients being made available) is more noticeable here – adopting a more normal step size reduces the function call count by around 30%, although again the impact on the final results is still less than 0.1%. This sensitivity explains why the Spellucci code uses the sophisticated approach that it does – since a model of the Hessian is being slowly constructed in quasi-Newton methods, it is important for efficiency that this is not corrupted by inaccurate information. Note also that the Spellucci quasi-Newton search with direct gradient calculations is clearly the fastest gradient-based method demonstrated of all those considered here.

It is clear from a study of these methods that computation of the gradients is key to how well they perform. Effort can be expended to do this highly accurately or, alternatively, more robust line searches can be used. In either case, the resulting searches are much more costly than where direct gradient information is available. It is these observations that have lead to the considerable interest in the development of adjoint and direct numerical differentiation schemes in many areas of analysis (and which are discussed in Chapter 4).

3.2.3 Noisy/Approximate Function Values

It will be clear from the previous subsections that gradient-based searches can rapidly identify optima in a fitness landscape with high precision. Although there is some variation between methods and also some subtleties in the fine workings of these searches, there is much experience on their use and readily available public domain codes. Of course, these methods only seek to follow the local gradient to the nearest optimum in the function, and so, if they are to be used where there are multiple optima, they must be restarted from several different starting points to try and find these alternative solutions. One great weakness in gradient-based searches, however, is their rather poor ability to deal with any noise in the objective function landscape.

Noise in computational methods is rather different from the kind found in ordinary experimental or in-service testing of products because it is repeatable in nature. When one carries out multiple wind tunnel tests on the same wing, each result will differ slightly from the last – this is a commonplace experience, and all practical experimenters will take steps to deal with the noise, usually by some form of averaging over multiple samples. With computational models, a rather different scenario arises: repeating a computation multiple times will (at least should) lead to the same results being returned each time. This does not mean that such results are free of noise, however: if the solver being used is iterative in nature or if it requires a discretization scheme, then both of these aspects will influence any results. When a small adjustment is made to the geometry being considered, then the convergence process or the discretization may well differ sufficiently for any changes in the results to be as much due to these sources of inaccuracy as to the altered geometry itself. If one then systematically plots the variation of some objective function for a fine grained series of geometry changes, the resulting curves appear to be contaminated with noise; see Figure 3.8 – the fact that the detail of such a noise landscape is repeatable does not mitigate the impact it can have on optimization processes, particularly those that make use of gradients, and most severely, if those gradients are obtained by finite differencing over short length scales.

To illustrate this fact, the results from the previous two tables may be compared with those in Table 3.3. Here, the same objective function has been used except that in each case a small (1%) random element has been added to the function (and where used, the gradient). The results in the table are averages taken over 100 runs of each search method with statistically independent noise landscapes in each case. It may be seen that there are two consequences: first, the methods take many more steps to converge (if they do converge), and secondly, the locations returned by the optimizers are often significantly in error compared to the actual optimum. These tests reveal that quasi-Newton methods have greater difficulty dealing with noise, in that not only do they return poorer results but also sometimes the

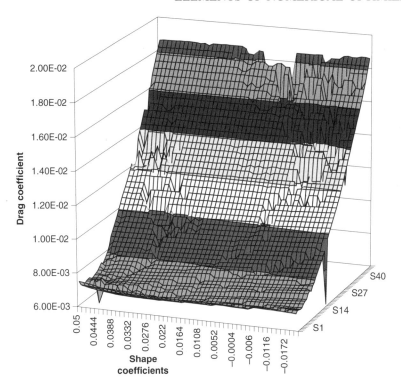

Figure 3.8 An objective function contaminated by discretization noise and the occasional failed calculation – airfoil section drag computed using a full potential code for variations in orthogonal shape functions (M = 0.80, t/c = 0.06, Cl = 0.4).

calculations become singular and fail. Note also that the standard deviations in the results in the table tend to be worse when using finite differences than for the directly evaluated gradients, although, interestingly, the mean value of the Polak–Ribiere conjugate gradient search due to Press et al. is in fact slightly better when using finite differences. Recall, however, that a rather coarse finite difference step length is being used here (0.01) and this in fact helps when dealing with a noisy function since then the chances of extreme gradients being calculated are reduced. If a much smaller (and more normal) step length of 1×10^{-7} is used, the results become much worse for the finite differencing schemes – that for the Press et al. Polak–Ribiere search drops to an average objective function value of -0.17595 (66.9% of the true value) with a standard deviation of 0.0886 (an order of magnitude worse) and an average search length of 1,083 calls. Clearly, when dealing with a function that is not strictly smooth, any gradient-based method must be used with extreme caution. If finite differencing is to be used, then step lengths must be adopted that are comparable with the noise content. Note that, again here, we also give results from the Gilbert and Nocedal conjugate gradient and Spellucci BFGS searches for comparison – they fare less well as they often fail to converge and give rather poorer results, though they retain their undoubted speed advantage.

Table 3.3 Average results of conjugate gradient and quasi-Newton searches on local optimum in the bump function of (3.2) with 1% noise

Method	Gradient Calculation	Source of Code	Final Objective Function Value (% of true optimal objective)	Standard Deviation in Final Objective Function Value	Number of Function Calls (gradient calls)
Fletcher–Reeves	Direct	Press et al.	−0.25564 (97.2)	0.0078	282 (128)
		Gilbert and Nocedal	−0.24292 (92.4) fails sometimes	0.0084	13 (13)
	Finite difference	Press et al.	−0.25265 (96.1)	0.0096	691 (0)
		Gilbert and Nocedal	−0.23763 (90.4) fails sometimes	0.0087	32 (0)
Polak–Ribiere	Direct	Press et al.	−0.25742 (97.9)	0.0077	455 (142)
		Gilbert and Nocedal	−0.24259 (92.3) fails sometimes	0.0088	13 (13)
	Finite difference	Press et al.	−0.26092 (99.2)	0.0100	628 (0)
		Gilbert and Nocedal	−0.23890 (90.9) fails sometimes	0.0090	34 (0)
Broyden–Fletcher–Goldfarb–Shanno	Direct	Press et al.	−0.24924 (94.8) fails sometimes	0.0095	143 (57)
		Spellucci	−0.07267 (27.6) fails most times	0.1016	32 (8)
	Finite difference	Press et al.	−0.24990 (95.1) fails sometimes	0.0116	269 (0)
		Spellucci	−0.01935 (7.4) fails always	0.0000	29 (0)

It is also the case that none of the codes used here has been specifically adapted for working in noisy landscapes. If the designer knows that the analysis methods in use have some kind of noise content, then great care should be taken when selecting and using search engines. Certainly, the methods that are fastest on clean data will rarely prove so effective when noise is present – they can however sometimes be adapted to cope in these circumstances (as here, for example, by using larger than normal step sizes when finite differencing).

3.2.4   Nongradient Algorithms

Because establishing the gradient of the objective function in all directions during a search can prove difficult, time consuming and sensitive to noise, a variety of search methods have been developed that work without explicit knowledge of gradients. These methods may be grouped in a variety of ways and are known by a range of names. Perhaps, the best term for them collectively is "zeroth order methods" (as opposed to first-order methods, which draw on the first differential of the function, etc.) although the term "direct search" is also often used. The most common distinction within these methods is between pattern/direct searches and stochastic/evolutionary algorithms. This distinction is both historical, in that pattern/direct search methods were developed first, and also functional since the stochastic and evolutionary methods are generally aimed at finding multiple optima in a landscape, while pattern/direct methods tend simply to be trying to find the location of the nearest optimum without recourse to gradient calculations. In either case, it will be clear that the problems of obtaining gradients by finite differencing in a noisy landscape are immediately overcome by simply not carrying out this step. It is also the case that these methods are often capable of working directly with discrete variable problems since at most stages all that they need to be able to do is to rank alternative designs. If, however, the problem at hand is not noisy, gradients can be found relatively easily and only a local search is required, it is always better to use gradient-based methods, such as the conjugate or quasi-Newton-based approaches already described. When there are modest amounts of noise or the gradients are modestly expensive to compute, no clear-cut guidance can be offered and experimentation with the various alternatives is then well worthwhile.

Pattern or Direct Search

Pattern (or direct) searches all involve some form of algorithm that seeks to improve a design based on samples in the vicinity of the current point. They work by making a trial step in some direction and seeing if the resulting design is better than that at the base point. This comparison does not require knowledge of the gradient or distance between the points, but simply that the two designs can be correctly ranked. Then, if the new design is an improvement, the step is captured and a further trial made, usually in the same direction but perhaps with a different step length. If the trial move does not improve the design, an alternative direction is chosen and a further trial made. If steps in all directions make the design worse, the step size is reduced and a further set of explorations made. Sometimes, the local trials are augmented with longer-range moves in an attempt to speed up the search process. Usually, these are based on analysis of recent successful samples. In this way, a series of gradually converging steps is made that tries to greedily capture improvements on the current situation. This process continues until some minimum length of step fails to produce an improvement in any direction and it is then assumed that the local optimum has been reached. A common feature of these searches is that the directions used in the local trial *span* the space being searched, that is, they comprise a set of directions that can be used to create a vector in any direction. In this way, they offer some guarantees of convergence, though often these are not as robust as for gradient-based methods.

In general, pattern searches have no way of accepting designs poorer than the current base point, even if moves in such a direction may ultimately prove to be profitable. It is this fact that makes these methods suitable only for finding the location of the nearest optimum and not for global searches over multimodal landscapes. If multiple optima are to be found with pattern searches, then they must be restarted from various locations so that multiple basins of attraction can be sampled. Usually, stochastic or evolutionary methods are better for such tasks, at least to begin with.

To see how pattern searches work, it is useful to review two well-accepted methods. The most well known methods are probably those due to Hooke and Jeeves (1960) and Nelder and Mead (1965). This latter search is also commonly referred to as the Simplex algorithm – not to be confused with the simplex method of linear programming. Both date from the 1960s and their continued use is testament to their inherent robustness in dealing with real problems and relative simplicity of programming. The approach due to Hooke and Jeeves is the earlier, so we begin with a description of their method.

The pattern search of Hooke and Jeeves begins at whatever point the user supplies and carries out an initial exploratory search by making a step of the given initial step size in the direction of increasing the first variable. If this improves the design, the base point is updated and the second variable is tested. If it fails, a negative step in the direction of the first variable is taken and again this is kept if it helps. If both fail, the second variable is tested, with any gains being kept, and so on, until all steps of this size have been tested and any gains made. This sequence of explorations gives a characteristic staircase pattern, aligned with the coordinate system; see the clusters of points in Figure 3.9.

When all coordinate directions have been tested, a pattern move is made. This involves changing all variables by the same amount as cumulated over the previous exploratory search (the diagonal moves in Figure 3.9). If this works, it is kept, while if not, it is abandoned (this happens twice in the figure, where it overshoots the downhill direction). In either case, a new exploratory search is made. At the end of the exploration, another pattern move is made; however, this time if the previous pattern move was successful, the next pattern move is built from a combination of the just completed exploratory steps and the previous pattern move, that is, the pattern move becomes bigger after each successful exploratory search/pattern move combination and slowly aligns itself to the most profitable direction of move. This allows the search to make larger and larger steps. Conversely, once a pattern move fails, the search resets and has to start accumulating pattern moves from scratch. Finally, if none of the exploratory steps yields an improvement, the step size is reduced by an arbitrary factor and the whole process restarted. This goes on until the minimum step size is reached. The only parameters needed are the initial and minimum step sizes, the reduction factor when exploratory moves fail and the total number of steps allowed before aborting the search. Clearly, a smaller final step size gives a more refined answer, while a smaller reduction factor tends to slow the search into a more careful process that is more capable of teasing out fine detail.

It is immediately apparent from studying the figure that the Hooke and Jeeves search will not initially align itself with the prevailing downhill direction. Rather, a number of pattern moves have to be made before the method adapts to the local topography. If, while

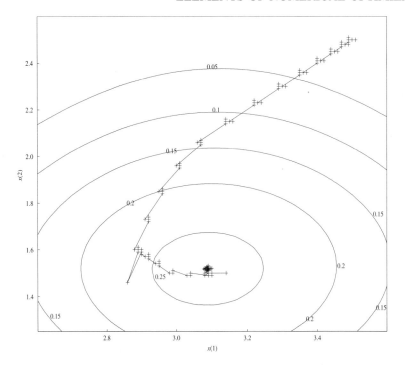

Figure 3.9 The Hooke and Jeeves pattern search in two dimensions.

Table 3.4   Results of pattern searches on local optimum in the bump function of (3.2) without and with 1% noise (using the code provided by Siddall in both cases)

Method	Noise	Final Objective Function Value (% of true optimal objective)	Number of Function Calls
Hooke and Jeeves	none	−0.26290 (100)	167
	1%	−0.22757 (86.6)	375
Nelder and Mead	none	−0.26290 (100)	181
	1%	−0.25409 (96.6)	149

these moves are made, the local landscape seen by the optimizer changes so that the ideal search direction swings around, as here, eventually the method starts to run uphill and a new set of explorations becomes necessary. Thus, it is almost inevitable that such a search will be slower than those that build models of the gradient landscape. Nonetheless, the search is robust and quite efficient when compared to other schemes, particularly on noisy landscapes or those with discontinuities; see Table 3.4, which shows results achieved from the version of the code provided in Siddall (1982).

The Nelder and Mead simplex search is based around the idea of the simplex, a geometrical shape that has $n + 1$ vertices in n dimensions (i.e., a triangle in two dimensions and a tetrahedron in three, etc). Provided a simplex encloses some volume (or area in two dimensions), then if one vertex is taken as the origin, the vectors connecting it to the other vertices span the vector space being searched (i.e., a combination of these vectors, suitably scaled, will reach any other point in the domain). The basic moves of the algorithm are as follows:

1. reflection – the vertex where the function is worst is reflected through the face opposite, but the volume of the simplex is maintained;

2. reflection and expansion – the worst vertex is again reflected through the face opposite but the point is placed further from the face being used for reflection than where it started with the volume of the simplex being increased;

3. contraction – a vertex is simply pulled toward the face opposite and the volume of the simplex reduced;

4. multiple contraction – a face is pulled toward the vertex opposite, again reducing the volume of the simplex.

To decide which step to take next, the algorithm compares the best, next to best and worst vertices of the simplex to a point opposite the worst point and outside the simplex (the trial point), labeled b, n, w and r, respectively in Figure 3.10, which illustrates the search in two dimensions (and where the simplexes are then triangles). The distance the trial point lies outside the simplex is controlled by a scaling parameter that can be user defined, but is commonly unity. The following tests are then applied in series:

1. if the objective function at the trial point lies between the best and the next to best, the trial point is accepted in lieu of the worst, that is, the simplex is reflected;

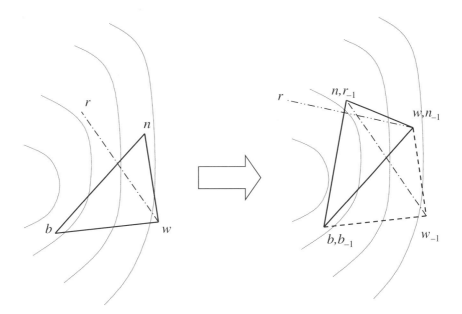

Figure 3.10 The Nelder and Mead simplex method.

2. if the trial point is better than the current best point, not only is the worst point replaced by the best but the simplex is expanded in this direction, that is, reflection and expansion – if the expansion fails, a simple reflection is used instead;

3. if the trial point is poorer than the current worst point, the simplex is contracted such that a new trial point is found along the line joining the current worst point and the existing trial point, if this new trial point is better than the worst simplex point, it is accepted, that is, contraction, while if it is not, all points bar the best are contracted, that is, multiple contraction;

4. otherwise, the trial point is at least better than the worst, so a new simplex is constructed on the basis of this point replacing the worst and in addition a contraction is also carried out.

This rather complex series of steps is set out in the original reference and many subsequent books. It is studied in some detail by Lagarias et al. (1998). The process terminates when the simplex is reduced to a user set minimum size or the average difference between the vertex objective function values is below a user-supplied tolerance. To apply the method, the user must supply the coordinates of the first simplex (usually by making a series of small deflections to a single starting point) and decide the parameters controlling the expansion and contraction ratios. Table 3.4 also shows results for using this routine on the bump function both with and without noise. Figure 3.11 illustrates the search path for the noise-free case.

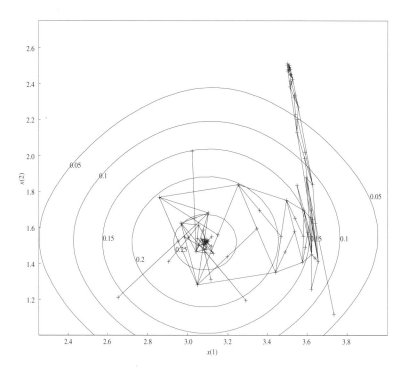

Figure 3.11 The Nelder and Mead simplex search in two dimensions.

The method is reasonably efficient and robust and not surprisingly therefore remains very popular.

This brief description of two of the oldest pattern or direct searches illustrates two aspects that are common to all such methods: first, there is a process for making cautious downhill movements by searching exhaustively across a set of directions that span the search space, typically with quite modest steps (sometimes referred to as "polling" the search space); and, secondly, there is some form of acceleration process that seeks to take larger steps when the previous few moves seem to justify this. The first of these is aimed at ensuring some form of convergence while the second tries to make the process more efficient by covering the ground with fewer steps. When designing a pattern search, these two issues must be borne in mind and suitable strategies implemented to allow the search to switch between them when needed.

Perhaps, the greatest area for innovation lies in the way that accelerated moves are made: in this phase, the algorithm is seeking to guess what the surface will be like some distance beyond what may be justified by a simple linear model using the current local exploration. A number of ideas have been proposed and some researchers refer to this part of pattern searches as "the oracle", indicating the attempt to see into the future! There are a number of themes that emerge in such exploration systems. One approach is to use as much of the previous history of the search as possible to build up some approximation that can used as a surrogate for the real function – this can then be analyzed to locate interesting places to try. It has one significant drawback – how far back should data be used in constructing this approximation? If the data used is too far from the current area of search, it may be misleading, and if it is too recent, it may add very little. The most common alternative is to scatter a small number of points forward from the current location and build an approximation that reaches into the search – clearly, for this to differ from the basic polling search process, it must use larger steps and fewer of them. Again, this is the drawback – if too few steps are used, the approximation will be poor, and if they are to close to the current design, the gains will be limited. Formal design of experiment (DoE) methods can help in planning such moves in that they help maximize the return on investment in such speculative calculations.

In any work on constructing oracle-like approximations, choice must also be made as to whether the approximation in use will interpolate the data used or alternatively a lower order regression model will be adopted. Adaptive step lengths can also be programmed in so that the algorithm becomes greedier the more successful it is. Provided the search is always able to fall back on the small step, exhaustive local polling searches using a basis that spans the space, it should be able to get (near) to the optimal location in the end. It is, of course, also the case that in much design work an improvement in design is the main goal, given reasonable amounts of computing effort, rather than a fully converged optimal design, which may be oversensitive to some of the uncertainties always present in computationally based performance prediction.

Although pattern searches remain popular, they do have drawbacks, often in terms of final convergence. For example, it has been demonstrated that the original Nelder and Mead method can be fooled by some search problems into collapsing its simplexes into degenerate forms that stall the search. In fact, all pattern search methods can be fooled by suitably chosen functions that are either discontinuous or have discontinuous gradients into terminating before finding a local optimum. There are a number of refinements now available that correct many of these issues, but the basic limitation remains. The work of

Kolda et al. (2003) shows how far provable convergence can be taken in various forms of pattern search, for example. Despite such limitations, pattern search methods often work better in practice than might be expected in theory. Moreover, a simple restart of a pattern search is often the easiest way to check for convergence. If the method is truly at an optimal location, the cost of such a restart is typically small compared to the cost of arriving at the optimum in the first place. Another approach is to take a small random walk in the vicinity of the final design and see if this helps at all – again, this is easy to program and is usually of low cost.

It will be apparent from all this that many variants of pattern/direct search exist and some effort has been directed toward establishing a taxonomy of methods (Kolda et al. 2003). One variant that the authors have found useful for problems with a small amount of multimodality is the Dynamic Hill Climbing (DHC) method of Yuret and de la Maza (1993) – this combines pattern search with an automated restart capability that seeks to identify fresh basins of attraction. Since these restarts are stochastic in nature, it bridges the division between pattern and stochastic methods that we consider next (and is thus a form of hybrid search; see later).

Stochastic and Evolutionary Algorithms

The terms stochastic and evolutionary are used to refer to a range of search algorithms that have two fundamental features in common. First, the methods all make use of random number sequences in some fashion, so that if a search is repeated from the same starting point with all parameters set the same, but with a different random number sequence, it will follow a different trajectory over the fitness landscape in the attempt to locate an optimal design. Secondly, the methods all seek to find globally, as opposed to locally, optimal designs. Thus, they are all capable of rejecting a design that is at some minimum (or maximum) of the objective function in the search for better designs, that is, they are capable of accepting poorer designs in the search for the best design.

These methods have a number of distinct origins and development histories but can, nevertheless, be broadly classified by a single taxonomy. The most well known methods are simulated annealing (SA), genetic algorithms (GA), evolution strategies (ES) and evolutionary programming (EP). We will outline each of these basic approaches in turn before drawing out their common threads and exploring their relative merits and weaknesses.

Simulated annealing optimization draws its inspiration from the process whereby crystalline structures take up minimum energy configurations if cooled sufficiently slowly from some high temperature. The basic idea is to allow the design to "vibrate" wildly to start with so that moves can be made in arbitrary directions and then to slowly "cool" the design so that the moves becomes less arbitrary and more downhill in focus until the design "solidifies", hopefully into some minimum energy state. The basic elements of this process are:

1. a scheme for generating (usually random) perturbations to the current design;

2. a temperature schedule that is monitored when taking decisions;

3. a model of the so-called Boltzmann probability distribution $P(E) = \exp(-dE/kT)$, where T is the temperature, dE is the change in energy and k is Boltzmann's constant.

At any point in the algorithm, the current objective function value is mapped to an energy level and compared to the energy level of a proposed change in the design produced using the perturbation scheme, that is, dE is calculated. If the perturbed design is an improvement on the current design, this move is accepted. If it is worse, it may still be accepted but this decision is made according to the Boltzmann probability, that is, a random number uniformly distributed between zero and one is drawn and, if $\exp(-dE/kT)$ is greater than this number, the design is still accepted. To begin with, the temperature T is set high so that almost all changes are kept. Then, as T is reduced, designs have increasingly to be an improvement to be used. In this way, the process starts as a simple random walk across the search space and ends as a greedy downhill search with random moves. This basic Metropolis algorithm is described by Kirkpatrick et al. (1983).

It is nowadays a commonly accepted practice that the temperature should be reduced in a logarithmic fashion though this can be done continuously or in a series of steps (the authors tend to use a fixed number of steps). Also, it is necessary to establish the value for the "Boltzmann" constant for any particular mapping of objective function to E. This is most readily achieved by making a few random samples across the search space to establish the kinds of variation seen when changes are made according to the scheme in use. The aim is that the initial value of kT should be significantly larger than the typical values of dE seen (typically, ten times more).

The most difficult aspect of setting up a SA scheme is to decide how to make the perturbations between designs. These need to be able to span the search space being used and can come from a variety of mechanisms. One very simple scheme is just to make random changes to the design variables. Many references describe much more elaborate approaches that have been tuned to problem-specific features of the landscape.

Genetic Algorithms are based on models of Darwinian evolution, that is, survival of the fittest. They have been discussed for at least 35 years in the literature but perhaps the most well known book on the subject is that by Goldberg (1989). The basic idea is that a pool of solutions, known as the population, is built and analyzed and then used to construct a new, and hopefully improved, pool by methods that mimic those of natural selection. Key among these are the ideas of fitness and crossover, that is, designs that are better than the average are more likely to contribute to the next generation than those that are below average (which provides the pressure for improvement) and that new designs are produced by combining ideas from multiple (usually two) "parents" in the population (which provides a mechanism for exploring the search space and taking successful ideas forward). This process is meant to mimic the natural process of crossover in the DNA of creatures that are successful in their environment. Here, the vector of design variables is treated as the genetic material of the design and it is this that is used when creating new designs. Various forms of encoding this information are used, of which the most popular are a simple binary mapping and the use of the real variables directly. There are numerous ways of modeling the selection of the fittest and for carrying out the "breeding" process of crossover. The simplest would be roulette wheel sampling whereby a design's chances of being used for breeding are random but directly proportional to fitness and one point binary crossover whereby a new design is created by taking a random length string of information from the leading end of parent one and simply filling the remaining gaps from the back end of parent two.

There are a number of further features that are found in GAs: mutation is commonly adopted, whereby small random changes are introduced into designs so that, in theory, all regions of the search space may be reached by the algorithm. Elitism is also normally used, whereby the best design encountered so far is guaranteed to survive to the next generation unaltered, so that the process cannot lose any gains that is has made (this, of course,

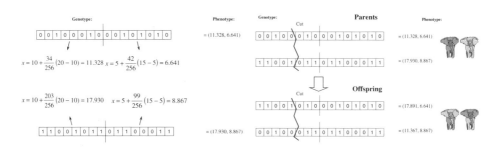

(a) Binary encoding of two designs – phenotype uses real numbers, for example, variables between 10.0 and 20.0 and 5.0 and 15.0 using eight binary digits for each variable.

(b) Cut points and results for crossover – here, crossover acts a single point in the genotype and all material before the cut point is exchanged in the offspring.

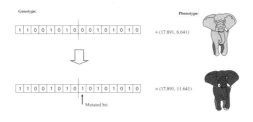

(c) Bit-flip mutation – a single bit in the genotype is flipped.

Figure 3.12   Genetic algorithm model and operators.

has no natural equivalent – it amounts to immortality until a creature is killed by a better one). Inversion or some form of positional mapping is also sometimes used to ensure that when crossover takes place the location of specific pieces of information in the genetic makeup can be matched to other, related aspects. The population pool in the method can vary in size during search and it can be updated gradually or all at one time. In addition, various methods have been proposed to prevent premature convergence (dominance) of the population by mediocre designs that arrive early in the search. Most common are various niche and resource sharing methods that make overly popular designs work harder for their place in the population. The literature in this field is vast but the interested reader can find most of the ideas that have proved useful by studying the first five sets of proceedings of the International Conference on Genetic Algorithms (Belew and Booker 1991; Forrest 1993; Grefenstette 1985, 1987; Schaffer 1989).

The basic processes involved can be illustrated by considering a simple "vanilla" GA applied to a two-dimensional problem that uses eight-bit binary encoding. In such a model each design is mapped into a binary string of 16 digits; see Figure 3.12a. The first eight bits are mapped to variable one such that 00000000 equates to the lower bound and 11111111 to the upper, with 256 values in total; and similarly for variable two. If we take two designs, these may be "crossed" using single-point crossover by randomly selecting a cut point between 1 and 15, say 5; see the upper part of Figure 3.12(b). Then, we create two "children" from the two "parents" by forming child one from bits 1–5 of parent one and 6–16 of parent two while child two is formed from the remaining material, that is, bits 1–5 of parent two and bits 6–16 of parent one; see the lower part of Figure 3.12(b). Mutation can then be added by simply flipping an arbitrary bit in arbitrary children according to some random number scheme, usually at low probability; see Figure 3.12(c). Provided that selection for parenthood is linked in some way to the quality of the designs being studied, this seemingly trivial process yields a powerful search engine capable of locating good designs in difficult multimodal search spaces. Nonetheless, a vanilla GA such as that just described is very far from representing the state of the art and will perform poorly when compared to the much more sophisticated methods now commonly available – its description here is given just to illustrate the basic process.

Evolution Strategies use evolving populations of solutions in much the same way as GAs. Their origins lie with a group of German researchers in the 1960s and 1970s (Back et al. 1991) (i.e., at about the same time as GAs). However, as initially conceived, they worked without the process of crossover and instead focused on a sophisticated evolving vector of mutation parameters. Also, they always use a real-valued encoding. The basic idea was to take a single parent design and make mutations to it that differed in the differing dimensions of the search. Then, the resulting child would either be adopted irrespective of any improvements or, alternatively, be compared to the parent and the best taken. In either case, the mutations adopted in producing the child were encoded into the child along with the design variables. These were stored as a list of standard deviations and also, sometimes, as correlations and were then themselves mutated in further operations, that is, the process evolved a mutation control vector that suited the landscape being searched at the same time as it evolved a solution to the problem itself. In later developments, a population-based formulation was introduced that does use a form of crossover and it is in this form that ESs are normally encountered today. They still maintain relatively complex mutation strategies as their hallmark, however. The resulting approach still takes one of two basic forms: with the mutated children automatically replacing their parents, a so-called (μ, λ)-ES, or with the next generation being selected from the combined parents and children, a $(\mu + \lambda)$-ES. In the latter case, if the best parent is better than the best child, it is commonly preserved unchanged (elitism). The best of the solutions is remembered at each pass to ensure that the final result is the best solution of all, irrespective of the selection mechanism.

Crossover can be discrete or intermediate on both the design vector and the mutation control vector. In some versions, the mutations introduced are correlated between the different search dimensions so that the search is better able to traverse long narrow valleys that are inclined to the search axes. The breeding of new children can be discrete and then individual design vector values are taken from either parent randomly or intermediate and the values are then the average of those of the parents. Separate controls for the breeding of the design vector and the mutation standard deviations are commonly available so that these can use different approaches.

In their simplest form, the mutations are controlled by vectors of standard deviations which evolve with the various population members. Each element in the design vector is changed by the addition of a random number with this standard deviation scaled to the range of the variable. The rate at which the standard deviations change is controlled by a parameter of the method, $\Delta\sigma$ such that $\sigma_i^{\text{NEXT}} = \sigma_i^{\text{PREV}} \exp(N(0, \Delta\sigma))$, where the σ_is are the individual standard deviations and $N(0, \Delta\sigma)$ is a random number with zero mean and $\Delta\sigma$ standard deviation. The σ_is are normally restricted to be less than some threshold (say 0.1) so that the changes to the design vector (in a nondimensional unit hypercube) do not simply push the vector to its extreme values. The initial values of the σ_is are user set, typically at 0.025. Since successful children will owe their success, at least in part, to their own vector of mutation controls, the process produces an adaptive mutation mechanism capable of rapidly traversing the design space where needed and also of teasing out fine detail near optimal locations. Because a population of designs is used, the method is capable of searching multiple areas of a landscape in parallel.

In EP algorithms (Fogel 1993), evolution is carried out by forming a mutated child from each parent in a population with mutation related to the objective function so that successful ideas are mutated less. The objective functions of the children are then calculated and a stochastic ranking process used to select the next parents from the combined set of parents and children. The best of the solutions is kept unchanged at each pass to ensure that only improvements are allowed. The mutation is controlled by a parameter that sets the order of the variation in the amount of mutation with ranking. A value of one gives a linear change, two a quadratic, and so on (only positive values being allowed). In all cases, the best parent is not mutated and almost all bits in the worst are changed.

The stochastic process for deciding which elements survive involves jousting each member against a tournament of other members selected at random (including possibly itself) with a score being allocated for the number of members in the tournament worse than the case being examined. Having scored all members of both parent and child generations, the best half are kept to form the next set of parents. The average number of solutions in the tournament is set by a second parameter. An elitist trap ensures that the best of all always survives.

In EP, the variables in the problem are typically discretized in binary arithmetic so that all ones (1111...) again would represent an upper bound and all zeros (0000...) the lower bound. The total number of binary digits used to model a given design vector (the word length) is the sum over each variable of this array – the longer the word length, the greater the possible precision but the larger the search space. The algorithm works by forming a given number of random guesses and then attempting to improve on them, maintaining a population of best guesses as the process continues.

A number of common threads can be drawn out from these four methods:

1. a population of parent designs is used to produce a population of children that inherit some of their characteristics (these populations may comprise only a single member as in SA);

2. the mechanisms by which new designs are produced are stochastic in nature so that multiple statistically independent runs of the search will give different results and therefore averaging will be needed when assessing a method;

3. if there are multiple parents, they may be combined to produce children that inherit some of their characteristics from each parent;

4. all children have some finite probability of undergoing a random mutation that allows the search to reach all parts of the search space;

5. the probability rates that control the process by which changes are made may be fixed, vary according to some preordained schedule or be adapted by the search itself depending on the local landscape;

6. a method is incorporated to record the best design seen so far, as the search can both cause the designs to improve and deteriorate;

7. some form of encoding may be used to provide the material that the search actually works with; and

8. a number of user set parameters must be controlled to use the methods.

Finally, it should be noted that no guarantees of ultimate convergence can be provided with evolutionary algorithms (EAs) and they are commonly slow at resolving precise optima, particularly when compared to gradient-based methods.

Figures 3.13(a–d) show the points evaluated by a variant of each search method on the bump test function with the best of generation/annealing temperature points linked by a line (the codes are from the Options search package[3]). In all cases, 500 evaluations have been allowed – such methods traditionally continue until told to stop rather than examining the current best solution, since they are always trying to find new basins of attraction that have yet to be explored. It is immediately apparent from the figures that all the methods search across the whole space and none get bogged down on the nearest local minimum, instead moving from local optimum to local optimum in the search for the globally best design. All correctly identify the best basin of attraction, and Table 3.5 shows how good the final values are when compared to the true optimum of −0.68002 in this unconstrained space, when averaged over 100 independent runs.

There is little to choose between the methods except that the ES method occasionally seems to get trapped in a slightly suboptimal basin. The performance of these methods on harder problems with more dimensions and more local optima is of course not easy to predict from such a simple trial. In almost all cases, careful tuning of a method to a landscape will significantly improve performance. Moreover, when dealing with global search methods like these, a trade-off must be made between speed of improvement in designs, quality of final design returned and robustness against varying random number sequences. Usually, a robust and fast improving process is most useful in the context of aerospace design, rather than one that occasionally does very well but often fails or a very slow and painstaking search that yields the best performance but only at very great computational expense. Another important feature is the extent to which parallel computation can be invoked. Clearly, with a population-based method that builds new generations in a stepwise fashion, an entire

---

[3]http://www.soton.ac.uk/~ajk/options/welcome.html

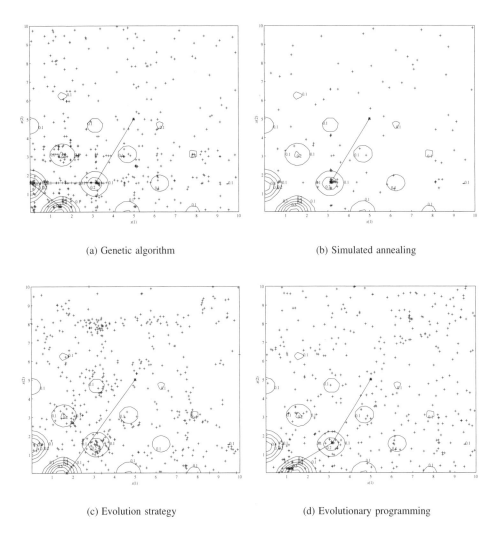

(a) Genetic algorithm   (b) Simulated annealing

(c) Evolution strategy   (d) Evolutionary programming

Figure 3.13   Evolutionary algorithm search traces on bump problem, (3.2), without constraints.

generation of calculations may be carried out simultaneously – such calculations can often readily be mapped onto the individual processors of a large cluster. The fact that SA does not readily allow this is one of the key drawbacks of the method.

One important common feature of the stochastic methods that make use of encoding schemes is that these can very easily be mapped onto discrete or integer optimization problems. For example, it is very easy to arrange for a binary encoding to map onto a list of items rather than to real numbers – in fact, this is probably a more natural use of such schemes. Even simulated annealing can be used in this way if a binary encoding scheme is used to generate the perturbations needed to move from design to design.

Table 3.5 Results of evolutionary searches on global optimization of the bump function of (3.2) averaged over 100 independent runs using 500 evaluations in each case (using the code provided by Keane in all cases)

Method	Average Result	% of True Optimum (−0.68002)	Standard Deviation in Final Objective Function Value
GA	−0.61606	90.6	0.0751
SA	−0.61896	91.0	0.1128
ES	−0.52065	76.6	0.1117
EP	−0.58528	86.1	0.0831

3.2.5 Termination and Convergence Aspects

In all search methods, some mechanism must be put in place to terminate the process. When engaged on local search, it is normal to terminate when the current design cannot be improved upon in the local vicinity, that is, an optimum has been located. For global search, no such simple test can be made, since there is always the possibility that a basin remains to be found that, if searched, will contain a better design. In the first case, the search is said to be converged, while in the second, it must be terminated using some more complex criterion. These two issues are, of course, related. Moreover, if the cost of evaluations is high, then it may well be necessary to terminate even a local search well before it is fully converged. We are thus always concerned with the rate of improvement of a search with each new evaluation.

In the classical gradient search literature, this rate of improvement is often described as being linear, quadratic or superlinear, depending on an analysis of the underlying algorithm. Formally, a search is said to be linear if, in the limit as the iteration count goes to infinity, the rate of improvement tends to some fixed constant that is less than unity, that is, if $|f(x_{i+1}) - f(x^*)|/|f(x_i) - f(x^*)| = \delta < 1$, where $f(x^*)$ is the final optimal objective function value. If the ratio tends in the limit to zero, the search is said to be superlinear, while if $|f(x_{i+1}) - f(x^*)|/|f(x_i) - f(x^*)|^2 = \delta$, the search is said to be quadratic. For example, the series 0.1, 0.01, 0.001, ... exhibits linear convergence, $1/i! = 1, 1/2, 1/6, 1/24, 1/125, ...$ is superlinear and 0.1, 0.01, 0.0001, 0.000000001, ... is quadratic.

Ideally, we would wish all searches to be superlinear or quadratic if possible. Newton's method is quadratically convergent if started sufficiently close to the optimum location. Conjugate gradient and quasi-Newton searches are always at least linear and can approach quadratic convergence. It is much more difficult to classify the speed of pattern and stochastic searches. Fortunately, when carrying out real design work, a steady series of improvements is usually considered enough! Nonetheless, faster searches are generally better than slower ones and so we normally seek a steep gradient in the search history. Note also that the accuracy with which an optimal design can be identified is limited to the square root of the machine precision (and not the full precision), so it is never worth searching beyond this limit – again, in real design work, such accuracy is rarely warranted since manufacturing constraints will limit the precision that can be achieved in any final product.

When carrying out global searches, where there are presumed to be several local optima, a range of other features must be considered when comparing search methods. Consider a graph that shows the variation of objective function with iteration count for a collection of

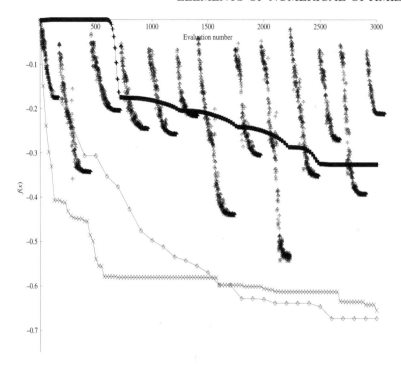

Figure 3.14   Search histories for a range of different search methods (+ – repeated Hooke and Jeeves pattern search with evolving penalty function, × – simulated annealing, arrows – dynamic hill climbing (Yuret and de la Maza 1993) and diamonds – genetic algorithm).

different search methods, such as Figure 3.14. A number of features can be observed in this plot. Obviously, the gradient tells us something about the convergence rate of the search at any point. However, this is nonuniform in these traces and clearly goes through various phases – this is to be expected if some of the time a search is attempting to identify new basins of attraction and sometimes it is rapidly converging to the best point in a newly located basin. Moreover, even if new basins can be regularly located, there will be no guarantee that each new basin will contain a design that improves on the best so far. Sometimes, by chance, or because the starting design is quite good, no basins beyond that first searched will contain better designs. Whatever the circumstances, the designer must make some decision on when to terminate a search and this cannot be based on local convergence criteria alone. The four most common approaches are:

1. to terminate the search after a fixed number of iterations;

2. to terminate the search after a fixed elapsed time;

3. to terminate the search after a fixed number of iterations have produced no further improvement in the global optimum; or

4. to terminate the search after a fixed number of distinct basins of attraction have been located and searched.

The last of these is only viable if the designer has some good understanding of the likely number of basins in the problem being dealt with, though this is often possible when dealing with variants on a familiar problem.

Given a termination criterion, it is then possible to try and rank different global search methods; see again Figure 3.14. Here, we see that one search exhibits very good initial speed but then slows down and is finally overtaken (simulated annealing), one that is slower but continues to steadily improve throughout (genetic algorithm), a third that works in cycles, with significant periods where it is searching inferior basins (dynamic hill climbing – a multistart method) and a fourth that appears to make no progress to begin with but which gets going some distance into the search (a Hooke and Jeeves pattern search with evolving penalty). The methods can be ranked in a number of ways:

1. by the steepness of the initial gradient (useful if the function evaluation cost is high and a simple improvement in design is all that is needed);

2. whether the ultimate design has a very good objective function (good if the function cost is low or if a specific and demanding performance target must be met); or

3. if the area below the trace curve is low (important when one is trying to balance time and quality in some way – N.B., it should be high for maximization, of course).

In addition, a number of other factors may be of interest, such as the volume of the search space traversed. Consider a narrow cylindrical pipe placed around the search path as it traverses the design space or small spheres placed around each new trial point – these represent the volume explored in some sense. The repeatability of the search will also be of interest if it is started from alternative initial designs or if used with differing random number sequences – this is particularly important for stochastic search methods. Ideally, a global search should explore the design space reasonably well within the limits of the available computing budget and also be reasonably robust to starting conditions and random number sequences. Depending on the circumstances, however, a search that is usually not very good at getting high-quality designs but occasionally does really well may be useful – this may be thought equivalent to having a temperamental star player on the team who can sometimes score in unexpected circumstances. Such players can be very useful at times even if they spend many games on the substitutes' bench! Again, we offer no clear-cut recommendation on how to rank search methods – we merely observe that many approaches are possible and the designer should keep an open mind as to how to make rational choices.

3.3 Constrained Optimization

In almost all practical design work, the problems being dealt with are subject to a variety of constraints. Such constraints are defined in many ways and restrict the freedom of the designer to arbitrarily choose dimensions, materials, and so on, that might otherwise lead to the best performance, lightest weight, lowest cost, and so on. It is also the case that many classical optimization problems are similarly constrained and this has led to the development of many ways of trying to satisfy constraints while carrying out optimization. More formally, we can recall our definition of the optimization problem as the requirement to minimize the function $f(\mathbf{x})$ subject to multiple inequality constraints $g_i(\mathbf{x}) \geq 0$ and equality constraints $h_j(\mathbf{x}) = 0$. Note that in many real design problems the number of constraints may be very large, for example, in stress analysis we may require that the Von Mises stress be less than some design stress at *every* node in a finite element mesh containing hundreds of thousands of elements.

A key issue in dealing with constraints during design is the tolerance with which they must be met. When examining the mathematical subtleties of a method's convergence, we may wish to satisfy constraints to within machine precision, but this is never required in practical design work. In many cases, slight violation of a constraint may be permissible if some other benefit is achieved by doing so – it is rare for anything in design work to be so completely black and white that a change in the last significant digit in a quantity renders a design unacceptable. Moreover, it is also arguable if equality constraints ever occur for continuous variables in design – rather, such constraints can usually be regarded as a tightly bound pair of inequality constraints and this is often how they must be treated in practice. Nonetheless, in this section on constraints, we deal with the formal problem and the methods that have been devised to deal with them – we simply urge the reader to keep in mind that in practical work things are never so simple.

3.3.1   Problem Transformations

The most common way of dealing with constraints is to attempt to recast the problem into one without constraints by some suitable transformation. The simplest example of this is where there is an algebraic equality constraint that may be used to eliminate one of the variables in the design vector. So, for example, consider trying to minimize the surface area of a rectangular box where the volume is to be held fixed. If the sides have lengths L, B and H, then it is clear that the total surface area A is $2(L \times B + L \times H + B \times H)$, while the volume V is just $L \times B \times H$. If the volume is given, then we may use this equality constraint to eliminate the variable H by setting it equal to $V/(L \times B)$ and so the objective function then becomes $2(L \times B + (L + B) \times V/(L \times B))$, that is, a function of two variables and the constraint limit (the value V), rather than three variables. Whenever such changes may be made, this should automatically be done before starting to carry out any optimization. Moreover, even when dealing with inequality constraints, such a transformation may still be possible: if an inequality is not active at the location of the optimum being sought, it can be ignored; if it is active, it can be replaced with an equality. Thus, if knowledge of which constraints will be active can be found, these constraints can be eliminated similarly, if simple enough.

Usually, however, the constraints being dealt with will result from some complex PDE solver that cannot be so easily substituted for. Even when this is the case, equality constraints can sometimes still usefully be removed by identifying a single variable that dominates the constraint and then solving an "inner loop" that attempts to find the value of the key design variable that satisfies the constraint by direct root searching. A classical example of this occurs in wing design – it is common in such problems to require that the wing produce a specified amount of lift while the angle of attack is one of the variables open to the designer. If the overall aim is to change wing shape so as to minimize drag while holding lift fixed, this can best be done by using two or three solves to identify the desired angle of attack for any given geometry rather than simply leaving the angle of attack as one of a number of design variables and attempting to impose a lift equality constraint on the search process. This substitution works because we know, *a priori*, that angle of attack directly controls lift and, moreover, that the relationship between the two is often very simple (approximately linear).

3.3.2 Lagrangian Multipliers

Normally, it is not possible to simply remove constraints by direct substitution; other methods must be adopted. The classical approach to the problem is to use Lagrangian multipliers. If we initially restrict our analysis to equality constraints, then we may substitute $f(\mathbf{x}) - \sum \lambda_j h_j(\mathbf{x})$ for the original constrained function. Here, the λ_j are the so-called Lagrangian multipliers and the modified function is referred to as the Lagrangian function (or simply the "Lagrangian" in much of the work in this field). If we now find a solution such that the gradients of the Lagrangian all go to zero at the same time as the constraints are satisfied, the final result will minimize the original problem (there are a number of further requirements such as the functions being differentiable, linear independence of constraint gradients, the Hessian of the Lagrangian being positive definite, and so on, the so-called Karush–Kuhn–Tucker (KKT) conditions, but we omit discussions of these here for clarity – see Boggs and Tolle (1989, 1995) for more complete details). The inclusion of the multipliers increases the problem dimensions, since they must be chosen as well as the design variables, but it also turns the optimization problem into the solution of a set of coupled (nonlinear) equations.

A simple example will illustrate the multiplier idea. Let the objective function being minimized $f(\mathbf{x})$ be just x^2 and the equality constraint be $x = 1$. It is obvious that the function has a minimum of 1 at $x = 1$ (no other value of x satisfies the constraint). The equivalent Lagrangian function with multiplier, $f'(\mathbf{x})$, is $x^2 - \lambda (x - 1)$, where the equality constraint has been set up such that it equals zero, as required in our formulation. Figure 3.15 shows

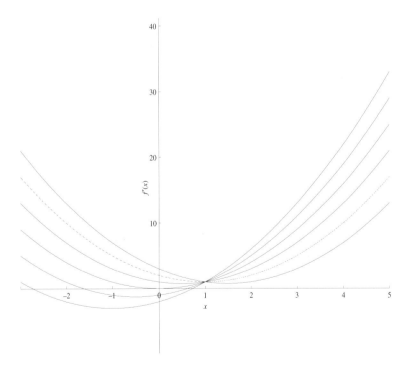

Figure 3.15 Plots of the Lagrangian function for various values of the Lagrangian multiplier (that shown dotted is for $\lambda = 2$).

a series of curves of this function for differing values of λ. Of these, the curve for $\lambda = 2$ is the one that has zero gradient (its minimum) at $x = 1$ and thus satisfies the constraint, that is, minimizing $x^2 - 2(x - 1)$ minimizes the original objective function and satisfies the equality constraint. When dealing with arbitrary functions, one must simultaneously seek zeros of all the gradients of $f'(\mathbf{x})$ and search over all the λs for values that meet the constraints.

Now, consider the situation where we have a single inequality constraint, for example, $x \geq 1$. It will be obvious that the solution to this problem is also at $x = 1$ and again the objective function is just 1. To deal with this problem using Lagrangian multipliers, we must introduce the idea of slack variables to convert the inequality constraint to an equality constraint. Thus, if we have constraints $g_i(\mathbf{x}) \geq 0$, we can write them as $g_i(\mathbf{x}) - a_i^2 = 0$, where a_i is a slack variable and then, proceeding as before, the Lagrangian becomes $f(\mathbf{x}) - \sum \lambda_j h_j(\mathbf{x}) - \sum \lambda_i [g_i(\mathbf{x}) - a_i^2]$. This of course introduces a further set of variables and so now we must, in principle, search over the xs, λs *and* the as. To obtain equations for the as, we just differentiate the Lagrangian with respect to these as and equate to zero, yielding $2\lambda_i a_i = 0$, so that at the minima either $\lambda_i = 0$ or $a_i = 0$. In the first case, the constraint has no effect (because it is not active at the minima), while in the second it has been reduced to a simple equality constraint. This is perhaps best seen as a rather involved mathematical way of saying that the inactive inequality constraints should be ignored and the active ones treated simply as equalities – hardly an astounding revelation. Nonetheless, in practice, it means that while searching for the Lagrangian multipliers that satisfy the constraints one must also work out which inequality constraints are active and only satisfy those that are – again something that is more difficult to do in practice than to write down. In our example, the inequality is always active and so the solution is as before and the curves of the Lagrangian remain the same.

Consider next our two-dimensional bumpy test function. Here, we have $(\cos^4 x_1 + \cos^4 x_2 - 2\cos^2 x_1 \cos^2 x_2)/(x_1^2 + 2x_2^2)$ subject to $x_1 x_2 > 0.75$ and $x_1 + x_2 < 15$. So, adding slack variables a_1 and a_2 and multipliers λ_1 and λ_2, the Lagrangian becomes $(\cos^4 x_1 + \cos^4 x_2 - 2\cos^2 x_1 \cos^2 x_2)/(x_1^2 + 2 x_2^2) - \lambda_1 (x_1 x_2 - 0.75 - a_1^2) - \lambda_2 (x_1 + x_2 - 15 - a_2^2)$. We can then study this new function for arbitrary values of the multipliers and slack variables. To use it to solve our optimization problem, we must identify which of the two inequality constraints will be active at the minimum being sought – let us arbitrarily deal with the product constraint and assume that it is active: remember in the interior space neither constraint is active and so the problem reverts to an unconstrained one as far as local search is concerned. To set the product constraint alone as active, we set a_1 and a_2 both to zero along with λ_2 and just search for λ_1, x and y. Remember that our requirement for a minimum is that the gradients of the Lagrangian with respect to x and y both be zero and that simultaneously the product constraint be satisfied as an equality, that is, three conditions for the three unknowns.

Figures 3.16(a–c) show plots of the gradients and the constraint line for three different values of λ_1 and it is clear that in the third of these, when λ_1 is equal to 0.5332, a solution of (1.60086, 0.46850) solves the system of equations (in the first figure the contours of zero gradient do not cross, while in the second they cross multiple times but away from the constraint line). Figures 3.17 and 3.18 show a plot of the original and Lagrangian functions, respectively, with this optimum position marked. It is clear that the minimum is correctly located and that in Lagrangian space it lies at a saddle point and *not* a minimum (i.e., a simple minimization of the augmented function will not locate the desired optimum – we

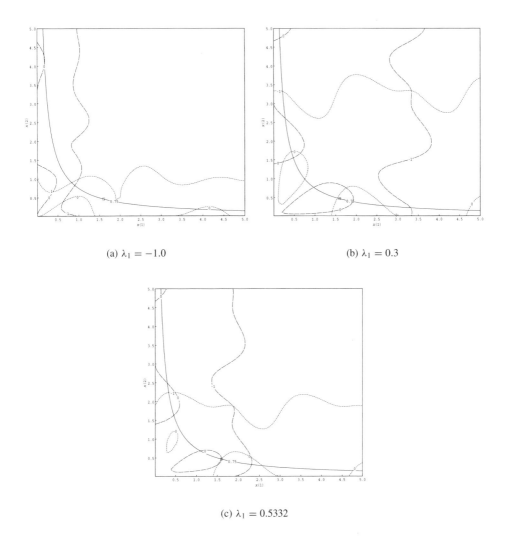

(a) $\lambda_1 = -1.0$ (b) $\lambda_1 = 0.3$

(c) $\lambda_1 = 0.5332$

Figure 3.16 Contour lines for gradients of the Lagrangian (dotted with respect to x_1, dashed with respect to x_2) and the equality constraint $x_1 x_2 = 0.75$. The square marker indicates the position of the constrained local minimum.

require only that the Hessian of the Lagrangian be positive definite, that is, a stationary point in the Lagrangian – this means that minimization of the Lagrangian will not solve the original problem and is a key weakness of methods that exploit the KKT conditions directly in this way). This process can be repeated for the other local optima on the product constraint and, by setting λ_1 to be zero and λ_2 nonzero, those on the summation constraint can also be found. The only other possibility is where the constraint lines cross so that both might be active at a minimum. Here, this lies outside the range of interest and so is not considered further.

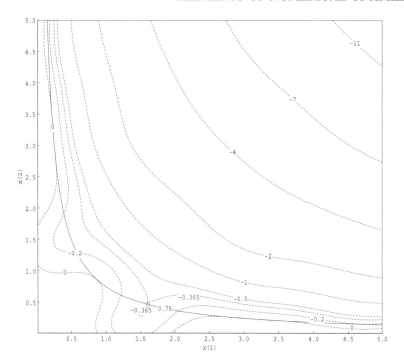

Figure 3.17 Contour lines of the Lagrangian (dotted) for $\lambda_1 = 0.5332$ and the equality constraint $x_1 x_2 = 0.75$. The square marker indicates the position of the constrained local minimum.

3.3.3 Feasible Directions Method

It will be clear that methods based on Lagrangian multipliers are designed to explore along active constraint boundaries. By contrast, the feasible directions method (Zoutendijk 1960) aims to stay as far away as possible from the boundaries formed by constraints. Clearly, this approach is only applicable to inequality constraints, but as has already been noted, these are usually in the majority in design work. The basic philosophy is to start at an active boundary and follow a "best" direction into the feasible region of the search space. There are two criteria used to select the best feasible direction. First is the obvious desire to improve the objective function as rapidly as possible, that is, to follow steepest descent downhill. This is, however, balanced with a desire to keep as far away as possible from the constraint boundaries. To do this, the quantity β is maximized subject to $-\mathbf{s}^T \mathrm{D} g_i(\{\mathbf{x}\}) + \theta_i \beta \leq 0$ for each active constraint, $\mathbf{s}^T \mathrm{D} f(\mathbf{x}) + \beta \leq 0$ and $|s_i| \leq 1$, where $\mathbf{s}$ is the vector that defines the direction to be taken and the θ_i are positive so-called push-off factors. The push-off factors determine how far the search will move away from the constraint boundaries. Again, these are problem dependent and are generally larger the more highly nonlinear the constraint is. The key to this approach is that the maximization problem thus defined is linear and so can be solved using linear programming methods which are extremely rapid and robust.

Once the search direction $\mathbf{s}$ has been fixed, we carry out a one-dimensional search in this direction either until a minimum has been found or a constraint has been violated. In the first case, we then revert to an unconstrained search as we are in the interior of the

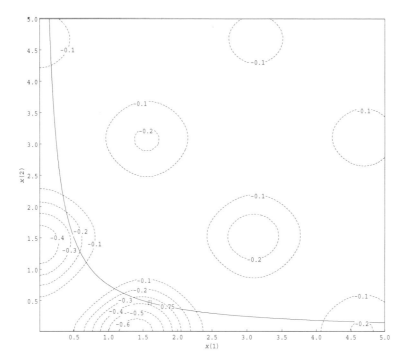

Figure 3.18   Contour lines of the original function and the constraint $x_1 x_2 = 0.75$. The square marker indicates the position of the constrained local minimum.

feasible region. In the second, we repeat the calculation of the feasible direction and iterate the procedure. This basic approach is used in the popular Conmin program.[4] In its basic form, it can become inefficient when searching into a shallow bowl on a constraint surface, since moving away from the boundary is then counterproductive.

3.3.4   Penalty Function Methods

It will be clear by now that there are no universally efficient and robust schemes for dealing with constraints in design search problems. Perhaps, not surprisingly, a large number of essentially heuristic approaches have therefore been developed that can be called upon depending on the case in hand. One such class revolves around the concept of penalty functions, which are somewhat related to Lagrangian multipliers. The simplest pure penalty approach that can be applied, the so-called one pass external function, is to just add a very large constant to the objective function value whenever any constraint is violated (or if we are maximizing to just subtract this): that is, $f_p(\mathbf{x}) = f(\mathbf{x}) + P$ if any constraint is violated, otherwise $f_p(\mathbf{x}) = f(\mathbf{x})$. Then, the penalized function is searched instead of the original objective. Provided the penalty added (P) is very much larger than the function being dealt with, this will create a severe cliff in the objective function landscape that will tend to make search methods reject infeasible designs. This approach is "external" because it is only applied in the infeasible regions of the search space and "one pass" because it is

---

[4] http://www.vrand.com

immediately severe enough to ensure rejection of any infeasible designs. Although simple to apply, the approach suffers from a number of drawbacks:

1. The slope of the objective function surface will not in general point toward feasible space in the infeasible region so that if the search starts from, or falls into, the infeasible region it will be unlikely to recover from this.

2. There is a severe discontinuity in the shape of the penalized objective at the constraint boundary and so any final design that lies on the boundary is very hard to converge to with precision, especially using efficient gradient-descent methods such as those already described (and, commonly, many optima in design are defined by constraint boundaries).

3. It takes no account of the number of constraint violations at any infeasible point.

Because of these limitations, a number of modifications have been proposed. First, a separate penalty may be applied for each violated constraint. Additionally, the penalty may be multiplied by the degree of violation of the constraint, and thirdly, some modification of the penalty may be made in the feasible region of the search near the boundary (the so-called interior space). None of these changes is as simple as might at first be supposed.

Consider first adding a penalty for each violated constraint: $f_p(\mathbf{x}) = f(\mathbf{x}) + mP$, where m is the number of violated constraints – this has the benefit of making the penalty more severe when multiple constraints are violated. Care must be taken, however, to ensure that the combined effect does not cause machine overflow. More importantly, it will be clear that this approach adds more cliffs to the landscape – now, there are cliffs along each constraint boundary so that the infeasible region may be riddled with them. The more discontinuities present in the objective function space, the harder it is to search, particularly with gradient-based methods.

Scaling the total penalty by multiplying by the degree of infeasibility is another possibility but can again lead to machine overflow. Now, $f_p(\mathbf{x}) = f(\mathbf{x}) + \sum P\langle|g_i(\mathbf{x})|\rangle + \sum P\langle|h_j(\mathbf{x})|\rangle$, where the angle brackets $\langle.\rangle$ are taken to be zero if the constraint is satisfied but return the argument value otherwise. In addition, if the desire is to cause the objective function surface to point back toward feasibility, then knowledge is required of how rapidly the constraint function varies in the infeasible region as compared to the objective function. If multiple infeasible constraints are to be dealt with in this fashion, they will need normalizing together so that their relative scales are appropriate. Consider dealing with a stress infeasibility in Pascals and a weight limit in metric tonnes – such elements will commonly be six orders of magnitude different before scaling. If individual constraint scaling is to be carried out, we need a separate P_i and P_j for each inequality and equality constraint, respectively: $f_p(\mathbf{x}) = f(\mathbf{x}) + \sum P_i\langle|g_i(\mathbf{x})|\rangle + \sum P_j\langle|h_j(\mathbf{x})|\rangle$. Finding appropriate values for all these penalties requires knowledge of the problem being dealt with, which may not be immediately obvious – in such cases, much trial and error may be needed before appropriate values are found.

Providing an interior component for a penalty function is even harder. The aim of such a function is, in some sense, to "warn" the search engine of an approaching constraint boundary so that action can be taken before the search stumbles over the cliff. Typically, this requires yet a further set of scaled penalties S_i, so that $f_p(\mathbf{x}) = f(\mathbf{x}) + \sum P_i\langle|g_i(\mathbf{x})|\rangle + \sum P_j\langle|h_j(\mathbf{x})|\rangle + \sum S_i/g_i^s(\mathbf{x})$, where the superscript s indicates a satisfied inequality constraint. Since the interior inequality constraint penalty goes to infinity at the boundary (where $g_i(\mathbf{x})$ goes to zero) and decreases as the constraint is increasingly satisfied (positive), this provides a shoulder on the feasible side of the function. However, in common with all interior penalties, this potentially changes the location of the true objective away from the boundary into the feasible region. Now, this may be desirable in design contexts, where a design that is on the brink of violating a constraint is normally highly undesirable, but again it introduces another complexity.

Thus far, all of the penalties mentioned have been defined only in terms of the design variables in use. It is also possible to make any penalty vary depending on the stage of the search, the idea being that penalties may usefully be weakened at the beginning of an optimization run when the search engine is exploring the landscape, provided that suitably severe penalties are applied at the end before the final optimum is returned. This leads to the idea of Sequential Unconstrained Minimization Techniques (SUMT), where a series of searches is carried out with ever more severe penalties, that is, the P_i, P_j and S_i all become functions of the search progress. Typically, the P values start small and increase while the S values start large and tend to zero. To see this, consider Figure 3.19 (after Siddall (1982) Fig. 6.27), where a single inequality constraint is shown in a two-dimensional problem and the final optimal design is defined by the constraint location. It is clear from the cross section that the penalized objective evolves as the constraints change so that initially a search approaching the boundary from either side sees only a gentle distortion to the search surface but finally this ends up looking identical to the cliff of a simple one pass exterior penalty. A number of variants on these themes have been proposed, and they are more fully explained by Siddall (1982) but it remains the case that the best choice will be problem specific and often therefore a matter of taste and experience.

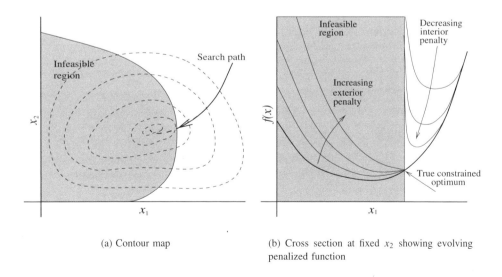

(a) Contour map

(b) Cross section at fixed x_2 showing evolving penalized function

Figure 3.19 Evolution of a penalty function in a Sequential Unconstrained Minimization Technique (SUMT).

3.3.5   Combined Lagrangian and Penalty Function Methods

Since many penalty functions introduce discontinuities which make search with efficient gradient descent methods difficult, a potentially attractive approach is to combine a penalty with the method of Lagrange multipliers. Again, we consider just the case of active constraints (i.e., equalities). Now, our modified function becomes $f(\mathbf{x}) - \sum \lambda_j h_j(\mathbf{x}) + c \sum h_j(\mathbf{x})^2$, where c is some large, but not infinite penalty – this model is therefore amenable to search by gradient-based methods provided the original objective and constraint functions are themselves smooth, that is, we have the smoothness of the Lagrangian approach combined with the applicability of unconstrained search tools of simple penalty methods. To proceed, we minimize this augmented function with arbitrary initial multipliers but then update the multipliers as $\lambda_j^{k+1} = \lambda_j^k - 2c_k h_j(\mathbf{x})^k$, where the superscript or subscript k indicates the major iteration count. Moreover, we also double the penalty c_k at every major iteration, although, it in fact only needs to be large enough to ensure that the Lagrangian has a minimum at the desired solution rather than a stationary point (N.B., if c_k becomes too large, the problem may become ill-conditioned). If this process is started sufficiently close to an optimum with sufficiently large initial penalty, a sequence of searches rapidly converges to the desired minimum using only unconstrained (potentially gradient based) methods. Now, while this still requires some experimentation, it can be simply used alongside any of the unconstrained searches already outlined. Table 3.6 shows the results achieved if we follow this approach on our bumpy test function with a Polak–Ribiere conjugate gradient method.

3.3.6   Sequential Quadratic Programming

As has already been noted, the minimum of our constrained function does not lie at a minimum of the Lagrangian, only at a stationary point where its Hessian is positive definite. The most obvious approach to exploiting this fact is simply to apply Newton's method to

Table 3.6   Results of a search with combined Lagrangian multipliers and penalty function on the bump function of (3.2)

Major Iteration (k)	x_1	x_2	$h(x)$
1	2.0	1.0	1.25
2	1.599091	0.4656781	−0.0053383
3	1.600862	0.4685018	0.0000067
4	1.600861	0.4684972	−0.0000011
5	1.600861	0.4684978	−0.0000001

Major Iteration (k)	λ	c	Penalized Lagrangian	$f(x)$
1	0.0	50	78.11924	0.005757
2	0.00028665	100	−0.36640314	0.36782957
3	0.5341084	200	−0.36497975	0.36497616
4	0.5327684	400	−0.36497975	0.36498033
5	0.5332084		−0.36497975	0.36497982

search for the appropriate solution to the equations of the multipliers and gradients, but this fails if the search is started too far from the optimum being sought. The minimum does, however, lie at a minimum of the Lagrangian in the subspace of vectors orthogonal to the gradients of the active constraints. This result is the basis for a class of methods known as Sequential Quadratic Programming (SQP) (Boggs and Tolle 1995). In SQP, a quadratic (convex) approximation to the Lagrangian with linear constraints is used, in some form, in an iterative search for the multiplier and variable values that satisfy the KKT conditions. At each step of this search, the quadratic optimization subproblem is efficiently solved using methods like those already discussed (typically quasi-Newton) and the values of the xs and λs updated until the correct solution is obtained. Ongoing research has provided many sophisticated codes in this class. These are all somewhat complex, and so a detailed exposition of their workings lies outside the scope of this text. Nonetheless, it will be clear that such methods have to allow for the fact that as the design variables change it may well be the case that the list of active inequality constraints also changes, so that some of λs then need setting to zero while others are made nonzero. SQP methods also assume that the functions and constraints being searched have continuous values and derivatives, and so on, and so will not work well when there are function discontinuities – a situation that is encountered all too often in design. Nonetheless, they remain among the most powerful tools for solving local constrained optimization problems that are continuous; see for example Lawrence and Tits (2001).

3.3.7 Chromosome Repair

So far, all of the constraint-handling mechanisms that have been described have been designed with traditional sequential optimizers in mind. Moreover, all have aimed at transforming the original objective function, either by eliminating constraints and thus changing the input vector to the objective calculation or by adding some terms to the function actually being searched. It is, however, possible to tackle constraints via the mechanism of repair – a solution is said to be repaired when the resulting objective function value is replaced by that of a nearby feasible solution. In the context of evolutionary computing, where such mechanisms are popular and there are some biological analogies, this is commonly referred to as "chromosome repair" since it is the chromosome that encodes the infeasible point that is being repaired. Of course, repairing a design can be a difficult and costly business and one that involves searching for a feasible (if not optimal) solution. The methods used to do this are essentially exactly the same as in any other search process except that now the objective is the feasibility of the solution rather than its quality and such a search is, by definition, not constrained further. Here, we assume that some mechanism is available to make the repair and restrict our discussion to what a search algorithm might do with a repaired solution. This depends very much on how the search itself works.

After repair, we have a modified objective function value and also the coordinates of the nearby feasible location where this applies. The simplest thing to do with this modified function is to use it to replace the original objective and continue searching. It will be immediately clear, however, that as one traverses down a vector normal to a constraint boundary the most likely neighboring repair location will not change and so the updated objective function would then become constant in some sense, that is, the objective function surface would be flat in the infeasible region and the search would have no way of knowing that a constraint had been violated. Now, this may not matter if a stochastic or population-based method is being used, since the next move will not be looking at the local gradient. Conversely, any downhill method is likely to become confused in such circumstances. In the evolutionary search literature, such an approach is termed *Baldwinian learning* (Baldwin 1896; Hinton and Nowlan 1987).

The alternative approach is to use both the function and variable values of the nearby feasible solution. This modifies the current design vector and so is not suitable for use in any sequential search process where the search is trying to unfold some strategy that guides the sequence of moves being taken: it would amount to having two people trying to steer a vehicle at the same time. This does not matter, however, in evolutionary or stochastic methods where the choice of the next design vector is anyway subject to random moves – the repair process can be seen as merely adding a further mutation to the design over and above that caused by the search algorithm itself. In the evolutionary search literature, this is termed *Lamarckian learning* after the early naturalist, Jean-Baptiste Lamarck (Gould 1980).

Since Baldwinian learning can confuse a gradient-based search and Lamarckian learning is incompatible with any form of directed search, it is no surprise that neither method finds much popularity in sequential search applications. When it comes to evolutionary search, they are more useful but still not simple to implement. Moreover, since the repair process may itself be expensive, most workers tend to prefer the Lamarckian approach, which keeps hold of all the information gleaned in making the repair. There is some evidence to show that such a process helps evolutionary methods when applied to any local improvement process (repair is just one of many ways that a design might be locally improved during an evolutionary search) (Ong and Keane 2004). Even so, the evidence is thus far not compelling enough to make the idea widely popular.

3.4   Metamodels and Response Surface Methods

One of the key factors in any optimization process is the cost of evaluating the objective function(s) and any constraints. In test work dealing with simple mathematical expressions this is not an issue, but in design activities, the function evaluation costs may well be measured in hours or even days and so steps must always be taken to minimize the number of function calls being used. Ideally, any mitigating strategies should work *irrespective* of the dimensions of the problem. One very important way of trying to do this is via the concept of response surface, surrogate or metamodels. In these, a small number of full function (or constraint) evaluations are used to seed a database/curve-fitting process, which is then interrogated in lieu of the full function whenever further information about the nature of the problem is needed by the optimization process.

At its simplest, such metamodeling is just the use of linear regression through existing calculations when results are needed for which the full simulations have not been run. Of course, it is unlikely that the original set of calculations will be sufficiently close to the desired optimal designs by chance, especially when working in many dimensions, so some kind of updating or refinement scheme is commonly used with such models. Updating enables the database and curve fit to be improved in an iterative way by selective addition of further full calculations. Figure 3.20 illustrates such a scheme. In this section, we introduce the basic ideas behind this way of working since they are key to many practical design

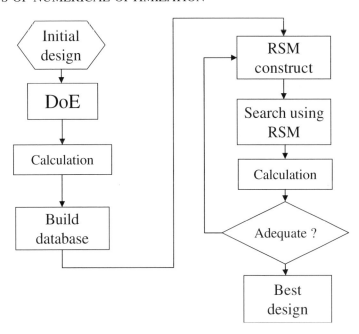

Figure 3.20 Simple serial response surface method with updating.

approaches – a much more extensive treatment is provided in Chapters 5, 6 and 7. We would note in passing that the idea of slowly building an internal model of the unknown function is similar in spirit to the gradient-based methods that build local approximations to the Hessian of a function – the main difference here is the explicit treatment of the model and the fact that in many cases they are built to span much wider regions of the search space.

There are a significant number of important decisions that a user must make when following a metamodel-based search procedure:

1. How should the initial database be seeded with design calculations – will this be a regular grid, random or should some more formal DoE approach be used?

2. Should the initial calculations be restricted to some local region of the search space surrounding the current best design or should they attempt to span a much larger range of possible configurations?

3. Can the initial set of calculations be run in parallel?

4. What kind of curve fit, metamodel or response surface should be used to represent the data found by expensive calculation?

5. What will we do if some of the calculations fail, either because of computing issues or because the underlying PDE cannot be solved for certain combinations of parameters?

6. Is it possible to obtain gradient information directly at reasonable cost via use of perturbation techniques or adjoint solvers and, if so, will these be used in building the metamodel?

7. Is one model enough or should multiple models be built with one for each function or constraint being dealt with?

8. If multiple models are to be built can they be correlated in some sense?

9. Should the model(s) strictly interpolate the data or is regression better?

10. Will the model(s) need some extensive tuning or training process to get the best from the data?

11. What methods should be used to search the resulting surface(s)?

12. Having searched the metamodel, should a fresh full calculation be performed and the database and curve fit updated?

13. If updates are to be performed, should any search on the surface(s) aim simply to find the best predicted design or should some allowance be made for improving the coverage of the database to reduce modeling errors, and so on?

14. If updates are to be performed, could multiple updates be run in parallel and, if so, how should these multiple points be selected?

15. If an update scheme is to be used, what criteria will be used to terminate the process?

Answering all these questions for any particular design problem so as to provide a robust and efficient search process requires considerable experience and insight.

3.4.1 Global versus Local Metamodels

At the time of writing, there are two very popular general metamodeling approaches. The first is to begin with a formal DoE spanning the whole search space, carry out these runs in parallel and then to build and tune a global response surface model using radial basis function or Kriging methods. This model is then updated at the most promising locations found by extensive searches on the metamodel, often using some formal criterion such as expectation of improvement to balance improvements and model uncertainty – this approach is illustrated in Figure 3.21 (Jones et al. 1998).

In the second approach, a local model is constructed around the current design location, probably using an orthogonal array or part-factorial design. The results are then fitted with a simple second-order (or even first-order) regression equation. This is searched to move in a downhill direction up to some amount determined by a trust region limit. The model is then updated with additional full calculations, discarding those results that lie furthest from the current location. The process is repeated until it converges on an optimum (Alexandrov et al. 1998b; Conn et al. 2000; Toropov et al. 1999). Such an approach can be provably convergent, at least to a local optimum of the full function, something which is not possible with global models.

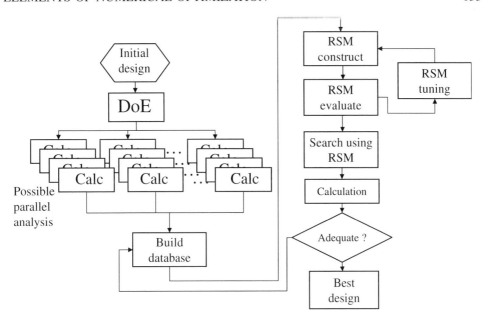

Figure 3.21 Improved response surface method calculation, allowing for parallel DoE-based calculations, RSM tuning and database update.

The choice between these two types of strategy depends largely on the goals of the designer. If an existing design is to be exploited and tuned to yield maximum performance without radical redesign, the trust region approach has much to commend it. Conversely, if the team is unsure where they might go and radical alternatives are required, the global strategy can be very effective. Moreover, as problem dimensions rise, the ability to construct any kind of accurate surrogate rapidly diminishes – one need only consider how the number of vertices of the search space bounding box rises with the power of the number of dimensions to realize just how little data a few hundred points represent when searching even a ten-dimensional search space.

3.4.2   Metamodeling Tools

Assuming that a surrogate approach is to be used, we next set out some of the most popular tools needed to construct a good-quality search process. We begin by noting that there are very many DoE methods available in the literature and the suitability of any particular one depends on both the problem to be modeled and also the class of curve fit to be used. Among the most popular are

1. pure random sequences (completely unbiased but fail to exploit the fact that we expect limited curvature in response surfaces, since we are commonly modeling physical systems);

2. full factorial designs (usually far too expensive in more than a few dimensions, also fail to study the interior of the design space well as they tend to focus on the extremes);

3. face centered cubic or central composite designs (cover more of the interior than full factorial but are even more expensive);

4. part-factorial designs (allow for limited computational budget but again do not cover the design space well);

5. orthogonal array designs, including Plackett and Burman, Box-Benken and Taguchi designs (reasonable coverage but designs tend to fix the number of trials to be used and also each variable is often restricted in the number of values it can take);

6. Sobol sequence and LPτ designs (allow arbitrary and extensible designs but have some limitations in coverage); and

7. Latin hypercubes of various kinds (not easily extensible but easy to design with good coverage over arbitrary dimensions).

Having run the DoE calculations, almost always in parallel, a curve fit of some kind must next be applied. In the majority of cases, the data available is not spread out on some regular grid and so the procedure used must allow for arbitrary point spacing. Those currently popular in the literature include

1. linear, quadratic and higher-order forms of polynomial regression;

2. cubic (or other forms of) splines, commonly assembled as a patchwork that spans the design space and having some form of smoothness between patches;

3. Shepard interpolating functions;

4. radial basis function models with various kernels (which are most commonly used to interpolate the data, but which can be set up to carry out regression in various ways and also can have secondary tuning parameters);

5. Kriging and its derivatives (which generally need the tuning of multiple hyperparameters that control curvature and possibly the degree of regression – this can be very expensive on large data sets in many dimensions);

6. support vector machines (which allow a number of differing models to be combined with key subsets of the data – the so-called support vectors – so that problems with many dimensions and large amounts of data can be handled with reasonable efficiency);

7. neural networks (which have a long history in function matching and prediction but which need extensive training and validation and which are also generally difficult to interpret); and

8. genetic programming (which can be used to design arbitrary functional forms to approximate given data, but which are again expensive to tune and validate).

Constructing such models always requires some computational effort and several of them require that the curve fit be tuned to match the data. This involves selecting various control or hyperparameters that govern things like the degree of regression and the amount of curvature. This can represent a real bottleneck in the search process since this is difficult to parallelize and can involve significant computing resources as the number of data points and dimensions rises – this is why the use of support vector approaches is of interest since they offer the prospect of affordable training on large data sets.

Typically, tuning (training) involves the use of a metric that measures the goodness of fit of the model combined with its ability to predict results at points where no data is available. For example, in Kriging, the so-called likelihood is maximized, while in neural networks it is common to reserve part of the data outside of the main training process to provide a validation set. Perhaps the most generally applicable metric is leave-out-one cross-validation, where a single element is removed from the training set, and its results predicted from the rest with the current hyperparameters. This is then repeated for all elements in the data and the average or variance used to provide a measure of quality. If this is used during training, it can, however, add to the computational expense.

Once an acceptable model is built, it should normally be refined by adding further expensive results in key locations. There is no hard and fast rule as to how much effort should be expended on the initial data set as compared to that on updating but, in the authors experience, a 50:50 split is often a safe basis to work on, although reserving as little as 5% of the available budget for update computations can sometimes work while there are other cases where using more than 75% of the effort on updating yields the best results. There are two main issues that bear on this choice: first, is the underlying function highly multimodal or not? – if it is, then more effort should go into updating; secondly, can updates be done in parallel or only serially? – if updates are only added one at a time while the initial DoE was done in parallel, then few updates will be possible. Interestingly, this latter point is more likely to be driven by the metric being used to judge the metamodel than the available hardware – to update in parallel, one needs to be able to identify groups of points that all merit further investigation at the same time – if the metric in use only offers a simplistic return, it may be difficult to decide where to add the extra points efficiently. This is where metrics such as expected improvement can help as they tend by their nature to be multimodal and so a suitable multistart hill climbing strategy can then often identify tens of update points for subsequent calculation from a single metamodel. The authors find Yuret's Dynamic Hill Climbing method ideal in this role (Yuret and de la Maza 1993), as are population-based searches which have niche forming capabilities (Goldberg 1989).

3.4.3 Simple RSM Examples

Figures 3.22, 3.23 and 3.24 show the evolution of a Kriging model on the bump problem already discussed as the number of data points is increased and the metamodel rebuilt and retuned. In this example, 100 points are used in the initial search and three sets of 10 updates are added to refine the surface. These are simply aimed at finding the best point in the search space and it can be seen that the area containing the global optimum is much better defined in the final result. Note that the area containing the optimum has to be in the original sample for the update policy to refine it – as this cannot usually be guaranteed, it is useful to include some kind of error metric in the update so as to continue sampling regions where little data is available, if such a strongly multimodal problem is to be dealt with in this way.

Figure 3.25 shows a trust region search carried on the bump function. In this search, a small local DoE of nine points is placed in the square region between (2,1) and (4,3) and then used to construct a linear plus squared term regression model. This is then searched within a unit square trust region based around the best point in the DoE. The best result

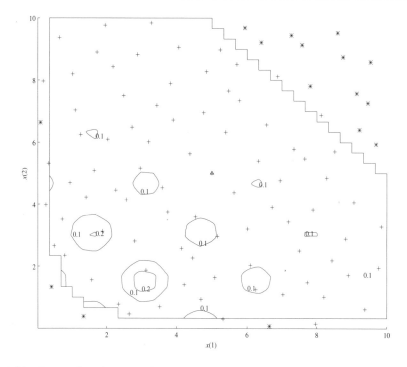

Figure 3.22 Bump function showing location of initial DoE of 100 LPτ points (points marked with an "*" are infeasible).

found in the search is then reevaluated using the true function and used to replace the oldest point in the DoE. This update process is carried out 15 times with a slowly decreasing trust region size, each time centered on the endpoint of the last search and, as can be seen from the figure, this converges to the nearby local optimum, using 24 ($= 9 + 15$) calls of the expensive function in all. Obviously, such an approach can become very confused if the trust region used encompasses many local optima and the search will then meander around aimlessly until the trust region size aligns with the local basin of attraction size. In many cases, particularly where there is little multimodality but significant noise, this proves to be a very effective means of proceeding.

3.5   Combined Approaches – Hybrid Searches, Metaheuristics

Given all the preceding ideas on how to perform optimization, it will be no surprise that many researchers have tried to combine the best aspects of these various methods into ever more powerful search engines. Before discussing such combined or "hybrid" searches, it is worth pointing out the "no free lunch" theorems of search (Wolpert and Macready 1997). These theorems essentially state that if a search method is improved for any one class of problem it will equivalently perform worse on some other class of problem and that, averaged over all classes of problems, it is impossible to outperform a random walk. Now,

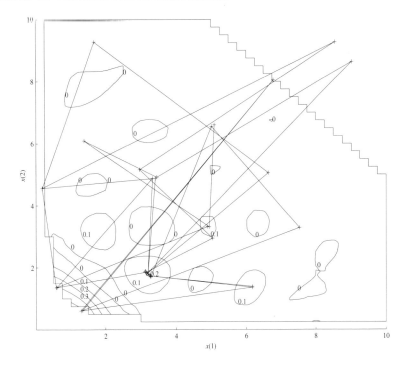

Figure 3.23 RSM of bump function showing search traces for three searches using Dynamic Hill Climbing (DHC) with parallel updates between them (straight lines between crosses link the minima located by each major iteration of the DHC search).

although this result might seem to make pointless much of what has gone before, it should be recalled that the problems being tackled in aerospace design all tend to share some common features and it is the exploitation of this key fact that allows systematic search to be performed. We know, for example, that small changes in the parameters of a design, in general, do not lead to random and violent changes in performance, i.e., that the models we use in design exhibit certain properties of smoothness and consistency, at least in broad terms. It is these features that systematic search exploit. Nonetheless, the no free lunch idea should caution us against any idea that we can make a perfect search engine – all we can in fact do is use our knowledge of the problem in hand to better adapt or tune our search to a particular problem, accepting that this will inevitably make the search less general purpose in nature, or, conversely, that by attempting to make a search treat a wider range of problems it will tend to work less well on any one task.

Despite this fundamental limit to what may be achieved, researchers continue to build interesting new methods and to try and ascertain what problems they may tackle well. Hybrid searches fit naturally into this approach to search. If one knows that certain features of a gradient-based method work well, but that its limitations for global search are an issue, it is natural to try and combine such methods with a global method such as a genetic algorithm. This leads to an approach currently popular in the evolutionary search community. Since evolutionary methods typically are built around a number of operators that are based on natural selection (such as mutation, crossover, selection and inversion), it is relatively easy

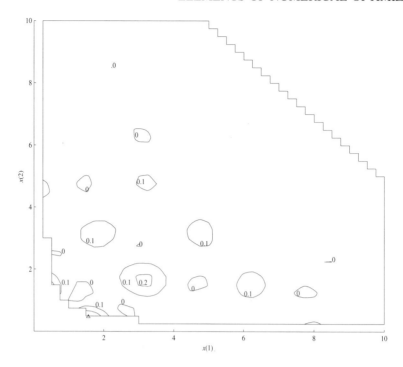

Figure 3.24 Final updated RSM of bump function after 30 updates – the best result found is marked by a triangle at (1.5659, 0.4792) where the estimated function value is 0.3602 and the true value is 0.3633, compare the exact solution of 0.3650 at (1.6009, 0.4685).

to add to this mix, ideas coming from the classical search literature such as gradient-based approaches. In such schemes, some (possibly all) members of a population are selected for improvement using, for example, a quasi-Newton method (which might possibly not be fully converged) and any gains used as another operator in the search mixture.

If another search is added into the inner workings of a main search engine, we think of the result as a hybrid search. Alternatively, we might simply choose to interleave, possibly partially converged, searches with one another by having some kind of control script that the combined process follows. This we would term a metaheuristic, especially if the logic followed makes decisions based on how the process has proceeded thus far. Of course, there are no real hard and fast boundaries between these two logics – the main difference is really in how they appear to the user – hybrid searches usually come boxed up and appear as a new search type to the designer while metaheuristics tend to appear as a scripting process where the user is fully exposed to, and can alter, the logic at play. In fact, the response surface methods already set out may be regarded as a special case of metaheuristics, spanning DoE-based search and whatever method is used to search the resulting metamodel.

To illustrate these ideas, we describe two approaches familiar to the authors – we would stress, however, that it is extremely straightforward to create new combinations of methods and the no free lunch theorems do not allow one to say if any one is better than another

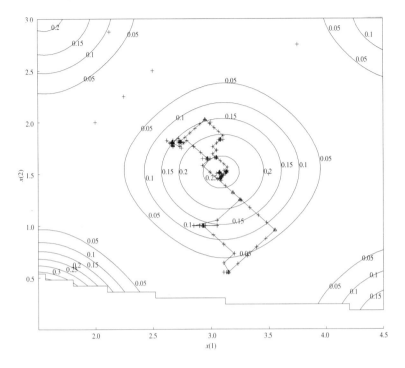

Figure 3.25 Trust region search with linear plus squared term regression model taken over nine points updated 15 times.

without first stating the problem class being tackled – here, we describe these combinations just to give a taste of what may be done. We would further note that combinations of optimizers can also be effective when dealing with multidisciplinary problems, a topic we return to in Chapter 9.

3.5.1 Glossy – A Hybrid Search Template

Glossy is a hybrid search template that allows for search time division between exploration and exploitation (Sóbester and Keane 2002). The basic idea is illustrated in Figure 3.26 – an initial population is divided between a global optimizer and a local one. Typically, the global method is a genetic algorithm and the local method is a gradient-based search such as the BFGS quasi-Newton method. The GA subpopulation is then iterated in the normal way for a set number of generations. At the same time, each individual in the BFGS population is improved using a traditional BFGS algorithm but only for as many downhill steps as the number of generations used in the GA. It is further assumed that each downhill search and each member of the GA is evaluated on its own dedicated processor so that all these process run concurrently and finish at the same time. Then, an exchange is carried out between the two subpopulations and at the same time the relative sizes of the subpopulations are updated. The searches are then started again and this procedure cycled

GLOSSY (GA-BFGS) We start with a **randomly generated population** (each circle represents one individual, the color of the circle indicates the objective value of the individual – the lighter the color, the better the objective value)...

● ○ ◐ ○ ● ○ ○ ●

...and **allocate the individuals** into two populations where they will undergo a sequence of local exploitation and a sequence of global exploration respectively. The population sizes (4 and 4 in this case) and the sequence lengths ($SL=2$ and $SG=4$ in our example) are set in advance.

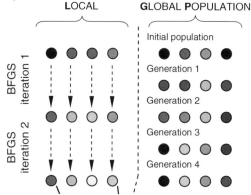

End of the first sequence. We perform the first reallocation step. **The populations will be resized according to the average objective value improvement per evaluation achieved during the last sequence.**

In this case the improvement calculated for the GA, was, say, three times higher than that achieved by BFGS. Thus, the ratio of the sizes of the two populations for the next sequence will be three. Therefore, the Local Population has to relinquish two individuals to the global one. Those two that **have achieved the least improvement** during the last sequence (individuals 1 and 4 in this case) will migrate. Those that have improved well locally are allowed to continue in the local population.

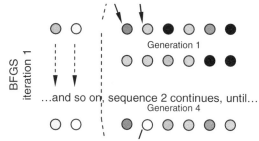

End of sequence 2. We perform the second reallocation step. Let us suppose that the efficiency of the GA has diminished slightly, so the performance ratio (and thus the population size ratio for the next sequence) is 5/3. Therefore, the Global Population now has to concede one individual to the global one. The individual with **the highest objective value** will migrate (the rationale being that the best individuals are likely to be in promising basins of attraction, i.e., they are worth improving locally)

...and sequence 3 begins.

Figure 3.26 The Glossy template (from Sóbester and Keane; reprinted by permission of the American Institute of Aeronautics and Astronautics, Inc.).

through until a suitable termination criterion is met. This strategy allows the GA to carry out global search and the BFGS to locally improve some designs, which can then be fed back to the GA to seed further global searches. At the same time, other globally improved points are handed back for local BFGS refinement. The changing size of the subpopulations allows the search to adapt to the progress made, becoming either more global or more local as the situation demands. Note that since the locally improved designs are returned directly to the GA population this is termed *meta-Lamarckian learning* in some parts of the literature.

3.5.2 Metaheuristics – Search Workflows

As already noted, metaheuristics are typically sequences of commands that invoke various search methods in an adaptive way depending on the progress of the search. They are typically written or adapted by the designer to suit the problem in hand using a general purpose workflow/scripting language such as Matlab® or Python, or the language provided in a commercial search tool such as the multidisciplinary optimization language (MDOL) system in iSIGHT® or even using operating system shell scripts, and so on. The key distinction over hybrid searches being the great flexibility this affords and its control by the designers themselves rather than via reliance on a programming team, etc.

The Geodise[5] toolkit is an example of this way of working. Geodise provides a series of commands in the Matlab environment that may be combined together however the user chooses to create a search workflow. It offers a range of individual search engines, DoE methods and metamodeling capabilities that can be used in concert with those already provided by the main Matlab libraries (or, indeed, any other third party library). Figure 3.27 provides a snippet of Matlab code that invokes this toolkit to carry out a sequence of gradient-based searches on a quadratic metamodel constructed by repeated sampling of the problem space. In this way, a simplistic trust region search can be constructed and the code given here is, in fact, that used to provide Figure 3.25 in the previous section.

It will be clear from examining this example that there is no easy distinction between a workflow language and the direct programming of an optimizer – the characteristic we look for when making the division is the use of existing multidimensional searches as building blocks *by the end users themselves*. Of course, it is a relatively trivial matter to subsequently box up such metacode to form new black-box methods for others to use. In fact, it is often the case that the distinction between classifications says as much about the expertise of the user as it does about the code – a skilled user who is happy to open up and change a Fortran or C code library might justifiably claim that the library provided the ingredients of a metaheuristic approach, while those less skilled would view any combinations worked in such a way as hybrids. We do not wish to labor this point – just to make clear that it is often worthwhile combining the search ingredients to hand in new ways to exploit any known features of the current problem, and that this will be most easily done using a metaheuristic scheme. This approach underpins much of the authors' current work in developing searches for the new problems that they encounter.

---

[5]http://www.geodise.org

```
% trust region search
%
% DoE_struct holds information for the actual function
% meta_struct holds information for the meta-model
% DoE_rslt holds the results from calls to the actual function
% meta_rslt holds the result from searches on the meta-model
% update_pt is used to hold the update points as they are calculated
% trust_sc is the scaling factor for reducing the trust region on each loop
% trust_wd is the current semi-width of the trust region
%
% define initial structures to define problem to be searched
%
DoE_struct=createbumpstruct;
meta_struct=createbumpstruct;
%
% initial DoE details
%
DoE_struct.OMETHD=2.8; DoE_struct.OLEVEL=0;
DoE_struct.VARS=[2,2]; DoE_struct.LVARS=[2,1];
DoE_struct.UVARS=[4,3]; DoE_struct.MC_TYPE=2;
%
% build initial DoE
%
DoE_struct.NITERS=9;
DoE_rslt = OptionsMatlab(DoE_struct);
n_data_pts = DoE_rslt.OBJTRC.NCALLS;
%
% set up trust region controls and initial trust region meta model
%
trust_sc=0.9; trust_wd=0.5;
meta_struct.OLEVEL=0; meta_struct.VARS=DoE_rslt.VARS;
meta_struct.LVARS=[2,1]; meta_struct.UVARS=[4,3];
%
% loop through trust region update as many times as desired
%
for (ii = 1:15)
%
% build and search regression meta model from the current data set
%
meta_struct.OBJMOD=3.3; meta_struct.OMETHD=1.8;
meta_struct.OVR_MAND=1; meta_struct.NITERS=500;
meta_rslt=OptionsMatlab(meta_struct,DoE_rslt);
%
% update trust region around best point but of reduced size
%
trust_wd=trust_wd*trust_sc;
meta_struct.LVARS=meta_rslt.VARS-trust_wd;
```

Figure 3.27   Matlab workflow for a trust region search using the Geodise toolkit
(http://www.geodise.org).

```
meta_struct.UVARS=meta_rslt.VARS+trust_wd;
%
% evaluate best point
%
DoE_struct.OMETHD=0.0;
DoE_struct.VARS=meta_rslt.VARS;
update_pt = OptionsMatlab(DoE_struct);
%
% add result to data set, eliminating the oldest point
%
DoE_rslt.OBJTRC.VARS(:,1)=[];
DoE_rslt.OBJTRC.VARS(:,n_data_pts)=update_pt.VARS;
DoE_rslt.OBJTRC.OBJFUN(1)=[];
DoE_rslt.OBJTRC.OBJFUN(n_data_pts)=update_pt.OBJFN;
%
%echo latest result and loop
%
update_pt.VARS update_pt.OBJFN
meta_struct.VARS=update_pt.VARS;
end
```

Figure 3.27 *Continued*

3.6   Multiobjective Optimization

Having set out a series of methods that may be used to identify single objectives, albeit ones limited by the action of constraints, we next turn to methods that directly address problems where there are multiple and conflicting goals. This is very common in design work, since engineering design almost always tensions cost against performance, even if there is only a single performance metric to be optimized. Usually, the design team will have a considerable number of goals and also some flexibility and/or uncertainty in how to interpret these.

Formally, when a search process has multiple conflicting goals, it is no longer possible to objectively define an optimal design without at the same time also specifying some form of weighting or preference scheme between them: recall that our fundamental definition of search requires the ability to rank competing designs. If we have two goals and some designs are better when measured against the first than the second, while others are the reverse, then it will not be possible to say which is best overall without making further choices. Sometimes, it is possible to directly specify a weighting (function) between the goals and so combine the figures of merit from the various objectives into a single number. When this is possible, our problem of course reduces directly to single value optimization and all the techniques already set out may be applied. Unfortunately, in many cases, such a weighting cannot be derived until such time as the best competing designs have been arrived at – this leads to the requirement for methods of identifying the possible candidates from which to select the so-called Pareto set, or front, already mentioned in earlier sections. We recall that this is the set of designs whose members are such that improving any one goal function of a member will cause some other goal function to deteriorate. However, before considering methods that work explicitly with multiple goals, we first consider formal schemes for identifying suitable weights that return our problem to that of single objective optimization.

We note also that, as when dealing with constraints, it is always wise to normalize all the goals being dealt with so that they have similar numerical magnitudes before starting any comparative work. This leads to better conditioned problems and prevents large numbers driving out good ones. We remember also that many design preferences are specified in the form of constraints, i.e., that characteristic A be greater than some limit or perhaps less than characteristic B, for example.

3.6.1 Multiobjective Weight Assignment Techniques

When a design team has multiple goals and the aim is to try and derive a single objective function that adequately represents all their intentions, the problem being tackled is often rather subjective. This is, however, an issue that is familiar to anyone who has ever purchased a house, an automobile, a boat, a computer, and so on: there will be a variety of different designs to choose from and a range of things that make up the ideal. These will often be in tension – the location of a house and its price are an obvious example – a property in a good location will command a high price. It is thus a very natural human activity to try and weigh up the pros and cons of any such purchase – we implicitly carry out some kind of balancing process. Moreover, our weightings often change as we learn more about the alternatives on offer. Multiobjective weight assignment techniques aim to provide a framework that renders this kind of process more repeatable and justifiable – they are, however, not amenable to very rigorous scientific analysis in the way that the equations of flow over a body are. Nonetheless, there is some evidence that a formalized approach leads to better designs (Sen and Yang 1998).

The basis of all weight assignment methods is an attempt to elicit information from the design team about their preferences. The most basic approach is direct weight assignment. In direct assignment, the team chooses a numerical scale, typically containing five or ten points, and then places adjective descriptions on this scale such as "extremely important", "important", "average", "unimportant" and "extremely unimportant". Each goal is then considered in turn against the adjectives and the appropriate weight chosen from the scale. This can be done by the team collectively, or each member asked to make a ranking and then average values taken. Often, the weighted goals, or perhaps their squares, are then simply added together to generate the required combined goal function. Although this may sound trivial, the fact that an explicit process is used tends to ensure that all team members views' are canvassed and that added thought is given to the ranking process. This approach tends to perform best when there are relatively few goals to be satisfied. It does not, however, give any guidance on what the function used to combine the individually weighted goals together should be, an issue we return to later on.

A slightly more complex way of eliciting and scoring preference information is the eigenvector method. In this approach, each pair of goals is ranked by stating a preference ratio, that is, if goal i is three times more important than goal j, we set the preference p_{ij} equal to three. Then, if goal j is twice as important as goal k, we get p_{jk} equal to two. Note that for consistency this would lead us to assume that p_{ik} was six, that is, three times two. However, if all pair-wise comparisons are made and there are many goals, such consistency is rarely achieved. In any case, having made all the pair-wise comparisons, a preference matrix $\mathbf{P}$ with a leading diagonal of ones can then be assembled. In the eigenvector scheme, we then seek the eigenvector $\mathbf{w}$ that satisfies $\mathbf{Pw} = \lambda_{\max}\mathbf{w}$, where $\lambda_{\max}$ is the maximum eigenvalue of the preference matrix. If the matrix is not completely consistent, it is still possible to seek a solution to this equation and also to gain a measure of any inconsistency by comparing the eigenvalue to the number of goals, since, for a self consistent matrix, the largest eigenvalue is always equal to the number of goals; see, for example, Saaty (1988). The eigenvector resulting from solution of this equation provides the weighting we seek between the goals. By way of example, consider a case where there are three goals and we assign scores to the matrix $\mathbf{P}$ as follows:

$$\mathbf{P} = \begin{bmatrix} 1 & 3 & 6 \\ 1/3 & 1 & 2 \\ 1/6 & 1/2 & 1 \end{bmatrix}$$

that is, goal one is three times as important as goal two and goal two is twice as important as goal three. It is easy to show that the largest eigenvalue of this matrix is three and the equivalent eigenvector, which produces the desired weighting, is $\{0.667, 0.222, 0.111\}^T$. The fact that the eigenvector is here equal to the number of goals indicates that the preference matrix is fully consistent. This approach works well for moderate numbers of goals (less than say 10) and copes well with moderate levels of inconsistency.

When the number of goals becomes large, the rapidly expanding number of pair-wise comparisons needed makes the eigenvector approach difficult to support. It is also difficult to maintain sufficient self consistency with very many pair-wise inputs. Consequently, a number of alternative schemes have been developed that permit incomplete information input and also iteration to improve consistency; see Sen and Yang (1998) for further details. There are also schemes that allow progressive articulation of preference information; see, for example, Benayoun et al. (1971).

3.6.2 Methods for Combining Goal Functions, Fuzzy Logic and Physical Programming

Even when the design team can assign suitable weights to each competing objective, this still leaves the issue of how the weighted terms should be combined. Most often, it is assumed that a simple sum is appropriate. However, there are many problems where such an approach would not lead to all possible Pareto designs being identified, even if every linear combination of weights were tried in turn. This is because, for some problems, the Pareto front is nonconvex; see, for example, Figure 3.28. The region in this figure between points A and B is such that, for any point on this part of the front, a simple weighted sum of the two goals can always be improved on by designs at either end *whatever* the chosen weights. To see this, we note that for any relative weighting between two design goals w all designs lying on the line of constant $wx + y$ in Figure 3.28 have the same combined objective when using a simple sum. For one particular design to be best for a given weight, the Pareto front at this point must be tangent to the equivalent weighting line – in the convex region, we cannot construct a tangent line that lies wholly to the left and below the Pareto front, that is, there is always some other design that, for that combination of weights, will lie outside of the convex region.

Since we know that all points on the Pareto front may be of interest to the designer (since each represents a solution that is nondominated), this fundamental limitation of a weighted sum of goal functions has led a number of researchers to consider more complex schemes. These range from the use of simple positive powers of each objective through to fuzzy logic and physical programming models. The basic idea is this: by combining our goals using some more complex function, the lines of constant combined objective in plots like

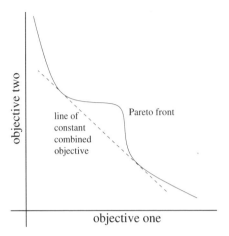

Figure 3.28   A Pareto front with a convex region and line indicating designs with equal combined objective under simple summation of weighted goals.

Figure 3.28 become curves. Provided the functions we use have sufficient curvature in this plot to touch the Pareto front tangentially at all points, given the correct choice of weights, the function in use can then overcome the problems of convexity. So, if we take the sum of squares for example, a contour of constant combined objective is then an ellipse – it will be obvious that such a shape will satisfy our tangency requirement for many convex Pareto fronts, but by no means all such possibilities. One can respond to this problem by using ever more complex functions, of course. However, they rapidly lose any physical meaning to the designer and this has led to the fuzzy logic and physical programming approaches, among others. In these, an attempt is made to capture the design team's subjective information and combine it in a nonlinear fashion *at the same time*, hopefully overcoming convexity while still retaining understanding of the formulation.

In the fuzzy logic approach (Ross 2004), a series of membership functions are generated that quantify the designer's linguistic expressions of preference. For example, if the designer believes that a goal must be at least five and preferably ten, but values over ten offer no further benefit, then a function such as that in Figure 3.29 might be used. Note that the shape of the function that translates the goal value to its utility is here piecewise linear but this is an arbitrary choice: a sigmoid function would just as well suffice. Similar functions are provided for all goals (and usually also all constraints), always mapping these into a scale from zero to one. The resulting utilities are then combined in some way. Typically, they might be summed and divided by the total number of functions, a product taken or simply the lowest overall used. In all cases, the resulting overall objective function still runs between zero and one and an optimization process can be used on the result. Note that this function will have many sharp ridges if piecewise linear membership functions are used or if a simple minimum selection process is used to combine them. Although it allows a more complex definition of the combined objective than simple weights, there is still no guarantee that all parts of the Pareto front will be reached by any subsequent search.

Physical programming (Messac 1996) is another method that seeks to use designer preferences to construct an aggregate goal function. It is again based on a series of linguistic statements that the design team may use to rank their preferences. Goals and constraints

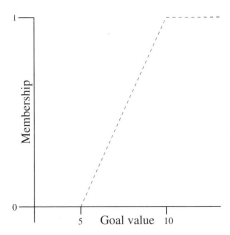

Figure 3.29 Simple fuzzy logic membership function.

are defined to be either hard (with rigid boundaries) or soft. When soft, the degree of acceptability is broken down between highly desirable, desirable, tolerable, undesirable, highly undesirable and unacceptable. Moreover, moving one goal across a given region is considered better than moving all the remaining goals across the next better region, for example, moving one across the tolerable region is better than moving all across the desirable region. In this way, the scheme aims to eliminate the worst features in a design. A principal motivation of the approach is to reduce the amount of time designers use studying aggregate goals with different sets of weightings, although designers may still wish to reassign what they mean by each of the linguistic definitions, of course.

In all cases, once a weighting and form of combined goal function has been decided, a single objective problem can then be tackled. If the outcome of a search with the chosen weights and function leads to designs that the team find inconsistent with their intuitive preferences, either the weight assignment process can then be iterated in light of these views or a different approach to constructing the combined goal function used. Sometimes, this may involve adding or removing preference information or goals. Commonly, such iteration is reduced when the more complex methods of weight assignment have been adopted, since the design team's preferences should then have been satisfied in as far as they have been articulated in the modeling.

3.6.3 Pareto Set Algorithms

Often, it is not possible or desirable to assign weights and combine functions before seeking promising designs – the designers may wish to "know what is out there" and actively seek a solution that is counterintuitive in the hope that this may lead to a gain in market share or a step forward in performance. In such circumstances, the Pareto solution set, or at least some approximation to it, must be sought.

The most obvious way to construct a Pareto set is to simply use a direct search to work through the various combinations of weights in a linear sum over the given objectives. Such an approach suffers from the obvious drawback that, when there are many objectives, there will be many weights to set and so building a reasonable Pareto set may be slow.

Also, it will not be clear, *a priori*, whether identifying designs that are optimal with one set of weights will be much harder than designs with other combinations – we already know that convex regions of the front cannot be found in this way – if a thorough survey over all possible weights is desired, this will further increase the cost of search. The desire to be able to efficiently construct Pareto sets, including designs lying in convex regions of the front, has lead to an entire subspecialization in the field of optimization research. It is also an area where much progress has been made using population-based methods, for the obvious reason that such methods work with sets of designs in the first place.

Perhaps, the best way to understand methods that aim to construct a Pareto set is to observe that, when constructing the front, the real goal is a well-balanced set of solutions, that is, one characterized by the members of the set being uniformly distributed along the front. Then, given that the designer chooses the number of members in the set, i.e., the granularity of the solution, it is the function of the search to place these points evenly along the front. Therefore, building Pareto fronts usually consists of two phases: first, a set of nondominated points must be found, and secondly, these must be spaced out evenly. One effective way of doing this is via an archive of nondominated points (Knowles and Corne 1999). As the search proceeds, if any design is identified that is currently nondominated, it is placed in an archive of fixed size – the size being defined *a priori* to reflect the granularity required. Then, as new nondominated points are found, these are added to the archive until such time as it is full. Existing points in the archive are pruned if the new points dominate them, of course. Thereafter, all new points are compared to the archived set and only added if either a) they dominate an existing point or b) they increase the uniformity of coverage of the set, that is, they reduce crowding and so spread out the solutions along the emerging Pareto front – with the crowded members being the target for pruning to maintain the set size. A number of crowding measures can be used in this role – for example, if e is the normalized Euclidean distance in the objective function space between members of the archive, then the authors find a useful crowding metric to be $1/\sum e$, where the sum is limited to the nearest 5% of solutions in the front (the normalization is achieved by dividing each objective function by the maximum range of function values seen in that direction).

The archive approach to locating the Pareto front additionally requires a way of generating new nondominated points as efficiently as possible. Since the archive represents a useful store of material about good designs, it is quite common in Pareto set construction to try to extract information from this set when seeking new points. The genetic algorithm operation of crossover forms a good way of doing this as does the mutation of existing nondominated points to form the starting points for downhill or other forms of searches. Another approach is to construct a pseudoobjective function and employ a single objective function optimizer to improve it. The aim of the pseudoobjective is to score the latest design indirectly by whether or not it is feasible, dominates the previous solution or is less crowded than the previous solution. Figure 3.30 is a chunk of pseudocode that implements such an approach and which can be used to identify Pareto fronts in the Options search package in conjunction with an evolutionary strategy single objective optimizer.[6]

An alternative, more direct solution is to treat the generation used in an evolutionary algorithm as the evolving Pareto set itself and then to rank each member in the set by the degree to which it dominates other members. By introducing some form of crowding or niche forming mechanism, such a search will tend to evolve an approximation to the Pareto front with reasonable coverage. In the nondominated sorting GA of Srinivas and Deb (1995),

---

[6]http://www.soton.ac.uk/~ajk/options/welcome.html

```
If(current solution is feasible)then
    If(current solution dominates previous solution)then
        Q=Q-1
    Else
        If(current solution is less crowded than previous solution)then
            Q=Q-1
        Else
            Q=Q+1
        Endif
    Endif
Else
    Q=Q+1
Endif
```

Figure 3.30 Pseudocode for simple objective function when constructing Pareto fronts (Q is the function minimized by the optimizer – this is set to zero at the beginning of a search and is nonstationary, and so, this code is only suitable for methods capable of searching such functions, for example, the evolution strategy approach; the crowding metric is $1/\sum e$, where the sum is limited to the nearest 5% of solutions).

first the nondominated individuals are collected, given an equal score and placed to one side. Then, the remaining individuals are processed to find those that dominate this new set to give a second tier of domination – these are again scored equally but now at reduced value and again placed to one side, and so on, until the entire population is sorted into a series of fronts of decreasing rank. Each rank has its score reduced according to a sharing mechanism that encourages diversity. The final score is then used to drive the selection of new points by the GA in the normal way. The Multi-Objective Genetic Algorithm (MOGA) method of Fonseca and Fleming (1995) is another variant on this approach although now the degree of dominance is used rather than a series of ranks assembled of equal dominance. Its performance is dependent on exactly how the sharing process that spaces points out along the front is implemented and how well this maps to the difficulty of finding points on any particular region of the front. In all cases, if a region is difficult to populate with nondominated designs, it is important that the weighting used when assembling new generations is sufficient to preserve these types of solutions in the population – something the archive approach does implicitly.

3.6.4 Nash Equilibria

An alternative way to trade off the goals in a multiobjective problem is via the mechanism of Nash equilibria (Habbal et al. 2003; Sefrioui and Periaux 2000). These are based on a game player model: each goal in the design problem is given over to a dedicated "player", who also has control of some subset of the design variables. Then, each player in turn tries to optimize their goal using their variables, but subject to the other variables in the problem being held fixed at the last values determined by the other players. As the players take turns to improve their own goals, the problem settles into an equilibrium state such that none can improve on their own target by unilaterally changing their own position. Usually, the variables assigned to a particular player are chosen to be those that most strongly correlate with their goal, although when a single variable strongly affects two goals, an arbitrary choice has to be made. In addition, a set of initial values has to be chosen to start the game off.

If carried out to complete convergence, playing a Nash game in this way will yield a result that lies on the Pareto front. Moreover, convergence to a nondominated point is usually very robust and quite rapid. The process can also be easily parallelized, since each player's search process can be handed to a dedicated processor and communication between players is only needed when each has found its latest optimal solution. Unfortunately, in many problems, different initial values will lead to different equilibria (i.e., different locations on the Pareto front), as will different combinations of assignments of the variables to different players. Nonetheless, such methods have found a good deal of use in economics and they are beginning to be adopted in aerospace design because of their speed and robustness.

3.7   Robustness

Thus far in this chapter, we have been concerned with tools that may be used to locate the optima in deterministic functions – the assumption being that such locations gave the ideal combinations of parameters that we seek in design. In fact, this is very often not so – as will by now have become clear, many such optima are defined by the actions of constraint boundaries. To develop a design that sits exactly on the edge of a hard constraint boundary is rarely what designers wish to do – this would lead to a degree of risk in the performance of the product that can rarely be justified by any marginal gain that might be attained. Rather, designers almost always wish their designs to exhibit robustness. Such robustness must cope with uncertainties coming from many sources: even today, our ability to predict product behavior at the design stage is by no means exact. Moreover, as-manufactured products always differ, sometimes substantially, from the nominal specifications. Finally, the in-service conditions experienced by products can rarely be foreseen with complete clarity. In consequence, some margin for uncertainty must be made. Thus, we say that a design whose nominal performance degrades only slightly in the face of uncertainty is robust, while one that has dramatic falloff in performance lacks robustness and is fragile. We return to this important topic in much more detail in Chapter 8 – here, we make some of the more obvious observations.

Given the relationship of robustness to changes in design performance, it will be obvious that robust designs will lie in smooth, flat regions of the search space. It is therefore often desirable that an optimizer can not only identify regions of high performance in the search space but also ones where plateaus of robustness may be found; see Figure 3.31. If the best design lies on the edge of the search space or where constraints are active, this will often mean that the way the objective function surface approaches the boundary will be important – a steep intersection with such a boundary will usually imply a fragile design (e.g., point (0,80) in Figure 3.31).

Some search methods are intrinsically better able to locate robust designs than others, while some may be adapted, albeit at extra cost, to yield robust designs. It is also worth noting that the way that targets are specified can also be used to influence this quality in design – if a constraint violation is embedded in the objective function rather than treated explicitly as a hard boundary, this will impact on the way the search proceeds – this can be an important way of managing the trade-off between nominal performance and robustness. In many senses, the desire for robustness can be seen as yet a further goal that designers must

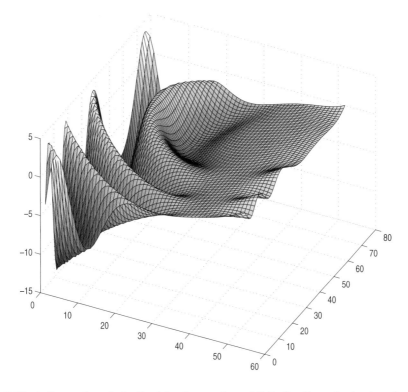

Figure 3.31 A fitness (maximization) landscape the exhibits fragile and robust regions, after Parmee (1996).

strive for and treated along with any others in a multiobjective framework. It is, however, often useful to treat robustness and uncertainty separately when carrying out search.

Perhaps, the most obvious way of dealing with robustness is to add artificial noise to the design vector or the resulting objective or constraints, or all of these, when calculating the performance of the design. This noise can be set to simulate uncertainty in the design so that the search engine must directly cope with this when attempting to find good configurations. Of course, this will mean that repeated calls by the search to the functions being dealt with will yield different responses – few search methods are able to cope with this directly, but it is perhaps the most direct way to simulate what actually happens in practice. Noise added to the design vector represents uncertainty in the as-manufactured product, whereas noise in the objective and constraint functions represents uncertainty in the operating conditions and the design team's representation of the relevant physics.

Some search engines, notably the stochastic search methods (usually with elitism switched off), can easily be set up to work with this kind of dynamic environment and, if direct searches are affordable, this way of working has much to commend it. Uncertainty generated by added noise can also be tolerated by many of the metamodeling approaches, provided that they include an element of regression in their model building capabilities. In either case, issues of search termination become more subtle and complex when the functions being searched are random. Still, using random functions in this way adds only modestly to the expense of the underlying search. Of course, to do this requires knowledge of how large any random perturbations in the design vector and modeling functions should be and this should be carefully assessed when specifying the problem being tackled – if the random elements included are unjustifiably large, then very pessimistic designs can result.

A more expensive but more exact way of proceeding is to attempt to sample the uncertainty statistics themselves at each point to be evaluated in the search space. To do this, the nominal design being evaluated is used to initiate a small random walk. Then the average, or perhaps worst design found during this walk is returned in place of the nominal design point. In this way, it is possible to make the search work on a pseudo mean or 95% confidence limit design so that any final optimal point will exhibit the desired degree of robustness. However, we have now replaced a single design evaluation with a random walk of perhaps 10–20 points (this will depend on the number of variables in play and the degree of accuracy required) – if this can be afforded it will allow significantly greater confidence to be placed in the final designs produced. An example of this kind of approach is given by Anthony et al. (2000). A more affordable approach is to combine these ideas – the functions and variables being treated as random during the search with random walks being added periodically so that a more precise understanding of the uncertainty is used from time to time.

The most expensive, but at the same time, most accurate way to deal with uncertainty in search is to use fully stochastic performance models throughout. Then, instead of working with deterministic quantities, the whole problem is dealt with in terms of the probability distribution functions (PDFs) of the inputs and outputs. This way of working is in its infancy at the time of writing, since stochastic PDE solvers are commonly orders of magnitude more expensive than their deterministic equivalents (at its most extreme each evaluation is replaced by a full, multidimensional, large scale Monte Carlo evaluation). It is also the case that only rarely do designers have sufficient information to construct the input variable PDFs for working in this way, especially if correlations arise between inputs (and they commonly do – and are even more commonly ignored in such work). Nonetheless, as stochastic PDE solvers become more mature it is inevitable that they will be increasingly used in optimization.

Part II

Sensitivity Analysis and Approximation Concepts

Having set out the background against which computational aerospace design takes place, we move on to consider in some depth a number of more advanced topics. These lie outside the scope of most of the texts that deal with aerospace analysis or design in a more traditional way. They all stem from a desire to exploit the highly sophisticated analysis tools now available to engineers, in a way suitable for teams seeking to produce the most competitive designs. One of the key aims of this book is to gather much of this material into a single volume for perhaps the first time. This work is inevitably somewhat mathematical and so some readers may wish to move directly to the case-studies of Part IV before returning to Parts II and III.

We begin the theoretical work by first making a more detailed study of the subject of sensitivity analysis. Sensitivity analysis is of key importance in computational approaches to design because designers obviously wish to know how best to improve their products. If sensitivity analysis can reveal efficiently which variables have most influence over key aspects of the design, and also in which direction to change them, this will significantly aid the design team's task.

We then move on to a deeper treatment of approximation concepts. As many of the analyses of interest to design teams are computationally expensive, finding effective means to reduce design cycle times by the use of approximations can make progress possible in what would otherwise be untenable circumstances. Such approximations can be constructed purely from the data derived by running analysis codes (so called "black-box" approaches) or by using knowledge of the physics being studied. We treat these two approaches separately, devoting a chapter to each. In general, black-box methods are simpler to apply but also usually less effective or efficient. As ever, the choices made will depend on the circumstances at hand – neither should be dismissed without some consideration as to their likely benefits.

4

Sensitivity Analysis

Sensitivity analysis is of fundamental importance to design based on computational approaches. It allows the use of gradient descent methods, reveals when optimal designs have been produced and indicates which variables are of most importance at any stage in the design process. Consider the discrete mathematical model of a physical system written in the form $\mathbf{R}(\mathbf{x}, \mathbf{w}) = 0$, where $\mathbf{x} = \{x_1, x_2, \ldots, x_p\}^T \in \mathbb{R}^p$ denotes a set of independent design variables and $\mathbf{w} \in \mathbb{R}^n$ is the vector of state variables.[1] In practice, the discrete form of the governing equations is commonly obtained by spatial and temporal discretization of a system of coupled partial differential equations (PDEs) obtained from continuum mechanics, although no such assumption is necessary in what follows. In the context of sensitivity analysis, one can either deal with the continuous or discrete form of the governing equations. However, for simplicity of presentation and for the sake of generality, much of the discussion in this chapter focuses on the discretized form of the equations.

Let an output quantity of interest be defined in the functional form $y = f(\mathbf{x}, \mathbf{w})$; for example, y could be an objective function that is to be minimized or a constraint to be satisfied in a design problem. Sensitivity analysis can then be formally defined as the task of computing the kth order partial derivatives of y with respect to the vector of independent variables $\mathbf{x}$.[2] In most practical problems, it is the first-order derivatives that are of interest, that is,

$$\nabla y = \left\{ \frac{\partial y}{\partial x_1}, \frac{\partial y}{\partial x_2}, \ldots, \frac{\partial y}{\partial x_p} \right\} \in \mathbb{R}^p. \tag{4.1}$$

Some applications of output sensitivities that arise in the context of design optimization are outlined below:

(a) Using a first-order Taylor series, changes in y can be quickly approximated when the vector of independent variables $\mathbf{x}$ is perturbed by small quantities. As discussed

---

[1] For example, in a structural design problem, the state vector $\mathbf{w}$ may contain the displacements at all the nodes of the finite element mesh.

[2] In some disciplines, the term *sensitivity analysis* is also used to loosely refer to techniques for estimating the degree of influence of each input variable on an output quantity. Such techniques generally employ global approaches like those described in Chapter 5 in the context of variable screening or importance analysis.

Computational Approaches for Aerospace Design: The Pursuit of Excellence.   A. J. Keane and P. B. Nair
© 2005 John Wiley & Sons, Ltd

later in Chapter 5, the output sensitivities can also be employed to construct surrogate models of the original function that are computationally cheaper to run.

(b) $-\nabla y$ and ∇y denote the direction of maximum decrease and increase in y, respectively. Hence, sensitivities can be employed in gradient-based search procedures such as those described in Chapter 3 to locate a local maximum or minimum of y.

(c) When $\mathbf{x}$ is modeled as a vector of zero-mean random variables with correlation matrix $\mathbf{C}$, then the first-order sensitivities can be employed to approximate the statistics of y as $\langle y \rangle = f(\langle \mathbf{x} \rangle, \mathbf{w})$, $\sigma_y = \nabla y \mathbf{C} \nabla y^T$, where $\langle \cdot \rangle$ denotes the expectation operator. A detailed discussion of the application of sensitivities to uncertainty analysis can be found in Chapter 8.

This chapter discusses various ways in which sensitivities may be calculated in an algebraic as well as a functional analysis setting. The advantages and drawbacks of different approaches are outlined and guidelines are provided for choosing the most appropriate technique for the problem at hand. Examples from the domains of structural analysis and fluid dynamics are used to illustrate the application of these general concepts in practice. We would, however, point out that the literature on sensitivity analysis is vast. The emphasis of this chapter is on the basic concepts underpinning various approaches to finding sensitivities. A more detailed account of specialized topics can be found in texts dedicated to the field, such as those by Choi and Kim (2005a,b), Mohammadi and Pironneau (2001), and so on.

4.1 Finite-difference Methods

A straightforward and commonly used approach for approximating derivatives involves the use of finite differences. Finite-difference (FD) approximations for a derivative are derived by truncating a Taylor series expansion of the function $y = f(x)$ about a point x. To outline the derivation of finite-difference approximations, consider the Taylor series expansion of a scalar function $f(x + \varepsilon)$

$$f(x + \varepsilon) = f(x) + \varepsilon \frac{df}{dx} + \frac{\varepsilon^2}{2!} \frac{d^2 f}{dx^2} + \frac{\varepsilon^3}{3!} \frac{d^3 f}{dx^3} + \cdots. \tag{4.2}$$

The forward finite-difference approximation for the first derivative can then be obtained by solving the preceding equation for $\frac{df}{dx}$, which gives

$$\frac{df}{dx} \approx \frac{f(x + \varepsilon) - f(x)}{\varepsilon} + \mathcal{O}(\varepsilon). \tag{4.3}$$

It is clear that the truncation error of the forward finite-difference approximation is $\mathcal{O}(\varepsilon)$ and hence this is a first-order approximation. To reduce the truncation error, an additional set of equations can be arrived at by writing down the Taylor series expansion of $f(x - \varepsilon)$ as follows:

$$f(x - \varepsilon) = f(x) - \varepsilon \frac{df}{dx} + \frac{\varepsilon^2}{2!} \frac{d^2 f}{dx^2} - \frac{\varepsilon^3}{3!} \frac{d^3 f}{dx^3} + \cdots. \tag{4.4}$$

Subtracting (4.4) from (4.2) and solving for $\frac{df}{dx}$ gives the following central difference approximation

$$\frac{df}{dx} = \frac{f(x + \varepsilon) - f(x - \varepsilon)}{2\varepsilon} + \mathcal{O}(\varepsilon^2). \tag{4.5}$$

It can be seen that the central difference scheme is second-order accurate since the truncation error is $\mathcal{O}(\varepsilon^2)$. However, this is achieved by using twice the number of function evaluations required by the forward difference approximation. It is similarly possible to construct higher-order finite-difference formulae.

Finite-difference approximations for the higher-order derivatives can be derived by nesting lower-order formulae. For example, the central difference approximation in (4.5) can be used to estimate the second-order derivative as follows:

$$\frac{\mathrm{d}^2 f}{\mathrm{d}x^2} = \frac{\mathrm{d}f(x+\varepsilon)/\mathrm{d}x - \mathrm{d}f(x-\varepsilon)/\mathrm{d}x}{2\varepsilon} + \mathcal{O}(\varepsilon^2). \tag{4.6}$$

Substituting central difference approximations for $\mathrm{d}f(x-\varepsilon)/\mathrm{d}x$ and $\mathrm{d}f(x+\varepsilon)/\mathrm{d}x$ into the preceding equation gives

$$\frac{\mathrm{d}^2 f}{\mathrm{d}x^2} = \frac{f(x+2\varepsilon) - 2f(x) + f(x-2\varepsilon)}{4\varepsilon^2} + \mathcal{O}(\varepsilon). \tag{4.7}$$

An attractive feature of finite-difference approximations is that the simulator computing y can be regarded as a black box and it is only necessary that it be run at a selected set of points. The computational cost depends on the order of the approximation and scales linearly with the number of design variables. In theory, the accuracy of the approximations should depend on the truncation error, that is, the approximation error can be made arbitrarily small by choosing a small enough step size. In practice, however, because of finite precision arithmetic (round-off errors), the accuracy of the approximations critically depends on the step size ε. More specifically, finite-difference approximations for the derivative can be inaccurate because of truncation or condition errors.

Truncation errors arise because of the neglected terms in the Taylor series expansion, whereas the condition error can be defined as the difference between the numerically evaluated value of the function and its "true" value. Consider, for example, the case when an iterative solver is employed to evaluate the function values. Here, if the iterative algorithm is not run to convergence, there will be an error in the function value (Haftka 1985). A similar situation arises when a perturbation in design variables leads to changes in mesh topology (and possibly the number of elements). For problems where the function values are computed by numerically approximating the solution to a system of PDEs, the error in the function value due to discretization errors may depend on the design variables. In order to negate the contaminating influence of discretization error, care needs to be taken to ensure that these errors are almost constant when the design variables are perturbed.

To illustrate the influence of condition errors, let the computed function value be given by $\hat{y} = y + \varepsilon_y y$, where ε_y is an error term. Then, the total error in the forward difference approximation due to truncation and condition errors is $\mathcal{O}(\varepsilon + \frac{\varepsilon_y}{\varepsilon})$.

Minimization of the combined truncation and condition errors is desirable to ensure accurate approximations for the derivatives. This leads to the familiar *step-size dilemma*. First, we need to choose a step size that minimizes the truncation error. At the same time, the step size chosen must not be so small as to cause significant subtractive cancellation errors.[3] In practice, it is often found that the approximation error decreases when ε is reduced until a critical value of the step size is reached, after which the error increases again.

---

[3]In the case of forward finite differences, cancellation error arises because of the term $f(x+\varepsilon) - f(x)$ in the numerator of (4.3) – it is difficult to accurately evaluate the difference between two terms that are similar in magnitude using *finite precision arithmetic*, particularly for small values of ε.

A number of approaches for choosing the step size optimally in order to trade off truncation error with the subtractive cancellation error can be found in the literature; see, for example, Brezillon et al. (1981). However, these approaches can lead to significant increase in the computational cost because of the need for additional function evaluations. The complex variable approach outlined next alleviates the limitations associated with finite-difference approximations, particularly when only the first-order derivatives are of interest.

4.2   Complex Variable Approach

The idea of using complex variables to derive differentiation formulae was proposed by Lyness and Trapp (1967). They developed a number of methods for sensitivity analysis, including a method for estimating the kth derivative of a function. More recently, this approach was revived by Squire and Trapp (1998), who presented a simple expression for reliably computing the first derivative. To illustrate the complex variable approach, consider an output function, $f(x)$, which is an analytic function of the scalar complex variable x, that is, the Cauchy–Riemann equations apply.[4] Under this assumption, y can be expanded in a Taylor series about a real point x as follows:

$$f(x + i\varepsilon) = f(x) + i\varepsilon \frac{df}{dx} - \frac{\varepsilon^2}{2!} \frac{d^2 f}{dx^2} - \frac{i\varepsilon^3}{3!} \frac{d^3 f}{dx^3} + \cdots . \tag{4.8}$$

Note that the above Taylor series is equivalent to considering a complex perturbation $i\varepsilon$ (where $i = \sqrt{-1}$) instead of a real perturbation as used in the conventional series given earlier in (4.2). Hence, this approach is also sometimes referred to as a complex-step approximation. The real and imaginary parts of (4.8) can be written as

$$\Re[f(x + i\varepsilon)] = f(x) - \frac{\varepsilon^2}{2!} \frac{d^2 f}{dx^2} + \mathcal{O}(\varepsilon^4), \tag{4.9}$$

$$\Im[f(x + i\varepsilon)] = \varepsilon \frac{df}{dx} - \frac{\varepsilon^3}{3!} \frac{d^3 f}{dx^3} + \mathcal{O}(\varepsilon^5). \tag{4.10}$$

Dividing (4.10) by ε gives the first-order derivative of f at x:

$$\frac{df}{dx} = \frac{\Im[f(x + i\varepsilon)]}{\varepsilon} + \mathcal{O}(\varepsilon^2). \tag{4.11}$$

Note that the above approximation gives an $\mathcal{O}(\varepsilon^2)$ estimate of the first derivative of y. In contrast to the finite-difference approximation (see (4.3)), the above estimate for the first derivative is not subject to subtractive cancellation error since it does not involve any difference operations. This is a major advantage of the complex-step approximation over the finite-difference approach since the "step-size" dilemma is effectively eliminated. In theory, one can use an arbitrarily small step-size ε without any significant loss of accuracy. In practice, ε is typically chosen to be around 10^{-8}.

---

[4]A function f of a complex variable $x = x_1 + ix_2$ is said to be analytic (i.e., differentiable in the complex plane) if the following Cauchy–Riemann equations hold:

$$\frac{\partial f_r}{\partial x_1} = \frac{\partial f_i}{\partial x_2}, \quad \frac{\partial f_r}{\partial x_2} = -\frac{\partial f_i}{\partial x_1},$$

where f_r and f_i are the real and imaginary parts of the function f, respectively.

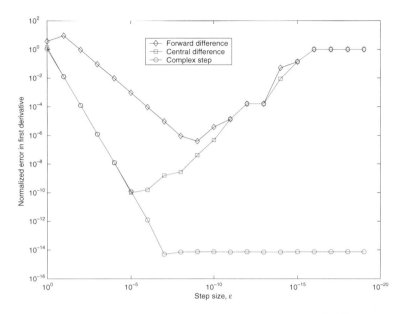

Figure 4.1 Comparison of normalized error in the first derivative ($|\frac{f'-\widehat{f}'}{f'}|$) computed using finite-difference methods and the complex-step approximation.

To illustrate the advantages of the complex variable approach over the finite difference technique, consider the simple one-dimensional function $y = x^2 - 10\cos(2\pi x)$. Figure 4.1 shows the dependence on the step size ε of the accuracy of the derivative approximations for the forward finite difference, central difference and the complex variable approaches. It can clearly be seen that as the step size is reduced, the accuracy of the finite-difference approximations degrades significantly because of subtractive cancellation errors. In contrast, for the complex variable approach, the accuracy is relatively insensitive to the step size.

It is worth noting here that using (4.9) and (4.10), an approximation for the second derivative can be written as

$$\frac{d^2 f}{dx^2} = \frac{2\,(f(x) - \Re[f(x + i\varepsilon)])}{\varepsilon^2} + \mathcal{O}(\varepsilon^2). \tag{4.12}$$

Unfortunately, the preceding equation is subject to cancellation errors due to the term in the numerator and hence the accuracy of the approximation will be sensitive to the step size ε.

4.2.1 Implementation Issues

To use the complex-step approach, we must have access to the source code of the simulator, which must, moreover, not already use complex arithmetic. For the complex variable approach to be attractive for such problems, it is desirable to have an automatic procedure for generating a modified source code that can furnish both the function value and its derivatives. This problem was studied in detail by Martins (2002) in his PhD dissertation. Apart from the obvious fact that one needs to have access to the source code, the process of generating a complex version of the original source code involves the following steps: (1) substitute all "real" type variable declarations with "complex" declarations, (2) define all functions and operators that are not defined for complex arguments, and (3) introduce a complex step for each independent variable and compute the derivative using (4.11).

The first step is relatively straightforward in programming languages such as Fortran, which support complex numbers as a standard data type. For codes written in Fortran, this involves simply substituting all occurrences of REAL*8 declarations with COMPLEX*16. Conversion of real data types to complex numbers can be achieved more elegantly if the programming language allows for intrinsic functions and operators to be overloaded. By leveraging the notion of operator overloading, far fewer changes are required to the original source, thereby leading to improved software maintainability. Fortunately, this capability is available in programming languages such as Fortran 90 and Matlab. However, care must be exercised in the second step since the complex-step approximation was derived under the assumption that the function f is analytic. It therefore becomes important to carefully examine the validity of this assumption when converting functions and operators to deal with complex arguments.

First, consider the case of relational operators such as "greater than" and "or", which are not defined for complex numbers. Here, the correct approach would be to only compare the real parts of the arguments and ignore the imaginary part. It is also worth noting that algorithms that employ conditional statements may result in discontinuities in their outputs. When the complex variable approach is used, the approximation for the derivative will be correct right up to the discontinuity. In theory, at the point of discontinuity, the derivative does not exist. However, if the function is defined at that point, the complex variable approach would still correctly estimate a one-sided derivative (Martins 2002).

Functions that accept complex arguments may be considered as an extension of their real counterpart. Note that the complex versions of the functions must also satisfy the Cauchy–Riemann equations for the complex-step approximation to be valid. Complex versions of most elementary operations and functions such as addition, multiplication and trigonometric operations are analytic in their arguments and can be used in a straightforward fashion. However, some can have singularities or branch cuts on which they are not analytic. As mentioned earlier, if the function is defined, the complex-step approximation would correctly approximate the one-sided derivative. The only standard complex function definition that is nonanalytic is the absolute value function or modulus operator (abs), that is, $|z| = \sqrt{x^2 + y^2}$. This function is not analytic, and hence the complex-step approximation cannot be employed to compute its derivative. Martins (2002) suggested the following analytic version of the function:

$$\text{abs}(x + \mathrm{i}y) = \begin{cases} -x - \mathrm{i}y & \forall x < 0 \\ +x + \mathrm{i}y & \forall x > 0 \end{cases}. \tag{4.13}$$

It should be noted that at $x = 0$ the derivative does not exist. However, by using the definition of the function for $x > 0$, it becomes possible to compute the function value at $x = 0$ and the correct right-hand-side derivative at that point.

Similar care needs to be exercised when applying the complex variable approach to certain trigonometric functions. For example, when a small step size is used, the complex variable approach will yield a spurious zero derivative for $\text{arcsin}(z)$ because of finite precision arithmetic. Martins (2002) suggested that in order to work around this problem the following linear approximation be used for such functions:

$$f(x + i\varepsilon) = f(x) + i\varepsilon\frac{\mathrm{d}f(x)}{\mathrm{d}x}. \tag{4.14}$$

Martins (2002) showed that when all functions are computed in this manner the complex variable approach is the same as the automatic or algorithmic differentiation approach to be presented later in Section 4.6.

In general, the complex variable approach requires roughly double the memory used by the primal solver. Numerical studies in the literature suggest that the run time required by the approach is around three times that required by the primal solver. The method has been applied to a number of problems including CFD (computational fluid dynamics), structural analysis and aeroelasticity; see, for example, Martins et al. (2004, 2003). Applications of the complex step approximation to some real-world problems are presented later in Section 4.9, where comparisons are made with other sensitivity analysis techniques.

4.3 Direct Methods

Direct methods (and the adjoint methods presented in the next section) present an efficient way to exactly compute sensitivities. However, these approaches may involve extremely detailed and cumbersome derivations and can only be applied when the analyst has a good understanding of the governing equations and the numerical scheme employed for its solution. To illustrate the direct approach, consider the discrete mathematical model of a physical system given in the form

$$\mathbf{R}(\mathbf{x}, \mathbf{w}(\mathbf{x})) = 0, \tag{4.15}$$

where $\mathbf{x} \in \mathbb{R}^p$ denotes a set of independent design variables and $\mathbf{w} \in \mathbb{R}^n$ is the vector of state variables. The residual $\mathbf{R}$ may explicitly depend on the set of design variables $\mathbf{x}$ while the state variable vector $\mathbf{w}$ is an implicit function of $\mathbf{x}$. In a majority of cases of practical interest in aerospace design, the system of algebraic equations in (4.15) is obtained via spatial and temporal discretization of a set of coupled PDEs governing the system under consideration.

Consider the case when it is desired to directly compute the derivative of the output quantity $y = f(\mathbf{x}, \mathbf{w})$ with respect to x_i, that is, it is necessary to compute

$$\frac{\mathrm{d}y}{\mathrm{d}x_i} = \frac{\partial f}{\partial x_i} + \frac{\partial f}{\partial \mathbf{w}}\frac{\mathrm{d}\mathbf{w}}{\mathrm{d}x_i}. \tag{4.16}$$

Now, differentiation of the governing equation (4.15) yields

$$\frac{\partial \mathbf{R}}{\partial \mathbf{w}}\frac{\mathrm{d}\mathbf{w}}{\mathrm{d}x_i} = -\frac{\partial \mathbf{R}}{\partial x_i}. \tag{4.17}$$

The sensitivity term $\mathrm{d}y/\mathrm{d}x_i$ can be readily computed by first solving (4.17) for $\mathrm{d}\mathbf{w}/\mathrm{d}x_i$ and substituting the result into (4.16). Note that (4.17) has to be solved as many times as the number of design variables in order to compute the gradient vector ∇y.

The basic steps involved in direct methods for sensitivity analysis will be illustrated next within the contexts of static structural analysis, eigenvalue problems and transient dynamic analysis.

4.3.1 Example: Static Structural Analysis

Consider the following system of linear algebraic equations governing the response of a linear structural system subjected to static loads:

$$\mathbf{K}(\mathbf{x})\mathbf{w}(\mathbf{x}) = \mathbf{f}(\mathbf{x}), \tag{4.18}$$

where $\mathbf{K} \in \mathbb{R}^{n \times n}$ denotes the structural stiffness matrix, $\mathbf{w} \in \mathbb{R}^n$ is the displacement vector and $\mathbf{f} \in \mathbb{R}^n$ is the force vector. $\mathbf{x} \in \mathbb{R}^p$ denotes the vector of design variables, which may include cross-sectional properties and geometric attributes.

To illustrate the direct sensitivity analysis approach, consider the problem of evaluating the first-order derivatives of the displacement vector with respect to the design variables at the point $\mathbf{x}^0$. Differentiating (4.18), we obtain the following system of linear algebraic equations for the derivative of the displacement vector:

$$\mathbf{K}(\mathbf{x}^0)\frac{d\mathbf{w}}{dx_i} + \frac{\partial\mathbf{K}}{\partial x_i}\mathbf{w}(\mathbf{x}^0) = \frac{d\mathbf{f}}{dx_i}. \tag{4.19}$$

Rearranging terms in the preceding equation gives

$$\mathbf{K}(\mathbf{x}^0)\frac{d\mathbf{w}}{dx_i} = \frac{d\mathbf{f}}{dx_i} - \frac{\partial\mathbf{K}}{\partial x_i}\mathbf{w}(\mathbf{x}^0). \tag{4.20}$$

It can be observed from (4.20) that computing the displacement sensitivity is equivalent to computing the response of the structure under the effect of a pseudoload vector $\frac{d\mathbf{f}}{dx_i} - \frac{\partial\mathbf{K}}{\partial x_i}\mathbf{w}(\mathbf{x}^0)$. Clearly, to compute this term, it is necessary to first solve the baseline static analysis problem $\mathbf{K}(\mathbf{x}^0)\mathbf{w}(\mathbf{x}^0) = \mathbf{f}(\mathbf{x}^0)$ to calculate $\mathbf{w}(\mathbf{x}^0)$. If a direct solver is used, then the factorized form of the stiffness matrix $\mathbf{K}(\mathbf{x}^0)$ is available as a by-product – this makes it possible to efficiently solve (4.20) using forward and backward substitutions. It can also be noted from (4.20) that it is necessary to compute the sensitivity of the global stiffness matrix with respect to x_i. In practice, this is often achieved by differentiating the element-level stiffness matrices with respect to x_i and assembling them using standard techniques.

In structural design problems with stress constraints, it may be necessary to compute the sensitivities of the stresses with respect to the design variables. The stresses are related to the displacements by $\sigma = \mathbf{S}\mathbf{w}$, where $\mathbf{S}$ is the stress-displacement matrix. Hence, the derivative of the stress can be computed as

$$\frac{d\sigma}{dx_i} = \frac{\partial\mathbf{S}}{\partial x_i}\mathbf{w}(\mathbf{x}^0) + \mathbf{S}(\mathbf{x}^0)\frac{d\mathbf{w}}{dx_i}. \tag{4.21}$$

A similar procedure can be employed for direct sensitivity analysis of nonlinear structural systems (Haftka and Gurdal 1992).

4.3.2 Example: Eigenvalue Problems

Consider the following generalized eigenvalue problem which governs the free-vibration response of a linear structural system:

$$\mathbf{K}(\mathbf{x})\boldsymbol{\phi}(\mathbf{x}) = \lambda(\mathbf{x})\mathbf{M}(\mathbf{x})\boldsymbol{\phi}(\mathbf{x}) \tag{4.22}$$

where $\mathbf{K}$ and $\mathbf{M} \in \mathbb{R}^{n \times n}$ denote the structural stiffness and mass matrices, respectively. λ and ϕ denote the eigenvalue and eigenvector and we assume that it is necessary to compute their sensitivities at the point $\mathbf{x}^0$. Let the eigenvalues and eigenvectors at this point be available by solving the baseline eigenvalue problem

$$\mathbf{K}(\mathbf{x}^0)\phi^0 = \lambda^0 \mathbf{M}(\mathbf{x}^0)\phi^0. \tag{4.23}$$

In the derivation that follows it is assumed that the eigenvalues of (4.23) are distinct. If the eigenvectors are mass normalized, then the following equations hold:

$$\phi_i^T \mathbf{K} \phi_j = \lambda_i \delta_{ij} \quad \text{and} \quad \phi_i^T \mathbf{M} \phi_j = \delta_{ij} \tag{4.24}$$

where δ_{ij} denotes the Kronecker delta function. Differentiating (4.22), we have

$$(\mathbf{K} - \lambda_i \mathbf{M}) \frac{\partial \phi_i}{\partial x_j} = \left[\lambda_i \frac{\partial \mathbf{M}}{\partial x_j} + \frac{\partial \lambda_i}{\partial x_j} \mathbf{M} - \frac{\partial \mathbf{K}}{\partial x_j} \right] \phi_i. \tag{4.25}$$

Since the matrices $\mathbf{K}$ and $\mathbf{M}$ are symmetric, it follows from (4.22) that $\phi_i^T (\mathbf{K} - \lambda_i \mathbf{M}) = 0$. Hence, premultiplying (4.25) by ϕ_i^T and using the normalization conditions in (4.24) gives the following expression for the eigenvalue derivative:

$$\frac{\partial \lambda_i}{\partial x_j} = \phi_i^T \left[\frac{\partial \mathbf{K}}{\partial x_j} - \lambda_i \frac{\partial \mathbf{M}}{\partial x_j} \right] \phi_i. \tag{4.26}$$

To compute the eigenvector derivative, it is necessary to solve (4.25). However, since the coefficient matrix of this equation is rank deficient, eigenvector sensitivity analysis is not a straightforward task (Fox and Kapoor 1968). For the sake of notational convenience, (4.25) can be rewritten as

$$[\mathbf{K} - \lambda_i \mathbf{M}] \frac{\partial \phi_i}{\partial x_j} = \mathbf{f}_i \quad \text{where} \quad \mathbf{f}_i = \left[\lambda_i \frac{\partial \mathbf{M}}{\partial x_j} + \frac{\partial \lambda_i}{\partial x_j} \mathbf{M} - \frac{\partial \mathbf{K}}{\partial x_j} \right] \phi_i. \tag{4.27}$$

It can be noted from (4.27) that computing the eigenvector derivative is analogous to computing the forced response of an undamped structure to harmonic excitation. Hence, a modal superimposition approach can be used to compute the eigenvector derivative. However, since all the eigenvectors of the baseline system are generally not available in practice, the modal summation approach will not give very accurate results. One way to overcome this difficulty is to employ Akgün's method (Akgün 1994), in which the accuracy of the modal superimposition approach is improved by including static correction terms for the eigenvectors that are not included in the summation. There are a large number of alternative methods in the literature for solving (4.27) particularly for cases when (4.23) has repeated eigenvalues (see, for example, Dailey (1989) and Zhang and Zerva (1997)) and for damped dynamic systems (Adhikari and Friswell 2001).

4.3.3 Example: Transient Dynamic Analysis

Consider the equations governing the dynamics of a viscously damped second-order system

$$\mathbf{M}\ddot{\mathbf{w}}(t) + \mathbf{C}\dot{\mathbf{w}}(t) + \mathbf{K}\mathbf{w}(t) = \mathbf{f}(t) \tag{4.28}$$

where $\mathbf{M}$, $\mathbf{C}$ and $\mathbf{K} \in \mathbb{R}^{n \times n}$ denote the mass, damping and stiffness matrices, respectively. $\mathbf{w}(t) \in \mathbb{R}^n$ denotes the displacement vector corresponding to the applied forcing $\mathbf{f} \in \mathbb{R}^n$. Here, the over-dots represent derivatives with respect to time $t \in [0, T]$.

The solution of (4.28) is sought with the following initial conditions:

$$\mathbf{w}(t) = \mathbf{u}_0, \qquad \dot{\mathbf{w}}(t) = \mathbf{v}_0 \tag{4.29}$$

and can be solved using a variety of temporal discretization schemes available in the literature (Hughes 1987). For illustration, consider the case when the Newmark family of time integration methods employing a predictor-corrector scheme is employed. Here, the velocity and displacement at step $k + 1$ are

$$\dot{\mathbf{w}}(k+1) = \bar{\mathbf{v}} + \gamma \Delta t \ddot{\mathbf{w}}(k+1),$$

$$\mathbf{w}(k+1) = \bar{\mathbf{u}} + \beta \Delta t^2 \ddot{\mathbf{w}}(k+1) \tag{4.30}$$

where $\bar{\mathbf{v}} = \dot{\mathbf{w}}(k) + (1 - \gamma)\Delta t \ddot{\mathbf{w}}(k)$ and $\bar{\mathbf{u}} = \mathbf{w}(k) + \Delta t \dot{\mathbf{w}}(k) + (\frac{1}{2} - \beta)\Delta t^2 \ddot{\mathbf{w}}(k)$ are the velocity and displacement predictors, respectively. β and γ are Newmark integration parameters. Using (4.30), (4.28) can be written in terms of the acceleration vector at step $k + 1$ as follows:

$$\left[\mathbf{M} + \gamma \Delta t \mathbf{C} + \beta \Delta t^2 \mathbf{K} \right] \ddot{\mathbf{w}}(k+1) = \mathbf{f}(k+1) - \mathbf{K}\bar{\mathbf{u}} - \mathbf{C}\bar{\mathbf{v}}. \tag{4.31}$$

Solution of (4.31) gives the value of $\ddot{\mathbf{w}}(k+1)$, which can be substituted into (4.30) to compute the velocity and displacement vectors at step $k + 1$. The Newmark time stepping scheme is unconditionally stable when $2\beta \geq \gamma \geq 1/2$. The accuracy is of second order only when $\gamma = 1/2$, and for $\gamma > 1/2$, the scheme is first-order accurate.

In the context of sensitivity analysis of the transient response of a dynamic system, there are two options. The first would be to directly differentiate (4.28) and subsequently apply a temporal discretization scheme. Alternatively, the temporally discretized equation in (4.31) can be differentiated. However, both approaches give rise to similar sensitivity equations of the form

$$\left[\mathbf{M} + \gamma \Delta t \mathbf{C} + \beta \Delta t^2 \mathbf{K} \right] \frac{\partial \ddot{\mathbf{w}}(k+1)}{\partial x_i} = \frac{\partial \mathbf{f}(k+1)}{\partial x_i} - \mathbf{K}\frac{\partial \bar{\mathbf{u}}}{\partial x_i} - \frac{\partial \mathbf{K}}{\partial x_i}\mathbf{w}(k+1) - \mathbf{C}\frac{\partial \bar{\mathbf{v}}}{\partial x_i}$$

$$- \frac{\partial \mathbf{C}}{\partial x_i}\dot{\mathbf{w}}(k+1) - \frac{\partial \mathbf{M}}{\partial x_i}\ddot{\mathbf{w}}(k+1), \tag{4.32}$$

where x_i denotes a design variable of interest. Note the implicit assumption in deriving (4.32) is that the time interval does not depend on the design variables.

Differentiation of (4.30) will give equations governing the sensitivities of the displacement and velocity vectors at step $k + 1$. Hence, the value of the acceleration sensitivity obtained by solving (4.32) can be substituted into these equations to obtain the derivatives of the displacement and velocity vectors.

The computational cost of this transient sensitivity analysis procedure will clearly depend on the values of the Newmark integration scheme parameters (β and γ), which dictate whether the time stepping is implicit or explicit. It can be seen from (4.30) that when $\beta = 0$ the scheme is implicit since the displacement and velocity vectors at step $k + 1$ are functions of the acceleration vector at step $k + 1$. For $\beta = 0$ and $\gamma = 1/2$, an explicit scheme is obtained. The explicit scheme is only conditionally stable and hence very small time steps are required to ensure numerical stability. Here, matrix factorizations are not required since the coefficient matrix in (4.31) is diagonal.

In the case of the implicit scheme, provided the time step is constant, the coefficient matrix in (4.31) and (4.32) are the same. Hence, the factored form of this coefficient matrix obtained from response analysis can be reused during sensitivity calculations to ensure computational efficiency. In contrast, for the case of the explicit scheme, calculating the right-hand side of (4.32) is more expensive compared to (4.31). Since no matrix factorizations are involved in the explicit scheme, the possibility of reusing factorizations does not exist. As a result, calculating the sensitivities using the explicit scheme becomes more expensive than response analysis.

In practice, the computational cost associated with transient analysis of large-scale structural systems is often kept tractable by adopting a reduced basis approach, that is, the displacement vector is approximated as

$$\mathbf{w}(t) = \mathbf{\Phi}\mathbf{q}(t) \tag{4.33}$$

where $\mathbf{\Phi} \in \mathbb{R}^{n \times m}$ is a matrix of basis vectors, which may include the free-vibration modes as well as additional Ritz vectors. $\mathbf{q}(t) \in \mathbb{R}^m$ is a vector containing the modal participation factors.

Using (4.33) and employing the Galerkin projection scheme, the governing equations in (4.28) can be rewritten as a reduced-order system of equations as follows:

$$\mathbf{M}_R\ddot{\mathbf{q}}(t) + \mathbf{C}_R\dot{\mathbf{q}}(t) + \mathbf{K}_R\mathbf{q}(t) = \mathbf{f}_R \tag{4.34}$$

where $\mathbf{M}_R = \mathbf{\Phi}^T\mathbf{M}\mathbf{\Phi}$, $\mathbf{C}_R = \mathbf{\Phi}^T\mathbf{C}\mathbf{\Phi}$, $\mathbf{K}_R = \mathbf{\Phi}^T\mathbf{K}\mathbf{\Phi} \in \mathbb{R}^{m \times m}$ and $\mathbf{f}_R = \mathbf{\Phi}^T\mathbf{f} \in \mathbb{R}^m$.

It is possible to apply the sensitivity analysis formulation developed for (4.28) directly to (4.34). If $m \ll n$, then the reduced basis approach will be computationally very efficient. However, the accuracy of the sensitivities will depend on the quality of the basis vectors. An issue that arises when differentiating (4.34) is whether the basis vectors should be considered to be constants or functions of the design variables; for a detailed discussion on this point, see Greene and Haftka (1989).

4.4 Adjoint Methods

Direct methods are not very efficient for problems with large numbers of design variables since (4.17) has to be solved as many times as the number of variables. For problems where the number of design variables (p) is greater than the number of outputs (the total number of objective and constraint functions), it becomes more efficient to use *adjoint* methods for computing the derivatives. The notion of adjoint sensitivity analysis has been in use for a long time and has been applied to compute the partial derivatives of outputs with respect to thousands of design variables at a cost comparable to a few function evaluations. Much of the earlier work on adjoint sensitivity analysis was conducted by nuclear engineers who continue to routinely apply such techniques to compute the derivatives of system responses such as the reactor temperature with respect to thousands of design variables (Cacuci 1981a,b). Adjoint methods have also been applied in the past to solve inverse problems in atmospheric and oceanographic research, where it is necessary to evaluate the derivatives of the residual norm indicating the discrepancies between numerical predictions and experimental observations with respect to the initial conditions and other unknown quantities. Adjoint equations have also been in use for a long time in the field of optimal control theory (Lions 1971).

In the context of design, it appears that the first application of adjoint methods was made by Pironneau (1974). In the field of aerospace design using CFD, the application of adjoint methods was pioneered by Jameson and his colleagues, who applied adjoint methods first to potential and inviscid flows (Jameson 1988, 2001; Jameson and Kim 2003). More recently, adjoint methods have been applied to fully viscous flows over a complete aircraft configuration by Reuther et al. (1999a,b). Jameson and Vassberg (2001) discuss the application of adjoint methods to an aircraft configuration design problem with over 4,000 variables. Adjoint sensitivity analysis formulations can be developed by either starting from the discretized governing equations or the original governing equations in continuous form.

4.4.1 Discrete Adjoint Formulation

The principle of duality or the Lagrange viewpoint can be employed for deriving discrete adjoint equations. To illustrate how the discrete adjoint formulation can be derived from a duality point of view, consider the case when only one design variable is present. Recall that in the direct method of sensitivity analysis applied to discrete systems of the form $\mathbf{R}(\mathbf{x}, \mathbf{w}(\mathbf{x})) = 0$ the following equations are used to compute the derivative of an output function y with respect to the variable x_i:

$$\frac{dy}{dx_i} = \frac{\partial f}{\partial x_i} + \frac{\partial f}{\partial \mathbf{w}} \frac{d\mathbf{w}}{dx_i}, \tag{4.35}$$

$$\frac{\partial \mathbf{R}}{\partial \mathbf{w}} \frac{d\mathbf{w}}{dx_i} = -\frac{\partial \mathbf{R}}{\partial x_i}. \tag{4.36}$$

The preceding equation can be compactly written in the form

$$\mathbf{L}\mathbf{u} = \mathbf{f} \tag{4.37}$$

where $\mathbf{L} = \partial \mathbf{R}/\partial \mathbf{w} \in \mathbb{R}^{n \times n}$, $\mathbf{u} = d\mathbf{w}/dx_i \in \mathbb{R}^n$ and $\mathbf{f} = -\partial \mathbf{R}/\partial x_i \in \mathbb{R}^n$. This equation is sometimes referred to as the *primal* since it arises from direct differentiation of the discrete governing equations. Similarly, (4.35) can be rewritten as $\frac{dy}{dx_i} = \frac{\partial f}{\partial x_i} + \mathbf{g}^T \mathbf{u}$, where $\mathbf{g}^T = \partial f/\partial \mathbf{w} \in \mathbb{R}^n$. It can clearly be seen that $\frac{dy}{dx_i}$ can also be computed from the equivalent equation

$$\frac{dy}{dx_i} = \frac{\partial f}{\partial x_i} + \mathbf{v}^T \mathbf{f} \tag{4.38}$$

where $\mathbf{v}$ is an adjoint vector that satisfies the equation

$$\mathbf{L}^T \mathbf{v} = \mathbf{g}. \tag{4.39}$$

The preceding equation can be viewed as the dual of (4.37) and the equivalence between them can be proved by showing that $\mathbf{g}^T \mathbf{u} = \mathbf{v}^T \mathbf{f}$ as follows:

$$\mathbf{v}^T \mathbf{f} = \mathbf{v}^T \mathbf{L}\mathbf{u} = (\mathbf{L}^T \mathbf{v})^T \mathbf{u} = \mathbf{g}^T \mathbf{u}. \tag{4.40}$$

In the adjoint approach, (4.39) is solved first for the adjoint vector $\mathbf{v}$, which can then be substituted into (4.38) to obtain the derivative dy/dx_i. Clearly, for the case with a single design variable and one output (i.e., single $\mathbf{f}$ and $\mathbf{g}$), the computational effort involved in computing sensitivities using either (4.37) or (4.39) is the same. Now, consider the case with p design variables and m outputs when it is required to compute the derivatives of the m outputs with respect to all the design variables. Here, there are p different $\mathbf{f}$ and m different $\mathbf{g}$. Hence, if the direct approach is used, (4.37) will have to be solved p times. In comparison, for the adjoint approach, (4.39) will have to be solved m times. Hence, the adjoint approach is to be preferred from a computational standpoint when $m < p$.

The discrete adjoint formulation can also be derived from a Lagrange viewpoint. The basic idea is to enforce satisfaction of the discrete governing equations via Lagrange multipliers. The Lagrange multipliers are interpreted here as adjoint variables. For example, when there is only one design variable, the augmented output function that enforces the governing equations via Lagrange multipliers can be written as $I(\mathbf{w}, x_i) = f(\mathbf{w}, x_i) - \boldsymbol{\lambda}^T \mathbf{R}(\mathbf{w}, x_i)$. Differentiating with respect to x_i, we obtain

$$\frac{dI(\mathbf{w}, x_i)}{dx_i} = \frac{\partial f}{\partial x_i} + \frac{\partial f}{\partial \mathbf{w}} \frac{d\mathbf{w}}{dx_i} - \boldsymbol{\lambda}^T \left(\frac{\partial \mathbf{R}}{\partial \mathbf{w}} \frac{d\mathbf{w}}{dx_i} - \frac{\partial \mathbf{R}}{\partial x_i} \right). \tag{4.41}$$

Now, if $\boldsymbol{\lambda}^T$ is chosen to satisfy the adjoint equation

$$\frac{\partial F}{\partial \mathbf{w}} - \boldsymbol{\lambda}^T \frac{\partial \mathbf{R}}{\partial \mathbf{w}} = 0 \tag{4.42}$$

then, $\frac{dI}{dx_i} = \frac{\partial f}{\partial x_i} + \boldsymbol{\lambda}^T \frac{\partial \mathbf{R}}{\partial x_i}$. It can be seen that (4.42) is similar to the equation derived earlier in (4.39) for the adjoint vector $\mathbf{v}$. Hence, the final form of the equations are the same irrespective of whether the starting point is the Lagrange multiplier approach or the duality viewpoint.

To illustrate some of the issues involved in applying the discrete adjoint approach, consider the problem defined earlier in Section 4.3.1 involving sensitivity analysis of a linear structural system subjected to static loads. Let the sensitivity of a stress constraint be of interest, that is, $f(\mathbf{x}) = \sigma(\mathbf{x}) - \sigma_{\max}$. Here, the adjoint equation can be written as

$$\mathbf{K}(\mathbf{x}^o)^T \mathbf{v} = \mathbf{g} \tag{4.43}$$

where $\mathbf{v}$ is an adjoint vector and $\mathbf{g}^T = \partial \sigma(\mathbf{x})/\partial \mathbf{w}$ denotes the gradient of the stress with respect to the displacement vector. Hence, the adjoint equation essentially involves computation of the structural response to the pseudoload vector $\mathbf{g}$. To compute the derivative of $f(\mathbf{x})$, (4.43) is first solved for $\mathbf{v}$ and the result is substituted into

$$\frac{df(\mathbf{x})}{dx_i} = \frac{\partial \sigma}{\partial x_i} + \mathbf{v}^T \mathbf{f}. \tag{4.44}$$

It may be noted that the discrete adjoint approach yields the same result as that obtained using the direct sensitivity analysis method. However, computation of the right-hand side of (4.43) requires access to the details of the finite element formulation since $\mathbf{g}$ is a function of the strain-displacement matrix. This makes implementation of the discrete adjoint approach difficult, especially when the source code of the finite element formulation is not available. An alternative approach based on the continuous adjoint method which circumvents this problem can be found in Akgün et al. (2001b).

4.4.2 Continuous Adjoint Formulation

The continuous adjoint formulation deals directly with the continuous form of the governing equations. To illustrate, consider a parameterized elliptic PDE of the form $\mathcal{L}(\mathbf{x})u(\mathbf{x}) = b(\mathbf{x})$, subject to the boundary condition

$$\mathcal{B}(\mathbf{x})u(\mathbf{x}) = c(\mathbf{x}) \tag{4.45}$$

where $\mathcal{L}$ and $\mathcal{B}$ are the domain and boundary operators, respectively. $b(\mathbf{x})$ and $c(\mathbf{x})$ are prescribed functions and $u(\mathbf{x})$ is the field variable. $\mathbf{x}$ is the set of independent design variables (not to be confused with the spatial coordinates). The governing PDE is to be solved over the bounded domain Ω with boundary $\partial\Omega$.

Let the sensitivity of the output functional of interest be written as[5]

$$\frac{\mathrm{d}y}{\mathrm{d}x_i} = \int_\Omega gu \, \mathrm{d}\Omega. \tag{4.46}$$

The functional (g, u) can also be written as (v, f), where v is the solution of the adjoint PDE, $\mathcal{L}^*v = g$, with additional adjoint boundary conditions. The adjoint operator $\mathcal{L}^*$ is defined by the identity $(v, \mathcal{L}u) = (\mathcal{L}^*v, u)$ which must hold for all functions v and u satisfying the boundary condition (4.45). In a similar fashion to the process used earlier in (4.40), the equivalence between the original and the adjoint versions of the problem can be shown as follows:

$$(v, f) = (v, \mathcal{L}u) = (\mathcal{L}^*v, u) = (g, u). \tag{4.47}$$

The basic idea of the continuous adjoint approach is to derive an adjoint PDE that governs the adjoint variables and to identify a suitable set of boundary conditions. Irrespective of the nature of the governing equations, the adjoint PDE always turns out to be linear. The adjoint PDE can then be discretized and solved. In principle, different numerical algorithms can be employed for the solution of the original PDE and its adjoint.

The adjoint PDEs can also be derived using the notion of duality, which is the natural extension of the notion of duality presented earlier in the context of the discrete adjoint formulation. To illustrate the steps involved in deriving an adjoint PDE, consider the one-dimensional convection-diffusion equation used by Giles and Pierce (2000) to illustrate the continuous adjoint approach:

$$\frac{\mathrm{d}u}{\mathrm{d}\xi} - c\frac{\mathrm{d}^2u}{\mathrm{d}\xi^2} = 0, \quad 0 < \xi < 1, \tag{4.48}$$

subject to the boundary conditions, $u(0) = u(1) = 0$. Using integration by parts, the inner product (v, f) can be simplified as

$$
\begin{aligned}
(v, \mathcal{L}u) &= \int_0^1 v\left(\frac{\mathrm{d}u}{\mathrm{d}\xi} - c\frac{\mathrm{d}^2u}{\mathrm{d}\xi^2}\right) \mathrm{d}\xi \\
&= \int_0^1 u\left(-\frac{\mathrm{d}u}{\mathrm{d}\xi} - c\frac{\mathrm{d}^2u}{\mathrm{d}\xi^2}\right) \mathrm{d}\xi + \left[vu - cv\frac{\mathrm{d}u}{\mathrm{d}\xi} + cu\frac{\mathrm{d}v}{\mathrm{d}\xi}\right]_0^1 \\
&= \int_0^1 u\left(-\frac{\mathrm{d}u}{\mathrm{d}\xi} - c\frac{\mathrm{d}^2u}{\mathrm{d}\xi^2}\right) \mathrm{d}\xi + \left[cv\frac{\mathrm{d}u}{\mathrm{d}\xi}\right]_0^1.
\end{aligned} \tag{4.49}
$$

---

[5]For example, if the output function is $\int_{\mathrm{d}\Omega} u^2 \, \mathrm{d}\Omega$, then $\mathrm{d}y/\mathrm{d}x_i = \int_{\mathrm{d}\Omega} gu \, \mathrm{d}\Omega$, where $g = 2\,\mathrm{d}u/\mathrm{d}x_i$.

It may be seen from (4.47) that the inner product $(v, \mathcal{L}u)$ must equal $(\mathcal{L}^*v, u)$. A careful examination of (4.49) suggests that for this relationship to hold true the adjoint operator needs to be defined as

$$\mathcal{L}^* = -\frac{d}{d\xi} - c\frac{d^2}{d\xi^2}. \tag{4.50}$$

Further, the adjoint boundary conditions need to be $v(0) = v(1) = 0$ in order to eliminate the boundary term in (4.49). It can be seen from the definition of the adjoint differential operator in (4.50) that there is a reversal in the direction of convection. This is a typical feature of many adjoint operators that arises because of sign changes when integrating by parts. For time-dependent PDEs, this implies a reversal of causality; see Giles and Pierce (2000).

Application of the continuous adjoint approach can be very cumbersome, particularly when dealing with three-dimensional turbulent flows. A major problem arises in the incorporation of boundary conditions on the adjoint variables since they are often nonphysical and nonintuitive. Overviews of progress made in the application of the continuous adjoint method to CFD can be found in Jameson (2001) and Jameson and Vassberg (2001). For an overview of progress made in application of adjoint methods to turbulent flows, see Le Moigne and Qin (2004) and the references therein.

4.4.3   Implementation Aspects

A number of researchers have applied the discrete adjoint approach to CFD problems, which involves solution of the following system of nonlinear differential equations:

$$\frac{d\mathbf{w}}{dt} + \mathbf{R}(\mathbf{x}, \mathbf{w}) = 0. \tag{4.51}$$

Let the sensitivity of an output function $y = f(\mathbf{x}, \mathbf{w})$ be of interest. Then, it follows that $dy/dx_i = \mathbf{g}^T \partial\mathbf{f}/\partial\mathbf{w} + \partial f/\partial x_i$, where $\mathbf{g}^T = d\mathbf{w}/dx_i$ and $\mathbf{f} = -\partial\mathbf{R}/\partial x_i$. Following the direct approach discussed earlier, $d\mathbf{w}/dx_i$ can be calculated by solving the following system of linear differential equations

$$\frac{d\mathbf{w}}{dt} + \mathbf{L}\frac{d\mathbf{w}}{dx_i} = \mathbf{f} \tag{4.52}$$

where $\mathbf{L} = \partial\mathbf{R}/\partial\mathbf{w}$. The adjoint counterpart of the above equation can be written as

$$-\frac{d\mathbf{v}}{dt} + \mathbf{L}^T\mathbf{v} = \mathbf{g}. \tag{4.53}$$

Note the reversal of time in the adjoint of the unsteady equation – in other words, (4.53) needs to be solved backward in time. The terminal conditions for the adjoint equation can be readily established once the solution of (4.51) has been computed. Further, since $\mathbf{L}$ and $\mathbf{L}^T$ have the same eigenvalues, the dynamics of the systems given by (4.52) and (4.53) are similar. As a consequence, the rate of convergence to the steady state (if it exists) will be similar when the same temporal discretization scheme is used to march these equations in time. Further, the convergence rate of (4.52) and (4.53) will be equal to the asymptotic convergence rate of the nonlinear flow solver, since $\mathbf{L}$ is the Jacobian of $\mathbf{R}(\mathbf{w})$.

The source term $\mathbf{f}$ in (4.52) can be computed using either automatic differentiation (AD) software (to be discussed later in Section 4.6) or the complex variable approach. In order to solve (4.53), the product term $\mathbf{L}^T\mathbf{v} = (\partial\mathbf{R}/\partial\mathbf{w})^T\mathbf{v}$ needs to be efficiently computed. In general, $\partial\mathbf{R}/\partial\mathbf{w}$ will be a large sparse matrix, and hence, to conserve computer memory, it is desirable to avoid storing it. A more efficient approach is to directly compute the product of this matrix with the vector $\mathbf{v}$. For the case when the Euler equations are approximated on an unstructured grid, the residual vector $\mathbf{R}(\mathbf{w})$ can be written as the sum of contributions from each edge of the grid, with each edge contributing only to the residuals at either end of the edge (Giles and Pierce 2000). In other words,

$$\mathbf{R}(\mathbf{w}) = \sum_e \mathbf{R}_e(\mathbf{w}). \qquad (4.54)$$

Linearization of (4.54) gives

$$\mathbf{L}\mathbf{w} = \sum_e \mathbf{L}_e\mathbf{w}, \quad \mathbf{L}_e = \frac{\partial\mathbf{R}_e}{\partial\mathbf{w}} \qquad (4.55)$$

where $\mathbf{L}_e$ is a sparse matrix whose only nonzero elements have row and column numbers matching one or other of the two nodes at either end of the edge. Therefore, $\mathbf{L}^T\mathbf{v} = \sum_e \mathbf{L}_e^T\mathbf{v}$.

As shown by Giles and Pierce (2000), in order to reduce memory requirements, the local products of the form $\mathbf{L}_e^T\mathbf{v}$ can be computed directly instead of first computing $\mathbf{L}_e$ and taking its product with $\mathbf{v}$. In principle, the matrix $\mathbf{L}_e$ can be computed using a direct approach. However, this process can be very cumbersome for the characteristic smoothing fluxes of the Euler equations and for the viscous fluxes in the Navier–Stokes equations. Hence, it has been suggested in the literature that it may be preferable in practice to employ instead AD tools or the complex variable approach. Note here that AD or the complex variable approach is only applied to evaluate the sensitivity of $\mathbf{R}_e(\mathbf{w})$ and not the flow field vector $\mathbf{w}$. In the case of the AD approach, the resulting code can be used to directly compute the product $\mathbf{L}_e^T\mathbf{w}$ in the direct mode or the term $\mathbf{L}_e^T\mathbf{v}$ in the reverse mode. For a detailed exposition of the issues involved in computing this term efficiently in the context of CFD using the Euler and Navier–Stokes equations, readers are referred to Giles et al. (2003).

An entirely different set of issues arises in the application of continuous adjoint methods. Recall that in the continuous approach the continuous adjoint PDE is derived first, which then needs to be discretized and solved to arrive at the sensitivities. The first obvious point is that as the mesh is refined the derivative estimates must converge to the correct value. Secondly, it is important that the finite-dimensional approximation for the function value and its gradient, evaluated by discretizing the adjoint problem, should be consistent. In theory, it is possible to discretize the adjoint PDE independent of the original PDE. Asymptotically, as the mesh is refined, the function value and its derivative should be consistent. However, for a given mesh, it is possible that because of inconsistencies between the function value and its derivatives, optimization algorithms may terminate at an erroneous stationary point. Another problem in the continuous adjoint approach arises when the derivatives of a so-called inadmissible function are to be computed (Arian and Salas 1999). Loosely speaking, an objective or constraint function is said to be inadmissible if it is not straightforwardly possible to identify the appropriate adjoint problem. A detailed analysis of continuous adjoint approach can be found in Lewis (1997).

In recent years, there has been a growing consensus that the discrete adjoint approach may be computationally more advantageous than the continuous formulation. A major factor in favor of the discrete approach is that it is much more straightforward to implement. It is worth remembering that the adjoint equations (and the boundary conditions) are a function of $f(\mathbf{x}, \mathbf{w})$, which could be either an objective function to be minimized or a constraint to be satisfied in a design problem. In other words, the adjoint problem is problem specific. This can be a limiting factor in practice. Typically, researchers who actively apply adjoint methods to aerodynamic design implement a library of adjoint formulations for the more commonly used objective and constraint functions encountered in practice (Campobasso et al. 2003).

4.5 Semianalytical Methods

Since the programming effort required to implement a direct or adjoint sensitivity analysis solver can be significant, resort is often made to semianalytical methods. Loosely speaking, semianalytical methods involve a hybridization of finite-difference (or complex variable) approximations with direct/adjoint methods (Adelman and Haftka 1986). To illustrate this concept, consider the case of static structural analysis, where the following equation is used to compute the derivatives of the displacements:

$$\mathbf{K}\frac{d\mathbf{w}}{dx_i} = -\frac{\partial \mathbf{K}}{\partial x_i}\mathbf{w} + \frac{\partial \mathbf{f}}{\partial x_i}. \tag{4.56}$$

It can be seen from this equation that the derivatives can be exactly computed only if the sensitivity of the stiffness matrix can be exactly calculated. For certain finite element formulations, this is readily possible provided the design parameterization is not overly complex, for example, when the design variables are the thicknesses of plates or shells, cross-sectional areas of truss members, and so on. However, in some cases, in order to reduce the programming effort, $\frac{\partial \mathbf{K}}{\partial x_i}$ can be approximated by finite differences or the complex variable method. This approach significantly reduces the programming effort, particularly when shape variables are involved (Van Keulen et al. in press).

4.6 Automatic Differentiation

As emphasized in previous sections, in most complex engineering design problems, the objective and constraint functions are computed using a computer code written in a high-level programming language such as Fortran, C, C++ or Java. Automatic differentiation is an enabling approach for sensitivity analysis when the *complete* source code of the simulator (including all libraries) is available. This approach is also known as computational differentiation or algorithmic differentiation in the literature, to indicate the fact that, in practice, manual intervention may often be required. The origins of AD can be traced back to the work of Beda (1959) in the former Soviet Union and Wengert (1964) in the United States.

The basic premise of the AD approach is that a computer program implementing a numerical algorithm can be decomposed into a *long* sequence of a limited set of elementary arithmetic operations such as additions, subtractions, multiplications, and calls to intrinsic functions (for example, sin(), cos(), exp(), etc.). AD involves the application of the chain rule of differential calculus to the original source code to create a new computer program that is capable of exactly computing the sensitivities of the outputs of interest with respect to the design variables. The main assumption invoked here is that the elemental functions are continuously differentiable on their open domains – more on this point will be presented in the subsections that follow.

It is worth noting here that AD is different from traditional symbolic differentiation that is applied to functions represented by a (typically compact) expression. The symbolic approach aims to arrive at an explicit expression for the derivative as a function of the input variables. This is a major limitation since the final expressions for the derivatives can be very lengthy and cumbersome to manage for most nontrivial functions with more than say three variables.

The attractive features of AD are that it potentially requires minimal human effort and that its computational efficiency can be greater than finite-difference approximations under a range of circumstances. Further, AD tools can be applied, in principle, to exactly compute the higher-order derivatives. In what follows, an overview of the forward and reverse mode of AD is presented and their relative merits are outlined. Much of the theoretical discussion in this section follows Griewank (1989).

4.6.1 Forward Mode

The first step in the forward mode of differentiation involves decomposing the computer program into elementary operations by introducing intermediate quantities. Given this decomposition, AD accumulates the derivatives of the elementary operations according to the chain rule to give the derivatives of the function. At the end of the sequence of computations, the augmented computer program computes both the function value and its derivatives.

Consider a function $y = f(\mathbf{x}) : \mathbb{R}^p \to \mathbb{R}$, which is the composition[6] of a finite set of *elementary* or *library* functions f_i that are continuous on their open domains. Let t_j be intermediate variables such that $t_1, t_2, \ldots, t_p$ correspond to the original input variables $x_1, x_2, \ldots, x_p$ and the last intermediate variable t_q is set to f. Further, let the first p elementary functions be $f_i = x_i$ for $i = 1, 2, \ldots, p$. A computer program for evaluating $f(x)$, which is the composition of $q - p$ elementary functions, can be written as follows.

> **Original Program**
> FOR $i = p + 1, p + 2, \ldots, q$
> $\quad t_i = f_i(t_j), \quad j \in \mathcal{J}_i$
> END FOR
> $y = t_q$

Each elementary function f_i depends on the already computed intermediate quantities, t_j, with j belonging to the index sets $\mathcal{J}_i \subset \{1, 2, \ldots, i - 1\}$ for $i = p + 1, p + 2, \ldots, q$.

To illustrate how the index sets are computed, it is instructive to visualize the original program in the form of a *graph*. Figure 4.2 shows the graph of an algorithm used to compute the product of four independent variables x_1, x_2, x_3, x_4. The graph representation of algorithms enables a rapid visualization of the dependencies between the intermediate variables. Since we assumed that the first-order derivatives of the elementary functions exist (i.e., $\nabla f_i = \partial f_i / \partial t_j$), application of the chain rule to the original computer program results in the following derivative code:

---

[6]Almost all scalar functions of practical interest can be written in this factorable form (Jackson and McCormick 1988). Hence, this representation is sufficiently general to represent any user-defined subroutine or computer program.

Forward Mode AD
FOR $i = 1, 2, \ldots, p$
$\quad \nabla t_i = e_i$
$\quad$ FOR $i = p + 1, p + 2, \ldots, q$
$\quad\quad t_i = f_i(t_j) \quad j \in \mathcal{J}_i$
$\quad\quad \nabla t_i = \sum_{j \in \mathcal{J}_i} \frac{\partial f_i}{\partial t_j} \nabla t_j$
$\quad$ END FOR
END FOR
$y = t_q$
$g = \nabla t_q$

where $e_i = (0, \ldots, 0, 1, 0, \ldots, 0)$ denotes the ith Cartesian basis vector in $\mathbb{R}^p$.

In summary, the forward mode starts out with the known derivatives $\nabla t_i, i = 1, 2, \ldots, p$. Subsequently, the derivatives of the intermediate variables $t_i, i = p + 1, p + 2, \ldots, q$ are propagated through the algorithm until $\partial y / \partial x$ is reached when the evaluation of y is completed. When the full Jacobian $\nabla f(\mathbf{x})$ is desired, ∇t_i should be set to be equal to the ith unit vector, that is, $\nabla t_i = e_i$. Hence, for a problem with p design variables, all derivative objects will be p-dimensional vectors. This will lead to a roughly n-fold increase in computational cost and memory requirements compared to evaluation of $f(\mathbf{x})$ alone.

In a number of applications, the full Jacobian may not be of direct interest. Rather, it may be necessary to compute the product of the Jacobian with a given vector $\mathbf{u} \in \mathbb{R}^p$, that is, the directional derivative of f in direction $\mathbf{u}$. Recall that this problem was discussed earlier in the context of the discrete adjoint approach (Section 4.4.3), where products of the form $(\partial \mathbf{R} / \partial \mathbf{w})^T \mathbf{u}$ were required. It can be easily established that if we set $\nabla t_i = u_i$, where $\mathbf{u} \in \mathbb{R}^p$ is a given vector, then the differentiated code will compute the inner product between $\nabla f(\mathbf{x})$ and $\mathbf{u}$. In this case, the derivative objectives will be scalar, which will reduce the computational cost to a modest multiple of that required to compute the function value. This approach will be more efficient than first computing $\nabla f(\mathbf{x})$ and taking its inner product with $\mathbf{u}$.

Assuming the time required for fetches and stores from and to memory is negligible, the computational effort required to evaluate $f(\mathbf{x})$ on a serial machine can be approximated as $\text{Work}(f) = \sum_{i=p+1}^{q} \text{Work}(f_i)$. Now, the computational effort involved in the direct AD mode is given by $\text{Work}(f, \nabla f) = \sum_{i=p+1}^{q} [\text{Work}(f_i, \nabla f_i) + pp_i(\text{mults} + \text{adds})]$, where the additional pp_i arithmetic operations arise from the steps involved in calculating ∇x_i.

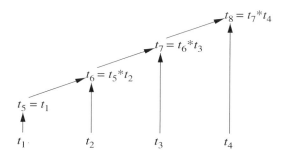

Figure 4.2 Graph of algorithm for computing the product $x_1 x_2 x_3 x_4$.

Clearly, the computational effort required for computing the derivatives using the forward mode increases linearly with the number of input variables (p). Hence, the computational cost of the forward mode is of the same order of magnitude as the divided (forward or backward) difference approach. One way to reduce the number of operations would be to exploit the fact that ∇t_j is a sparse vector. However, it is generally felt that the resulting increase in implementation complexity may not be justifiable, except for some special cases. It can also be noted that the differentiated code will require extra memory to store the derivatives of the intermediate variables. One way to reduce this memory requirement would be to run through the basic loop p times, each run evaluating the derivative only with respect to one independent variable. This approach would require only twice the memory of the original code; however, this reduction is achieved at the expense of increased computational effort. An alternative approach, referred to as the reverse mode, which is computationally more efficient than the forward mode, is described next.

4.6.2 Reverse Mode

The reverse mode calculates the function value first before beginning a backward pass to compute the derivatives. This mode is also based on the chain rule and is mathematically similar to the adjoint approach presented earlier in Section 4.4. In short, the reverse mode of AD computes the derivative of the final output with respect to an intermediate quantity, which can be thought of as an adjoint quantity. In order to propagate adjoints, the program flow needs to be reversed and, further, any intermediate quantity that affects the final output needs to be stored or recomputed. Recall that the forward mode of AD associates a derivative object ∇t_i with every intermediate variable t_i. In contrast, the reverse mode of AD associates with each intermediate variable t_i the scalar derivative $\bar{t}_i = \frac{\partial t_q}{\partial t_i}$.

Note that $\bar{t}_i$ is the partial derivative of the final output with respect to the ith intermediate variable t_i. The new set of variables $\bar{t}_i$ can be interpreted as adjoint quantities similar to those encountered earlier in the context of adjoint sensitivity analysis. Now, by definition, we have $\bar{t}_q = 1$ and $\frac{\partial f(x)}{\partial t_i} = \bar{t}_i$, for $i = 1, 2, \ldots, p$. Application of the chain rule yields

$$\bar{t}_j = \sum_{i \in \mathcal{I}_j} \frac{\partial f_i}{\partial t_j} \bar{t}_i,$$

where $\mathcal{I}_j = \{i \le q : j \in \mathcal{J}_i\}$. It can be seen that $\bar{t}_j$ can be computed directly once the values of $\bar{t}_i$ for $i > j$ are known. Applying this idea to the computer program described earlier, the following augmented code results. If the vector $\bar{g}$ is set to zero and $\gamma = 1$, then the output g is the gradient vector ∇f. Otherwise, the output becomes $g = \bar{g} + \gamma \nabla f$.

Reverse Mode AD
FOR $i = p + 1, p + 2, \ldots, q$
    $t_i = f_i(t_j), \quad j \in \mathcal{J}_i$ **Forward Sweep**
    $\bar{t}_i = 0$
END FOR
$y = t_q \quad , \quad \bar{t}_q = \gamma$
$\{\bar{t}_i\}_{i=1}^p = \bar{g}$
FOR $i = q, q - 1, \ldots, p + 1$
    $\bar{t}_j = \bar{t}_j + \frac{\partial f_i}{\partial t_j} \bar{t}_i$ for all $j \in \mathcal{J}_i$ **Reverse Sweep**
END FOR
$g = \{\bar{t}_i\}_{i=1}^p$

Under the assumption that the computational cost of evaluating every elementary function and its derivative (f_i and ∇f_i) is independent of p, it has been shown that *"The evaluation of a gradient by reverse accumulation never requires more than five times the effort of evaluating the underlying function by itself (Griewank 1989)"*. In other words, the theoretical upper bound on the computational cost of the reverse mode of AD is five times the computer time required for one function evaluation. Studies in the literature suggest that this upper bound is rather pessimistic – in practice, much higher computational efficiency can be achieved. Since the computational cost of the reverse mode does not depend on the number of independent variables, it is preferable to the forward mode. The drawback of the reverse mode is that it is necessary to store all the intermediate variables as well as the computational graph of the complete algorithm in order to execute the reverse sweep. As a consequence, the memory requirements increase dramatically – in CFD applied to flow over complex three-dimensional geometries, the memory requirements may well be prohibitive. However, with careful manual tuning, it is possible in some cases to significantly reduce the memory requirements; see Section 4.10.1 for a brief study that illustrates that in some cases the efficiency and memory requirements of reverse mode AD can be comparable to hand-coded adjoint codes.

4.6.3 AD Software Tools and Implementation Aspects

Comprehensive lists of AD software can be found on-line at various sites.[7] To use these software tools, the user has only to specify the independent and dependent variables and pass the set of routines that are invoked to compute the dependent variables, given the independent variables. Let the original source code be denoted by $y = SC(\mathbf{x})$, where $\mathbf{x}$ and y denote the independent and dependent variables, respectively. Then the AD preprocessor generates a new source code of the form $[y, \nabla y] = SC'(\mathbf{x})$. The modified source code must be compiled and linked to appropriate AD libraries to create an executable. This is capable of furnishing the function value and its derivatives at any given point of interest.

The two main approaches for implementing AD are (1) source code transformation and (2) derived data types and operator overloading. In the first approach, the original source code is processed with a parser and a (very much) larger code generated by including additional lines for calculating derivatives. AD packages implementing this approach include ADIFOR, TAMC, DAFOR, GRESS, Odysse and PADRE2 for Fortran and ADIC for C/C++. By contrast, in the second approach, AD is implemented by creating a new data structure containing both the function value and its derivative. The notion of overloading is then leveraged to redefine the existing operators so that the function value and its derivatives can be simultaneously computed. For instance, the multiplication operation $x * y$ can be overloaded to perform both $x * y$ and $dx * y + x * dy$. In languages such as C and C++, this involves simply replacing a standard data type such as float with dfloat, which is a new derivative data type. Clearly, this approach leads to minimal changes to the structure and flow of the original source code compared to the source code transformation approach. Most existing software tools for AD make use of the former approach, however. Packages that use the derived data-type approach include ADO1, ADOL-F, IMAS and OPTIMA90 for Fortran, and ADOL-C for C/C++ codes.

---

[7]See `http://www-unix.mcs.anl.gov/autodiff/AD_Tools/index.html#contents` or `http://www.autodiff.org` for details of the AD codes mentioned here.

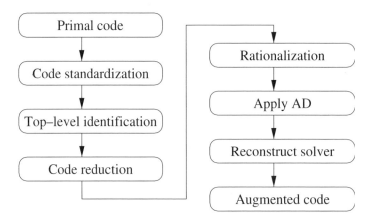

Figure 4.3   Steps involved in preparing code for automatic differentiation (Cusdin and Müller 2003).

A number of case studies that discuss in detail the issues involved in applying AD to various problems can be found in the literature. Cusdin and Müller (2003) presented an excellent overview of the steps involved in applying a variety of AD tools to CFD codes. Figure 4.3 shows their suggested process chart. The first task in most applications is to restructure the original code to enforce Fortran 77 or ANSI C standards prior to invoking AD tools. The second step in Figure 4.3 involves identifying the top-level routine, that is, the routine from which all functions that affect the dependent variables are called. In the third step of code reduction, portions of the code that conduct pre- and postprocessing via manipulation of input and output data files are identified. These routines are not submitted to the AD software. The code is hence restructured such that all the input data (apart from the independent variables) are read outside the functions to be differentiated. Similar changes are made to the postprocessing routines.

The rationalization step is primarily concerned with replacing noncontinuous flow constructs such as goto and continue with if and do statements. Recall the underlying assumption of AD that the function to be differentiated is composed of elemental functions that are continuously differentiable on their open domains. For computer codes containing subroutines that satisfy this condition, AD tools will work without any manual intervention. However, it may so happen that at a point on the boundary of an open domain, the function may be continuous but its gradient may jump to a finite value or even infinity. This becomes an important issue when the original code contains branches, some kink functions (e.g., abs) or inverse functions (e.g., sqrt, arctan, or arccos). Care needs to be exercised to handle such cases correctly.

AD tools such as ADIFOR allow the detection of possible nondifferentiable points by generating the derivative code with *exception handling* enabled. Tadjouddine et al. (2005) discuss a case where it was necessary to rewrite some of the code (portions involving the use of operators such as arccos, abs, sin and tan and certain branching constructs) to handle the nondifferentiability issue. Note that the rewritten code was equivalent to the original code and calculates the same function values; however, differentiability of the functions was guaranteed around the vicinity of their arguments. Once the issue of nondifferentiability is taken care of, via manual intervention if necessary, the final derivative code can be generated.

This involves defining the independent and dependent variables, which allows the AD tool to parse the original computer code to generate an augmented derivative code, at least in principle.

Prior to using the derivative code, validation studies should be carried out by comparing the results with finite-difference approximations. If the results obtained are found to be satisfactory, attention can then be focused on improving the computational efficiency and reducing memory requirements. In most complex application problems, it may be possible to significantly improve the generated code via manual tuning. This step becomes a necessity when the derivative code requires more memory than is available to the analyst on his/her computer. In the satellite structure dynamics application discussed in the case studies, it was found that naive application of ADIFOR 3.0 to the LAPACK routine zgesv (a solver for complex linear algebraic equations) resulted in very inefficient code and, further, the results were inconsistent with finite-difference approximations. Hence, it was decided to hand-code the derivative of this function. This led to a derivative code using reverse mode AD which was around seven times faster than the forward finite-difference approximation.

In summary, restructuring/rewriting of the original code and manual tuning of the derivative code are often required when AD tools are applied to sensitivity analysis of complex computer codes. An example illustrating the computational gains to be achieved via tuning the primal and the derivative code will be presented later in Section 4.10.1, while the application of ADIFOR to a structural dynamics problem is described in the first case study. For more detailed discussion of the theoretical aspects of AD, the interested reader is referred to the book by Griewank (2000) and a recent conference proceedings edited by Corliss et al. (2001).

4.7   Mesh Sensitivities for Complex Geometries

In many practical design problems, the geometry of the artifact under consideration must often be manipulated to meet design objectives. In order to compute the sensitivities of the objective and constraint functions when using discretized PDE solvers, the analyst has to evaluate the derivative of the mesh (coordinates) with respect to the design variables. To illustrate how these terms arise, let us write the discrete mathematical model describing flow over an airfoil as follows:

$$\mathbf{R}(\mathbf{x}, \mathcal{M}(\mathbf{x}), \mathbf{w}) = 0, \tag{4.57}$$

where $\mathbf{x} \in \mathbb{R}^p$ denotes a set of independent design variables, $\mathcal{M}(\mathbf{x})$ denotes the coordinates of the mesh used to discretize the computational domain, and $\mathbf{w} \in \mathbb{R}^n$ is the vector of state variables.

Application of the chain rule to (4.57) results in

$$\frac{\partial \mathbf{R}}{\partial \mathbf{w}} \frac{d\mathbf{w}}{dx_i} + \frac{\partial \mathbf{R}}{\partial \mathcal{M}} \frac{\partial \mathcal{M}}{\partial x_i} + \frac{\partial \mathbf{R}}{\partial x_i} = 0. \tag{4.58}$$

It can be clearly seen that to solve for $\frac{d\mathbf{w}}{dx_i}$ it is necessary to compute the mesh sensitivity term $\frac{\partial \mathcal{M}}{\partial x_i}$, that is,

$$\frac{\partial \mathcal{M}}{\partial x_i} = \frac{\partial \mathcal{M}}{\partial \mathcal{M}_s} \frac{\partial \mathcal{M}_s}{\partial \boldsymbol{\xi}} \frac{\partial \boldsymbol{\xi}}{\partial x_i}, \tag{4.59}$$

where $\frac{\partial \mathcal{M}}{\partial \mathcal{M}_s}$ is the sensitivity of the mesh coordinates with respect to the surface mesh, $\frac{\partial \mathcal{M}_s}{\partial \boldsymbol{\xi}}$ is the sensitivity of the surface mesh with respect to the shape vectors of the airfoil, and $\frac{\partial \boldsymbol{\xi}}{\partial x_i}$ is the sensitivity of the shape vectors with respect to the design variables.

In practice, it is difficult to compute the terms in (4.59) since many off-the-shelf mesh generators and geometry modeling tools do not have sensitivity analysis capabilities. For certain two-dimensional airfoil geometry representations, it is possible to apply direct methods to compute the mesh sensitivities (Sadrehaghighi et al. 1995). Alternatively, if the source code for the mesh generator and geometry modeling software is available, the complex variable or AD approaches described earlier can be applied to compute the mesh sensitivities.

Samareh (2001a) proposed an approach that simplifies the process of computing the mesh sensitivities in the context of aerodynamic design. This approach employs a free-form deformation algorithm similar to those used in the motion picture industry for digital animation. The basic idea is to parameterize the *changes* in the geometry rather than the geometry itself. For example, a B-spline net can be used to describe the changes in the baseline geometry. The net is placed around the baseline mesh and the changes at each point on the surface are obtained by interpolation from the changes in the B-spline net. The control points in the net may be directly used as the design variables, or they may be grouped further into design variables such as camber, thickness and twist. The main advantage of this parameterization approach is that it is very flexible in scope and, further, meshes developed in the preliminary design stages for the purpose of analysis can be readily used in subsequent stages. It can, however, lead to modified geometries that can no longer be described by the original elements in the initial CAD model.

Given such a parameterization strategy, the computational mesh can be deformed when the surface geometry (driven by the design variables) is changed during the course of the design process. For the case of inviscid flows, mesh deformation can be carried out using a spring analogy. Here, the edges of the mesh are treated as tension springs and the following equilibrium equation is solved using a Jacobi iteration procedure:

$$\sum_j K_{ij}(\Delta z_i - \Delta z_j) = 0, \tag{4.60}$$

where Δz_i and Δz_j denote the changes in the coordinates of nodes i and j from the baseline mesh to the deformed mesh due to design variable perturbations, respectively. K_{ij} are the spring stiffnesses, which are calculated as $K_{ij} = 1/l_{ij}^2$, where l_{ij} is the length of the edge connecting node i to node j.

To circumvent the possibility of mesh entanglement, the design perturbations are applied in an incremental fashion analogous to standard procedures for large deflection analysis of structures subjected to static loads (Hughes 1987). Studies in the literature suggest that this approach works well for Euler-based designs, provided the initial mesh does not contain stretched or deformed cells. For viscous flows with boundary layers, the approach is not adequate and can easily lead to crossing mesh lines and negative volumes. To alleviate these problems, the nodes near the viscous surfaces can be shifted by interpolating the changes in the coordinates at the boundaries of the nearest surface triangles or edge. It may also be necessary to introduce a smoothing procedure to ensure that away from the surface the mesh movement reverts to that of the procedure used for inviscid flows. More detailed expositions of mesh deformation strategies for viscous flows can be found in the literature, for example, see Nielsen (1998).

Studies conducted in the literature suggest that mesh deformation strategies work well for most two-dimensional problems. However, the fundamental limitation of this approach is that a baseline mesh is morphed to accommodate the perturbed geometry – no additional mesh points are inserted or removed. Hence, the mesh quality may be poor when the baseline geometry is changed radically. This problem becomes more critical when mesh deformation algorithms are applied to three-dimensional geometries. One way around this is to periodically update the baseline mesh during design iterations. Development of more reliable mesh deformation strategies continues to be an area of active research. There is also a wealth of literature on shape sensitivity analysis in the context of structural optimization (Choi and Kim 2005a,b).

4.8   Sensitivity of Optima to Problem Parameters

In this section, we consider the problem of computing the sensitivities of an optimum solution with respect to problem parameters appearing in the optimization formulation (Sobieszczanksi-Sobieski et al. 1982). To illustrate, consider the nonlinear programming problem:

$$\begin{aligned} \textbf{Minimize}: \quad & f(\mathbf{x}, \theta) \\ \textbf{Subject to}: \quad & g_i(\mathbf{x}, \theta) \leq 0, \quad i = 1, 2, \ldots, m \end{aligned} \tag{4.61}$$

where $f(\mathbf{x}, \theta)$ and $g_i(\mathbf{x}, \theta)$ denote the objective and constraint functions, respectively. $\mathbf{x} \in \mathbb{R}^p$ is a vector of design variables and θ is a (fixed) parameter appearing within the objective and constraint functions (also referred to as a problem parameter). Let $\mathbf{x}^*$ denote the optimum solution of (4.61) and let $f^* := f(\mathbf{x}^*, \theta)$ denote the value of the objective function at this point.

Here, we are interested in calculating the derivatives of the optimum solution $\mathbf{x}^*$ and the corresponding objective function value f^* with respect to the problem parameter θ. For example, in a structural optimization problem, θ may denote the loading with respect to which the structural members are sized. These sensitivity terms will be useful for estimating new optima that may result from perturbing θ. As we shall see later in Chapter 9, the optimum sensitivities are also required in the implementation of some bilevel architectures for multidisciplinary design optimization.

The optimum solution $\mathbf{x}^*$ satisfies the Karush–Kuhn–Tucker (KKT) conditions

$$\nabla f(\mathbf{x}^*, \theta) + \sum_{i \in \mathcal{J}} \lambda_i g_i(\mathbf{x}^*, \theta) = 0 \tag{4.62}$$

and

$$g_i(\mathbf{x}^*, \theta) = 0, \quad \lambda_i > 0, \quad i \in \mathcal{J}, \tag{4.63}$$

where λ_i denote the Lagrange multipliers (which are an implicit function of θ) and $\mathcal{J}$ contains the set of all active constraints. Let m' denote the total number of active constraints, that is, $m' = \dim(\mathcal{J}) \leq m$. Suppose that for small perturbations in θ the set of active constraints do not change, that is, the KKT conditions remain valid. Then, differentiating (4.62) and

(4.63) with respect to θ gives

$$\sum_{k=1}^{p} \left[\frac{\partial^2 f}{\partial x_i \partial x_k} + \sum_{j \in \mathcal{J}} \lambda_j \frac{\partial^2 g_i}{\partial x_i \partial x_k} \right] \frac{\partial x_k^*}{\partial \theta} + \sum_{j \in \mathcal{J}} \frac{\partial \lambda_j}{\partial \theta} \frac{\partial g_j}{\partial x_i}$$

$$+ \frac{\partial^2 f}{\partial x_i \partial \theta} + \sum_{j \in \mathcal{J}} \lambda_j \frac{\partial^2 g_j}{\partial x_i \partial \theta} = 0 \tag{4.64}$$

and

$$\frac{\partial g_j}{\partial \theta} + \sum_{i=1}^{p} \frac{\partial g_j}{\partial x_i} \frac{\partial x_i^*}{\partial \theta} = 0, \tag{4.65}$$

where $i = 1, 2, \ldots, p$ and $j \in \mathcal{J}$.

Rearranging (4.64) and (4.65) in matrix form, we have

$$\begin{bmatrix} \mathbf{P} & \mathbf{Q} \\ \mathbf{Q}^T & \mathbf{O} \end{bmatrix} \begin{bmatrix} \dfrac{d\mathbf{x}^*}{d\theta} \\[2ex] \dfrac{\partial \boldsymbol{\lambda}}{\partial \theta} \end{bmatrix} = \begin{bmatrix} \mathbf{p} \\ \mathbf{q} \end{bmatrix}, \tag{4.66}$$

where $\mathbf{P} \in \mathbb{R}^{p \times p}, \mathbf{Q} \in \mathbb{R}^{p \times m'}, \mathbf{p} \in \mathbb{R}^p, \mathbf{q} \in \mathbb{R}^{m'}, \partial \mathbf{x}^*/\partial \theta = \{\partial x_1^*/\partial \theta, \partial x_2^*/\partial \theta, \ldots, \partial x_p^*/\partial \theta\}^T$ $\in \mathbb{R}^p$ and $\partial \boldsymbol{\lambda}/\partial \theta = \{\partial \lambda_1/\partial \theta, \partial \lambda_2/\partial \theta, \ldots, \partial \lambda_{m'}/\partial \theta\}^T \in \mathbb{R}^{m'}$. $\mathbf{O} \in \mathbb{R}^{n \times m'}$ is a matrix of zeros. The elements of $\mathbf{P}, \mathbf{Q}, \mathbf{p}$ and $\mathbf{q}$ are given by

$$P_{ij} = \frac{\partial^2 f}{\partial x_i \partial x_j} + \sum_{k \in \mathcal{J}} \lambda_k \frac{\partial^2 g_k}{\partial x_i \partial x_j}, \quad Q_{ij} = \frac{\partial g_j}{\partial x_i}, \quad j \in \mathcal{J}, \tag{4.67}$$

$$p_i = \frac{\partial^2 f}{\partial x_i \partial \theta} + \sum_{k \in \mathcal{J}} \lambda_k \frac{\partial^2 g_k}{\partial x_i \partial \theta} \quad \text{and} \quad q_i = \frac{\partial g_i}{\partial \theta}, \quad i \in \mathcal{J}. \tag{4.68}$$

Solution of the $(p + m') \times (p + m')$ matrix system of equations (4.66) gives the sensitivity of $\mathbf{x}^*$ with respect to θ. Clearly, to ensure a unique solution, the coefficient matrix of (4.66) must be nonsingular. The sensitivity of f^* can subsequently be calculated as

$$\frac{df^*}{d\theta} = \frac{\partial f^*}{\partial \theta} + \sum_{i=1}^{p} \frac{\partial f^*}{\partial x_i^*} \frac{\partial x_i^*}{\partial \theta}. \tag{4.69}$$

When the sensitivities with respect to multiple problem parameters are of interest, only the right-hand side of (4.66) needs to be updated. Given the factored form of the coefficient matrix, sensitivities with respect to many parameters can be efficiently computed provided p and m' are not too large. It may also be noted from (4.67) and (4.68) that the second-order derivatives of the objective and constraints are required to calculate $\mathbf{P}$ and $\mathbf{p}$. Hence, sensitivity analysis of the optimum is feasible only if these terms are easy to compute. Rao (1996) discuss an alternative formulation that does not involve second-order derivatives; however, this involves interpreting θ as an additional optimization variable, and hence, this is computationally attractive only when the derivatives with respect to a few problem parameters are of interest.

4.9 Sensitivity Analysis of Coupled Systems

The various approaches outlined so far in this chapter are applicable to sensitivity analysis problems encountered in single-discipline studies. In this section, we examine how the techniques discussed earlier can be applied to multidisciplinary sensitivity analysis following on from the ideas discussed in Section 2.3. By way of example, consider a coupled system involving two disciplines (or subsystems), whose governing equations can be written as

$$\mathbf{R}_1(\mathbf{x}, \mathbf{w}_1, \mathbf{w}_2) = 0 \tag{4.70}$$

and

$$\mathbf{R}_2(\mathbf{x}, \mathbf{w}_2, \mathbf{w}_1) = 0 \tag{4.71}$$

where $\mathbf{x}$ denotes the total set of design variables for both disciplines, and $\mathbf{w}_1$ and $\mathbf{w}_2$ denote the state variables for disciplines 1 and 2, respectively.

The governing equations considered here are typical of those arising from integrated aerodynamics-structures analysis. Because of the nature of the coupling between (4.70) and (4.71), an iterative scheme is required for carrying out multidisciplinary analysis; see Sections 2.3.2 and 9.1.1. If such a coupled solver is available, then the approaches outlined in this chapter can be readily applied. Even so, care must be exercised since incomplete convergence of the iterations may lead to highly inaccurate results when finite-difference methods are used; see the discussion in Section 4.1.

In many scenarios when a tightly coupled solver is not available, it becomes more practical (as well as computationally more efficient) to synthesize the multidisciplinary sensitivities from the disciplinary sensitivities. The problem of computing multidisciplinary (system) sensitivities from disciplinary (subsystem) sensitivities was extensively studied by Sobieszczanski-Sobieski (1990a,b), who presented an approach based on global sensitivity equations (GSEs). This approach has subsequently been applied by a number of researchers to complex problems; see, for example, Giunta (2000), Maute et al. (2001), and Martins et al. (2005).

To illustrate the application of GSEs to multidisciplinary sensitivity analysis, let the sensitivities of the output function $y = f(\mathbf{x}, \mathbf{w}_1, \mathbf{w}_2)$ be of interest. The derivative of this function with respect to a design variable x_i can be written as

$$\frac{dy}{dx_i} = \frac{\partial f}{\partial x_i} + \frac{\partial f}{\partial \mathbf{w}_1} \frac{d\mathbf{w}_1}{dx_i} + \frac{\partial f}{\partial \mathbf{w}_2} \frac{d\mathbf{w}_2}{dx_i}. \tag{4.72}$$

Differentiating (4.70) and (4.71) with respect to x_i gives

$$\frac{\partial \mathbf{R}_1}{\partial \mathbf{w}_1} \frac{d\mathbf{w}_1}{dx_i} + \frac{\partial \mathbf{R}_1}{\partial \mathbf{w}_2} \frac{d\mathbf{w}_2}{dx_i} = -\frac{\partial \mathbf{R}_1}{\partial x_i} \tag{4.73}$$

$$\frac{\partial \mathbf{R}_2}{\partial \mathbf{w}_1} \frac{d\mathbf{w}_1}{dx_i} + \frac{\partial \mathbf{R}_2}{\partial \mathbf{w}_2} \frac{d\mathbf{w}_2}{dx_i} = -\frac{\partial \mathbf{R}_2}{\partial x_i}. \tag{4.74}$$

becomes

$$\begin{bmatrix} \dfrac{\partial \mathbf{R}_1}{\partial \mathbf{w}_1} & \dfrac{\partial \mathbf{R}_1}{\partial \mathbf{w}_2} \\[2ex] \dfrac{\partial \mathbf{R}_2}{\partial \mathbf{w}_1} & \dfrac{\partial \mathbf{R}_2}{\partial \mathbf{w}_2} \end{bmatrix} \begin{bmatrix} \dfrac{d\mathbf{w}_1}{dx_i} \\[2ex] \dfrac{d\mathbf{w}_2}{dx_i} \end{bmatrix} = - \begin{bmatrix} \dfrac{\partial \mathbf{R}_1}{\partial x_i} \\[2ex] \dfrac{\partial \mathbf{R}_2}{\partial x_i} \end{bmatrix}. \tag{4.75}$$

Sobieszczanski-Sobieski (1990b) also presented an alternative approach, referred to as GSE2, in which instead of directly differentiating the governing equations in (4.70) and

(4.71), the state variable is differentiated. We first rewrite (4.70) and (4.71) as follows: $\mathbf{w}_1 = \mathcal{R}_1(\mathbf{x}, \mathbf{w}_2)$ and $\mathbf{w}_2 = \mathcal{R}_2(\mathbf{x}, \mathbf{w}_1)$. Differentiating these versions of (4.70) and (4.71) with respect to x_i gives

$$\frac{d\mathbf{w}_1}{dx_i} = \frac{\partial \mathcal{R}_1}{\partial x_i} + \frac{\partial \mathcal{R}_1}{\partial \mathbf{w}_2}\frac{d\mathbf{w}_2}{dx_i} \tag{4.76}$$

and

$$\frac{d\mathbf{w}_2}{dx_i} = \frac{\partial \mathcal{R}_2}{\partial x_i} + \frac{\partial \mathcal{R}_2}{\partial \mathbf{w}_1}\frac{d\mathbf{w}_1}{dx_i}. \tag{4.77}$$

The preceding equations can be compactly written in matrix form as

$$\begin{bmatrix} \mathbf{I} & -\dfrac{\partial \mathcal{R}_1}{\partial \mathbf{w}_2} \\ -\dfrac{\partial \mathcal{R}_2}{\partial \mathbf{w}_1} & \mathbf{I} \end{bmatrix} \begin{bmatrix} \dfrac{d\mathbf{w}_1}{dx_i} \\ \dfrac{d\mathbf{w}_2}{dx_i} \end{bmatrix} = \begin{bmatrix} \dfrac{\partial \mathcal{R}_1}{\partial x_i} \\ \dfrac{\partial \mathcal{R}_2}{\partial x_i} \end{bmatrix}. \tag{4.78}$$

Note that the derivatives of $\mathcal{R}_1$ and $\mathcal{R}_2$ appearing in the coefficient matrix and the right hand side are local derivatives.

It worth noting that for some problems GSE2 may be more efficient than GSE1, since the dimension of the coefficient matrix of (4.78) may be smaller than that of (4.75). This is true for aerodynamics-structures interaction problems where only the surface aerodynamic pressures affect the structural analysis and only the surface structural deformations affect the aerodynamic analysis. Hence, the matrices $\partial \mathcal{R}_1/\partial \mathbf{w}_2$ and $\partial \mathcal{R}_2/\partial \mathbf{w}_1$ turn out to be very sparse. Note that this observation holds true only for coupled systems where the state variables of one discipline depend only on a subset of the state variables of the other discipline. In the limit when the systems are fully coupled, the computational cost incurred by GSE1 and GSE2 are similar.

Martins (2002) proposed adjoint versions of the GSE approaches. For example, the adjoint version of GSE1 can be written as

$$\begin{bmatrix} \dfrac{\partial \mathbf{R}_1}{\partial \mathbf{w}_1} & \dfrac{\partial \mathbf{R}_1}{\partial \mathbf{w}_2} \\ \dfrac{\partial \mathbf{R}_2}{\partial \mathbf{w}_1} & \dfrac{\partial \mathbf{R}_2}{\partial \mathbf{w}_2} \end{bmatrix}^T \begin{bmatrix} \lambda_1 \\ \lambda_2 \end{bmatrix} = -\begin{bmatrix} \dfrac{\partial f}{\partial \mathbf{w}_1} \\ \dfrac{\partial f}{\partial \mathbf{w}_2} \end{bmatrix}, \tag{4.79}$$

where λ_1 and λ_2 are appropriately defined adjoint vectors corresponding to both subsystems. Similarly, the adjoint version of GSE2 can be written as

$$\begin{bmatrix} \mathbf{I} & -\dfrac{\partial \mathcal{R}_1}{\partial \mathbf{w}_2} \\ -\dfrac{\partial \mathcal{R}_2}{\partial \mathbf{w}_1} & \mathbf{I} \end{bmatrix}^T \begin{bmatrix} \lambda_1 \\ \lambda_2 \end{bmatrix} = \begin{bmatrix} \dfrac{\partial f}{\partial \mathbf{w}_1} \\ \dfrac{\partial f}{\partial \mathbf{w}_2} \end{bmatrix}. \tag{4.80}$$

Note that because of the significant size of the coefficient matrix of (4.79) in many real-world applications, the computational cost of multidisciplinary sensitivity analysis may be prohibitive. Martins (2002) proposed a *lagged coupled adjoint method* for alleviating this computational obstacle. The basic idea is to first rewrite (4.79) as

$$\begin{bmatrix} \dfrac{\partial \mathbf{R}_1}{\partial \mathbf{w}_1}^T & \dfrac{\partial \mathbf{R}_2}{\partial \mathbf{w}_1}^T \\ \dfrac{\partial \mathbf{R}_1}{\partial \mathbf{w}_2}^T & \dfrac{\partial \mathbf{R}_2}{\partial \mathbf{w}_2}^T \end{bmatrix} \begin{bmatrix} \lambda_1 \\ \lambda_2 \end{bmatrix} = -\begin{bmatrix} \dfrac{\partial f}{\partial \mathbf{w}_1} \\ \dfrac{\partial f}{\partial \mathbf{w}_2} \end{bmatrix}. \tag{4.81}$$

In the lagged coupled adjoint method, the preceding equation is solved using a fixed point iterative scheme as follows:

$$\left[\frac{\partial \mathbf{R}_1}{\partial \mathbf{w}_1}\right]^T \boldsymbol{\lambda}_1^{k+1} = \frac{\partial f}{\partial \mathbf{w}_1} - \frac{\partial \mathbf{R}_2}{\partial \mathbf{w}_1}^T \boldsymbol{\lambda}_2^k, \tag{4.82}$$

$$\left[\frac{\partial \mathbf{R}_2}{\partial \mathbf{w}_2}\right]^T \boldsymbol{\lambda}_2^{k+1} = \frac{\partial f}{\partial \mathbf{w}_2} - \frac{\partial \mathbf{R}_1}{\partial \mathbf{w}_2}^T \boldsymbol{\lambda}_1^k \tag{4.83}$$

where k denotes the iteration index. Note that here the adjoint vector in one discipline is calculated using the adjoint vector of the other discipline, obtained at the previous iteration. The main advantage of this scheme is that the coefficient matrices to be inverted are much smaller compared to (4.79). If the fixed point iterations converge, the final result should match those obtained by solving (4.79) directly. For further details of the adjoint approach to global sensitivity analysis, see Martins et al. (2005).

4.10 Comparison of Sensitivity Analysis Techniques

Given the variety of approaches available for sensitivity analysis, it is important to conduct a careful examination of the problem under consideration before deciding on the most appropriate one to adopt. In general, if the source code of the simulator is not available, the analyst has to resort to finite-difference approximations. Finite differencing is, however, the method of last resort since the analyst has to experiment with various values of the step size in order to estimate the derivatives with reasonable accuracy. Moreover, the best accuracy that can be achieved if any noise is present may be rather limited in precision. This may influence the choice of any search engine to those that work well with limited precision gradients.

Most structural analysis programs that offer sensitivity analysis capabilities use the direct method, implemented with semianalytical approximations. Direct methods are usually preferred in structural problems since the number of output functions (for example, stresses at various elements) often exceeds the number of design variables. Examples of commercial structural analysis tools that have a sensitivity analysis capability include Nastran®[8] and Genesis®.[9] Semianalytical methods are primarily used because of their ease of implementation, since no element-level sensitivities are required. In comparison, significant programming effort is required to implement the analytical direct sensitivity approach. Johnson (1997) discusses the motivation for implementing adjoint methods in Nastran and also presents some structural dynamics case studies.

In a number of instances, it is possible to develop a sensitivity analysis capability even when the source of the structural analysis software is not available. In the case of commercial FEA packages for structural analysis which allow the use of user subroutines and/or access to the internal databases, it may be possible to implement direct and adjoint methods. For example, the DMAP language in Nastran allows for the manipulation of stiffness matrices, stresses and displacements by the user without access to the source code. The basic idea is to compute the term $-\frac{\partial \mathbf{K}}{\partial x_i}\mathbf{w}(\mathbf{x}^o) + \frac{\partial \mathbf{f}}{\partial x_i}$ after baseline analysis and subsequently employ this term as a pseudoload vector to compute the derivative of the displacement vector; see

---

[8]http://www.mscsoftware.com/products/
[9]http://www.vrand.com/Genesis.html

(4.19). A detailed exposition of the issues involved in incorporating a sensitivity analysis capability within commercial FEA software is presented by Choi and Duan (2000).

If the analysis source code is available, then the user has a much larger number of options for sensitivity analysis. Before deciding on the appropriate technique for the problem at hand, the key points worth considering are: (1) the ratio of the number of design variables to the number of output quantities whose sensitivities are required, (2) programming effort, (3) the computational cost and (4) memory requirements.

If the number of design variables is greater than the number of output quantities (the total number of objective and constraint functions), then the adjoint approach becomes attractive, otherwise the direct approach is preferable. In general, hand-coded direct and adjoint formulations can be very efficient compared to other methods for most problems of practical interest. The only major drawback of this approach is the significant amount of programming effort required. The analyst implementing the sensitivity analysis algorithms also needs to have a good understanding of the governing equations and the numerical scheme employed for its solution. Finally, hand-written codes are prone to errors, especially as the problem complexity increases.

Automatic differentiation is a suitable choice for applications where high accuracy is required and it is not possible to invest significant staff effort in programming. This is in contrast to direct and adjoint sensitivity analysis techniques, which require significant manual effort and which are prone to computer bugs. Further, AD gives exact results for the derivatives. On the downside, the automatically generated code may be difficult to maintain and can require significant computer memory to run. In practice, manual tuning and restructuring of the original as well as the derivative code may be required to optimize the performance.

The memory requirements of the complex variable approach are independent of the number of design variables. In contrast, when AD is used, the memory requirements scale linearly with respect to the number of design variables. This characteristic can lead to significant difficulties for problems with large numbers of design variables.

It is generally argued that the complex variable approach results in more maintainable code compared to AD. In the case of AD, when new design variables are added, the original source code needs to be reprocessed to generate a modified source code that includes the additional design variables. Further, since the modified source is automatically generated by a parser, it may not be easy to read and modify. However, with the emergence of AD tools employing derived data types, this problem may disappear in the near future. A detailed review of the options available for structural sensitivity analysis is given in Van Keulen et al. (in press).

Two brief case studies are presented next to illustrate the relative merits of the various approaches to sensitivity analysis described in this chapter.

4.10.1   Case Study: Aerodynamic Sensitivity Analysis

Cusdin and Müller (2003) conducted a detailed study on two CFD codes to compare a variety of AD tools with the complex variable approach and hand-coded sensitivity analysis. The first CFD code is the 2D Euler solver Eusoldo, which uses a finite volume technique on unstructured grids. The second code is the node-centered edge-based Navier–Stokes code Hydra for which a linearized and adjoint version has been derived using hand-written code. Hydra was initially developed by researchers at the Rolls-Royce University Technology

Table 4.1   Comparison of performance of various sensitivity analysis techniques applied to Eusoldo. Time is given for the primal in seconds and for the sensitivity codes as a factor of the primal. The light figures are based on an unmodified primal, and bold ones refer to run times based on a primal code tuned for improved AD performance. Hand sensitivity code is not modified. Reproduced from (Cusdin and Müller 2003) by permission of J.D. Müller and P. Cusdin

	Itanium 2	Xeon	Athlon
Primal	58.2 (**52.9**)	49.9 (**47.2**)	65.2 (**64.8**)
Hand	2.43	2.33	2.24
Complex	10.78 (**10.49**)	12.93 (**11.39**)	9.78 (**8.09**)
ADIFOR 2.0	(**2.23**)	(**2.52**)	(**2.39**)
TAF fwd	4.56 (**2.22**)	2.99 (**2.23**)	2.86 (**2.10**)
Tap. fwd	1.87 (**1.96**)	2.58 (**2.32**)	2.17 (**2.17**)
Hand	2.23	2.65	2.35
TAF rev	8.16 (**2.16**)	4.09 (**2.92**)	3.77 (**2.65**)
Tap. rev	11.55 (**8.28**)	12.02 (**3.79**)	8.82 (**3.14**)

Table 4.2   Comparison of performance of various sensitivity analysis techniques applied to Hydra. Reproduced from (Cusdin and Müller 2003) by permission of J.D. Müller and P. Cusdin

	Itanium 2	Xeon	Athlon
Primal	134.2 (**124.8**)	315.0 (**310.4**)	521.8 (**516.4**)
Hand	1.87	1.20	1.13
ADIFOR 2.0	3.72 (**3.41**)	1.71 (**1.62**)	1.68 (**1.66**)
TAF fwd	3.79 (**3.37**)	1.25 (**1.21**)	1.18 (**1.17**)
Tap. fwd	3.71 (**1.91**)	2.14 (**2.10**)	1.89 (**1.87**)
Hand	1.77	1.10	1.16
TAF rev	3.43 (**3.35**)	1.25 (**1.21**)	1.26 (**1.22**)
Tap. rev	20.96 (**3.49**)	11.64 (**1.99**)	10.16 (**1.62**)

Center in CFD at the University of Oxford on behalf of Rolls-Royce Plc. The code continues to be actively developed by Rolls-Royce, working with its University Technology Centers, for a wide range of aerospace, marine and industrial applications. The four AD tools used in this study were ADIFOR 2.0, Tapenade, TAMC and TAF.[10]

Tables 4.1 and 4.2 summarize the performance of the various sensitivity analysis techniques used in the study on various computer architectures. The timings shown in bold typeface are those obtained after manual tuning of the primal and the derivative code. It can be observed that the results obtained go against the preconceived notion that codes generated

---

[10]See `http://www-unix.mcs.anl.gov/autodiff/AD\_Tools/index.html\#contents` or `http://www.autodiff.org` for details of the AD codes mentioned here.

Table 4.3 Memory performance of the primal, tangent and adjoint flux functions in Hydra. Memory is given in MB for the primal code and for the sensitivity code as a factor of the primal. The light figures refer to the unmodified demand and the bold figures refer to the code tuned for CPU performance. Code generated by reverse mode Tapenade shows a considerable improvement. Reproduced from (Cusdin and Müller 2003) by permission of J.D. Müller and P. Cusdin

	Linear	**Adjoint**
Primal	343	343
Hand	1.27 (**1.27**)	1.27 (**1.27**)
ADIFOR	1.62 (**1.62**)	
TAF	1.27 (**1.27**)	1.27 (**1.27**)
Tapenade	1.62 (**1.62**)	1.91 (**1.62**)

by AD will be inferior to a hand-written sensitivity code. In particular, for Eusoldo, the performance of the tuned TAF derivative code (in reverse mode) is faster than the hand-written code on the Itanium 2 processor. Similarly, the performance of the tuned derivative code obtained using Tapenade is faster than the hand-written sensitivity code. For the case of Eusoldo, it was found that the efficiency of the complex variable approach is poor compared to other techniques.

Cusdin and Müller (2003) also compared the memory requirements of various sensitivity analysis techniques – see Table 4.3 for a summary. It can be seen that TAF-generated linear and adjoint code requires 27% more memory than the primal, which is the same as the hand-written sensitivity code. In comparison, the complex variable approach effectively doubles the memory requirements.

4.10.2 Case Study: Aerostructural Sensitivity Analysis

This section summarizes a case study conducted on the aerostructural design of a transonic jet by Martins (2002) to compare various sensitivity analysis techniques. Figure 4.4 shows the aircraft geometry, the flow solution and its associated mesh, and the primary structure inside the wing. This is a multidisciplinary coupled problem that is solved within an aerostructural design framework consisting of an aerodynamic analysis and design module (which includes a geometry engine and a mesh perturbation algorithm), a linear finite element structural solver, an aerostructural coupling procedure together with various postprocessing tools.

The aerodynamic analysis and design module used SYN107-MB, which is a parallel multiblock solver for the Euler and the Reynolds-averaged Navier–Stokes equations. This solver also includes a sensitivity analysis capability based on adjoint methods (Alonso et al. 2003). Structural analysis is carried out using the FESMEH code developed by Holden (1999). Multidisciplinary coupling is handled using a linearly consistent and conservative approach. A centralized database containing jig shape, pressure distributions and displacements is employed to facilitate data transfer between the CFD and structural analyses.

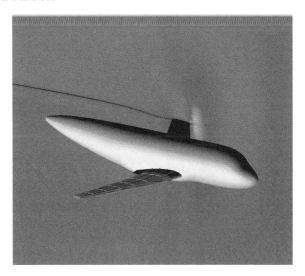

Figure 4.4   Aerostructural wing problem. Reproduced by permission of J.R.R.A. Martins.

Table 4.4   Comparison of the accuracy and computational requirements of various sensitivity analysis methods applied to the aerostructural problem (Martins 2002)

Method	Sample Sensitivity	Wall Time	Memory
Complex	-30.0497600	1.00	1.00
ADIFOR	-30.0497600	2.33	8.09
Analytic	-30.0497600	0.58	2.42
FD	-30.0497243	0.88	0.72

The wing structure model shown in Figure 4.4 is constructed using a wing box with four spars evenly distributed from 15 to 80% of the chord. Ribs are distributed evenly along the span at every eighth of the semispan. A total of 640 finite elements were used in the structural model with the appropriate thicknesses of the spar caps, shear webs, and skins being chosen on the basis of the expected loads for this design.

A comparison of the complex variable method and AD with analytical and finite-difference approaches is presented in Table 4.4. It can be seen that all methods except the finite-difference approach give very accurate results and that even the FD scheme is accurate to six significant figures. In terms of CPU time, the analytical approach is the most efficient; however, a significant quantity of human input is required to implement it. The complex variable approach and AD are comparatively easier to set up. For the problem under consideration, the complex variable approach seems to provide a good trade-off between implementation effort and efficiency if high accuracy gradients are required. It can also be noted that for this problem the complex variable approach is more efficient than ADIFOR. Of course, such efficiency ratios are problem dependent.

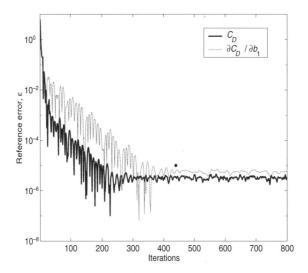

Figure 4.5 Convergence of drag coefficient (C_D) and its first derivative ($\partial C_D / \partial x_i$). Reproduced by permission of J.R.R.A. Martins.

Figure 4.5 depicts the rate of convergence of the drag coefficient (C_D) and its first-order derivative. It can be seen that the convergence of the derivative lags behind C_D. A comparison of the accuracies of the forward finite difference and complex-step approximations as a function of the step size reveals that using the complex variable approach high accuracy can be achieved with step sizes smaller than 10^{-5}. In comparison, finite-difference approximations are much less accurate, even when an optimum step size is used.

5

General Approximation Concepts and Surrogates

The computational cost associated with the use of high-fidelity simulation models poses a serious impediment to the successful application of formal optimization algorithms in engineering design. Even though advances in computing hardware and numerical algorithms have reduced costs by orders of magnitude over the last few decades, the fidelity with which engineers desire to model engineering systems has also increased considerably. For example, in the 1970s, aerodynamic design was routinely carried out using potential flow methods, which on modern computers take only a few minutes to run. However, over the last decade, much focus has been placed on leveraging Navier–Stokes solvers along with turbulence models in the design process. Each evaluation of such high-fidelity models may take significant compute time for complex geometries.

In many engineering design problems, thousands of function evaluations may be required to locate a near-optimal solution. Therefore, when expensive high-fidelity simulation models are used for predicting design improvements, the naive application of optimization algorithms can lead to exorbitant computational requirements. In the 1970s, Schmidt and his coworkers (Schmit and Farshi 1974; Schmit and Miura 1976) pioneered the concept of *sequential approximate optimization* to improve the efficiency of structural optimization. The basic idea was to analyze an initial design to generate data that could be used to construct approximations of the objective and constraint functions. The optimization algorithm is then applied to minimize this approximate model subject to the approximated constraints. Appropriate move limits[1] were applied to account for the fact that the approximations may be valid only in the neighborhood of the initial design. After solving this approximate optimization problem, the models are updated and the process is continued until suitable convergence criteria are met.

Subsequently, more general approximation techniques have been developed for improving the efficiency of the design process in areas other than structural analysis. It is now increasingly commonplace to employ computationally cheap approximation models in lieu

---

[1]Move limits can be interpreted as the magnitude of permissible modifications to an initial design.

Computational Approaches for Aerospace Design: The Pursuit of Excellence.   A. J. Keane and P. B. Nair
© 2005 John Wiley & Sons, Ltd

of exact calculations to design complex systems. Another motivation for employing approximation models in design arises from the fact that in many practical cases commercial-off-the-shelf analysis codes are used instead of in-house codes. This can lead to significant programming complexities in interfacing analysis codes that were originally intended to work in stand-alone mode and for which the source code is not available to the optimization routines. By constructing approximation models of the analysis codes, programming complexities can be significantly reduced by linking the optimization routines to the approximation models in lieu of the original codes. It is also worth noting that approximation techniques have applications in design space exploration, visualization, screening, nondeterministic analysis and code debugging, that is, any application where the analysis model has to be evaluated repeatedly for different input values.

This chapter describes a range of approaches for building approximation models of high-fidelity analysis codes. Here, the high-fidelity analysis model is represented by the functional relationship $y = f(\mathbf{x})$, where $\mathbf{x} \in \mathbb{R}^p$ is the vector of inputs to the simulation code and y is a scalar output.[2] The objective of approximation is to construct a model $\widehat{y} = \widehat{f}(\mathbf{x}, \boldsymbol{\alpha}) \approx f(\mathbf{x})$, that is computationally cheaper to evaluate than the high-fidelity code. $\boldsymbol{\alpha}$ is a vector of undetermined parameters, which is estimated either by employing a black-box or a physics-based approach. In what follows, we shall interchangeably refer to approximation models as *surrogates* or *metamodels*.

There are a wide range of techniques for constructing models from observational data; see, for example, the texts by Vapnik (1998), Bishop (1995), Schölkopf and Smola (2001) and Hastie et al. (2001) for excellent expositions of the problem of learning functional relationships from data. This chapter introduces a selection of techniques that appear to be promising for applications in engineering design. Before delving into algorithmic aspects of surrogate modeling techniques, we first review the steps involved in black-box modeling from a classical statistics point of view. Subsequently, we present an overview of black-box modeling from the perspective of statistical learning theory. Finally, we revisit an ever important question that often arises in practice – should we use interpolation or regression techniques to construct surrogate models when the observational data is generated using computer models?

In black-box approaches, the intention is to build a surrogate without using any domain-specific knowledge of the analysis code, that is, the code is considered to be a computational module that cannot be intrusively modified. A typical black-box approach involves running the analysis code at a number of preselected inputs to generate a set of input–output data.[3] Subsequently, a surrogate model is trained to learn the input–output mapping by minimizing an appropriate loss function. This nonintrusive approach to surrogate construction has a number of practical advantages as well as theoretical limitations, which are examined in later sections of this chapter.

In contrast, physics-based approaches exploit to some degree either the continuous or discrete form of the governing equations solved by the analysis code. As a consequence, physics-based approaches may require intrusive modifications to the analysis code, since generality is sacrificed for the sake of improved efficiency and accuracy – the next chapter considers such domain-specific approximation concepts in detail.

---

[2]In general, y may contain a vector of outputs of interest such as the objective and constraint functions in a design problem. However, for simplicity of presentation, the primary focus of this chapter is on the multiple input–single output case.

[3]Note that we will refer to the input–output dataset interchangeably as the training dataset or observational data.

5.1   Local Approximations

We begin by considering classical local approximations before moving onto more general schemes. The term "local" is used here to refer to approximations that give accurate predictions only in the vicinity of the design point around which they are built. Perturbation series are one example of a local approximation; for example, the scalar equation $y = (x + \delta x)^{-1}$ can be approximated by the binomial series

$$\widehat{y} = \left[1 - (\delta x/x) + (\delta x/x)^2 - (\delta x/x)^3 + \cdots \right] x^{-1}. \tag{5.1}$$

Clearly, the above series will converge only when $|\delta x/x| < 1$.

A matrix generalization of the scalar binomial series for the equation $\mathbf{Y} = (\mathbf{X} + \Delta\mathbf{X})^{-1}$, where $\mathbf{Y}, \mathbf{X}, \Delta\mathbf{X} \in \mathbb{R}^{n \times n}$, can be written as

$$\widehat{\mathbf{Y}} = (1 - (\mathbf{X}^{-1}\Delta\mathbf{X}) + (\mathbf{X}^{-1}\Delta\mathbf{X})^2 - (\mathbf{X}^{-1}\Delta\mathbf{X})^3 + \cdots)\mathbf{X}^{-1}. \tag{5.2}$$

The series in (5.2) is useful in the context of static structural reanalysis, stochastic mechanics and matrix preconditioning; we shall cover some of these applications in more detail in the next chapter. Note that (5.2) is also referred to as the Neumann series in the numerical analysis literature. For this series, it is straightforward to establish that convergence can be guaranteed only when $||\mathbf{X}^{-1}\Delta\mathbf{X}|| < 1$ or the spectral radius of the matrix $\mathbf{X}^{-1}\Delta\mathbf{X}$ is less than one. This limited radius of convergence is a feature shared by most local approximation techniques and because of which they are useful only for small perturbations.

Perturbation methods can be applied to a wide class of problems such as reanalysis of structural systems subjected to static and dynamic loads (Stewart and Sun 1990). For example, given the solution of the linear algebraic system of equations $\mathbf{A}\mathbf{w} = \mathbf{f}$, perturbation methods can be applied to approximate the solution to the perturbed equation $(\mathbf{A} + \Delta\mathbf{A})(\mathbf{w} + \Delta\mathbf{w}) = \mathbf{f} + \Delta\mathbf{f}$. An account of such methods for static and eigenvalue reanalysis will be presented in the next chapter. We continue our discussion of local methods here by examining the Taylor series approximation and showing how its accuracy can be improved through the use of intervening variables.

5.1.1   Taylor Series Approximation

Taylor series approximation of response quantities typically involves the use of a model of the form

$$\widehat{y}(\mathbf{x}) = f(\mathbf{x}^0) + \sum_{i=1}^{p}(x_i - x_i^0)\frac{\partial f(\mathbf{x}^0)}{\partial \mathbf{x}_i} + \frac{1}{2}\sum_{i=1}^{p}\sum_{j=1}^{p}(x_i - x_i^0)(x_j - x_j^0)\frac{\partial^2 f(\mathbf{x}^0)}{\partial x_i \partial x_j} + \cdots,$$

$$\tag{5.3}$$

where $\mathbf{x} = \{x_1, x_2, \ldots, x_p\}$ denotes the design variables and $\mathbf{x}^0 = \{x_1^0, x_2^0, \ldots, x_p^0\}$ is the point around which the Taylor series is expanded.

Using sensitivity analysis techniques such as those described in Chapter 4, the linear form of this approximation model can often be constructed very easily. However, computation of the second-order derivatives tends to be expensive and so these terms are rarely used in practice.

Given a Taylor series approximation, the response quantities of interest can be rapidly approximated when the input variables are perturbed. The major drawback of the Taylor series is that it has a small radius of convergence, similar to the perturbation series discussed earlier. As a consequence, the approximations can be very poor for large perturbations in the design variables (Storaasli and Sobieszczanski-Sobieski 1974). We next outline two approaches for improving the accuracy of the Taylor series approximation: intervening variables and multipoint approximations.

5.1.2 Intervening Variables

Early efforts in the field of structural optimization focused on developing alternative strategies for improving the accuracy of the Taylor series. One powerful concept in this area is the notion of intervening variables, proposed by Schmidt and coworkers (Mills-Curran et al. 1983; Schmit and Farshi 1974; Schmit and Miura 1976). To illustrate this concept, consider the case of approximating the stresses within a truss structure subjected to static loads, as a function of member cross-sectional areas (A). A standard result from basic engineering mechanics is that the stress level in any member of a statically determinate structure is inversely proportional to its cross-sectional area. This linear relationship between stress and the reciprocal of A does not hold for statically indeterminate structures. Even so, numerical studies in the literature indicate that by expanding the stress levels in statically indeterminate structures using a Taylor series in terms of $1/A$ gives more accurate approximations compared to a Taylor series expansion in terms of A itself. Here, $1/A$ can be interpreted as an intervening variable that approximately linearizes the stress.

The basic idea of this approach is to replace the original design variable vector $\mathbf{x}$ by the intervening variable vector $\boldsymbol{\xi}$, such that the function to be approximated, which is originally nonlinear in $\mathbf{x}$, turns out to be linear in $\boldsymbol{\xi}$. In practice, it is usually not possible to define intervening variables that exactly linearize the function of interest. Nonetheless, numerical studies have shown that the reciprocals of sizing variables are good intervening variables for approximating the stress levels in many structures.

To illustrate the idea more generally, consider a set of m intervening variables defined as $\xi_i = \xi_i(\mathbf{x})$, $\forall i = 1, 2, \ldots, m$. The first-order Taylor series expansion of the original function $f(\mathbf{x})$ in terms of these intervening variables can then be written as

$$\widehat{y}_I = f(\mathbf{x}^0) + \sum_{i=1}^{m} (\xi_i(\mathbf{x}) - \xi_i(\mathbf{x}^0)) \left(\frac{\partial f(\mathbf{x})}{\partial \xi_i} \right)_{\xi^0} \tag{5.4}$$

where $\boldsymbol{\xi}^0 = \{\xi_1(\mathbf{x}^0), \xi_2(\mathbf{x}^0), \ldots, \xi_m(\mathbf{x}^0)\}$. For the special case of reciprocal approximation, the intervening variables become $\xi_i = 1/x_i$, $i = 1, 2, \ldots, p$. Hence, (5.4) can be rewritten in terms of the original variables as

$$\widehat{y}_R = f(\mathbf{x}^0) + \sum_{i=1}^{p} (x_i - x_i^0) \frac{x_i^0}{x_i} \left(\frac{\partial f(\mathbf{x}^0)}{\partial x_i} \right). \tag{5.5}$$

A similar idea was proposed by Canfield (1990) for eigenvalue approximation using the Rayleigh quotient. In that approach, the numerator and denominator of the Rayleigh quotient (potential energy and kinetic energy, respectively) are considered to be intervening variables, which are expanded separately using Taylor series in terms of the design variables of interest. It was shown that the intervening variable approach leads to significantly better approximations for the eigenvalues compared to the conventional Taylor series expression.

Another related notion is that of *conservative approximations* that are suitable for interior and extended interior penalty function formulations. The key idea is to hybridize the Taylor series and the reciprocal approximation so as to generate conservative approximations.[4] To illustrate this idea, let us first subtract the reciprocal approximation from the linear Taylor series approximation, which gives

$$\widehat{y} - \widehat{y}_R = \sum_{i=1}^{p} \frac{(x_i - x_i^0)^2}{x_i} \left(\frac{\partial f(\mathbf{x}^0)}{\partial x_i} \right). \tag{5.6}$$

It can be seen that the sign of each term in the sum is determined by the sign of $(\frac{\partial f(\mathbf{x}^0)}{\partial x_i})x_i$. In other words, when this term is negative, the reciprocal approximation is greater than the first-order Taylor series approximation, and vice versa when the term is positive. Now, consider the case when we need to approximate the constraint function $g(\mathbf{x}) \geq 0$. Here, to arrive at a conservative approximation, i.e., one that is smallest, we can hybridize the linear and reciprocal approximation as follows:

$$\widehat{g}(\mathbf{x}) = g(\mathbf{x}^0) + \sum_{i=1}^{p} G_i(x_i - x_i^0) \left(\frac{\partial g}{\partial x_i} \right)_{\mathbf{x}^0} \tag{5.7}$$

where $G_i = 1$, if $(\frac{\partial g(\mathbf{x})}{\partial x_i})x_i \leq 0$, otherwise $G_i = \frac{x_i^0}{x_i}$.

5.2 Multipoint Approximations

Multipoint approximation techniques were conceived in the 1990s to employ data from multiple design points to improve the range of applicability of the Taylor series. In their simplest form, the function and gradient values at two design points are used to construct the so-called two-point approximation. To illustrate the two-point method, let the function value and its gradient at the design points $\mathbf{x}^{(1)} = \{x_1^{(1)}, x_2^{(1)}, \ldots, x_p^{(1)}\}$ and $\mathbf{x}^{(2)} = \{x_1^{(2)}, x_2^{(2)}, \ldots, x_p^{(2)}\}$ be given; that is, $f(\mathbf{x}^{(1)})$, $\nabla f(\mathbf{x}^{(1)})$, $f(\mathbf{x}^{(2)})$, and $\nabla f(\mathbf{x}^{(2)})$.

Fadel et al. (1990) introduced a two-point exponential approximation (TPEA) method, which involves intervening variables in terms of exponentials, that is, a first-order Taylor series approximation of $f(\mathbf{x})$ is derived around the point $\mathbf{x}^{(2)}$ in terms of the intervening variables $\xi_i = x_i^{q_i}, i = 1, 2, \ldots, p$. The first-order Taylor series approximation in terms of these intervening variables can be written as

$$\widehat{f}(\mathbf{x}) = f(\mathbf{x}^{(2)}) + \sum_{i=1}^{p} \left(\frac{(x_i^{(2)})^{1-q_i}}{q_i} \right) \frac{\partial f(\mathbf{x}^{(2)})}{\partial x_i} \left(x_i^{q_i} - (x_i^{(2)})^{q_i} \right). \tag{5.8}$$

The exponents $q_i, i = 1, 2, \ldots, p$ are computed by matching the derivative of the above approximation with the actual derivatives at $\mathbf{x}^{(1)}$, that is, $\nabla \widehat{f}(\mathbf{x}^{(1)}) = \nabla f(\mathbf{x}^{(1)})$. After some algebraic manipulations, a closed form expression for q_i can then be written as

$$q_i = 1 + \left(\ln \left[\frac{\partial f(\mathbf{x}^{(1)})}{\partial x_i} \bigg/ \frac{\partial f(\mathbf{x}^{(2)})}{\partial x_i} \right] \bigg/ \ln(x_i^{(1)}/x_i^{(2)}) \right). \tag{5.9}$$

---

[4]The approximations are not conservative in an absolute sense. The only guarantee is that the predicted values will be conservative compared to the Taylor series and the reciprocal approximation.

The basic two-point approximation technique has evolved considerably since its conception. Notably, Wang and Grandhi (1995) developed two-point adaptive nonlinear approximations (TANA), which also allow for the possibility of including additional information to improve accuracy. In the TANA-1 formulation of Wang and Grandhi (1995), the same intervening variables as in TPEA are used. However, the approximation is assumed to have the form

$$\widehat{f}(\mathbf{x}) = f(\mathbf{x}^{(1)}) + \sum_{i=1}^{p} \left(\frac{(x_i^{(2)})^{1-q_i}}{q_i} \right) \frac{\partial f(\mathbf{x}^{(2)})}{\partial x_i} \left(x_i^{q_i} - (x_i^{(2)})^{q_i} \right) + \varepsilon_1, \qquad (5.10)$$

where ε_1 is the residue of the first-order Taylor series approximation in terms of the intervening variables ξ_i. Note that in contrast to the TPEA method the Taylor series is constructed around the point $\mathbf{x}^{(1)}$. The exponents q_i are subsequently computed by matching the derivatives of $\widehat{f}$ to the exact derivatives at the point $\mathbf{x}^{(2)}$. The correction term ε_1 can be computed in closed form by enforcing the condition $\widehat{f}(\mathbf{x}^{(1)}) = f(\mathbf{x}^{(1)})$.

In the TANA-2 formulation of Wang and Grandhi (1995), a truncated second-order Taylor series approximation around the point $\mathbf{x}^{(2)}$ is employed. Assuming that the Hessian matrix is a multiple of the identity matrix, that is, $\varepsilon_2 I \in \mathbb{R}^{p \times p}$, the TANA-2 approximation can be written as

$$\widehat{f}(\mathbf{x}) = f(\mathbf{x}^{(2)}) + \sum_{i=1}^{p} \left(\frac{(x_i^{(2)})^{1-q_i}}{q_i} \right) \frac{\partial f(\mathbf{x}^{(2)})}{\partial x_i} \left(x_i^{q_i} - (x_i^{(2)})^{q_i} \right)$$

$$+ \frac{\varepsilon_2}{2} \sum_{i=1}^{p} \left(x_i^{q_i} - (x_i^{(2)})^{q_i} \right)^2. \qquad (5.11)$$

In order to compute the undetermined constants $q_i, i = 1, 2, \ldots, p$ and ε_2, we need at least $p + 1$ equations. p equations can be obtained by matching the derivatives of the approximation with the actual values at the point $\mathbf{x}^{(1)}$, that is, $\partial \widehat{f}(\mathbf{x}^{(1)})/\partial x_i = \partial f(\mathbf{x}^{(1)})/\partial x_i$, $i = 1, 2, \ldots, p$. An additional equation can be derived by enforcing $\widehat{f}(\mathbf{x}^{(1)}) = f(\mathbf{x}^{(1)})$. Solving these $p + 1$ equations, the undetermined constants in (5.11) can be computed. Numerical studies have shown that the TANA-2 formulation gives better accuracy compared to other two-point approximations.

In order to further improve the computational efficiency of TANA-2, Xu and Grandhi (1998) proposed an alternative formulation referred to as TANA-3. Here, it is assumed that the Hessian matrix can be approximated by $\varepsilon_3(\mathbf{x})I$, where

$$\varepsilon_3(\mathbf{x}) = \frac{H}{\left[\sum_{i=1}^{p} \left(x_i^{q_i} - (x_i^{(1)})^{q_i} \right)^2 + \sum_{i=1}^{p} \left(x_i^{q_i} - (x_i^{(2)})^{q_i} \right)^2 \right]}, \qquad (5.12)$$

and where $q_i, i = 1, 2, \ldots, p$ and H are constants that can be computed by enforcing the conditions $\widehat{f}(\mathbf{x}^{(1)}) = f(\mathbf{x}^{(1)})$ and $\partial \widehat{f}(\mathbf{x}^{(1)})/\partial x_i = \partial f(\mathbf{x}^{(1)})/\partial x_i$, $i = 1, 2, \ldots, p$. After some algebra, it turns out that q_i can be computed using the expression in (5.9) and the constant H is

$$H = 2 \left[f(\mathbf{x}^{(1)}) - f(\mathbf{x}^{(2)}) - \sum_{i=1}^{p} \left(\frac{(x_i^{(2)})^{1-q_i}}{q_i} \right) \frac{\partial f(\mathbf{x}^{(2)})}{\partial x_i} \left((x_i^{(1)})^{q_i} - (x_i^{(2)})^{q_i} \right) \right]. \qquad (5.13)$$

It can be noted from the expression for q_i in (5.9) that the two-point approximation is not defined when either of the following conditions hold:

$$\frac{\partial f(\mathbf{x}^{(1)})}{\partial x_i} \bigg/ \frac{\partial f(\mathbf{x}^{(2)})}{\partial x_i} \leq 0 \quad \text{or} \quad (x_i^{(1)}/x_i^{(2)}) \leq 0. \tag{5.14}$$

In order to ensure that two-point approximations can be computed in a numerically stable fashion, special treatment is necessary when the above conditions hold. Xu and Grandhi (1998) suggested that the value of 1 or -1 should be assigned if the expression for q_i is not valid. In order to avoid numerical instabilities when $(x_i^{(1)}/x_i^{(2)}) \approx 0$, it was suggested that suitable upper and lower bounds on q_i should also be imposed. It is also possible to construct three-point approximations using procedures similar to those discussed here; see, for example, Guo et al. (2001) and the references therein.

5.3 Black-box Modeling: a Statistical Perspective

A common feature of the local approximation techniques discussed in the previous sections is that they require sensitivity information. There are a number of practical applications where such data may not be readily available to the analyst. This situation commonly arises when using software tools without a sensitivity analysis capability (and for which the source code is not available and where finite differencing is not reliable or affordable). In the remainder of this chapter, we examine more general approximation techniques that can be employed to construct global surrogate models. The possibility of using sensitivity information with these models (if available) to improve approximation quality will also be explored.

We consider again a deterministic computer code that takes as input the vector $\mathbf{x} \in \mathbb{R}^p$ and returns a scalar output $y(\mathbf{x}) \in \mathbb{R}$. Further, for a given set of n input vectors $\mathbf{X} = \{\mathbf{x}^{(1)}, \mathbf{x}^{(2)}, \ldots, \mathbf{x}^{(n)}\} \in \mathbb{R}^{p \times n}$, the corresponding output values $\mathbf{y} = \{y^{(1)}, y^{(2)}, \ldots, y^{(n)}\} \in \mathbb{R}^n$ are assumed to be available. To simplify notation, we shall also denote the observational data by the set of vectors, $\mathbf{z}_i, i = 1, 2, \ldots, n$, where the data vector $\mathbf{z}$ is defined as $\mathbf{z} = [\mathbf{x}, y]$. This training data is often obtained in practice by applying a design of experiments technique (we cover this topic in more detail in Section 5.8) to decide the design points at which the simulation code should be run. Given the *training data*, the approximation problem reduces to prediction of the output $y(\mathbf{x})$ given a new design point $\mathbf{x}$.

In general, black-box surrogate modeling is an iterative procedure involving the following steps: (1) data generation, (2) model-structure selection, (3) parameter estimation and (4) model validation. Techniques in this field can be classified into different categories depending on how these steps are carried out. To begin with, we adopt a classical statistics perspective on the steps involved along with a brief discussion on how these four steps can be brought together in an iterative process. The objective is to highlight how existing procedures in the statistics literature can be leveraged to construct surrogate models for engineering design applications. Some of the deficiencies in classical statistics and the pitfalls associated with naive application of existing statistical procedures to surrogate model construction are also discussed – for a more detailed account of the issues discussed here, see Chatfield (1995) and Hosking et al. (1997).

5.3.1 Data Generation

To construct a surrogate model, we first need a set of observational or training data. A straightforward approach would be to generate data by running the analysis code at a large number of design points, which are chosen randomly. However, when the computational budget is limited, it may not be possible to generate a large training dataset. Further, the quality of the surrogate model crucially depends on the location of the training points. For example, in the context of design optimization, we are interested in designing or planning a sequence of analysis code runs that will enable us to construct a surrogate model with maximum predictive capability within the allotted computational budget. Moreover, we wish the surrogate to be most accurate in those regions of the search space that the optimizer will sample.[5] This leads to two conflicting objectives in the data generation phase: (1) how to decide on a set of points where the analysis code should be run so as to maximize the predictive capability of the surrogate model and (2) how to minimize the number of runs of the analysis code. Clearly, a rational approach for addressing these issues must take into account the structure of the surrogate model employed to fit the data.

In the classical statistics literature, this problem has been extensively studied for over 70 years in the context of designing physical experiments (Fisher 1935). Physical experiments are difficult to set up and, in some applications, generation of data may take substantial calendar time. This is similar to the situation faced in computational engineering where substantial effort is required to create and mesh a complex engineering system and generation of results for each prototype is computationally expensive. However, there are fundamental differences between physical experiments and computer experiments. For example, physical experiments can contain significant random noise, whereas computer models employed in design applications are usually deterministic. Further, most existing techniques in the design of experiments literature are tailored to construct linear and quadratic models. This can be primarily attributed to the fact that results from physical experiments are usually contaminated by a significant level of noise and more complex models may overfit this noise. Even so, some of the techniques from the design of experiments literature have recently found applications in the data generation phase of surrogate modeling. We discuss the issues involved in design of experiments for design optimization applications in more depth in Section 5.8.

5.3.2 Model-structure Selection

The next step in surrogate construction involves specifying the structure of the model to be used. In statistical regression techniques, it is assumed that the relationship between the input $\mathbf{x}$ and the output y is observed with error. Hence, a model of the form $\widehat{y} = \widehat{f}(\mathbf{x}, \boldsymbol{\alpha}) + e$ is often employed, where e is a zero-mean error term with some probability distribution.

In practice, one may use a linear, quadratic or a generalized nonlinear model to represent $\widehat{f}(\mathbf{x}, \boldsymbol{\alpha})$, that is, either a parametric or nonparametric model can be used to represent the input–output relationship. A model is said to be *parametric*, if after the undetermined set of parameters are estimated using the training dataset, the training data is no longer needed when making predictions at new points, for example, polynomial models. An attractive feature of parametric models is that they are readily interpretable and can hence provide useful insights into the nature of the input–output relationship. However, when the true input–output relationship is complex, parametric models can be of limited utility. Hence, these methods are best applied to approximate the input–output relationship over a small

---

[5]Ultimately, we wish our predictive capability to support design improvement – depending on the search strategy in use, this will require predictions mainly in regions where we expect to find promising designs, rather than completely uniformly in the search space.

region of the design space. In contrast, *nonparametric* models require the training dataset to make predictions even after the set of undetermined parameters $\boldsymbol{\alpha}$ has been estimated from the data. Kernel methods such as radial basis function approximations, support vector machines and Gaussian process modeling (discussed later in this chapter) are commonly encountered examples of nonparametric models.

In classical statistics, model-structure selection continues to be a largely subjective step. In practice, a suitable structure for the function $\widehat{f}(\mathbf{x}, \boldsymbol{\alpha})$ is decided by inspecting the data or from experience gained by analysis of similar datasets.

5.3.3 Parameter Estimation

After choosing a model structure that is believed to be appropriate for the problem at hand, we need to estimate the set of unknown model parameters $\boldsymbol{\alpha}$ by minimizing a suitable error or loss function.[6] Parameter estimation techniques can be broadly classified into different categories on the basis of which loss function is employed and the algorithm employed to minimize the function.

In classical statistics, $\boldsymbol{\alpha}$ is typically estimated by the maximum likelihood approach. Let $y = f(\mathbf{x}) + e$ denote the true underlying relationship, where e is a noise term. Then, in a statistical approach, we specify a functional form for the conditional probability density $p(y|\mathbf{x})$ in terms of the parameter vector $\boldsymbol{\alpha}$. Maximization of the conditional density gives an estimate of the value of $\boldsymbol{\alpha}$ that is *most likely* to have generated the given data. The likelihood of a dataset $(\mathbf{x}^{(1)}, y^{(1)}), \ldots, (\mathbf{x}^{(n)}, y^{(n)})$, given a parameterized model of the form $\widehat{f}(\mathbf{x}, \boldsymbol{\alpha})$, is given by

$$p(\{\mathbf{x}^{(1)}, \ldots, \mathbf{x}^{(n)}\}, \{y^{(1)}, \ldots, y^{(n)}\}|\boldsymbol{\alpha}) = \prod_{i=1}^{n} p(\mathbf{x}^{(i)}, y^{(i)}|\boldsymbol{\alpha}) = \prod_{i=1}^{n} p(y^{(i)}|\mathbf{x}^{(i)}, \boldsymbol{\alpha}) p(\mathbf{x}^{(i)}).$$

$$(5.15)$$

In practice, it is preferable to convert the products into sums by taking the logarithm of the preceding equation. Further, the terms containing $p(\mathbf{x}^{(i)})$ can be neglected since they are independent of $\boldsymbol{\alpha}$. This gives the following function which can be maximized to estimate $\boldsymbol{\alpha}$:

$$\sum_{i=1}^{n} \ln p(y^{(i)}|\mathbf{x}^{(i)}, \boldsymbol{\alpha}). \qquad (5.16)$$

If we assume further that $p(y^{(i)}|\mathbf{x}^{(i)})$ is Gaussian (i.e., the outputs are corrupted by Gaussian noise), then the maximum likelihood loss function (in minimization form) reduces to the well-known least-squares error function

$$\sum_{i=1}^{n} \left(y^{(i)} - \widehat{f}(\mathbf{x}^{(i)}, \boldsymbol{\alpha}) \right)^2. \qquad (5.17)$$

The use of maximum likelihood estimation to compute $\boldsymbol{\alpha}$ has been justified in the classical statistics literature by asymptotic arguments. In other words, it can be shown that the estimates for $\boldsymbol{\alpha}$ are consistent and efficient as the size of the observational dataset $n \to \infty$. No theoretical justification exists that this is the best way to estimate the parameter

---

[6]We shall refer to error functions as loss functions interchangeably, as loss function is a term commonly used in the machine and statistical learning theory literature.

vector α when only a finite (and, in practice, perhaps small) set of observational data is available.

Another important component in parameter estimation involves representing the accuracy of α in probabilistic terms. Note that in traditional statistical inference the undetermined parameter vector α is regarded as fixed and does not have a probability distribution. So, to quantify the accuracy of α in probabilistic terms, it is assumed that the observational data is sampled from a probabilistic generator. Likelihood theory provides an asymptotic large sample approximation for the confidence region of the parameter estimates and for predictions made using the models. It is important to note that the approximations for the confidence regions are computed under the assumption that the specified model structure for $\hat{f}(\mathbf{x}, \alpha)$ is correct. For example, if we use a linear function to model the output of an analysis code, then the usual confidence intervals provided by likelihood theory are meaningful only if the actual relationship between the inputs and output is linear. If the actual relationship is nonlinear, these confidence intervals may be misleading.

Often, in practice, it is observed that the accuracy of the model can deteriorate because of points that do not fit the overall trends in the data, also referred to as outliers. Attempts to resolve this problem have led to the emergence of robust statistics (Huber 1981), which is a body of methods for constructing models that are less susceptible to outliers.

5.3.4 Model Assessment

After the model parameters have been estimated, some diagnostic checks need to be performed to judge the suitability of the model for optimization or uncertainty analysis studies. The inadequacy of a statistical model may arise from many causes. For example, the phenomenon of over-fitting may lead to poor models. This generally happens when the specified model structure for $\hat{f}(\mathbf{x}, \alpha)$ has too many parameters, which in turn leads to a model that reproduces the noise in the data rather than the true underlying relationship between $\mathbf{x}$ and y. The converse situation is underfitting, which arises when the model structure is too simple and is not capable of representing the true patterns in the data. A statistical model can also be inadequate when an inappropriate model structure is chosen. Model-assessment strategies are hence required to check whether a given model is adequate. Ideally, model assessment should lead to objective criteria for comparing different models, thereby providing guidelines for selecting the best model.

In classical statistics, goodness-of-fit tests and diagnostic plots are employed to check the adequacy of the model structure chosen for the problem at hand. An example is the t-statistic, whose distribution is used to assess model adequacy (Myers and Montgomery 1995). An extreme value of this statistic can be used as an indicator of the fact that the model structure is wrong or something very unusual has occurred. In certain cases, it may be possible to use the t-statistic to come up with a way of modifying the model structure to improve accuracy. It is important to note here that the t-statistic is computed under the assumption that the model structure is correct.

Subjective judgments of model adequacy can also be made by using diagnostic plots. For example, if the model structure is correctly specified for regression problems with noise, the residuals $y^{(i)} - \hat{f}(\mathbf{x}^{(i)}, \alpha)$ will be approximately independently distributed. A plot of residuals against fitted values can hence be used to check model adequacy. Diagnostic plots can also be used to identify unusual data points or outliers.

There are also a number of simulation based techniques in the literature such as the bootstrap, which enables the adequacy of the model to be assessed more accurately when only a finite number of data points are available (Efron 1981; Hastie et al. 2001). We discuss model-assessment procedures for a number of surrogate modeling techniques in more detail later on in this chapter.

5.3.5 Iterative Model-building Procedures

Conceptually, it should be possible to develop an iterative approach for building models from data by combining the steps of model-structure selection, parameter estimation and model assessment. However, there are two bottlenecks to automating this iterative procedure since model-structure selection and model assessment are largely subjective in nature. The difficulty in automating model selection can be alleviated to some extent by restricting ourselves to choosing the best model from a sequence of nested candidate models, that is, we choose from the sequence of models $\mathcal{M}_1(\boldsymbol{\alpha}^1), \mathcal{M}_2(\boldsymbol{\alpha}^2), \ldots, \mathcal{M}_m(\boldsymbol{\alpha}^m)$, where every element of $\boldsymbol{\alpha}^i$ is also included in $\boldsymbol{\alpha}^{i+1}$.

Once the best model is chosen from this nested sequence, predictions can be made under the assumption that this model is "correct". However, this procedure has a number of fundamental problems. Take, for example, the case of stepwise regression, which is widely used for deciding which elements of the vector $\mathbf{x}$ should appear in the final model. It is possible here that none of the models in the nested sequence is capable of adequately modeling the data. Further, because of random noise, it is possible that variables that are unrelated to the output may appear in the final model. The fact that several models have been estimated and tested on the same dataset is not taken into account when deriving the statistical properties of the parameter estimates in the final model. As a result, the confidence intervals for the estimated regression coefficients can be highly biased. This problem led Miller (1983) to recommend that, in practice, it is often better to use all of the available variables rather than a stepwise regression procedure.

Since the 1970s, a number of alternative frameworks for comparison of nonnested models have been developed to address difficulties associated with the classical frequentist approach. Notable among them are Akaike's information criterion and Rissanen's principle of minimum description length (MDL) (Hansen and Yu 2001). We shall employ these criteria later on in Section 5.5.3 to construct parsimonious radial basis function approximations.

5.3.6 Perspectives from Statistical Learning Theory

The major drawback of classical frequentist approaches arises from the implicit assumption that the assumed model structure is "correct" and the problem is just to estimate the model parameters from the data. In practice, the ultimate objective in model building is to arrive at an appropriate model structure that is capable of describing the observational data. Even though a number of diagnostic checks are available in the literature to check model adequacy, model-structure selection is still largely a subjective process whose success depends to a great extent on the experience of the analyst. Further, asymptotic arguments (i.e., those assuming the number of training points $n \to \infty$) are commonly used to justify the procedures employed for parameter estimation and model assessment. Existing procedures in the classical frequentist statistics literature cannot be rigorously justified for practical modeling problems where the size of the observational dataset is finite. A more detailed exposition of the inherent flaws in existing procedures for model building in the classical literature has been presented by Chatfield (1995).

This deficiency of classical frequentist approaches motivated Vapnik and Chervonenkis (1971) to develop an alternative statistical framework for constructing models from data. The body of methods developed to address this issue forms the backbone of *statistical learning theory* – see the article by Poggio et al. (2004) for a concise overview and the review paper by Evgeniou et al. (2000) for a more detailed account. Statistical learning theory provides insights into the problem of learning from a finite set of observational data and provides pointers on how models with good generalization capability can be constructed. It is presumed that the real model structure is truly unknown and that the aim is to select the best possible model from a given set of models. In contrast to classical statistics where asymptotic arguments are used, the criteria used for comparing models in statistical learning theory are based on finite-sample statistics.

Before delving into a brief account of some basic results in learning theory, let us define a few of the fundamental notions used in that area. First, assume that the observational data, $z_i, i = 1, 2, \ldots, n$ are n realizations of a random process drawn from the distribution $\mathcal{P}(z)$. Then, from a decision-theoretic perspective, the best model for representing the input–output relationship minimizes the following function:

$$R(\alpha) = \int Q(z, \alpha) \, d\mathcal{P}(z), \tag{5.18}$$

where $Q(z, \alpha)$ is an appropriate loss function. In more general terms, $Q(z, \alpha)$ is a (negative) utility measure, the minimization of which leads to an optimal decision consistent with the risk preferences defined by the utility measure (Hosking et al. 1997). Hence $R(\alpha)$ is also referred to as the true risk or the expected loss.

In practice, however, $R(\alpha)$ cannot be computed since the statistical properties of the observational data $\mathcal{P}(z)$ are unknown. Hence, one has to resort to a finite-sample approximation $R_{\text{emp}}(\alpha)$, also referred to as the empirical risk, that is,

$$R_{\text{emp}}(\alpha, n) = \frac{1}{n} \sum_{i=1}^{n} Q(z_i, \alpha). \tag{5.19}$$

In most model-fitting procedures, parameter estimation is carried out by minimizing R_{emp}, which is a finite-sample approximation of R. However, ideally, we wish to minimize the true risk R. This is because R_{emp} can be a gross underestimate of R (particularly for small n) and hence the model constructed by minimizing R_{emp} may have poor generalization properties. Statistical learning theory studies the conditions under which minimization of R_{emp} also leads to minimization of the true risk. The basic premise is that the true risk for a given model is fixed but unknown since the probability measure of the observational data $\mathcal{P}(z)$ is fixed but unknown. Since R_{emp} is a random variable obtained by sampling, the aim is to construct confidence regions for $R(\alpha)$ given $R_{\text{emp}}(\alpha, n)$.

Vapnik and Chervonenkis (1971) derived a series of probability bounds for R given a model with average empirical loss R_{emp}. In a significant departure from existing results in classical frequentist statistics, these bounds do not assume that the chosen model is correct and further no asymptotic arguments are invoked. It was shown that the size of the bounds is largely determined by the ratio of a measure of model complexity (which is now referred to as the Vapnik–Chervonenkis (VC) dimension in the statistical learning theory literature) to the number of data points. Let $\widehat{f}(\mathbf{x}, \boldsymbol{\alpha})$ be an approximation model having VC dimension h. Then, with probability $1 - \eta$, the following upper bound holds for function approximation:

$$R \leq \frac{R_{\text{emp}}}{(1 - c\sqrt{\varepsilon})_+}, \quad \text{where} \quad \varepsilon = a_1 \frac{h[\log(a_2 n / h) + 1] - \log(\eta/4)}{n}, \tag{5.20}$$

and c, a_1 and a_2 are constants usually set to one (Cherkassky and Mulier 1998).

Loosely speaking, the VC dimension of a model can be thought of as the maximum number of data points that it can fit exactly. For example, in the case of linear regression, the VC dimension is related to the number of terms, since n terms can fit n data points exactly. In fact, the VC dimension is much more general than the total number of free parameters in a model and is able to indicate more clearly the ability of models to fit data. The VC dimension can be defined for a wide class of models, including neural networks, classification and regression trees and radial basis functions. The concepts of VC dimension and confidence bounds for various loss functions employed in regression and classification problems are explained in more detail in the texts by Vapnik (1998), Cristianini and Shawe-Taylor (2000) and Schölkopf and Smola (2001).

Theoretical results in statistical learning rigorously support the intuition that minimizing the empirical risk alone may not lead to models with good generalization capability – it is also necessary to control a measure of model complexity (or the VC dimension) to avoid overfitting. Even though bounds on R_{emp} in terms of the VC dimension are often very loose, they provide a good criterion for comparing models. This criterion forms the basis of Vapnik's *structural risk minimization* approach, in which a nested sequence of models with increasing VC dimensions $h_1 < h_2 < \cdots$ are constructed and the model with the smallest value of the upper bound is selected.

5.3.7 Interpolation versus Regression

In the domain of computational engineering design, we primarily rely on *deterministic* computer models to predict measures of design qualities given the values of the inputs. So, let us assume that an input–output dataset is made available by running the models at a number of inputs. Subsequently, it is required to choose a parameterized flexible metamodel to approximate the input–output mapping. An important question that arises in this context is *should the metamodel structure be chosen to be flexible enough to exactly interpolate the training data or is some form of regression desirable?*

This question of whether one should employ interpolation or regression techniques to approximate the output of a deterministic computer model has been the subject of much debate in recent years. Since no random measurement errors exist in the dataset, the usual measures of uncertainty derived from least-square residuals do not have any obvious statistically meaningful interpretation. In spite of this, many papers in the literature continue to use linear or quadratic response surface methods and model validation methodology developed for noisy data to construct surrogate models for computationally produced data. Strictly speaking, the approximation errors of response surface models when applied to datasets generated by deterministic computer models can be entirely attributed to the inadequacy of the assumed model structure.

Since there is no random noise in observational data generated by running deterministic computer models, the application of regression techniques cannot be rigorously justified. Often, regression techniques are employed in the literature for dealing with computer models under the pretext that in the presence of numerical errors in the simulations the surrogate models must not be allowed to interpolate the data. For example, take the case when the analysis code is a CFD solver, which is not run to full convergence. Here, the drag values calculated by the computer model will have some error. Hence, it makes sense not to use an interpolating surrogate model for the drag values. However, the error in drag is reproducible if the solver is run again for the same inputs. In other words, numerical errors arising in such problems do not have a direct analogy with random noise. Hence, the use of regression techniques must be treated with caution since it is statistically questionable to treat numerical errors as random quantities that are independently and identically distributed. Certainly, there is a need for a more rigorous framework for surrogate modeling in the presence of numerical errors since existing statistical approaches are not strictly applicable to this class of problems.

It should be noted that, in the preceding discussion, we are not suggesting that regression techniques should not be applied to design optimization. Rather, one should exercise caution in interpreting the results of model significance based on least-square residuals, F-statistics, and so on, since they do not have any statistical meaning for deterministic experiments. Further, it is also important to note that polynomial models cannot capture highly nonlinear variations in the response being modeled, which consequently restricts the class of problems to which they can be successfully applied. A more detailed discussion of the statistical pitfalls associated with the application of response surface methodology to deterministic computer experiments can be found in Sacks et al. (1989) and Simpson et al. (1997).

In summary, interpolation techniques are more appropriate from a theoretical point of view when training datasets are obtained by running a deterministic computer model. However, this does not mean that, for a given problem, an interpolation model ($\mathcal{M}_{\text{int}}$) will work better than a regression model ($\mathcal{M}_{\text{reg}}$). In fact, what is more important in practice is the generalization performance of the surrogate model in the context of the process we intend to deploy – in other words, the generalization of $\mathcal{M}_{\text{int}}$ and $\mathcal{M}_{\text{reg}}$ should be compared and contrasted. We recommend that for any problem the user should always consider a number of surrogate models, $\mathcal{M}_1, \mathcal{M}_2, \ldots$, constructed using techniques such as those discussed in this chapter. Then, a range of criteria should be employed to help decide on the surrogate model most appropriate for the problem at hand – see also the discussion in Section 5.10.

5.4   Generalized Linear Models

In this section, two popular approaches for constructing surrogate models are examined – polynomial response surface models and radial basis functions. Both techniques can be treated as instances of generalized linear models, which have a long history in the statistics and function approximation literature. A generalized linear model can be expressed in the following form:

$$\widehat{y}(\mathbf{x}) = \sum_{i=1}^{m} \alpha_i \phi_i(\mathbf{x}), \tag{5.21}$$

where $\phi_i, i = 1, 2, \ldots, m$ denote a set of fixed basis functions[7] and $\alpha_i, i = 1, 2, \ldots, m$ are the undetermined coefficients in the approximation.

---

[7]As an aside, a generalized nonlinear model uses a tunable basis function, that is, $\phi := \phi(\mathbf{x}, \boldsymbol{\theta})$, where $\boldsymbol{\theta}$ is a set of hyperparameters. A neural network with sigmoid transfer functions (Bishop 1995) belongs to this class of models.

5.4.1   Response Surface Methods: Polynomial Models

Response surface methodology (RSM) involves the application of design of experiments techniques, regression analysis and analysis of variance techniques to plan and interpret data from physical experiments (Box and Draper 1987; Myers and Montgomery 1995). The origin of RSM dates back to the early twentieth century when agricultural and biological scientists required tools for modeling data from physical experiments. Since physical experiments are subject to measurement errors, least-square regression techniques were employed under the assumption that the measurement errors for each data point are independent and identically distributed. For experiments with significant measurement errors and when data generation is expensive and time consuming, it was often found that rather simple linear and quadratic models worked best. Further, the accuracy of the response surface can be quantified by examining the statistics of the least-square residuals.

Over the last decade, RSM has been increasingly applied to construct polynomial approximations of computer models. To illustrate RSM, consider as before a problem where $\{\mathbf{x}^{(i)}, y^{(i)}, i = 1, 2, \ldots, n\}$ denote the training dataset, $\mathbf{x} \in \mathbb{R}^p$ is the input vector and $y \in \mathbb{R}$ is the output. A quadratic response surface model for y can be written as

$$\widehat{y} = c_0 + \sum_{1 \leq j \leq p} c_j x_j + \sum_{1 \leq j \leq p, k > j} c_{p-1+j+k} x_j x_k, \tag{5.22}$$

where c_0, c_j, and $c_{p-1+j+k}$ are the undetermined coefficients in the model. Let $\mathbf{c} = \{c_0, c_1, \ldots, c_{m-1}\} \in \mathbb{R}^m$ denote the vector of undetermined coefficients in the quadratic response surface model, where $m = (p + 1)(p + 2)/2$. The polynomial model can be compactly written as $\widehat{y} = \mathbf{c}^T \bar{\mathbf{x}}$, where $\bar{\mathbf{x}} = \{1, x_1, x_2, \ldots, x_1^2, x_1 x_2, x_1 x_3, \ldots, x_p^2\}$ can be interpreted as a vector of basis functions.

Substituting the observational data into the model, the undetermined coefficient vector $\mathbf{c}$ can be estimated by solving the matrix system of equations $\mathbf{Ac} = \mathbf{y}$, where $\mathbf{y} = \{y^{(1)}, y^{(2)}, \ldots, y^{(n)}\}^T \in \mathbb{R}^n$ and

$$\mathbf{A} = \begin{bmatrix} 1 & x_1^{(1)} & x_2^{(1)} & \cdots & (x_p^{(1)})^2 \\ \cdot & \cdot & \cdot & \cdots & \cdot \\ \cdot & \cdot & \cdot & \cdots & \cdot \\ \cdot & \cdot & \cdot & \cdots & \cdot \\ 1 & x_1^{(n)} & x_2^{(n)} & \cdots & (x_p^{(n)})^2 \end{bmatrix} \in \mathbb{R}^{n \times m}. \tag{5.23}$$

Assume that the noise in the data is zero-mean Gaussian with standard deviation σ. Further, let the noise at each point be uncorrelated with the noise at other points. Then, an unbiased estimate for the variance of the noise is

$$\widehat{\sigma}^2 = \frac{\mathbf{y}^T \mathbf{y} - \mathbf{c}^T \mathbf{A}^T \mathbf{y}}{n - m}, \tag{5.24}$$

where the numerator is the sum of the square of errors in the model.

Analysis of variance can be applied to estimate a measure of uncertainty in the computed coefficients of the response surface model. The t-statistic for c_{j-1} can be defined as

$$t = \frac{c_{j-1}}{\sqrt{\widehat{\sigma}^2 (\mathbf{A}^T \mathbf{A})_{jj}^{-1}}}, \tag{5.25}$$

where $(\mathbf{A}^T \mathbf{A})_{jj}$ is the term in the jth diagonal of the matrix $\mathbf{A}^T \mathbf{A}$.

The t-statistic can be computed for all the coefficients in the polynomial model. This statistic can be interpreted as an estimate of the reciprocal of the standard deviation of each coefficient as a fraction of its value. Hence, coefficients that have not been accurately estimated will have a low t-statistic. This allows for the possibility of dropping coefficients with low t-statistic to improve the accuracy of the response surface model; see (Box and Draper 1987; Myers and Montgomery 1995) for a detailed description of this approach.

The negative impact of outliers in the observational data on the final model can also be alleviated by employing a weighted least-squares minimization technique, which involves the solution of the modified system of normal equations

$$\mathbf{A}^T \mathbf{W} \mathbf{A} \mathbf{c} = \mathbf{A}^T \mathbf{W} \mathbf{y}, \tag{5.26}$$

where $\mathbf{W} \in \mathbb{R}^{n \times n}$ is a diagonal matrix containing weights that must be assigned to each data point. There are a number of recipes in the literature for choosing the weight of each point; see, for example, Myers and Montgomery (1995).

In practice, the matrix $\mathbf{A}$ may be poorly conditioned, which may cause numerical instabilities to arise when the normal equations are directly solved to compute the RSM coefficients. Hence, it is preferable to employ the singular value or QR decomposition scheme to estimate $\mathbf{c}$. Further, to ensure robust estimates for the coefficients, the number of training points must be chosen to be greater than the number of undetermined coefficients in the model, that is, $n > m$ (if $n = m$, the model simply interpolates the data, of course). Because of the requirement of using more than $(p + 1)(p + 2)/2$ points to fit a quadratic model, RSM is computationally expensive to apply for problems with more than, say, ten design variables.

A number of applications of RSM to design optimization can be found in the literature; see, for example, the dissertations of Balabanov (1997); Giunta (1997); Venter (1998) and the references therein. An excellent overview of the theoretical aspects of RSM has been presented by Venter et al. (1998). Balabanov et al. (1999) presented a reasonable design space approach that was applied to design of a High Speed Civil Transport configuration involving 25 variables. Papila and Haftka (2000) presented an iteratively reweighted least-squares fitting approach that can be used to repair outliers when applying RSM.

As pointed out earlier, a polynomial approximation can be interpreted as a parametric model. Even though such models are highly interpretable, they are of limited utility when modeling complex input–output relationships. For example, quadratic response surfaces are incapable of capturing multiple extrema of the original function. Hence, polynomial models are useful only when it is desired to approximate an input–output relationship over a small region of the design space, for example, in the vicinity of the current design. Fortunately, the drawbacks associated with parametric models can be alleviated to some extent via the use of nonparametric models that employ more flexible basis functions – we shall focus on this class of modeling techniques in the remainder of the chapter.

5.4.2 Neural Network Approximations

We begin by mentioning neural networks, which are based on highly idealized models of brain structures where multiple connected elements within the model represent the neurons of the brain in some sense. When building neural networks, the topology of the interconnected neuron models has to be fixed and there are many alternatives mentioned in the literature.

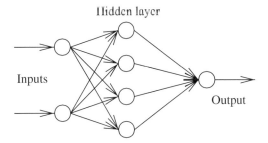

Figure 5.1 A feed-forward neural network with one hidden layer – the input neurons scale the inputs, the output neuron sums contributions from the hidden layer and the hidden layer applies the activation functions.

The most popular configuration for function modeling appears to be a feed-forward system with an input layer, one hidden layer and an output layer; see Figure 5.1. This corresponds to a model of the form

$$\widehat{y}(\mathbf{x}) = \sum_{i=1}^{m} \alpha_i \phi(a_i), \quad \text{where } a_i = \sum_{j=1}^{p} w_{ij} x_j + \beta_j, \tag{5.27}$$

and where α, w and β are undetermined parameters representing the weights and bias terms of the network. $\phi(x)$ is a transfer function that is typically chosen to be the sigmoid, that is, $(1 + e^{-x})^{-1}$ or the tanh function. The total number of undetermined parameters is $(p + 2)m$, where m and p denote the number of neurons in the hidden layer and the number of inputs, respectively.

Since $\widehat{y}(\mathbf{x})$ is a nonlinear function of the undetermined parameters, feed-forward neural networks belong to the class of generalized nonlinear models. Their predictive capabilities have much in common with Radial Basis Function (RBF) models and in many cases RBFs are now used in preference since they are rather simpler to analyze. See Bishop (1995) for a detailed exposition of neural networks and numerical schemes for estimating their parameters.

5.4.3 Radial Basis Function Approximations

Radial basis function (RBF) approximations also belong to the class of generalized linear models. They differ from standard polynomial response surface models in the choice of basis functions. In particular, a radial basis function of the form $K(||\mathbf{x} - \mathbf{x}_c||)$ is used here, where $\mathbf{x}_c$ is commonly referred to as the center of the basis function (Powell 1987). Consider the case when the training data is generated by running a deterministic computer model. Here, an interpolating RBF approximation of the form

$$\widehat{y}(\mathbf{x}) = \sum_{i=1}^{n} \alpha_i K(||\mathbf{x} - \mathbf{x}^{(i)}||) \tag{5.28}$$

Table 5.1 Examples of Radial basis functions of the form $K(||\mathbf{x} - \mathbf{x}_c||)$, where $\mathbf{x}_c$ denotes the center. In the case of Gaussian, multiquadric and inverse multiquadric RBFs, θ denotes the shape parameter

Linear splines	$||\mathbf{x} - \mathbf{x}_c||$
Thin plate splines	$||\mathbf{x} - \mathbf{x}_c||^k \ln ||\mathbf{x} - \mathbf{x}_c|| \quad k \in [2, 4, \ldots]$
Cubic splines	$||\mathbf{x} - \mathbf{x}_c||^3$
Gaussian	$\exp\left(-\frac{||\mathbf{x}-\mathbf{x}_c||^2}{\theta}\right)$
Multiquadrics	$\left(1 + \frac{||\mathbf{x}-\mathbf{x}_c||^2}{\theta}\right)^{1/2}$
Inverse multiquadrics	$\left(1 + \frac{||\mathbf{x}-\mathbf{x}_c||^2}{\theta}\right)^{-1/2}$

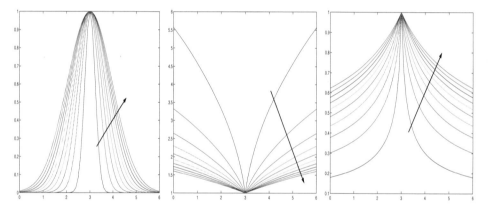

Figure 5.2 Shapes of Gaussian, multiquadric and inverse multiquadric RBFs as a function of the shape parameter. The arrow is shown in the direction of increasing θ.

is of interest,[8] where $K(||\mathbf{x} - \mathbf{x}^{(i)}||) : \mathbb{R}^p \to \mathbb{R}$ is a RBF and $\boldsymbol{\alpha} = \{\alpha_1, \alpha_2, \ldots, \alpha_n\}^T \in \mathbb{R}^n$ denotes the vector of undetermined weights.

Typical choices for the RBF include linear splines, cubic splines, multiquadrics, thin-plate splines, and Gaussian functions. The structure of some commonly used RBFs and their parameterizations are shown in Table 5.1. In the case of Gaussian, multiquadrics and inverse multiquadrics, a shape parameter (θ) is used to control the domain of influence of the RBF. Figure 5.2 shows how the shape of these RBFs (and hence the smoothness of the approximation) can be controlled by varying θ.

Note that the basis function K is also sometimes referred to as a *kernel* in the machine learning literature. Hence, methods for classification, function approximation and density

---

[8]As an aside, the representer theorem in approximation theory states that, under very general conditions, the solution to the minimization problem $\min_{f \in \mathcal{H}_k} \left[\sum_{i=1}^n Q(y^{(i)}, f(\mathbf{x}^{(i)})) + \lambda ||f||_{\mathcal{H}_k}\right]$ has the form $f(\mathbf{x}) = \sum_{i=1}^n \alpha_i K(\mathbf{x}, \mathbf{x}^{(i)})$, where Q is a loss function, λ is a regularization parameter and $\mathcal{H}_k$ denotes a reproducing kernel Hilbert space (Wahba 1990). Since any RBF can be expressed as $K(\mathbf{x}, \mathbf{x}^{(i)})$, this result provides a theoretical justification for employing the model structure in (5.28) to solve function approximation problems.

estimation that employ approximations of the form given in (5.28) are also referred to as kernel methods (Vapnik 1998). Theoretical analysis of the universal approximation characteristics of a wide class of kernels can be found in the literature; see, for example, Liao et al. (2003) and the references therein. The results derived there suggest that by using suitable kernels in a generalized linear model it becomes possible to approximate any function to an arbitrary degree of accuracy. The kernels that are capable of universal approximation include RBFs as well as models employing tunable kernels such as feed-forward neural networks. Franke (1982) compared a variety of interpolation techniques on two-dimensional problems where a scattered set of observational data is available. It was shown that multiquadrics give best results compared to other techniques considered in the study. This observation has led to the widespread application of multiquadrics to model complex two- and three-dimensional surfaces in geodesy, image processing and natural resource modeling (Carr et al. 1997).

Given a suitable RBF, the weight vector can be computed by solving the linear algebraic system of equations $\mathbf{K\alpha} = \mathbf{y}$, where $\mathbf{y} = \{y^{(1)}, y^{(2)}, \ldots, y^{(n)}\}^T \in \mathbb{R}^n$ denotes the vector of outputs and $\mathbf{K} \in \mathbb{R}^{n \times n}$ denotes the Gram matrix formed using the training inputs, that is, the ijth element of $\mathbf{K}$ is computed as $K(||\mathbf{x}^{(i)} - \mathbf{x}^{(j)}||)$.

Michelli (1986) showed that nonsingularity of the Gram matrix $\mathbf{K}$ can be theoretically guaranteed for a class of RBFs only when the set of input vectors in the training dataset are distinct. Specifically, nonsingularity can be guaranteed only if positive-definite kernels (such as Gaussian and inverse multiquadric) are used in (5.28).[9] However, for extreme values of θ (as the basis functions become flatter) in these kernels, $\mathbf{K}$ can become highly ill-conditioned.[10] Wang (2004) proved a monotonic relationship between the condition number of $\mathbf{K}$ and the shape parameter for multiquadrics. In practice, it is often observed that high accuracy is achieved only at the verge of numerical instability.

If the RBF is not positive definite, a polynomial term $\mathcal{P}$ needs to be appended to (5.28) along with some constraints. In other words, if K is a conditionally positive-definite basis function of order q, then to ensure an unique solution for the weight vector, the original RBF approximation in (5.28) is rewritten as

$$\widehat{y}(\mathbf{x}) = \sum_{i=1}^{n} \alpha_i K(||\mathbf{x} - \mathbf{x}^{(i)}||) + \mathcal{P}_{q-1}(\mathbf{x}), \qquad (5.29)$$

where $\mathcal{P}_{q-1}$ is a polynomial of order $q - 1$. To simplify notation, let us represent the polynomial term as a generalized linear model of the form $\mathcal{P}_{q-1} = \sum_{i=1}^{n_q} c_i \varphi_i(\mathbf{x})$, where φ_i is a basis function and n_q is the number of undetermined coefficients c_i. The following homogeneous constraint equations are imposed to ensure that a square system of equations result for $\boldsymbol{\alpha} = \{\alpha_1, \alpha_2, \ldots, \alpha_n\}$ and $\mathbf{c} = \{c_1, c_2, \ldots, c_{n_q}\}$:

$$\sum_{i=1}^{n} \alpha_i \varphi_k(\mathbf{x}^{(i)}) = 0, \quad k = 1, 2, \ldots, n_q. \qquad (5.30)$$

Substituting the training data points into (5.29) and (5.30), it follows that the weight vector can be computed by solving an $(n + n_q) \times (n + n_q)$ linear algebraic system of equations

---

[9]If K is a positive-definite kernel, then the Gram matrix $\mathbf{K}$ is symmetric positive definite. Such basis functions are also commonly referred to as Mercer kernels in learning theory (Vapnik 1998).

[10]A matrix is said to be ill-conditioned if the logarithm of its condition number is greater than the machine precision. The global condition number of a square matrix can be defined as the ratio of the largest to smallest eigenvalue.

of the form $\mathbf{A}\widetilde{\boldsymbol{\alpha}} = \mathbf{b}$, where

$$\mathbf{A} = \begin{bmatrix} \mathbf{K} & \mathbf{P} \\ \mathbf{P}^T & \mathbf{0} \end{bmatrix}, \quad \widetilde{\boldsymbol{\alpha}} = \{\boldsymbol{\alpha}, \mathbf{c}\}^T, \quad \mathbf{b} = \{\mathbf{y}, \mathbf{0}\}^T, \qquad (5.31)$$

and $\mathbf{P} \in \mathbb{R}^{n \times n_q}$ is a matrix that arises from substitution of the input vectors in the training dataset into the polynomial term $\mathcal{P}$. In the case of multiquadrics (which are conditionally positive definite of order one), a constant term can be used instead of a full-order polynomial. Here, the coefficient matrix $\mathbf{A}$ becomes

$$\mathbf{A} = \begin{bmatrix} \mathbf{K} & \mathbf{1} \\ \mathbf{1}^T & 0 \end{bmatrix} \in \mathbb{R}^{(n+1) \times (n+1)}, \qquad (5.32)$$

where $\mathbf{1} \in \mathbb{R}^n$ is a column vector of ones.

The RBF interpolation approach can be applied to datasets arising from deterministic computer experiments. However, when the training dataset is corrupted by random noise, a regression technique becomes necessary. A regularization approach can be employed to construct RBF approximations that attempt to filter noise in the output (Poggio and Girosi 1990). Assuming that the output vector $\mathbf{y}$ is corrupted by zero-mean Gaussian noise and a positive-definite kernel is used, RBF regression essentially involves solving $(\mathbf{K} + \lambda \mathbf{I})\boldsymbol{\alpha} = \mathbf{y}$, to compute the weight vector, where $\mathbf{I} \in \mathbb{R}^{n \times n}$ is the identity matrix and λ is a regularization parameter that controls the amount of regression. Ideally, λ should be set to the variance of any noise in the output.

For problems with multiple outputs, different weight vectors are necessary to model each output of interest. Fortunately, the weight vector corresponding to each output can be efficiently computed once the matrix $\mathbf{K}$ is decomposed. This is possible since only the right-hand side of the system of equations $\mathbf{K}\boldsymbol{\alpha} = \mathbf{y}$ is changed for each output of interest. Hence, the RBF approximation technique is particularly efficient for design problems where it is necessary to construct surrogates for the constraint functions as well as the objective. Note that construction of RBF approximations requires $\mathcal{O}(n^2)$ memory and $\mathcal{O}(n^3)$ operations since $\mathbf{K}$ is a dense symmetric matrix. For a typical dataset with 100 training points, 10 inputs, and five outputs, surrogate model construction using linear splines generally takes a few seconds on a modern workstation. When dealing with computationally expensive problems that cost more than a few minutes of CPU time per function evaluation, this training cost is generally negligible. However, when a few thousand training points are used, the computational cost and memory requirements can become prohibitive. Such costs are increased when there is also the need to establish suitable values for θ and λ when using kernels with tunable parameters – see Section 5.4.4 for a discussion of techniques for selecting the values of θ and λ for a given dataset. We also present later in Section 5.5.2 a greedy approach for constructing RBF approximations, which significantly reduces the computational cost and memory requirements when dealing with large datasets. An alternative approach based on fast multipole expansions was proposed by Cherrie et al. (2002) to construct RBF approximations for very large datasets. However, this approach can be used only when multiquadrics are employed as basis functions.

5.4.4 Hermite Interpolation using Radial Basis Functions

In a number of application areas, it is possible to efficiently compute the sensitivities of the objective and constraint functions; see Chapter 4 for a detailed exposition of such methods.

For these problems, it can make sense to use sensitivity data to construct surrogate models as they can be more accurate than those built using function values only. In the context of generalized linear models, the output sensitivities can be readily incorporated into the surrogate model using the idea of Hermite interpolation (Zhongmin 1992). To illustrate the idea of Hermite interpolation, let us denote the training dataset by $\{\mathbf{x}^{(i)}, y(\mathbf{x}^{(i)}), \nabla y(\mathbf{x}^{(i)})\}$, $i = 1, 2, \ldots, n$, where $\nabla y(\mathbf{x}) = \{\partial y/\partial x_1, \partial y/\partial x_2, \ldots, \partial y/\partial x_p\} \in \mathbb{R}^p$ denotes the first derivative of the output $y(\mathbf{x})$ with respect to the components of the input vector. Then, a Hermite interpolant can be written in terms of a set of RBFs as follows:

$$\widehat{y}(\mathbf{x}) = \sum_{i=1}^{n} \alpha_i K(||\mathbf{x} - \mathbf{x}^{(i)}||) + \sum_{i=1}^{n} \sum_{j=1}^{p} \alpha_{ij} \frac{\partial K(||\mathbf{x} - \mathbf{x}^{(i)}||)}{\partial x_j}, \tag{5.33}$$

where α_i and α_{ij}, $i = 1, 2, \ldots, n$, $j = 1, 2, \ldots, p$ are undetermined weights.

Since the training dataset contains $y(\mathbf{x}) \in \mathbb{R}$ and $\nabla y(\mathbf{x}) \in \mathbb{R}^p$ at n points, we need a total of $n(p + 1)$ linear algebraic equations to ensure a unique solution. The first set of n equations using the function values at the points $\mathbf{x}^{(i)}, i = 1, 2, \ldots, n$ can be written as

$$\sum_{i=1}^{n} \alpha_i K(||\mathbf{x}^{(k)} - \mathbf{x}^{(i)}||) + \sum_{i=1}^{n} \sum_{j=1}^{p} \alpha_{ij} \frac{\partial K(||\mathbf{x}^{(k)} - \mathbf{x}^{(i)}||)}{\partial x_j} = y(\mathbf{x}^{(k)}), \tag{5.34}$$

where $k = 1, 2, \ldots, n$.

An additional set of np equations can be derived by matching the surrogate derivatives with the derivative information available in the training dataset as follows:

$$\nabla \widehat{y}(\mathbf{x}^{(i)}) = \nabla y(\mathbf{x}^{(i)}), \quad i = 1, 2, \ldots, n. \tag{5.35}$$

Note that to apply (5.35) the Hermite interpolant in (5.33) needs to be differentiated with respect to x_l, which yields

$$\frac{\partial \widehat{y}(\mathbf{x})}{\partial x_l} = \sum_{i=1}^{n} \alpha_i \frac{\partial K(||\mathbf{x} - \mathbf{x}^{(i)}||)}{\partial x_l} + \sum_{i=1}^{n} \sum_{j=1}^{p} \alpha_{ij} \frac{\partial^2 K(||\mathbf{x} - \mathbf{x}^{(i)}||)}{\partial x_l \partial x_j}. \tag{5.36}$$

It can be noted from the preceding equation that in order to construct a Hermite interpolation model, the kernel K must be differentiable at least twice. The system of linear algebraic equations arising from (5.34) and (5.35) can be compactly written as $\mathbf{K}_g \alpha_g = \mathbf{y}_g$, where

$$\alpha_g = \{\alpha_1, \alpha_{11}, \alpha_{12}, \ldots, \alpha_{1p}, \ldots, \alpha_n, \alpha_{n1}, \alpha_{n2}, \ldots, \alpha_{np}\}^T \in \mathbb{R}^{n(p+1)}, \tag{5.37}$$

and

$$\mathbf{y}_g = \left\{ y(\mathbf{x}^{(1)}), \frac{\partial y(\mathbf{x}^{(1)})}{\partial x_1}, \frac{\partial y(\mathbf{x}^{(1)})}{\partial x_2}, \ldots, \frac{\partial y(\mathbf{x}^{(1)})}{\partial x_p}, \ldots \right.$$

$$\left. \ldots, y(\mathbf{x}^{(n)}), \frac{\partial y(\mathbf{x}^{(n)})}{\partial x_1}, \frac{\partial y(\mathbf{x}^{(n)})}{\partial x_2}, \ldots, \frac{\partial y(\mathbf{x}^{(n)})}{\partial x_p} \right\}^T \in \mathbb{R}^{n(p+1)}. \tag{5.38}$$

The coefficient matrix $\mathbf{K} \in \mathbb{R}^{n(p+1) \times n(p+1)}$ can be written in partitioned form as follows

$$\mathbf{K}_g = \begin{bmatrix} \mathbf{\Phi}_{11} & \mathbf{\Phi}_{12} & \cdots & \mathbf{\Phi}_{1n} \\ \mathbf{\Phi}_{21} & \mathbf{\Phi}_{22} & \cdots & \mathbf{\Phi}_{2n} \\ \cdots & \cdots & \cdots & \cdots \\ \mathbf{\Phi}_{n1} & \mathbf{\Phi}_{n2} & \cdots & \mathbf{\Phi}_{nn} \end{bmatrix}, \tag{5.39}$$

where

$$
\Phi_{ij} = \begin{bmatrix}
K(||\mathbf{x}^{(i)} - \mathbf{x}^{(j)}||) & \dfrac{\partial K(||\mathbf{x}^{(i)} - \mathbf{x}^{(j)}||)}{\partial x_1} & \cdots & \dfrac{\partial K(||\mathbf{x}^{(i)} - \mathbf{x}^{(j)}||)}{\partial x_p} \\
\dfrac{\partial K(||\mathbf{x}^{(i)} - \mathbf{x}^{(j)}||)}{\partial x_1} & \dfrac{\partial^2 K(||\mathbf{x}^{(i)} - \mathbf{x}^{(j)}||)}{\partial x_1^2} & \cdots & \dfrac{\partial^2 K(||\mathbf{x}^{(i)} - \mathbf{x}^{(j)}||)}{\partial x_1 \partial x_p} \\
\cdots & \cdots & \cdots & \cdots \\
\dfrac{\partial K(||\mathbf{x}^{(i)} - \mathbf{x}^{(j)}||)}{\partial x_p} & \dfrac{\partial^2 K(||\mathbf{x}^{(i)} - \mathbf{x}^{(j)}||)}{\partial x_p \partial x_1} & \cdots & \dfrac{\partial^2 K(||\mathbf{x}^{(i)} - \mathbf{x}^{(j)}||)}{\partial x_p^2}
\end{bmatrix}. \tag{5.40}
$$

This set of equations can be solved to compute the undetermined weights of the Hermite RBF interpolant. Conditions governing the nonsingularity of $\mathbf{K}_g$ can be found in Zhongmin (1992). It may be noted that in comparison to the standard RBF approximation technique presented earlier, the size of the system of equations to be solved in the Hermite interpolation approach is a function of p since $\mathbf{K}_g \in \mathbb{R}^{n(p+1) \times n(p+1)}$. As a result, the computational cost and memory requirements of Hermite interpolation become significant when the number of training points and design variables is increased. One way to reduce the computational effort and memory requirements is to use sensitivity information only at those training points that are of particular interest. For example, if the surrogate model is to be used for design studies, then it may make sense to use the sensitivity information only at those points in the training dataset that correspond to promising designs. Let $\mathcal{J}$ denote the set of indices of those points in the training dataset that are of particular interest. Then, a Hermite interpolant that employs output sensitivities only at those points in $\mathcal{J}$ can be written as

$$
\widehat{y}(\mathbf{x}) = \sum_{i=1}^{n} \alpha_i K(||\mathbf{x} - \mathbf{x}^{(i)}||) + \sum_{i \in \mathcal{J}} \sum_{j=1}^{p} \alpha_{ij} \frac{\partial K(||\mathbf{x} - \mathbf{x}^{(i)}||)}{\partial x_j}. \tag{5.41}
$$

The undetermined weights in the approximation can be computed on the lines of the procedure outlined earlier. If the cardinality of $\mathcal{J}$ is k, then an $(n + kp) \times (n + kp)$ system of equations needs to be solved to calculate the weights in (5.41).

On similar lines, it is possible to rewrite the Hermite interpolant such that the derivatives with respect to only important variables are considered. Again, let $\mathcal{J}$ denote the indices of the design variables whose derivatives are to be considered. Then, the Hermite interpolant can be rewritten as

$$
\widehat{y}(\mathbf{x}) = \sum_{i=1}^{n} \alpha_i K(||\mathbf{x} - \mathbf{x}^{(i)}||) + \sum_{i=1}^{n} \sum_{j \in \mathcal{J}} \alpha_{ij} \frac{\partial K(||\mathbf{x} - \mathbf{x}^{(i)}||)}{\partial x_j}. \tag{5.42}
$$

A comparison of the prediction accuracies of the Hermite and standard RBF interpolation techniques for a simple one-dimensional problem is shown in Figure 5.3. It can be clearly seen that the Hermite interpolant captures the true trends of the underlying function more accurately than the simple data only model.

5.4.5 Tuning RBF Shape and Regularization Parameters

Recall that Gaussian, multiquadric and inverse multiquadric RBFs are defined in terms of a shape parameter θ that needs to be selected by the user (see Table 5.1). Irrespective of the chosen value of θ, the resulting model is always an interpolant. However, the generalization

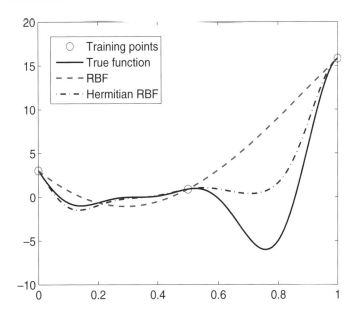

Figure 5.3 Comparison of Hermite RBF with standard RBF. A Gaussian kernel is used for both cases.

performance (and the degree of smoothness) of RBF approximations can depend to a significant extent on the value of the shape parameter. Hence, it is of interest to examine techniques for estimating the optimum value of the shape parameter in order to ensure that the model performs well on unseen data. These techniques can also be applied to the problem of RBF regression, where the regularization parameter λ, which controls the degree of regression, needs to be set.

In what follows, we use the notation introduced earlier in Section 5.3 for the sake of generality, since the methods presented here for tuning the RBF shape and regularization parameters can also be applied to a variety of other approximation techniques. Let $Q(\mathbf{z}, \boldsymbol{\alpha}_n)$ denote the function that minimizes the empirical risk

$$R_{\text{emp}}(\boldsymbol{\alpha}) = \frac{1}{n} \sum_{i=1}^{n} Q(\mathbf{z}_i, \boldsymbol{\alpha}), \tag{5.43}$$

where $\mathbf{z}_i = [\mathbf{x}^{(i)}, y^{(i)}]$ denotes the ith training point. For the case of the standard RBF approximation presented earlier in (5.28), $Q(\mathbf{z}_i, \boldsymbol{\alpha}) = (y^{(i)} - \sum_{j=1}^{n} \alpha_j K(\|\mathbf{x}^{(i)} - \mathbf{x}^{(j)}\|))^2$ is the loss or prediction error at the ith training point. $\boldsymbol{\alpha}_n$ denotes the vector of weights which are computed by solving the linear algebraic system of equations $\mathbf{K}\boldsymbol{\alpha} = \mathbf{y}$. If the RBF approximation is an interpolant, then $Q(\mathbf{z}, \boldsymbol{\alpha}_n) = 0$, that is, the prediction error on the training dataset is zero.

As discussed earlier in the context of statistical learning theory, a good approximation model should minimize the true risk (generalization error) $R(\boldsymbol{\alpha}_n) = \int Q(\mathbf{z}, \boldsymbol{\alpha}_n) \, d\mathcal{P}(\mathbf{z})$, where $\mathcal{P}(\mathbf{z})$ is the joint statistical distribution of $\mathbf{x}$ and y. Since $\mathcal{P}(\mathbf{z})$ is unknown, the true risk cannot be computed. In theory, it is possible to numerically approximate $R(\boldsymbol{\alpha}_n)$ by creating a large set of validation points. This can provide a criterion that can be minimized to estimate the RBF shape parameters. However, in many practical cases, where the training dataset is generated by running a computationally expensive computer model, it is infeasible to generate a large separate validation dataset. This motivates the development of computationally more efficient alternatives for approximating the true risk.

The leave-one-out procedure presents one way to estimate the accuracy of the model $Q(\mathbf{z}, \boldsymbol{\alpha}_n)$ without the need for creating additional data. This procedure essentially involves leaving the kth training point out and minimizing R_{emp} to obtain the model $Q(\mathbf{z}, \boldsymbol{\alpha}_{n-1})$. The model can then be run to predict the loss (measure of prediction error) at the point that was deleted from the training dataset. We denote the loss at the kth point, which was not used to construct the model, as $Q(\mathbf{z}_k, \boldsymbol{\alpha}_{n-1}|\mathbf{z}_k)$. Repeating this process n times leads to the following leave-one-out estimator:

$$\mathcal{L}(\mathbf{z}_1, \mathbf{z}_2, \ldots, \mathbf{z}_n) = \sum_{i=1}^{n} Q(\mathbf{z}_i, \boldsymbol{\alpha}_{n-1}|\mathbf{z}_i). \tag{5.44}$$

It can be shown that the leave-one-out estimator is almost unbiased, that is,

$$\left\langle \frac{\mathcal{L}(\mathbf{z}_1, \mathbf{z}_2, \ldots, \mathbf{z}_n)}{n} \right\rangle = \langle R(\boldsymbol{\alpha}_n) \rangle, \tag{5.45}$$

where $\langle \cdot \rangle$ denotes the expectation operator.

In other words, on an average, over all possible datasets drawn from the distribution $\mathcal{P}(\mathbf{z})$, the mean of the leave-one-out error equals the mean value of true risk when $n \to \infty$. Even though the leave-one-out error is an unbiased estimator of true risk, its variance can be high because the n training datasets created by deleting each point in turn are similar to each other. A biased estimator with lower variance can be obtained using a k-fold cross-validation procedure. The basic idea here is to divide the training dataset randomly into k subsets. The approximation model is then constructed using $k-1$ subsets and the remainder data is used to estimate the model error. In a similar fashion to the leave-one-out procedure, the k-fold cross-validation process is repeated k times to arrive at an estimate of model accuracy. The best value of k for a given problem depends on the nature of the underlying input–output relationship and the number of training points. Hastie et al. (2001) recommend that, in practice, a five- or 10-fold cross-validation procedure gives a good compromise.

From a practical viewpoint, computing the leave-one-out estimator can be expensive since, each time a training point is deleted, a new surrogate model needs to be constructed. For the case of RBF approximations, it is possible to use some shortcuts from numerical linear algebra to efficiently compute the leave-one-out error. To illustrate, let the QR decomposition of the Gram matrix $\mathbf{K}$ be available such that $\mathbf{K} = \mathbf{QR}$, where $\mathbf{Q} \in \mathbb{R}^{n \times n}$ is an orthogonal matrix and $\mathbf{R} \in \mathbb{R}^{n \times n}$ is an upper triangular matrix. Then, the weight vector $\boldsymbol{\alpha}$ can be computed by solving the upper triangular system of equations $\mathbf{R}\boldsymbol{\alpha} = \mathbf{Q}^T \mathbf{y}$. Further, the leave-one-out error can be computed cheaply as follows:

$$\mathcal{L} = (1/n) \sum_{k=1}^{n} \left(\left(y^{(k)} - \widehat{y}(\mathbf{x}^{(k)}) \right) / \left(1 - \mathbf{Q}(k, :)\mathbf{Q}(k, :)^T \right) \right)^2, \tag{5.46}$$

where $\mathbf{Q}(k, :)$ denotes the kth row of $\mathbf{Q}$ and $\widehat{y}(\mathbf{x}^{(k)})$ is the prediction made by the surrogate model constructed using all the n points.

The RBF shape and regularization parameters can be tuned by directly minimizing the leave-one-out estimator or the k-fold cross-validation error measure. However, this is computationally expensive particularly when the RBF shape parameters are to be tuned, because each time the shape parameter is perturbed, the Gram matrix **K** needs to be decomposed again. However, for the case when only the regularization parameter λ is to be tuned, it is possible to efficiently compute the leave-one-out error provided a singular value decomposition of **K** is available (Golub and Van Loan 1996). An alternative approach for tuning the shape parameter based on maximum likelihood estimation is described later in Section 5.6.2.

5.5 Sparse Approximation Techniques

We next look at some algorithms for constructing generalized linear models, in which a finite subset of basis functions are used to approximate the output of interest. Again, we consider the problem of constructing an approximation model using observational data $\mathcal{D} := (\mathbf{x}^{(i)}, y(\mathbf{x}^{(i)}))$, $i = 1, 2, \ldots, n$, where $\mathbf{x} \in \mathbb{R}^p$ denotes the input vector and $y \in \mathbb{R}$ denotes the target to be approximated. Here, the focus is on sparse models of the form $\widehat{y}(\mathbf{x}) = \sum_{j=1}^{m} \alpha_j K(\mathbf{x}, \tilde{\mathbf{x}}^{(j)})$, where $\{\tilde{\mathbf{x}}^{(j)}\}_{j=1}^{m} \subseteq \mathcal{D}$.[11] Further, $K(\mathbf{x}, \mathbf{x}^{(i)})$ is a Mercer kernel (positive-definite basis function).

5.5.1 The Support Vector Machine

The support vector machine (SVM) derived from Vapnik–Chervonenkis theory (Vapnik 1998) has emerged as a popular technique for function approximation, classification and density estimation in recent years. Here, we present a brief and simplified account of the basic steps involved in the SVM algorithm in the context of function approximation. A detailed account of the theoretical and computational aspects of SVMs can be found in the texts by Vapnik (1998), Cristianini and Shawe-Taylor (2000) and Schölkopf and Smola (2001).

In contrast to polynomial models and RBF approximations based on least-squares minimization, SVM regression typically involves minimization of the ε-insensitive loss function $\sum_{i=1}^{n} |y^{(i)} - \widehat{y}(\mathbf{x}^{(i)}, \boldsymbol{\alpha})|_{\varepsilon}$, where

$$|y^{(i)} - \widehat{y}(\mathbf{x}^{(i)}, \boldsymbol{\alpha})|_{\varepsilon} = \begin{cases} 0 & \text{if } |y^{(i)} - \widehat{y}(\mathbf{x}^{(i)}, \boldsymbol{\alpha})| < \varepsilon, \\ |y^{(i)} - \widehat{y}(\mathbf{x}^{(i)}, \boldsymbol{\alpha})| - \varepsilon & \text{otherwise.} \end{cases} \tag{5.47}$$

It can be seen that the ε-insensitive function applies a linear penalty only when the absolute value of the prediction error is greater than ε – otherwise, the loss at that point is set to zero. Note that ε is a user-defined parameter that should be specified on the basis of the expected level of error/noise in the outputs.

The model structure chosen in SVMs is a generalized linear model of the form $\widehat{y}(\mathbf{x}) = \alpha_0 + \sum_{i=1}^{n} \alpha_i K(\mathbf{x}, \mathbf{x}^{(i)})$, where K is a Mercer kernel. The principle of structural risk minimization (Vapnik 1998) states that to create a model that generalizes well the empirical risk as well as a measure of model complexity should be simultaneously minimized. In the present context, the L_2 norm of the weight vector $\boldsymbol{\alpha} = \{\alpha_0, \alpha_1, \ldots, \alpha_n\}^T \in \mathbb{R}^{n+1}$ can be used as a measure of model complexity. Hence, $\boldsymbol{\alpha}$ can be computed by minimizing the cost function $1/2\|\boldsymbol{\alpha}\|_2^2 + \mathcal{C} \sum_{i=1}^{n} |y^{(i)} - \widehat{y}(\mathbf{x}^{(i)})|_{\varepsilon}$, where $\mathcal{C}$ is a weight factor that dictates the trade-off between the empirical risk and model complexity. Instead of directly minimizing

---

[11]This is similar to RBF interpolators discussed in the previous section, where $m = n$. However, since here only m basis functions are used (where $m < n$), the error over the training dataset will be nonzero. Hence, the model can be interpreted as a sparse approximate regressor.

this cost function, the undetermined weight vector can be computed by solving the following constrained convex quadratic programming problem:

Minimize: $\frac{1}{2}\alpha^T\alpha$

Subject to: $y^{(i)} - \alpha_0 - \sum_{j=1}^{n} \alpha_j K(\mathbf{x}^{(i)}, \mathbf{x}^{(j)}) \leq \varepsilon, \quad i = 1, 2, \ldots, n,$ (5.48)

$\alpha_0 + \sum_{j=1}^{n} \alpha_j K(\mathbf{x}^{(i)}, \mathbf{x}^{(j)}) - y^{(i)} \leq \varepsilon, \quad i = 1, 2, \ldots, n.$

The two constraints given above ensure that the prediction error of the optimal model lies in the interval $[-\varepsilon, \varepsilon]$. However, to account for the possibility that the model may not precisely meet this condition, the constraints can be relaxed by introducing slack variables $\mathcal{E}_i$, $\mathcal{E}_i^*$, $i = 1, 2, \ldots, n$, which leads to the following reformulation of the optimization problem:

Minimize: $\frac{1}{2}\alpha^T\alpha + \mathcal{C}\sum_{i=1}^{n}(\mathcal{E}_i + \mathcal{E}_i^*)$

Subject to: $y^{(i)} - \alpha_0 - \sum_{j=1}^{n} \alpha_j K(\mathbf{x}^{(i)}, \mathbf{x}^{(j)}) \leq \varepsilon + \mathcal{E}_i,$

$\alpha_0 + \sum_{j=1}^{n} \alpha_j K(\mathbf{x}^{(i)}, \mathbf{x}^{(j)}) - y^{(i)} \leq \varepsilon + \mathcal{E}_i^*,$ (5.49)

$\mathcal{E}_i, \mathcal{E}_i^* \geq 0, \quad i = 1, 2, \ldots, n.$

As before, the user-defined parameter $\mathcal{C}$ dictates how the empirical risk is traded off with model complexity. For example, the empirical risk of the final model will decrease when the value of $\mathcal{C}$ is increased, while the model complexity term will grow.

The Lagrangian of the above optimization problem can be written as

$$\mathcal{L}(\alpha, \mathcal{E}, \mathcal{E}^*; \tilde{\alpha}, \tilde{\alpha}^*, \beta, \beta^*) = \frac{1}{2}\alpha^T\alpha + \mathcal{C}\sum_{i=1}^{n}(\mathcal{E}_i + \mathcal{E}_i^*)$$

$$- \sum_{i=1}^{n} \tilde{\alpha}_i \left(y^{(i)} - \alpha_0 - \sum_{j=1}^{n} \alpha_j K(\mathbf{x}^{(i)}, \mathbf{x}^{(j)}) - \varepsilon - \mathcal{E}_i\right)$$

$$- \sum_{i=1}^{n} \tilde{\alpha}_i^* \left(\alpha_0 + \sum_{j=1}^{n} \alpha_j K(\mathbf{x}^{(i)}, \mathbf{x}^{(j)}) - y^{(i)} - \varepsilon - \mathcal{E}_i^*\right)$$

$$- \sum_{i=1}^{n} (\beta_i \mathcal{E}_i + \beta_i^* \mathcal{E}_i^*). \tag{5.50}$$

The saddle point of $\mathcal{L}$ can be found by minimizing with respect to the primal variables $\alpha, \mathcal{E}, \mathcal{E}^*$ and maximizing with respect to the Lagrange multipliers $\tilde{\alpha}, \tilde{\alpha}^*, \tilde{\beta}, \tilde{\beta}^*$. Setting the corresponding derivatives to zero and eliminating the primal variables, we obtain the following expression for the final approximation:

$$\widehat{y}(\mathbf{x}) = \sum_{i=1}^{n} (\tilde{\alpha}_i - \tilde{\alpha}_i^*) K(\mathbf{x}, \mathbf{x}^{(i)}) \tag{5.51}$$

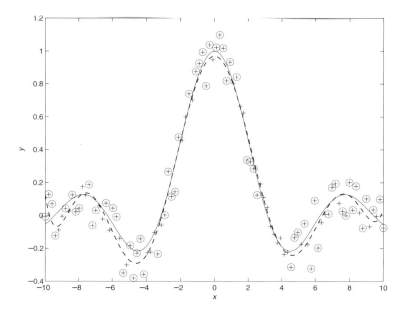

Figure 5.4 SVM approximation of the sinc function using 100 noisy samples. The solid line represents the true sinc function and the dashed line is the SVM approximation. The support vectors are circled. A Gaussian RBF with $\theta = 1$ is used.

and the optimization problem in the dual space (with unknowns $\tilde{\boldsymbol{\alpha}}$ and $\tilde{\boldsymbol{\alpha}}^*$) becomes:

$$
\begin{aligned}
\textbf{Maximize:} \quad & -\sum_{i=1}^{n} \varepsilon(\tilde{\alpha}_i + \tilde{\alpha}_i^*) + \sum_{i=1}^{n} y^{(i)}(\tilde{\alpha}_i - \tilde{\alpha}_i^*) \\
& -\frac{1}{2}\sum_{i,j=1}^{n}(\tilde{\alpha}_i - \tilde{\alpha}_i^*)(\tilde{\alpha}_j - \tilde{\alpha}_j^*)K(\mathbf{x}^{(i)}, \mathbf{x}^{(j)})
\end{aligned}
\tag{5.52}
$$

$$
\textbf{Subject to:} \quad \sum_{i=1}^{n}(\tilde{\alpha}_i - \tilde{\alpha}_i^*) = 0,
$$
$$
0 \leq \tilde{\alpha}_i, \tilde{\alpha}_i^* \leq \mathcal{C}, \; i = 1, 2, \dots, n.
$$

The quadratic programming problem arising in the SVM formulation is guaranteed to be convex if K is a Mercer kernel. Because of the nature of the constraints in the dual space, it is generally found that, at the optima, for a subset of data points, $\tilde{\alpha}_i - \tilde{\alpha}_i^* = 0$. Hence, it follows from (5.51) that only some of the kernels (basis functions) appear in the final model. The data points associated with these kernels are called the *support vectors*. Figure 5.4 shows the approximation constructed using a SVM for a dataset generating by adding Gaussian noise $\mathcal{N}(0, 0.1)$[12] to the sinc function $(\sin(x)/x)$.

Since training SVMs involves the solution of a quadratic programming problem, they are computationally more expensive than standard RBF approximations. Further, cross-validation procedures similar to those discussed in Section 5.5.3 need to be applied to estimate appropriate values of the user-defined parameters ε and $\mathcal{C}$. A detailed account of numerical schemes for efficiently training SVMs and estimating the user-defined parameters can be found in the text by Schölkopf and Smola (2001). A comparison of SVMs with a

---

[12]Note that we denote a Gaussian random variable with mean μ and variance σ^2 using the compact notation $\mathcal{N}(\mu, \sigma^2)$.

number of other surrogate modeling techniques can be found in Clarke et al. (to appear). This study suggests that SVMs are very promising for modeling data from computer experiments. However, much work remains to be undertaken to evaluate the potential of SVMs for modeling data from computer codes involving iterative computations that are terminated before complete convergence (for example, CFD codes). Here, the ε-insensitive loss function is expected to be particularly attractive for ensuring that the surrogate model filters out the errors in the output values.

5.5.2 Greedy Approximations

Next, we outline a greedy approach for constructing sparse approximate interpolators at a computational cost lower than SVM and RBF approximations. To illustrate the idea of greedy approximations, consider the RBF interpolation technique where the weight vector is computed by solving the dense linear system $\mathbf{K}\alpha = \mathbf{y}$, which requires $\mathcal{O}(n^2)$ memory and $\mathcal{O}(n^3)$ operations. The key idea behind greedy approximations is to reduce the memory requirements and computational costs by solving instead the overdetermined least-squares problem, $\min \|\mathbf{K}^m \alpha^m - \mathbf{y}\|_2$, where $\mathbf{K}^m \in \mathbb{R}^{n \times m}$ is a rectangular matrix formed by choosing a subset of the n columns of the original Gram matrix and $\alpha^m \in \mathbb{R}^n$ denotes the truncated weight vector. Note that since $\mathbf{K}$ is symmetric positive definite when Mercer kernels are used, $\mathbf{K}^m$ has full column rank. This approach intrinsically leads to a *sparse* model since $m < n$ and hence only a subset of the training data is employed in the final model. In addition, this approach requires only $\mathcal{O}(mn)$ memory and $\mathcal{O}(m^3)$ operations for computing the weight vector. Since the empirical/training error of the sparse approximation $\|\mathbf{K}^m \alpha^m - \mathbf{y}\| = \varepsilon > 0$ for $m \ll n$, such a model can also be referred to as a sparse approximate interpolator. Natarajan (1999) presented a theoretical justification for regularization via sparse approximate interpolation. It was shown that the interpolation error (ε) and noise in the target vector $\mathbf{y}$ will tend to cancel out if ε is chosen *a priori* to be the noise intensity.

The problem statement considered here can be stated as follows: *Given a dictionary of basis functions* $\mathcal{D} := \{\phi_1(\mathbf{x}), \phi_2(\mathbf{x}), \ldots, \phi_n(\mathbf{x})\}$, *a training dataset* $\{\mathbf{x}^{(i)}, y^{(i)}\}$, $i = 1, 2, \ldots, n$, *and* $\varepsilon > 0$, *find the smallest subset of m functions* ϕ_i, $i = 1, 2, \ldots, m$ *from* $\mathcal{D}$ *and constants* α_i, $i = 1, 2, \ldots, m$, *if they exist, such that* $\sum_{i=1}^{n} (\sum_{j=1}^{m} \alpha_j \phi_j(\mathbf{x}^{(i)}) - y^{(i)})^2 \leq \varepsilon$.

Clearly, choosing m basis functions from a dictionary with n elements involves combinatorial search over an $^n C_m$ space. As shown by Natarajan (1995), this problem is NP-hard, and hence recourse has to be made to suboptimal greedy search schemes to ensure computational efficiency. A number of papers devoted to the solution of a similar class of NP-hard approximation problems can be found in the signal processing, machine learning and numerical linear algebra literature. Such algorithms have applications in sparse approximate interpolation (Natarajan 1995), pattern recognition (Nair et al. 2003), signal compression, and computation of sparse approximate inverse preconditioners (Grote and Huckle 1997).

Here, we focus on the sequential forward greedy algorithm proposed by Nair et al. (2003) since numerical studies have shown that this approach is computationally cheaper and has modest memory requirements compared to other greedy algorithms in the literature. We consider the case when all the basis functions in the dictionary $\mathcal{D}$ are positive-definite RBFs – extension to positive semidefinite functions is readily possible provided

---

Algorithm 5.1: Template of a general greedy algorithm for function approximation

---

Inputs: A dictionary of basis functions $\mathcal{D} := \{\phi_1, \phi_2, \ldots, \phi_n\}$, the training dataset $\{\mathbf{x}^{(i)}, y^{(i)}\}, i = 1, 2, \ldots, n$, and tolerance for residual ε.

Set $k = 0$, $\boldsymbol{\alpha}^0 = 0$, $\mathbf{r}^0 = \mathbf{y}$, and $\mathcal{I}^k = [\ ]$

while $\|\mathbf{r}^k\| < \varepsilon$, **do**

1. $k \leftarrow k + 1$.
2. Find $i_k = arg \max_{j \notin \mathcal{I}^{k-1}} J_j$
3. $\mathcal{I}^k \leftarrow [\mathcal{I}^{k-1}, i_k]$
4. Compute $\mathbf{K}(:, i_k) = \{\phi_{i_k}(\mathbf{x}^{(1)}), \phi_{i_k}(\mathbf{x}^{(2)}), \ldots, \phi_{i_k}(\mathbf{x}^{(n)})\}^T \in \mathbb{R}^n$
5. Update weight vector $\boldsymbol{\alpha}$
6. Update residual $\mathbf{r}^k$

end

---

a polynomial of appropriate order is appended to the approximation as discussed earlier in Section 5.4.2.

The basic structure of a greedy algorithm to solve the above problem is shown in Algorithm 5.1. Here, $\mathcal{I}^k$ denotes the set of cardinality k, which contains the indices of the basis functions chosen from the dictionary $\mathcal{D}$ at iteration k. $\mathbf{K}(:, i_k)$ denotes the i_kth column of the Gram matrix $\mathbf{K}$. It can be noted that the core of the algorithm consists of two key elements – (1) selection of an index i_k (or basis function) at each iteration in step 2 by finding the basis function in the dictionary that leads to a maximum value of the criterion J, and (2) updating the weight vector $\boldsymbol{\alpha}$ and the residual $\mathbf{r} = \mathbf{y} - \mathbf{K}\boldsymbol{\alpha}$ to reflect the fact that a new basis function has been appended to the approximation.

A number of criteria (J) can be formulated to greedily select a basis function from $\mathcal{D}$ at each iteration. One simple and cheap way to compute i_k is to search for the point in the training dataset where the residual error is highest. This point can then be used as a center for the new basis function. Once a new basis function has been appended to the model, the weight vector $\boldsymbol{\alpha}$ and the residual $\mathbf{r}$ can be updated using gradient descent techniques such as those presented by Mallat and Zhang (1993); Nair et al. (2003); Schaback and Wendland (2000). In order to ensure numerical stability and fast convergence of the residual error, a QR factorization scheme is preferable to update $\boldsymbol{\alpha}$ and $\mathbf{r}$.

The goal of QR factorization is to decompose the matrix $\mathbf{K}$ into the product of an orthogonal matrix $\mathbf{Q}$ and an upper triangular matrix $\mathbf{R}$, that is, $\mathbf{K} = \mathbf{QR}$. A greedy variant of the QR factorization algorithm is summarized in Algorithm 5.2. Algorithm 5.2 can be interpreted as a thin incremental QR factorization scheme where the pivot is chosen greedily with respect to $\mathbf{y}$. More specifically, in step 2, we search for the maximum absolute value of the residual $\mathbf{r}$ and subsequently, the RBF centered at the point with highest residual is chosen as the next basis function.

In step 6 of Algorithm 5.2, the operation $\mathbf{q}^k \leftarrow [\mathbf{Q}^{k-1} \perp \mathbf{K}(:, i_k)]$ denotes the result of orthogonalizing a column vector $\mathbf{K}(:, i_k)$ with respect to the columns of the orthogonal matrix $\mathbf{Q}^{k-1} \in \mathbb{R}^{n \times (k-1)}$. It is to be noted here that a naive application of the classical

---

Algorithm 5.2: Greedy QR factorization scheme for sparse approximate interpolation using RBFs.

Inputs: A dictionary of basis functions $\mathcal{D} := \{\phi_1, \phi_2, \ldots, \phi_n\}$, the training dataset $\{\mathbf{x}^{(i)}, y^{(i)}\}, i = 1, 2, \ldots, n$, and tolerance for residual ε.

Set $k = 0$, $\boldsymbol{\alpha}^0 = 0$, $\mathbf{r}^0 = \mathbf{y}$, and $\mathcal{I}^k = [\ ]$

while $||\mathbf{r}^k|| < \varepsilon$, **do**

1. $k \leftarrow k + 1$.

2. Find $i_k = arg \max\limits_{j \notin \mathcal{I}^{k-1}} |\mathbf{r}_j^{k-1}|$

3. $\mathcal{I}^k \leftarrow [\mathcal{I}^{k-1}, i_k]$

4. Compute $\mathbf{K}(:, i_k) = \{\phi_{i_k}(\mathbf{x}^{(1)}), \phi_{i_k}(\mathbf{x}^{(2)}), \ldots, \phi_{i_k}(\mathbf{x}^{(n)})\}^T \in \mathbb{R}^n$

5. If $\mathbf{K}(:, i_k)$ numerically lies in the span of $\mathbf{Q}^{k-1}$ **break**.

6. $\mathbf{q}^k \leftarrow [\mathbf{Q}^{k-1} \perp \mathbf{K}(:, i_k)] \quad \in \mathbb{R}^n$

7. $\mathbf{Q}^k \leftarrow [\mathbf{Q}^{k-1}, q^k] \quad \in \mathbb{R}^{n \times k}$

8. $\mathbf{R}^k \leftarrow \begin{bmatrix} \mathbf{R}^{k-1} & (\mathbf{Q}^{k-1})^T \mathbf{K}(:, i_k) \\ 0 & (\mathbf{q}^k)^T \mathbf{K}(:, i_k) \end{bmatrix} \in \mathbb{R}^{k \times k}$

9. $\mathbf{r}^k \leftarrow \mathbf{r}^{k-1} - (\mathbf{q}^k)^T \mathbf{y} q^k$

end

Compute $\boldsymbol{\alpha}^k$ by solving $\mathbf{R}^k \boldsymbol{\alpha}^k = \mathbf{Q}^{k^T} \mathbf{y}$.

Gram–Schmidt algorithm can lead to numerical instabilities, particularly when the Gram matrix is highly ill-conditioned. It is preferable in practice to use the algorithm of Daniel et al. (1976) to update the thin QR factorization when a new basis function is selected from the dictionary. This implementation of updating the QR decomposition is based on the use of elementary two-by-two reflection matrices and the Gram–Schmidt process with reorthogonalization, thereby ensuring efficiency and numerical stability.

Step 5 of Algorithm 5.2 is included to address the concern that at later stages of the iterations it is possible that the new column $\mathbf{K}(:, i_k)$ may numerically lie in span $\{\mathbf{Q}^{k-1}\}$. To circumvent this problem, it is necessary to monitor the reciprocal condition number of the matrix $\tilde{\mathbf{Q}} := [\mathbf{Q}^{k-1}, \mathbf{K}(:, i_k)/||\mathbf{K}(:, i_k)||]$, which can be efficiently computed since the columns of $\mathbf{Q}^{k-1}$ are orthonormal (Reichel and Gragg 1990). In practice, the iterations can be terminated if the reciprocal condition number is lower than a specified threshold.

Assuming that the greedy iterative scheme terminates at the mth iteration, the weight vector can then be found by solving the upper triangular system of equations

$$\mathbf{R}^m \boldsymbol{\alpha}^m = (\mathbf{Q}^m)^T \mathbf{y}, \tag{5.53}$$

where $\mathbf{R}^m \in \mathbb{R}^{m \times m}$ and $\mathbf{Q}^m \in \mathbb{R}^{n \times m}$.

It can be noted from Algorithm 5.2 that the computational cost and memory requirements of the greedy algorithm scale as $\mathcal{O}(nm)$. Hence, this approach can be applied to large datasets. The cost savings compared to the standard RBF approach becomes particularly remarkable for cases when models with good generalization can be constructed using a few basis functions. We next consider in detail the issue of how to terminate the greedy algorithm using estimates of the generalization error.

5.5.3 Stopping Criteria: Empirical Risk versus Complexity Trade-off

From the description of the greedy algorithm in the previous section, it is clear that the user is required to specify the termination criterion, which could be either the tolerance on the residual (ε) or the final number of basis functions (m). This is a general problem in deciding metamodel structure. Consider the case when no regularization is employed. Then, if the specified tolerance on the residual (which is equivalent to the empirical error) is smaller than the noise intensity, the final model will tend to *overfit* the data. To illustrate this phenomena, the typical trends in the generalization error and training error as a function of the number of basis functions are shown in Figure 5.5 for a model problem. It can be noted that as complexity is increased the training error falls but generalization error reaches a minimum before becoming worse. In other words, there is an optimum level of complexity beyond which the model will tend to overfit the noise (here, the use of 10 basis functions seems an appropriate selection).

As discussed earlier in Section 5.3.6, the problem of overfitting can be overcome by trading off the empirical risk with model complexity. This observation is embodied in Vapnik's structural risk minimization principle. See also the "Null" hypothesis used in classical polynomial regression (Ralston and Rabinowitz 2001). A more traditional approach to this problem employs a regularization parameter to improve generalization. In the present context, the greedy algorithms can be applied to the regularized Gram matrix $\mathbf{K} + \lambda\mathbf{I}$, where λ is the regularization parameter. Alternatively, a shrinkage parameter (Friedman 2001) can be used. As a result, the empirical error is guaranteed to converge to a nonzero value. It is possible to enforce regularization via sparsity, that is, by finding the optimum level of sparsity that leads to the best generalization performance. This approach was adopted in

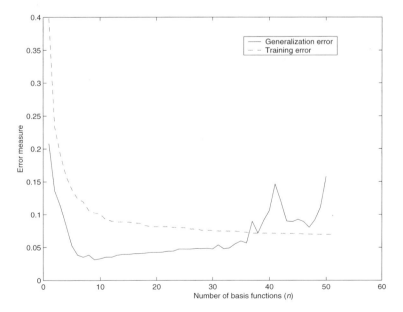

Figure 5.5 Trends in the generalization error when the number of basis functions are increased in the absence of regularization.

Vincent and Bengio (2001), where a validation dataset (or cross-validation) was employed to decide the optimum level of sparsity; see also Keane (2004).

There are a number of alternative model selection criteria in the literature that can be applied to decide when to terminate the greedy algorithm. Rissanen's principle of MDL is one such enabling technique that has been employed in the past for constructing parsimonious models. The basic idea is to choose the model that gives the shortest description of the data. From an information theoretic point of view, this translates into minimizing a code length of the form $-\ln L(\mathcal{D}|\boldsymbol{\alpha}) + \mathfrak{C}(\boldsymbol{\alpha})$, where $L(\cdot|\boldsymbol{\alpha})$ represents the data likelihood, given the model parameters ($\boldsymbol{\alpha}$), while the second term is a measure of complexity of the model in question. Numerous ways of describing this code length exist in the literature; see Hansen and Yu (2001) for a more detailed exposition of the MDL principle. Of particular interest is the two-stage MDL criterion, where the model parameters $\boldsymbol{\alpha}$ are selected and $L(\boldsymbol{\alpha})$ approximated in the first stage. This is followed by a second stage that involves computing the description length of the data, that is, $-\ln f(\mathcal{D}|\boldsymbol{\alpha})$.

In the present context of greedily minimizing the least-squares error function, applying the two-stage MDL criterion yields the following approximation to the code length at the kth stage (where k is the number of basis functions chosen from the dictionary):

$$\frac{n}{2}\ln(\mathbf{r}^T\mathbf{r}) + \frac{k}{2}\ln n. \tag{5.54}$$

An alternative to the MDL criterion is the Akaike information criterion (AIC) suitably modified for use in small samples (Sugiura 1978):

$$\frac{n}{2}\ln(\mathbf{r}^T\mathbf{r}) + \frac{k}{2}\frac{1+k/n}{1-(k+2)/n}. \tag{5.55}$$

It can be clearly seen from (5.54) and (5.55) that the first term of both criteria (a measure of empirical error) will tend to decrease as more basis functions are chosen from the dictionary. In contrast, the second term, which represents the model complexity, increases at a linear rate. As a consequence, when either MDL or AIC is used as the termination criterion, the sparsity level for which they reach a minimum represents an optimal trade-off between empirical risk and model complexity.

It can also be shown that the leave-one-out error can be computed cheaply at each iteration of the greedy algorithm as follows:

$$LOO(\mathbf{r}, k) = \frac{1}{n}\sum_{i=1}^{n}\left(\frac{\mathbf{r}(i)}{(1 - \mathbf{Q}^k(i, :)\mathbf{Q}^k(i, :)^T)}\right)^2 \tag{5.56}$$

where $\mathbf{Q}^k(i, :)$ denotes the ith row of the orthogonal matrix Q^k and $\mathbf{r}(i)$ denotes the ith element of the residual error vector (which is the prediction error at the ith training point). The leave-one-out error can also be used as a termination criterion for the greedy algorithm. Further, for the case when the dictionary contains RBFs with shape parameters, they can be tuned by minimizing the leave-one-out error.

Figure 5.6 shows how the greedy approximation technique filters noise for a one-dimensional test function; c.f. Figure 5.4. For this problem, all the stopping criteria indicate that good generalization performance is predicted when using eight Gaussian kernels. It is clear from the figure that the sparse model approximates the "true" sinc function closely. For a more detailed study of the performance of greedy algorithms, see Cotter et al. (1999); Nair et al. (2003).

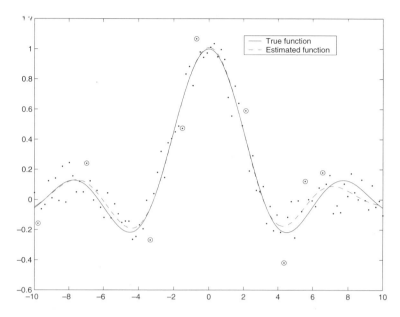

Figure 5.6 The centers of the basis functions (circled dots) chosen by the AIC criteria for approximating the sinc(x) function using 100 samples corrupted with Gaussian noise $\mathcal{N}(0, 0.1)$.

5.6 Gaussian Process Interpolation and Regression

We present next a statistically rigorous approach for constructing surrogate models of deterministic computer codes based on Gaussian stochastic process models. The underlying foundations of this approach were originally developed in the 1960s in the field of geostatistics, where this method is referred to as *Kriging*[13] (Cressie 1993). Sacks et al. (1989) popularized this approach in their paper on design and analysis of computer experiments (DACE). The book by Santner et al. (2003) on DACE presents a detailed description of work in this area. In the optimization literature, Gaussian process models are often referred to as Kriging or DACE models.

The stochastic process modeling approach has also attracted much interest in a section of the neural computing community that is involved in applying Bayesian formalisms to function approximation and classification. In the neural computing literature, this approach is referred to as Gaussian process regression. Much of the intensive research activity in this area resulted from the seminal work of Neal (1996), who showed that a large class of neural network models will converge to Gaussian process priors over functions in the limit of an infinite number of hidden units in the network. For an excellent discussion of the connections between neural networks and Gaussian process models, the reader is referred to MacKay (2003).

---

[13]The term "Kriging" was coined after South African mining engineer D. G. Krige, who developed methods for inferring ore grade distributions from sample grades in the 1950s (Matheron 1963).

5.6.1 Basic Formulation

We discuss the basic formulation of Gaussian process modeling from a Bayesian perspective (Currin et al. 1991; Neal 1998). In Bayesian modeling theory, a Gaussian process prior is generally placed on the unknown computer model. This may seem to be at odds with the reality that the computer model is indeed deterministic. However, when the underlying compute model is computationally expensive, we are treating its output $y(\mathbf{x})$ as random in the sense that it is an unknown quantity until the model is run, and it will not be a trivial exercise to learn the true value of $y(\mathbf{x})$; see, for example, Kennedy and O'Hagan (2000) for a discussion of this viewpoint. This convenient fiction provides us a useful statistical framework for assessing the accuracy of the metamodel. In essence, using a Gaussian process prior, the application of Bayesian inferencing leads to the posterior process that is also Gaussian. The mean of the posterior process can be used for prediction purposes, and its variance may be interpreted as a measure of uncertainty involved in making the prediction.

The model structure typically used in stochastic process approximation of the relationship $y = f(\mathbf{x})$ can be compactly written as

$$Y(\mathbf{x}) = \beta + Z(\mathbf{x}), \tag{5.57}$$

where β is an unknown hyperparameter to be estimated from the data and $Z(\mathbf{x})$ is a Gaussian stochastic process with zero mean and covariance given by

$$\text{Cov}\left(Z(\mathbf{x}, \mathbf{x}')\right) = \Gamma(\mathbf{x}, \mathbf{x}') = \sigma_z^2 R(\mathbf{x}, \mathbf{x}'). \tag{5.58}$$

In other words, the observed outputs of the simulation code $\{y^{(1)}, y^{(2)}, \ldots, y^{(n)}\}$ are assumed to be realizations of a Gaussian random field with mean β and covariance Γ. Here, $R(\cdot, \cdot)$ is a parameterized correlation function that can be tuned to the training dataset and σ_z^2 is the so-called process variance. It is possible to replace the hyperparameter β with a linear or quadratic polynomial in $\mathbf{x}$. However, for a sufficiently flexible choice of the correlation function, a constant β is often found to be sufficient for modeling highly complex input–output relationships.

For mathematical convenience, a commonly used correlation function is the stationary family which obeys the *product correlation rule*,[14] that is,

$$R(\mathbf{x}, \mathbf{x}') = \prod_{j=1}^{p} \mathfrak{r}(x_j - x_j'), \tag{5.59}$$

where $\mathfrak{r}$ is a one-dimensional positive-definite function. The simplest possible correlation function is perhaps the nonnegative linear function given below:

$$\mathfrak{r}(d) = 1 - \frac{1}{\theta}|d|, \qquad |d| < \theta,$$

$$= 0, \qquad |d| \geq \theta,$$

---

[14]A correlation function obeys the product correlation rule if it can be represented as the product of one-dimensional functions. These correlation functions are attractive for modeling multidimensional functions since they allow certain multidimensional integrals of the approximation to be written as products of one-dimensional integrals; see Section 5.9 and Chapter 8, which deals with uncertainty analysis using Gaussian process models.

where $\theta > 0$ is an unknown hyperparameter to be estimated from the data. (Mitchell et al. 1990) proposed a nonnegative cubic correlation function of the form

$$\mathfrak{r}(d) = 1 - 6\left(\frac{d}{\theta}\right)^2 + 6\left(\frac{|d|}{\theta}\right)^3, \qquad |d| < \frac{\theta}{2},$$

$$= 2\left(1 - \frac{|d|}{\theta}\right)^3, \qquad \frac{\theta}{2} \le |d| < \theta,$$

$$= 0, \qquad |d| \ge \theta.$$

A number of other choices for the correlation function (including nonstationary kernels and recipes for creating new kernels) can be found in the literature; see, for example, MacKay (2003). In most applications in the literature, the following Gaussian function is used:

$$R(\mathbf{x}, \mathbf{x}') = \prod_{j=1}^{p} \exp(-\theta_j |\mathbf{x}_j - \mathbf{x}'_j|^{m_j}), \tag{5.60}$$

where $\theta_j \ge 0$ and $0 < m_j \le 2$ are undetermined hyperparameters. Note that for $m_j = 2$ all sample paths of this process are infinitely differentiable. This value can hence be used in applications where it is believed that the underlying function being modeled is infinitely differentiable. If there is no basis for this assumption, then it may be preferable to tune the parameters m_j to the data – this allows for the possibility of modeling functions that may be discontinuous. The hyperparameters θ_j control the degree of nonlinearity in the model. For example, small values of θ_j indicate that the output is a smooth function of the jth variable, and large values indicate highly nonlinear behavior.

For the case when the observational dataset is corrupted by noise, an additional term has to be appended to the covariance function to model noise, that is,

$$\Gamma(\mathbf{x}, \mathbf{x}') = \sigma_z^2 R(\mathbf{x}, \mathbf{x}') + \mathcal{R}_n, \tag{5.61}$$

where $\mathcal{R}_n$ is the noise model. When the noise is output dependent, we can set $\mathcal{R}_n = \delta_{ii'}\theta_{p+1}$, where $\delta_{ii'}$ is the Kronecker delta function and θ_{p+1} is an unknown hyperparameter to be estimated from the data. When the noise is input dependent, a parameterized spatial correlation function has to be employed. In short, the implementation aspects of Gaussian process modeling do not change much for noisy datasets. The only difference is that additional hyperparameters have to be introduced to model noise. Further, the resulting model will not be an interpolator since an attempt is made to filter noise in the data. In the derivation that follows, we shall assume that the noise is output dependent and subsequently consider interpolating Gaussian process models as a special case.

In Gaussian process modeling, prior knowledge about the unknown function $y = f(\mathbf{x})$ is represented by a Gaussian process with parameterized mean and correlation functions, that is, any finite set of observations $y^{(1)}, y^{(2)}, \dots, y^{(n)}$ has a multivariate Gaussian distribution. Notice that the prior does not make use of the training dataset that is available to us. However, the specified mean and correlation functions can be used to dictate some of the properties that we desire in our approximation; for example, the Gaussian correlation function in (5.60) with $m_j = 2$ can be used to express the prior belief that the underlying function is infinitely differentiable. The prior can be now be updated using the training dataset.

Let y^* denote the unknown value of the output at the point $\mathbf{x}^*$. By assumption, $y^{(1)}, y^{(2)}, \ldots, y^{(n)}, y^*$ have a joint Gaussian distribution with specified covariance structure, that is,

$$\begin{bmatrix} \mathbf{y} \\ y^* \end{bmatrix} \sim \mathcal{N} \left(\begin{bmatrix} \mathbf{1}_{n+1}\beta \end{bmatrix}, \begin{bmatrix} \mathbf{\Gamma} & \boldsymbol{\gamma}(\mathbf{x}^*) \\ \boldsymbol{\gamma}(\mathbf{x}^*)^T & \Gamma(\mathbf{x}^*, \mathbf{x}^*) \end{bmatrix} \right), \tag{5.62}$$

where $\mathbf{y} = \{y^{(1)}, y^{(2)}, \ldots, y^{(n)}\}^T \in \mathbb{R}^n$ is the vector of outputs in the training dataset. $\mathbf{\Gamma} \in \mathbb{R}^{n \times n}$ is a symmetric positive-definite matrix computed using the training points, that is, the ijth element of $\mathbf{\Gamma}$ is given by $\Gamma(\mathbf{x}^{(i)}, \mathbf{x}^{(j)})$. $\boldsymbol{\gamma}(\mathbf{x}^*) = \sigma_z^2 \{R(\mathbf{x}^*, \mathbf{x}^{(1)}), R(\mathbf{x}^*, \mathbf{x}^{(2)}), \ldots, R(\mathbf{x}^*, \mathbf{x}^{(n)})\}^T \in \mathbb{R}^n$ and $\mathbf{1}_{n+1} = \{1, 1, \ldots, 1\}^T \in \mathbb{R}^{n+1}$.

We are ultimately interested in deriving an expression for the distribution of the unknown output at the point $\mathbf{x}^*$ in terms of the training dataset. Using (5.62), the conditional distribution of y^* given $\mathbf{y}$ (the posterior) is also Gaussian and can be written as

$$y^* | \mathbf{y} \sim \mathcal{N} \left(\beta + \boldsymbol{\gamma}(\mathbf{x}^*)^T \mathbf{\Gamma}^{-1}(\mathbf{y} - \mathbf{1}\beta), \Gamma(\mathbf{x}^*, \mathbf{x}^*) - \boldsymbol{\gamma}(\mathbf{x}^*)^T \mathbf{\Gamma}^{-1} \boldsymbol{\gamma}(\mathbf{x}^*) \right). \tag{5.63}$$

In summary, when a Gaussian process prior over functions is used, the posterior process is also Gaussian, that is, $Y(\mathbf{x}) | \mathbf{y} \sim \mathcal{N}(\widehat{y}(\mathbf{x}), \sigma_z^2 C(\mathbf{x}, \mathbf{x}'))$. Using (5.63), the posterior mean and covariance can be written as

$$\widehat{y}(\mathbf{x}) = \beta + \boldsymbol{\gamma}(\mathbf{x})^T \mathbf{\Gamma}^{-1} (\mathbf{y} - \mathbf{1}\beta) \tag{5.64}$$

$$\text{and} \quad C(\mathbf{x}, \mathbf{x}') = \Gamma(\mathbf{x}, \mathbf{x}') - \boldsymbol{\gamma}(\mathbf{x})^T \mathbf{\Gamma}^{-1} \boldsymbol{\gamma}(\mathbf{x}'). \tag{5.65}$$

For the case of interpolating Gaussian process models, the covariance function is given by $\Gamma(\mathbf{x}, \mathbf{x}') = \sigma_z^2 R(\mathbf{x}, \mathbf{x}')$. Hence, the posterior mean and covariance can be rewritten as

$$\widehat{y}(\mathbf{x}) = \beta + \mathbf{r}(\mathbf{x})^T \mathbf{R}^{-1} (\mathbf{y} - \mathbf{1}\beta) \tag{5.66}$$

$$\text{and} \quad C(\mathbf{x}, \mathbf{x}') = \sigma_z^2 \left(R(\mathbf{x}, \mathbf{x}') - \mathbf{r}(\mathbf{x})^T \mathbf{R}^{-1} \mathbf{r}(\mathbf{x}') \right). \tag{5.67}$$

Here, $\mathbf{R}$[15] is the correlation matrix whose ijth element is computed as $R(\mathbf{x}^{(i)}, \mathbf{x}^{(j)})$. $\mathbf{r}(\mathbf{x}) = \{R(\mathbf{x}, \mathbf{x}^{(1)}), R(\mathbf{x}, \mathbf{x}^{(2)}), \ldots, R(\mathbf{x}, \mathbf{x}^{(n)})\}^T \in \mathbb{R}^n$ is the correlation between $\mathbf{x}$ and all points in the training dataset and $\mathbf{1} = \{1, 1, \ldots, 1\}^T \in \mathbb{R}^n$. Note that the Gaussian process model can also be derived using a different line of approach that involves computing the best linear unbiased estimator. However, this leads to a slightly different expression for $C(\mathbf{x}, \mathbf{x}')$; see Sacks et al. (1989) for details.

It is clear from the equations for the posterior process that the Bayesian inferencing approach ultimately leads to an approximation of the computer model as a multidimensional Gaussian random field. The randomness due to (5.65) or (5.67) can be interpreted as an estimate of the uncertainty involved in making predictions at a new point $\mathbf{x}$ using a limited amount of training data – namely, the posterior variance $\sigma^2(\mathbf{x}) = C(\mathbf{x}, \mathbf{x}) = \sigma_z^2(1 - \mathbf{r}(\mathbf{x})^T \mathbf{R}^{-1} \mathbf{r}(\mathbf{x}'))$ provides an estimate of the uncertainty involved in predicting the output at $\mathbf{x}$.

Note that when the ith training point is substituted into (5.66), we have $\widehat{y}(\mathbf{x}^{(i)}) = y^{(i)}$ since $\mathbf{r}(\mathbf{x}^{(i)})$ is the ith column of the correlation matrix $\mathbf{R}$. Hence, the Gaussian process model is an interpolator irrespective of the correlation function being used. The

---

[15]Notice that for the case of output dependent noise, $\mathbf{\Gamma} = \sigma_z^2 \mathbf{R} + \theta_{p+1} \mathbf{I}$, where $\mathbf{I}$ denotes the identity matrix and θ_{p+1} is the hyperparameter used to model noise.

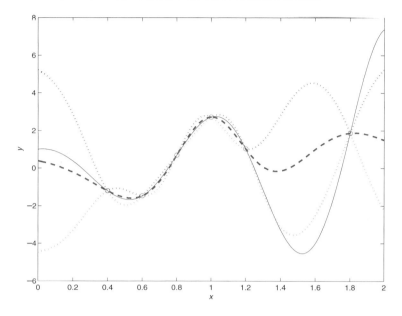

Figure 5.7 Posterior mean (dashed) and error ($\pm 3\sigma$ – dotted) for a smooth function predicted using Gaussian process interpolation. The true function is shown by the solid curve and the training points are indicated by circles. It can be seen that the posterior variance is zero at the training points and increases in between them.

posterior variance $\sigma^2(\mathbf{x})$ will be zero for all the training points. In general, the posterior variance will increase as the point $\mathbf{x}$ is moved further away from the set of training points. If the covariance function in (5.61) is used to handle output noise, then the model will regress the training data. Further, the posterior variance at the training points will be nonzero. Figures 5.7 and 5.8 show the posterior mean and error ($\pm 3\sigma$) computed using Gaussian process interpolation and regression models for one-dimensional test problems.

In practice, for the sake of computational efficiency, it is preferable to compute the Cholesky decomposition of $\mathbf{R}$. This allows the posterior mean to be computed using a vector–vector product, that is, $\widehat{y}(\mathbf{x}) = \beta + \mathfrak{r}(\mathbf{x})^T \mathbf{w}$, where $\mathbf{w} = \mathbf{R}^{-1}(\mathbf{y} - \mathbf{1}\beta)$. However, the computation of the variance (or error bar) of the posterior process requires a forward and backward substitution.

It is instructive to rewrite the posterior mean in (5.66) as follows:

$$\widehat{y}(\mathbf{x}) = \beta + \sum_{i=1}^{n} w_i R(\mathbf{x}, \mathbf{x}^{(i)}), \qquad (5.68)$$

where w_i denotes the ith component of the vector $\mathbf{w}$. It can be noted from this equation that the final expression for the posterior mean is similar to the model structure used in generalized linear models such as RBF approximations; see (5.21). The main difference is that in the Bayesian approach a parameterized correlation function is used (this is equivalent to employing positive-definite basis functions in a generalized linear model). Since the basis

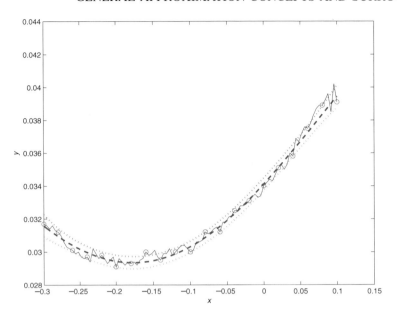

Figure 5.8 Posterior mean (dashed) and error ($\pm 3\sigma$ – dotted) for a noisy function predicted using Gaussian process regression. The true function is shown by the solid curve and the training points are indicated by circles. It can be observed that the model attempts to filter noise in the data and that the posterior variance is nonzero at the training points. The dataset used was generated from partially converged CFD runs on a simplified airfoil problem with a single design variable (Forrester 2004).

functions are tuned via iterative optimization of a statistical criterion, a Gaussian process model can be thought of as a generalized nonlinear model.[16]

It is of interest to note that by choosing new design points that maximize $\sigma^2(\mathbf{x})$, the accuracy of the metamodel can be improved. This criterion will lead to sampling those regions of the design space where the current metamodel is likely to be most inaccurate – techniques for planning computer experiments based on this idea will be presented later in Section 5.8.

5.6.2 Maximum Likelihood Estimation

We next turn our attention to the problem of estimating the unknown parameters in the Gaussian process model. Let the covariance function Γ be parameterized in terms of the vector $\boldsymbol{\theta} = \{\theta_1, \theta_2, \ldots, \theta_p\}$ (along with additional parameters if a noise model is used); see, for example, (5.60). The other hyperparameters to be estimated are β and σ_z^2. In order to compute these unknown hyperparameters, a maximum likelihood estimation (MLE) procedure can be employed – this corresponds to an empirical Bayesian treatment of the interpolation/regression problem. Alternative techniques such as Markov chain Monte Carlo

---

[16]Loosely speaking, in a generalized nonlinear model, the parameter estimation problem requires the application of nonlinear optimization techniques. This is also the case for the Gaussian process interpolation/regression technique since an iterative maximum likelihood estimation procedure is needed to estimate the hyperparameters within the correlation function.

simulation, which allow a more complete Bayesian treatment of the hyperparameters, exist; however, they require significant computational effort; see Rasmussen (1996) for details.

As discussed previously in Section 5.3.3, the MLE approach leads to those values of the undetermined parameters that are most likely to have generated the training dataset. For the case of Gaussian process interpolation,[17] since we assume that the observed outputs have a Gaussian distribution, the likelihood function is

$$(2\pi)^{-n/2}(\sigma_z^2)^{-n/2}|\mathbf{R}|^{-1/2}\exp\left(-\frac{1}{2\sigma_z^2}(\mathbf{y}-\mathbf{1}\beta)^T\mathbf{R}^{-1}(\mathbf{y}-\mathbf{1}\beta)\right). \tag{5.69}$$

Hence, the negative log-likelihood function to be minimized becomes

$$L(\boldsymbol{\theta},\beta,\sigma_z^2) = -\frac{1}{2}\left[n\ln(2\pi)+n\ln\sigma_z^2+\ln|\mathbf{R}|+\frac{1}{2\sigma_z^2}(\mathbf{y}-\mathbf{1}\beta)^T\mathbf{R}^{-1}(\mathbf{y}-\mathbf{1}\beta)\right]. \tag{5.70}$$

In order to obtain the maximum likelihood estimate of β, we first differentiate L with respect to β, which gives

$$\frac{\partial L(\boldsymbol{\theta},\beta,\sigma_z^2)}{\partial\beta} = \frac{1}{2\sigma_z^2}(\mathbf{1}^T\mathbf{R}^{-1}\mathbf{y}-\mathbf{1}^T\mathbf{R}^{-1}\mathbf{1}\beta). \tag{5.71}$$

Setting $\frac{\partial L}{\partial\beta}=0$ gives the following estimate for β

$$\widehat{\beta} = \frac{\mathbf{1}^T\mathbf{R}^{-1}\mathbf{y}}{\mathbf{1}^T\mathbf{R}^{-1}\mathbf{1}}. \tag{5.72}$$

Similarly, the maximum likelihood estimate of σ_z^2 can be computed as

$$\widehat{\sigma_z^2} = \frac{1}{n}(\mathbf{y}-\mathbf{1}\beta)^T\mathbf{R}^{-1}(\mathbf{y}-\mathbf{1}\beta). \tag{5.73}$$

Now, differentiating L with respect to θ_j gives

$$\frac{\partial L(\boldsymbol{\theta})}{\partial\theta_j} = -\frac{1}{2}\left[\mathrm{Tr}\left(\mathbf{R}^{-1}\frac{\partial\mathbf{R}}{\partial\theta_j}\right)-(\mathbf{y}-\mathbf{1}\beta)^T\mathbf{R}^{-1}\frac{\partial\mathbf{R}}{\partial\theta_j}\mathbf{R}^{-1}(\mathbf{y}-\mathbf{1}\beta)\right], \tag{5.74}$$

where Tr denotes the trace of a matrix. Note that in the above derivation we have used the following identity: $\partial|\mathbf{R}|/\partial\theta_j=\mathrm{Tr}(\mathbf{R}^{-1}\partial\mathbf{R}/\partial\theta_j)$.

Clearly, setting the preceding equation to zero does not lead to an analytical solution for $\boldsymbol{\theta}$. Hence, an iterative optimization procedure has to be employed to minimize L as a function of $\boldsymbol{\theta}$. For a given value of $\boldsymbol{\theta}$, estimates of β and σ_z^2 can be obtained using (5.72) and (5.73), respectively. These computed values of β and σ_z^2 can be substituted into (5.70) to calculate the log-likelihood function. In principle, the maximum likelihood estimate of $\boldsymbol{\theta}$ can

---

[17]For Gaussian process regression, a similar procedure can be used where the likelihood function is given by $(2\pi)^{-n/2}|\boldsymbol{\Gamma}|^{-1/2}\exp\left(-(\mathbf{y}-\mathbf{1}\beta)^T\boldsymbol{\Gamma}^{-1}(\mathbf{y}-\mathbf{1}\beta)\right)$. Here, the negative log-likelihood function to be minimized is given by $-n/2\ln(2\pi)-1/2\ln|\boldsymbol{\Gamma}|-1/2(\mathbf{y}-\mathbf{1}\beta)^T\boldsymbol{\Gamma}^{-1}(\mathbf{y}-\mathbf{1}\beta)$. It can be noted that the first term is independent of the hyperparameters and can hence be ignored. The second term can be interpreted as a measure of model complexity, while the last term is a weighted measure of training error since that is the only term that depends on the outputs. Hence, minimization of the negative log likelihood is equivalent to trading off model complexity with empirical risk – in contrast to SVMs where user-specified weighting parameters are used, the weighting parameters automatically arise in the case of Gaussian process modeling.

be estimated using any standard optimization technique. When a gradient-based optimization technique is used, the derivatives can be computed using (5.74). However, in practice, it is often found that the MLE landscape is highly multimodal and, further, ridges of constant values may lead to significant difficulties in convergence when gradient-based methods are used. Hence, it is preferable to employ stochastic global search techniques to locate a good starting point for a gradient-based optimizer. Sacks et al. (1989) offer the view that even if the MLE iterations are not fully converged the resulting parameter estimates lead to good surrogate models with useful error bars. There is also a growing body of numerical evidence to suggest that, for some cases, if the constant trend function β is replaced with a linear or quadratic polynomial, $L(\boldsymbol{\theta})$ can be less multimodal and hence easier to optimize using gradient-based techniques (Martin and Simpson 2004).

From a numerical point of view, for some datasets, the correlation matrix **R** may be ill-conditioned, particularly when two or more training points lie close to each other. To circumvent this difficulty, a small term (say 10^{-6}) can be added to the diagonal elements of **R**. Note that the resulting model would no longer be a perfect interpolant. An alternative approach would be to employ a singular value decomposition of **R** – this is, however, computationally expensive for large datasets.

Since computing $L(\boldsymbol{\theta})$ and its gradients generally involves computing and decomposing a dense $n \times n$ correlation matrix (requiring $\mathcal{O}(n^3)$ operations) at each iteration, MLE can be prohibitively expensive even for moderately sized data, for example, say a few thousand points. This is shown in Figure 5.9, where the training time is shown for a dataset with 10 variables. Unfortunately, other schemes such as Markov chain Monte Carlo sampling for approximating the predictive distribution are also plagued by similar problems involving training costs that escalate with dataset size. Recent work to address this issue has led to data parallel MLE techniques, which are discussed later in Section 5.7.

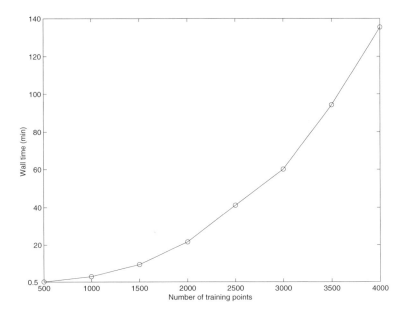

Figure 5.9 Computational cost of MLE as a function of data size for a model problem with 10 variables.

5.6.3 Incorporating Sensitivity Information

Next, consider the case when both the target value $y(\mathbf{x})$ and its sensitivities can be efficiently computed by the computer code. Then, it may be useful to exploit the sensitivity information to construct a more accurate Gaussian process model – this is similar in spirit to the Hermite RBF interpolation technique presented earlier. Following the standard Gaussian process formulation, the covariance function can be written as

$$\mathrm{Cov}\left[y(\mathbf{x}^{(i)}), y(\mathbf{x}^{(j)})\right] = \sigma_g^2 R(\mathbf{x}^{(i)}, \mathbf{x}^{(j)}), \tag{5.75}$$

where σ_g^2 is the process variance term to be estimated from the data and R is a correlation function. Now, to incorporate correlations between the function value and its derivatives into the model, the covariance function can be modified as follows (Morris et al. 1993):

$$\mathrm{Cov}\left[y(\mathbf{x}^{(i)}), \frac{\partial y(\mathbf{x}^{(j)})}{\partial x_k}\right] = \sigma_g^2 \frac{\partial R(\mathbf{x}^{(i)}, \mathbf{x}^{(j)})}{\partial x_k}, \tag{5.76}$$

$$\mathrm{Cov}\left[\frac{\partial y(\mathbf{x}^{(i)})}{\partial x_k}, y(\mathbf{x}^{(j)})\right] = -\sigma_g^2 \frac{\partial R(\mathbf{x}^{(i)}, \mathbf{x}^{(j)})}{\partial x_k}, \tag{5.77}$$

$$\mathrm{Cov}\left[\frac{\partial y(\mathbf{x}^{(i)})}{\partial x_k}, \frac{\partial y(\mathbf{x}^{(j)})}{\partial x_l}\right] = -\sigma_g^2 \frac{\partial^2 R(\mathbf{x}^{(i)}, \mathbf{x}^{(j)})}{\partial x_k \partial x_l}. \tag{5.78}$$

Using the modified function, the expression for the posterior mean $\widehat{y}$ becomes

$$\widehat{y}(\mathbf{x}) = \widehat{\beta}_g + \mathbf{r}_g^T \mathbf{R}_g^{-1}\left(\mathbf{y}_g - \widehat{\beta}_g \mathbf{f}_g\right), \tag{5.79}$$

where

$$\mathbf{y}_g = \left\{y(\mathbf{x}^{(1)}), \ldots, y(\mathbf{x}^{(n)}), \frac{\partial y(\mathbf{x}^{(1)})}{\partial x_1}, \frac{\partial y(\mathbf{x}^{(1)})}{\partial x_2}, \ldots, \frac{\partial y(\mathbf{x}^{(n)})}{\partial x_p}\right\}^T \in \mathbb{R}^{n(p+1)}, \tag{5.80}$$

and $\mathbf{f}_g = \{1, 1, \ldots, 1, 0, 0, \ldots, 0\}^T \in \mathbb{R}^{n(p+1)}$ is a vector containing n ones and np zeros. $\mathbf{R}_g \in \mathbb{R}^{n(p+1) \times n(p+1)}$ is the correlation matrix computed using (5.75), (5.76), (5.77) and (5.78), and β_g is an undetermined hyperparameter.

In a similar fashion to the standard formulation of Gaussian process interpolation, a wide family of parameterized correlation functions can be used in such models. However, because of the definition of the modified covariance function in (5.75), (5.76), (5.77) and (5.78), R must be differentiable at least twice. This condition is satisfied by the correlation function in (5.60). Let the correlation function R be parameterized in terms of $\boldsymbol{\theta} = \{\theta_1, \theta_2, \ldots, \theta_p\} \in \mathbb{R}^p$. Then, the set of hyperparameters, $\boldsymbol{\theta}$, β_g and σ_g^2 can be estimated by maximizing the following log-likelihood function:

$$L_g(\boldsymbol{\theta}, \beta_g, \sigma_g^2) = -n(p+1) \ln \sigma_g^2 - \ln |\mathbf{R}_g| - \frac{1}{\sigma_g^2}(\mathbf{y}_g - \beta_g \mathbf{f}_g)^T \mathbf{R}_g^{-1}(\mathbf{y}_g - \beta_g \mathbf{f}_g). \tag{5.81}$$

Note that when $\boldsymbol{\theta}$ is kept fixed maximization of the log-likelihood function leads to the following estimates for β_g and σ_g:

$$\widehat{\beta}_g = \frac{\mathbf{f}_g^T \mathbf{R}_g^{-1} \mathbf{y}_g}{\mathbf{f}_g^T \mathbf{R}_g^{-1} \mathbf{f}_g}, \tag{5.82}$$

$$\widehat{\sigma}_g^2 = \frac{1}{n(p+1)}(\mathbf{y}_g - \beta_g \mathbf{f}_g)^T \mathbf{R}_g^{-1}(\mathbf{y}_g - \beta_g \mathbf{f}_g). \tag{5.83}$$

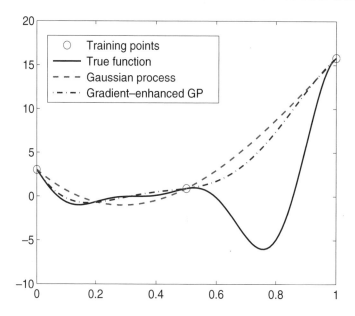

Figure 5.10 Comparison of gradient-enhanced Gaussian process interpolation model with standard formulation.

Substitution of (5.82) and (5.83) into (5.81) leads to a log-likelihood function in terms of θ, which can once again be maximized using a nonlinear optimization technique. Note that since the dimension of the correlation matrix ($\mathbf{R}_g$) is significantly higher compared to the standard Gaussian process model MLE of the hyperparameters will be computationally more expensive. Further, the memory requirements will also increase significantly since the dimension of $\mathbf{R}_g$ is $n(p+1) \times n(p+1)$. As a consequence, it may not be feasible to apply gradient-enhanced Gaussian process models to problems with large number of variables even for modestly sized training datasets. It is possible, in theory, to employ ideas such as those discussed earlier in the context of Hermite RBF interpolation to reduce the computational cost and memory requirements. However, these ideas remain to be fully tested.

A comparison of the prediction accuracies of the standard Gaussian process model with the gradient-enhanced version for a one-dimensional test problem is shown in Figure 5.10. More detailed studies on gradient-enhanced Gaussian process models can be found in Chung and Alonso (2002) and Leary et al. (2004b).

5.6.4 Assessment and Refinement of Gaussian Process Models

After constructing a Gaussian process model, we need to perform validation studies to assess how well the approximate model agrees with the true model. A brute force approach would be to run the true model at a number of additional points (testing points) and check how well the approximate model correlates at these points. In most practical applications, generation of additional testing data is computationally expensive. This motivates the development of alternative model diagnostic measures that can be evaluated more cheaply.

Consider the case of Gaussian process modeling where the application of (5.66) and (5.67) to a new design point $\mathbf{x}$ gives both the expected value of $y(\mathbf{x})$ as well as an estimate

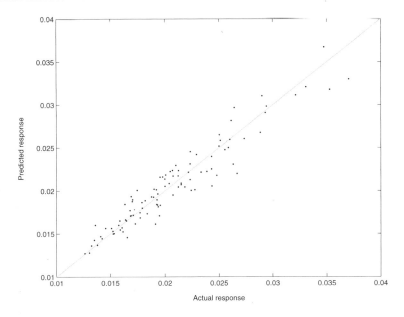

Figure 5.11 Plot of leave-one-out prediction versus true value at all training points.

of the prediction error in terms of the posterior variance $\sigma^2(\mathbf{x}) = C(\mathbf{x}, \mathbf{x})$. To compute the accuracy of the posterior mean, a leave-one-out cross-validation procedure similar to that discussed earlier in the context of RBF approximations can be employed. This involves leaving the ith training point out and computing the posterior mean at the point $\widehat{y}_{-i}(\mathbf{x})$. The accuracy of the posterior mean can then be examined by plotting this quantity versus the true function value $y^{(i)}$ for all the training points; see, for example, Figure 5.11. Meckesheimer et al. (2002) noted that the leave-one-out procedure may significantly underestimate the actual prediction error – it was suggested that a k-fold cross-validation scheme (with $k = 0.1n$ or $\sqrt{n}$) may give a better indication of generalization ability.

The accuracy of the posterior variance $\sigma^2(\mathbf{x})$ (error estimate) depends on the validity of the assumption made in creating the metamodel, namely, that a Gaussian process prior is appropriate for the black-box simulation code. Techniques for metamodel diagnostics available in the literature can be applied to check the validity of this assumption and the accuracy of the metamodel. One such measure is the "standardized cross-validated residual" (SCVR) (Jones et al. 1998) defined as

$$\text{SCVR}_i = \frac{y^{(i)} - \widehat{y}_{-i}(\mathbf{x}^{(i)})}{\sigma_{-i}^2(\mathbf{x}^{(i)})}, \quad i = 1, 2, \ldots, n, \tag{5.84}$$

where $\widehat{y}_{-i}(\mathbf{x})$ and $\sigma_{-i}^2(\mathbf{x})$ denote the mean and variance of the metamodel prediction at a point $\mathbf{x}$ without using the ith training point. SCVR_i can be computed for all the training points by removing the contribution of the corresponding point from the correlation matrix $\mathbf{R}$. In the cross-validation procedure, it is generally assumed (for computational reasons) that the MLE estimates for the hyperparameters do not change when one training point is removed from the training set. Mitchell and Morris (1992) showed that the prediction error at the deleted point can be efficiently computed using the equation $y^{(i)} - \widehat{y}_{-i}(\mathbf{x}^{(i)}) = q_i(g_i - \beta w_i)$, where

g_i and w_i are the ith elements of the vectors $\mathbf{g} = \mathbf{R}^{-1}\mathbf{y}$ and $\mathbf{w} = \mathbf{R}^{-1}\mathbf{1}$, respectively. q_i is the ith element of the vector $\mathbf{q}$, which is defined as the inverse of the diagonal of $\mathbf{R}^{-1}$. The above equation allows the cross-validation error to be efficiently computed.

If the Gaussian process prior is appropriate for the problem under consideration, $SCVR_i$ will roughly lie in the interval $[-3, +3]$. This implies that given the posterior predictions of the metamodel at a new design point $\mathbf{x}$ the actual output value lies in the interval $[\widehat{y}(\mathbf{x}) - 3\sigma(\mathbf{x}), \widehat{y}(\mathbf{x}) + 3\sigma(\mathbf{x})]$ with a high level of confidence.

A measure of generalization error of the metamodel can be defined using the statistics of the covariance term $C(\mathbf{x}, \mathbf{x}')$. Assuming that the vectors $\mathbf{x}$ and $\mathbf{x}'$ have a prescribed probability distribution $\mathcal{P}(\mathbf{x})$, the generalization error can be defined as

$$\Delta_{ge} = \sigma_z^2 \int_\chi \left(R(\mathbf{x}, \mathbf{x}') - \mathbf{r}(\mathbf{x})^T \mathbf{R}^{-1}\mathbf{r}(\mathbf{x}') \right) d\mathcal{P}(\mathbf{x}) \, d\mathcal{P}(\mathbf{x}'). \tag{5.85}$$

Note that when $\mathcal{P}(\mathbf{x})$ corresponds to the joint probability density function (pdf) of multidimensional uncorrelated Gaussian random variables and the correlation function obeys the product correlation rule the above integral can be evaluated analytically. This is because the multidimensional integral can be expressed as product of one-dimensional integrals. Alternatively, Monte Carlo simulation procedures or design of experiment techniques can be employed to numerically approximate Δ_{ge}. It is also worth noting that the expression for Δ_{ge} can be used during the metamodel construction phase to decide whether more design points should be added to improve the accuracy.

The assessment criteria presented here can also sometimes provide useful guidelines on how to improve the existing model. There are at least three potential avenues for improving a Gaussian process surrogate model: (1) changing the correlation function, (2) introducing intermediate variables and (3) using a polynomial model instead of the constant term β in (5.57). Clearly, the optimal covariance function for modeling any input–output relationship is data dependent. However, computational experience accumulated in the literature suggests that the parameterized correlation function in (5.60) offers sufficient flexibility for modeling smooth and highly nonlinear functions. The metrics presented in this section for assessing model accuracy can be employed to check whether an alternative correlation function (such as a linear or cubic spline) is more appropriate for the problem at hand. Bayesian model comparison techniques may also be employed to compare models constructed using different covariance functions. For example, the likelihood values can be compared to select an appropriate correlation function – the interested reader is referred to MacKay (2003) for an excellent overview of the Bayesian approach to model comparison. The idea of using intermediate variables to improve the accuracy of Gaussian process models is discussed in Jones et al. (1998) and Snelson et al. (2004).

5.7 Data Parallel Modeling

As mentioned earlier, a common problem with most kernel methods (such as RBF approximations and Gaussian process modeling) is their poor scaling with data size. For example, typical training times for such methods scale as $\mathcal{O}(n^3)$, where n is the number of training points, along with a storage requirement that scales as $\mathcal{O}(n^2)$. (An exception is the greedy algorithm presented earlier for constructing parsimonious RBF approximations). Clearly, surrogate model construction for large datasets rapidly becomes computationally intractable. Recent research has focused on these problems intensively and current state-of-the-art algorithms have managed to bring down training times closer to $\mathcal{O}(n^2)$ (Joachims 1999; Schölkopf and Smola 2001) through various innovative heuristics. In the context of Gaussian process regression, researchers have attempted to tackle the computational cost issue through Skilling's approximations (Gibbs and MacKay 1997), on-line learning techniques based on sequential model updates (Csató and Opper 2001), and model sparsification strategies such as those discussed earlier in Section 5.5.2.

A popular approach to scaling machine learning algorithms is that of Divide And Conquer (DAC). In general, DAC is used in the context of *metalearning*, which involves the generation of an ensemble of global models over random partitions of data. The basic rationale behind the idea is that computational and storage complexity can be reduced by partitioning the original data into smaller manageable chunks and then applying a base learner to each of the partitions in sequential or parallel mode. Prediction at a new point is achieved by suitably aggregating the outputs of all the models. Recent examples of such metalearning approaches include parallel mixtures of SVMs (Collobert et al. 2002) and Bayesian Committee Machines (Tresp 2000).

Here, we discuss how the Gaussian process modeling technique can be reformulated using a DAC philosophy in order to improve scalability following Choudhury et al. (2002). In doing this, the meta- or global-learning approach is abandoned in favor of a local learning strategy. In contrast to metalearning, local learning strategies construct models for spatially localized partitions of the data. The motivation for this arises from the fact that local learning techniques are known to generally perform better than metalearning techniques, which suffer from high variance and poor generalization.

In what follows, the local learning problem is addressed explicitly from the point of view of the covariance function, which is the fundamental building block of Gaussian stochastic processes. Specifically, it is shown how compactly supported covariance functions defined over spatially localized clusters naturally lead to a data parallel scheme for modeling large-scale data – this new covariance function also allows the MLE problem to be decomposed into smaller decoupled subproblems. This leads to a significant speedup in the training phase of Gaussian process models. Further, this approach also leads to sparse[18] yet effective predictive models. Since equally sized or balanced partitioning of data is crucial to the efficiency of such data parallel approaches, we next present a simple clustering scheme, which outputs clusters of (nearly) equal sizes. Finally, although the data parallel approach is presented here in the context of Gaussian process models, we stress that this strategy may also be readily applied to other kernel methods such as RBFs and SVMs.

5.7.1 Data Parallel Local Learning

To illustrate the data parallel approach, assume the existence of m disjoint and spatially localized subsets of the training data, say $\mathcal{C}_1, \mathcal{C}_2, \ldots \mathcal{C}_m$. Given such a partitioning, the following covariance model can be used:

$$\tilde{R}(\mathbf{x}^{(i)}, \mathbf{x}^{(j)}; \mathfrak{F}(\mathbf{x}^{(i)}), \mathfrak{F}(\mathbf{x}^{(j)}), \boldsymbol{\theta}) = \delta_{\mathfrak{F}(\mathbf{x}^i)\mathfrak{F}(\mathbf{x}^{(j)})} \, R(\mathbf{x}^{(i)}, \mathbf{x}^{(j)}; \boldsymbol{\theta}), \tag{5.86}$$

where $\mathbf{x}^{(i)}, \mathbf{x}^{(j)} \in \mathbb{R}^p$ are input vectors, δ_{ij} is the Kronecker delta function, $\boldsymbol{\theta}$ is a set of hyperparameters and $\mathfrak{F} : \mathfrak{F}(\mathbf{x}) \longmapsto \{1, 2, \ldots, m\}$ is the assignment function, which maps

---

[18]The term *sparse* is used here loosely to denote models that only use a subset of the training dataset to make predictions.

the input point $\mathbf{x}$ to one of m available clusters.[19] Then, the covariance function in (5.86) can be immediately written for cluster i as

$$\tilde{R}(\mathbf{x}^{(1)}, \mathbf{x}^{(2)}; \mathfrak{F}(\cdot), \boldsymbol{\theta}) = R(\mathbf{x}^{(1)}, \mathbf{x}^{(2)}; \boldsymbol{\theta}_i), \ \forall \ \mathfrak{F}(\mathbf{x}^{(1)}) = \mathfrak{F}(\mathbf{x}^{(2)}) = i$$

$$= 0 \qquad \qquad \text{otherwise} \qquad \qquad (5.87)$$

where $\boldsymbol{\theta}_i$ denotes the set of hyperparameters for the local model trained on the ith cluster. This is readily identified as an example of a compactly supported covariance function popularly used in the RBF literature (Schaback and Wendland 2000). Consider the case when $m = 2$, that is, when the data has been partitioned into two disjoint spatially localized subsets. Then, using (5.87), the correlation matrix can be written in block diagonal form as follows:

$$\mathbf{R} = \begin{pmatrix} \mathbf{R}_{11} & \mathbf{0} \\ \mathbf{0} & \mathbf{R}_{22} \end{pmatrix} \qquad \qquad (5.88)$$

where $\mathbf{R}_{ii} \in \mathbb{R}^{n_i \times n_i}$ contains correlation terms explicitly from the ith cluster, which consists of n_i points. Since, in this case, the determinant of the correlation matrix $\mathbf{R}$ can be written as the product of determinants of the blocks $\mathbf{R}_{11}$ and $\mathbf{R}_{22}$, the log-likelihood function defined earlier in (5.70) can be split into individual log likelihoods for the two partitions, that is, $L(\boldsymbol{\theta}) = L_1(\boldsymbol{\theta}_1) + L_2(\boldsymbol{\theta}_2)$.

It is important to note at this point that while the conventional Gaussian process model uses a single set of hyperparameters to model the full data the data parallel version uses different sets of hyperparameters to model the covariance structures of different regions of the data. Hence, maximizing the overall log likelihood for the data is equivalent to maximizing the log likelihood for each partition, since the individual log-likelihood functions depend on separate sets of hyperparameters. It may seem at first sight that the proposed compactly supported covariance function would result in a less accurate oversimplified global model. However, numerical studies by Choudhury et al. (2002) do not seem to bear out this observation. Indeed, it can be argued that given enough flexibility in the choice of the local covariance models any performance degradation can be readily controlled.

From the preceding discussion, it is clear that the use of a compactly supported covariance function naturally leads to a data parallel local learning approach, and hence provides a means to handle large datasets. Having made this connection, we are now confronted with the problem of partitioning or clustering the data into subsets. From a data parallel point of view, the subset sizes chosen need to be (nearly) equal.[20] What is needed is an algorithm that creates clusters that are localized in the data space. Local models can then be built on each of these clusters. Finally, the ensemble of local models can be used in place of the global model.

The basic approach suggested here is set out as a pseudocode in Algorithm 5.3, where $\mathcal{C}_1, \mathcal{C}_2, \ldots, \mathcal{C}_m$ denote the m clusters or partitions of data and $\mathcal{M}_1, \mathcal{M}_2, \ldots, \mathcal{M}_m$ denote

---

[19]For example, $\mathfrak{F}(\mathbf{x})$ could be the nearest neighbor operator, where the neighbors considered are the cluster centers.

[20]This is primarily because of efficiency reasons and also to reduce the waiting times when operating in a multiprocessor scenario.

Algorithm 5.3: An outline of the data parallel approach to surrogate modeling.

| **Inputs** | : Training Data $\mathcal{D}$ |
| | : Number of partitions desired m |

Begin Modeling

| **Step 1** | : $[\mathcal{C}_1, \mathcal{C}_2, \ldots, \mathcal{C}_m] :=$ **Partition**$(\mathcal{D}, m)$ |
| **Step 2** | : **For** $i = 1, \ldots, m$ |

$$M_i := \textbf{CreateLocalModel}(\mathcal{C}_i)$$

        end For

$$\mathcal{M} = \bigcup_{i=1}^{m} M_i$$

End Modeling

| **Prediction** | : Given a new point $\mathbf{x}$ |
| | $y(\mathbf{x}) = \textbf{Aggregate}(\mathcal{M}, \mathbf{x})$ |

---

the corresponding local prediction models. As is clear from Algorithm 5.3, a complete description of the data parallel approach requires the specification of three functions, that is, Partition(), CreateLocalModel() and Aggregate().

5.7.2 Data Partitioning

Although clustering techniques abound in the literature, (expectation maximization and k-means being two of the more popular ones), few of them seem to offer direct control on the number of points to be held in each cluster. One way to generate spatially balanced clusters would be to solve the nonlinear programming problem

$$\underset{\mathbf{c}_1, \mathbf{c}_2, \ldots, \mathbf{c}_m}{\textbf{Minimize}} : \sum_{i=1}^{m} (W_i - W_{\text{des}})^2, \tag{5.89}$$

where $\mathbf{c}_i$ and W_i denote the center and number of points in the ith cluster, respectively, and W_{des} is the desired number of points per cluster.

Note that since W_i can only be expressed in terms of indicator functions the resulting objective function is nondifferentiable. As a result, one has to resort to less-efficient nongradient search techniques. In this section, we present a simple geometry motivated clustering algorithm that attempts to solve (5.89) greedily (Choudhury et al. 2002). We refer to this method as "GeoClust".

In the GeoClust technique, a greedy strategy is employed to incrementally update the cluster centers. The pseudocode of the algorithm is presented in Algorithm 5.4. Consider the case when data is to be partitioned into three clusters $\mathcal{C}_1, \mathcal{C}_2, \mathcal{C}_3$ with cluster centers $\mathbf{c}_1, \mathbf{c}_2$ and $\mathbf{c}_3$, respectively. GeoClust is an iterative scheme that updates cluster centers until each cluster contains equal or near-equal instances. During the course of clustering, each cluster center $\mathbf{c}_i$ moves toward every other cluster $\mathbf{c}_j$ according to the fraction $(\frac{W_j}{W_i} - 1)$, where W_i, W_j denote the strengths (number of points) of the ith and the jth clusters, respectively. Thus, at any iteration, cluster center $\mathbf{c}_1$ moves toward cluster center $\mathbf{c}_2$ by the fraction

---

Algorithm 5.4: GeoClust

Inputs   : Data $\mathcal{D}$; Number of partitions desired m
                   : Maximum number of iterations $Maxiter$
                   : learning parameter γ
Initialize m random centers, $\mathbf{c}_i^0, i = 1, 2, \dots, m$
For t = 1 : $Maxiter$
  • Assign each data point to the cluster nearest to it
  • Compute number of points in each cluster: $W_1, W_2, \dots, W_m$
  • For each cluster $\mathcal{C}_i$, Update the cluster center as follows:
    ○ Compute $\delta\mathbf{v}_i = \sum_{j=1, j \neq i}^{m} (\frac{W_j}{W_i} - 1)(\mathbf{c}_j^{t-1} - \mathbf{c}_i^{t-1})$
    ○ Update center $\mathbf{c}_i^t = \mathbf{c}_i^{t-1} + \gamma\delta\mathbf{v}_i$
  • End
End

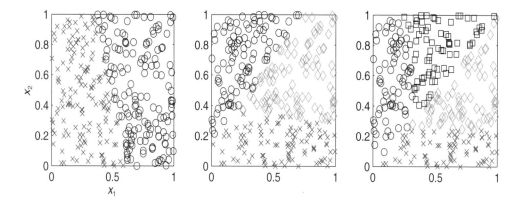

Figure 5.12 The partitions obtained by GeoClust for a toy dataset when the desired number of partitions are increased from 2 to 4 (left to right)

$(\frac{W_2}{W_1} - 1)$ along the direction of the vector $(\mathbf{c}_2 - \mathbf{c}_1)$ and by a fraction $(\frac{W_3}{W_1} - 1)$ along the direction $(\mathbf{c}_3 - \mathbf{c}_1)$ toward cluster center $\mathbf{c}_3$. At the end of each iteration, the cluster strengths W_1, W_2, W_3 are reassessed. The process continues until the increments become zero or very small. The iterative process is assisted by a relaxation/learning factor γ, which is typically set to a low value. In our experiments, we have found 10^{-2} to be a reasonable choice when the data is normalized to lie within [0, 1]. Note that as the cluster strengths become similar the relative movement of the centers toward each other slows down and finally ceases. Hence, the algorithm converges when the clusters have equal strengths.[21] Figure 5.12 illustrates how GeoClust partitions a toy 2-D dataset into two, three and four clusters respectively.

---

[21]When the number of instances is not exactly divisible by the number of clusters, the cluster strengths settle to nearly equal values.

5.7.3 Prediction with Multiple Models

Since our desired computing paradigm is essentially data parallel, we generate multiple models for explaining the given data. Given a new input point, the task of predicting its output therefore must take into account the outcomes from the multiple models in some reasonable way. Intuitively, such an aggregation strategy should also depend on the clustering technique used.

Consider Gaussian process interpolation using the compactly supported covariance model proposed in (5.86). Given a new point $\mathbf{x}$, the posterior mean is given by

$$\widehat{y}(\mathbf{x}) = \beta + \mathbf{r}(\mathbf{x})^T \mathbf{R}^{-1} (\mathbf{y} - \mathbf{1}\beta) \qquad (5.90)$$

where $\mathbf{r}(\mathbf{x}) = \left[\delta_{\mathfrak{F}(\mathbf{x})\mathfrak{F}(\mathbf{x}^{(1)})} R(\mathbf{x}, \mathbf{x}^{(1)}; \boldsymbol{\theta}), \ldots, \delta_{\mathfrak{F}(\mathbf{x})\mathfrak{F}(\mathbf{x}^{(n)})} R(\mathbf{x}, \mathbf{x}^{(n)}; \boldsymbol{\theta}) \right]$. It is easy to see that this prediction strategy amounts to a hard assignment strategy – given a new point, determine the cluster center nearest to it and then use the corresponding local model for prediction. Mathematically, (5.90) is now equivalent to

$$\textbf{Determine Cluster} : i = \mathfrak{F}(\mathbf{x}) ; \quad \mathfrak{F}(\cdot) \in \{1, \ldots, m\} \qquad (5.91)$$

$$\textbf{Predict} : \widehat{y}(\mathbf{x}) = \widehat{y}(\mathbf{x}) = \beta + \mathbf{r}(\mathbf{x}, \boldsymbol{\theta}_i)^T \mathbf{R}_{i,\theta}^{-1} (\mathbf{y} - \mathbf{1}\beta) \qquad (5.92)$$

where the subscript i is used indicate that the data points from the ith cluster and the corresponding set of hyperparameters, are to be used, respectively. Alternatively, a sum weighted on distances from all cluster centers could be used.

5.7.4 Computational Aspects

As noted earlier, the key computational bottlenecks involved in applying the conventional Gaussian process modeling approach to a large dataset are related to storing the dense correlation matrix ($\mathcal{O}(n^2)$ memory) and computing its factorization ($\mathcal{O}(n^3)$ operations). The computational advantages of the local learning strategy over a global Gaussian process modeling approach are significant and straightforward to evaluate. Specifically, for n data points and m clusters, a sequential data parallel version of the algorithm requires training time of $m\mathcal{O}((\frac{n}{m})^3)$ only, as compared to $\mathcal{O}(n^3)$ for the global Gaussian process model. Further, the attendant memory requirements are reduced from $\mathcal{O}(n^2)$ for the conventional case to $m\mathcal{O}((\frac{n}{m})^2)$. The parallel version of the algorithm offers even more advantages. Typically, in such a scenario, one has access to at least as many processors as the number of clusters specified[22]. The overall training time now reduces to $\mathcal{O}((\frac{n}{m})^3)$ and the memory requirements fall to $\mathcal{O}((\frac{n}{m})^2)$ per processor involved. The computer time required by the data parallel approach as a function of n is shown in Figure 5.13. It can be clearly seen that the proposed approach in both its sequential and parallelized form is much faster than the conventional Gaussian process modeling (see Figure 5.9).

It is clear from the preceding discussion that the efficiency of the algorithm increases with increasing number of clusters m. Further, it has been shown elsewhere that the use of compactly supported kernels produces more stable solutions (Schaback, 1998). However, it is important to note that increasing the number of clusters beyond a certain limit actually leads to a deterioration in the generalization capability of the resulting model. This results

---

[22]Conversely, it is reasonable to assume that the number of clusters are prespecified by the user according to the number of processors available.

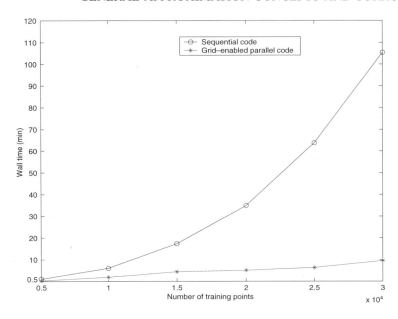

Figure 5.13 A comparison of computational speedups obtained using the parallel and the sequential versions of the data parallel local learning approach. The computational cost of conventional Gaussian process modeling is shown in Figure 5.9.

from an ever reducing number of points being available for modeling each cluster.[23] Moreover, unless some form of weighted averaging is used, discontinuities may arise at cluster boundaries.

5.8   Design of Experiments (DoE)

In the derivation of the surrogate modeling techniques covered in this chapter, we have assumed that a set of observational data is already available. Clearly, the location of the points in this set will have a significant impact on the accuracy of any surrogate model. In this section, we examine the problem of how to generate training data that leads to approximation models that generalize well.

The subject of design of experiments (DoE) has a long history and was originally studied in the context of designing *physical* experiments (Fisher 1935), such as those encountered in agriculture yield analysis and experimental chemistry. A distinguishing feature of physical experiments is the presence of random error, that is, the outcome of the experiment may not be repeatable. The objective of DoE in such a context is to generate data that can be used to fit a regression model (usually a linear response surface model) that reliably predicts the true trends of the input–output relationship. Classical DoE techniques include central composite design, Box–Behnken design, and full- and fractional-factorial designs. A common feature of these techniques is that the sample points are placed at the extremes of the parameter space

---

[23]From the point of view of compactly supported covariance functions, poor generalization arises when the support radius is reduced below a certain threshold.

to alleviate the contaminating influence of measurement noise. There is also a wide class of DoE techniques based on optimality criteria suitable for cases when a linear or quadratic polynomial response surface model is employed. In summary, classical DoE techniques are appropriate if the data is corrupted by noise, which make it necessary to employ regression techniques to filter that noise. A number of excellent texts on DoE techniques for physical experiments can be found in the literature; see, for example, Montgomery (1997).

In contrast to physical experiments, observations made using computer experiments are not subject to random errors. Hence, to extract maximum information about the underlying input–output relationship, the sample points chosen should fill the design space in an optimal sense. It makes little sense to employ classical DoE techniques, which place points at the extremes of the design space. A detailed explanation of this issue can be found in Sacks et al. (1989). The focus of this section is on DoE techniques that are suitable for computer experiments. The problem statement we consider can be stated as follows:

Problem: *Given a bounded design space* $\Omega_{\mathbf{x}}$, *generate a set of samples* $\mathbf{x}^{(1)}, \mathbf{x}^{(2)}, \ldots, \mathbf{x}^{(n)}$ *at which to run a computer model* $y = f(\mathbf{x})$ *such that a surrogate model constructed using the input–output dataset* $\{\mathbf{x}^{(i)}, y^{(i)}, i = 1, 2, \ldots, n\}$ *generalizes well.*

Clearly, a DoE technique appropriate for solving the above problem must take into consideration the geometry of the design space $\Omega_{\mathbf{x}}$ and the type of surrogate model that will be fitted to the data. Most of the work in this field assume that $\Omega_{\mathbf{x}}$ is a box-shaped domain – this occurs when bounds on each design variable are of the form $x_i^l \leq x_i \leq x_i^u$. However, in many practical design problems with general nonlinear constraints, the domain $\Omega_{\mathbf{x}}$ may be highly irregular. For such cases, one can apply any DoE technique to generate points on a box domain that encloses $\Omega_{\mathbf{x}}$ and subsequently delete points that lie outside $\Omega_{\mathbf{x}}$. However, this often leads to destruction of the space filling nature of the initial design. An approach based on optimality criteria that can overcome this problem is presented later in this section. It is also possible to formulate DoE techniques that create n samples in one shot, or alternatively a stage-wise (or model updating) procedure can be employed. In the stage-wise procedure, at stage $k + 1$, the trends of the input–output relationship observed from data generated at stage k are exploited. This type of approach can be very powerful when the objective is to minimize or maximize $f(\mathbf{x})$ – stage-wise approaches for optimization are presented in Chapter 7.

We discuss next DoE techniques that attempt to create space-filling designs without using any information about the type of surrogate model being employed. Finally, DoE techniques based on optimality criteria tailored to Gaussian process models are presented – this approach also allows for the possibility of treating cases when the design space $\Omega_{\mathbf{x}}$ is irregular. Our discussion is largely based on the review papers by Koehler and Owen (1996) and Giunta et al. (2003). For a detailed account of DoE techniques appropriate for computer experiments, the interested reader is referred to the recent book by Santner et al. (2003).

Before proceeding further, some of the terminology commonly used in the DoE literature is explained.

1. *Factors:* The design variables that are of interest. For simplicity, we assume throughout this section that all factors are normalized to lie in the interval [0, 1].

2. *Levels:* The number of possible values each design variable can take. In practice, each continuous variable is usually allowed to take only a finite set of values.

3. *Full factorial design:* The most basic experimental design is the so-called full factorial design, which involves selecting all the corners of the hypercube and possibly sets of

interior points. For a problem with p design variables and k levels (such as 0 and 1), this results in a total of k^p points. Clearly, this is unaffordable for computationally expensive problems when p and k are large.

4. *Design matrix:* The matrix $\mathbf{X} = [\mathbf{x}^{(1)}, \mathbf{x}^{(2)}, \ldots, \mathbf{x}^{(n)}]^T \in \mathbb{R}^{n \times p}$ containing all the inputs values is referred to as the design matrix.

5.8.1   Monte Carlo Techniques

Monte Carlo techniques are perhaps the simplest of all DoE methods, wherein the basic idea is to use a random number generator to sample the design space $\Omega_{\mathbf{x}}$. Hence, the essential ingredient of a Monte Carlo sampling procedure is the numerical algorithm employed for generating random numbers. When $\Omega_{\mathbf{x}}$ is box shaped, Monte Carlo techniques are straightforward to implement. For irregular design spaces, implementation may be difficult depending on how the geometry of $\Omega_{\mathbf{x}}$ is characterized. The implementation details also depend on the assumed joint probability distribution of the input vector $\mathbf{x}$. In practice, the major disadvantage of the Monte Carlo technique is that the points generated may not be space filling, since large areas of the design space may be left unexplored while others are sampled over densely. They can only really be justified where it is believed that the correlation length in the underlying problem being sampled is very small.

Stratified variants of the Monte Carlo technique ensure that the points generated are more uniformly distributed in $\Omega_{\mathbf{x}}$. In stratified sampling, the design space is divided into bins of equal probability and at least one point is placed within a bin. For example, if each design variable has two intervals, then the total number of bins in $\Omega_{\mathbf{x}}$ is 2^p. This approach suffers from the same drawbacks as full factorial designs. We note here that the topic of Monte Carlo sampling is vast, and for a detailed overview the reader may consult texts such as Sobol (1994) and Evans and Swartz (2000).

5.8.2   Latin Hypercube Sampling

The Latin hypercube sampling (LHS) technique was proposed by McKay et al. (1979) as an alternative to Monte Carlo techniques for designing computer experiments. The basic idea is to divide the range of each design variable into n bins of equal probability – this leads to a total of n^p bins in the design space. Subsequently, n samples are generated such that, for each design variable, no two values should lie in a bin. In other words, when a one-dimensional projection of the design is taken, there will be one and only one sample in each bin.

The LHS algorithm generates samples in a box-shaped domain as follows:

$$x_j^{(i)} = \frac{\pi_j^{(i)} + U_j^i}{n}, \quad \forall 1 \le j \le p, \quad 1 \le i \le n, \tag{5.93}$$

where n is the number of samples, $U \in [0, 1]$ is a uniform random number and π is an independent random permutation of the sequence of integers $0, 1, \ldots, k-1$. The subscript denotes the design variable number and superscript in brackets denotes the sample number. It should be noted that design matrix produced using LHS is not repeatable because of the random number generator used for calculating U and π.

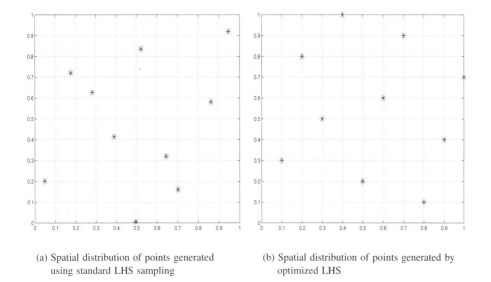

(a) Spatial distribution of points generated
using standard LHS sampling

(b) Spatial distribution of points generated by
optimized LHS

Figure 5.14 Comparison of spatial distribution of points obtained using the standard LHS algorithm and the optimized version.

Points generated by the standard LHS sampling algorithm are shown in Figure 5.14(a). It can be seen that if one-dimensional projections are taken along x_1 or x_2, there will be exactly one sample in each bin.

The space-filling characteristics of designs produced using LHS are not guaranteed to be good all the time (a diagonal arrangement in Figure 5.14(a) would also be possible, for example). This has motivated the development of optimal LHS designs, which give a more uniform coverage of the design space. One such approach is due to Audze and Eglais (1977), wherein an n^k grid is first generated. Subsequently, the points are placed such that no two points lie along the same grid line and the metric

$$\sum_{i=1}^{n} \sum_{j=i+1}^{n} \frac{1}{d_{ij}^2}, \qquad (5.94)$$

is minimized. Here, d_{ij} is the Euclidean distance between points i and j. The design matrix obtained using this approach for $p = 2$ and $n = 10$ is shown in Figure 5.14(b). It can be noted that the optimized approach leads to points that are more uniformly spread out in the design space.

A simple modification of the LHS algorithm involves setting $U = 0.5$. Since this setting ensures that each point is placed at the center of its bin, the resulting algorithm is referred to as *lattice sampling*. A similar idea of placing each point at the center of its bin is used when constructing *uniform designs*. In addition to this, uniformity is imposed in the p-dimensional design space. Hence, uniform designs are generally more space filling compared to LHS. A review of the theoretical aspects and applications of uniform designs has been presented by Fang et al. (2000).

5.8.3 Sampling using Orthogonal Arrays

An orthogonal array can be defined by the matrix $\mathbf{O}_A \in \mathbb{R}^{n \times p}$, where each element is an integer number between 0 and $q - 1$ (Hedayat et al. 1999). $\mathbf{O}_A$ is said to have *strength* t, if in every $n \times t$ submatrix, all the q^t possible rows appear the same $\mathcal{G}$ number of times. Hence, $n = \mathcal{G}q^t$. This means that in a t-dimensional projection all the points will be uniformly spaced. Hence, orthogonal arrays can be seen as a generalization of LHS sampling whose one-dimensional projections are uniformly spaced. $\mathcal{G}$ is termed the *index* of the array, which denotes the number of points that occur in each bin when a t-dimensional projection of the samples is taken.

An orthogonal array can be compactly denoted by $\mathbf{O}_A(n, p, k, t)$, where n is the total number of samples, p is the number of input variables, k is the number of bins over which each input variable is discretized and t is the strength of the array. Two examples of orthogonal arrays are shown below:

$$
\mathbf{O}_A(4, 3, 2, 2) = \begin{bmatrix} 0 & 0 & 0 \\ 1 & 0 & 1 \\ 1 & 1 & 0 \\ 0 & 1 & 1 \end{bmatrix} \quad \text{and} \quad \mathbf{O}_A(8, 3, 2, 2) = \begin{bmatrix} 0 & 0 & 0 \\ 0 & 0 & 1 \\ 0 & 1 & 0 \\ 1 & 0 & 0 \\ 0 & 1 & 1 \\ 1 & 1 & 0 \\ 1 & 0 & 1 \\ 1 & 1 & 1 \end{bmatrix}.
$$

Since $n = \mathcal{G}k^t$, it follows that $\mathcal{G} = 1$ for $\mathbf{O}_A(4, 3, 2, 2)$ and $\mathcal{G} = 2$ for $\mathbf{O}_A(8, 3, 2, 2)$.

Randomized orthogonal arrays can be used to plan computer experiments (Koehler and Owen 1996). Similar to the LHS algorithm, this can be achieved as follows:

$$
x_j^{(i)} = \frac{\pi_j(A_j^{(i)}) + U_j^i}{k}, \quad \forall 1 \leq j \leq p, \quad 1 \leq i \leq n.
$$

Here, $\pi_j(A_j^{(i)})$ denotes the ijth element of $\mathbf{O}_A$.

5.8.4 Minimum Discrepancy Sequences

The *discrepancy* of a design matrix can be defined as a measure of how much the distribution of the points deviates from an ideal uniform distribution. Minimum discrepancy sequences were originally conceived to develop space-filling points, primarily for the purpose of efficient numerical integration of multidimensional functions. Some popular examples of minimum discrepancy sequences are Hammersley, Halton, Sobol and Faure sequences. A mathematical description of these sequences is beyond the scope of this text and the interested reader is referred to more specialized works; see, for example, Niederreiter (1992) and Sobol (1994).

Minimum discrepancy sequences are also commonly referred to as quasi-Monte Carlo methods since these methods generate a deterministic sequence of points. This makes sense for incremental model building. Figure 5.15(a) shows the distribution of the first ten points from the Sobol sequence for a two-dimensional problem. Even though these points are not uniformly distributed, it can be seen that once the first 100 points from the Sobol sequence are drawn (see Figure 5.15(b)), the design space is filled quite uniformly.

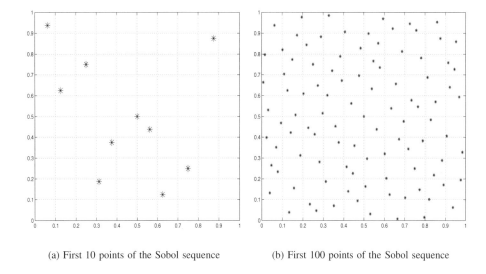

(a) First 10 points of the Sobol sequence   (b) First 100 points of the Sobol sequence

Figure 5.15 Spatial distribution of points generated from the Sobol sequence.

5.8.5 DoE Using Optimality Criteria

A plan for conducting experiments can also be decided by maximizing or minimizing a suitable figure of merit. One possible criterion to minimize is the distance measure discussed earlier in the context of optimum LHS. We discuss here some optimality criteria that can be used to construct designs for Gaussian process models. Clearly, a good DoE approach should lead to points that minimize the posterior variance. This means that the design matrix should be generated such that the following criterion is minimized:

$$\int_{\Omega_x} \left(1 - \mathbf{r}(\mathbf{x})^T \mathbf{R}^{-1} \mathbf{r}(\mathbf{x})\right) d\mathbf{x}.$$

A DoE approach that employs this criterion is discussed in Sacks et al. (1989). An attractive feature of this approach is that it is possible to apply additional constraints that ensure that all the samples in the design matrix $\mathbf{X}$ lie within the design space Ω_x. This makes optimality criteria–based DoE approaches suitable for generating points in irregular design spaces. The major disadvantage is that the computational effort can be significant compared to the other DoE techniques discussed earlier.

An alternative model free criterion that is geometrically very appealing is to maximize the minimum distance between any two points in the design matrix (Johnson et al. 1990). This approach should, in principle, lead to a set of samples that are uniformly distributed in an optimal sense. Note, however, that when n is small, most of the points will end up at the extremities of the design space and the interior will only be filled in as n is increased. Currin et al. (1991) discuss a criterion based on entropy measure that involves finding the design matrix such that the determinant of the correlation matrix $\mathbf{R}$ is maximized.

The optimality criteria–based approaches to DoE are generally implemented by starting off with a design matrix $\mathbf{X}$ obtained via LHS sampling. Subsequently, the values for each column of $\mathbf{X}$ are permutated so as to minimize/maximize some figure of merit. This combinatorial optimization problem is usually solved using simulated annealing or a genetic algorithm.

5.8.6 Recommendations for Optimization Studies

Since design optimization studies are iterative in nature, a sequential or stage-wise DoE approach is perhaps the most attractive. The basic idea is to generate an initial design matrix to decide the location of the points at which the computer model must be run. A baseline surrogate model can then be constructed using the data thus generated. Subsequently, promising designs can be identified by optimizing the surrogate model in lieu of the original computer model. If the performance of the designs identified using the surrogate model is not satisfactory, the original design matrix can be augmented by a selection of points explored during surrogate optimization. By running the computer code on these new points, additional data can be generated for updating the baseline surrogate model. This process can be carried out in a stage-wise fashion until a design with satisfactory performance is reached. We cover this important topic of stage-wise DoE and surrogate-assisted optimization in more detail in Chapter 7. Since the initial design matrix will be eventually updated in this approach, the choice of the DoE technique used to generate the initial design matrix is not that critical. Our experience suggests that good performance can be achieved provided a space-filling DoE technique is used at the first stage.

As pointed out earlier in this section, the DoE method must work synergistically with the surrogate modeling technique in use to ensure good generalization. It has been noted in previous studies (Meckesheimer et al. 2001) that Gaussian process models can encounter difficulties when some factorial designs and central composite designs are used. In general, space-filling designs such as LHS, uniform designs and minimum discrepancy sequences work best with nonparametric models such as RBFs and Gaussian processes.

Simpson et al. (2001) conducted an extensive study using datasets from computer experiments to compare how different DoE techniques (LHS, Hammersley sequences, orthogonal arrays and uniform designs) work in conjunction with a selection of surrogate modeling techniques commonly used in engineering practice (quadratic polynomials, Kriging, RBF approximations with linear splines and multivariate adaptive regression splines). The accuracy of the final model was judged using the maximum error, $\max\{|y(\mathbf{x}^{(i)}) - \widehat{y}(\mathbf{x}^{(i)})|\}$, and the mean-square error, $(1/l)\sum_{i=1}^{l}(y(\mathbf{x}^{(i)}) - \widehat{y}(\mathbf{x}^{(i)}))^2$. It was found that uniform designs and Hammersley sampling sequences result in more accurate surrogate models compared to standard LHS as regards to the mean-square error. Orthogonal arrays tend to result in models that have lowest value of maximum error, which can be attributed to the fact that points generated using orthogonal arrays are placed at the corners of the design space. Simpson et al. (2001) made the recommendation that models with low values of root mean-square error should be preferred to models with low maximum error. This was because the maximum error can always be reduced by employing a sequential model updating strategy.

5.9 Visualization and Screening

Once a surrogate model has been constructed, it becomes possible to obtain information on how the input variables affect the output quantities of interest. This has two immediate applications, which we examine in this section: (1) visualization of the design space and (2) variable screening to rank the variables according to their importance.

5.9.1 Design Space Visualization

Design space visualization is concerned with the generation of graphical plots that aid the analyst to gain insights into the design space under consideration. This may involve, among other things, understanding the key functional relationships between the inputs and outputs of interest, such as the structure of any nonlinearity and the interactions between variables. Scatter plots are perhaps the most straight forward approach for design space visualization. However, when there are many variables, this approach may not lead to any useful insights.

The main effect of a design variable x_i can be computed by integrating out the effects of other variables as follows:

$$M(x_i) = \frac{1}{V} \int_{\Omega_x} y(\mathbf{x}) \prod_{j \neq i} dx_j, \tag{5.95}$$

where V is the volume of the design space Ω_x over which integration is carried out. Similarly, the interaction effects between two variables x_i and x_j can be computed as

$$M(x_i, x_j) = \frac{1}{V^2} \int_{\Omega_x} y(\mathbf{x}) \prod_{k \neq i, j} dx_k \tag{5.96}$$

If we have a surrogate model for the output of interest, say $\widehat{y}$, then the main and interaction effects can be computed very efficiently. For certain polynomial models, the multidimensional integrals appearing in the above expressions can be evaluated analytically. For Gaussian process models, analytical integration is possible only if a correlation function obeying the product correlation rule is employed. Otherwise, DoE techniques can be used to numerically approximate the integrals defining the main and interaction effects. Graphical plots of these effects can provide useful insights into the input–output relationship. For example, if the output is relatively insensitive to a particular design variable, say x_i, then the main effect $M(x_i)$ will tend to be very flat.

Another approach to design space visualization that becomes possible when surrogate models are available is the hierarchical axis technique (HAT) plot. These can be useful when the analyst wishes to visualize multidimensional functions with between one and six variables; see Figure 16.2. This shows a four-dimensional problem and also includes an equality constraint, an inequality constraint and the positions of points in the DoE used to construct the data. An excellent overview of the state of the art in design space visualization and techniques for visualizing multidimensional functions can be found in the dissertation of Holden (2004).

5.9.2 Variable Screening

The objective of variable screening is to obtain as much information on how the input variables affect the design using as few runs of the high-fidelity code as possible. In many cases, only a few of the variables are responsible for most of the variation in the response, whereas most of the other variables contribute little. An estimate of the relative ranking of the importance of the design variables may then make it possible to reduce the number of design variables being considered, as well as the number of training points required to construct an accurate surrogate. As mentioned earlier in this section, one way to achieve this is to compare the main effect of each variable calculated using (5.95).

Myers and Montgomery (1995) discuss a number of approaches based on factorial designs that can be used in conjunction with polynomial response surface models to screen problems. This, however, can be computationally very expensive for studies with large numbers of design variables. More efficient alternatives based on fractional factorial and Plackett–Burman designs exist – in these, an assumption is again made that the output can be well approximated by polynomial models.

Welch et al. (1992) presented an approach that uses the hyperparameters in Gaussian process models $(\theta_1, \theta_2, \ldots, \theta_p)$ to screen out variables of little importance. In this approach, the response is first evaluated at an n point DoE where the variables are all normalized over the range [0, 1]. Subsequently, MLE of the hyperparameters is carried out and the relative importance of the ith variable is judged by examining the value of θ_i. For example, when the Gaussian correlation function in (5.60) in used, a small value of θ_i suggests that the ith variable does not significantly affect the output. This approach was applied to construct approximation models for two problems with 20 variables using only 30–50 runs of the computer code.

Morris (1991) suggested that the statistics of the following "elementary effect" can be used to estimate the importance of the ith variable:

$$\mathfrak{I}_i = y\left(x_1, \ldots, x_i + \delta, \ldots, x_p\right) - y(\mathbf{x}) \tag{5.97}$$

The mean and variance of $\mathfrak{I}_i$ can be calculated by randomly selecting points in the design space $\Omega_{\mathbf{x}}$. If the ith variable does not effect the output significantly, then the mean and variance of $\mathfrak{I}_i$ will be small. A large mean together with small variance implies a linear effect, independent of the other variables, whereas a large variance implies either a nonlinear independent effect or interactions with other variables. In practice, the statistics of $\mathfrak{I}_i$ can be efficiently approximated by using a surrogate model in lieu of the original function $y(\mathbf{x})$.

A comparison of various screening methods can be found in Trocine and Malone (2000) and Leary et al. (2002). In general, it is difficult to recommend which is the best technique to use since results can be highly problem dependent. Nonetheless, if screening reveals a wide variation in the effects of different factors, then reducing the problem size can greatly simplify computational studies. Simpson et al. (2004) note that the interactions between screening methods and optimization need to be carefully studied since variables that are not deemed to be important during initial studies may become important later during optimization. Hence, it may be necessary to revisit variables that were initially screened out.

5.10   Black-box Surrogate Modeling in Practice

A major drawback of black-box modeling techniques is that a large number of training points may be needed to construct an accurate surrogate. This is particularly true for problems with a large number of input variables and highly nonlinear input–output relationships. This in turn leads to a significant increase in the computational cost, because of the requirement of running the analysis code at a large number of design points. Further, the training times of many surrogate modeling techniques tend to increase significantly with the number of training points.

A number of recent studies in the literature have examined strategies to circumvent this *curse of dimensionality* that arises from the fact that the number of hypercubes required to fill out a compact region of a p-dimensional space grows exponentially with p. Balabanov et al. (1999) proposed a reasonable design space approach that can be used in conjunction with variable-fidelity models to circumvent the curse of dimensionality. Another idea that has attracted some attention has been the use of output sensitivity information from the analysis code to improve the fidelity of the surrogate, for example, the Hermite interpolation and gradient-enhanced Gaussian process modeling techniques discussed in this chapter. However, in general, it is difficult to construct a global surrogate model that accurately captures the input–output relationship over the entire design space. This has motivated a number of researchers to use local surrogate models that are valid only over a small region of the design space (Ong et al. 2003; Toropov et al. 1993); see Chapter 7. Even though promising results have been reported for some problems, this fundamental difficulty associated with the curse of dimensionality is not expected to disappear. On the positive side, black-box modeling techniques are mature for real-life applications with a moderate number of design variables $\mathcal{O}(10 - 20)$. This is clear from the wide body of applications published in the open literature over the last decade. An overview of the state of the art in surrogate modeling can be found in the summary of a panel discussion on approximation methods held at the 9th *AIAA/ISSMO Conference on Multidisciplinary Analysis & Optimization* (Simpson et al. 2004).

In practice, it is often the case that any observational dataset needs to be transformed (or preprocessed) to extract the best performance from a given surrogate modeling technique. It is always sensible in practice to normalize the input vectors to avoid ill-conditioned matrices – this is particularly the case for RBF approximations and Gaussian process modeling. The usual practice is to normalize all the design variables to lie in the interval $[0, 1]$ or $[-1, 1]$. In some cases, the performance of the surrogate model can be improved by applying a transformation to the input variables – this is similar in spirit to the intervening variable idea discussed earlier in this chapter. As a result, the input–output relationship is essentially *warped* to aid the surrogate model to generalize better. A discussion on the issue of variable transformation in the context of Gaussian process modeling can be found in Jones et al. (1998) and Snelson et al. (2004). Sóbester et al. (2004) presented an approach based on genetic programming that attempts to automatically estimate the best variable transformation for a given dataset and surrogate modeling technique. The applicability of this approach to problems with many variables is yet to be established.

It is impossible to provide simple guidelines on the best surrogate modeling technique given a set of observational data. We recommend that when choosing a surrogate model most appropriate for the problem at hand the user should definitely consider constructing a sequence of models, $\mathcal{M}_1, \mathcal{M}_2, \ldots$, using different techniques. Subsequently, a set of objective criteria can be employed to choose the (approximately) best model. If a validation dataset is available, then the performance of all models on this dataset can be compared by examining the root mean-square error and/or maximum error. For cases where limited data is available, the leave-one-out estimator or the k-fold cross-validation procedure discussed earlier in Section 5.4.4 can be employed. These criteria makes sense if the aim is to create a surrogate model that closely *emulates* the true input–output relationship over the design space.

In the context of optimization, however, the objective is to create a surrogate model that accurately captures the *trends* of the relationship between $\mathbf{x}$ and y, and in particular, over those regions where high performance designs lie. The mean-square error may not be a very good indicator of how well such trends are captured – it is possible that in some cases a model with high mean-square error may predict the location of the minima/maxima better than a model with lower mean-square error. If a surrogate model predicts the trends of the input–output relationship accurately, then, given two design points $\mathbf{x}^{(1)}$ and $\mathbf{x}^{(2)}$, the surrogate model should be able to predict correctly if $f(\mathbf{x}^{(1)}) > f(\mathbf{x}^{(2)})$ or vice versa – the absolute value of the surrogate model predictions at $\mathbf{x}^{(1)}$ and $\mathbf{x}^{(2)}$ are not very important in this context.

For example, consider the case when it is required to minimize $f(\mathbf{x})$ via minimization of a surrogate model. Let a surrogate model $\widehat{y} = \widehat{f}_{ij}(\mathbf{x})$ be constructed by deleting two points from the training dataset (say point i and j). Then, the following error measure can be used to evaluate the predictive capability of the surrogate model

$$\mathcal{L}_{ij} = \mathcal{I}[f(\mathbf{x}^{(i)}) < f(\mathbf{x}^{(j)})] - \mathcal{I}[\widehat{f}_{ij}(\mathbf{x}^{(i)}) < \widehat{f}_{ij}(\mathbf{x}^{(j)})], \qquad (5.98)$$

where $\mathcal{I}[.]$ denotes the indicator function, which returns the value 1 if its argument is true, otherwise it returns 0. Repeating this leave-two-out process for a random selection of points i and j, the expected value of $\mathcal{L}_{ij}$ can be estimated. This value can be used to compare the suitability of models for optimization purposes.

Finally, we note that during optimization studies it is a standard practice to augment black-box models as improved designs are obtained. This leads to the important topic of surrogate model updating, which we discuss more fully in Chapter 7.

6

Physics-based Approximations

The approximation techniques covered in the previous chapter were general in scope and can be applied to a variety of black-box simulation codes, since no assumptions were made regarding the governing equations being solved. In this chapter, we examine physics-based approximation concepts, which require a deeper understanding of the governing equations and the numerical methods employed for their solution. The methods described here attempt to circumvent the curse of dimensionality associated with black-box surrogate modeling by exploiting domain-specific knowledge. More specifically, the focus is on leveraging knowledge of equations governing the problem physics to construct a "customized" metamodel. We begin by examining how variable-fidelity models can be employed to construct such surrogate models. Subsequently, we examine in detail reduced basis methods for constructing physics-based approximations of linear and nonlinear systems. In most of the methods discussed here, the generality and ease of implementation of black-box approaches is sacrificed in order to achieve improved accuracy. A notable advantage of most of the physics-based approaches discussed is that they do not suffer from the curse of dimensionality and can hence be readily applied to problems with greater numbers of design variables.

We would also note that the construction of the low- fidelity, approximate semiempirical models, which are commonly used in concept design (see, for example, Bradley (2004)), is not considered here. Many concept design teams use customized tools based on a mixture of empirical expressions, look-up tables and other experimental results to predict design performance. No general advice can be given on the construction of such tools, although we do consider how such methods may be used alongside high-fidelity PDE solvers.

6.1 Surrogate Modeling using Variable-fidelity Models

In this section, we discuss techniques for fusing data from simulation models of varying fidelity to construct a surrogate model. These techniques are useful for problems where a number of simulation models are available (each modeling the underlying system with a different level of accuracy). Some simple strategies based on local approximations are outlined first and subsequently a more general approach based on global surrogate modeling is presented.

Computational Approaches for Aerospace Design: The Pursuit of Excellence.   A. J. Keane and P. B. Nair
© 2005 John Wiley & Sons, Ltd

For simplicity of presentation, we consider the case when two analysis models are available, $f_{lo}(\mathbf{x})$ and $f_{hi}(\mathbf{x})$, where $\mathbf{x} \in \mathbb{R}^p$ denotes the vector of design variables and the fidelity of f_{hi} is greater than that of f_{lo}. In a number of problems of practical interest, it is possible to create a hierarchy of variable-fidelity models and the methods discussed here can be applied to such cases. Some examples of situations where variable-fidelity models arise are: (1) the model f_{hi} uses a finer mesh to solve the governing equations compared to f_{lo}; (2) the governing equations being solved by model f_{hi} contain more detailed aspects of the problem physics compared to f_{lo}, for example, Navier–Stokes versus Euler equations; (3) the high-fidelity model f_{hi} is evaluated using a fully converged run whereas the low-fidelity model f_{lo} is based on a partially converged run – such variable-fidelity models can be readily constructed for CFD or CSM problems where an iterative solver is employed for analysis; and (4) f_{lo} is a semiempirical approximation model constructed for conceptual design studies. In all these cases, the low-fidelity model f_{lo} can be viewed as a global physics-based approximation of f_{hi}. The methods presented in this section attempt to correct this global approximation so as to ensure better agreement with the high-fidelity model, that is, they attempt to "fuse" the models together. Note that we shall henceforth refer to the corrected low-fidelity model by $\widehat{f}$.

6.1.1 Zero-order Scaling

The simplest approach for model fusion involves the use of a scale factor based on the values of $f_{hi}(\mathbf{x})$ and $f_{lo}(\mathbf{x})$ at a single point, say $\mathbf{x}^0$. This approach, which is also referred to as the local-global approximation strategy (Haftka 1991), involves the use of the following scale factor:

$$A(\mathbf{x}) = \frac{f_{hi}(\mathbf{x})}{f_{lo}(\mathbf{x})}. \tag{6.1}$$

The scaled (corrected) low-fidelity model can then be written as

$$\widehat{f}(\mathbf{x}) = A(\mathbf{x}^0) f_{lo}(\mathbf{x}). \tag{6.2}$$

Similarly, if we define an additive scale factor of the form $B(\mathbf{x}) = f_{hi}(\mathbf{x}) - f_{lo}(\mathbf{x})$, then the scaled low-fidelity model can be written as

$$\widehat{f}(\mathbf{x}) = f_{lo}(\mathbf{x}) + B(\mathbf{x}^0). \tag{6.3}$$

The key idea of both the multiplicative and additive scaling approaches is to correct the low-fidelity model so that it agrees with the high-fidelity model at the point $\mathbf{x}^0$, that is, $\widehat{f}(\mathbf{x}^0) = f_{hi}(\mathbf{x}^0)$. The scaled low-fidelity model can then be used as a surrogate for f_{hi} in the vicinity of $\mathbf{x}^0$. It is also possible to define subtractive and division-based scaling factors of the form $C(\mathbf{x}) = f_{lo} - f_{hi}$ and $D(\mathbf{x}) = f_{lo}/f_{hi}$.

6.1.2 First-order Scaling

In first-order scaling techniques (Haftka 1991), the idea is to use first-order Taylor series approximations for the scale factors $A(\mathbf{x})$ and $B(\mathbf{x})$. This leads to the following expression for the scaled low-fidelity model when the multiplicative approach is used:

$$\widehat{f}(\mathbf{x}) = \left[A(\mathbf{x}^0) + \nabla A(\mathbf{x})^T (\mathbf{x} - \mathbf{x}^0) \right] f_{lo}(\mathbf{x}) \tag{6.4}$$

where

$$\nabla A(\mathbf{x}^0) = \frac{1}{f_{lo}(\mathbf{x}^0)} \nabla f_{hi}(\mathbf{x}^0) - \frac{f_{hi}(\mathbf{x}^0)}{f_{lo}^2(\mathbf{x}^0)} \nabla f_{lo}(\mathbf{x}^0). \tag{6.5}$$

Similarly, when the additive approach is used, the scaled low-fidelity model can be written as $\widehat{f}(\mathbf{x}) = f_{lo}(\mathbf{x}) + B(\mathbf{x}^0) + \nabla B(\mathbf{x})^T (\mathbf{x} - \mathbf{x}^0)$ where $\nabla B(\mathbf{x}^0) = \nabla f_{hi}(\mathbf{x}^0) - \nabla f_{lo}(\mathbf{x}^0)$.

It may be noted from these expressions that the first-order scaling approach can be applied only when the first-order sensitivities of f_{hi} and f_{lo} are available. Further, this approach also ensures that both the function value and the sensitivities computed using the scaled low-fidelity model agree with those calculated using the high-fidelity model at the point $\mathbf{x}^0$. As a consequence, first-order scaling approaches will have improved accuracy compared to zero-order approaches in the vicinity of $\mathbf{x}^0$.

6.1.3 Second-order Scaling

A straightforward extension of the first-order scaling approach can be formulated by using second-order Taylor series expansions of the scaling factors $A(\mathbf{x})$ and $B(\mathbf{x})$ (Eldred et al. 2004). For example, in the multiplicative approach, a second-order Taylor series expansion of the scaling function $A(\mathbf{x})$ defined in (6.1) can be used to arrive at the following expression for the scaled low-fidelity model:

$$\widehat{f}(\mathbf{x}) = \left[A(\mathbf{x}^0) + \nabla A(\mathbf{x}^0)^T (\mathbf{x} - \mathbf{x}^0) + \frac{1}{2}(\mathbf{x} - \mathbf{x}^0)^T \nabla^2 A(\mathbf{x}^0)(\mathbf{x} - \mathbf{x}^0) \right] f_{lo}(\mathbf{x}) \tag{6.6}$$

where

$$\nabla^2 A(\mathbf{x}^0) = \frac{1}{f_{lo}(\mathbf{x}^0)} \nabla^2 f_{hi}(\mathbf{x}^0) - \frac{f_{hi}(\mathbf{x}_c)}{f_{lo}^2(\mathbf{x}_c)} \nabla^2 f_{lo}(\mathbf{x}^0) + 2 \frac{f_{hi}(\mathbf{x})}{f_{lo}^3(\mathbf{x})} \nabla f_{lo}(\mathbf{x}^0) \nabla f_{lo}^T(\mathbf{x}^0)$$

$$- \frac{1}{f_{lo}^2(\mathbf{x}^0)} \left(\nabla f_{lo}(\mathbf{x}^0) \nabla f_{hi}^T(\mathbf{x}^0) + \nabla f_{hi}(\mathbf{x}^0) \nabla f_{lo}^T(\mathbf{x}^0) \right). \tag{6.7}$$

Similarly, in the additive approach, the scaled low-fidelity model is written in terms of the second-order Taylor series expansion of the scaling function $B(\mathbf{x})$ as

$$\widehat{f}(\mathbf{x}) = f_{lo}(\mathbf{x}) + B(\mathbf{x}^0) + \nabla B(\mathbf{x}^0)^T (\mathbf{x} - \mathbf{x}^0) + \frac{1}{2}(\mathbf{x} - \mathbf{x}^0)^T \nabla^2 B(\mathbf{x}^0)(\mathbf{x} - \mathbf{x}^0) \tag{6.8}$$

where $\nabla^2 B(\mathbf{x}^0) = \nabla^2 f_{hi}(\mathbf{x}^0) - \nabla^2 f_{lo}(\mathbf{x}^0)$.

Clearly, second-order scaling techniques can be implemented only when the second-order sensitivities of f_{lo} and f_{hi} are available. Hence, this approach is not practical for problems where the second-order sensitivities cannot be efficiently computed. Eldred et al. (2004) proposed the application of quasi-Newton or Gauss–Newton approximations for the second-order derivatives which make use of first-order sensitivity information. In these approaches, approximations for the Hessians of f_{hi} and f_{lo} are sequentially built up during the course of the design iterations in much the same way that quasi-second-order search methods build Hessian matrices (see Chapter 3).

6.1.4 Multipoint Corrections

It is also possible to develop a hybrid approach that uses a combination of additive and multiplicative corrections in order to improve the global approximation characteristics of the scaled low-fidelity model (Eldred et al. 2004). Here, the scaled low-fidelity model is written as a weighted combination of (6.2) and (6.3) as follows:

$$\widehat{f}(\mathbf{x}) = \gamma \left(f_{lo}(\mathbf{x}) + B(\mathbf{x}^0) \right) + (1 - \gamma) A(\mathbf{x}^0) f_{lo}(\mathbf{x}), \qquad (6.9)$$

where γ is an undetermined parameter. It can be noted from (6.9) that $\widehat{f}(\mathbf{x}^0) = f_{hi}(\mathbf{x}^0)$. The parameter γ can be computed by imposing an additional matching condition at a different design point, say $\mathbf{x}^{(1)}$. For example, $\mathbf{x}^{(1)}$ could be a previously evaluated design point in the vicinity of $\mathbf{x}^0$. Hence, using the condition $\widehat{f}(\mathbf{x}^{(1)}) = f_{hi}(\mathbf{x}^{(1)})$, it follows that γ can be computed as

$$\gamma = \frac{f_{hi}(\mathbf{x}^{(1)}) - A(\mathbf{x}^0) f_{lo}(\mathbf{x}^{(1)})}{f_{lo}(\mathbf{x}^{(1)}) + B(\mathbf{x}^0) - A(\mathbf{x}^0) f_{lo}(\mathbf{x}^{(1)})}. \qquad (6.10)$$

In a similar fashion, it is possible to derive first- and second-order multipoint correction schemes to ensure that the gradients of $\widehat{f}$ agree with those computed using the high-fidelity model at a chosen set of points. Alternatively, the two-point approximation techniques discussed earlier in Section 5.2 can be used to derive more accurate model correction factors.

6.1.5 Global Scaling using Surrogate Models

The scaling approaches discussed so far are based on correcting the low-fidelity model so that it shows better agreement with the high-fidelity model in the vicinity of a single (or a few) point(s). Hence, the corrected model will only be locally accurate – the example discussed next illustrates this point numerically. It is possible to formulate scaling schemes that improve the global accuracy of the corrected low-fidelity model by employing black-box surrogate modeling techniques such as those discussed previously in Chapter 5. To illustrate this idea, consider the case when a set of observational data $\{\mathbf{x}^{(i)}, f_{lo}(\mathbf{x}^{(i)}), f_{hi}(\mathbf{x}^{(i)})\}$, $i = 1, 2, \ldots, m$ is available. Then, a global surrogate model of the multiplicative scaling factor (say $\widehat{A}(\mathbf{x})$) can be constructed by using the training dataset $\{\mathbf{x}^{(i)}, f_{hi}(\mathbf{x}^{(i)})/f_{lo}(\mathbf{x}^{(i)})\}$, $i = 1, 2, \ldots, m$, where $\mathbf{x}^{(i)}$ denotes the input vector and $A(\mathbf{x}) = f_{hi}(\mathbf{x})/f_{lo}(\mathbf{x})$ is the scalar output to be approximated. Finally, a globally corrected low-fidelity model can be written as $\widehat{f}(\mathbf{x}) = \widehat{A}(\mathbf{x}) f_{lo}(\mathbf{x})$, which can be used as a global surrogate of the high-fidelity model.

If the first-order sensitivities of f_{hi} and f_{lo} are available, then it may be worthwhile to use the Hermite RBF interpolation or gradient-enhanced Gaussian process modeling approaches discussed previously in Chapter 5. Similar procedures can be devised for the cases when the additive, subtractive or division scaling approaches are used. The global scaling approaches can lead to highly accurate and efficient surrogate models, particularly when f_{lo} can be evaluated very cheaply. The accuracy of the surrogate model constructed using this approach can in many cases be better than that constructed by using only the high-fidelity training dataset $\{\mathbf{x}^{(i)}, f_{hi}(\mathbf{x}^{(i)})\}$, $i = 1, 2, \ldots, m$. This is because the scaling factors may be better behaved and hence easier to approximate (Keane 2003; Leary et al. 2003a). The additive variant is applied in the case study described in Chapter 12.

A general Bayesian approach for surrogate modeling that employs data from k variable-fidelity models was proposed by Kennedy and O'Hagan (2000). The basic idea is to assume that a correlation exists between $f_{lo}(\mathbf{x})$ and $f_{hi}(\mathbf{x})$. Then, a training dataset of the form $\{\widetilde{\mathbf{x}}^{(i)}, f_{hi}(\mathbf{x}^{(i)})\}$, $i = 1, 2, \ldots, m$ can be used to construct a surrogate model for f_{hi}, where the modified input vector is defined as $\widetilde{\mathbf{x}}^{(i)} = \{x_1, x_2, \ldots, x_p, f_{lo}(\mathbf{x}^{(i)})\} \in \mathbb{R}^{p+1}$. In other words, the low fidelity model output becomes an input for the surrogate model. It is possible to apply the Gaussian process approach or any of the techniques discussed in Chapter 5 to this augmented training dataset to construct a surrogate; see, for example, El-Beltagy (2004).

Note that in all the techniques discussed in this section the low-fidelity model needs to be evaluated at any new point $\mathbf{x}$ before predicting $\widehat{f}$. Hence, for these methods to be useful in practice, it is important that f_{lo} is computationally cheap to run.

6.1.6   An Example

To illustrate some of the features of scaled low-fidelity models, we consider a test problem where the high- and low-fidelity models are represented by the one-dimensional algebraic functions $f_{hi}(x) = x^3 - 2x^2 - \sin(2\pi x)$ and $f_{lo}(x) = (x - 0.1)^3 - 2(x + 0.1)^2 - \sin(2\pi x + 0.5) + 0.1$, respectively. Plots of f_{hi} and f_{lo} are shown in Figure 6.1. It can be observed that the qualitative behavior of f_{lo} is similar to f_{hi}. However, there are significant disparities in the outputs predicted by the models, particularly for $x > 0.8$.

Consider the case when it is of interest to scale the low-fidelity model so as to ensure agreement with the high-fidelity model at $x = 1$. The corrected low-fidelity models obtained using zero-, first- and second-order *multiplicative* scaling factors are shown in Figure 6.2. It can be seen that the zero-order correction only ensures that $\widehat{f}$ agrees with f_{hi} at $x = 1$. In contrast, the first- and second-order additive scaling approaches ensure that $\widehat{f}$ matches the derivatives of f_{hi} at $x = 1$. However, since all these correction schemes are local in nature, the corrected low-fidelity models do not agree well with the high-fidelity model at points further away from $x = 1$. In particular, the second-order correction tends to deviate significantly from f_{hi} for values of x further away from 1.

The corrected models constructed using *additive* scaling factors are shown in Figure 6.3. The trends are similar to those obtained using the multiplicative approach in the vicinity of $x = 1$. Here, however, the second-order correction tends to deviate significantly from f_{hi} at points far away from $x = 1$. As a consequence, appropriate move limits need to be

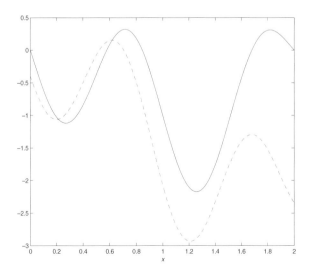

Figure 6.1   Plots of f_{hi} (solid line) and $f_{lo}(x)$ (dashed line) for the model problem.

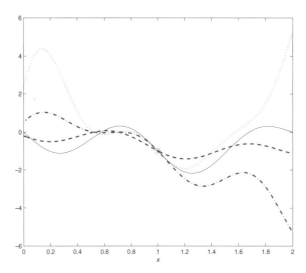

Figure 6.2 Plots of the corrected low-fidelity model $\widehat{f}$ constructed using zero- (dashed line), first- (dash-dot line) and second-order (dotted line) multiplicative scaling factors. The high-fidelity model f_{hi} is depicted by a solid line and the star denotes the point at which the low-fidelity model is corrected.

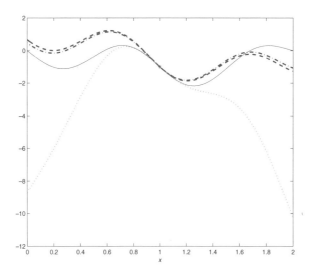

Figure 6.3 Plots of the corrected low-fidelity model $\widehat{f}$ constructed using zero- (dashed line), first- (dash-dot line) and second-order (dotted line) additive scaling factors. The high-fidelity model f_{hi} is depicted by a solid line and the star denotes the point at which the low-fidelity model is corrected.

applied if such corrected low fidelity models are to be used in design optimization studies. Alternatively, global approximations of the multiplicative or additive correction factors could be employed to ensure that $\widehat{f}$ agreed with f_{hi} over a larger region of the design space.

6.2 An Introduction to Reduced Basis Methods

We next consider reduced basis methods, which aim to build models with fewer unknowns than the original high-fidelity model. To illustrate reduced basis methods, consider the discrete mathematical model of a physical system written in the form

$$\mathbf{R}(\mathbf{w}, \mathbf{x}) = 0, \tag{6.11}$$

where $\mathbf{w} \in \mathbb{R}^n$ denotes the discretized vector of field variables, $\mathbf{x} \in \mathbb{R}^p$ denotes the vector of design variables and $\mathbf{R} : \mathbb{R}^n \times \mathbb{R}^p \to \mathbb{R}^n$. Note that $\mathbf{w}$ is an implicit function of the design variables. The above form of discrete equations is usually obtained in practice by spatial and temporal discretization of PDEs governing the system response. Hence, (6.11) is representative of a wide class of problems encountered in structural analysis, fluid mechanics, heat transfer, etc.

The main idea used in reduced basis methods is to approximate the field variable vector $\mathbf{w}$ as

$$\widehat{\mathbf{w}} = c_1 \boldsymbol{\psi}_1 + c_2 \boldsymbol{\psi}_2 + \cdots + c_m \boldsymbol{\psi}_m = \boldsymbol{\Psi} \mathbf{c}, \tag{6.12}$$

where $\boldsymbol{\Psi} = \{\boldsymbol{\psi}_1, \boldsymbol{\psi}_2, \ldots, \boldsymbol{\psi}_m\} \in \mathbb{R}^{n \times m}$ denotes a matrix of *known* basis vectors and $\mathbf{c} = \{c_1, c_2, \ldots, c_m\}^T \in \mathbb{R}^m$ is a vector of *undetermined* coefficients.

The key assumption made in reduced basis methods is that the response vector lies in the subspace spanned by a set of basis vectors. This allows the original problem with n unknowns to be recast as a problem with m unknowns. If $m \ll n$, it becomes possible to approximate $\mathbf{w}$ very efficiently by solving a reduced-order problem. It is also generally the case that the undetermined coefficients in (6.12) are estimated so that the governing equations are satisfied in some sense (e.g., by minimizing the integral of an appropriate error norm evaluated over the design space) – this ensures that the reduced basis approximation is global in nature. Since the model parameters are estimated using the governing equations, reduced basis methods also belong to the class of physics-based approximation techniques.[1]

Clearly, the main ingredients of reduced basis methods are the choice of basis vectors and the numerical scheme employed to compute the vector of undetermined coefficients. The accuracy and convergence rate (as the number of basis vectors is increased) will depend on the choice of basis vectors and the parameter estimation scheme. In what follows, we examine the choices associated with these ingredients in greater detail.

6.2.1 Choice of Basis Vectors

Ideally, a good set of basis vectors should be easy to compute as well as guaranteed to be linearly independent. For the general form of discrete governing equations specified in (6.11), there are three commonly used options for the basis vectors. To illustrate this, consider the case when there is only one design variable, say x.

---

[1]We note here that many numerical procedures for solving PDEs such as the finite element method can also be interpreted as a reduced basis method. This is because an approximate solution is sought in a subspace spanned by a set of known basis functions, thereby reducing the original infinite-dimensional problem into a numerically tractable finite-dimensional problem.

1. *The Lagrange subspace* is formed by using the exact solutions of the discrete governing equations calculated at various values of the design variable, say $x^{(1)}$, $x^{(2)}$, ..., $x^{(m)}$, that is,

$$\Psi = \text{span} \left\{ \mathbf{w}(x^{(1)}), \mathbf{w}(x^{(2)}), \ldots, \mathbf{w}(x^{(m)}) \right\} \in \mathbb{R}^{n \times m}. \quad (6.13)$$

2. *The Hermite subspace* is formed by combining the Lagrange subspace with the first-order derivatives of the field variable vector $\mathbf{w}$ evaluated at $x^{(1)}$, $x^{(2)}$, ..., $x^{(m)}$, that is,

$$\Psi = \text{span} \left\{ \mathbf{w}(x^{(1)}), \frac{\partial \mathbf{w}(x^{(1)})}{\partial x}, \ldots, \mathbf{w}(x^{(m)}), \frac{\partial \mathbf{w}(x^{(m)})}{\partial x} \right\} \in \mathbb{R}^{n \times 2m}. \quad (6.14)$$

3. *The Taylor subspace* is written in terms of the derivatives of the field variable vector $\mathbf{w}$ at the point x^*, that is,

$$\Psi = \text{span} \left\{ \mathbf{w}(x^*), \frac{\partial \mathbf{w}(x^*)}{\partial x}, \frac{\partial^2 \mathbf{w}(x^*)}{\partial x^2}, \ldots \frac{\partial^{m-1} \mathbf{w}(x^*)}{\partial x^{m-1}} \right\} \in \mathbb{R}^{n \times m}. \quad (6.15)$$

It is clear from the above that basis vectors spanning the Hermite subspace can be computed only when the first-order sensitivities of $\mathbf{w}$ are available. Further, the dimension of the Hermite subspace grows as $m(p + 1)$, where p is the number of design variables. The Taylor subspace may not be very useful since, for most problems of practical interest, the higher-order derivatives cannot be efficiently computed. The Lagrange subspace appears to be the most attractive from a practical viewpoint since the basis vectors spanning this subspace can be easily computed. The only point of concern with all three subspaces is the possibility that some of the basis vectors could be linearly dependent. However, as shown later in Section 6.5, this difficulty can be easily overcome by employing the singular value decomposition scheme.

Another important point worth noting here is that the quality of the basis vectors spanning the Lagrange, Hermite and Taylor subspaces will depend on the choice of design parameter values, $x^{(1)}, x^{(2)}, \ldots, x^{(m)}$. Hence, it is of interest to develop strategies for optimally choosing these points such that a rich set of basis vectors can be obtained. Unfortunately, there are few guidelines available in the literature for designing such computer experiments. Given this lack of specialized techniques, it makes sense to use a space-filling design of experiments technique for deciding the parameter values at which the full-order model should be solved (see Section 5.9).

Reduced basis methods using the subspaces discussed here have been applied in the past to solve a wide range of problems, including structural mechanics and fluid flow; see, for example, Balmes (1996); Ito and Ravindran (1998); Nagy (1979); Noor (1981); Peterson (1989); Prud'homme et al. (2002). In the context of design, the first application of the Lagrange subspace was perhaps made by Fox and Miura (1971).

Now, consider the special case when the governing discrete equations in (6.11) are linear and can be written in the form $\mathbf{Kw} = \mathbf{f}$. Then, basis vectors spanning the Krylov subspace can be used to approximate $\mathbf{w}$:

$$\mathcal{K}_m(\mathbf{K}, \mathbf{r}_0) = \text{span} \left\{ \mathbf{r}_0, \mathbf{Kr}_0, \ldots, \mathbf{K}^{m-1} \mathbf{r}_0 \right\}. \quad (6.16)$$

Here, $\mathbf{r}_0 = \mathbf{Kw}_0 - \mathbf{f} \in \mathbb{R}^n$ is the residual error vector when $\mathbf{w}_0 \in \mathbb{R}^n$ is used as an initial guess.

The application of Krylov subspace methods for iterative solution of linear systems is an active area of research in numerical linear algebra, with a rich history spanning more than 50 years; see Saad and Van der Vorst (2000) for a historical overview of progress made in the last century. There are a number of specialized texts on this topic; see, for example, Saad (1996). Some of the better-known methods in this class are Arnoldi's method, the Lanczos method, and the Generalized Minimal Residual (GMRES) method. We discuss the Krylov subspace and its theoretical properties in more detail in the next section.

It may be noted from (6.16) that the basis vectors spanning the Krylov subspace are functions of the design variable vector $\mathbf{x}$. To illustrate this point, let us first write the governing linear algebraic equations in the parametric form $\mathbf{K}(\mathbf{x})\mathbf{w}(\mathbf{x}) = \mathbf{f}(\mathbf{x})$. Then, the ith basis vector can be expressed in parametric form as $[\mathbf{K}(\mathbf{x})]^{i-1}\{\mathbf{K}(\mathbf{x})\mathbf{w}_0 - \mathbf{f}(\mathbf{x})\}$. Hence, for any given value of $\mathbf{x}$, the basis vectors have to be computed first by carrying out a series of matrix-vector multiplications before the reduced basis approximation can be constructed. In contrast, basis vectors spanning the Lagrange, Hermite and Taylor subspaces are independent of $\mathbf{x}$. Here, the basis vectors are computed in the preprocessing phase and they are used for approximating the solution to (6.11) for any given value of $\mathbf{x}$. Even though the preprocessing phase may incur significant computational effort since the high-fidelity model has to be run at a selected number of design points, reduced basis approximations of $\mathbf{w}$ can be efficiently computed across the design space once the basis vectors are *fixed*. In practice, it is often the case that parameterized basis vectors give more accurate results than nonparametric systems. However, if the reduced basis model is to be evaluated at a large number of design points, then the latter approach would be computationally more efficient.

6.2.2   Schemes for Computing Undetermined Coefficients

Given an appropriate set of linearly independent basis vectors $\mathbf{\Psi}$, what is needed next is a numerical scheme for computing the vector of undetermined coefficients $\mathbf{c}$. A key objective is to estimate the coefficients so that the discrete governing equations are satisfied in some sense – this is precisely why we refer to this class of methods as physics-based approximations. In contrast, in the black-box approaches discussed in Chapter 5, the undetermined coefficients in the surrogate model are computed without regard to the governing equations. We outline next the Bubnov–Galerkin and Petrov–Galerkin schemes, which can be employed to estimate the undetermined coefficients in (6.12).

- *Bubnov–Galerkin scheme:* In the Bubnov–Galerkin scheme, $\mathbf{c}$ is computed such that the residual error vector obtained by substituting $\widehat{\mathbf{w}} = \mathbf{\Psi}\mathbf{c}$ into the governing equations is orthogonal to the approximating subspace. For the discrete governing equations in (6.11), this means that the following conditions are imposed:

$$\mathbf{R}(\mathbf{\Psi}\mathbf{c}, \mathbf{x}) \perp \boldsymbol{\psi}_i, \quad \text{i.e.,} \quad \boldsymbol{\psi}_i^T \mathbf{R}(\mathbf{\Psi}\mathbf{c}, \mathbf{x}) = 0, \quad i = 1, 2, \ldots, m. \qquad (6.17)$$

- *Petrov–Galerkin scheme:* In the Petrov–Galerkin scheme, $\mathbf{c}$ is computed by directly minimizing the L_2 norm of the residual error, that is, $\|\mathbf{R}(\mathbf{\Psi}\mathbf{c}, \mathbf{x})\|_2$.

It can be noted that when the governing equations are nonlinear the residual error vector is a nonlinear function of $\mathbf{c}$. Hence, nonlinear least-squares minimization techniques have to be employed to estimate the undetermined coefficients. However, when the governing equations are linear, $\mathbf{c}$ can be efficiently computed by solving a reduced-order linear problem.

We discuss the theoretical properties of the Bubnov–Galerkin and Petrov–Galerkin schemes, particularly for the case of linear systems, in more detail in the next section.

6.3 Reduced Basis Methods for Linear Static Reanalysis

In this section, we discuss reduced basis methods for *re*analysis of linear structures subjected to static loads. Even though our focus is on structural systems, the methods discussed here can be applied to any problem where discretization of the governing PDEs leads to a system of linear algebraic equations. Such approaches are important as many design improvement methods, including optimization, involve the solution of a sequence of modified baseline systems.

Consider the matrix system of equations for static equilibrium of a modified linear structural system

$$(\mathbf{K}_0 + \Delta\mathbf{K})(\mathbf{w}_0 + \Delta\mathbf{w}) = \mathbf{f}_0 + \Delta\mathbf{f}, \tag{6.18}$$

where $\mathbf{K}_0$ and $\Delta\mathbf{K} \in \mathbb{R}^{n\times n}$ denote the baseline stiffness matrix and its perturbation, respectively. $\mathbf{f}_0$ and $\Delta\mathbf{f} \in \mathbb{R}^n$ denote the force vector applied to the baseline structure and its perturbation, respectively. $\mathbf{w}_0 = \mathbf{K}_0^{-1}\mathbf{f}_0 \in \mathbb{R}^n$ is the displacement of the baseline structure corresponding to the force vector $\mathbf{f}_0 \in \mathbb{R}^n$. $\Delta\mathbf{w} \in \mathbb{R}^n$ denotes the perturbation of the displacement vector. Note the implicit assumption made here that the stiffness matrix and force vector are functions of the vector of design variables $\mathbf{x}$, that is, $\mathbf{K} \equiv \mathbf{K}(\mathbf{x})$ and $\mathbf{f} \equiv \mathbf{f}(\mathbf{x})$. The baseline stiffness matrix and force vector are computed at $\mathbf{x}^0$ and the perturbed system in (6.18) corresponds to $\mathbf{x} = \mathbf{x}^0 + \Delta\mathbf{x}$.

The static reanalysis problem involves computing $\Delta\mathbf{w}$ using the data available from analysis of the baseline structure, without explicitly solving (6.18) in its exact form. For example, if a direct method such as LU or Cholesky factorization is applied to analyze the baseline structure by solving $\mathbf{K}_0\mathbf{w}_0 = \mathbf{f}_0$, then as a by-product of this analysis, we have access to the baseline displacement vector $\mathbf{w}_0$ as well as the factored form of the baseline stiffness matrix $\mathbf{K}_0$. Henceforth, to simplify the notation, we denote the perturbed stiffness matrix, force vector and displacement vector by $\mathbf{K} = \mathbf{K}_0 + \Delta\mathbf{K}$, $\mathbf{f} = \mathbf{f}_0 + \Delta\mathbf{f}$ and $\mathbf{w} = \mathbf{w}_0 + \Delta\mathbf{w}$, respectively.

It is possible to apply the local approximation techniques (including intervening variables and two-point approximations) discussed in Chapter 5 to approximate the displacement vector. However, as mentioned earlier, the accuracy of local approximations can be poor, particularly for large changes in the design variables. Global surrogate models can be used to circumvent this problem – however, since $\mathbf{w}$ is a n-dimensional vector, this approach can be computationally prohibitive for systems with large numbers of degrees of freedom (dof) and design variables. This motivates the application of reduced basis methods to the static reanalysis problem.

Kirsch (1991) proposed a reduced basis technique for static reanalysis of structural systems. It was shown that the terms of a binomial series expansion (also referred to as the Neumann series) can be used as basis vectors for accurately approximating the static response for large changes in the design variables. The Neumann series for the solution of (6.18) can be written as

$$\mathbf{w} = \sum_{i=0}^{\infty} (-1)^i (\mathbf{K}_0^{-1}\Delta\mathbf{K})^i \mathbf{K}_0^{-1}\mathbf{f}. \tag{6.19}$$

Note that the Neumann series will converge only when $\|\mathbf{K}_0^{-1}\Delta\mathbf{K}\| < 1$ or $\rho(\mathbf{K}_0^{-1}\Delta\mathbf{K}) < 1$, where ρ denotes the spectral radius. Hence, this series can be interpreted as a local approximation method with a limited radius of convergence. Since the technique proposed by Kirsch combines this local approximation series with the global characteristics of the reduced basis method, it was dubbed the "combined approximations" (CA) technique. The CA technique attempts to improve the accuracy of (6.19) by approximating the perturbed displacement vector as

$$\widehat{\mathbf{w}} = c_1\boldsymbol{\psi}_1 + c_2\boldsymbol{\psi}_2 + \cdots + c_m\boldsymbol{\psi}_m = \boldsymbol{\Psi}\mathbf{c}, \tag{6.20}$$

where $\boldsymbol{\psi}_i = (\mathbf{K}_0^{-1}\Delta\mathbf{K})^{i-1}\mathbf{K}_0^{-1}\mathbf{f} \in \mathbb{R}^n$ is the ith basis vector. $\boldsymbol{\Psi} = [\boldsymbol{\psi}_1, \boldsymbol{\psi}_2, \ldots, \boldsymbol{\psi}_m] \in \mathbb{R}^{n\times m}$ and $\mathbf{c} = \{c_1, c_2, \ldots, c_m\}^T \in \mathbb{R}^m$ denote the matrix of basis vectors and the vector of undetermined coefficients, respectively.

Over the last decade, a number of studies on the application of the CA technique as well as its extensions have been reported in the literature. Examples include static sensitivity analysis (Kirsch 1994), reanalysis of structures subjected to topological modifications (Chen and Yang 2004; Kirsch and Liu 1997; Kirsch and Papalambros 2001), structural optimization (Kirsch 1997), and damage-tolerant design (Garcelon et al. 2000). A recent analysis by Nair (2002a) has shown that the CA method can be viewed as a preconditioned Krylov subspace method. Since Krylov methods have a long history in numerical linear algebra, we have adopted this perspective when describing CA techniques.

We begin our discussion of reduced basis methods by focusing on the case when the structure is modified parametrically, that is, the dimensions of $\mathbf{K}_0$ and $\mathbf{K}$ are the same. In the final part of this section, we cover techniques that can be applied to cases when the structure is modified topologically – that is, the total number of dof of the modified structure may be greater or less than that of the baseline structure.

6.3.1 Choice of Basis Vectors

In the context of the solution of (6.18), Krylov methods seek to approximate the displacement vector $\mathbf{w}$ using the affine subspace $\mathbf{w}_0 + \mathcal{K}_m(\mathbf{K}, \mathbf{r}_0)$, where $\mathcal{K}_m$ denotes the Krylov subspace of order m defined in (6.16). The following theorem[2] can be used to justify the application of the Krylov subspace to approximate $\mathbf{w}$:

Theorem: *If the minimal polynomial of a square matrix $\mathbf{K}$ has degree q,[3] then the solution to $\mathbf{Kw} = \mathbf{f}$ lies in the Krylov subspace $\mathcal{K}_q(\mathbf{K}, \mathbf{f})$.*

Hence, the displacement of the modified structure can be approximated using basis vectors spanning the Krylov subspace. However, for a general matrix $\mathbf{K}$, the degree of the minimal polynomial could be rather high. As a consequence, a large number of basis vectors may be required to achieve accurate approximations. As shown next, preconditioning schemes can be employed to improve the quality of these basis vectors.

Consider the case when the matrix $\mathbf{M}$ is used as a left preconditioner for (6.18), which gives the transformed equation $\mathbf{MKw} = \mathbf{Mf}$. The key idea here is to choose the preconditioner $\mathbf{M}$ such that the degree of the minimal polynomial of the transformed coefficient matrix $\mathbf{MK}$ is lower than that of the original coefficient matrix $\mathbf{K}$. In other words, we

---

[2]For a proof, the interested reader is referred to the expository article by Ipsen and Meyer (1998).

[3]For a diagonalizable matrix, q is the total number of distinct eigenvalues. As a check, consider the case when $\mathbf{K} = \alpha\mathbf{I}$, where $\mathbf{I}$ is the identity matrix. Then $q = 1$ and hence only one basis vector is required to compute the exact solution.

wish the matrix $\mathbf{MK}$ to have a small number of distinct eigenvalues to ensure that accurate approximations can be obtained using a small number of basis vectors. One way to achieve this is to set $\mathbf{M} = \mathbf{K}_0^{-1}$.[4] This preconditioning scheme assumes that the decomposed form of the baseline stiffness matrix is available. This is easily satisfied if a direct solver is used for analyzing the baseline structure. The preconditioned matrix system of equations for static equilibrium can hence be rewritten as

$$\mathbf{K}_0^{-1}(\mathbf{K}_0 + \Delta \mathbf{K})\mathbf{w} = \mathbf{K}_0^{-1}(\mathbf{f}_0 + \Delta \mathbf{f}) \Rightarrow (\mathbf{I} + \mathbf{K}_0^{-1}\Delta\mathbf{K})\mathbf{w} = \mathbf{w}_0 + \mathbf{K}_0^{-1}\Delta\mathbf{f}. \qquad (6.21)$$

Note that the solutions of (6.18) and (6.21) are the same. If the zero vector is used as an initial guess for constructing a Krylov subspace for the solution of (6.21), then $\mathbf{r}_0 = \mathbf{K}_0^{-1}\mathbf{f}$. The left preconditioned Krylov subspace may hence be written as

$$\mathcal{K}_m(\mathbf{K}_0^{-1}\Delta\mathbf{K}, \mathbf{K}_0^{-1}\mathbf{f}) = \text{span}\left\{\mathbf{K}_0^{-1}\mathbf{f}, \ldots, (\mathbf{K}_0^{-1}\Delta\mathbf{K})^{m-1}\mathbf{K}_0^{-1}\mathbf{f}\right\}. \qquad (6.22)$$

A comparison of (6.19) and (6.22) clearly shows that the m terms of the Neumann series are identical to the basis vectors spanning the left preconditioned Krylov subspace. This fact was used by Nair (2002a) to show that Kirsch's CA technique is equivalent to a preconditioned Krylov subspace method with the preconditioner $\mathbf{K}_0^{-1}$.

Since the ith basis vector is given by the recursive sequence $\boldsymbol{\psi}_i = \mathbf{K}_0^{-1}\Delta\mathbf{K}\boldsymbol{\psi}_{i-1}$, it can be efficiently computed using the factored form of the baseline stiffness matrix $\mathbf{K}_0$. To illustrate, consider the case when the Cholesky decomposition of $\mathbf{K}_0$ is available, that is, $\mathbf{K}_0 = \mathbf{U}_0^T\mathbf{U}_0$, where $\mathbf{U}_0$ is an upper triangular matrix. Then, we first solve the lower triangular system of equations, $\mathbf{U}_0^T\mathbf{t} = \Delta\mathbf{K}\boldsymbol{\psi}_{i-1}$, by forward substitution. Subsequently, the upper triangular system of equations, $\mathbf{U}_0\boldsymbol{\psi}_i = \mathbf{t}$, can be solved by backward substitution to arrive at $\boldsymbol{\psi}_i$.

An alternative choice of basis vectors was proposed by Kirsch and Liu (1997) for cases when $\text{Rank}(\Delta\mathbf{K}) = s \ll n$. Here, the perturbed stiffness matrix can be written as $\mathbf{K} = \mathbf{K}_0 + \mathbf{K}_1 + \mathbf{K}_2 + \cdots + \mathbf{K}_s$, where $\mathbf{K}_i \in \mathbb{R}^{n \times n}$ is a rank one matrix. For the case when $\Delta\mathbf{f} = 0$, the displacement of the baseline structure $\mathbf{w}_0$ is chosen as the first basis vector. The subsequent s basis vectors are computed as $\boldsymbol{\psi}_i = \mathbf{K}_0^{-1}\mathbf{K}_{i-1}\mathbf{w}_o, i = 2, 3, \ldots s + 1$. Numerical studies have shown that this choice of basis vectors gives exact results. Akgün et al. (2001a) showed that, for this choice of basis vectors, the reduced basis approximation is equivalent to the Sherman–Morrison–Woodbury formula

$$\left(\mathbf{K}_0 + \mathbf{U}\mathbf{V}^T\right)^{-1} = \mathbf{K}_0^{-1} - \mathbf{K}_0^{-1}\mathbf{U}\left(\mathbf{I} + \mathbf{V}^T\mathbf{K}_0^{-1}\mathbf{U}\right)^{-1}\mathbf{V}^T\mathbf{K}_0^{-1}, \qquad (6.23)$$

where, $\mathbf{U}, \mathbf{V} \in \mathbb{R}^{n \times s}$. Since the preceding equation involves the decomposition of the matrix $(\mathbf{I} + \mathbf{V}^T\mathbf{K}_0^{-1}\mathbf{U}) \in \mathbb{R}^{s \times s}$, it is useful only when $s \ll n$. In practice, for high-rank changes in the stiffness matrix, it is more efficient to use basis vectors spanning the preconditioned Krylov subspace to approximate the displacement vector.

For the sake of numerical stability, it is preferable that the basis vectors are made orthogonal to each other or the stiffness matrix. Kirsch and Papalambros (2001) used the

---

[4]This choice of preconditioner can be motivated by writing $\mathbf{MK} = \mathbf{K}_0^{-1}(\mathbf{K}_0 + \Delta\mathbf{K}) = \mathbf{I} + \mathbf{K}_0^{-1}\Delta\mathbf{K}$, where $\mathbf{I}$ is the identity matrix. Clearly, for small perturbations in the design variables, all the eigenvalues of the matrix $\mathbf{I} + \mathbf{K}_0^{-1}\Delta\mathbf{K}$ will be clustered around unity. Hence, from a numerical viewpoint, the degree of the minimal polynomial of the transformed coefficient matrix will be small. In practice, even for the case of large perturbations in the design variables, the degree of the minimal polynomial $q \ll n$. As a result, we can expect to achieve good approximations using a small number of basis vectors.

classical Gram–Schmidt algorithm to orthogonalize the basis vectors to **K**. In the linear algebra literature, orthogonalization of basis vectors spanning the Krylov subspace is traditionally carried out using Arnoldi's method, which is a version of the Gram–Schmidt orthogonalization procedure tailored to the Krylov subspace (Saad 1996). This is because, if the Krylov subspace is directly computed first, then because of finite precision arithmetic, the terms in the series $\mathbf{r}_0, \mathbf{B}\mathbf{r}_0, \ldots, \mathbf{B}^{m-1}\mathbf{r}_0$ tend to approach the direction of the dominant eigenvector of the matrix **B**, and hence the vectors can become linearly dependent. From a numerical standpoint, it is not advisable to compute the basis vectors first and orthogonalize them later as an afterthought.

6.3.2   Bubnov–Galerkin and Petrov–Galerkin Schemes

The undetermined coefficients in the reduced basis approximation can be computed either using the Bubnov–Galerkin or the Petrov–Galerkin scheme. In the case of the Bubnov–Galerkin scheme (which is also sometimes referred to as an orthogonal projection scheme), the residual error is orthogonalized with respect to the approximating subspace $\mathcal{K}$ as follows:

$$\mathbf{K}\boldsymbol{\Psi}\mathbf{c} - \mathbf{f} \perp \boldsymbol{\psi}_i \quad \Rightarrow \quad \boldsymbol{\psi}_i^T (\mathbf{K}\boldsymbol{\Psi}\mathbf{c} - \mathbf{f}) = 0, \quad i = 1, 2, \ldots, m. \tag{6.24}$$

Applying the above orthogonality conditions, we arrive at the following reduced-order $m \times m$ system of equations for the coefficient vector $\mathbf{c}$:

$$\boldsymbol{\Psi}^T \mathbf{K} \boldsymbol{\Psi} \mathbf{c} = \boldsymbol{\Psi}^T \mathbf{f}. \tag{6.25}$$

This equation can be efficiently solved for $\mathbf{c}$ and subsequently the perturbed displacement vector can be approximated as $\widehat{\mathbf{w}} = \boldsymbol{\Psi}\mathbf{c}$. For the case when the coefficient matrix is Hermitian positive definite,[5] the Galerkin scheme ensures that the **K**-norm[6] of the error converges as the number of basis vectors is increased.

The Petrov–Galerkin scheme is an oblique projection scheme in which the residual error vector is made orthogonal to the subspace $\mathcal{L}$. Typically, $\mathcal{L} = \text{span}\{\mathbf{K}\boldsymbol{\psi}_1, \mathbf{K}\boldsymbol{\psi}_2, \ldots, \mathbf{K}\boldsymbol{\psi}_m\}$. This gives

$$\mathbf{K}\boldsymbol{\Psi}\mathbf{c} - \mathbf{f} \perp \mathbf{K}\boldsymbol{\psi}_i \quad \Rightarrow \quad \boldsymbol{\psi}_i^T \mathbf{K}^T (\mathbf{K}\boldsymbol{\Psi}\mathbf{c} - \mathbf{f}) = 0, \quad i = 1, 2, \ldots, m. \tag{6.26}$$

Applying the above m orthogonality constraints, we arrive at the following reduced-order $m \times m$ system of equations for $\mathbf{c}$:

$$\boldsymbol{\Psi}^T \mathbf{K}^T \mathbf{K} \boldsymbol{\Psi} \mathbf{c} = \boldsymbol{\Psi}^T \mathbf{K}^T \mathbf{f}. \tag{6.27}$$

Note that the Petrov–Galerkin scheme is equivalent to minimization of the L_2 norm of the residual error vector $\mathbf{K}\boldsymbol{\Psi}\mathbf{c} - \mathbf{f}$.[7] Hence, it follows that $||\mathbf{K}\boldsymbol{\Psi}\mathbf{c} - \mathbf{f}||_2$ will converge when the

---

[5]**K** is a Hermitian matrix if $\mathbf{K}^* = \mathbf{K}$, where the superscript * is used to denote the complex conjugate transpose. For the case of self-adjoint systems, the stiffness matrix **K** is always symmetric positive definite and therefore Hermitian.

[6]The **K**-norm of the error can be defined as $\{\widehat{\mathbf{w}} - \mathbf{w}\}^T \mathbf{K}\{\widehat{\mathbf{w}} - \mathbf{w}\}$. Note that the result regarding convergence of this norm assumes that the basis vectors are mutually orthogonal.

[7]The interested reader can verify this statement by defining the residual error function $(\mathbf{K}\boldsymbol{\Psi}\mathbf{c} - \mathbf{f})^T (\mathbf{K}\boldsymbol{\Psi}\mathbf{c} - \mathbf{f})$, which is to be minimized with respect to $\mathbf{c}$. Taking the derivative of this error function with respect to $\mathbf{c}$ and setting it equal to zero will lead to the same $m \times m$ reduced-order system of equations as in (6.27).

number of basis vectors is increased. This result holds for general matrices and hence the Petrov–Galerkin scheme is to be preferred for non-Hermitian problems.[8]

In summary, the coefficient vector $\mathbf{c}$ can be efficiently computed by solving the system of reduced-order equations in (6.25) or (6.27). Subsequently, (6.20) can be used to approximate the static response as $\widehat{\mathbf{w}} = \boldsymbol{\Psi}\mathbf{c}$. The approaches discussed so far can be readily applied to approximate the displacement vector when the baseline structure is parametrically modified.

6.3.3 Topologically Modified Structures

We consider next the case when design modifications lead to changes in topology. For example, in topology optimization of structures, a selection of elements and joints may be added or deleted during the iterations. If topological modifications do not lead to a change in the total number of dof, the approaches presented earlier for parametrically modified structures can be directly employed. However, if modifications lead to an increase or decrease in the total number of dof, the reduced basis formulations presented earlier cannot be directly applied.

First, consider the case when the total number of dof is decreased as a consequence of topological modifications. Let the baseline and modified structure have n and n_1 dof, respectively, where $n_1 < n$. Further, let $\mathbf{K}_M \in \mathbb{R}^{n_1 \times n_1}$ and $\mathbf{f}_M \in \mathbb{R}^{n_1}$ denote the perturbed stiffness matrix and force vector, respectively. The analysis equations of the modified structure are hence given by $\mathbf{K}_M \mathbf{w}_M = \mathbf{f}_M$. The perturbed governing equations can also be written in the form $\mathbf{K}\mathbf{w} = \mathbf{f}$, where

$$\mathbf{K} = \mathbf{K}_0 + \Delta\mathbf{K} = \begin{bmatrix} \mathbf{K}_M & \mathbf{0} \\ \mathbf{0} & \mathbf{0} \end{bmatrix} \in \mathbb{R}^{n \times n} \quad \text{and} \quad \mathbf{f} = \mathbf{f}_0 + \Delta\mathbf{f} = \begin{bmatrix} \mathbf{f}_M \\ \mathbf{0} \end{bmatrix} \in \mathbb{R}^n. \quad (6.28)$$

In some cases, deletion of structural elements and joints may lead to a singular modified stiffness matrix (a conditionally stable structure) which prevents exact reanalysis being carried out. Kirsch and Papalambros (2001) proposed to directly solve the perturbed equations using a reduced basis approach by defining

$$\Delta\mathbf{K} = \mathbf{K}_0 - \begin{bmatrix} \mathbf{K}_M & \mathbf{0} \\ \mathbf{0} & \mathbf{0} \end{bmatrix} \in \mathbb{R}^{n \times n}. \quad (6.29)$$

The definition of $\Delta\mathbf{K}$ in (6.29) enables the application of the approaches discussed earlier for parametrically modified structures to approximate the modified displacement vector. In other words, the matrix $\mathbf{K}_0^{-1}$ is used as a preconditioner for solving the perturbed problem $(\mathbf{K}_0 + \Delta\mathbf{K})\mathbf{w} = \mathbf{f}$, where $\Delta\mathbf{K}$ and $\mathbf{f}$ are given by (6.29) and (6.28), respectively. Kirsch and Papalambros (2001) note that even if the perturbed stiffness matrix is singular the

---

[8]Non-Hermitian coefficient matrices arise, for example, in the context of dynamic reanalysis of structural systems in the frequency domain. Here, we need to solve the system of equations $(\mathbf{H}_0(\omega) + \Delta\mathbf{H}(\omega))\mathbf{w}(\omega) = \mathbf{f}$, where $\mathbf{H}_0 = \mathbf{K}_0 - \omega^2 \mathbf{M}_0 + i\omega \mathbf{C}_0$ denotes the dynamic stiffness matrix of the baseline structure. Here, $\mathbf{M}_0$ and $\mathbf{C}_0$ denote the baseline mass and damping matrices, respectively. ω is the frequency of excitation and $i = \sqrt{-1}$. $\Delta\mathbf{H}$ is the perturbation in the dynamic stiffness matrix. It can be easily verified that $\mathbf{H}^* \neq \mathbf{H}$ and hence the dynamic stiffness matrix is non-Hermitian – more specifically, $\mathbf{H}$ is a complex symmetric matrix. Another point worth noting is that the transpose operations in the Bubnov–Galerkin and Petrov–Galerkin projection schemes need to be replaced by the complex conjugate transpose when the coefficient matrix and basis vectors are complex. This is because, by definition, two complex vectors $\mathbf{r}_1$ and $\mathbf{r}_2$ are orthogonal to each other if and only if $\mathbf{r}_1^* \mathbf{r}_2 = 0$. A discussion on reduced basis methods for frequency response reanalysis can be found in Bouazzouni et al. (1997) and Nair and Keane (2001).

reduced-order coefficient matrix obtained by using m basis vectors to approximate $\mathbf{w}$ will be nonsingular provided $m < n$.

Now, consider the case when topological modifications lead to an increase in the total number of dof. This situation can arise, for example, when additional elements and joints are added to the baseline structure. Let n and n_2 denote the total number of dof of the baseline and modified structures, respectively, where $n_2 > n$. One way to approach this problem would be to rewrite the equations governing static equilibrium of the perturbed system in the form $(\mathbf{K}_M + \Delta \mathbf{K})\mathbf{w} = \mathbf{f}$, where $\mathbf{K}_M \in \mathbb{R}^{n_2 \times n_2}$ is the so-called modified initial stiffness matrix, whose factored form can be cheaply computed by updating the factorization of the baseline stiffness matrix $\mathbf{K}_0$. This would allow for the possibility of using $\mathbf{K}_M^{-1}$ as a preconditioner in the spirit of the approach described earlier for static reanalysis of parametrically modified structures. A discussion of this approach to reanalysis can be found in Kirsch and Papalambros (2001). Here, we discuss a simpler approach proposed by Chen and Yang (2004), which is motivated by ideas taken from model condensation procedures.

We first write the equations of the perturbed system in partitioned form as

$$
\begin{bmatrix} \mathbf{K}_0 + \Delta \mathbf{K}_{11} & \Delta \mathbf{K}_{12} \\ \Delta \mathbf{K}_{21} & \Delta \mathbf{K}_{22} \end{bmatrix} \begin{bmatrix} \mathbf{w}_1 \\ \mathbf{w}_2 \end{bmatrix} = \begin{bmatrix} \mathbf{f}_1 \\ \mathbf{f}_2 \end{bmatrix}, \tag{6.30}
$$

where $\mathbf{K}_0 \in \mathbb{R}^{n \times n}$ is the baseline stiffness matrix. $\Delta \mathbf{K}_{11} \in \mathbb{R}^{n \times n}$, $\Delta \mathbf{K}_{12} \in \mathbb{R}^{n \times (n_2 - n)}$, $\Delta \mathbf{K}_{21} \in \mathbb{R}^{(n_2 - n) \times n}$ and $\Delta \mathbf{K}_{22} \in \mathbb{R}^{(n_2 - n) \times (n_2 - n)}$ are perturbation matrices. $\mathbf{f}_1 \in \mathbb{R}^n$ and $\mathbf{f}_2 \in \mathbb{R}^{n_2 - n}$ are partitions of the perturbed force vector.

It follows from (6.30) that

$$
(\mathbf{K}_0 + \Delta \mathbf{K}_{11})\mathbf{w}_1 + \Delta \mathbf{K}_{12}\mathbf{w}_2 = \mathbf{f}_1 \tag{6.31}
$$

$$
\text{and} \quad \Delta \mathbf{K}_{21}\mathbf{w}_1 + \Delta \mathbf{K}_{22}\mathbf{w}_2 = \mathbf{f}_2. \tag{6.32}
$$

After some algebraic manipulations, we have

$$
\left[\mathbf{K}_0 + \Delta \mathbf{K}_{11} - \Delta \mathbf{K}_{12}\Delta \mathbf{K}_{22}^{-1}\Delta \mathbf{K}_{21} \right] \mathbf{w}_1 = \mathbf{f}_1 - \Delta \mathbf{K}_{12}\Delta \mathbf{K}_{22}^{-1}\mathbf{f}_2 \tag{6.33}
$$

and

$$
\Delta \mathbf{K}_{22}\mathbf{w}_2 = (\mathbf{f}_2 - \Delta \mathbf{K}_{21}\mathbf{w}_1). \tag{6.34}
$$

To proceed further, we note that (6.33) can be written in the form $(\mathbf{K}_0 + \Delta \mathbf{K})\mathbf{w} = \mathbf{f}$, where $\Delta \mathbf{K} = \Delta \mathbf{K}_{11} - \Delta \mathbf{K}_{12}\Delta \mathbf{K}_{22}^{-1}\Delta \mathbf{K}_{21}$ and $\mathbf{f} = \mathbf{f}_1 - \Delta \mathbf{K}_{12}\Delta \mathbf{K}_{22}^{-1}\mathbf{f}_2$. Hence, the reduced basis formulation developed earlier in this section for static reanalysis of parametrically modified structures can be readily applied.[9] We can use the factored form of the baseline stiffness matrix $\mathbf{K}_0$ as a preconditioner to solve (6.33) for $\mathbf{w}_1 \in \mathbb{R}^n$. Subsequently, the displacements at the other degrees of freedom can be computed by solving (6.34) for $\mathbf{w}_2 \in \mathbb{R}^{n_2 - n}$. It can be noted from (6.33) and (6.34) that we also need to factorize the matrix $\Delta \mathbf{K}_{22}$. Fortunately, since $\Delta \mathbf{K}_{22} \in \mathbb{R}^{(n_2 - n) \times (n_2 - n)}$, the factored form of this matrix can be efficiently computed when $n_2 - n$ is small.

---

[9]As an aside, it is worth noting that the coefficient matrix of (6.33) can be referred to as the *Schur complement* matrix associated with $\mathbf{w}_1$ (Saad 1996). Schur complements also arise in domain decomposition schemes, where partitioned systems of equations are solved using preconditioned iterative methods.

6.3.4   Implementation Issues

The CA techniques discussed in this section were originally developed by Kirsch and his colleagues (see, for example, Kirsch (1991, 2003); Kirsch and Liu (1997); Kirsch and Papalambros (2001)) apparently independent of developments in the field of numerical linear algebra on Krylov methods. We deliberately adopted the viewpoint of Krylov methods and matrix preconditioning in our exposition since these approaches have a long history and a rigorous mathematical underpinning. The connections between CA techniques and Krylov methods have an important ramification on the practical issue of integrating static reanalysis techniques with numerical optimization algorithms. The success of the CA technique taken together with the observations made here suggests that it may be worthwhile to invest computational effort in constructing a good preconditioner. Since a large number of repeated analysis will be carried out during any optimization iterations, a good preconditioning scheme may lead to a significant improvement in the overall efficiency of the optimization process. In practice, a direct method can be employed to construct a preconditioner by decomposing a baseline stiffness matrix. During the course of the optimization iterations, structural analysis may then be carried out using preconditioned Krylov solvers. It is also possible to periodically update/refresh the preconditioner during the optimization iterations. We recommend that instead of directly implementing the reduced basis formulation the user should consider using an existing Krylov solver that takes as input the factored form of the preconditioner $\mathbf{M}$. A number of robust software libraries implementing preconditioned Krylov methods can be found in the public domain.[10] We make this recommendation primarily for the purpose of ensuring numerical stability and efficiency as well as ease of implementation.

It will be clear from the approaches discussed in this section that given a particular set of basis vectors and a projection scheme the accuracy of the reduced basis method clearly depends on the perturbation matrix $\Delta \mathbf{K}$. Further, since the methods discussed here have a rigorous mathematical underpinning, nearly exact results can be obtained within the limitations imposed by finite precision arithmetic, provided a suitable number of basis vectors is used. Because of these factors, reduced basis methods do not suffer from the curse of dimensionality encountered in black-box approaches to surrogate modeling.

It is to be noted that reduced basis methods can also be applied to nonlinear problems since most numerical methods for analyzing nonlinear systems solve a sequence of linear problems (Kirsch 2003). Here, during the solution of each linear subproblem, a reduced basis method can be employed to expedite the calculations. Note that reduced basis methods employing the Krylov subspace are commonly used in the literature to solve nonlinear algebraic equations (Fokkema et al. 1998). Since a detailed account of Krylov methods is beyond the scope of this book, we have opted for a simplified description to outline some of the key ideas. We highly recommend the text by Saad (1996) for a rigorous exposition of Krylov methods. We believe that a good understanding of the theoretical aspects of Krylov methods and projection schemes is helpful in gaining a broader perspective on reduced basis methods.

---

[10]See, for example, `http://www.netlib.org`.

6.4   Reduced Basis Methods for Reanalysis of Eigenvalue Problems

In this section, we discuss how reduced basis methods can be applied to reanalysis of eigenproblems. The free-vibration undamped natural frequencies and mode shapes of a linear self-adjoint structural system can be computed by solving the generalized symmetric algebraic eigenproblem $\mathbf{K}_0 \boldsymbol{\phi}^0 = \lambda^0 \mathbf{M}_0 \boldsymbol{\phi}^0$, where $\mathbf{K}_0, \mathbf{M}_0 \in \mathbb{R}^{n \times n}$ are the structural stiffness and mass matrices, respectively. The system matrices can be considered to be a general function of the design variable vector denoted by $\mathbf{x} = \{x_1, x_2, \ldots, x_p\} \in \mathbb{R}^p$. $\lambda^0 \in \mathbb{R}$ and $\boldsymbol{\phi}^0 \in \mathbb{R}^n$ denote an eigenvalue and eigenvector, respectively.

Consider the case when the design variables are perturbed by $\Delta \mathbf{x}$. Let $\Delta \mathbf{K}$ and $\Delta \mathbf{M}$ denote the corresponding perturbations in the stiffness and mass matrices, respectively. The perturbed eigenproblem can be written as

$$(\mathbf{K}_0 + \Delta \mathbf{K})(\boldsymbol{\phi}^0 + \Delta \boldsymbol{\phi}) = (\lambda^0 + \Delta \lambda)(\mathbf{M}_0 + \Delta \mathbf{M})(\boldsymbol{\phi}^0 + \Delta \boldsymbol{\phi}), \tag{6.35}$$

where $\Delta \lambda$ and $\Delta \boldsymbol{\phi}$ denote the eigenvalue and eigenvector perturbation, respectively. This can be written in compact form as

$$\mathbf{K} \boldsymbol{\phi} = \lambda \mathbf{M} \boldsymbol{\phi}. \tag{6.36}$$

Often, it is found that, even for small to moderate perturbations in the stiffness and mass matrices, significant alterations in the modal characteristics of the structure may occur. Hence, an exact reanalysis becomes necessary to compute the perturbed eigenparameters with sufficient accuracy. The objective of approximate reanalysis procedures is to compute the perturbed eigenparameters using the results of exact analysis for the baseline system, without recourse to solving (6.36) in its exact form.

Note that a first-order approximation for the modified eigenvalues and eigenvectors can be computed using the sensitivity expressions derived earlier in Chapter 4. However, this local approximation based on first-order sensitivities can be highly inaccurate even for moderate perturbations in the design variables. The other alternative is to use the eigenvectors of the baseline structure as starting vectors in an iterative eigensolution procedure such as the Lanczos or subspace iteration method – provided sufficient number of iterations are carried out, convergence to the exact values of the perturbed eigenvalues and eigenvectors is possible. In this section, however, we primarily focus on approximate techniques which can be computationally much cheaper.

6.4.1   Improved First-order Approximation

One way to approximate the solution of a perturbed eigenproblem is to use the baseline eigenvector and the first-order approximation term as basis vectors for Ritz analysis of the perturbed problem (Murthy and Haftka 1988; Nair et al. 1998a). An assumption is made here that the ith eigenvector of the perturbed system can be approximated in the subspace spanned by $\boldsymbol{\phi}_i^0$ and $\Delta \boldsymbol{\phi}_i$, where $\Delta \boldsymbol{\phi}_i$ is a first-order approximation of the eigenvector perturbation. Hence, an approximation for the perturbed eigenvector can be written as

$$\widehat{\boldsymbol{\phi}}_i = \zeta_1 \boldsymbol{\phi}_i^0 + \zeta_2 \Delta \boldsymbol{\phi}_i, \tag{6.37}$$

where ζ_1 and ζ_2 are the undetermined scalar quantities in the approximate representation of the perturbed eigenvector. The assumption implicit in this proposition is that even for moderate to large perturbations in the structural parameters the first-order approximation yields a $\Delta\phi_i$ vector, which usually gives a reasonable indication of the likely change of the baseline eigenvector, although the magnitude or even direction of change may be erroneous.

The approximation (6.37) can be expressed in matrix form as $\widehat{\phi}_i = \mathbf{T}\zeta$, where $\mathbf{T} = [\phi_i^0, \Delta\phi_i] \in \mathbb{R}^{n \times 2}$ and $\zeta = \{\zeta_1, \zeta_2\}^T \in \mathbb{R}^2$. Substituting this approximation into (6.36) and applying the Bubnov–Galerkin projection scheme, the resulting set of equations can be expressed as

$$\mathbf{K}_T \zeta = \lambda \mathbf{M}_T \zeta, \qquad (6.38)$$

where $\mathbf{K}_T = \mathbf{T}^T \mathbf{K} \mathbf{T} \in \mathbb{R}^{2 \times 2}$ and $\mathbf{M}_T = \mathbf{T}^T \mathbf{M} \mathbf{T} \in \mathbb{R}^{2 \times 2}$. Hence the original $n \times n$ eigensystem is represented by a reduced 2×2 eigensystem for each eigenmode to be approximated.

The solution of the reduced-order eigenproblem in (6.38) leads to two possible values for the approximate eigenvalue of mode i, say $\widehat{\lambda}_i^{\min}$ and $\widehat{\lambda}_i^{\max}$, and the coefficients in the reduced basis ζ_1 and ζ_2. Because of the inclusion principle, $\widehat{\lambda}_i^{\min} \leq \lambda_i^{\mathrm{rqa0}} \leq \widehat{\lambda}_i^{\max}$, where $\lambda_i^{\mathrm{rqa0}}$ is the zero-order Rayleigh quotient approximation defined as

$$\lambda_i^{\mathrm{rqa0}} = \frac{\phi_i^{0^T} \mathbf{K} \phi_i^0}{\phi_i^{0^T} \mathbf{M} \phi_i^0}. \qquad (6.39)$$

Clearly, for the fundamental mode of the perturbed system, $\widehat{\lambda}_i^{\min}$ is the best approximation. For the higher modes, the eigenvalue that is closest to the higher-order eigenvalue perturbation derived in Eldred et al. (1992) can be chosen as the best approximation for that mode. The expression for the higher-order eigenvalue perturbation is

$$\widehat{\lambda}_i = \lambda_i^o + \frac{\phi_i^{o^T} \left(\Delta\mathbf{K} - \lambda_i^o \Delta\mathbf{M}\right) \left(\phi_i^o + \Delta\phi_i\right)}{\phi_i^{o^T} \left(\mathbf{M}^o + \Delta\mathbf{M}\right) \left(\phi_i^o + \Delta\phi_i\right)}. \qquad (6.40)$$

After the best approximation for the eigenvalue has been chosen, the corresponding eigenvector approximation can be readily computed as $\widehat{\phi}_i = \mathbf{T}\zeta$. Note that it is possible to improve the accuracy of results by orthogonalizing the basis vectors ϕ_i^0 and $\Delta\phi_i$ to the approximate eigenvector of mode $i - 1$, that is, $\widehat{\phi}_{i-1}$.

The approximation procedure considered here could also be interpreted as an improved Rayleigh quotient approximation procedure with one free parameter, that is, $\frac{\zeta_2}{\zeta_1}$. In fact, if the reanalysis problem was set up as finding the stationary values of a Rayleigh quotient for each eigenmode, an identical set of equations of the form derived earlier would result. The eigenvalue approximation for mode i can be rewritten as

$$\widehat{\lambda}_i = \frac{\{\zeta_1\phi_i^0 + \zeta_2\Delta\phi_i\}^T \mathbf{K}\{\zeta_1\phi_i^0 + \zeta_2\Delta\phi_i\}}{\{\zeta_1\phi_i^0 + \zeta_2\Delta\phi_i\}^T \mathbf{M}\{\zeta_1\phi_i^0 + \zeta_2\Delta\phi_i\}}. \qquad (6.41)$$

It can be seen that, for $\zeta_1 = \zeta_2$, (6.41) reduces to the first-order Rayleigh quotient approximation for the perturbed eigenvalue and the first-order Taylor series for the perturbed eigenvector. The first-order Rayleigh quotient is given by

$$\lambda_i^{\mathrm{rqa1}} = \frac{\{\phi_i^0 + \Delta\phi_i\}^T \mathbf{K}\{\phi_i^0 + \Delta\phi_i\}}{\{\phi_i^0 + \Delta\phi_i\}^T \mathbf{M}\{\phi_i^0 + \Delta\phi_i\}}. \qquad (6.42)$$

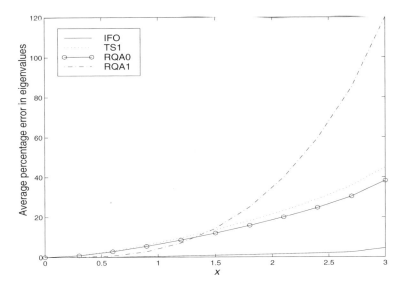

Figure 6.4   Comparison of average approximation errors in first four eigenvalues using different methods for a model problem with one design variable.

If $\Delta\boldsymbol{\phi}_i$ is considered to be a very small quantity or the value of ζ_2 is taken as zero, (6.41) reduces to the zero-order Rayleigh quotient approximation defined in (6.39). The assumption made in this case is that the mode shapes are invariant to parametric perturbations.

A comparison of the accuracies of the improved first-order approximation method (IFO) with the first-order Taylor series approximation (TS1), zero-order Rayleigh quotient (RQA0) and first-order Rayleigh quotient (RQA1) for a model problem is shown in Figure 6.4. It can be seen that the improved first-order method gives significantly better results compared to the other methods. More detailed comparison studies can be found in Nair et al. (1998a). The results obtained there suggest that the improved first-order method gives highly accurate approximations for the fundamental mode, particularly when the rank of $\Delta\mathbf{K}$ and $\Delta\mathbf{M}$ is small. However, the approximations for the higher modes can be poor for modest design changes.

6.4.2   Global Reduced Basis Procedures

A major drawback of the improved first-order approximation method is that since an independent set of reduced-order eigenproblems is solved the approximations can be poor for the higher modes because of *mode crossing*.[11] We outline next a global reduced basis procedure that alleviates this difficulty by simultaneously approximating all the eigenvalues and eigenvectors of interest by constructing and solving a single reduced-order eigenvalue problem.

---

[11] Mode crossing is a phenomenon, because of which, the character of the ith mode of the baseline structure may not correspond to the ith mode of the modified structure. For example, consider a structure where the second mode is a bending-torsion mode. Then, in the modified structure, it could so happen that the second mode is a pure bending mode. Hence, the second eigenvector of the baseline structure and its first-order perturbation will not be good basis vectors for approximating the second mode of the modified structure.

Consider the case when it is desired to approximate the lowest $\widehat{m}$ eigenvalues of the perturbed eigenvalue problem. The main idea used in global reduced basis methods is to approximate the eigenvector of the modified structure as $\widehat{\boldsymbol{\phi}} = \boldsymbol{\Psi}\mathbf{c}$, where $\boldsymbol{\Psi} \in \mathbb{R}^{n \times m}$ denotes a matrix of basis vectors and $\mathbf{c} \in \mathbb{R}^m$ is a vector of undetermined coefficients. Note that in practice, to ensure good approximations, it is necessary to choose $m > \widehat{m}$. Substituting this approximation into the perturbed eigenvalue problem in (6.36) gives the residual error vector $\mathbf{K}\boldsymbol{\Psi}\mathbf{c} - \lambda \mathbf{M}\boldsymbol{\Psi}\mathbf{c}$. Now, let us apply the Bubnov–Galerkin condition $\mathbf{K}\boldsymbol{\Psi}\mathbf{c} - \lambda \mathbf{M}\boldsymbol{\Psi}\mathbf{c} \perp \boldsymbol{\Psi}$. This results in the reduced-order eigenproblem

$$\mathbf{K}_R\mathbf{c} = \widehat{\lambda}\mathbf{M}_R\mathbf{c}, \quad \text{where} \quad \mathbf{K}_R = \boldsymbol{\Psi}^T\mathbf{K}\boldsymbol{\Psi}, \mathbf{M}_R = \boldsymbol{\Psi}^T\mathbf{M}\boldsymbol{\Psi} \in \mathbb{R}^{m \times m}. \tag{6.43}$$

The lowest $\widehat{m}$ eigenvalues of (6.43) can be interpreted as approximations (Ritz values) for the lowest eigenvalues of the perturbed eigenvalue problem (6.36). Further, an approximation for the ith eigenvector of the modified structure can be computed as $\widehat{\boldsymbol{\phi}}_i = \boldsymbol{\Psi}\mathbf{c}_i$, where $\mathbf{c}_i$ is the ith eigenvector of (6.43).

In a fashion similar to the reduced basis methods discussed previously in the context of static structural reanalysis, the accuracy of the approximation depends on the quality of the basis vectors. One obvious choice is to use the first m baseline eigenvectors as basis vectors for approximating the eigenparameters of the modified system, that is, $\boldsymbol{\Psi} = \{\boldsymbol{\phi}_1^0, \boldsymbol{\phi}_2^0, \ldots, \boldsymbol{\phi}_m^0\} \in \mathbb{R}^{n \times m}$. However, this set of basis vectors may not give good approximations when parametric modifications lead to significant alterations in the modal characteristics. In the spirit of the improved first-order method presented earlier, we could use $\boldsymbol{\Psi} = \{\boldsymbol{\phi}_1^0, \Delta\boldsymbol{\phi}_1, \boldsymbol{\phi}_2^0, \Delta\boldsymbol{\phi}_2, \ldots, \boldsymbol{\phi}_{\widehat{m}}^0, \Delta\boldsymbol{\phi}_{\widehat{m}}\} \in \mathbb{R}^{n \times 2\widehat{m}}$. This, however, leads to total of $2\widehat{m}$ basis vectors and hence the computational cost involved in solving (6.43) increases. It is possible here to employ the singular value decomposition procedure discussed in the next section to extract a smaller set of dominant basis vectors from this subspace.

We outline next how ideas from the subspace iteration procedure can be used to enrich the basis vectors. In the subspace iteration procedure, given an initial set of basis vectors, say $\boldsymbol{\Psi}^i$, they are updated as follows:

(1) solve $\mathbf{K}\widetilde{\boldsymbol{\Psi}}^{i+1} = \mathbf{M}\boldsymbol{\Psi}^i$;

(2) substitute $\boldsymbol{\Psi} = \widetilde{\boldsymbol{\Psi}}^{i+1}$ in (6.43) and solve the reduced-order eigenproblem for the eigenvalues $(\widehat{\lambda}_1, \widehat{\lambda}_2, \ldots, \widehat{\lambda}_{\widehat{m}})$ and eigenvectors $(\mathcal{E} = \{\mathbf{c}_1, \mathbf{c}_2, \ldots, \mathbf{c}_{\widehat{m}}\})$;

(3) set $\boldsymbol{\Psi}^{i+1} = \widetilde{\boldsymbol{\Psi}}^{i+1}\mathcal{E}$.

The iterations are terminated when the eigenvalue approximations have converged within specified tolerance levels and the Sturm sequence check is passed (Bathe 1996).

It can be noted that the first step involves the solution of the linear algebraic system of equations $(\mathbf{K}_0 + \Delta\mathbf{K})\widetilde{\boldsymbol{\Psi}}^{i+1} = (\mathbf{M}_0 + \Delta\mathbf{M})\boldsymbol{\Psi}^i$, which is similar to that encountered in Section 5.3 in the context of static reanalysis of parametrically modified structures. Hence, the matrix $\mathbf{K}_0^{-1}$ can be used as a preconditioner to efficiently solve this system of equations. Provided an appropriate convergence tolerance is applied, the modified subspace iteration procedure should converge to the exact eigenvalues and eigenvectors of the perturbed system. If only an approximation is desired, then one iteration should suffice. Detailed numerical studies on the performance of the modified subspace iteration procedure can be found in Kirsch and Bogomolni (2004). That paper also discusses the application of inverse iteration schemes for eigenvalue and eigenvector reanalysis of modified structures.

It is also worth pointing out that in this section we have focused on symmetric eigen-problems. However, dynamic analysis of non-self-adjoint systems leads to nonsymmetric eigenproblems where approximations have to be obtained for the right and left eigenvectors; see, for example, Murthy and Haftka (1988). Further, free-vibration analysis of nonproportionally damped systems leads to a polynomial eigenproblem. In principle, reduced basis methods can be applied to reanalysis of such eigenproblems – however, little work has been carried out on this topic. We end this section by noting two important contributions to exact eigenvalue reanalysis from the applied mathematics community: (1) the algorithm of Carey et al. (1994) based on the Lanczos method and (2) a reduced basis approach based on the generalized Krylov subspace proposed by Zhang et al. (1998).

6.5   Reduced Basis Methods for Nonlinear Problems

We now turn our attention to reduced basis methods for nonlinear problems. Before proceeding further, we point out that it is possible to apply the reduced basis methods discussed earlier to nonlinear problems provided a suitably *rich* set of basis vectors is available. This should be clear from Section 6.2, where a general discrete governing equation is used to introduce the idea of reduced basis approximations. Also, most numerical algorithms for analyzing nonlinear systems are based on the solution of a sequence of linear subproblems. This opens up the possibility of employing the reduced basis methods developed for linear systems to speed up the solution of the linear subproblems encountered during nonlinear analysis. In this section, we discuss a general approach that uses the proper orthogonal decomposition procedure to derive basis vectors appropriate for design problems.

6.5.1   Proper Orthogonal Decomposition

Proper orthogonal decomposition (POD) involves computing an optimal set of bases that best describes (or reconstructs) a given multidimensional dataset. The POD procedure is also referred to as principal component analysis and Karhunen-Lòeve expansion in the literature. There is a long history behind the application of the POD to areas such as image processing, signal analysis and data compression. This approach has been applied by a number of researchers to understand the important dynamical structures or coherent structures seen in fluid flows (Berkooz et al. 1993; Sirovich 1987). More recently, the POD procedure has been applied to construct reduced-order models of dynamical systems (e.g., fluid flows described by the Navier–Stokes equations) in order to efficiently design optimal control strategies.

In the literature, the POD procedure is usually described within the framework of calculus of variations applied to multidimensional spatio-temporal datasets. However, here we describe it for steady-state problems from the perspective of singular value decomposition (SVD). Even though both perspectives are equivalent, we believe that the SVD perspective is more straightforward to understand. Again, we consider systems governed by general discrete governing equations of the form $\mathbf{R}(\mathbf{w}, \mathbf{x}) = 0$, where $\mathbf{w} \in \mathbb{R}^n$ denotes the discretized vector of field variables and $\mathbf{x} \in \mathbb{R}^p$ denotes the vector of design variables. As discussed earlier in Section 6.2, the Lagrange subspace can be used to approximate $\mathbf{w}$, that is, the subspace spanned by solutions of the high-fidelity model at the design points $\mathbf{x}^{(1)}, \mathbf{x}^{(2)}, \ldots, \mathbf{x}^{(\widehat{m})}$, that is, $\boldsymbol{\Psi} = \mathrm{span}\{\mathbf{w}(\mathbf{x}^{(1)}), \mathbf{w}(\mathbf{x}^{(2)}), \ldots, \mathbf{w}(\mathbf{x}^{(\widehat{m})})\} \in \mathbb{R}^{n \times \widehat{m}}$. In POD, each solution vector is referred to as a snapshot.[12]

The objective of POD is to extract a set of m dominant basis vectors from the subspace spanned by the snapshots so that the solution can be approximated as

$$\widehat{\mathbf{w}} = \langle \mathbf{w} \rangle + \sum_{i=1}^{m} c_i \boldsymbol{\psi}_i, \tag{6.44}$$

where $\langle \mathbf{w} \rangle = (1/\widehat{m}) \sum_{i=1}^{\widehat{m}} \mathbf{w}(\mathbf{x}^{(i)})$ is the mean of the snapshots, $\boldsymbol{\psi}_i$, $i = 1, 2, \ldots, m$ denote a set of basis vectors and c_i, $i = 1, 2, \ldots, m$ are undetermined coefficients. We first define the set of *modified* snapshots obtained by subtracting $\langle \mathbf{w} \rangle$ from $\mathbf{w}(\mathbf{x}^{(i)})$,

$$\widetilde{\mathbf{w}}^{(i)} = \mathbf{w}(\mathbf{x}^{(i)}) - \langle \mathbf{w} \rangle, \quad i = 1, 2, \ldots, \widehat{m}. \tag{6.45}$$

Let $\mathbf{A}$ denote the matrix whose columns are the modified snapshots, that is, $\mathbf{A} = [\widetilde{\mathbf{w}}^{(1)}, \widetilde{\mathbf{w}}^{(2)}, \ldots, \widetilde{\mathbf{w}}^{(\widehat{m})}]^T \in \mathbb{R}^{\widehat{m} \times n}$. The SVD of $\mathbf{A}$ can be written as

$$\mathbf{A} = \mathbf{U}\boldsymbol{\Sigma}\mathbf{V}^T, \tag{6.46}$$

where $\mathbf{U} \in \mathbb{R}^{\widehat{m} \times \widehat{m}}$ and $\mathbf{V} \in \mathbb{R}^{n \times n}$ are orthogonal matrices (also referred to as left and right singular vectors). $\boldsymbol{\Sigma} \in \mathbb{R}^{\widehat{m} \times n}$ is a diagonal matrix whose diagonal elements $\boldsymbol{\Sigma}_{ii}$ consist of $q = \min(\widehat{m}, n)$ nonnegative numbers σ_i arranged in decreasing order, that is, $\sigma_1 \geq \sigma_2 \geq \cdots \geq \sigma_q \geq 0$. σ_i are referred to as the singular values of $\mathbf{A}$.[13]

The columns of $\mathbf{V}$ are referred to as the proper orthogonal modes or empirical modes of the system. In other words, we can use the basis vectors $\boldsymbol{\psi}_i = \mathbf{V}(:, i) \in \mathbb{R}^n$ to approximate $\mathbf{w}$. These basis vectors are orthonormal, that is, $\boldsymbol{\psi}_i^T \boldsymbol{\psi}_j = \delta_{ij}$, where δ_{ij} denotes the Kronecker delta operator. The number of dominant modes are those corresponding to the largest m singular values. The energy (or variance in the data) captured by the first m modes can be computed as

$$E(m) = \frac{\sum_{i=1}^{m} \sigma_i^2}{\sum_{i=1}^{\widehat{m}} \sigma_i^2}. \tag{6.47}$$

For most datasets, the singular values decay rapidly, and hence only a small number of basis vectors is required to represent the snapshots. Figure 6.5 shows the largest 20 singular values obtained for a problem involving analysis of fluid flow over a circular cylinder with time-dependent angular velocity. The snapshot matrix was generated here by using the velocity fields at different instants of time. Here, only six basis vectors are required to capture 99.9% of the energy in the flow.

It is worth noting that instead of the SVD approach the proper orthogonal modes can also be computed by solving for the largest m eigenvalues and corresponding eigenvectors of the matrix $\mathbf{K} = \mathbf{A}^T\mathbf{A} \in \mathbb{R}^{n \times n}$. Hence, it follows that $\mathbf{V}$ is the matrix of eigenvectors of $\mathbf{K}$ and σ_i^2, $i = 1, 2, \ldots, \widehat{m}$ are its largest eigenvalues.

For cases where $\widehat{m} \ll n$, instead of using the SVD procedure, it is computationally more efficient to use the "method of snapshots" proposed by Sirovich (1987). In that approach,

---

[12]In the present context, the snapshots are solutions obtained for different values of the design variables. In the case of spatio-temporal datasets, each snapshot corresponds to the solution obtained at a particular instant in time.

[13]Note that σ_i are also the singular values of $\mathbf{A}^T$. Further, the rank of $\mathbf{A}$ equals the total number of nonzero singular values. Hence, the number of numerically significant singular values gives an estimate of the effective dimension of the Lagrange subspace (or set of snapshots).

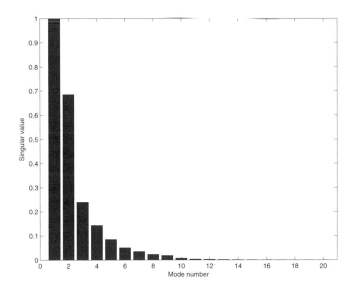

Figure 6.5 Decay of singular values of the snapshot matrix $\mathbf{A}$ for flow over a circular cylinder subjected to rotational excitations (see for example Ravi et al. 2002).

we first solve the $\widehat{m} \times \widehat{m}$ eigenvalue problem $\mathbf{AA}^T\mathbf{U} = \Lambda\mathbf{U}$. After computing $\mathbf{U}$, premultiplying (6.46) by $\mathbf{U}^T$ gives $\mathbf{U}^T\mathbf{A} = \Sigma\mathbf{V}^T$. Then, the first m rows of $\mathbf{U}^T\mathbf{A}$ normalized to unit magnitude become the proper orthogonal modes.

The POD procedure is general in scope and can be applied to prune any subspace by extracting a finite set of dominant basis vectors. For example, the POD procedure can also be applied to extract a set of orthogonal basis vectors from the Hermite subspace defined earlier in (6.14). This approach can be attractive for problems where the first-order sensitivities of the field variable $\mathbf{w}$ can be efficiently computed.

6.5.2 Reduced-order Modeling

We next look at how reduced-order models can be constructed using the basis vectors generated via POD. It is worth noting that it is possible to employ the proper orthogonal modes for Galerkin approximation of the continuous form of the governing equations. A number of researchers have applied this idea to the time-dependent Navier–Stokes equations; see, for example, Ravindran (1999) and the references therein. In these papers, the Navier–Stokes equations are reduced to a system of nonlinear ordinary differential equations (ODEs) with m states. Considering the fact that conventional spatial discretization of the Navier–Stokes equations may well result in the order of ten million unknowns for complex flows, the reduced-order model offers significant computational cost savings.

Consider the case when the reduced basis approximation procedure is to be applied to the discrete form of the governing equations. As before, we write the reduced basis approximation as $\widehat{\mathbf{w}} = c_1\boldsymbol{\psi}_1 + c_2\boldsymbol{\psi}_2 + \cdots + c_m\boldsymbol{\psi}_m = \boldsymbol{\Psi}\mathbf{c}$. Let us substitute this approximation into the discrete governing equations $\mathbf{R}(\mathbf{w}, \mathbf{x}) = 0$. The residual can then be written in terms of the undetermined coefficients in the reduced basis approximation as $\mathbf{R}(\boldsymbol{\Psi}\mathbf{c}, \mathbf{x})$. The vector of undetermined coefficients **c** can subsequently be computed by minimizing $||\mathbf{R}(\mathbf{\Psi c, x})||$ using a nonlinear least-squares minimization technique.

LeGresley and Alonso (2000) applied POD-based reduced-order models to efficiently approximate subsonic and transonic flows over airfoils. A Lagrange subspace was first created by running an Euler solver for various geometries. Subsequently, the POD procedure was employed to generate a set of orthogonal basis vectors. These basis vectors were then used for reduced basis approximation of the conserved flow variables. It was shown that the reduced-order models constructed using this set of basis vectors can exactly reproduce the solutions corresponding to geometries used to create the Lagrange subspace, that is, the reduced basis approximation is an interpolant.

We now illustrate the POD-based reduced-order modeling approach on a simplified version of the airfoil problem (LeGresley and Alonso 2003). Consider the case when we seek to approximate the pressure distribution (C_p) as a function of the Mach number (M). To employ the reduced basis approach, a set of snapshots first needs to be created by running the full-order flow solver at various Mach numbers. The C_p distributions for a set of Mach numbers for the RAE2822 airfoil are shown in Figure 6.6. The C_p distributions predicted using the reduced-order model for $M = 0.37$ and $M = 0.67$ (whose solutions are not included in the set of snapshots) are shown in Figure 6.7. It can be noted that for subsonic flow conditions the set of snapshots used is sufficiently rich to approximate the C_p distribution with high accuracy. However, at $M = 0.67$, a shock forms on the upper surface of the airfoil and its location is difficult to approximate accurately using the available snapshots. In other words, the basis vectors are not rich enough to ensure accurate approximations when parametric perturbations lead to shifts in the location of shocks. This is a typical limitation of reduced basis methods, namely, the accuracy can be poor if the system behavior does not lie in the subspace spanned by the basis vectors.

The C_p distribution obtained by direct subspace projection is also shown in Figure 6.7 for $M = 0.67$. To illustrate how direct subspace projection is carried out, let us suppose that $\mathbf{w}_e$ denotes the exact solution obtained by running the full-order flow solver. Then, $\mathbf{w}_e$ can be approximately represented using the POD basis as $\mathbf{w}_e \approx \widehat{\mathbf{w}}_e = \langle \mathbf{w} \rangle + \sum_{i=1}^{m} c_i \boldsymbol{\psi}_i$, where by orthonormality of the basis vectors we have $c_i = \boldsymbol{\psi}_i^T \mathbf{w}_e - \boldsymbol{\psi}_i^T \langle \mathbf{w} \rangle$.[14] $\widehat{\mathbf{w}}_e$ is referred to as the subspace projection and $||\mathbf{w}_e - \widehat{\mathbf{w}}_e||$ is a measure of the ability of the POD subspace to accurately represent the true solution $\mathbf{w}_e$. Clearly, this error measure can be estimated only if we know the true solution and hence it is not useful in practice. However, from a theoretical perspective, if $||\mathbf{w}_e - \widehat{\mathbf{w}}_e||$ is high, then the reduced-order model will be fundamentally incapable of giving good approximations because of the poor quality of the approximating subspace.

More recently, LeGresley and Alonso (2003) proposed a domain decomposition scheme to improve the accuracy of POD-based reduced-order models. The key idea is to use the POD modes for global approximation of the flow field and to use a standard finite volume scheme in localized regions where shocks occur. The improvement in accuracy achieved by this hybrid procedure for $M = 0.67$ is shown in Figure 6.8. It can be seen that it now becomes possible to closely approximate the location of the shock.

It is expected that by combining POD-based reduced-order models with conventional spatial discretization schemes efficient multiscale algorithms for analyzing parameter dependent PDEs can be developed. The idea of multiscale modeling is to use (at least) two

---

[14]To derive this expression, premultiply both sides of the subspace approximation by $\boldsymbol{\psi}_i^T$ and use the fact that $\boldsymbol{\psi}_i^T \boldsymbol{\psi}_j = \delta_{ij}$.

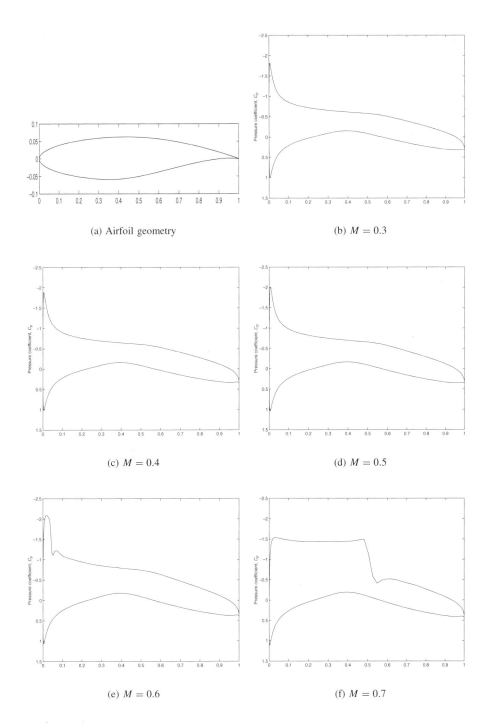

Figure 6.6   Snapshots of C_p distributions at various Mach numbers. Reproduced from (LeGresley and Alonso 2003) by permission of Patrick LeGresley and Juan J. Alonso.

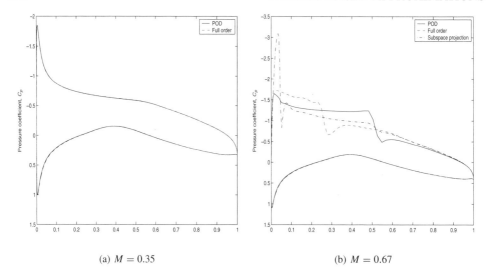

(a) $M = 0.35$ (b) $M = 0.67$

Figure 6.7   Comparison of approximations for C_p obtained at $M = 0.35$ and $M = 0.67$. Reproduced from (LeGresley and Alonso 2003) by permission of Patrick LeGresley and Juan J. Alonso.

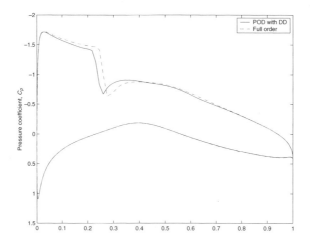

Figure 6.8   Approximation for C_p obtained using domain decomposition scheme at $M = 0.67$. Reproduced from (LeGresley and Alonso 2003) by permission of Patrick LeGresley and Juan J. Alonso.

models – a coarse model to represent the global features of the solution and a detailed model to capture the local features. The work by LeGresley and Alonso (2003) presents a major step in that direction. The main issues involved in designing a multiscale scheme are (1) the numerical method employed to identify the localized regions where a full-order model should be used and (2) the approach used for handling the coupling between the two models in order to arrive at consistent values for the field variables in the interface region.

We refer the interested reader to the dissertation of Lucia (2001), which proposes numerical schemes based on combining reduced-order and full-order models for flows with moving shocks.

Finally, it is to be noted that the computational cost of the POD procedure may become prohibitive for problems with a large number of snapshots. For such problems, the dominant basis vectors can be computed using centroidal Voronoi tessellations; see, for example, Du et al. (1999). This approach appears to be promising since it does not involve the solution of an eigenvalue problem and can also be efficiently parallelized. Du and Gunzburger (2002) present an approach for model reduction that hybridizes centroidal Voronoi tessellations with POD.

6.5.3 Concluding Remarks

We conclude by mentioning the possibility of constructing physics-based surrogate models by extending the black-box surrogate modeling approaches discussed in Chapter 5. To illustrate, consider a parameterized two-dimensional steady-state problem where the governing PDEs can be expressed as $\mathcal{L}w = f$ and the boundary conditions by $\mathcal{B}w = g$, where $\mathcal{L}$ and $\mathcal{B}$ denote parameterized differential operators in space. Let us denote the spatial coordinates by η and ζ. Then, it follows that the field variable w is a function of η, ζ and the vector of design variables $\mathbf{x}$, that is, $w \equiv w(\eta, \zeta, \mathbf{x}) : \mathbb{R}^{p+2} \to \mathbb{R}$. The key idea of the physics-based surrogate modeling approach is to postulate a model for w in the form $\widehat{w}(\eta, \zeta, \mathbf{x}, \boldsymbol{\alpha})$, where $\boldsymbol{\alpha}$ is a vector of undetermined coefficients. For example, this model could be a linear combination of radial basis functions or a feed-forward neural network. The undetermined coefficients in the surrogate model can be computed by minimizing $\int ||\mathcal{L}\widehat{w} - f|| \, d\eta \, d\zeta \, d\mathbf{x}$ subject to the constraint $\int ||\mathcal{B}\widehat{w} - g|| \, d\eta \, d\zeta \, d\mathbf{x} \leq \varepsilon$, where ε is a small constant. Computation of $\boldsymbol{\alpha}$ by solving this optimization problem will ensure that the surrogate model satisfies the governing equations in some sense. Similarly, it is also possible to derive error functions using the discrete form of the governing equations.

The physics-based surrogate modeling approach can be interpreted as a generalization of conventional PDE solution techniques to solve parameterized operator problems. The most attractive feature of this approach is that, in theory, it becomes possible to construct surrogate models without using any observational data, relying instead purely on the equations being solved. The interested reader is referred to Nair (2002b) for a detailed exposition of continuous and discrete formulations for physics-based surrogate modeling. Even though encouraging results have been obtained for some simple test problems, further work is required to evaluate the applicability of this approach to complex problems.

Part III

Frameworks for Design Space Exploration

In this third part of the book, we examine a number of practical issues that must be addressed if the best results are to be obtained from the tools already described.

In many cases, this involves defining relatively sophisticated workflows or metaheuristics to control the process of design exploration, and so we begin by discussing how approximation techniques such as those described in Chapters 5 and 6 can be leveraged to improve the computational efficiency of design optimization algorithms. The key idea, already outlined in Section 3.4, is to develop algorithms that make use of computationally cheap surrogates to carry out the bulk of the function evaluations, while the computationally expensive high-fidelity model is run sparingly to check and control the progress of optimization. Necessarily, the approaches adopted must take into account the types of surrogate available and also the aims of the designer. Sometimes relatively inefficient methods may be preferred because they are simpler to set up or manage. It must also be remembered that even the most complex approaches are still subject to the "no free lunch" theorems, in that, by adopting highly sophisticated approaches, one is commonly sacrificing generality of application.

Having discussed various ways of managing surrogates, we move on to set out the major issue of uncertainty in design. Since all engineering products are subject to some uncertainties in manufacture as well as in operating conditions, their actual performance will, of necessity, be equally subject to uncertainties. This naturally leads to stochastic treatment of the design problem and Chapter 8 reviews some of the basic issues involved and how they may be tackled. Such stochastic treatments are still relatively rarely applied in industrial design offices and this is a rapidly advancing field of research. Consequently, the chapter can only be an introduction to the area.

The last chapter in Part III discusses a number of frameworks that can be used to orchestrate design search over multiple disciplines. Since such work is by no means routine, even in the largest aerospace companies, the best approaches to take are yet to be clearly established. We fully anticipate that this aspect of the book will be the fastest to date since so much active research is taking place in this area.

Finally, it should also be noted that, although we cover most of the approaches currently in use for the topics discussed in this part of the book, the relevant literature is extremely large and no doubt we have omitted some schemes that are of real benefit in some situations – the reader is strongly encouraged to refer to the primary literature for the latest contributions and developments.

7

Managing Surrogate Models in Optimization

The idea of using approximation models in optimization is not new and is implicit in most gradient-based mathematical programming algorithms. For example, most nonlinear optimization techniques use line searches on a polynomial approximation of the objective function to compute the next iterate; see Section 3.1. Further details on such techniques can be found in any standard text on optimization theory and are not covered in depth here. Instead, our primary focus is on how general approximation techniques can be exploited to accelerate a variety of optimization methods, including gradient-based and evolutionary algorithms. A key theme throughout this discussion is the distinction between building accurate surrogates and the improvement of a design by optimization: in design optimization one is not concerned with the accuracy of any intermediate predictions, but only with the ability to ultimately reach good quality designs.

Apparently, the idea of using approximation techniques to speed up design optimization procedures was first studied by Schmit and his coworkers in the 1970s; see, for example, Schmit and Farshi (1974) and Schmit and Miura (1976). The motivation for this work arose from the fact that the computer hardware available in the 1970s was incapable of designing systems of even modest complexity. In the widely acknowledged first published work on structural optimization using mathematical programming techniques, Schmit (1960) notes that nearly 30 min were required by the computer available to the author to solve a three-bar truss design problem. Because of the dramatic improvements in computing power and numerical methods now available, the same problem can currently be solved in fractions of a second. However, the complexity of systems that engineers wish to study has grown in concert with computer technology. As a consequence, the idea of using approximate models to reduce the computational burden associated with design optimization is still relevant and continues to be an area of active research in the engineering design and applied mathematics communities.

Computational Approaches for Aerospace Design: The Pursuit of Excellence.   A. J. Keane and P. B. Nair
© 2005 John Wiley & Sons, Ltd

To provide an overview of the issues involved in surrogate-assisted optimization, consider a general nonlinear programming problem of the form

$$
\begin{aligned}
\text{Minimize}_{\mathbf{x}} : \quad & f(\mathbf{x}) \\
\text{Subject to} : \quad & g_i(\mathbf{x}) < 0, \quad i = 1, 2, \ldots, m \\
& h_i(\mathbf{x}) = 0, \quad i = 1, 2, \ldots, q \\
& \mathbf{x}_l \leq \mathbf{x} \leq \mathbf{x}_u,
\end{aligned}
\tag{7.1}
$$

where $\mathbf{x} \in \mathbb{R}^p$ is the vector of design variables and $\mathbf{x}_l$ and $\mathbf{x}_u$ are vectors of lower and upper bounds, respectively. $f(\mathbf{x})$ is the objective function to be minimized, while $g_i(\mathbf{x})$ and $h_i(\mathbf{x})$ denote inequality and equality constraints, respectively.

Our focus is on the case where the evaluation of $f(\mathbf{x})$, $g(\mathbf{x})$ and $h(\mathbf{x})$ are computationally expensive, and the aim is to obtain a near optimal solution on a *limited* computational budget. This normally rules out the direct application of an optimization algorithm to solve (7.1). Intuitively, the most straightforward way to tackle the computational cost issue is to use a strategy that employs computationally cheap surrogate models for the objective and constraint functions to solve an approximation to (7.1). In other words, we replace $f(\mathbf{x})$, $g_i(\mathbf{x})$ and $h_i(\mathbf{x})$ with surrogate models that are computationally cheaper to run (say, $\widehat{f}$, $\widehat{g}_i$ and $\widehat{h}_i$) and solve the following problem instead:

$$
\begin{aligned}
\text{Minimize}_{\mathbf{x}} : \quad & \widehat{f}(\mathbf{x}) \\
\text{Subject to} : \quad & \widehat{g}_i(\mathbf{x}) < 0, \quad i = 1, 2, \ldots, m \\
& \widehat{h}_i(\mathbf{x}) = 0, \quad i = 1, 2, \ldots, q \\
& \mathbf{x}_l \leq \mathbf{x} \leq \mathbf{x}_u.
\end{aligned}
\tag{7.2}
$$

Let $\widehat{\mathbf{x}}^*$ denote the optimal solution obtained by solving (7.2) when $\mathbf{x}^0$ is used as the initial guess. Since approximate models for the objective and constraint functions are used in (7.2), in general, $\widehat{\mathbf{x}}^* \neq \mathbf{x}^*$, where $\mathbf{x}^*$ is the "true" optimum obtained by solving (7.1). The actual performance of $\widehat{\mathbf{x}}^*$ can be checked by calculating $f(\widehat{\mathbf{x}}^*)$, $g_i(\widehat{\mathbf{x}}^*)$ and $h_i(\widehat{\mathbf{x}}^*)$. There are three possible outcomes at this point:

1. $\widehat{\mathbf{x}}^*$ is a satisfactory solution;

2. $\widehat{\mathbf{x}}^*$ is an improvement over $\mathbf{x}^0$, however, the solution needs to be improved further to satisfy design requirements; or

3. $\widehat{\mathbf{x}}^*$ is a poorer solution than the initial guess.

In the first scenario, the analyst may decide to accept $\widehat{\mathbf{x}}^*$ as the final solution. A rational approach to tackle the second and third scenarios would be to update the surrogate models using the objective and constraint function values at the point $\widehat{\mathbf{x}}^*$ (and perhaps additional data points) and solve (7.2) again using the updated models – these steps can be carried out in an iterative fashion; see, for example, Figure 3.21 and the surrounding discussion. In many design problems, this simple strategy may give good solutions, provided the surrogate models are updated appropriately and sufficient iterations are carried out. However, there is no theoretical guarantee that this iterative process will converge to the optima of the original nonlinear programming problem (7.1).

A variety of approaches for integrating surrogate models with optimization algorithms have been published in the literature, particularly over the last decade. Since these strategies are essentially centered around tackling the problem of managing the interplay between low- and high-fidelity models, they are often referred to as approximation model management frameworks (AMMF), model management frameworks (MMF) or surrogate management frameworks (SMF). Clearly, the design of a surrogate-assisted optimization strategy has to take into account the optimization algorithm being used (e.g., gradient, nongradient, stochastic, etc.) and the type of surrogate model (e.g., local, global or physics-based surrogate models). Some surrogate-assisted search algorithms are carefully designed so that the convergence properties of the underlying optimization algorithm in use can be inherited – this is the case for the trust-region methods described in the next section and also surrogate-assisted pattern search algorithms (Booker et al. 1998). Many other strategies used in the literature focus on generating near optimal solutions on a limited computational budget without much consideration of the issue of guaranteeing convergence to the optima of the original problem. It is reassuring if theoretical guarantees of convergence can be made – however, since such results are usually arrived at by assuming that the number of evaluations of the high-fidelity model tends to infinity, theoretical convergence analysis is a topic that is largely sidestepped by many engineering designers: they will often settle for a process that gives rapid improvements from a baseline design even if it may subsequently stall.

In this chapter, we outline a number of strategies for surrogate-assisted optimization that we consider mature enough for real-life applications. Some examples are also provided to illustrate the performance of the algorithms discussed.

7.1 Trust-region Methods

To illustrate the basic idea used in trust-region methods, it is instructive to first consider the mechanisms employed by standard nonlinear programming techniques to search for a new iterate. Consider the unconstrained minimization problem $\min f(\mathbf{x})$, where $\mathbf{x} \in \mathbb{R}^p$. Let $\mathbf{x}^0$ be an initial guess for the minimum of $f(\mathbf{x})$. Then, a gradient-based approach computes the next iterate as $\mathbf{x}^{k+1} = \mathbf{x}^k + \lambda \Delta \mathbf{x}$, where λ is the step length and $\Delta \mathbf{x}$ is the direction along which $f(\mathbf{x})$ decreases.[1] Once the direction of descent $\Delta \mathbf{x}$ is fixed, the problem reduces to computing the scalar λ such that $f(\mathbf{x}^{k+1}) < f(\mathbf{x}^k)$. The standard way to achieve this is to solve the one-dimensional minimization problem

$$\min_{\lambda} f(\mathbf{x}^k + \lambda \Delta \mathbf{x}) \tag{7.3}$$

which is referred to as line search since the direction of descent is fixed.

It must be clear from the above discussion that to improve the computational efficiency of optimization algorithms we can replace the objective function $f(\mathbf{x})$ with a computationally cheap surrogate model during line searches. This idea will work well provided the surrogate $\widehat{f}(\mathbf{x})$ is able to predict an iterate $\mathbf{x}^{k+1}$ that is an improvement over $\mathbf{x}^k$, that is, $\widehat{f}(\mathbf{x}^{k+1}) < \widehat{f}(\mathbf{x}^k) \Rightarrow f(\mathbf{x}^{k+1}) < f(\mathbf{x}^k)$. However, in many cases, the approximation is valid only in the vicinity of $\mathbf{x}^k$, that is, $\widehat{f}(\mathbf{x}^k + \lambda \Delta \mathbf{x})$ is a good approximation of the original function only when λ varies in an interval of limited but unknown width. In order to ensure that the surrogate always works within its range of validity, we need to impose an additional

---

[1]Optimization algorithms differ in the choice of how $\Delta \mathbf{x}$ is computed. For example, in the steepest descent method, $\Delta \mathbf{x} = -\nabla f(\mathbf{x}^0)$, see Section 3.2.

constraint, that is, the following bound-constrained optimization problem is solved:

$$\min_{\mathbf{s}} \widehat{f}(\mathbf{x}^0 + \mathbf{s}) \text{ subject to: } ||\mathbf{s}|| \leq \delta. \tag{7.4}$$

Here, $\mathbf{s}$ is the step, that is, $\lambda \Delta \mathbf{x}$ in a line search, and δ can be interpreted as a *move limit*, that is, the magnitude of permissible modifications to $\mathbf{x}^k$. For a general approximation model, it is difficult, if not impossible, to decide on an appropriate value for δ *a priori*. Even though heuristic recipes can be employed, they may compromise the ability of the optimization scheme to converge to the minima of the original objective function $f(\mathbf{x})$. This has motivated the development of *trust-region* methods for nonlinear optimization, which provide a rigorous basis for adapting the move limit δ so that convergence to a minimum of $f(\mathbf{x})$ can be theoretically guaranteed. Here, δ is interpreted as the trust-region radius since it indicates the region over which the surrogate can be *trusted* to be a faithful approximation of $f(\mathbf{x})$. Note that in classical trust-region methods the focus is on the case where $\widehat{f}(\mathbf{x})$ is a polynomial (usually a second-order Taylor series approximation with an approximate Hessian); see, for example, Dennis and Schnabel (1983). In this section, we set out how ideas from classical trust-region methods can be employed for solving computationally expensive optimization problems when general approximation models are employed.

7.1.1 Unconstrained Problems

The extension of the classical trust-region algorithm to surrogate-assisted optimization was proposed by Alexandrov et al. (1998b) and that paper forms the basis for much of the material covered in this section. The approach is similar to the development of the MARS scheme proposed by Toropov and Alvarez (1998). The main steps of the algorithm for the case of bound-constrained problems are shown in Algorithm 7.1, which are discussed in more detail in what follows.

The inputs for Algorithm 7.1 are the initial guess $\mathbf{x}^0$ and the initial value of the trust-region radius δ_k. Further, at each iteration, the following zero- and first-order consistency conditions are imposed on the surrogate model

$$\widehat{f_k}(\mathbf{x}^k) = f(\mathbf{x}^k) \text{ and } \nabla \widehat{f_k}(\mathbf{x}^k) = \nabla f(\mathbf{x}^k). \tag{7.5}$$

In other words, it is ensured that the function value and gradient computed using the surrogate model agree with those calculated using the high-fidelity model at the current iterate. This ensures that the surrogate is at least first-order accurate in the vicinity of the initial guess. As we shall see later, these conditions are necessary to guarantee convergence.

In step (3), which can be interpreted as the trust-region subproblem, the surrogate model $\widehat{f_k}(\mathbf{x}^k + \mathbf{s})$ is minimized subject to the constraint $||\mathbf{s}|| \leq \delta_k$ to predict the *prospective* step $\mathbf{s}^k$. The trust-region constraint $||\mathbf{s}|| \leq \delta_k$ ensures that we work within the range of validity of the surrogate. In practice, the L_∞ norm can be used for this constraint, since this reduces the norm-based constraint to simple bound constraints of the form $-\delta_k \mathbf{1} \leq \mathbf{s} \leq \delta_k \mathbf{1}$, where $\mathbf{1} = \{1, 1, \ldots, 1\} \in \mathbb{R}^p$. Using the L_∞ norm for the trust-region constraint makes it possible to implement Algorithm 7.1 using existing optimization routines capable of handling bound constraints; for example, a bound-constrained quasi-Newton method can be used to estimate $\mathbf{s}^k$. Further, a fully converged solution to the trust-region subproblem is not necessary. In order to guarantee convergence, only the "fraction of Cauchy decrease" condition needs to

---

Algorithm 7.1: Trust-region framework for solving computationally expensive uncon
strained optimization problems using surrogates.

Require: Initial guess $\mathbf{x}^0 \in \mathbb{R}^p$ and initial value of trust-region radius $\delta_o > 0$

Ensure: $\widehat{f}_k(\mathbf{x}^k) = f(\mathbf{x}^k)$ and $\nabla \widehat{f}_k(\mathbf{x}^k) = \nabla f(\mathbf{x}^k)$ {Ensure that the surrogate model
   meets zero- and first-order consistency conditions}

1:    $k = 0$
2:    **while** convergence criteria not met **do**
3:        Compute $\mathbf{s}^k$ by minimizing $\widehat{f}_k(\mathbf{x}^k + \mathbf{s})$ subject to the constraint $||\mathbf{s}|| \leq \delta_k$ {Solve trust
      region subproblem to estimate prospective step}
4:        Compute $\rho_k = \left(f(\mathbf{x}^k) - f(\mathbf{x}^k + \mathbf{s}^k) \right) / \left(\widehat{f}_k(\mathbf{x}^k) - \widehat{f}_k(\mathbf{x}^k + \mathbf{s}^k) \right)$ {Compute ratio of
      actual to predicted improvement in objective function}
5:        **if** $\rho_k \leq 0$ **then** {Surrogate is inaccurate – reject step and shrink δ to improve
      surrogate accuracy}
6:           $\delta_{k+1} \leftarrow 0.25\delta_k$
7:           $\mathbf{x}^{k+1} \leftarrow \mathbf{x}^k$
8:        **else if** $0 < \rho_k < 0.25$ **then** {Surrogate is marginally accurate – accept the step
      but shrink δ to improve surrogate accuracy}
9:           $\mathbf{x}^{k+1} \leftarrow \mathbf{x}^k + \mathbf{s}^k$
10:         $\delta_{k+1} \leftarrow 0.25\delta_k$
11:       **else if** $0.25 < \rho_k < 0.75$ **then** {Surrogate is moderately accurate – accept the
      step and keep δ unchanged}
12:         $\mathbf{x}^{k+1} \leftarrow \mathbf{x}^k + \mathbf{s}^k$
13:         $\delta_{k+1} \leftarrow \delta_k$
14:       **else if** $\rho_k \geq 0.75$ and $||\mathbf{s}^k|| < \delta_k$ **then** {Surrogate is accurate and the step lies
      within trust-region bounds – accept the step and keep δ unchanged}
15:         $\mathbf{x}^{k+1} \leftarrow \mathbf{x}^k + \mathbf{s}^k$
16:         $\delta_{k+1} \leftarrow \delta_k$
17:       **else if** $\rho_k \geq 0.75$ and $||\mathbf{s}^k|| = \delta_k$ **then** {Surrogate is accurate and the step lies on
      the trust-region bounds – accept the step and increase δ by a factor of two}
18:         $\mathbf{x}^{k+1} \leftarrow \mathbf{x}^k + \mathbf{s}^k$
19:         $\delta_{k+1} \leftarrow 2\delta_k$
20:      **end if**
21:      Construct updated surrogate $\widehat{f}_{k+1}$ satisfying the zero- and first-order consistency con
      ditions at $\mathbf{x}^{k+1}$.
22:      $k \leftarrow k + 1$
23: **end while**

---

be satisfied in the following form: there exist $\beta > 0$ and $C > 0$, independent of k, for which
the step $\mathbf{s}^k$ satisfies

$$f(\mathbf{x}^k) - \widehat{f}(\mathbf{x}^k + \mathbf{s}^k) > \beta ||\nabla f(\mathbf{x}^k)|| \min \left(\delta_k, \frac{||\nabla f(\mathbf{x}^k)||}{C} \right). \tag{7.6}$$

Once the step $\mathbf{s}^k$ is computed by minimizing the surrogate model, the high-fidelity model
is evaluated at $\mathbf{x}^k + \mathbf{s}^k$ to calculate ρ_k, which is a measure of the actual versus predicted
improvement (decrease) in the objective function – see step (4). Depending on the value of
ρ_k, a decision is made whether to accept the prospective step. For example, if $\rho_k \leq 0$, then

the step $\mathbf{s}^k$ has led to an *increase* in the objective function value compared to the initial guess. This suggests that the surrogate model is not capable of predicting improvements in the true objective function. Hence, we reject the step and shrink the trust-region radius δ_k to improve our chances of generating a better step at the next iteration. When $\rho_k > 0$, the step $\mathbf{s}^k$ has led to a reduction in the true objective function. Hence, we accept the step. Depending on the actual value of ρ_k, we either reduce δ_k, keep it unchanged or increase it – see steps (8–16), which are based on the recommendations made by Giunta and Eldred (2000).

Finally, once appropriate decisions are made regarding the current step and changes to the trust-region radius, the surrogate model is updated if necessary. If the step is rejected, we reduce the trust-region radius and reuse the surrogate without update. However, if the step is accepted, the surrogate model $\widehat{f_k}$ needs to be rebuilt so that the zero- and first-order conditions in (7.5) are satisfied at $\mathbf{x}^{k+1} = \mathbf{x}^k + \mathbf{s}^k$. The way $\widehat{f_k}$ is updated will depend on the approximation technique in use. For example, when a Hermite RBF interpolant is used (see Section 5.4.4), we append the point $(\mathbf{x}^{k+1}, f(\mathbf{x}^{k+1}), \nabla f(\mathbf{x}^{k+1}))$ to the baseline training dataset to ensure that the updated surrogate matches the high-fidelity function value and its gradient at $\mathbf{x}^{k+1}$.

In summary, the basic idea of the trust-region approach is to adaptively increase or decrease δ_k at each iteration, depending on how well $\widehat{f}(\mathbf{x})$ predicts improvements in the true objective $f(\mathbf{x})$. In practice, the algorithm can be terminated when either $f(\mathbf{x}^{k+1}) - f(\mathbf{x}^k) \le \varepsilon_1$ or $||\mathbf{x}^{k+1} - \mathbf{x}^k|| \le \varepsilon_2$, where ε_1 and ε_2 are user-specified tolerances.

In order to ensure convergence[2] of trust-region frameworks employing general surrogate models, Alexandrov et al. (1998b) showed that the zero- and first-order consistency conditions in (7.5) have to be imposed at the initial guess used to solve each trust-region subproblem; see step (3) in Algorithm 7.1. Further, each iterate should satisfy the fraction of Cauchy decrease condition given in (7.6). If these two assumptions are satisfied, then it is possible to show that

$$\lim_{k \to \infty} \inf ||\nabla f(\mathbf{x}^k)|| = 0. \tag{7.7}$$

It can be readily verified that a number of the surrogate modeling techniques described in Chapter 5 satisfy the zero- and first-order consistency conditions by construction: these include second-order Taylor series, two-point approximations, Hermite RBF interpolants and gradient-enhanced Gaussian process models. If a variable-fidelity approach is employed, where $\widehat{f}$ is a global physics-based or empirical model such as those covered in Section 6.1, then a first-order scaling approach needs to be employed to ensure that the zero- and first-order consistency conditions are satisfied.

7.1.2   Extension to Constrained Problems

The trust-region algorithm can be readily extended to the case with general nonlinear constraints by using an augmented Lagrangian approach to convert the original problem in (7.1) into an unconstrained problem with additional variables as described in Chapter 3; see also, Alexandrov et al. (2000) and Rodriguez et al. (1998). Here, we discuss an approach that can be easily implemented in conjunction with a sequential quadratic programming (SQP) algorithm (Gill et al. 1981).

---

[2]Here, convergence is defined as the mathematical assurance that the iterates produced by an algorithm, started from an arbitrary location, will converge to a stationary point or local optima of the original high-fidelity optimization problem and not that the overall best optimum will be found.

In the case of constrained nonlinear programming problems of the form (7.1), the prospective step $\mathbf{s}^k$ can be computed by solving the following trust-region subproblem:

$$
\begin{aligned}
\text{Minimize} : \quad & \widehat{f}(\mathbf{x}^k + \mathbf{s}) \\
\text{Subject to} : \quad & \widehat{g}_i(\mathbf{x}^k + \mathbf{s}) < 0, \quad i = 1, 2, \ldots, m \\
& \widehat{h}_i(\mathbf{x}^k + \mathbf{s}) = 0, \quad i = 1, 2, \ldots, q \\
& ||\mathbf{s}||_\infty \leq \delta_k.
\end{aligned}
\tag{7.8}
$$

Note that here the surrogates for the objective and constraint functions are updated at each iteration – however, we do not explicitly indicate the dependency of $\widehat{f}$, $\widehat{g}$ and $\widehat{h}$ on k for simplicity of notation. The approximate optimization problem in (7.8) can be readily solved using an SQP routine such as FFSQP (Lawrence and Tits 2001), which is capable of handling inequality and equality constraints. However, note that this approach leads to the requirement of explicitly constructing separate surrogate models for the objective and each of the constraint functions.[3] In contrast, if an augmented Lagrangian approach is employed, then we end up solving an unconstrained optimization problem, albeit with a larger number of variables.[4]

Further, the ratio of predicted versus actual decrease in the objective function (see step (4) in Algorithm 7.1) is replaced with

$$
\rho_k = \frac{L(\mathbf{x}^k) - L(\mathbf{x}^k + \mathbf{s}^k)}{\widehat{L}(\mathbf{x}^k) - \widehat{L}(\mathbf{x}^k + \mathbf{s}^k)},
\tag{7.9}
$$

where $L(\mathbf{x})$ is the augmented objective function obtained using the L_1 penalty function, that is,

$$
L(\mathbf{x}) = f(\mathbf{x}) + \mathcal{P} \sum_{i=1}^{m} \max(0, g_i(\mathbf{x})) + \mathcal{P} \sum_{i=1}^{q} |h_i(\mathbf{x})|.
\tag{7.10}
$$

Here, $\mathcal{P} > 0$ is a penalty parameter. In practice, this parameter can be increased by an order of magnitude at each iteration (whenever the prospective step is accepted) to ensure that the final solution satisfies all the constraints.

The decisions made regarding acceptance of prospective step and adaptation of trust-region radius based on the computed value of ρ_k can then proceed on similar lines to those carried out for the unconstrained case in Algorithm 7.1.

7.2 The Space Mapping Approach

Next, we present a simplified account of space mapping algorithms, which were originally developed in the context of tackling computationally expensive microwave circuit design problems (Bandler et al. 1994). The central ideas used make this an attractive approach for tackling engineering design problems using variable-fidelity models. They are closely related to the model fusion approaches previously discussed in Section 6.1. The main difference between the approaches is that whereas in model fusion the results of the low-fidelity model

---

[3]For some problems, the various surrogates can be linked since they may well be correlated; see, for example, the description of co-Kriging in Cressie (1993).

[4]Whether or not a single but more complex surrogate is simpler to construct and update than a series of models, one for each function, will depend on the problem being considered.

are scaled in some way to agree with those of the high-fidelity model, in space mapping a distortion is applied to the *inputs* to the low-fidelity model to cause its maxima and minima to align in *space* with those of the high-fidelity code. So, in some sense, the space mapping idea is akin to drawing the low-fidelity model on a rubberized sheet, which can then be distorted to agree topologically with the high-fidelity results.[5] N.B.: in the space mapping literature, the low- and high-fidelity models are referred to as *coarse* and *fine* models, respectively.

7.2.1 Mapping Functions

To illustrate the main idea used in space mapping approaches, consider the unconstrained minimization problem

$$\mathbf{x}_{hi}^* = \arg \min_{\mathbf{x}_{hi}} f_{hi}(\mathbf{x}_{hi}), \tag{7.11}$$

where $f_{hi}(\mathbf{x}_{hi}) : \mathbb{R}^p \to \mathbb{R}$ is the high-fidelity objective function, which is computationally expensive to evaluate, $\mathbf{x}_{hi} \in \mathbb{R}^p$ is the vector of design variables and the subscript indicates that the variables are defined in the *high-fidelity space*.

Following Bakr et al. (2000), we then assume that a computationally cheap low-fidelity model $f_{lo}(\mathbf{x}_{lo})$ is available, which can be used as a surrogate of the high-fidelity model. The key idea of space mapping is to define a mapping function $\mathbf{x}_{lo} = \mathbf{p}(\mathbf{x}_{hi}) \in \mathbb{R}^p$ such that $f_{lo}(\mathbf{p}(\mathbf{x}_{hi}))$ is an improved approximation of the high-fidelity model over a local or global region of the design space. We can also interpret $\mathbf{p}(\mathbf{x}_{hi})$ as encompassing a set of intervening variables that are introduced to improve the accuracy of the coarse model. Hence, the space mapping approach also has some parallels with the approximation techniques based on intervening variables described in Chapter 5: both approaches apply a transformation to the design variables to improve accuracy.

We begin the space mapping process by solving

$$\mathbf{x}_{lo}^* = \arg \min_{\mathbf{x}_{lo}} f_{lo}(\mathbf{x}_{lo}) \tag{7.12}$$

to establish the argument that minimizes the low-fidelity model. The mapping function $\mathbf{p}(\mathbf{x}_{hi})$ is then computed by minimizing the difference between $f_{lo}(\mathbf{p}(\mathbf{x}_{hi}))$ and $f_{hi}(\mathbf{x}_{hi})$ either at a series of points in an iterative scheme or over a finite volume of the design space. This involves the solution of an optimization problem of the form

$$\mathbf{p}(\mathbf{x}_{hi}) = \arg \min_{\mathbf{p}(\mathbf{x}_{hi})} || f_{hi}(\mathbf{x}_{hi}) - f_{lo}(\mathbf{p}(\mathbf{x}_{hi})) ||. \tag{7.13}$$

If the mapping function is computed by solving the preceding minimization problem and the distance metric $|| f_{hi}(\mathbf{x}_{hi}) - f_{lo}(\mathbf{p}(\mathbf{x}_{hi})) ||$ is close to zero, then $f_{lo}(\mathbf{p}(\mathbf{x}_{hi})) \approx f_{hi}(\mathbf{x}_{hi})$. The process of defining the mapping $\mathbf{p}(\mathbf{x}_{hi})$ is referred to as parameter estimation in the space mapping literature.

A number of parameter estimation schemes have been proposed – they all involve two stages: first, the functional form of the mapping must be chosen and, secondly, any undetermined coefficients in the mapping must be found using data coming from the available low- and high-fidelity models. In the original version of space mapping (Bandler et al. 1995),

---

[5] Note that in space mapping it is often implicitly assumed that the ranges of the high- and low-fidelity functions are the same so that their relative magnitudes do not need scaling – if they are not, then it may be necessary to adopt a hybrid approach where both space mapping and some form of scaling or fusion are adopted.

the mapping function was computed by minimizing the difference between $f_{lo}(\mathbf{p}(\mathbf{x}_{hi}))$ and $f_{hi}(\mathbf{x}_{hi})$ at several points $\mathbf{x}_{hi} + \delta\mathbf{x}_i$, that is, the following problem is solved:

$$\mathbf{p}(\mathbf{x}_{hi}) = \arg\min_{\mathbf{p}(\mathbf{x}_{hi})} \sum_i f_{hi}(\mathbf{x}_{hi} + \delta\mathbf{x}_i) - f_{lo}(\mathbf{p}(\mathbf{x}_{hi} + \delta\mathbf{x}_i))^2. \qquad (7.14)$$

Having established the required mapping, the location of the minimum in the high-fidelity space is estimated by $\mathbf{x}_{hi}^* = \mathbf{p}^{-1}(\mathbf{x}_{lo}^*)$. This value of $\mathbf{x}_{hi}$ is then substituted into the high-fidelity code and a check made to see if it satisfies requirements. If not, the new high-fidelity point is added to the dataset and a revised space mapping created and so on. Note that if $\mathbf{p}(\mathbf{x}_{hi})$ is a simple function its inverse may be readily found; if a more complex mapping is used, then finding $\mathbf{p}^{-1}(\mathbf{x}_{lo}^*)$ will itself require a solution scheme. In the original approach proposed by Bandler et al. (1995), Broyden's method for nonlinear equations (Broyden 1965) was used to find $\mathbf{x}_{hi}^*$.

Later, Bakr et al. (1998) introduced a trust-region approach for solving (7.11) in which the corrected low-fidelity model is used as a surrogate for the high-fidelity model – this approach essentially follows the steps outlined in Algorithm 7.1. In this revised "aggressive space mapping" approach, the mapping is built up over a series of points using quasi-Newton iteration $\mathbf{x}_{hi}^{(k+1)} = \mathbf{x}_{hi}^{(k)} + \mathbf{h}^{(k)}$, where $\mathbf{h}^{(k)}$ is the solution of $\mathbf{B}^{(k)}\mathbf{h}^{(k)} = \mathbf{x}_{lo}^* - \mathbf{p}^{(k)}(\mathbf{x}_{hi}^{(k)})$. Here, $\mathbf{B}^{(k)}$ is an approximation of the Jacobian $\partial\mathbf{p}^{(k)}/\partial\mathbf{x}_{hi}$ evaluated at $\mathbf{x}_{hi}^{(k)}$, which can be iteratively updated using Broyden's update as

$$\mathbf{B}^{(k+1)} = \mathbf{B}^{(k)} + \frac{\mathbf{p}^{(k+1)}(\mathbf{x}_{hi}^{(k+1)}) - \mathbf{p}^{(k)}(\mathbf{x}_{hi}^{(k)}) - \mathbf{B}^{(k)}\mathbf{h}^{(k)}}{(\mathbf{h}^{(k)})^T\mathbf{h}^{(k)}}(\mathbf{h}^{(k)})^T. \qquad (7.15)$$

$\mathbf{p}^{(k+1)}(\mathbf{x}_{hi}^{(k+1)})$ is computed by solving the parameter estimation problem defined in (7.13) at the point $\mathbf{x}_{hi}^{(k+1)}$. The update process starts with $\mathbf{B}^{(0)} = \mathbf{I}$, where $\mathbf{I}$ is the identity matrix. To account for the fact that the corrected low-fidelity model $f_{lo}(\mathbf{p}(\mathbf{x}_{hi}))$ may agree with $f_{hi}(\mathbf{x}_{hi})$ only in the close vicinity of the current iterate, a trust-region framework of the form outlined earlier in Algorithm 7.1 is employed. The basic steps involved in the local space mapping based trust-region framework are outlined in Algorithm 7.2.

To illustrate this process, consider a simple one-dimensional problem where $f_{hi}(\mathbf{x}_{hi}) = \sin((x_{hi}^2 - 1)/2)$, $f_{lo}(\mathbf{x}_{lo}) = \sin(x_{lo})$ and we wish to minimize f_{hi} in the range $x_{hi} = 2$ to 3.5, starting from an initial guess of $x_{hi} = 2.7$. Of course, here, $\mathbf{x}_{lo}^* = -1.5708$. In this model, a first-order mapping function $\mathbf{p}(\mathbf{x}_{hi}) = ax_{hi} + b$ is adopted. We begin by setting the gradient of the mapping function to unity and finding an initial value for b that causes $f_{hi}(\mathbf{x}_{hi})_{x_{hi}=2.7} =$

Algorithm 7.2: Outline of a local space mapping based optimization algorithm.

Require: $\mathbf{x}_{hi}^{(0)} \in \mathbb{R}^p$, $\delta_0 > 0$, $\mathbf{B}^{(0)} = \mathbf{I} \in \mathbb{R}^{p\times p}$, $\mathbf{p}^{(0)}(\mathbf{x}_{hi}^{(0)}) = \mathbf{x}_{hi}^{(0)}$, $k = 0$

1: **while** convergence criteria not met **do**

2:     $\mathbf{x}_{hi}^{(k+1)} = \arg\min_{\mathbf{x}_{hi}} f_{lo}(\mathbf{p}^{(k)}(\mathbf{x}_{hi}^{(k)}))$ subject to $||\mathbf{x}_{hi} - \mathbf{x}_{hi}^{(k)}|| \le \delta_k$

3:     evaluate $f_{hi}(\mathbf{x}_{hi})$

4:     perform a parameter estimation to obtain $\mathbf{p}^{(k+1)}(\mathbf{x}_{hi}^{(k+1)})$

5:     update δ, $\mathbf{B}$

6:     $k \leftarrow k + 1$

7: **end while**

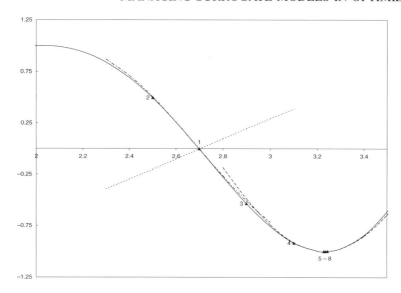

Figure 7.1 A one-dimensional space mapping (solid line f_{hi}, dotted line f_{lo}^1, chained line f_{lo}^2 and dashed line f_{lo}^8; triangles indicate where f_{hi} is evaluated).

$f_{lo}(x_{hi} + b)_{x_{hi}=2.7}$, here $b = -2.7034$. Figure 7.1 illustrates this problem – the dotted line shows the initial space mapped low-fidelity function with these parameter values, and obviously, it is a poor predictor. We next seek the minimum of this function subject to a move limit, here taken to be fixed at 0.2, and so the next iterate at which we evaluate the high-fidelity model is $x_{hi} = 2.5$. Given the value of f_{hi} at this point, it is possible to provide an estimate for the gradient of the space mapping and to construct a second surrogate; here, we get $p(x_{hi}) = -2.65x_{hi} + 7.1516$ (notice that the parameter estimation process is trivial since we can directly invert the low-fidelity function). This gives the chained line in Figure 7.1. which is clearly much improved. Minimizing the surrogate and staying within the same move limit leads to the next iterate of $x_{hi} = 2.9$. The parameters of the space mapping are then based on the functions values of x_{hi} at 2.7 and 2.9 and a further improvement in the values for a and b can be obtained. This process can be continued until at iteration eight the process converges with the final model (shown dotted in the figure) of $p(x_{hi}) = -3.2288x_{hi} + 0.5706$.

7.2.2 Global Space Mapping

To provide a space mapping that works globally requires a fully flexible parameterized model structure for the mapping function $p(x_{hi})$. To do this, one could employ any black-box surrogate modeling approach such as those discussed in Chapter 5. In the space mapping literature, a feed-forward neural network model with p inputs and p outputs is often used to parameterize the mapping function (Bandler et al. 1999). Let α denote the vector composed of the undetermined parameters in the specified model structure for the mapping, that is, $p(x_{hi}) \equiv p(x_{hi}, \alpha)$. Further, let $x_{hi}^{(i)}, i = 1, 2, \ldots$ denote the points at which we have evaluated the fine model. Then, α can be estimated by solving the following nonlinear optimization problem:

$$\min_{\boldsymbol{\alpha}} \sum_i \left(f_{\text{hi}}(\mathbf{x}_{\text{hi}}^{(i)}) - f_{\text{lo}}(\mathfrak{p}(\mathbf{x}_{\text{hi}}^{(i)}, \boldsymbol{\alpha})) \right)^2 . \qquad (7.16)$$

Since the above objective function is the empirical loss function, it is a good practice to append a regularization term to ensure that the estimated mapping function leads to good predictions at new points. The Huber norm is often employed instead of the least-squares error measure to ensure good generalization (Bandler et al. 1999). In a similar fashion to the local approach, the corrected model resulting from the global space mapping approach can be used as a surrogate for accelerating optimization algorithms.

7.3   Surrogate-assisted Optimization using Global Models

In this section, we revisit the issue of how global black-box surrogate models can be applied to optimize computationally expensive functions. A major attraction of the approaches presented is that the sensitivities of the high-fidelity model outputs are not required – this is in contrast to trust-region methods where sensitivity information is required to satisfy the zero- and first-order consistency conditions (thereby theoretically guaranteeing global convergence). Our primary focus in this section is on statistical approaches which can be used when Gaussian process models (also known as Kriging and DACE models) are employed as surrogates in numerical optimization. Before focusing exclusively on Gaussian process models, however, we first outline the basic steps involved in stage-wise unconstrained optimization of computationally expensive functions using global surrogates:

1. generate an initial design matrix using a space-filling Design of Experiments (DoE) technique and run the high-fidelity model at these points to create the observational (training) dataset $\mathcal{D}^0 \equiv \{\mathbf{x}^{(i)}, f(\mathbf{x}^{(i)})\}, i = 1, 2, \ldots, n_0$;

2. construct a surrogate model $\widehat{f}(\mathbf{x})$ using $\mathcal{D}^0$;

3. apply an optimization algorithm or search technique to $\widehat{f}(\mathbf{x})$ to locate promising design points;

4. verify the performance of design points identified in the previous step by running the high-fidelity model $f(\mathbf{x})$; and finally,

5. check if stopping criteria are met – if they are met stop, else refine the dataset $\mathcal{D}^0$ with the additional points generated at step 3 and go back to step 2.

The construction of a good global surrogate model requires an experimental design that is space filling so as to capture the essential trends of the function being modeled. Hence, new points generated in a stage-wise fashion should be placed in those regions of the design space that have not yet been fully explored in order to improve the quality of the surrogate.[6] In contrast, the goal of optimization is to generate points that lead to improvements in the objective function, that is, new points should be placed only where high-quality performance is predicted. Hence, when developing a strategy for combining

---

[6]In the machine learning literature, the problem of selecting data points to improve an existing model is referred to as *active learning*; see, for example, MacKay (1992).

global surrogate models and DoE techniques with optimization algorithms, it is necessary to balance the conflicting concerns of good experimental design (exploration) with those of optimization (exploitation).

To make this idea more concrete, take the case when we wish to solve the unconstrained minimization problem: min $f(\mathbf{x})$. Then, from the perspective of optimization, that is, to find good solutions, it is obvious that we should minimize $\widehat{f}(\mathbf{x})$. However, from the perspective of DoE, minimization of $\widehat{f}(\mathbf{x})$ will not ensure that the iterates generated will improve the accuracy of the baseline surrogate over the active range of $\mathbf{x}$. Torczon and Trosset (1998) proposed an approach where the conflicting concerns arising from optimization and DoE are accommodated by minimizing the aggregate *merit function*

$$\widehat{f}(\mathbf{x}) - \rho d_{\min}(\mathbf{x}), \quad \text{where} \quad d_{\min}(\mathbf{x}) = \min_i ||\mathbf{x} - \mathbf{x}^{(i)}||_2 \quad \text{and} \quad \rho \geq 0. \qquad (7.17)$$

The first term in the merit function ensures that $\widehat{f}(\mathbf{x})$ is minimized whereas the second term ensures that the distance of the new iterate to all the points in the baseline training dataset is maximized.[7] The second term in (7.17) can be interpreted as an adaptation of the *maximin* criterion used in DoE to generate space-filling designs (Johnson et al. 1990). ρ is a user-defined parameter that has to be appropriately selected since the orders of magnitude of $\widehat{f}(\mathbf{x})$ and $d_{\min}(\mathbf{x})$ can be different unless these terms are suitably normalized. The value of ρ also dictates how much emphasis is placed on converging to a minimum of $\widehat{f}(\mathbf{x})$ versus learning more about the underlying function in areas of the design space that have not yet been explored. In practice, it is preferable to start with a large value of ρ and reduce it across the optimization iterations to ensure eventual convergence to a minimum of the original high-fidelity function.

It is worth noting here that the approach based on minimizing the aggregate merit function in (7.17) can be applied irrespective of the type of global surrogate modeling strategy being used; for example, polynomial response surface models, RBF approximations or Gaussian process modeling approaches could be used. We now move on to describe statistical criteria that can be used to balance the concerns of DoE with those of optimization when Gaussian process models are used as surrogates.[8] The key ideas are first illustrated for unconstrained problems and subsequently extended to problems with general nonlinear constraints.

7.3.1 The Expected Improvement Criterion

Suppose that an initial set of training data is made available by running the high-fidelity model at the points generated by a space-filling DoE technique, that is, $\mathcal{D}^0 \equiv \{\mathbf{x}^{(i)}, f(\mathbf{x}^{(i)})\}$, $i = 1, 2, \ldots, n_0$. This dataset can be used to construct a Gaussian process model to approximate the input–output relationship (see Section 5.6 for details). Recollect that, in this approach, we make predictions at a new point not contained in the training dataset using the posterior (Gaussian) distribution $F(\mathbf{x}) \sim \mathcal{N}(\widehat{f}(\mathbf{x}), \sigma^2(\mathbf{x}))$. The posterior mean $\widehat{f}(\mathbf{x})$ can be used as a surrogate for the original high-fidelity objective function $f(\mathbf{x})$, while the posterior variance $\sigma^2(\mathbf{x})$ gives an estimate of the uncertainty involved in making predictions using a finite set of input–output data.

---

[7]It can be noted from (7.17) that minimization of $-d_{\min}(\mathbf{x})$ is equivalent to maximizing the minimum distance between $\mathbf{x}$ and the points in the training dataset. Hence, this term ensures that the iterates will not be closely clustered in the design space.

[8]Similar statistical criteria can be constructed for RBF models; see Sóbester et al. (2005).

From the viewpoint of DoE, to improve the accuracy of the baseline surrogate, we should augment the dataset $\mathcal{D}^0$ with additional points, say $\tilde{\mathbf{x}}$, at which the posterior variance is high, that is, by solving the optimization problem

$$\tilde{\mathbf{x}} = \arg \max_{\mathbf{x}} \sigma^2(\mathbf{x}). \qquad (7.18)$$

However, from the perspective of finding iterates that lead to reductions in the objective function, we need to minimize $\widehat{f}(\mathbf{x})$. The expected improvement criterion (Jones et al. 1998; Schonlau 1997) discussed next enables the posterior mean and variance of the Gaussian process model to be simultaneously used to balance the concerns of optimization with those of DoE.

Let $f_{\min} = \min(f^{(1)}, f^{(2)}, \ldots, f^{(n_0)})$ denote the objective function value of the best point in the baseline training dataset. Then, the improvement we are likely to make at a new point, say $\mathbf{x}$, is given by $I(\mathbf{x}) = \max(f_{\min} - F(\mathbf{x}), 0)$. Since $F(\mathbf{x})$ is a random variable, given $\mathbf{x}$, we can define the expected improvement as

$$\langle I(\mathbf{x}) \rangle = \langle \max(f_{\min} - F(\mathbf{x}), 0) \rangle, \qquad (7.19)$$

where $\langle \cdot \rangle$ denotes the expectation operator.[9] Using the fact that $F(\mathbf{x})$ is Gaussian, the expected improvement can be written in closed form as

$$\langle I(\mathbf{x}) \rangle = (f_{\min} - \widehat{f}(\mathbf{x})) \Phi\left(\frac{f_{\min} - \widehat{f}(\mathbf{x})}{\sigma(\mathbf{x})}\right) + \sigma(\mathbf{x}) \phi\left(\frac{f_{\min} - \widehat{f}(\mathbf{x})}{\sigma(\mathbf{x})}\right), \qquad (7.20)$$

where $\Phi()$ and $\phi()$ denote the standard normal density and distribution functions, respectively.[10] It may be noted that since the model interpolates the data $I(\mathbf{x}) = 0$ at all the points in the training dataset.

In summary, to minimize the original objective function $f(\mathbf{x})$, we first construct a baseline Gaussian process model using n_0 points generated by applying a space-filling DoE technique. Subsequently, a new iterate is generated by maximizing the expected improvement criterion, that is,

$$\mathbf{x}^{(k+1)} = \arg \max_{\mathbf{x}} \langle I(\mathbf{x}) \rangle. \qquad (7.21)$$

The high-fidelity model is then evaluated at $\mathbf{x}^{(k+1)}$ and the result $(\mathbf{x}^{(k+1)}, f(\mathbf{x}^{(k+1)}))$ is augmented to the baseline training dataset, that is, $\mathcal{D}^{k+1} \leftarrow \mathcal{D}^k \cup (\mathbf{x}^{(k+1)}, y(\mathbf{x}^{(k+1)}))$. The augmented dataset is used to update the surrogate model; a discussion on how this can be achieved efficiently is presented later in Section 7.3.5. The updated surrogate is then used to solve (7.21) for the next iterate. This iterative process is continued until specified convergence criteria are met (for example, when the expected improvement is less than 1% of best function value obtained so far).

Figure 7.2 depicts the iterates obtained by maximizing the expected improvement criterion for a one-dimensional test function. Here, the baseline training dataset consists of three points and the figure shows the locations of the next seven iterates. It can be seen that the

---

[9]If f is a function of random variables, say θ, then $\langle f(\theta) \rangle = \int f(\theta) \mathcal{P}(\theta) \, d\theta$, where $\mathcal{P}(\theta)$ is the joint probability density function of θ. Hence $\langle I(\mathbf{x}) \rangle = \int \max(f_{\min} - F(\mathbf{x}), 0) \mathcal{P}(F) \, dF$, where $\mathcal{P}(F)$ is a Gaussian distribution with mean $\widehat{f}$ and variance σ^2.

[10]Note that $\langle I(\mathbf{x}) \rangle$ can be computed using the erf function, which is available in many numerical computation libraries; see, for example, http://www.netlib.org

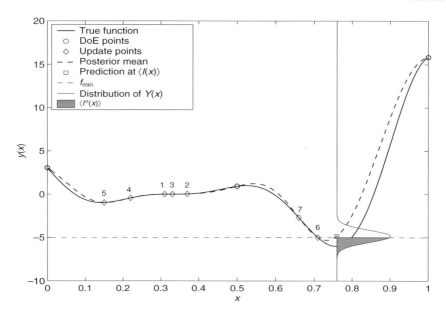

Figure 7.2 Update process for a simple test function driven by the expected improvement criterion, also illustrating the probability of improvement. Reproduced from (Forrester 2004) by permission of Alexander I. S. Forrester.

first five iterates converge toward a local optima. However, the subsequent two iterates lead the search to the vicinity of the global optimum. The iterate obtained at the eighth update, for which the posterior distribution is depicted, is the global optimum of this test function.

In some design problems, particularly those involving manipulation of geometrical design variables, it is possible that the high-fidelity model may crash (fail) at certain points in the design space (Booker et al. 1998). Even though a sensible geometry parameterization scheme can alleviate such problems, this situation can be almost inevitable in many design problems. Hence, some precautionary measures are required to handle the situation when the design point obtained by maximizing $\langle I(\mathbf{x}) \rangle$, say $\mathbf{x}^{(k+1)}$, leads to a failure of the simulator to return the objective function value. Forrester (2004) suggests that in such situations, a *dummy* point should be created as $(\mathbf{x}^{(k+1)}, \widehat{f}(\mathbf{x}^{(k+1)}))$ and this point should be augmented to the training dataset, that is, a point based on the existing surrogate rather than the true function. This strategy will ensure that $\langle I(\mathbf{x}^{(k+1)}) \rangle = 0$ and hence the search process will not revisit this point during subsequent iterations.

Jones et al. (1998) conjecture that, for continuous functions, algorithms based on maximization of the expected improvement criterion converge to the global optimum. This is based on the argument that if the number of updates $k \to \infty$ then the design space will be sampled at all locations and hence the global optima will eventually be located. However, in practice, because of an inevitably finite computational budget, the iterations are usually terminated when the number of points in the training dataset is greater than say a few thousand (or even a few hundred). Further, the computational cost of updating the surrogate model and the memory requirements grow with the number of training points as $\mathcal{O}(n^3)$ and $\mathcal{O}(n^2)$, respectively. Care also needs to be exercised to circumvent numerical problems that may arise due to ill conditioning of the correlation matrix — this may happen when the iterate $\mathbf{x}^{(k+1)}$ generated by solving (7.21) is close to any of the points in $\mathcal{D}^k$ in the sense of the Euclidean distance measure. One way to avoid this problem is to use a Gaussian process regression model instead of an interpolator. Another point worth noting is that the landscape of the expected improvement criterion can be highly multimodal. Hence, stochastic search techniques may be required to locate the maxima. Alternatively, the branch-and-bound algorithm which uses bounds on $\langle I(\mathbf{x}) \rangle$ can be employed (Jones et al. 1998).

7.3.2 The Generalized Expected Improvement Criterion

The expected improvement criterion works well provided the size of the initial DoE is sufficient to reliably estimate the hyperparameters of the Gaussian process model. However, since we only maximize the expectation (first moment) of the improvement, which corresponds to an average case analysis, the search may be too local. Schonlau (1997) proposed a generalized improvement criterion to circumvent this problem. To illustrate this idea, let us write the improvement we are likely to make at a new point $\mathbf{x}$ in terms of an integer l as follows:

$$I^l(\mathbf{x}) = \begin{cases} \left[f_{\min} - F(\mathbf{x}) \right]^l & \text{if } F(\mathbf{x}) < f_{\min} \\ 0 & \text{otherwise.} \end{cases} \tag{7.22}$$

Clearly, for $l = 1$, applying the expectation operation to (7.22), we recover the expected improvement criterion defined earlier in (7.20). Let us now apply the expectation operation to (7.22) for $l = 0$, which gives

$$\langle I^0(\mathbf{x}) \rangle = P\left[F(\mathbf{x}) < f_{\min} \right] = P\left[\frac{F(\mathbf{x}) - \widehat{f}(\mathbf{x})}{\sigma(\mathbf{x})} < \frac{f_{\min} - \widehat{f}(\mathbf{x})}{\sigma(\mathbf{x})} \right]$$

$$= \Phi\left(\frac{f_{\min} - \widehat{f}(\mathbf{x})}{\sigma(\mathbf{x})} \right). \tag{7.23}$$

Hence, for $l = 0$, the generalized expected improvement criterion reduces to the probability of improvement criterion, that is, $\langle I^0(\mathbf{x}) \rangle = P[F(\mathbf{x}) < f_{\min}]$.

For $l = 1, 2, \ldots$, Schonlau (1997) showed that the generalized expected improvement can be written as

$$\langle I^l(\mathbf{x}) \rangle = \sigma^l \sum_{k=0}^{l} (-1)^k \binom{l}{k} u^{l-k} T_k, \tag{7.24}$$

where $T_0 = \Phi(u)$, $T_1 = -\phi(u)$ and T_k for $k > 1$ can be computed recursively as $T_k = -u^{k-1}\phi(u) + (k - 1)T_{k-2}$, and $u = (f_{\min} - \widehat{f}(\mathbf{x}))/\sigma(\mathbf{x})$.

Note that the integer l is used to control the exploratory nature of the search process, that is, larger values of l make the search more global.

7.3.3 The Weighted Expected Improvement Criterion

Sóbester et al. (2005) introduced the weighted expected improvement criterion, which can be defined as

$$\langle I_w(\mathbf{x}) \rangle = \begin{cases} w(f_{\min} - \widehat{f}(\mathbf{x}))\Phi\left(\dfrac{f_{\min} - \widehat{f}(\mathbf{x})}{\sigma(\mathbf{x})} \right) + (1 - w)\sigma(\mathbf{x})\phi\left(\dfrac{f_{\min} - \widehat{f}(\mathbf{x})}{\sigma(\mathbf{x})} \right), & \sigma^2 > 0 \\ 0, & \sigma^2 = 0 \end{cases}$$

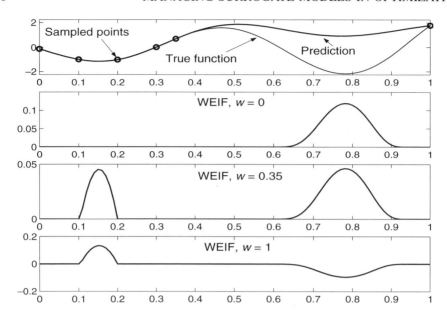

Figure 7.3 Influence of weighting parameters on the expected improvement criterion (Sóbester 2004).

where $w \in [0, 1]$ is a weighting factor that can be used to trade-off local exploitation with global exploration, that is, $w = 1$ emphasizes local exploitation whereas $w = 0$ emphasizes global exploration. Note that $w = 0.5$ gives $\langle I_w(\mathbf{x}) \rangle = 0.5 \langle I(\mathbf{x}) \rangle$. The influence of the weight parameter w is graphically depicted in Figure 7.3 for a one-dimensional test function. Detailed numerical studies on this criterion including strategies for adapting the parameter w across the optimization iterations can be found in the dissertations of Sóbester (2004) and Forrester (2004).

7.3.4 Extension to Constrained Problems

A straightforward way to extend these approaches to constrained problems would be to employ a penalty function approach, for example, the L_1 penalty function used earlier in the context of trust-region methods in Section 7.1.2. We describe here an alternative approach in which the generalized expected improvement criterion in (7.22) is rewritten by taking the constraints into account (Schonlau 1997). To illustrate, consider a nonlinear programming problem where we wish to minimize $f(\mathbf{x})$ subject to inequality constraints of the form $g_i(\mathbf{x}) \leq 0$, $i = 1, 2, \ldots, m$. As before, let us denote the Gaussian process surrogate model for the objective function by $F(\mathbf{x}) \sim \mathcal{N}(\widehat{f}(\mathbf{x}), \sigma^2(\mathbf{x}))$. Further, let the Gaussian process surrogate model for the ith constraint be given by $G_i(\mathbf{x}) \sim \mathcal{N}(\widehat{g_i}(\mathbf{x}), \sigma_i^2(\mathbf{x}))$. In other words, we assume that separate Gaussian process models for the objective and constraint functions are constructed using a baseline training dataset. Then, the generalized expected improvement subject to the constraints can be defined as

$$I_c^l(\mathbf{x}) = \begin{cases} \left[f_{\min} - F(\mathbf{x}) \right]^l & \text{if } F(\mathbf{x}) < f_{\min} \quad \text{and} \quad G_i(\mathbf{x}) \leq 0 \text{ for } i = 1, \ldots, m \\ 0 & \text{otherwise.} \end{cases} \tag{7.25}$$

Here, f_{min} is defined as the minimum *feasible* value of the objective among all the points in the training dataset.

The expected value of I_c^l can be written as a multidimensional integral of the form

$$\langle I_c^l(\mathbf{x}) \rangle = \int \cdots \int (f_{min} - F(\mathbf{x}))^l \mathcal{P}(F, G_1, G_2, \ldots, G_m) \, \mathrm{d}F \, \mathrm{d}G_1 \, \mathrm{d}G_2 \cdots \mathrm{d}G_m, \quad (7.26)$$

where $\mathcal{P}(F, G_1, G_2, \ldots, G_m)$ denotes the joint probability density function of F, G_1, $G_2, \ldots, G_m$. Even though from the posterior distributions we know the mean and variance of these quantities, we have no information about correlations between them. Hence, to proceed further, we invoke the assumption that the posterior distributions for the objective and constraint functions are statistically independent. Under this assumption, the preceding multidimensional integral simplifies to

$$\langle I_c^l(\mathbf{x}) \rangle = \langle I^l(\mathbf{x}) \rangle P[G_1 \leq 0] P[G_2 \leq 0] \cdots P[G_m \leq 0]. \quad (7.27)$$

In other words, the generalized expected improvement defined earlier in (7.22) for the unconstrained case is multiplied by the probability that each constraint is satisfied.[11] The probability of constraint satisfaction can be easily computed since, given $\mathbf{x}$, G_i is a Gaussian random variable with known mean and variance, that is, $P[G_i \leq 0] = \Phi(\widehat{g}_i(\mathbf{x})/\sigma_i(\mathbf{x}))$.

7.3.5 Correlation Matrix Updating

When the training dataset is augmented, it is possible to efficiently update the Gaussian process model in an incremental fashion. To illustrate, let the Cholesky factorization of the correlation matrix at iteration k be given by $\mathbf{R}^k = \mathbf{L}^k (\mathbf{L}^k)^T \in \mathbb{R}^{n_k \times n_k}$, where $\mathbf{L}^k \in \mathbb{R}^{n_k \times n_k}$ is a lower triangular matrix and n_k is the number of training points used to construct the surrogate at the kth iteration. If one additional data point, say $\mathbf{x}^{(n_k+1)}$, is added to the training dataset, then the updated correlation matrix can be written as

$$\mathbf{R}^{k+1} = \begin{bmatrix} \mathbf{R}^k & \mathbf{r} \\ \mathbf{r}^T & R(\mathbf{x}^{(n_k+1)}, \mathbf{x}^{(n_k+1)}) \end{bmatrix} \in \mathbb{R}^{(n_k+1) \times (n_k+1)}, \quad (7.28)$$

where $\mathbf{r} \in \mathbb{R}^{n_k}$ is the correlation between the new point and all the points in the training dataset and $R(\cdot, \cdot)$ is the correlation function.

The Cholesky factorization of $\mathbf{R}^{k+1}$ can be written as $\mathbf{L}^{k+1} (\mathbf{L}^{k+1})^T$, where

$$\mathbf{L}^{k+1} = \begin{bmatrix} \mathbf{L}^k & 0 \\ \mathbf{z}^T & d \end{bmatrix}. \quad (7.29)$$

$\mathbf{z} \in \mathbb{R}^{n_k}$ can be computed by solving the lower triangular system of equations $\mathbf{L}^k \mathbf{z} = \mathbf{r}$ and $d = \sqrt{R(\mathbf{x}^{(k+1)}, \mathbf{x}^{(k+1)}) - \mathbf{z}^T \mathbf{z}}$.

Since the major computational cost in updating the Cholesky factorization of the correlation matrix arises from the solution of a lower triangular system of equations, only $\mathcal{O}(n_k^2/2)$ operations are involved. Clearly, this update procedure is useful only when the hyperparameters of the Gaussian process model are not being updated. Managing the update cost in global approximation schemes of this kind becomes an important concern as the number

---

[11] As noted by Sasena et al. (2002), by maximizing $\prod_{i=1}^{m} P[G_i \leq 0]$, it is possible to generate an initial feasible design point even when none of the points in the training dataset used to construct the surrogate are feasible. This step may be required in some cases to determine f_{min} since in (7.25) we implicitly assume that at least one point in the training dataset is feasible.

of data points rises. In many cases, it is computationally more efficient to identify multiple updates when searching the improvement functions and compute the equivalent true function values in parallel to produce an effective process. Such multiple updates can then be added to the surrogate in one go and the hyperparameters retuned only once. Note that the approach outlined here can also be applied to update RBF models when new points are appended to the training dataset.

7.4   Managing Surrogate Models in Evolutionary Algorithms

In this section, we present strategies for managing surrogate models in EAs such as those discussed earlier in Section 3.2. Since EAs make use of probabilistic recombination operators, controlling the step size of design changes (to control the accuracy of approximate fitness predictions) is not as straightforward as in gradient-based optimization algorithms. This difficulty becomes particularly severe when local approximation models with a limited range of validity are employed during evolutionary search. In principle, global surrogate models could be employed to circumvent this problem. However, in practice, because of the curse of dimensionality, global models become increasingly difficult to construct for problems with large numbers of variables.

In what follows, we first describe some strategies for employing surrogate models in conjunction with standard EAs. Subsequently, we present a hybrid EA that makes use of local surrogate models and an SQP optimizer employing the trust-region framework to manage the interplay between the surrogate and the high-fidelity model (Ong et al. 2003). Numerical studies on two test functions are carried out to illustrate the performance of the algorithms described. We conclude with an outline of ongoing research work focusing on surrogate-assisted EAs.

7.4.1   Using Global Surrogate Models in Standard EAs

A template of an EA employing global surrogate models based on the approach suggested by Ratle (2001) is shown in Algorithm 7.3. The basic idea is to run the EA on the surrogate model with recourse made to the high-fidelity model periodically to generate additional data points for updating the surrogate. Clearly, for this approach to work well, precautions have to be taken to ensure sufficient diversity in the population. Some studies on test functions to evaluate the performance of this strategy are carried out later on in this section.

Many variants on this basic idea are possible. For example, consider the case when a Gaussian process surrogate model is used. Here, for each individual in an EA generation, the posterior variance can be calculated using a Gaussian process model. On the basis of this value, which provides an error bar on the approximation, a decision can be made as to whether the fitness of the individual under consideration should be evaluated using the surrogate or the original high-fidelity model. Another possibility is to employ statistical measures such as the expected improvement criteria discussed earlier in Section 7.3 as the fitness function driving evolutionary search. This idea makes sense since these surfaces can sometimes be more multimodal than the original objective function. The idea of using global physics-based approximations (reduced basis methods) to accelerate EAs was studied by Nair et al. (1998b) and Nair and Keane (2001). For more general approximation models, Ratle (2001) examined a strategy for integrating evolutionary search with Kriging models.

Algorithm 7.3: Template of a surrogate-assisted evolutionary optimization strategy based on global surrogate models.

Require: A database containing a population of designs (*Optional: upload a historical database if one exists*)

1: Construct surrogate model $\widehat{f}(\mathbf{x})$ using all points in the database.
2: Set fitness function := surrogate model
3: **while** computational budget not exhausted **do**
4:     **if** fitness function := high-fidelity model **then**
5:         Evaluate fitness of all individuals using high-fidelity model
6:         Update database with any new designs generated using the exact model
7:         Update surrogate model using all designs in the database
8:     **else**
9:         Evaluate fitness of all individuals using surrogate model
10:    **end if**
11:    **if** search over $\widehat{f}(\mathbf{x})$ has converged **then**
12:        fitness function := high-fidelity model
13:    **else**
14:        fitness function := surrogate model
15:    **end if**
16:    Apply standard EA operators to create a new population
17: **end while**

This problem was studied by El-Beltagy et al. (1999), where it is argued that the issue of balancing the concerns of optimization with that of DoE/exploration must be addressed. Numerical studies were presented for certain pathological cases to show that the idea of constructing an accurate global surrogate model is fraught with difficulties due to the curse of dimensionality. Liang et al. (2000) proposed a strategy for coupling EAs with local search and quadratic response surface methods. However, when working with multimodal problems, the accuracy of quadratic models may become questionable. Jin et al. (2003) presented a framework for coupling EAs and neural network-based surrogate models. This approach uses both the expensive and approximate models throughout the search, with an empirical criterion to decide the frequency at which each model should be used. Karakasis et al. (2003) discuss the application of RBF approximations with EAs to aerodynamic design.

Computational experience accumulated over studies in the literature (see also Section 7.4.3) suggests that when global surrogates are used in conjunction with standard EAs convergence to a false optima is very likely for problems with a large number of variables due to inaccuracies in the surrogate model. Nonetheless, they may still be of use in design studies where any improvement is worthwhile and a scheme that is simple to implement is required. An alternative approach is to adopt a local surrogate-assisted hybrid EA, based on a combination of evolutionary search with gradient-based optimization. This circumvents some of the limitations associated with global surrogates.

7.4.2   Local Surrogate-assisted Hybrid EAs

Hybrid EAs use a combination of global and local search to balance the goals of exploring and exploiting promising regions of the design space. We discuss here a surrogate-assisted hybrid EA that employs a feasible SQP optimization algorithm for local search – in this approach, each individual in the EA population is used as a local search starting point. This makes it easier to design surrogate-assisted EAs since we can employ a trust-region framework for interleaving use of the exact models for the objective and constraint functions with computational cheap surrogate models during local search.

In the approach described, local surrogate models are constructed using data points that lie in the vicinity of an initial guess. This local learning technique is an instance of the transductive inference paradigm, which has been the focus of recent research in statistical learning theory (Chapelle et al. 1999; Vapnik 1998). Traditionally, surrogate models are constructed using inductive inference, which involves using a training dataset to estimate a functional dependency and then using the computed model to predict the outputs at the points of interest. However, when constructing surrogate models for optimization, we are specifically interested in ensuring that the models predict the objective and constraint function values accurately at the sequence of iterates generated during the search – how well the model performs at other points in the design space is of no concern in this specific context. Transductive inference offers an elegant solution to this problem by directly estimating the outputs at the point of interest in one step; the reader is referred to Chapter 8 of Vapnik (1998) for a detailed theoretical analysis of its superior generalization capabilities over standard inductive inference.

Transduction can be implemented, for example, by constructing RBF approximations using data points in the local neighborhood of an optimization iterate. In other words, instead of constructing a global surrogate model, a local model is created on the fly whenever the objective and constraint functions must be estimated at a design point during local search. This idea of constructing local models is similar in spirit to trust-region searches and also the moving least-squares approximation technique (Levin 1998). Localized training data can be readily selected from a search engine database containing previous iterates, which is continuously updated as the search progresses. Further, the algorithm outlined here can be efficiently parallelized on grid computing architectures since it does not compromise on the intrinsic parallelism offered by EAs.

The basic steps of the search strategy are outlined in Algorithm 7.4. In the first step, we initialize a database using a population of designs, either randomly or using DoE techniques such as Latin hypercube sampling. All the design points thus generated and the associated exact values of the objective and constraint functions are archived in the database that will be used later for constructing local surrogate models. Alternatively, one could use a database containing the results of a previous search on the problem or a combination of the two.

Subsequently, a hybrid EA is employed, where for each nonduplicated design point or chromosome in the population, a local search is conducted using surrogates. Each individual in an EA generation is used as an initial guess for local search in the spirit of Lamarckian learning. A trust-region framework is employed to manage the interplay between the original objective and constraint functions and computationally cheap surrogate models during the local search. The local search strategy demonstrated here embeds the feasible SQP optimizer within a trust-region framework. However, instead of adopting an augmented Lagrangian approach, the objective and constraint functions are handled separately using the approach of Giunta and Eldred (2000). More specifically, during local search for each chromosome in an EA generation, we solve a sequence of trust-region subproblems of the form

$$\text{Minimize}: \quad \widehat{f}^k(\mathbf{x} + \mathbf{x}_c^k)$$

$$\text{Subject to}: \quad \widehat{g}_i^k(\mathbf{x} + \mathbf{x}_c^k) \leq 0, i = 1, 2, \ldots, m \tag{7.30}$$
$$||\mathbf{x}||_\infty \leq \delta_k$$

Algorithm 7.4: Template of an algorithm for integrating local surrogate models with hybrid EAs for optimization of computationally expensive problems.

Require: A database containing a population of designs (*Optional: upload a historical database if one exists*)

    while computational budget not exhausted **do**

        Evaluate all individuals in the population using the high-fidelity fitness function.

        for Each non-duplicated individual in the population **do**

            Apply trust-region enabled feasible SQP solver to each individual in the population by interleaving the exact and local surrogate models for the objective and constraint functions.

            Update the database with any new design points generated during the trust-region iterations and their exact objective and constraint function values.

            Replace the individuals in the population with the locally improved solution in the spirit of Lamarckian learning.

        end for

        Apply standard EA operators to create a new population.

    end while

---

where $k = 0, 1, 2, \ldots, k_{\max}$. $\mathbf{x}_c^k$ and δ_k are the starting point and the trust-region radius used for local search at iteration k, respectively.

For each subproblem (i.e., during each trust-region iteration), surrogate models of the objective and constraint functions, namely, $\widehat{f}^k(\mathbf{x})$ and $\widehat{g}_i^k(\mathbf{x})$ are created dynamically. The n nearest neighbors of the initial guess, $\mathbf{x}_c^k$, are first extracted from the archived database of design points evaluated so far using the exact analysis code. The criterion used to determine the similarity between design points is the simple Euclidean distance metric. These points are then used to construct local surrogate models of the objective and constraint functions. It is worth noting here that care has to be taken to ensure that repetitions do not occur in the training dataset, since this may lead to a singular Gram matrix.

The surrogate models thus created are used to facilitate the necessary objective and constraint function estimations in the local searches. During local search, we initialize the trust region δ using the minimum and maximum values of the design points used to construct the surrogate model. We find this initialization strategy to work well for the problems we have worked with in the past. After each iteration, the trust-region radius δ_k is updated on the basis of a measure that indicates the accuracy of the surrogate model at the kth local optimum, $\mathbf{x}^{k*}$. After computing the exact values of the objective and constraint functions at this point, the figure of merit, ρ^k, is calculated as

$$\rho^k = \min(\rho_f^k, \rho_{g_i}^k), \text{ for } i = 1, 2, \ldots, p \tag{7.31}$$

where

$$\rho_f^k = \frac{f(\mathbf{x}_c^k) - f(\mathbf{x}^{k*})}{\widehat{f}(\mathbf{x}_c^k) - \widehat{f}(\mathbf{x}^{k*})} \text{ and } \rho_{g_i}^k = \frac{g_i(\mathbf{x}_c^k) - g_i(\mathbf{x}^{k*})}{\widehat{g}_i(\mathbf{x}_c^k) - \widehat{g}_i(\mathbf{x}^{k*})}. \tag{7.32}$$

The above equations provide a measure of the actual versus predicted change in the objective and constraint function values at the kth local optimum. The value of ρ^k is then used to update the trust-region radius using steps (5-20) outlined earlier in Algorithm 7.1.

The exact values of the objective and constraint functions at the optimal solution of the kth subproblem are combined with the n nearest neighboring design points to generate a new surrogate model for the next iteration. In addition, the initial guess for the $(k + 1)$th iteration within each local search is determined by

$$\mathbf{x}_c^{k+1} = \mathbf{x}^{k*}, \ \text{if} \ \rho^k > 0$$

$$= \mathbf{x}_c^k, \ \text{if} \ \rho^k \leq 0. \tag{7.33}$$

The trust-region iterations (for each chromosome) are terminated when $k \geq k_{max}$, where k_{max} is the maximum number of trust-region iterations set *a priori* by the user. At the end of k_{max} trust-region iterations for a chromosome, the exact fitness of the locally optimized design point is determined. Provided it is found to be better than that of the initial guess, then Lamarckian learning proceeds. Lamarckian learning forces the genotype to reflect the result of improvement by placing the locally improved individual back into the population to compete for reproductive opportunities. In addition, the locally optimized design point and its corresponding objective and constraint function values are added to the database. This process of hybrid EA search is continued until the computational budget is exhausted or a user-specified termination criterion is met.

Note that apart from the parameters used in standard EAs the surrogate-assisted search algorithm has only two additional user-specified parameters, k_{max} and n. Later in this section, we present experimental studies to investigate the effect of these parameters on convergence trends. Moreover, in the local surrogate-based EA, it is relatively straightforward to achieve parallelism, since local search for each individual in an EA generation can be conducted independently. To ensure load balancing, we only need to specify that the number of trust-region iterations be kept the same for each individual, since we evaluate the expensive code at most once per iteration.

7.4.3 Numerical Studies on Test Functions

To illustrate these ideas, we present some brief numerical results obtained from implementing various surrogate-assisted optimization strategies within a standard binary coded genetic algorithm (GA). Here, we have employed a population size of 50, uniform crossover and mutation operators at probabilities 0.6 and 0.01, respectively. A linear ranking algorithm is used for selection. The codes implementing the objective and constraint functions were wrapped on NetSolve servers (Casanova and Dongarra 1998) running Red Hat Linux on Pentium processors. The GA code is linked to the NetSolve client library so that the objective function and constraint evaluation modules, and the local search routines can be invoked remotely.

The feasible SQP implementation used is the FFSQP code developed by Lawrence and Tits (2001). When started from an infeasible point (one that violates at least one of the linear or nonlinear constraints), FFSQP first carries out an optimization in which it minimizes the maximum of the constraint violations. Subsequently, it minimizes the objective function, while maintaining feasibility of the iterates.

A pair of benchmark problems commonly used in the global optimization literature are used to illustrate the behavior of the surrogate-assisted EAs. These tests show how the approaches discussed can reduce computational costs when used on more time consuming problems. The first example considered is unconstrained minimization of the Rastrigin function, defined as

$$\textbf{Minimize}: \ \ 10d + \sum_{i=1}^{d}(x_i^2 - 10\cos(2\pi x_i)) \tag{7.34}$$

$$\textbf{Subject to}: \ -5.12 \leq x_i \leq 5.12, i = 1, 2, \ldots, d.$$

The second problem considered is maximization of the bump test function repeated below and already described in Chapter 3. Its main purpose is to test how methods cope with optima that occur hard up against the constraint boundaries commonly found in engineering design.

$$\textbf{Maximize} : \text{abs} \left(\sum_{i=1}^{d} \cos^4(x_i) - 2 \prod_{i=1}^{d} \cos^2(x_i) \right) \Big/ \sqrt{\sum_{i=1}^{d} i x_i^2}$$

$$\textbf{Subject to} : \qquad \prod_{i=1}^{d} x_i > 0.75 \qquad\qquad (7.35)$$

$$\sum_{i=1}^{d} x_i < 15d/2$$

$$0 \le x_i \le 10, i = 1, 2, \ldots, d.$$

Figure 7.4 shows both the Rastrigin and bump functions for $d = 2$. Twenty-dimensional ($d = 20$) versions of the test functions are used for the numerical studies. The Rastrigin function has a unique global optimum at which the function value is zero. For the bump test function, even though the global optimum is not precisely known for $d = 20$, a value of 0.8 can be obtained after around 100,000 function evaluations using a conventional, directly applied GA.

A RBF approximation with linear splines was used to construct the local surrogate models used for all the numerical studies. The averaged convergence trends of the surrogate-assisted EAs as a function of the total number of function evaluations are shown in Figures 7.5, 7.6 and 7.7. The results presented have been averaged over 20 runs for each test function. Also shown in the figures are averaged convergence trends obtained using a standard GA and a global surrogate modeling strategy. The algorithm based on global surrogate models follows the approach outlined in Algorithm 7.3 and also uses a RBF surrogate model.

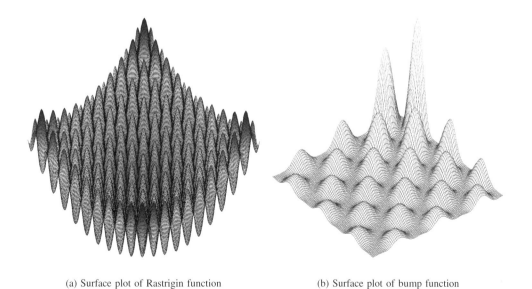

(a) Surface plot of Rastrigin function   (b) Surface plot of bump function

Figure 7.4 Two-dimensional versions of test functions used.

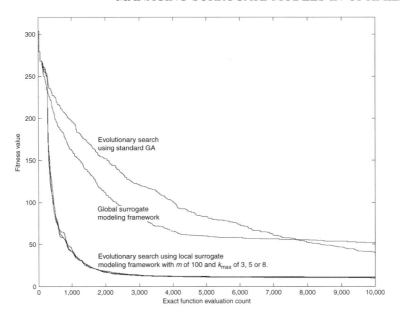

Figure 7.5   Averaged convergence trends for $n = 100$ and various values of k_{max} (3, 5 and 8) compared with standard GA and global surrogate modeling framework for the 20D Rastrigin function.

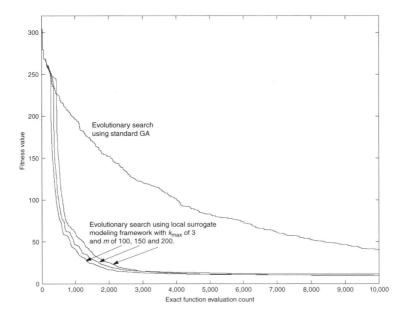

Figure 7.6   Averaged convergence trends for $k_{max} = 3$ and various values of n (100, 150 and 200) compared with standard GA for the 20D Rastrigin function.

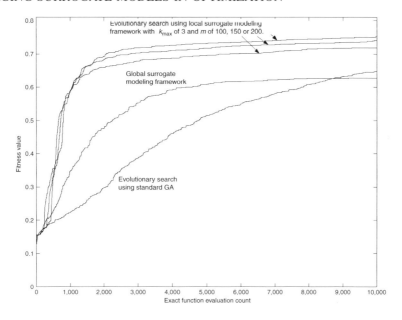

Figure 7.7 Averaged convergence trends for $k_{max} = 3$ and various values of n (100, 150 and 200) compared with standard GA for the 20D bump function.

The results obtained for the test functions show that the global surrogate framework displays early sign of stalling. This is consistent with previous studies in the literature (El-Beltagy et al. 1999; Ratle 2001), which suggest that when global surrogate models are applied to high-dimensional and multimodal test functions, the search generally tends to stall early on. Such an effect is a result of the curse of dimensionality, which often leads to early convergence at false global optima of the surrogate model. In contrast, the results obtained using the local surrogate modeling approach clearly demonstrate that solutions close to the global optima can be obtained on a limited computational budget. As surrogates are used only for local searches, that is, as the exact model is used for all analyzes conducted at the EA level, the chances for convergence to false global optima are greatly reduced. In addition, the use of the trust-region framework maintains convergence close to the local optima of the original problem during the SQP steps.

For these benchmark problems, it is also possible to study the effect of increasing the maximum number of trust-region iterations and the number of nearest neighbors (employed to construct the local surrogate model) on the convergence behavior; see again Figures 7.5, 7.6 and 7.7. At least two observations can be made from the convergence trends. First, it appears that there is not much to be gained by increasing k_{max} beyond three. Secondly, it appears that smaller values of n generally lead to faster convergence during the early stages of search, but there is a tendency to stall in later stages. This suggests the possibility of adaptively selecting n during the search. Setting n to be a fixed proportion of the available results in the growing database of computed results appears to work reasonably well for a wide class of problems. We note that Ong et al. (2004) proposed an extension of the local surrogate modeling approach described here, using Hermite RBF approximations to improve the convergence rate of the local surrogate-assisted search strategy. A hierarchical approach that combines global and surrogate models to accelerate EAs has been presented by Zhou et al. (2004).

7.5 Concluding Remarks

In this chapter, we have presented an outline of several strategies for integrating surrogate models in optimization algorithms. Surrogate-assisted optimization continues to be an area of active research and new approaches regularly appear in the open literature. For example, a novel approach was proposed by Forrester (2004) for a class of problems which employs iterative solvers for evaluating the objective and constraint functions. Surrogate-assisted optimization algorithms for multiobjective problems have also started to appear recently: see, for example, Emmerich and Naujoks (2004). Surrogate-assisted optimization algorithms can be also be applied to robust design – this topic is discussed further in the next chapter. A number of surrogate-assisted approaches are used in the case studies presented in Part IV.

8

Design in the Presence of Uncertainty

In this chapter, we focus on the design of engineering systems in the presence of uncertainty. The motivation for this arises from the fact that some degree of uncertainty in characterizing any real engineering system is inevitable. When carrying out analysis and design of a structural system, deterministic characterization of the system properties and its environment may not be desirable for many reasons, including uncertainty in material properties due to statistically inhomogeneous microstructure, variations in nominal geometry due to manufacturing tolerances, uncertainty in loading due to the nondeterministic nature of the operating environment, and so on. In aerodynamic wing design, uncertainties may arise from changes in flight speed, variability in nominal geometry arising from manufacturing tolerances, damage processes, icing, and so on.

A common practice in engineering is to assume nominal values for any uncertain parameters when carrying out design studies. In fact, this assumption of determinism has been made implicitly in most of the formulations discussed so far in this book. Unfortunately, the deterministic approach generally leads to a final design whose performance may degrade significantly because of perturbations arising from uncertainties. This problem is exacerbated when dealing with designs that have been heavily optimized, because an optimum solution usually tends to lie at the extrema of the objective function or on a constraint boundary – small perturbations may then lead to either significant changes in performance or violation of the design constraints. As a consequence, an optimum design obtained without taking uncertainty into account can potentially be a high risk solution that is likely to violate design requirements or, in the worst case, fail when a physical prototype is built and tested.

One way of safeguarding against uncertainty is to employ more stringent constraints than would ideally be imposed. For example, when designing a structural system, to ensure that the final design does not fail, design constraints are often imposed on the stresses in the form $\sigma(\mathbf{x}) \leq \sigma_{\max}$, where $\sigma_{\max}$ is the failure stress of the material in use and $\mathbf{x}$ is a set of design variables. To account for uncertainties (e.g., those arising from inaccuracies in the analysis model, loading conditions, constitutive laws, manufacturing tolerances, in-service degradation, and so on.), the constraints may be rewritten as $F_s \sigma(\mathbf{x}) \le \sigma_{max}$, where F_s is the so-called *factor of safety*. Typical values for the factor range from 1.2 to 3; the actual value is often decided on the basis of prior experience with the material being used and similar design concepts. Clearly, the new optimum design becomes more conservative as we increase F_s, since the optimum is moved further away from the original constraint boundary $\sigma(\mathbf{x}) - \sigma_{max} = 0$. In the context of aerospace structural design, it is not desirable to use a high factor of safety since a conservative design will entail a weight penalty.[1] Further, with the increasing use of new materials and radical aerospace design concepts for which no prior design experience (and limited experimental data) exists, it is not straightforward to decide on an appropriate value of F_s.

In this chapter, we examine a number of approaches that can be employed to rationally accommodate uncertainties in the design process to arrive at robust designs. This is in contrast to approaches based on factors of safety where parameter uncertainty is not explicitly incorporated into the design formulation and uncertainties arising from all sources are lumped into a single parameter. In other words, in the factor of safety approach, uncertainty considerations enter the design formulation through the backdoor.

There is no universally accepted definition for what constitutes a robust design. Loosely speaking, a design is said to be robust if its performance does not degrade significantly because of perturbations arising from uncertainties. At this stage, we are deliberately vague about the permissible degree of degradation in nominal performance since this depends on the design specifications, which are problem specific. Nonetheless, this working definition suffices for motivating the approaches to robust design we describe. More broadly, the overall goal of robust design is to produce systems that perform as intended over their life cycles in the face of uncertainty. A related notion is that of reliability-based design, where the intention is to design a system with a specified probability of failure, which is of course invariably small.

The idea of robust design can be traced back at least as far as the fundamental observation made decades ago by the Japanese engineer Genichi Taguchi that one should design a product in such a way as to make its performance insensitive to variation in variables beyond the control of the designer. Taguchi is widely regarded as the founding father of robust design methods since he made significant contributions to the algorithmic and philosophical foundations of this topic (Taguchi 1987). The methodology proposed by Taguchi has been widely used in many engineering sectors over the last few decades and a number of case studies demonstrating its benefits can be found in the literature; see, for example, the text by Phadke (1989). We discuss Taguchi methods in more detail in Section 8.4.

Figure 8.1 illustrates the notion of robust design from a probabilistic viewpoint. Here, the horizontal axis represents changes in the design variable while the vertical axis represents the resulting measure of design performance, denoted by $f(x)$, which should be as small as possible. The probabilistic variability in the performance of two candidate designs are shown in the presence of uncertainty in the nominal value of x. It may be noted that the design with better nominal performance suffers from higher variability compared to the alternative design, which has marginally worse nominal performance. Hence, the latter design is more robust and may therefore be preferable. In design, it often makes sense to accept a small loss in nominal performance for improved robustness. We refer the reader to Section 1.6.5 for a more detailed discussion of the issues involved in trading off nominal performance with robustness.

---

[1] The Federal Aviation Authority (FAA) airworthiness certification requires a factor of safety equal to 1.5 for man-rated aircraft structures; however, non-man-rated systems such as missiles may use safety factors as low as 1.02–1.03 (Zang et al. 2002).

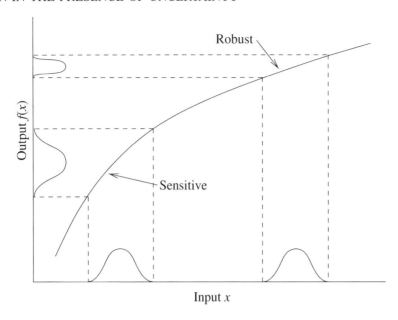

Figure 8.1 Graphical illustration of the notion of robust design (uncertainties are illustrated by typical input and output probability density functions).

On the basis of the original idea proposed by Taguchi, the parameters in any design problem may be classified into two groups: (1) *control* factors composed of the inputs that the designer is free to manipulate and (2) *noise* factors, which are the inputs that are difficult or expensive to control. For example, in an airfoil design problem, the control factors are the design variables used to parameterize the airfoil geometry, whereas the noise factors may include the operational Mach number and geometric uncertainty in manufacture. A robust design problem may involve, for example, designing the airfoil geometry such that the aerodynamic performance does not degrade significantly because of perturbations in these quantities – see the case study in Chapter 11.

Design optimization in the presence of uncertainty essentially involves three steps: (1) identification, modeling and representation of uncertainties so as to ultimately translate available data (and possibly expert opinion) into mathematical models based on probability theory or nonprobabilistic approaches, (2) propagating uncertainties through any computer models to quantify their impact on system performance, and (3) formulation and solution of an optimization problem with appropriate objective and constraint functions that ensure the optimum solution obtained is robust against uncertainties. We cover the choices associated with these steps in greater detail in what follows. We begin with an overview of approaches for modeling and representing parameter uncertainty. Subsequently, we describe a range of approaches for efficiently propagating uncertainty through computational models. Finally, we discuss a number of optimization formulations that can accommodate models of parameter uncertainty to arrive at robust designs. Some of the formulations described in this chapter are based on Chapters 5 and 7. The reader may wish to read those chapters first before delving into the contents that follow.

We now describe the notation used in this chapter. We denote the control factors (design variables) by $\mathbf{x} = \{x_1, x_2, \ldots, x_p\} \in \mathcal{X}$ and the noise factors by $\boldsymbol{\xi} = \{\xi_1, \xi_2, \ldots, \xi_{p_\xi}\} \in \mathcal{E}$. $\mathcal{X}$ denotes the space in which the design variables lie, for example, a box-shaped region in $\mathbb{R}^p$, where p is the total number of design variables. $\mathcal{E}$ is the space in which the uncertain parameters lie. For example, when the elements of $\boldsymbol{\xi}$ are uniformly distributed random variables or intervals, then $\mathcal{E}$ is also a box-shaped region in $\mathbb{R}^{p_\xi}$, where p_ξ is the total number of uncertain variables. Note that in some cases the design variables may also contain uncertainty, which may arise, for example, because of manufacturing tolerances. In such cases, we assume that $\mathbf{x}$ denotes the mean value of the design variables and the uncertainty in $\mathbf{x}$ (perturbations in $\mathbf{x}$) are modeled using a separate set of variables, which are placed in the vector $\boldsymbol{\xi}$. This makes the notation used here sufficiently general for mathematically formulating a wide class of design problems in the presence of uncertainty.

8.1 Uncertainty Modeling and Representation

Uncertainties can be broadly classified into two groups to distinguish between their origin: *aleatory* and *epistemic* (Oberkampf et al. 2004). Aleatory uncertainty can be defined as the inherent variation associated with the physical system under consideration or its operating environment. In contrast, epistemic uncertainty arises because of any lack of knowledge or information in any phase or activity of the modeling process. Hence, epistemic uncertainty is not an inherent property of the system under consideration since, given sufficient data, this uncertainty could be removed. Uncertainties can also be classified into *parametric* and *model-structure* uncertainties. The former arises from quantities that can be described parametrically, whereas the latter are due to the uncertainty arising from the inadequacy of the computational models being used to predict design performance.

In recent years, there has been significant debate on the philosophical and theoretical foundations of uncertainty representation schemes. In 2002, Sandia National Laboratories hosted the "Epistemic Uncertainty Workshop", which invited researchers to solve and discuss a set of Challenge problems. The outcomes of this workshop are summarized in a special issue of Reliability Engineering and System Safety (Vol. 85, 2004). This issue contains a number of papers that discuss various uncertainty analysis techniques including, probability theory, Dempster–Shafer evidence theory, random sets, possibility theory, fuzzy sets, polynomial chaos expansions and information-gap models. It appears to be the general consensus that in some situations of practical interest nonprobabilistic approaches may be required for uncertainty quantification, particularly when faced with limited data.

The total prediction error associated with a computational model used for describing a physical phenomena or predicting the performance of a design artifact may be decomposed as

$$\varepsilon = \varepsilon_h + \varepsilon_m + \varepsilon_d, \tag{8.1}$$

where ε_h is the error arising from numerical approximation of the original governing equations, ε_m is the error arising from the assumed model structure, and ε_d is the error arising from uncertainty in the data used as inputs. The problem of reducing ε_h has been studied in much detail in the computational mechanics literature and has led to increasing sophistication in the numerical algorithms for solving PDEs. ε_m can be reduced by using a more detailed model structure, for example, in fluid flow problems, a large eddy simulation can be used instead of a Reynolds-averaged Navier–Stokes solver. The contribution of ε_d to the total prediction error can be quantified by employing an uncertainty model based on probability theory or nonprobabilistic approaches and subsequently propagating this through the computational model. Estimation and control of error in computational simulations is an area which has seen increased research activity in recent years; see, for example, Oden et al. (2005).

In this section, we present a brief summary of a selection of approaches that may be used for representing parameter uncertainty. We first describe probabilistic modeling, which continues to be the most popular approach for representing parameter uncertainty because of its universality. Subsequently, we mention a number of nonprobabilistic approaches that have started to receive much attention in recent years. The problem of how to propagate various types of input uncertainties through high-fidelity computational models to derive uncertainty models for the outputs of interest is discussed in the following section.

8.1.1 Probabilistic Approaches

Probabilistic approaches have a long history in engineering design for representing parameter uncertainty. Three types of probabilistic models are commonly used to represent parameter uncertainty in design: random variable, random field and time-dependent stochastic processes. Probabilistic models are appropriate when sufficient experimental or field data exists on the quantities of interest. To illustrate, consider an uncertain scalar ξ. Here, given a sufficient number of realizations of this scalar, kernel density estimation techniques can be employed to fit a probability density function (pdf) $\mathcal{P}(\xi)$ to the data (see Izenman (1991) for a review); Figure 8.2(a). Alternatively, the structure of the pdf can be assumed (e.g., Gaussian, log-normal, beta, etc.) and its parameters can be estimated from the data. The validity of the assumed statistical model can be checked by goodness-of-fit tests such as chi-square or Kolmogorov–Smirnov tests (Kanji 1999).

When the uncertain parameter of interest is a spatially varying quantity (for example, the constitutive matrix of a random heterogeneous solid), then a random field model becomes appropriate. An example random field model is the Gaussian process surrogate described earlier in Section 5.6. The idea of random field model can be extended to represent uncertain parameters that vary as a function of both space and time. The interested reader is referred to the text by VanMarcke (1983) for a detailed exposition of random field modeling techniques.

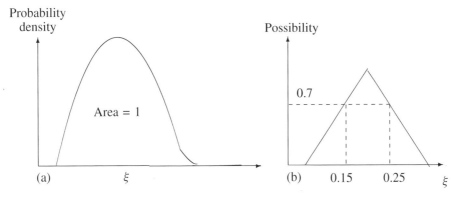

Figure 8.2 Examples of uncertainty modeling using probability theory (a) and fuzzy set theory (b).

However, in practice, it is often the case that sufficient data may not exist for accurately estimating the joint pdf of the uncertain parameters. In such situations, it is necessary to invoke simplifying assumptions or perhaps solicit expert opinion. For example, when modeling uncertainties arising from manufacturing tolerances on an engineering component, a Gaussian model is often used. The use of a Gaussian uncertainty model can sometimes be justified on the basis of data/experience generated from previous studies or in some cases by virtue of the central-limit theorem.[2] In practice, the Gaussian assumption is often made for the sake of mathematical convenience since a Gaussian distribution can be specified uniquely by its first two moments. However, when representing uncertainty in the Young's modulus of a component say, a Gaussian model cannot be justified. This is because the support of a Gaussian distribution is $[-\infty, +\infty]$, so with nonzero probability, the Young's modulus can turn out to be negative. This is of course not physically permissible. Hence, recourse has to be made to nonnegative distributions such as the log-normal or uniform distributions.

The designer must exercise care when insufficient data exists to accurately estimate the distributions of the uncertain parameters. In general, even if the assumed structure of the joint pdf is wrong, the first two statistical moments of an output quantity may not have significant error. However, the statistics of the extremes can be grossly erroneous if poor assumptions are made regarding the structure of the input distributions. A case in point is reliability analysis where it is necessary to compute the probability that a system may fail. The probability of failure is dictated by the tails of the pdf of the failure criterion, which are highly sensitive to the assumed statistical model for the inputs. The results for failure probability must be interpreted with caution if insufficient data exists to justify the choice of the joint pdf of the input uncertainties.[3]

In the approaches discussed in this chapter, we assume that the reader has an understanding of basic probability theory such as the notion of continuous random variables, random fields, statistical moments, distribution functions, and so on. For a tutorial overview of probabilistic and nonprobabilistic approaches to uncertainty representation, the reader may consult Helton et al. (2004). A detailed exposition of probability theory can be found in many standard texts dedicated to this topic; see, for example, Loève (1977).

8.1.2 Nonprobabilistic Approaches

It must be clear from the previous discussion that probability theory may be inappropriate when insufficient field data exists for accurately estimating the probability distributions of the uncertain inputs. We discuss here nonprobabilistic approaches, which can be seen as alternatives to probability theory. Nonprobabilistic approaches that have attracted some attention by the engineering design community in recent years include evidence theory, possibility theory, interval analysis and convex modeling. Here, we focus on interval analysis and convex modeling since evidence theory and possibility theory are advanced topics that lie beyond the scope of this book.[4]

---

[2]The central-limit theorem states that if $\xi_1, \xi_2, \ldots, \xi_m$ are mutually independent random variables with finite mean and variance (with possibly different distributions), the sum $S_m = \sum_{i=1}^{m} \xi_i$ tends to have a Gaussian distribution if no single variable contributes significantly to S_m, as $m \to \infty$.

[3]When dealing with multiple uncertain inputs, it is also common practice to ignore any correlations between the inputs – this is often far from true and such correlations can have profound consequences for system reliability.

[4]The number of publications discussing the application of evidence theory and possibility theory to engineering design is few compared to those dealing with applications of probability theory; however, in recent years, a

The simplest nonprobabilistic approach is perhaps the interval representation of uncertainties. The idea here is to represent the uncertain parameter ξ by the interval $[\xi^-, \xi^+]$, where ξ^- and ξ^+ denote the lower and upper bound, respectively. The interval bounds are then propagated through the analysis models to arrive at bounds on the output variables of interest, which can, of course, be conservatively wide. Note that all values within the interval are equally likely – this is in contrast to probabilistic representations where the extremes occur with much lower frequency than the average value. Interval methods have been applied to represent uncertainty in engineering systems by a number of researchers; see, for example, Rao and Chen (1998).

Ben-Haim and Elishakoff (1990) proposed a more general approach referred to as convex modeling, where all uncertain parameters are represented by convex sets. A few examples of convex models are given below:

- The uncertain function has envelope bounds

$$\xi^-(t) \le \xi(t) \le \xi^+(t). \tag{8.2}$$

- The uncertain function has an integral square bound

$$\int \xi^2(t)\, dt \le \alpha. \tag{8.3}$$

- The Fourier coefficients of the uncertain function belong to an ellipsoidal set

$$\mathbf{a}^T \mathbf{W} \mathbf{a} \le \alpha, \tag{8.4}$$

where $\mathbf{a}$ is the vector of Fourier coefficients of $\xi(t)$ and $\mathbf{W}$ is a weighting matrix.

Figure 8.3 shows the distinctions between an interval model of uncertainty and an ellipsoidal convex model for a two-dimensional case. The rectangular box defines the region

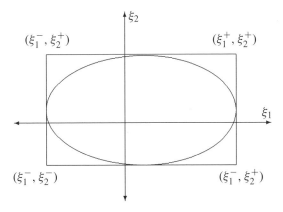

Figure 8.3 Uncertainty modeling using an interval model and ellipsoidal convex set for a problem with two variables.

number of research groups have started to examine the application of nonprobabilistic approaches to design; see, for example, Agarwal et al. (2004); Nikolaidis et al. (2004a); Rao and Cao (2002); Venter and Haftka (1999).

in which the uncertain variables lie for the interval model, that is, $[\xi_1^-, \xi_1^+] \times [\xi_2^-, \xi_2^+]$, whereas the ellipse shows the domain within which the uncertain parameters can lie when an ellipsoidal convex model is used.

Note that in the preceding convex models we have defined the uncertain parameter ξ as a function of the independent variable t (e.g., time or a spatial coordinate) for the sake of generality. For the case when ξ is a scalar, the envelope bounds reduce to an interval. Hence, the convex modeling approach includes interval representations as a special case, while allowing for more complex representations of uncertainty. It can be noted that a convex model defines an infinite set of values that an uncertain variable can take. A central-limit type theorem can be proved to justify the use of convex models to represent uncertainty. To illustrate, consider the set of functions

$$S_m = \left\{ \xi : \xi = \frac{1}{m} \sum_{i=1}^m \upsilon_i \quad \forall \, \upsilon_i \in \Upsilon, i = 1, 2, \ldots, m \right\}. \tag{8.5}$$

Arstein's theorem shows that regardless of the structure of the set Υ the set of all functions ξ formed by superimposition of elementary functions drawn from Υ will be convex as $m \to \infty$ (Ben-Haim and Elishakoff 1990).

More recently, Ben-Haim (2001) proposed information-gap (info-gap) theory as a rational approach for decision making in the presence of parameter uncertainty. In this approach, uncertainty is represented by convex sets. Two examples of info-gap models for the case when $\boldsymbol{\xi}$ is a vector are

$$\mathcal{E}(\alpha, \widetilde{\boldsymbol{\xi}}) = \{\boldsymbol{\xi} : \|\boldsymbol{\xi} - \widetilde{\boldsymbol{\xi}}\| \leq \alpha\}, \quad \alpha > 0 \tag{8.6}$$

$$\text{and} \quad \mathcal{E}(\alpha, \widetilde{\boldsymbol{\xi}}) = \{\boldsymbol{\xi} : (\boldsymbol{\xi} - \widetilde{\boldsymbol{\xi}})^T \mathbf{W} (\boldsymbol{\xi} - \widetilde{\boldsymbol{\xi}}) \leq \alpha\}, \quad \alpha > 0, \tag{8.7}$$

where $\mathbf{W}$ is a weighting matrix, α is a scalar parameter indicating the magnitude of uncertainty, and $\widetilde{\boldsymbol{\xi}}$ can be interpreted as the nominal value of $\boldsymbol{\xi}$.

It can be noted from (8.6) and (8.7) that an info-gap model is a family of nested sets. For a given value of α, the set $\mathcal{E}(\alpha, \widetilde{\boldsymbol{\xi}})$ represents the variability of $\boldsymbol{\xi}$ around the center point $\widetilde{\boldsymbol{\xi}}$. Clearly, the support of this set (the domain over which $\boldsymbol{\xi}$ varies) increases as α is increased. Because of this reason, α is referred to as the *uncertainty parameter*, which expresses the information gap between what is known ($\widetilde{\boldsymbol{\xi}}$ and the structure of the set) and what needs to be known for an ideal solution (the exact value of $\boldsymbol{\xi}$). It is worth noting that info-gap theory is general in scope since it is concerned with making decisions in the presence of severe uncertainty rather than just representing and propagating uncertainty through a model. We discuss the application of info-gap decision theory to robust design later in Section 8.8.

Fuzzy set theory proposed by Zadeh (1965) also provides a powerful approach for modeling parameter (and linguistic) uncertainty based on inexact or incomplete knowledge. In standard set theory, an element either lies inside a set or outside it – in the jargon of fuzzy set theory, this is a *crisp* set since the degree of membership of a point is either 1 if it is inside the set or 0 if outside the set. In contrast, in fuzzy modeling, uncertainty is represented using sets with *fuzzy* boundaries. In other words, a membership function is associated with a fuzzy set, which indicates the degree of membership of a given point. The degree of membership can vary from 0 to 1. A triangular fuzzy membership function is shown in Figure 8.2(b) for the sake of illustration. The vertical axis represents the possibility, which varies between 0 and 1, while the horizontal axis shows the values of the uncertain variable.

For illustration, a dotted horizontal line is depicted to indicate how the membership function can be interpreted. Here, with possibility 0.7, the uncertain variable lies between 0.15 and 0.25. Hence, at a given possibility level (also referred to as an α-cut), ξ is an interval variable. Because of this, fuzzy modeling can be seen as a generalization of interval analysis. As discussed earlier in Chapter 3, fuzzy set theory can also be applied to multiobjective design. The text by Ross (2004) gives a detailed exposition of fuzzy set theory.

8.2   Uncertainty Propagation

In this section, we describe a number of approaches for propagating uncertainty through computational models. These approaches have some parallels with the approximation concepts discussed previously in Chapters 5 and 6, where we distinguish between black-box and physics-based algorithms. Similarly, uncertainty propagation can be carried out using black-box approaches (which include sensitivity-based methods, local approximations and global surrogate modeling) and physics-based approximations such as subspace projection schemes.

An ultimate goal in uncertainty analysis is to develop general-purpose methods that can be routinely applied to complex systems with uncertain parameters. It appears to be the general consensus that, in the short term, progress can be made by developing *nonintrusive* uncertainty analysis techniques that exploit existing deterministic computer models. These approaches consider the computer as a black-box that returns function values given an input vector. We discuss a number of approaches that satisfy this requirement. We also present a brief overview of physics-based approaches that involve intrusive modifications to the analysis code or, in some cases, the development of new stochastic solvers.

When probabilistic models with specified statistics are used to represent uncertainty, the uncertainty propagation problem essentially involves computing the statistical moments of the output and, in some cases, its complete probability distribution. In contrast, when interval or convex models are used to represent uncertainty, then the uncertainty propagation problem usually involves computing bounds on the outputs of interest to calculate the least- and most-favorable responses.

8.2.1   Simulation Methods

Simulation methods are nonintrusive and can be applied to virtually any uncertainty propagation problem. The most popular simulation method is the Monte Carlo simulation (MCS) technique. Given the joint pdf of the uncertain parameters, the MCS technique can be applied to compute the complete statistics of the response quantities of interest with an arbitrary level of accuracy, provided sufficient number of samples is used. This approach is hence often used as a benchmark for evaluating the performance of new uncertainty analysis techniques. To motivate simulation methods, consider the multidimensional integral given below:

$$I = \langle \phi(\boldsymbol{\xi}) \rangle = \int_{\mathcal{E}} \phi(\boldsymbol{\xi}) \mathcal{P}(\boldsymbol{\xi}) \, d\boldsymbol{\xi}, \tag{8.8}$$

where $\phi(\boldsymbol{\xi})$ is a function calculated by running an expensive computer model. The above integral arises, for example, when computing the kth statistical moment of an output function $f(\boldsymbol{\xi})$, in which case we set $\phi(\boldsymbol{\xi}) = f^k(\boldsymbol{\xi})$.

Multidimensional integrals of this form can rarely be evaluated analytically, except for special cases. The basic idea of simulation techniques is to numerically approximate the above multidimensional integral. In Monte Carlo integration, $\phi(\boldsymbol{\xi})$ is evaluated at various points generated by drawing samples from the distribution $\mathcal{P}(\boldsymbol{\xi})$, say $\boldsymbol{\xi}^{(1)}, \boldsymbol{\xi}^{(2)}, \ldots, \boldsymbol{\xi}^{(m)}$. Subsequently, the integral is approximated by an ensemble average of the realizations of ϕ, that is,

$$\langle \phi(\boldsymbol{\xi}) \rangle \approx \widetilde{\phi} = \frac{1}{m} \sum_{i=1}^{m} \phi(\boldsymbol{\xi}^{(i)}). \tag{8.9}$$

$\widetilde{\phi}$ is referred to as the Monte Carlo *estimate* of I, that is, an approximation to $\langle \phi(\boldsymbol{\xi}) \rangle$. The strong law of large numbers states that, if ϕ is integrable over $\mathcal{E}$, then $\widetilde{\phi} \to I$ almost surely as $m \to \infty$. The variance of the Monte Carlo estimate is given by

$$\mathrm{Var}(\widetilde{\phi}) = \frac{\sigma_\phi^2}{m}, \quad \text{where} \quad \sigma_\phi^2 = \frac{1}{(m-1)} \sum_{i=1}^{m} (\phi(\boldsymbol{\xi}^{(i)}) - \widetilde{\phi})^2 \tag{8.10}$$

is the sample estimate of the variance of $\phi(\boldsymbol{\xi})$. The variance computed using (8.10) can be used to judge the accuracy of the Monte Carlo estimate. It follows from (8.10) that the standard error of $\widetilde{\phi}$ is given by $\sigma_\phi/\sqrt{m}$. Note that the error estimate is independent of the dimension of $\boldsymbol{\xi}$.

The major drawback of the MCS technique is that to satisfy accuracy requirements for complex systems it may be required to use a very large sample size. Since the convergence rate of the Monte Carlo estimate is $\mathcal{O}(1/\sqrt{m})$, to improve accuracy by one decimal place, around 100 times more samples will be required. Clearly, this will be computationally prohibitive when the runtime of the computer model used to calculate $\phi(\boldsymbol{\xi})$ is high. It is, however, easy to parallelize the MCS technique since $\phi(\boldsymbol{\xi}^{(i)})$ for different i can be calculated independently of each other.

To address the computational cost issue, a number of alternative approaches have been proposed that are less expensive than MCS while maintaining comparable accuracy in the computed statistics. For example, the convergence rate of the MCS technique can be improved by employing Latin hypercube sampling and minimum discrepancy sequences, as discussed previously in Section 5.8. A significant development in this area was the extension of the importance sampling techniques originally proposed by Marshall (1956) to structural reliability analysis; see, for example, Ang et al. (1992) and Wu (1994). To illustrate, consider the multidimensional integral (8.8). The idea of importance sampling is to approximate the integral I using points generated by sampling a new distribution that concentrates the points at those regions where $\phi(\boldsymbol{\xi})$ is large, since they make significant contributions to the value of the integral. This is achieved by rewriting the integral (8.8) as

$$I = \int_{\mathcal{E}} \phi(\boldsymbol{\xi}) \frac{\mathcal{P}(\boldsymbol{\xi})}{\widetilde{\mathcal{P}}(\boldsymbol{\xi})} \widetilde{\mathcal{P}}(\boldsymbol{\xi}) \, d\boldsymbol{\xi}, \tag{8.11}$$

where $\widetilde{\mathcal{P}}(\boldsymbol{\xi})$ is the importance sampling function. Samples are drawn from this distribution and I is approximated by the Monte Carlo estimate

$$I \approx \widetilde{\phi}_{\mathrm{imp}} = \frac{1}{m} \sum_{i=1}^{m} \phi(\boldsymbol{\xi}^i) \frac{\mathcal{P}(\boldsymbol{\xi}^{(i)})}{\widetilde{\mathcal{P}}(\boldsymbol{\xi}^{(i)})}. \tag{8.12}$$

Given a set of observations $\psi(\boldsymbol{\xi}^{(i)})$, $i = 1, 2, \ldots, m$, it can be noted that the estimate $\tilde{\phi}_{imp}$ (and its variance) will depend on the choice of the sampling function $\widetilde{\mathcal{P}}$. The fundamental idea used in importance sampling is to choose $\widetilde{\mathcal{P}}$ such that the variance of $\tilde{\phi}_{imp}$ is small, since this will ensure that $\tilde{\phi}_{imp}$ converges rapidly to I. Developing estimators with minimum variance is the central theme of variance reduction techniques; for a comprehensive survey of techniques used to evaluate multidimensional integrals in statistics, see Evans and Swartz (1995). Note that the problem of designing a good sampling function is challenging since it is problem dependent; see Hammersley and Handscomb (1964) for a theoretical analysis.

O'Hagan (1987) argues that the importance sampling scheme violates the *likelihood principle* since, given the same information (i.e., the observations $\phi(\boldsymbol{\xi}^{(1)}), \phi(\boldsymbol{\xi}^{(2)}), \ldots, \phi(\boldsymbol{\xi}^{(m)}))$, we obtain different estimates for $\langle\phi(\boldsymbol{\xi})\rangle$, depending on the choice of the sampling function (which in theory may be arbitrary). In practice, the user must exercise care since an arbitrary choice of sampling function may lead to an increase in the variance of the estimator (8.12). Another weakness of the MCS technique is that it wastes information since the realizations of the inputs $\boldsymbol{\xi}^{(1)}, \boldsymbol{\xi}^{(2)}, \ldots, \boldsymbol{\xi}^{(m)}$ are not used in the final estimate. The Bayesian Monte Carlo technique described in Section 8.2.5 aims to address these limitations by using realizations of $\boldsymbol{\xi}$ and the corresponding values of $\phi(\boldsymbol{\xi})$ to estimate $\langle\phi(\boldsymbol{\xi})\rangle$.

8.2.2 Taylor Series Approximations

We next look at how Taylor series expansions can be employed to approximate the statistical moments of the outputs of interest. Consider the second-order Taylor series approximation of the function $\phi(\boldsymbol{\xi})$ at the point $\boldsymbol{\xi}^0$:

$$\widehat{\phi}(\boldsymbol{\xi}) = \phi(\boldsymbol{\xi}^0) + \sum_{i=1}^{p_\xi} \frac{\partial\phi}{\partial\xi_i}(\xi_i - \xi_i^0) + \frac{1}{2}\sum_{i=1}^{p_\xi}\sum_{j=1}^{p_\xi}\frac{\partial^2\phi}{\partial\xi_i\partial\xi_j}(\xi_i - \xi_i^0)(\xi_j - \xi_j^0). \tag{8.13}$$

For simplicity, suppose that the elements of $\boldsymbol{\xi}$ are uncorrelated random variables, that is, $\langle\boldsymbol{\xi}\boldsymbol{\xi}^T\rangle$ is a diagonal matrix. Note that this is not usually the case and hence it may be necessary to apply a transformation first; see, for example, Melchers (1999). To illustrate one such transformation, let $\boldsymbol{\theta} \in \mathbb{R}^q$ denote a set of correlated random variables with covariance matrix $\mathbf{C} = \langle\boldsymbol{\theta}\boldsymbol{\theta}^T\rangle \in \mathbb{R}^{q\times q}$. Then $\boldsymbol{\xi} = \mathbf{T}^T\boldsymbol{\theta} \in \mathbb{R}^{p_\xi}$ denotes a set of uncorrelated random variables, where $\mathbf{T} \in \mathbb{R}^{q\times p_\xi}$ is an orthogonal matrix containing the eigenvectors corresponding to the largest p_ξ eigenvalues of $\mathbf{C}$ and $p_\xi \leq q$. This procedure can hence also be employed to reduce the number of random variables from q to p_ξ.

After some algebraic manipulations, the mean and standard deviation of $\phi(\boldsymbol{\xi})$ can be approximated using (8.13) as follows:

$$\widehat{\mu}_\phi = \widehat{\phi}(\boldsymbol{\xi}^0) + \frac{1}{2}\sum_{i=1}^{p_\xi}\frac{\partial^2\phi}{\partial\xi_i^2}\sigma_{\xi_i}^2, \tag{8.14}$$

$$\widehat{\sigma}_\phi = \sqrt{\sum_{i=1}^{p_\xi}\left(\frac{\partial\phi}{\partial\xi_i}\right)^2\sigma_{\xi_i}^2 + \frac{1}{2}\sum_{i=1}^{p_\xi}\sum_{j=1}^{p_\xi}\left(\frac{\partial^2\phi}{\partial\xi_i\partial\xi_j}\right)^2\sigma_{\xi_i}^2\sigma_{\xi_j}^2}. \tag{8.15}$$

Neglecting the second-order terms in (8.14) and (8.15) results in first-order approximations for the mean and standard deviation of $\phi(\boldsymbol{\xi})$. It may also be noted from the preceding equations that knowledge of the first two statistical moments of the input uncertainties is required to approximate the mean and standard deviation of $\phi(\boldsymbol{\xi})$. This is in contrast to simulation methods, which require the joint pdf of $\boldsymbol{\xi}$. On the downside, sensitivity-based approximations are local in nature and hence their accuracy can be poor when the coefficients of variation of the inputs are increased. As discussed later, this difficulty can be overcome to some extent by employing global surrogate models.

Cao et al. (2003) proposed a variance reduction technique that uses sensitivity information to accelerate the MCS technique. To illustrate, consider the multidimensional integral (8.8). The first-order Taylor series expansion of $\phi(\boldsymbol{\xi})$ about $\langle \boldsymbol{\xi} \rangle$ can be written as

$$\widehat{\phi}(\boldsymbol{\xi}) = \phi(\langle \boldsymbol{\xi} \rangle) + \sum_{i=1}^{p_{\xi}} \frac{\partial \phi}{\partial \xi_i} (\xi_i - \langle \xi_i \rangle). \tag{8.16}$$

The following sensitivity-enhanced estimate can be used to approximate $\langle \phi(\boldsymbol{\xi}) \rangle$

$$\langle \phi(\boldsymbol{\xi}) \rangle \approx \widetilde{\phi}_g = \phi(\langle \boldsymbol{\xi} \rangle) + \frac{1}{m} \sum_{i=1}^{m} \left(\phi(\boldsymbol{\xi}^{(i)}) - \widehat{\phi}(\boldsymbol{\xi}^{(i)}) \right). \tag{8.17}$$

It may be noted from (8.16) that as $m \to \infty$, $(1/m) \sum_{i=1}^{m} \widehat{\phi}(\boldsymbol{\xi}^{(i)}) \to \phi(\langle \boldsymbol{\xi} \rangle)$ since

$$\sum_{i=1}^{p_{\xi}} \frac{\partial \phi}{\partial \xi_i} \int_{\mathcal{E}} (\xi_i - \langle \xi_i \rangle) \, \mathcal{P}(\boldsymbol{\xi}) \, d\boldsymbol{\xi} = 0. \tag{8.18}$$

Hence, $\widetilde{\phi}_g$ will converge to the Monte Carlo estimate (8.9). The analysis by Cao et al. (2003) shows that the variance of $\widetilde{\phi}_g$ is smaller than the variance of the standard Monte Carlo estimate $\widetilde{\phi}$ defined in (8.9). In practice, (8.17) generally results in an order of magnitude improvement in accuracy. Therefore, it always makes sense to exploit sensitivity information, if available, to accelerate simulation techniques in this way.

8.2.3 Laplace Approximation

An approach commonly used in Bayesian statistics to approximate multidimensional integrals is the Laplace approximation (Evans and Swartz 1995):[5]

$$I = \int \exp[-\phi(\boldsymbol{\xi})] \, d\boldsymbol{\xi} \approx \exp[-\phi(\boldsymbol{\xi}^*)] (2\pi)^{p_{\xi}/2} |\mathbf{H}|^{-1/2}, \tag{8.19}$$

where $\boldsymbol{\xi}^* = \arg \min \phi(\boldsymbol{\xi})$ and $\mathbf{H} = \nabla^2 \phi(\boldsymbol{\xi}^*)$.

The Laplace approximation is derived by using a second-order Taylor series expansion of $\phi(\boldsymbol{\xi})$ around $\boldsymbol{\xi}^*$. Noting that by definition $\partial \phi(\boldsymbol{\xi}^*)/\partial \xi_i = 0$, the integrand can be expressed as

$$\exp[-\phi(\boldsymbol{\xi})] \approx \exp\left[-\phi(\boldsymbol{\xi}^*) - \frac{1}{2} (\boldsymbol{\xi} - \boldsymbol{\xi}^*)^T \mathbf{H} (\boldsymbol{\xi} - \boldsymbol{\xi}^*) \right]. \tag{8.20}$$

Substituting (8.20) into (8.19), we have

$$I \approx \exp[-\phi(\boldsymbol{\xi}^*)] \int \exp\left[-\frac{1}{2} (\boldsymbol{\xi} - \boldsymbol{\xi}^*)^T \mathbf{H} (\boldsymbol{\xi} - \boldsymbol{\xi}^*) \right] d\boldsymbol{\xi}$$

$$= \exp[-\phi(\boldsymbol{\xi}^*)] (2\pi)^{p_{\xi}/2} |\mathbf{H}|^{-1/2}. \tag{8.21}$$

---

[5]This approach is also referred to as the saddle-point approximation by physicists (MacKay 2003).

In the above derivation, we have used the standard result for the normalization constant of a Gaussian distribution with mean $\boldsymbol{\xi}^*$ and covariance matrix $\mathbf{H}$. Note that the accuracy of the Laplace approximation depends on how much the region around $\boldsymbol{\xi}^*$ contributes to the value of the integral. The accuracy may be poor if $\phi(\boldsymbol{\xi})$ has multiple maxima; for further details of the theoretical properties of this approximation, see Evans and Swartz (1995).

In order to evaluate the multidimensional integral for the kth statistical moment of a function $f(\boldsymbol{\xi})$ using (8.19), we replace $\phi(\boldsymbol{\xi})$ by $-\log f^k(\boldsymbol{\xi}) - \log \mathcal{P}(\boldsymbol{\xi})$, where $\mathcal{P}(\boldsymbol{\xi})$ denotes the joint pdf of $\boldsymbol{\xi}$. However, note here that to implement the Laplace approximation we need to find the minimum of $\phi(\boldsymbol{\xi})$ as well as its Hessian. This makes the Laplace approximation computationally expensive to implement when $\phi(\boldsymbol{\xi})$ is defined by a high-fidelity model. However, this approximation can be efficiently calculated when a computational cheap surrogate model is used in lieu of $\phi(\boldsymbol{\xi})$.

8.2.4 Reliability Analysis

Reliability analysis is essentially concerned with calculating the probability that a system may fail given a statistical model of the uncertain parameters affecting its response. To illustrate, consider the case when the criterion governing failure of the system under consideration can be written as $g(\boldsymbol{\xi}) \leq 0$, where $\boldsymbol{\xi} \in \mathbb{R}^{p_\xi}$ is a set of random variables whose statistics are known.[6] Note that the function $g(\boldsymbol{\xi})$ is sometimes referred to as the *safety margin*. The equation $g(\boldsymbol{\xi}) = 0$ is commonly referred to as the *limit state function* or the equation of the failure surface. Figure 8.4 depicts how the limit state function partitions the uncertainty space into the failure domain and the safe domain.

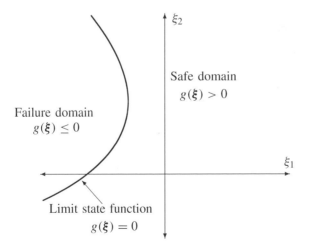

Figure 8.4 Limit state function in reliability analysis. The region where $g(\boldsymbol{\xi}) \leq 0$ is the failure domain whereas the region where $g(\boldsymbol{\xi}) > 0$ is the safe domain.

---

[6]Here, for the sake of simplicity, we consider the case when a single criterion governs failure. However, in general, complex systems may have multiple modes of failure (possibly time-dependant) that need to be taken into account when carrying out reliability assessment studies; see, for example, Melchers (1999).

The central focus of reliability analysis is the development of efficient numerical methods for calculating the probability that $g(\boldsymbol{\xi}) \leq 0$, that is,

$$P_f = P\left[g(\boldsymbol{\xi}) \leq 0\right] = \int_{g(\boldsymbol{\xi}) \leq 0} \mathcal{P}(\boldsymbol{\xi})\, \mathrm{d}\boldsymbol{\xi}, \tag{8.22}$$

where $\mathcal{P}(\boldsymbol{\xi})$ is the joint pdf of $\boldsymbol{\xi}$. Note that the above multidimensional integral needs to be evaluated over the failure domain and hence it represents the volume in $\mathbb{R}^{p_\xi}$ over which $g(\boldsymbol{\xi}) \leq 0$. The reliability of the system under consideration is then given by $(1 - P_f)$ or $100(1 - P_f)\%$.

The integral in (8.22) is difficult to evaluate in practice for the following reasons: (1) the domain of integration $g(\boldsymbol{\xi}) \leq 0$ is usually not available in analytical form and is often computed by running a computationally expensive computer model, (2) in practice, p_ξ can be large and hence direct numerical evaluation of the multidimensional integral is often computationally prohibitive, (3) in many situations, only the first two statistical moments of $\boldsymbol{\xi}$ are available and hence assumptions have to be made regarding the structure of the joint pdf $\mathcal{P}(\boldsymbol{\xi})$.

Before proceeding further, we first describe how simulation methods can be applied to approximate P_f. Here, given a realization of $\boldsymbol{\xi}$ drawn from $\mathcal{P}(\boldsymbol{\xi})$, say $\boldsymbol{\xi}^{(i)}$, we can check whether the system has failed or not by calculating $g(\boldsymbol{\xi}^{(i)})$ – this can be interpreted as a binary process whose outcome can be flagged as either 1 (failed) or 0 (safe). The probability of failure can hence be approximated by the Monte Carlo estimate

$$P_f \approx \frac{1}{m} \sum_{i=1}^{m} \mathcal{I}\left[g(\boldsymbol{\xi}^{(i)}) \leq 0\right], \tag{8.23}$$

where $\mathcal{I}$ is the indicator function which returns the value 1 if $g(\boldsymbol{\xi}^{(i)}) \leq 0$ and 0 if the converse is true.

Even though simulation schemes are easy to set up for reliability analysis, they can be computationally expensive, particularly when each evaluation of the limit state function involves running a high-fidelity model. Further, a large number of samples may be required to accurately estimate P_f. For example, to estimate a probability of failure of the order of magnitude 10^{-6}, clearly more than one million samples will be required to ensure accurate approximations. Specifically, with $100(1 - \nu)\%$ confidence, the error bound for the Monte Carlo estimate (8.23) is given by

$$z_{1-\nu/2}\sqrt{\frac{P_f(1 - P_f)}{m}}, \tag{8.24}$$

where $z_{1-\nu/2}$ denotes the $(1 - \nu/2)$ quantile of the standard normal distribution.

We next present an outline of the central ideas used in first- and second-order reliability methods (referred to as FORM and SORM, respectively) that enable the integral in (8.22) to be efficiently approximated; for a survey of recent developments, see Rackwitz (2001). Our discussion is based on the paper by Langley (1999), who showed that a number of reliability methods can be interpreted as special cases of an asymptotic formulation, which we discuss first. The asymptotic formulation is based on the Laplace approximation discussed in the earlier section. The integral (8.22) is first rewritten as

$$P_f = \int_{g(\boldsymbol{\xi}) \leq 0} \exp[-\phi(\boldsymbol{\xi})] \, d\boldsymbol{\xi}, \tag{8.25}$$

where $\phi(\boldsymbol{\xi}) = -\ln[\mathcal{P}(\boldsymbol{\xi})]$ is the negative of the log likelihood function.

Langley (1999) showed that an asymptotic approximation for (8.25) can be written as

$$P_f \approx (2\pi)^{p_\xi/2} |\mathbf{X}|^{-1/2} \mathcal{P}(\boldsymbol{\xi}^*) \exp(\beta^2/2) \Phi(-\beta), \tag{8.26}$$

where $\mathbf{X} \in \mathbb{R}^{p_\xi \times p_\xi}$ is a matrix whose ijth element is given by

$$X_{ij} = \frac{\partial^2 \phi}{\partial \xi_i \partial \xi_j} + \frac{|\nabla \phi|}{|\nabla g|} \frac{\partial^2 g}{\partial \xi_i \partial \xi_j} \tag{8.27}$$

$$\text{and} \quad \beta = \frac{|\nabla \phi|}{|\nabla g|} \sigma, \quad \text{where} \quad \sigma^2 = \sum_{i=1}^{p_\xi} \sum_{j=1}^{p_\xi} (\mathbf{X}^{-1})_{ij} \frac{\partial g}{\partial \xi_i} \frac{\partial g}{\partial \xi_j} \tag{8.28}$$

and Φ is the cumulative normal distribution function. Note that all the terms in the preceding equations are evaluated at the point $\boldsymbol{\xi}^*$, which is defined as

$$\boldsymbol{\xi}^* = \arg \min \phi(\boldsymbol{\xi}) \quad \text{subject to:} \quad g(\boldsymbol{\xi}) = 0. \tag{8.29}$$

It is worth noting that the approximation for P_f in (8.26) is asymptotic, that is, the approximation error tends to zero as $\beta \to \infty$. The major computational effort associated with this approximation involves solving the constrained minimization problem (8.29) and calculation of the sensitivities of the safety margin g. Note that (8.29) can be solved for $\boldsymbol{\xi}^*$ using specialized algorithms in the reliability analysis literature (Melchers 1999) or alternatively, surrogate-assisted optimization algorithms such as those discussed previously in Chapter 7 can be employed. We now proceed to briefly discuss how the asymptotic approximation reduces to FORM and SORM.

In general, $\mathcal{P}(\boldsymbol{\xi})$ will be non-Gaussian and hence the variables $\boldsymbol{\xi}$ are often transformed into a set of uncorrelated Gaussian random variables with zero mean and unit variance. The Rosenblatt transformation (Rosenblatt 1952) is used here and the algorithmic aspects of this transformation can be found in any standard text on reliability analysis (Madsen et al. 1986; Melchers 1999). Here, for simplicity of presentation, we assume that the variables $\boldsymbol{\xi}$ are uncorrelated Gaussian random variables with zero mean and unit variance, i.e., $\mathcal{P}(\boldsymbol{\xi}) = (1/(2\pi)^{p_\xi/2}) \exp(-\|\boldsymbol{\xi}\|_2^2/2)$. Hence, the negative log likelihood function can be written as

$$\phi(\boldsymbol{\xi}) = -\ln[\mathcal{P}(\boldsymbol{\xi})] = \ln\left[(2\pi)^{p_\xi/2}\right] + \frac{1}{2}\|\boldsymbol{\xi}\|_2^2, \tag{8.30}$$

where $\|\cdot\|_2$ denotes the L_2 norm.

In the case of FORM, it is assumed that the contributions of the second-order derivatives of the safety margin are negligible and hence the approximation in (8.26) becomes

$$P_f \approx \Phi(-\beta) \quad \text{where} \quad \beta = \left[\sum_{i=1}^{p_\xi} \xi_i^{*2}\right]^{1/2}. \tag{8.31}$$

It can be noted from (8.29) that, when ϕ is defined by (8.30), $\boldsymbol{\xi}^*$ is computed by minimizing $||\boldsymbol{\xi}||_2$ subject to the constraint $g(\boldsymbol{\xi}) = 0$. From a geometric viewpoint, $\boldsymbol{\xi}^*$ denotes the point on the limit state function that lies closest to the origin. Hence, $\boldsymbol{\xi}^*$ is commonly referred to as the most probable point of failure (MPP).

Now, consider the case when second-order effects are to be included in reliability analysis. Then, for the case when the variables $\boldsymbol{\xi}$ are uncorrelated Gaussian random variables with zero mean and unit variance (i.e., ϕ is given by (8.30)), the asymptotic approximation (8.26) becomes

$$P_f \approx \Phi(-\beta)|\mathbf{X}|^{-1/2}, \tag{8.32}$$

where β is given by (8.28) and the ijth element of the matrix $\mathbf{X}$ is given by

$$X_{ij} = \delta_{ij} + \left[\sum_{k=1}^{p_\xi} \xi_k^{*2} \right]^{1/2} \frac{\partial^2 g}{\partial \xi_i \partial \xi_j} \frac{1}{|\nabla g|}. \tag{8.33}$$

The approximation in (8.32) is similar to the SORM result in Madsen et al. (1986). It is worth noting that there are a number of alternative second-order methods for reliability analysis; see, for example, Adhikari (2004). The interested reader is referred to Langley (1999) for a detailed exposition on how the asymptotic approximation (8.26) reduces to well-known results yielded by various SORM approximations.

In summary, the main steps carried out by FORM and SORM are (1) transform $\boldsymbol{\xi}$ to a set of uncorrelated Gaussian random variables, (2) estimate the MPP of failure by solving a constrained optimization problem, (3) compute first- or second-order approximation of the safety margin g at this point, and (4) estimate P_f using the approximated limit state function. The accuracy of the approximation for P_f can be improved further by using importance sampling schemes (Ang et al. 1992; Wu 1994). For example, a standard Gaussian pdf centered at the MPP of failure can be used as the sampling function, since the region surrounding this point is expected to make significant contributions to the integral for P_f.

In the context of reliability-based design optimization, most researchers make use of FORM for computational reasons since computing the second-order derivatives of the safety margin may well be impossible or computationally intractable in many situations. This difficulty can be overcome by using a quadratic response surface model; see Section 5.4.1. Also note that, at each function evaluation, evaluating the performance of a design (approximating the probability of failure) using say FORM involves solving a minimization problem to calculate the MPP of failure. Hence, a reliability-based design optimization formulation may incur computational cost roughly one order of magnitude higher than an equivalent deterministic formulation. We revisit the topic of reliability-based design optimization later in Section 8.7.

8.2.5 Uncertainty Analysis using Surrogate Models: the Bayesian Monte Carlo Technique

We next address some of the issues involved in uncertainty analysis using a surrogate model of the black-box simulation code. Much of the work employing this line of approach makes use of polynomial response surface models; see, for example, Haldar and Mahadevan (2000). The approach involves first constructing a polynomial model of the simulation code (generally a quadratic function of the input variables), which is then used as a surrogate of the original computationally expensive model during MCS. It has been demonstrated on a number of problem domains that this method can give reasonable approximations using a fraction of the computational effort required for a full-fledged MCS using the simulation code. However, as discussed earlier in Chapter 5, it may be preferable in many cases to use a more flexible surrogate model, particularly when dealing with complex systems.

Here, we focus on the Bayesian Monte Carlo technique where the Gaussian process model described previously in Section 5.6 is used as a surrogate of the original analysis code (Rasmussen and Ghahramani 2003). To illustrate, consider the case when it is required to evaluate the multidimensional integral (8.8) to calculate $\langle \phi(\boldsymbol{\xi}) \rangle$. Recollect that Gaussian process modeling involves running the analysis code at a chosen set of design points to generate the training dataset $\{\boldsymbol{\xi}^{(i)}, \phi(\boldsymbol{\xi}^{(i)})\}$, $i = 1, 2, \ldots, m$.[7] Subsequently, a maximum likelihood estimation technique is applied to estimate the hyperparameters within the correlation function $R(\boldsymbol{\xi}, \boldsymbol{\xi}')$. This process ultimately leads to an approximation of the black-box simulation code as a multidimensional Gaussian random function since the posterior distribution is given by $\mathfrak{P}(\boldsymbol{\xi}) \sim \mathcal{N}(\widehat{\phi}(\boldsymbol{\xi}), C_{\phi}(\boldsymbol{\xi}, \boldsymbol{\xi}'))$, where $\widehat{\phi}(\boldsymbol{\xi})$ and $C_{\phi}(\boldsymbol{\xi}, \boldsymbol{\xi}')$ denote the posterior mean and covariance, respectively – see (5.66) and (5.67) in Section 5.6.1. Since the surrogate is itself a random function,[8] its statistical moments have to be described in terms of random variables. More specifically, the mean of the posterior distribution $\mathfrak{P}(\boldsymbol{\xi})$ is also a Gaussian random variable, that is,

$$K = \int_{\mathcal{E}} \mathfrak{P}(\boldsymbol{\xi}) \mathcal{P}(\boldsymbol{\xi}) \, \mathrm{d}\boldsymbol{\xi} \sim \mathcal{N}\left(\langle K \rangle, \sigma_K^2\right), \tag{8.34}$$

where

$$\langle K \rangle = \int_{\mathcal{E}} \widehat{\phi}(\boldsymbol{\xi}) \mathcal{P}(\boldsymbol{\xi}) \, \mathrm{d}\boldsymbol{\xi} \quad \text{and} \quad \sigma_K^2 = \int_{\mathcal{E}} \int_{\mathcal{E}} C_{\phi}(\boldsymbol{\xi}, \boldsymbol{\xi}') \mathcal{P}(\boldsymbol{\xi}) \mathcal{P}(\boldsymbol{\xi}') \, \mathrm{d}\boldsymbol{\xi} \, \mathrm{d}\boldsymbol{\xi}'. \tag{8.35}$$

Here, $\langle K \rangle$ can be interpreted as an estimate of $\langle \phi(\boldsymbol{\xi}) \rangle$, while σ_K^2 is a probabilistic error bar which is analogous to the variance of the Monte Carlo estimate in (8.10).

Note that the multidimensional integrals in (8.35) can be evaluated analytically when the input variables are Gaussian random variables and the correlation function $R(\boldsymbol{\xi}, \boldsymbol{\xi}')$ obeys the product correlation rule (O'Hagan et al. 1999). A detailed exposition on the analytical evaluation of similar multidimensional integrals that arise when surrogate models are employed for uncertainty propagation can be found in Chen et al. (2004). Fortunately, since predicting the output at a new point $\boldsymbol{\xi}$ using the metamodel is computationally very cheap (only matrix vector and vector products are involved), simulation techniques can be readily applied to compute the statistics of interest. Further, it is also possible to employ the first-order derivatives of $\widehat{\phi}$ and C_{ϕ} with respect to $\boldsymbol{\xi}$ to accelerate the convergence of simulation schemes using the approach described earlier in Section 8.2.2.

In the context of reliability analysis, it is of interest to compute integrals of the form

$$I(f^+) = P\left[\mathfrak{P}(\boldsymbol{\xi}) \le f^+\right] = \int_{\mathcal{E}} \mathcal{I}\left[\mathfrak{P}(\boldsymbol{\xi}) \le f^+\right] \mathcal{P}(\boldsymbol{\xi}) \, \mathrm{d}\boldsymbol{\xi}, \tag{8.36}$$

---

[7]Note that, in general, ϕ is a function of the design variables $\mathbf{x}$ as well as $\boldsymbol{\xi}$; however, we suppress the dependency of ϕ on $\mathbf{x}$ for simplicity of presentation.

[8]Note that the surrogate is not random because of input uncertainty; rather, the posterior mean and variance indicate the uncertainty involved in predicting $\phi(\boldsymbol{\xi})$ using a surrogate that is constructed using a finite number of training points.

where $\mathcal{I}$ is the indicator function. Since $\mathfrak{P}(\boldsymbol{\xi})$ is a Gaussian process, $I(f^+)$ is also random. The posterior mean of (8.36) can be computed as

$$\langle I(f^+) \rangle = \int_{\mathcal{E}} P\left[\mathfrak{P}(\boldsymbol{\xi}) \leq f^+\right] \mathcal{P}(\boldsymbol{\xi}) \, d\boldsymbol{\xi} \approx \frac{1}{m} \sum_{i=1}^{m} P\left[\mathfrak{P}(\boldsymbol{\xi}^{(i)}) \leq f^+\right]. \tag{8.37}$$

Here, $P[\mathfrak{P}(\boldsymbol{\xi}) \leq f^+]$ denotes the probability that the surrogate model output at the point $\boldsymbol{\xi}$ is less than or equal to f^+. This can be computed using the cumulative normal distribution function since, given $\boldsymbol{\xi}$, $\mathfrak{P}$ is a Gaussian random variable with mean $\widehat{\phi}(\boldsymbol{\xi})$ and variance $C(\boldsymbol{\xi}, \boldsymbol{\xi})$. However, the preceding integral cannot be evaluated analytically and hence a simulation procedure needs to be applied.

An approximate upper bound on $I(f^+)$ can be calculated using the Monte Carlo estimate

$$\frac{1}{m} \sum_{i=1}^{m} \mathcal{I}\left[\widehat{\phi}(\boldsymbol{\xi}^{(i)}) + 3C_\phi(\boldsymbol{\xi}^{(i)}, \boldsymbol{\xi}^{(i)}) \leq f^+\right]. \tag{8.38}$$

However, before using this bound, the analyst should verify that the standardized cross-validated residual (SCVR) values[9] for most training points roughly lie in the interval $[-3, +3]$. If this statistical test is successful, then the error bounds computed for the statistical moments are expected to be reasonably accurate (Nair et al. 2001). Similarly, an approximate lower bound on $I(f^+)$ can also be calculated. The bounds provide an idea of the accuracy of the approximations obtained using the surrogate.

An alternative approximation for $I(f^+)$ can be derived using the Monte Carlo estimate

$$\langle I(f^+) \rangle \approx \frac{1}{m} \sum_{i=1}^{m} \mathcal{I}\left[P[\mathfrak{P}(\boldsymbol{\xi}^{(i)}) \leq f^+] \geq 0.5\right]. \tag{8.39}$$

Here, for each realization of $\boldsymbol{\xi}$, we compute $P[\mathfrak{P}(\boldsymbol{\xi}^{(i)}) \leq f^+]$, which gives the probability of the surrogate output being less than the specified critical value f^+. If this probability is greater than 0.5, the indicator function $\mathcal{I}$ returns the value 1; otherwise the indicator function returns 0. The accuracy of this estimate will depend on how well the posterior variance $C(\boldsymbol{\xi}^{(i)}, \boldsymbol{\xi}^{(i)})$ predicts the errors we are likely to make at a candidate point $\boldsymbol{\xi}^{(i)}$. If the posterior variance is high, then it may be necessary to generate new points for updating the baseline Gaussian process model by solving the optimization problem (7.18) defined earlier in Section 7.3.1.

8.2.6 Polynomial Chaos Expansions

The idea of polynomial chaos (PC) representations of stochastic processes was introduced by Wiener (1938) as a generalization of Fourier series expansion. More specifically, Wiener used multidimensional Hermite polynomials as basis functions for representing stochastic processes. The basic idea is to project the process under consideration onto a stochastic subspace spanned by a set of complete orthogonal random polynomials.

To illustrate the process of constructing PC expansions, let $\kappa_i(\theta)$, $i = 1, 2, \ldots, \infty$ denote a set of polynomials that form an orthogonal basis in the L_2 space of random variables. Note that here we use the symbol θ to indicate the dependence of any quantity on a random

---

[9]See Section 5.6.4 for a discussion on model diagnostic measures for validating Gaussian process models.

dimension. It can be shown that a general second-order stochastic process (i.e., a process with finite variance) $h(\theta)$ can be represented as

$$h(\theta) = c_0 \gamma_0 + \sum_{i_1=1}^{\infty} c_{i_1} \gamma_1(\xi_{i_1}) + \sum_{i_1=1}^{\infty} \sum_{i_2=1}^{i_1} c_{i_1 i_2} \gamma_2(\xi_{i_1}, \xi_{i_2})$$

$$+ \sum_{i_1=1}^{\infty} \sum_{i_2=1}^{i_1} \sum_{i_3=1}^{i_2} c_{i_1 i_2 i_3} \gamma_3(\xi_{i_1}, \xi_{i_2}, \xi_{i_3}) + \cdots, \tag{8.40}$$

where $\gamma_k(\xi_{i_1}, \xi_{i_2}, \ldots, \xi_{i_k})$ denotes the generalized PC of order k, which is a tensor product of one-dimensional basis functions $\kappa_i, i = 1, 2, \ldots, k$.

In the original work of Wiener, γ_p is chosen to be a multidimensional Hermite polynomial in terms of a set of uncorrelated Gaussian random variables $\xi_1, \xi_2, \ldots$, which have zero mean and unit variance. For example, if Hermite polynomials are used, a second-order two-dimensional PC expansion of $h(\theta)$ can be written as

$$h(\theta) = h_0 + h_1 \xi_1 + h_2 \xi_2 + h_3(\xi_1^2 - 1) + h_4 \xi_1 \xi_2 + h_5(\xi_2^2 - 1). \tag{8.41}$$

It can be seen from the above equation that the first term of the PC expansion represents the mean value of $h(\theta)$ since ξ_1 and ξ_2 are uncorrelated Gaussian random variables with zero mean and unit variance. Another point worth noting here is that the number of terms in the expansion grows very quickly with the dimension of $\boldsymbol{\xi}$ and the order of the expansion.

More recently, Xiu and Karniadakis (2002) proposed a generalized PC approach that employs basis functions from the Askey family of orthogonal polynomials. The Hermite chaos expansion appears as a special case in this generalized approach, which is referred to as Wiener–Askey chaos. The motivation for this generalization arises from the observation that the convergence of Hermite chaos expansions can be far from optimal for non-Gaussian inputs. In such cases, the convergence rate can be improved by replacing Hermite polynomials with other orthogonal polynomials that better represent the input. For example, when the elements of $\boldsymbol{\xi}$ have a uniform distribution, then an expansion in Legendre basis functions converges faster compared to Hermite polynomials.

For notational convenience, (8.40) can be rewritten as

$$h(\theta) = \sum_{i=0}^{\infty} h_i \varphi_i(\boldsymbol{\xi}), \tag{8.42}$$

where there is a one-to-one correspondence between the functions $\gamma_p(\xi_{i_1}, \xi_{i_2}, \ldots, \xi_{i_p})$ and $\varphi_i(\boldsymbol{\xi})$. Also note here, that $\varphi_0 = 1$ and $\langle \varphi_i \rangle = 0$ for $i > 0$. Since $\varphi_i(\boldsymbol{\xi}), i = 0, 1, 2, \ldots, \infty$ form an orthogonal basis in the L_2 space of random variables

$$\langle \varphi_i(\boldsymbol{\xi}) \varphi_j(\boldsymbol{\xi}) \rangle = \langle \varphi_i^2(\boldsymbol{\xi}) \rangle \delta_{ij}, \tag{8.43}$$

where δ_{ij} is the Kronecker delta operator and $\langle \cdot \rangle$ is defined as

$$\langle f(\boldsymbol{\xi}) g(\boldsymbol{\xi}) \rangle = \int f(\boldsymbol{\xi}) g(\boldsymbol{\xi}) W(\boldsymbol{\xi}) \, d\boldsymbol{\xi}. \tag{8.44}$$

Here, $W(\boldsymbol{\xi})$ is the weight function corresponding to the PC basis. The weight function is chosen to correspond to the distribution of the elements of $\boldsymbol{\xi}$ (Xiu and Karniadakis 2002).

For example, when Hermite polynomials are used as basis functions, the weight function is given by the multidimensional Gaussian distribution.

Cameron and Martin (1947) proved that the Hermite chaos expansion converges in a mean-square sense for any second-order stochastic process when the number of terms is increased. This result also holds for the case when Weiner–Askey chaos expansions are used (Xiu and Karniadakis 2002).

As discussed in the next section, PC expansions can be used to solve a wide class of stochastic operator problems, both algebraic and differential. However, these approaches are intrusive in scope since knowledge of the governing equations is required for their implementation. A nonintrusive PC expansion scheme is also possible, since by virtue of the orthogonality properties of the basis functions, the coefficients h_i that appear in (8.42) can be written as

$$h_i = \frac{\langle h(\theta)\varphi_i \rangle}{\langle \varphi_i^2 \rangle}. \tag{8.45}$$

The term in the denominator is easy to calculate and is tabulated in Ghanem and Spanos (1991). The numerator can be replaced by its Monte Carlo estimate. This involves computing the ensemble average of $h(\theta)\varphi_i$, which can be calculated by generating a number of realizations of the uncertain parameters and running the computer model at these points (Reagan et al. 2003). Note that when Hermite polynomials are used the final form of the approximation is similar to a polynomial response surface model.

8.2.7 Physics-based Uncertainty Propagation

The basic idea of physics-based methods is to incorporate uncertainty directly into the governing equations of the system under consideration. For example, when a probabilistic model is used to represent uncertainty, then, in many instances, the uncertainty propagation problem can be reduced to solving a set of stochastic partial differential equations governing the system response. A major contribution in this direction was made by Ghanem and Spanos (1991), who developed a methodology based on PC expansions that can be applied to solve a wide class of problems in computational stochastic mechanics.

To illustrate the chaos projection scheme, consider the case when the system under consideration can be analyzed by solving a system of linear random algebraic equations $\mathbf{K}(\boldsymbol{\xi})\mathbf{w}(\boldsymbol{\xi}) = \mathbf{f}(\boldsymbol{\xi})$ for the process $\mathbf{w}(\boldsymbol{\xi}) \in \mathbb{R}^n$, where $\mathbf{K}(\boldsymbol{\xi}) \in \mathbb{R}^{n \times n}$ and $\mathbf{f}(\boldsymbol{\xi}) \in \mathbb{R}^n$. Suppose that $\mathbf{K}$ is a second-order process. Then, it follows from the Cameron–Martin theorem that $\mathbf{K}$ admits the chaos decomposition

$$\mathbf{K}(\boldsymbol{\xi}) = \sum_{i=1}^{P_1} \mathbf{K}_i \varphi_i(\boldsymbol{\xi}), \tag{8.46}$$

where $\mathbf{K}_i \in \mathbb{R}^{n \times n}$ are deterministic matrices and φ_i denote the PC basis functions. Similarly, the right-hand-side can be expanded as $\mathbf{f}(\boldsymbol{\xi}) = \sum_{i=1}^{P_2} \mathbf{f}_i \varphi_i(\boldsymbol{\xi})$.

The basic idea of the chaos projection scheme is to represent the unknown solution process $\mathbf{w}$ by a PC expansion, that is,

$$\mathbf{w}(\boldsymbol{\xi}) \approx \widehat{\mathbf{w}}(\boldsymbol{\xi}) = \sum_{i=1}^{P_3} \mathbf{w}_i \varphi_i(\boldsymbol{\xi}), \tag{8.47}$$

where $\mathbf{w}_i \in \mathbb{R}^n$, $i = 1, 2, \ldots, P_3$ denote a set of undetermined vectors. The undetermined vectors are computed by employing a stochastic Bubnov–Galerkin projection scheme in which the residual error is made orthogonal to the approximating space of PC basis functions, that is,

$$\mathbf{K}(\boldsymbol{\xi}) \left[\sum_{i=1}^{P_3} \mathbf{w}_i \varphi_i(\boldsymbol{\xi}) \right] - \mathbf{f}(\boldsymbol{\xi}) \perp \varphi_k(\boldsymbol{\xi}), \quad k = 1, 2, \ldots, P_3. \tag{8.48}$$

Using PC expansions for $\mathbf{K}$, $\mathbf{f}$, and the definition of orthogonality between random functions (which follows from the inner product defined in (8.44)), the preceding equation reduces to

$$\sum_{i=1}^{P_1} \sum_{j=1}^{P_3} \mathbf{K}_i \mathbf{w}_j \langle \varphi_i(\boldsymbol{\xi}) \varphi_j(\boldsymbol{\xi}) \varphi_k(\boldsymbol{\xi}) \rangle - \sum_{i=1}^{P_2} \mathbf{f}_i \langle \varphi_i(\boldsymbol{\xi}) \varphi_k(\boldsymbol{\xi}) \rangle = 0, \quad \forall \ k = 1, 2, \ldots, P_3. \tag{8.49}$$

Rearranging terms in (8.49), we finally end up with a deterministic matrix system of equations of size $nP_3 \times nP_3$, which is in fact the total number of unknowns in (8.47). Similar procedures can be devised for the case when the chaos decomposition scheme is to be applied to solve the stochastic governing equations in their continuous form. In general, the computational effort associated with solving the final deterministic equations resulting from Bubnov–Galerkin projection can be significantly higher than solving a deterministic version of the governing equations (since the number of unknowns is a multiple of the length of the solution vector $\mathbf{w}$). Even so, this approach can be significantly faster than the MCS technique for problems with a modest number of random variables. For a more detailed description of the chaos projection scheme and its applications, see Nair (2004) and the references therein.

An alternative projection scheme can be formulated for efficiently solving linear stochastic PDEs using a stochastic version of the Krylov subspace described earlier in Section 6.3.1 (Nair 2001; Nair and Keane 2002b). Here, when solving a system of linear random algebraic equations of the form $\mathbf{K}(\boldsymbol{\xi})\mathbf{w}(\boldsymbol{\xi}) = \mathbf{f}(\boldsymbol{\xi})$, the solution process $\mathbf{w}$ is approximated as follows:

$$\widehat{\mathbf{w}}(\boldsymbol{\xi}) = c_1 \boldsymbol{\psi}_1(\boldsymbol{\xi}) + c_2 \boldsymbol{\psi}_2(\boldsymbol{\xi}) + \cdots + c_m \boldsymbol{\psi}_m(\boldsymbol{\xi}) = \boldsymbol{\Psi}(\boldsymbol{\xi})\mathbf{c}, \tag{8.50}$$

where $\mathbf{c} = \{c_1, c_2, \ldots, c_m\}^T \in \mathbb{R}^m$ denotes a set of undetermined coefficients and $\boldsymbol{\Psi}(\boldsymbol{\xi}) = [\boldsymbol{\psi}_1, \boldsymbol{\psi}_2, \ldots, \boldsymbol{\psi}_m] \in \mathbb{R}^{n \times m}$ is a matrix of basis vectors spanning the stochastic Krylov subspace

$$\mathcal{K}_m(\mathbf{K}(\boldsymbol{\xi}), \mathbf{f}(\boldsymbol{\xi})) = \text{span} \left\{ \mathbf{f}(\boldsymbol{\xi}), \mathbf{K}(\boldsymbol{\xi})\mathbf{f}(\boldsymbol{\xi}), \mathbf{K}(\boldsymbol{\xi})^2 \mathbf{f}(\boldsymbol{\xi}), \ldots, \mathbf{K}(\boldsymbol{\xi})^{m-1} \mathbf{f}(\boldsymbol{\xi}) \right\}. \tag{8.51}$$

The undetermined coefficients c_i can be computed by the Bubnov–Galerkin scheme, which involves imposing the orthogonality condition

$$\mathbf{K}(\boldsymbol{\xi})\boldsymbol{\Psi}(\boldsymbol{\xi})\mathbf{c} - \mathbf{f}(\boldsymbol{\xi}) \perp \boldsymbol{\psi}_i(\boldsymbol{\xi}), \quad \forall \ i = 1, 2, \ldots, m. \tag{8.52}$$

Applying the above condition[10] and using PC expansions for $\mathbf{K}$ and $\mathbf{f}$, we have the following reduced-order $m \times m$ system of equations for the undetermined coefficients

$$\left\langle \boldsymbol{\Psi}(\boldsymbol{\xi})^T \sum_{i=1}^{P_1} \mathbf{K}_i \varphi_i(\boldsymbol{\xi}) \boldsymbol{\Psi}(\boldsymbol{\xi}) \right\rangle \mathbf{c} = \sum_{i=1}^{P_2} \langle \varphi_i(\boldsymbol{\xi}) \boldsymbol{\Psi}(\boldsymbol{\xi})^T \mathbf{f}_i \rangle. \tag{8.53}$$

---

[10]Note that here we use the definition that two random vectors $\mathbf{r}_1$ and $\mathbf{r}_2$ are said to be orthogonal to each other if $\langle \mathbf{r}_1^T \mathbf{r}_2 \rangle = 0$.

In practice, the matrix $\langle \mathbf{K}(\boldsymbol{\xi}) \rangle^{-1}$ is used as a preconditioner to ensure accurate approximations using a small number of basis vectors. Numerical studies have shown that the stochastic Krylov method gives results of comparable accuracy to chaos projection schemes at a fraction of the computational effort. A detailed exposition of the theoretical and computational aspects of stochastic subspace projection schemes can be found in Nair (2004).

8.2.8 Output Bounds and Envelopes

When a nonprobabilistic model is used to represent uncertainties, alternative techniques are required to propagate uncertainty. For example, when all the uncertain input parameters are represented by intervals, then we seek to estimate the intervals (bounds) of the output quantities of interest. In theory, interval algebra can then be used to propagate uncertainties. This involves substituting all elementary operations (such as addition, subtraction, multiplication, etc.) by their interval extensions (Neumaier 1990). However, in practice, this can lead to bounds that are overly conservative and hence useless for engineering design purposes. A number of approaches that circumvent this difficulty in the context of solving *interval* linear algebraic equations (which arise when interval/fuzzy models are used to represent parameter uncertainty) can be found in the literature; see, for example, Rao and Chen (1998) and Muhanna and Mullen (2001).

For a general nonlinear function, it may be preferable in practice to estimate its upper and lower bounds using optimization techniques. These bounds represent the most-favorable and least-favorable values of the outputs. The phrase "antioptimization" is used to refer to the process of calculating the least-favorable response (Ben-Haim and Elishakoff 1990). To illustrate, in order to compute the upper and lower bounds on $f(\mathbf{x}, \boldsymbol{\xi})$ when $\boldsymbol{\xi}$ is represented by a convex model say $\mathcal{E}$, the following two optimization problems need to be solved.

$$f^{-}(\mathbf{x}) = \min_{\boldsymbol{\xi} \in \mathcal{E}} f(\mathbf{x}, \boldsymbol{\xi}) \quad \text{and} \quad f^{+}(\mathbf{x}) = \max_{\boldsymbol{\xi} \in \mathcal{E}} f(\mathbf{x}, \boldsymbol{\xi}). \tag{8.54}$$

Clearly, solving the preceding optimization problems repetitively for different values of $\mathbf{x}$ to calculate the bounding envelope of $f(\mathbf{x}, \boldsymbol{\xi})$ is computationally expensive. To reduce computational cost, surrogate-assisted algorithms can be used; see Chapter 7.

When the elements of $\boldsymbol{\xi}$ are modeled by fuzzy sets with appropriate membership functions, the uncertainty propagation problem involves computing the membership functions of the outputs of interest. This essentially involves solving a sequence of optimization problems of the form (8.54) for various α-cuts of the input membership functions. This is because, for fixed α (possibility level), the elements of $\boldsymbol{\xi}$ are interval variables. Langley (1999) proposed computational schemes for calculating a possibilistic measure of safety. The basic idea is to calculate the maximum amount of uncertainty that a given design can tolerate before it fails. We revisit this idea later in Section 8.8, where we discuss robust design approaches based on info-gap theory.

8.3 Taguchi Methods

Taguchi methods[TM] are based on the fundamental observation that *quality*[11] is a characteristic that must be designed into the product, rather than a requirement that is imposed only

---

[11]For the purpose of our discussion, quality can be interpreted as a measure of variability in design performance in the presence of uncertainty. We make this idea more concrete later in this section.

at the production stage as an on-line quality control exercise.[12] There is a huge amount of literature on Taguchi methods and they continue to be used by many industries for product design; see, for example, the text by Phadke (1989), and for a tutorial overview, see Sanchez (1994) and Parks (2001).

Taguchi methods divide the design process into three stages:

1. *Systems design:* This stage is concerned with activities such as construction and validation of a functional model to predict design performance, identifying the feasible regions of the control and noise factors, and so on.

2. *Parameter design:* Parameter design focuses on selecting appropriate values for the control factors by optimizing a figure of merit that embodies the designer's notion of product quality.

3. *Tolerance design:* The final stage involves a more detailed study of the designs identified in Stage 2 to decide on the tolerances that can be permitted for the control factors, which eventually form targets for on-line quality control.

In this section, we present an outline of Taguchi methods for robust design, with an emphasis on those elements that can be potentially exploited when applying formal optimization techniques to robust design. Our discussion focuses on the parameter design stage. To illustrate, let $y(\mathbf{x}, \boldsymbol{\xi})$ denote the performance measure of a design. As before, we denote the control factors (design variables) by $\mathbf{x}$ while $\boldsymbol{\xi}$ refers to the noise factors. Taguchi suggested that the optimum design should be selected by minimizing a mean-square deviation (MSD) measure of performance. Three types of objective functions (also sometimes referred to as Taguchi loss functions), that provide a measure of product quality can be defined, depending on the design intent (Trosset 1996).

1. If quality is measured by how close y comes to a target value y_{target}, then

$$\text{MSD}(\mathbf{x}) = \left\langle \left(y(\mathbf{x}, \boldsymbol{\xi}) - y_{\text{target}} \right)^2 \right\rangle \approx \frac{1}{m} \sum_{i=1}^{m} \left(y(\mathbf{x}, \boldsymbol{\xi}^{(i)}) - y_{\text{target}} \right)^2. \qquad (8.55)$$

2. If quality is measured by how small y is, then[13]

$$\text{MSD}(\mathbf{x}) = \left\langle y^2(\mathbf{x}, \boldsymbol{\xi}) \right\rangle \approx \frac{1}{m} \sum_{i=1}^{m} y^2(\mathbf{x}, \boldsymbol{\xi}^{(i)}). \qquad (8.56)$$

3. If quality is measured by how large y is, then

$$\text{MSD}(\mathbf{x}) = \left\langle \frac{1}{y^2(\mathbf{x}, \boldsymbol{\xi})} \right\rangle \approx \frac{1}{m} \sum_{i=1}^{m} \frac{1}{y^2(\mathbf{x}, \boldsymbol{\xi}^{(i)})}. \qquad (8.57)$$

---

[12]Taguchi methods and Robust Design are registered trademarks of the American Supplier Institute for Taguchi methodology.

[13]It may be noted that $\langle y^2 \rangle = \langle y \rangle^2 + \sigma_y^2$, where σ_y is the standard deviation of y. Hence, the mean square value can be interpreted as an aggregate merit function which ensures that the mean and standard deviation of y is simultaneously minimized.

Monte Carlo estimates of MSD are also given in the preceding equations, where m is the sample size and $\boldsymbol{\xi}^{(i)}$ denotes the ith realization of the noise factors $\boldsymbol{\xi}$ drawn from the joint pdf $\mathcal{P}(\boldsymbol{\xi})$.

Parameter design in Taguchi methods essentially involves finding the optimal value of the control factors $\mathbf{x}$ by minimizing MSD$(\mathbf{x})$. In practice, the objective function is chosen to be the signal-to-noise ratio, $-10\log_{10}$ MSD$(\mathbf{x})$. In the original approach of Taguchi, this minimization problem is solved using techniques from the DoE literature with the control and noise factors treated separately. Specifically, an orthogonal array is used to arrive at settings for the control factors – this is referred to as the "control array" or "inner array". For each setting of the control factors $\mathbf{x}$, a second orthogonal array is created for deciding settings for the noise factors $\boldsymbol{\xi}$ – the second array is referred to as the "noise array" or "outer array". In other words, for each value of $\mathbf{x}$ in the inner array, the signal-to-noise ratio is computed using the outer array. Subsequently, the data thus generated is analyzed by standard analysis of variance techniques to identify values of $\mathbf{x}$ that result in robust performance.

It must be clear from the above description that Taguchi methods essentially use a one-shot DoE approach. This is in contrast to the surrogate-assisted optimization strategies discussed in Chapter 7 where the search for the optimum is carried out in a sequential stage-wise fashion. It should, however, be noted that Taguchi's approach was originally developed with the understanding that in some situations the design performance can be evaluated only using physical experiments, which can be time consuming. Hence, only a small set of candidate designs can be evaluated within the allocated time frame. When computer models are used to predict design performance, Taguchi's approach to robust design based on separate inner and outer arrays can be highly inefficient. A detailed account of the controversies surrounding Taguchi's methods can be found in the panel discussion edited by Nair (1992).

8.4 The Welch–Sacks Method

Welch et al. (1990) and Welch and Sacks (1991) suggested a more efficient alternative to Taguchi's approach where, instead of using separate inner and outer arrays, the control and noise parameters are grouped together. In other words, we define the new variable $\widetilde{\mathbf{x}} = \{\mathbf{x}, \boldsymbol{\xi}\} \in \mathbb{R}^{p+p_\xi}$ and a DoE matrix is then created for $\widetilde{\mathbf{x}}$. Clearly, this approach, which uses a single experimental array for both control and noise factors, will require far fewer evaluations of the analysis code. Running the analysis model at the points in the DoE matrix, we ultimately end up with the training dataset $\{\widetilde{\mathbf{x}}^{(i)}, y(\widetilde{\mathbf{x}}^{(i)})\}, i = 1, 2, \ldots, m$. This training dataset can be used to construct a surrogate model $\widehat{y}(\widetilde{\mathbf{x}})$ from which MSD$(\mathbf{x})$ can be approximated (alternatively, a surrogate model can be directly constructed for MSD).[14] The resulting approximation for the signal-to-noise ratio function can then be minimized to efficiently determine the optimum solution. This is again a one-shot surrogate-assisted optimization strategy, since the baseline surrogate is not updated.

As discussed in the introduction to Chapter 7, one-shot approaches to surrogate-assisted optimization are fraught with practical difficulties, particularly when the accuracy of the surrogate is poor. It is hence desirable to develop alternative strategies that enable the surrogate

---

[14]It is worth noting that the approach discussed here can be extended to cases with multiple performance criteria by defining a general loss function of the form $l[y_1(\mathbf{x}, \boldsymbol{\xi}), y_2(\mathbf{x}, \boldsymbol{\xi}), \ldots, y_q(\mathbf{x}, \boldsymbol{\xi})]$, where $y_1, y_2, \ldots, y_q$ denote indicators of design performance. A robust solution can then be found by minimizing the expected loss function (Trosset 1996). An alternative approach based on the solution of a compromise decision support problem has been presented by Chen et al. (1996).

model to be updated in a stage-wise fashion. In principle, it is possible to devise criteria that enable the judicious selection of additional points to improve the surrogate. To briefly illustrate one such possibility for surrogate-assisted robust design using Gaussian process models, consider the case when the original *deterministic* design problem involves solving

$$\mathbf{x}^* = \arg \min_{\mathbf{x}} y(\mathbf{x}, \langle \boldsymbol{\xi} \rangle), \qquad (8.58)$$

where we have assumed that the noise factors are set to their nominal values. It follows from (8.56) that Taguchi's approach involves solving the following reformulation of the original optimization problem

$$\mathbf{x}^* = \arg \min_{\mathbf{x}} \int_{\mathcal{E}} y^2(\mathbf{x}, \boldsymbol{\xi}) \mathcal{P}(\boldsymbol{\xi}) \, d\boldsymbol{\xi}. \qquad (8.59)$$

To simplify notation, let us denote $y^2(\mathbf{x}, \boldsymbol{\xi})$ by $\phi(\mathbf{x}, \boldsymbol{\xi})$. Let a Gaussian process model for $\phi(\mathbf{x}, \boldsymbol{\xi})$ be constructed using the training dataset $\{\widetilde{\mathbf{x}}^{(i)}, y^2(\widetilde{\mathbf{x}}^{(i)})\}$, $i = 1, 2, \ldots, m$. Recollect that Gaussian process modeling results in a posterior distribution of the form $\mathfrak{P}(\mathbf{x}, \boldsymbol{\xi}) \sim \mathcal{N}(\widehat{\phi}(\mathbf{x}, \boldsymbol{\xi}), C_\phi^2(\mathbf{x}, \boldsymbol{\xi}))$ – see Section 8.2.5. If the posterior distribution is used as a surrogate for $\phi(\mathbf{x}, \boldsymbol{\xi})$, then the optimization problem (8.59) becomes

$$\mathbf{x}^* = \arg \min_{\mathbf{x}} \int_{\mathcal{E}} \mathfrak{P}(\mathbf{x}, \boldsymbol{\xi}) \mathcal{P}(\boldsymbol{\xi}) \, d\boldsymbol{\xi}. \qquad (8.60)$$

Since $\mathfrak{P}(\mathbf{x}, \boldsymbol{\xi})$ is a Gaussian random field, the above objective function is a random variable given $\mathbf{x}$. This follows from the same line of reasoning used earlier in Section 8.2.5 to describe the application of Gaussian process models to uncertainty propagation. Specifically, it is possible to compute probabilistic error bounds on the integral. Note that these error bounds have nothing to do with randomness in $\boldsymbol{\xi}$, rather they indicate an estimate of the approximation error in the surrogate model (which is inevitable since a finite number of training points is used). If we denote the integral in (8.60) by $\widehat{I}(\mathbf{x})$, then its mean (μ_I) and standard deviation (σ_I) can be computed using the approach discussed in Section 8.2.5; see (8.35).

An approximate optimization problem can be solved for $\mathbf{x}^*$ by minimizing $\mu_I(\mathbf{x})$. Since $\sigma_I(\mathbf{x})$ can be interpreted as an estimate of the error in $\widehat{I}(\mathbf{x})$, it can be used to decide additional locations in the design space at which the computer model must be run to best update the surrogate. Alternatively, on the lines of the idea discussed earlier in Section 7.3, the new objective function $\mu_I(\mathbf{x}) - \rho \sigma_I(\mathbf{x})$ can be minimized, where $\rho \geq 0$ is a user-specified parameter that controls the balance between minimizing the actual objective (local exploitation) and selecting points that may potentially improve the surrogate accuracy (global exploration). Other implementation aspects of this approach will be similar to those discussed in Section 7.3.

8.5 Design for Six Sigma

In this section, we discuss some alternative formulations for robust design. To begin with, consider unconstrained minimization of the function $f(\mathbf{x}, \boldsymbol{\xi})$, where $\mathbf{x} \in \mathbb{R}^p$ is the vector of design variables and $\boldsymbol{\xi} \in \mathbb{R}^{p_\xi}$ is a set of random variables with joint pdf $\mathcal{P}(\boldsymbol{\xi})$. As discussed in previous sections, Taguchi methods or the Welch–Sacks method can be applied to solve this problem. An alternative approach would be to pose the robust design problem as a multiobjective optimization problem involving simultaneous minimization of the mean of

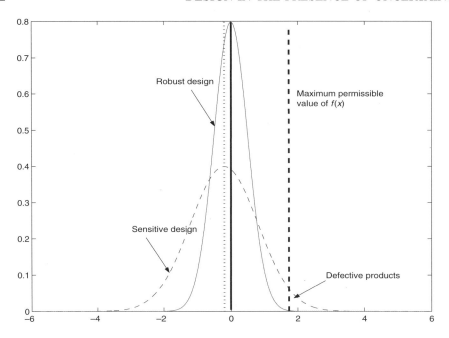

Figure 8.5 Illustration of the characteristics of a robust design obtained by simultaneously minimizing nominal performance and its variance. The pdfs of the objective function $f(x)$ for the robust and a nonrobust design are shown by the solid and dashed curves, respectively. The solid vertical line corresponds to the nominal performance of the robust design while the vertical dotted line shows that of the nonrobust design, which has better nominal performance. The dashed vertical line represents the maximum permissible value of $f(x)$. It can be noted that even though the nominal performance of the nonrobust design is well within the design requirement perturbations arising from uncertainties may lead to violation of the specified bound on $f(x)$. In contrast, the robust solution satisfies the design requirements and has a smaller probability of exceeding the maximum permissible value of $f(x)$.

the objective $\langle f(\mathbf{x}, \boldsymbol{\xi}) \rangle$ and its variance $\langle (f(\mathbf{x}, \boldsymbol{\xi}) - \langle f(\mathbf{x}, \boldsymbol{\xi}) \rangle)^2 \rangle$; see, for example, Chen et al. (2000). The overall goal of this formulation (or in fact any robust design procedure) is sketched out in Figure 8.5.

We could minimize the aggregate objective function $\mu_f(\mathbf{x}) + \mathcal{W}\sigma_f(\mathbf{x})$, where $\mathcal{W}$ is a user-specified weighting parameter, and μ_f and σ_f denote the mean and standard deviation of $f(\mathbf{x}, \boldsymbol{\xi})$. Clearly, the parameter $\mathcal{W}$ determines the trade-off between nominal performance and robustness. Design for six sigma[15] essentially involves maintaining six standard deviations of performance variation with the specified acceptable limits (Koch et al. 2004). More specifically, given a design specification of the form, $f^- \leq f(\mathbf{x}, \boldsymbol{\xi}) \leq f^+$, $\forall \boldsymbol{\xi} \in \mathcal{E}$, this approach seeks to ensure that $\mu_f \pm 6\sigma_f$ lies within f^- and f^+. If f has a Gaussian distribution, then design for six sigma essentially involves specifying a target reliability level of 99.9999998% (i.e., the probability of failure cannot exceed 2×10^{-7}). Of course, in many

---

[15]It is worth noting here that design for six sigma is essentially a quality improvement process and philosophy; more detailed overviews can be found in texts dedicated to this topic (Pyzdek 2000). However, here we primarily focus on optimization formulations which can be employed to implement the central aim of design for six sigma.

problems of practical interest, $f(\mathbf{x}, \boldsymbol{\xi})$ will be non-Gaussian – here, design for six sigma may be interpreted as an approach that optimizes for worst-case scenarios in some sense.

To illustrate how this idea can be extended to constrained design problems, consider the case when we seek to minimize $f(\mathbf{x}, \boldsymbol{\xi})$ subject to inequality constraints of the form $g_i(\mathbf{x}, \boldsymbol{\xi}) \leq 0$. To ensure that the objective and constraint functions lie within $\pm 6\sigma$ limits, the original optimization problem may be reformulated as follows:[16]

$$
\begin{aligned}
\textbf{Minimize:} \quad & \mu_f(\mathbf{x}) + \mathcal{W}\sigma_f(\mathbf{x}) \\
\textbf{Subject to:} \quad & \mu_{g_i}(\mathbf{x}) + \mathcal{W}\sigma_{g_i}(\mathbf{x}) \leq 0,
\end{aligned}
\tag{8.61}
$$

where μ_{g_i} and σ_{g_i} denote the mean and standard deviation of the ith constraint function, respectively.

Note that to calculate the objective and constraint functions in (8.61) the mean and standard deviation of $f(\mathbf{x}, \boldsymbol{\xi})$ and $g(\mathbf{x}, \boldsymbol{\xi})$ have to be calculated. This can be carried out using any of the approaches discussed previously in Section 8.2; see also Koch et al. (2004) and Eldred et al. (2002).

8.6   Decision-theoretic Formulations

In this section, we present an overview of decision-theoretic formulations for robust design (Huyse 2001; Trosset et al. 2003). To illustrate, consider the problem where we wish to minimize $f(\mathbf{x}, \boldsymbol{\xi})$, where $\mathbf{x} \in \mathcal{X}$ denotes the decision variables controlled by the designer and $\boldsymbol{\xi} \in \mathcal{E}$ represents the uncertain variables beyond the control of the designer. The (unattainable) goal is to find $\mathbf{x}^* \in \mathcal{X}$ such that for every $\boldsymbol{\xi} \in \mathcal{E}$

$$
f(\mathbf{x}^*, \boldsymbol{\xi}) \leq f(\mathbf{x}, \boldsymbol{\xi}) \quad \forall \, \mathbf{x} \in \mathcal{X}.
\tag{8.62}
$$

The above problem of finding $\mathbf{x}^* \in \mathcal{X}$ that simultaneously minimizes $f(\mathbf{x}, \boldsymbol{\xi})$ for each $\boldsymbol{\xi} \in \mathcal{E}$ is the central problem of statistical decision theory (Pratt et al. 1996). Clearly, the condition (8.62) is too stringent and hence finding $\mathbf{x}^*$ is impossible unless it is relaxed. Two approaches are commonly used in this context. The first involves the solution of the following minimax optimization problem

$$
\mathbf{x}^* = \min_{\mathbf{x} \in \mathcal{X}} \phi(\mathbf{x}), \quad \text{where} \quad \phi(\mathbf{x}) = \max_{\boldsymbol{\xi} \in \mathcal{E}} f(\mathbf{x}, \boldsymbol{\xi}).
\tag{8.63}
$$

The second approach, motivated by Bayes' principle, involves solving

$$
\mathbf{x}^* = \min_{\mathbf{x} \in \mathcal{X}} \phi(\mathbf{x}), \quad \text{where} \quad \phi(\mathbf{x}) = \int_{\mathcal{E}} f(\mathbf{x}, \boldsymbol{\xi}) \mathcal{P}(\boldsymbol{\xi}) \, d\boldsymbol{\xi},
\tag{8.64}
$$

and $\mathcal{P}(\boldsymbol{\xi})$ denotes the pdf of the uncertain parameters. In this approach, $\mathbf{x}^*$ is referred to as the Bayes' decision and $\phi(\mathbf{x}^*)$ is referred to as the Bayes' risk.[17]

Clearly, the minimax principle is conservative since it optimizes for the worst-case scenario.[18] In contrast, the Bayes principle is essentially concerned with the average-case

---

[16]Many other variants of this general idea are possible; see, for example, Koch et al. (2004). Another possibility is to use the Taguchi loss function defined earlier in (8.56) as the objective function to be minimized.

[17]This notion has also been discussed earlier in Section 5.3.6 in the context of surrogate modeling via empirical risk minimization.

[18]As an aside, it is worth noting here that the minimax formulation can also be used in cases where an interval or convex model is used to represent the uncertain parameters.

performance since only the mean objective is minimized. Huyse (2001) conducted detailed numerical studies on the application of the latter approach to robust design of airfoil shapes against uncertainty in the Mach number. The objective was to minimize the drag coefficient (C_d) subject to a lower bound on the lift coefficient (C_l), given a statistical model for uncertainty in the Mach number (M). A second-order Taylor expansion of C_d with respect to M around its nominal value was used to efficiently compute the objective function $\int_{\mathcal{E}} C_d(\mathbf{x}, M) \mathcal{P}(M) \, dM$. Numerical studies suggest that this approach gives improved designs compared to multipoint techniques where a linear combination of C_d values at a fixed set of Mach numbers is minimized.

Trosset et al. (2003) suggested that the robust design formulation based on Bayes' principle can be efficiently implemented using a surrogate-assisted optimization strategy involving the following steps: (1) evaluate the original objective $f(\mathbf{x}, \boldsymbol{\xi})$ at a set of design sites by varying both $\mathbf{x}$ and $\boldsymbol{\xi}$, (2) construct a computationally cheap surrogate model $\widehat{f}(\mathbf{x}, \boldsymbol{\xi})$, (3) use the surrogate model in lieu of the original objective to arrive at the approximate objective $\widehat{\phi}(\mathbf{x})$, and (4) minimize the approximation $\widehat{\phi}(\mathbf{x})$.

In principle, it is possible to update the surrogate model using the new design points generated by solving the approximate problem. Note here that we are interested in minimizing $\phi(\mathbf{x})$ and not the original objective function $f(\mathbf{x}, \boldsymbol{\xi})$. When a Gaussian process model is used as a surrogate, it is possible to derive error bars for the approximate objective function $\widehat{\phi}(\mathbf{x})$. These error bars (variance) can subsequently be used to formulate a surrogate-assisted optimization strategy on the lines of that discussed earlier in the context of the Welch–Sacks method.

8.7 Reliability-based Optimization

In this section, we outline reliability-based design optimization formulations that take into account higher-order statistics of the objective and constraint functions. The goal of reliability-based optimization is to minimize an objective function defined in terms of the nominal performance and its variability subject to a target reliability on the constraint functions. In other words, an original inequality constraint of the form $g \leq 0$ is replaced with a reliability constraint of the form $P[g \leq 0] \leq R_g$, where R_g is the specified reliability target. The probability that the inequality constraint is not satisfied is hence given by $1 - R_g$.

Consider the case when we seek to minimize the objective function $f(\mathbf{x}, \boldsymbol{\xi})$ subject to inequality constraints of the form $g(\mathbf{x}, \boldsymbol{\xi}) \leq 0$. As before, $\mathbf{x} \in \mathbb{R}^p$ and $\boldsymbol{\xi} \in \mathbb{R}^{p_\xi}$ denote the design variables and noise factors, respectively. The reliability-based design optimization formulation can be formally stated as

$$\begin{aligned} \underset{\mathbf{x}}{\text{Minimize}:} \quad & l[f(\mathbf{x}, \boldsymbol{\xi})] \\ \text{Subject to}: \quad & P[g_i(\mathbf{x}, \boldsymbol{\xi})] \leq R_{g_i}, \end{aligned} \tag{8.65}$$

where $l[f(\mathbf{x}, \boldsymbol{\xi})]$ is the new objective function that ensures that the mean and variance of $f(\mathbf{x}, \boldsymbol{\xi})$ are simultaneously minimized, and R_{g_i} is the reliability target for the ith constraint. For example, $l[f] = \mu_f(\mathbf{x}) + \mathcal{W}\sigma_f(\mathbf{x})$, where μ_f and σ_f denote the mean and standard deviation of $f(\mathbf{x}, \boldsymbol{\xi})$, respectively, and $\mathcal{W}$ is a user-specified weighting parameter. Alternative objective functions can be formulated such as the Taguchi quality measures defined earlier in Section 8.3 or we could maximize the probability that $f(\mathbf{x}, \boldsymbol{\xi})$ is less than a specified upper bound, that is, $P[f(\mathbf{x}, \boldsymbol{\xi}) \leq f^+]$.

It may be noted from (8.65) that at each function evaluation (i.e., at each given $\mathbf{x}$) we need to calculate the probability of every constraint value at this point being greater or less than specified threshold values.[19] These probabilities are dictated by the tails of the probability distributions of the constraint functions. This is in contrast to the robust design formulations discussed in earlier sections where it is necessary to calculate only the first two statistical moments of f and g. Hence, the reliability-based optimization formulation is useful in practice only if the tails of the constraint distributions can be estimated accurately and efficiently.

The accuracy with which the constraints in (8.65) can be evaluated depends on the fidelity of the joint pdf of $\boldsymbol{\xi}$ and the numerical scheme employed for reliability analysis. If limited data is available for specification of the joint pdf $\mathcal{P}(\boldsymbol{\xi})$, then significant errors may arise in the constraints, particularly when the target reliability is very close to 1 (i.e., a small probability of infeasibility is allowed). It may be preferable in such situations to use a robust design formulation such as those described in previous sections. Another potential source of error arises from the numerical scheme employed for reliability analysis. In practice, FORM and SORM discussed earlier in Section 8.2.4 are often used since the MCS technique can be prohibitively expensive. The approximations for the constraints calculated using FORM and SORM can be poor when $g(\mathbf{x}, \boldsymbol{\xi})$ is a highly nonlinear function of the uncertain variables and p_ξ is large; see Thacker et al. (2001) for details.

From a computational viewpoint, when FORM or SORM approximations are used, it is necessary to solve a minimization problem to compute the MPP of failure at each function evaluation; see (8.29). Further, when a gradient-based nonlinear programming technique is used, it is also necessary to compute the sensitivities of the probability of constraint infeasibility. Further details on practical aspects of reliability-based design optimization can be found in the literature; see, for example, Du and Chen (2000), Gumbert et al. (2003) and Youn and Choi (2004).

8.8   Robust Design using Information-gap Theory

In this section, we briefly outline how info-gap decision theory (Ben-Haim 2001) can be applied to produce robust designs. This approach is nonprobabilistic since the uncertain parameters $\boldsymbol{\xi}$ are represented by an info-gap model of the form $\mathcal{E}(\alpha, \widetilde{\boldsymbol{\xi}})$, where α is the uncertainty parameter and $\widetilde{\boldsymbol{\xi}}$ denotes the center point (or nominal value) of $\boldsymbol{\xi}$; see (8.6) and (8.7) for examples of info-gap models. As discussed previously in Section 8.2.2, info-gap models are appropriate for problems with severe uncertainty where limited data is available to accurately estimate the joint pdf of $\boldsymbol{\xi}$.

In info-gap decision theory, the robustness of a design is defined as the maximum amount of uncertainty that it can tolerate before failure, that is,

$$\alpha_{\max}(\mathbf{x}) = \max\{\alpha : \text{ minimal requirements are always satisfied}\}, \qquad (8.66)$$

where $\alpha_{\max}(\mathbf{x})$ is referred to as the *immunity* to failure. For an unconstrained optimization problem, $\alpha_{\max}$ can hence be computed by solving

$$\alpha_{\max}(\mathbf{x}) = \max\{\alpha : \max_{\boldsymbol{\xi} \in \mathcal{E}(\alpha, \widetilde{\boldsymbol{\xi}})} f(\mathbf{x}, \boldsymbol{\xi}) \leq f^+\}, \qquad (8.67)$$

where f^+ is the highest permissible value of the objective function.

---

[19]If we set the objective function in (8.65) to $P[f(\mathbf{x}, \boldsymbol{\xi}) \leq f^+]$, then the probability of $f(\mathbf{x}, \boldsymbol{\xi})$ exceeding a threshold value needs to be computed as well.

For the general nonlinear programming problem where we seek to minimize $f(\mathbf{x}, \boldsymbol{\xi})$ subject to inequality constraints of the form $g_i(\mathbf{x}, \boldsymbol{\xi}) \leq 0$, the immunity to failure of a candidate design may be defined as

$$\alpha_{\max}(\mathbf{x}) = \max\{\alpha : \max_{\boldsymbol{\xi} \in \mathcal{E}(\alpha, \tilde{\boldsymbol{\xi}})} f(\mathbf{x}, \boldsymbol{\xi}) \leq f^+ \text{ subject to: } g_i(\mathbf{x}, \boldsymbol{\xi}) \leq 0\}. \qquad (8.68)$$

The basic idea is to find a robust design by varying $\mathbf{x}$ such that the immunity to failure is maximized. In other words, we wish to search for the design that can tolerate the largest amount of uncertainty and yet satisfy all the specified design requirements. This can be posed as a two-level optimization problem. In the first level, $\mathbf{x}$ is varied to maximize $\alpha_{\max}(\mathbf{x})$, and in the inner loop, the optimization subproblem in (8.68) is solved by varying $\boldsymbol{\xi}$ and α. This idea was applied by Hemez and Ben-Haim (2004) to optimization problems under uncertainty arising in model updating using test data.

It is worth noting that the computational expense associated with this formulation is comparable to reliability-based design optimization since a two-level problem is solved. However, it is possible to reduce this cost by employing surrogate-assisted optimization strategies such as those discussed in this chapter and earlier in Chapter 7.

8.9   Evolutionary Algorithms for Robust Design

In the case of evolutionary algorithms, it is possible to directly incorporate robustness considerations into the search process. One example is the noisy phenotype scheme for use in GA optimization proposed by Tsutsui and Ghosh (1997), which they refer to as Genetic Algorithms with Robust Solution Searching Schemes or GAs/RS³ in short. They introduced a Simple Evaluation Model (SEM) for finding robust solutions in conjunction with a GA. The only difference between this robust search scheme and the traditional GA lies in the evaluation component, where a random noise vector, δ, is added to the genotype before fitness evaluation. Hence, this approach is applicable only when uncertainty is restricted to components of the design vector $\mathbf{x}$. This robust search scheme generally operates on the basis that individuals whose fitness suffers because of uncertainty would most likely fail to reproduce in the selection process, while robust individuals would be more likely to survive across the GA generations.

Tsutsui and Ghosh (1999) later reported the Multiple Evaluation Method (MEM). In contrast to SEM, the MEM constructs k new intermediate chromosomes by adding several random vector noises δ_i (where $i = 1, 2, \ldots, k$), to the original chromosome and the fitnesses of the resulting k perturbed individuals are evaluated. Subsequently, the perceived fitness of an individual can be evaluated in different ways. In the Standard SEM, the perceived fitness value of an individual equals the fitness corresponding to the perturbed chromosome. In Average MEM, the perceived fitness value is given by the average fitness of all k perturbed individuals. If the perceived fitness is taken as the worst among the k perturbed individuals, then we have the Worst MEM. Note that worst would represent the minimum on a maximization optimization problem or vice versa for a minimization problem. Hence, the Average MEM may be considered as a more conservative variant of the SEM with the Worst MEM being the most conservative. The case study of Chapter 10 illustrates some of these approaches.

Arnold and Beyer (2002) also reported the study of an $(1 + 1)$ Evolution Strategy with isotropic normal mutations under Category II uncertainty. Their analysis presupposes that besides evaluating the fitness of the perturbed individual the true fitness of the parent individual should also be considered. The perceived fitness value of an individual is then taken as the worst or average of the parent and perturbed individual.

8.10   Concluding Remarks

The development of efficient strategies for designing complex systems in the presence of uncertainty is an area of intense research activity. In particular, much of the focus of ongoing research is on improving the computational efficiency of optimization strategies that explicitly deal with parameter uncertainty and the development of nonprobabilistic approaches that can cope with limited data. Other directions of research include the extension of uncertainty propagation and robust design methods to multidisciplinary systems (Allen and Maute 2004; DeLaurentis and Mavris 2000; Du and Chen 2002). For a summary of the current state of the art in this area, we recommend that the reader consult the CRC Engineering Design Reliability Handbook edited by Nikolaidis et al. (2004b).

The objective of this chapter was to provide the reader with an overview of some well-established techniques used in this area and highlight a selection of recent research directions. Even though our exposition is not exhaustive, we hope that this has provided the reader with a good starting point from which they may pursue an in-depth study of topics that are of interest. Numerical studies dealing with design optimization of space structures under uncertainty can be found in Section 10.6. An additional study that demonstrates the benefits offered by robust design methods for airfoil design is described in Chapter 11.

9

Architectures for Multidisciplinary Optimization

The optimization algorithms described in Chapters 3 and 7 can be applied to solve a wide class of engineering design problems since few restrictive assumptions were made regarding how the objective and constraint functions might be calculated. In this chapter, we focus on optimization of coupled systems, that is, problems where the objective and constraint function evaluations require iterations between two or more disciplinary analysis codes. Such design problems are "multidisciplinary" in nature since evaluating the performance of a candidate design involves executing (or iterating between) a suite of analysis codes representing different disciplinary perspectives. Perhaps, the best example of a multidisciplinary design scenario is aircraft design, where the designer has to consider the interactions between many disciplines including structures, aerodynamics, propulsion and flight control, when carrying out performance calculations and trade-off studies. A simplified overview of the coupling and types of data transferred between four disciplines that are important drivers in aircraft performance analysis is set out in Figure 9.1. An overview of the issues involved in aircraft design is given in Chapter 1. In this chapter, we recapitulate some of the key problems that lead to the need for specialized optimization algorithms/workflows to efficiently tackle multidisciplinary design problems.

A traditional practice in aerospace design is to employ a sequential staggered approach to solve multidisciplinary design problems, since large corporations are often divided along disciplinary lines into different divisions (possibly at separate geographical locations), each focusing on a different aspect of vehicle performance, for example, structures, aerodynamics, propulsion and flight control groups. This organizational structure has been put in place primarily to address the challenges arising from the ever-increasing complexity of modern aerospace vehicles, which calls for expertise in a wide variety of specialized fields. By appropriate division of labor among the specialist disciplinary groups, it becomes possible to solve complex design problems via the strategy of divide-and-conquer.

To illustrate the sequential approach to multidisciplinary design, consider the problem of integrated aerodynamics-structures optimization of aircraft wings. Here, the simplest approach is to first carry out aerodynamic shape optimization to decide optimum values for the design variables describing the wing planform geometry – structural considerations

Computational Approaches for Aerospace Design: The Pursuit of Excellence.   A. J. Keane and P. B. Nair
© 2005 John Wiley & Sons, Ltd

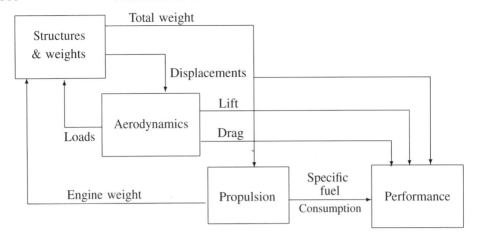

Figure 9.1 Graphical outline of interactions between structures, aerodynamics, propulsion and performance in aircraft design.

are not directly entered into the aerodynamic design formulation but are accounted for by imposing bound constraints on the airfoil thickness at several stations along the span of the wing. Subsequently, given the aerodynamic loading, the structural team can be given the task of designing the internal wing geometry (by varying the dimensions of the internal wing box structure, sizes and location of ribs, stiffeners, spars, etc.), to meet structural strength requirements.

Clearly, this sequential approach neglects the possibility of fully exploiting the coupling between aerodynamics and structures at the design stage to conduct trade-off studies, and hence convergence to a suboptimal solution is likely. For example, in the supersonic business jet design problem considered by Martins et al. (2004), the sequential approach leads to a design with a range of 7,137 nautical miles and structural weight of 44,772 lbs. In contrast, an integrated approach that takes the coupling in the problem physics into account produces a design with a range of 7,361 nautical miles and structural weight of 43,761 lbs. This case study serves to illustrate the potential gains[1] that may be achieved by adopting an integrated multidisciplinary perspective in aircraft design. Additionally, when a sequential approach is used to design multidisciplinary systems, it is possible that during the final stages of the design process potentially costly design modifications may become necessary to meet performance targets. Another undesirable feature of the sequential approach is that some disciplines will have to wait for data and design decisions generated by others before they are in a position to conduct their own discipline-specific design studies.

Multidisciplinary design optimization (MDO) is an enabling methodology for the design of complex systems, the physics of which involve couplings between various interacting disciplines/phenomena. The underlying focus of MDO methodology is to develop formal

---

[1]The actual gains achieved by a multidisciplinary approach will be problem dependent. In general, a multidisciplinary approach that exploits the coupling in the problem physics will always result in better designs than the sequential approach for systems with a high degree of coupling between the disciplines. As discussed in Chapter 1, the idea of adopting a multidisciplinary approach to aircraft design has been around for a long time and can be traced at least as far back as the research work that led to the first powered flight by the Wright brothers. However, it is only because of the dramatic advances made in computer hardware and numerical methods over the last decade that a truly multidisciplinary approach to design using high-fidelity models has started to look more realistic.

procedures for exploiting the synergistic effects of the coupling in the problem physics at every stage of the design process. One advantage offered by this methodology is the achievement of calendar-time compression via concurrency in the product design and development cycle. It also allows for trade-off studies between performance, manufacturing, maintainability, economics, and life cycle issues to be conducted at all stages of the design process. Hence, the adoption of MDO methodology is expected to lead to not only better designs but also the establishment of a more physically meaningful design practice as compared to the sequential approach in which the synergistic effects of coupling are not dealt with explicitly. Summarizing the white paper prepared by the AIAA MDO Technical Committee in 1991, Giesing and Barthelemy (1998) state that

"MDO provides a human-centered environment 1) for the design of complex systems, where conflicting technical and economic requirements must be rationally balanced, 2) that compresses the design cycle by enabling a concurrent engineering process where all the disciplines are considered early in the design process, while there remains much design freedom and key trades can be effected for an overall system optimum, 3) that is adaptive as various analysis/simulation capabilities can be inserted as the design progresses and the team of designers tailor their tool to the need of the moment, and 4) that contains a number of generic tools that permit the integration, of the various analysis capabilities, together with their sensitivity analyses and that supports a number of decision-making problem formulations".

The potential for MDO algorithms in the aerospace vehicle design process is significant and hardly any formal justification is required for their application. However, in spite of extensive research work on this topic, MDO is yet to find widespread acceptability in industrial design practice. This can be attributed to organizational and computational challenges beyond those encountered in single-discipline optimization (Sobieszczanski and Haftka 1997). Organizational challenges arise because of the existing organizational structure of large aerospace corporations, where the different disciplinary groups expect to have some degree of autonomy and decision-making power in the design process. It is also commonly the case that each discipline tends to work with their own set of legacy computer codes that have been developed over many years and which they often wish to maintain as their own. Computational challenges arise from the increasing push toward the use of high-fidelity computational models to predict design improvements. In summary, the prevailing organizational scenario and computational cost considerations rule out the direct application of an optimizer coupled to a large monolithic multidisciplinary analysis code carrying out objective and constraint evaluations, except perhaps for conceptual design studies.[2] Software integration and establishment of common standards for interdisciplinary data transfer are also challenging. Even though, as discussed earlier in Chapter 2, significant advances have been made in alleviating some of these difficulties, many challenges remain to be overcome.

We next describe a model problem to illustrate some of the key issues involved in MDO and define the terminology used in the remainder of this chapter. Subsequently, we move on to discuss various architectures for MDO. We use the term "architecture" to refer to how

---

[2]A number of software systems for carrying out conceptual aircraft design studies use an MDO approach, for example, ACSYNT (Vanderplaats 1976) and FLight OPtimisation System (FLOPS) (McCullers 1984). Here, because of the simplicity of the disciplinary analysis codes (which are usually semiempirical or based on low-fidelity physics), software integration is straightforward and, further, the objective and constraint function evaluations are computationally cheap. We do not focus on this type of approach in this chapter. Instead, our focus is on the case when high-fidelity disciplinary analysis codes involving, for example, CSM and CFD are employed to conduct preliminary or detailed design studies.

the multidisciplinary system is decomposed and the optimization formulation employed to meet design requirements. More broadly, MDO methods can be classified, on the basis of the design architecture or formulation, the optimization algorithms used for design space search, the methods used for constructing approximation models to accelerate the design process, and the framework employed for managing any variable-fidelity models. We do not aim for an exhaustive exposition in this chapter since MDO algorithm development is an area of active research and new approaches continue to be proposed in the literature. Moreover, research work focusing on theoretical analysis of MDO algorithms has only recently been initiated. Instead, we primarily focus on describing the fundamental ideas used in designing MDO architectures and the computational properties of a few of the more promising approaches.

9.1 Preliminaries

To set the stage for the exposition of MDO algorithms, we first describe a model problem involving integrated aerodynamics-structures optimization. Clearly, the MDO architectures described here can be formulated to deal with more general multidisciplinary problems with an arbitrary number of disciplines. However, we have chosen to describe various approaches using a two-discipline problem to ensure clarity of presentation. We first discuss the nomenclature and terminology used in MDO using the model problem to make the ideas that follow more concrete.[3]

9.1.1 A Model Problem

Consider the coupled system involving two disciplines (or subsystems), namely, aerodynamics and structures, shown in Figure 9.2. The coupling shown here is typical of aerodynamics-structures interaction encountered in steady-state aeroelastic analysis of flexible aircraft wings. The variables in Figure 9.2 and potential choices for them are described in detail below:

- x_m denotes the vector of *multidisciplinary design variables*, that is, variables that are common to (shared by) both disciplines. For the model problem, x_m are the design variables used to parameterize the wing surface geometry such as the wing taper, sweep, aspect ratio, and so on.

- x_1 and x_2 are the vectors of *disciplinary design variables* corresponding to the aerodynamics and structures disciplines, respectively; for example, x_1 may include airfoil geometry at various locations along the wing span outside the wing box, whereas x_2 may include wing box structural dimensions and skin thicknesses. Note that x_1 and x_2 are disjoint sets.

- w_1 and w_2 are the internal *state variables* computed during aerodynamic and structural analysis, respectively. In other words, these are the outputs obtained by numerically solving appropriate conservation laws governing the discipline-specific problem

---

[3]A more general notation applicable to problems with an arbitrary number of disciplines is presented by Cramer et al. (1994). However, the notation used here is sufficiently flexible to illustrate the conceptual ideas used in MDO formulations.

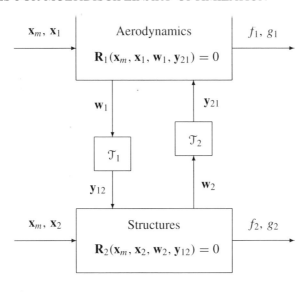

Figure 9.2 Schematic representation of a model MDO problem involving integrated aerodynamics-structures analysis.

physics. Here, $\mathbf{w}_1$ contains the flow variables being solved for by the aerodynamics solver and $\mathbf{w}_2$ contains the structural displacements.

- f_1 and f_2 are the objective functions arising from aerodynamic and structural considerations, respectively; for example, f_1 could be the aerodynamic drag, whereas f_2 could be the structural weight.

- g_1 and g_2 are the constraint functions corresponding to aerodynamics and structures, respectively; for example, g_1 may include limits on the angle of attack and lift coefficient, whereas g_2 may include constraints ensuring that the wing structure does not fail because of the stress distributions induced by aerodynamic loading and other external forces acting on it.

- $\mathbf{y}_{12}$ is the vector of coupling variables computed by the aerodynamics discipline that is required to carry out structural analysis, that is, the aerodynamic loading on the wing. Note that the loading is computed from the aerodynamic pressure distribution that forms a part of the aerodynamic state vector $\mathbf{w}_1$. Symbolically, this involves a transformation of the form $\mathbf{y}_{12} = \mathcal{T}_1(\mathbf{x}_m, \mathbf{x}_1, \mathbf{x}_2, \mathbf{w}_1)$.

- $\mathbf{y}_{21}$ denotes the vector of coupling variables computed by the structures discipline that is required to carry out aerodynamic analysis, that is, the deformed shape of the wing, which is calculated using the structural displacement vector $\mathbf{w}_2$ via a transformation of the form $\mathbf{y}_{21} = \mathcal{T}_2(\mathbf{x}_m, \mathbf{x}_1, \mathbf{x}_2, \mathbf{w}_2)$. Note that the deformed shape of the wing is required to construct the mesh for solving the PDEs governing aerodynamics.

The problem of constructing the transformations $\mathcal{T}_1$ and $\mathcal{T}_2$ that facilitate interdisciplinary data transfer for integrated aerodynamics-structures interaction studies has been extensively studied in the literature; see, for example, Maute et al. (2001) and the references therein. Note that these transformations usually involve curve fits to the spatial distributions of the state variables $\mathbf{w}_1$ and $\mathbf{w}_2$ since the meshes at the interface between aerodynamics and structures may not share a common set of nodes. For example, the transformation $\mathbf{y}_{12} = \mathcal{T}_1(\mathbf{w}_1)$ may involve calculating the aerodynamic loads on the nodes of the structural mesh using the pressure distributions obtained from flow analysis which uses a different mesh. Here, $\mathcal{T}_1$ is normally constructed using a consistent and conservative approach (Farhat et al. 1998).

Note that the dimensions of the coupling variables are usually much smaller than those of the state variables, that is, $\dim(\mathbf{y}_{12}) < \dim(\mathbf{w}_1)$ and $\dim(\mathbf{y}_{21}) < \dim(\mathbf{w}_2)$. Hence, the transformations $\mathcal{T}_1$ and $\mathcal{T}_2$ may also facilitate data compression. For example, instead of transferring the aerodynamic loading at all nodes of the structural mesh, a spline or RBF approximation can be fitted to the spatial distribution of the aerodynamic loading and the coefficients of this approximation can be passed instead to the structures discipline. The *coupling bandwidth* of the multidisciplinary system refers to the dimensions of $\mathbf{y}_{12}$ and $\mathbf{y}_{21}$. When the coupling bandwidth is high, we refer to the multidisciplinary system as tightly coupled. As we shall see later, tightly coupled systems can be difficult to analyze and optimize efficiently.

We say that *single-discipline feasibility* for aerodynamics is achieved when we have successfully solved for the flow state variables $\mathbf{w}_1$ given the design variables $(\mathbf{x}_m, \mathbf{x}_1)$ and the deformed wing shape $\mathbf{y}_{21}$. Similarly, single-discipline feasibility for structures is achieved when we have successfully solved for the structural displacements $\mathbf{w}_2$ given the design variables $(\mathbf{x}_m, \mathbf{x}_2)$ and the aerodynamic loading $\mathbf{y}_{12}$. Note that single-discipline feasibility has nothing to do with the disciplinary constraints being satisfied.

The state equations for the two disciplines can be mathematically expressed as a simultaneous system of equations as follows:

$$\mathbf{R}_1(\mathbf{x}_m, \mathbf{x}_1, \mathbf{w}_1, \mathbf{y}_{21}(\mathbf{w}_2)) = 0, \tag{9.1}$$

$$\mathbf{R}_2(\mathbf{x}_m, \mathbf{x}_2, \mathbf{w}_2, \mathbf{y}_{12}(\mathbf{w}_1)) = 0. \tag{9.2}$$

The discrete governing equations in (9.1) and (9.2) are obtained in practice by applying an appropriate numerical discretization scheme to the conservation laws (or PDEs) governing the discipline-specific physics. When aerodynamic analysis is carried out using an Euler or Navier–Stokes solver, $\mathbf{R}_1$ is a system of nonlinear algebraic equations. However, structural analysis is usually carried out using linear finite element analysis and hence, in that case, $\mathbf{R}_2$ denotes a system of linear algebraic equations of the form $\mathbf{K}\mathbf{w}_2 = \mathbf{f}$, where $\mathbf{K}$ is the structural stiffness matrix and $\mathbf{f}$ is the force vector, which is a function of $\mathbf{y}_{12}$. However, if a detailed structural optimization of the wing structure is to be carried out, then we have to study panel buckling and, in that case, the structural governing equations will also be nonlinear. Note that in (9.1) and (9.2) we have explicitly indicated the dependence of the coupling variables on the state variables.

Sometimes, it is notationally convenient to employ an equivalent representation of the governing equations (9.1) and (9.2) in the form

$$\mathbf{a}_1 = \mathbf{A}_1(\mathbf{x}_m, \mathbf{x}_1, \mathbf{a}_2), \tag{9.3}$$

$$\mathbf{a}_2 = \mathbf{A}_2(\mathbf{x}_m, \mathbf{x}_2, \mathbf{a}_1), \tag{9.4}$$

where $\mathbf{a}_i$ are the coupling variables from the ith discipline. The coupling in the governing equations (9.3) and (9.4) implies that $\mathbf{a}_i$ are implicit functions of all the design variables, that is, $\mathbf{a}_1 = \mathbf{a}_1(\mathbf{x}_m, \mathbf{x}_1, \mathbf{x}_2)$ and $\mathbf{a}_2 = \mathbf{a}_2(\mathbf{x}_m, \mathbf{x}_1, \mathbf{x}_2)$.

As we shall see later, the alternative representation of the governing equations given by (9.3) and (9.4) makes the formulation of some MDO architectures notationally convenient. Note that, for convenience, we sometimes compactly denote the vector containing all the design variables by $\mathbf{x} = \{\mathbf{x}_m, \mathbf{x}_1, \mathbf{x}_2\} \in \mathbb{R}^p$.

9.1.2 Multidisciplinary Analysis

Multidisciplinary analysis is the process whereby an iterative scheme is applied to the disciplinary governing equations to establish equilibrium or *multidisciplinary feasibility*. An iterative scheme is required here because of the nature of the coupling between the disciplinary analysis equations as indicated in (9.1) and (9.2). It can be noted that even if both the disciplinary governing equations are linear an iterative scheme will still be required to carry out multidisciplinary analysis. A brief description of approaches to multidisciplinary analysis has already been provided in Section 2.3. Here, we make some of the ideas and the algorithms more concrete in the context of the model problem under consideration.

One way to arrive at a multidisciplinary feasible solution for given values of $(\mathbf{x}_m, \mathbf{x}_1, \mathbf{x}_2)$ is to iterate between both the disciplines using an initial guess for either $\mathbf{y}_{12}$ or $\mathbf{y}_{21}$. The steps involved are outlined in Algorithm 9.1 for the case when we start with an initial guess for $\mathbf{y}_{21}$. Note that this algorithm is sequential since structural analysis can be carried out only after the coupling variables computed from aerodynamic analysis become available; see step (3). The iterations can be terminated when $||\mathbf{y}_{12}^{k+1} - \mathbf{y}_{12}^k|| \le \varepsilon$ and $||\mathbf{y}_{21}^{k+1} - \mathbf{y}_{21}^k|| \le \varepsilon$, where ε is a user-specified tolerance parameter. This sequential iterative process can be interpreted as a generalized Gauss–Seidel scheme because of its parallels with the Gauss–Seidel algorithm for solving linear algebraic equations (Arian 1998).

In order to eliminate the couplings between the disciplinary analysis modules, the original system can be decomposed as shown in Figure 9.3. For given values of $\mathbf{x}_m$, $\mathbf{x}_1$ and $\mathbf{x}_2$, multidisciplinary analysis now requires an initial guess for both $\mathbf{y}_{12}$ and $\mathbf{y}_{21}$. Iterations involving aerodynamic and structural analysis converge to a multidisciplinary feasible solution when the *interdisciplinary* or *coupling compatibility constraints* are satisfied, that is, when $||\mathbf{y}_{12}^{k+1} - \mathbf{y}_{12}^k|| = 0$ and $||\mathbf{y}_{21}^{k+1} - \mathbf{y}_{21}^k|| = 0$. Here, $\mathbf{y}_{12}^k$ and $\mathbf{y}_{21}^k$ are the values of the coupling variables used at the current iteration, and $\mathbf{y}_{12}^{k+1}$ and $\mathbf{y}_{21}^{k+1}$ are the new values of coupling variables after disciplinary analysis using the current values of $\mathbf{y}_{21}$ and $\mathbf{y}_{12}$, respectively. The steps involved are outlined in Algorithm 9.2. It can be noted that steps (2–3)

Algorithm 9.1: Generalized Gauss–Seidel scheme for multidisciplinary analysis.

Require: $\mathbf{x}_m$, $\mathbf{x}_1$, $\mathbf{x}_2$, $\mathbf{y}_{21}^0$, $k = 0$

  1: **while** convergence criteria not satisfied **do**
  2:    Solve for $\mathbf{w}_1^k$ the aerodynamics governing equation $\mathbf{R}_1(\mathbf{x}_m, \mathbf{x}_1, \mathbf{w}_1, \mathbf{y}_{21}^k) = 0$
  3:    $\mathbf{y}_{12}^k \leftarrow \mathcal{T}_1(\mathbf{w}_1^k)$
  4:    Solve for $\mathbf{w}_2^k$ the structural governing equation $\mathbf{R}_2(\mathbf{x}_m, \mathbf{x}_2, \mathbf{w}_2, \mathbf{y}_{12}^k) = 0$
  5:    $\mathbf{y}_{21}^{k+1} \leftarrow \mathcal{T}_2(\mathbf{w}_2^k)$
  6:    $k \leftarrow k + 1$
  7: **end while**

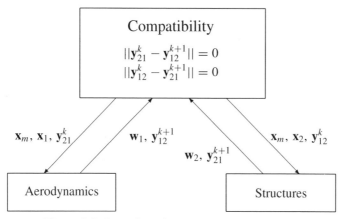

Figure 9.3 Data flow for distributed multidisciplinary analysis.

Algorithm 9.2: Generalized Jacobi scheme for multidisciplinary analysis.

Require: $\mathbf{x}_m$, $\mathbf{x}_1$, $\mathbf{x}_2$, $\mathbf{y}_{21}^0$, $\mathbf{y}_{12}^0$, $k = 0$

1: **while** convergence criteria not satisfied **do**
2:    Solve for $\mathbf{w}_1^k$ the aerodynamics governing equation $\mathbf{R}_1(\mathbf{x}_m, \mathbf{x}_1, \mathbf{w}_1, \mathbf{y}_{21}^k) = 0$
3:    $\mathbf{y}_{12}^{k+1} \leftarrow \mathcal{T}_1(\mathbf{w}_1^k)$
4:    Solve for $\mathbf{w}_2^k$ the structural governing equation $\mathbf{R}_2(\mathbf{x}_m, \mathbf{x}_2, \mathbf{w}_2, \mathbf{y}_{12}^k) = 0$
5:    $\mathbf{y}_{21}^{k+1} \leftarrow \mathcal{T}_2(\mathbf{w}_2^k)$
6:    $k \leftarrow k + 1$
7: **end while**

---

and steps (4–5) are independent of each other and hence they can be conducted in parallel. Hence, this decomposition of the original coupled system makes it possible to concurrently analyze a design (and thus perform distributed optimization as we shall see in later sections). As discussed in Arian (1998), this parallel iterative scheme for multidisciplinary analysis can be interpreted as a generalized Jacobi approach. Further, it was also shown that the generalized Jacobi scheme will take more iterations than the generalized Gauss–Seidel scheme to converge to a multidisciplinary feasible solution.

It is worth noting that the generalized Gauss–Seidel and Jacobi iteration schemes may not always converge, particularly for tightly coupled systems; a theoretical analysis of conditions under which convergence can be guaranteed can be found in Arian (1998) and Ortega and Rheinboldt (1970). As discussed earlier in Section 2.3, a more robust solution procedure can be devised by employing a Newton scheme to solve the coupled governing equations in (9.1) and (9.2). Here, given an initial guess for $\mathbf{w}_1$ and $\mathbf{w}_2$ (say $\mathbf{w}_1^k$ and $\mathbf{w}_2^k$), a linear system of equations of the following form is solved at each iteration:

$$\begin{bmatrix} \dfrac{\partial \mathbf{R}_1}{\partial \mathbf{w}_1} & \dfrac{\partial \mathbf{R}_1}{\partial \mathbf{w}_2} \\[2mm] \dfrac{\partial \mathbf{R}_2}{\partial \mathbf{w}_1} & \dfrac{\partial \mathbf{R}_2}{\partial \mathbf{w}_2} \end{bmatrix} \begin{pmatrix} \Delta \mathbf{w}_1 \\ \Delta \mathbf{w}_2 \end{pmatrix} = - \begin{pmatrix} \mathbf{R}_1\left(\mathbf{w}_1^k, \mathbf{y}_{21}(\mathbf{w}_2^k)\right) \\ \mathbf{R}_2\left(\mathbf{w}_2^k, \mathbf{y}_{12}(\mathbf{w}_1^k)\right) \end{pmatrix}. \tag{9.5}$$

Given the solution of (9.5), the state variables are updated as

$$\mathbf{w}_1^{k+1} = \mathbf{w}_1^k + \Delta\mathbf{w}_1 \text{ and } \mathbf{w}_2^{k+1} = \mathbf{w}_2^k + \Delta\mathbf{w}_2. \tag{9.6}$$

Newton's method converges to a stationary point given a *sufficiently close* initial guess if $||\mathbf{J}^{-1}||$ is bounded and $||\mathbf{J}(\mathbf{u}) - \mathbf{J}(\mathbf{v})|| \leq c||\mathbf{u} - \mathbf{v}||$, where $\mathbf{J}$ is the Jacobian matrix (i.e., the coefficient matrix in (9.5)) and c is a constant.[4] Hence, Newton's method is not globally convergent since convergence to a stationary point cannot be guaranteed from an arbitrary initial guess.

Global convergence can be ensured by employing a damped Newton scheme, where as before, we first solve (9.5) for $\Delta\mathbf{w}_1$ and $\Delta\mathbf{w}_2$. However, instead of using (9.6), the state variables are now updated as

$$\mathbf{w}_1^{k+1} = \mathbf{w}_1^k + \alpha^k\Delta\mathbf{w}_1 \text{ and } \mathbf{w}_2^{k+1} = \mathbf{w}_2^k + \alpha^k\Delta\mathbf{w}_2, \tag{9.7}$$

where the scalar α^k is computed by solving the following one-dimensional minimization (line search) problem

$$\alpha^k = \arg\min_\alpha ||\mathbf{R}_1(\mathbf{w}_1^k + \alpha\Delta\mathbf{w}_1)||_2^2 + ||\mathbf{R}_2(\mathbf{w}_2^k + \alpha\Delta\mathbf{w}_2)||_2^2. \tag{9.8}$$

Clearly, we recover the standard Newton method if we set $\alpha^k = 1$ in (9.7).

A full Newton approach may not be attractive from a practical viewpoint, since the off-diagonal terms of the Jacobian matrix ($\partial\mathbf{R}_1/\partial\mathbf{w}_2$ and $\partial\mathbf{R}_2/\partial\mathbf{w}_1$) are not easy to compute. Hence, recourse has to be made to inexact Newton methods that employ approximations to the off-diagonal terms of the Jacobian matrix. For example, if an iterative method such as GMRES (Saad 1996) is used to solve (9.5), then the derivatives appearing in the off-diagonal terms need not be explicitly computed since these methods only make use of matrix-vector products. Hence the off-diagonal terms can be approximated by finite differences. To accelerate computations, the factored form of the diagonal blocks can be used to precondition (9.5); see, for example, Kim et al. (2003).

It can be noted that the Newton methods discussed previously require disciplinary analysis codes that are capable of returning the residual vector given an approximation to the state variables. Unfortunately, many stand-alone disciplinary analysis codes do not have this capability. Hence modifications to the source code of the analysis modules may become necessary. An alternative approach referred to as the *multilevel* Newton method was proposed by Kim et al. (2003) to ensure that multidisciplinary analysis can be carried out without intrusive modifications to the disciplinary analysis source codes. In this approach, we first rewrite the governing equations in (9.1) and (9.2) in the following equivalent form:

$$\mathbf{w}_1 = \mathcal{R}_1(\mathbf{w}_2) \text{ and } \mathbf{w}_2 = \mathcal{R}_2(\mathbf{w}_1). \tag{9.9}$$

The first equation in (9.9) implies that given the structural state variables $\mathbf{w}_2$ we carry out a fully converged aerodynamic analysis to calculate $\mathbf{w}_1$. Similarly, the second equation implies that given the aerodynamic state variable $\mathbf{w}_1$ we calculate the structural displacement vector $\mathbf{w}_2$. In other words, single-discipline feasibility is achieved for both aerodynamics and structures. Note that the representation of the governing equations in (9.9) avoids dealing with residuals.[5] The motivation for this is to develop an alternative formulation for multidisciplinary analysis that can work with black-box disciplinary analysis codes.

---

[4]The second condition states that the derivatives of $\mathbf{R}_1$ and $\mathbf{R}_2$ are Lipschitz continuous.

[5]This idea was used earlier to derive an alternative set of global sensitivity equations referred to as GSE2 in Section 4.9.

Using (9.9), the Newton correction terms can be computed by solving the following linear algebraic system of equations:

$$
\begin{bmatrix} \mathbf{I} & -\dfrac{\partial \mathcal{R}_1}{\partial \mathbf{w}_2} \\[2mm] -\dfrac{\partial \mathcal{R}_2}{\partial \mathbf{w}_1} & \mathbf{I} \end{bmatrix} \begin{pmatrix} \Delta \mathbf{w}_1 \\ \Delta \mathbf{w}_2 \end{pmatrix} = - \begin{pmatrix} \mathbf{w}_1^k - \mathcal{R}_1(\mathbf{w}_2) \\ \mathbf{w}_2^k - \mathcal{R}_2(\mathbf{w}_1) \end{pmatrix}. \tag{9.10}
$$

When solving (9.10) for $\Delta \mathbf{w}_1$ and $\Delta \mathbf{w}_2$, terms of the form

$$(\partial \mathcal{R}_i / \partial \mathbf{w}_i)^{-1} (\partial \mathcal{R}_j / \partial \mathbf{w}_j)^{-1} \quad \text{for } i, j = 1, 2,$$

arise, which, however, can be calculated by finite-difference approximations. On the lines of the procedure outlined earlier using (9.7) and (9.8), it is also possible to formulate a damped version of the multilevel Newton scheme.

As discussed earlier in Section 2.3, irrespective of the type of numerical scheme employed for multidisciplinary analysis, it is possible that the final solution may depend on the initial guess. This is because a general nonlinear system can have multiple equilibrium points. Even though such situations are rarely encountered in aerodynamics-structures inter-action studies, the designer must exercise care when carrying out multidisciplinary analysis to ensure that the most physically meaningful solution is selected. This problem becomes increasingly severe as further analysis disciplines are considered.

9.2 Fully Integrated Optimization (FIO)

Much of the early research work in MDO focused on system optimization approaches, also known as fully integrated optimization (FIO), multidisciplinary feasible optimization or nested analysis and design (NAND). The idea is to directly couple an optimizer to the multidisciplinary analysis code, often a not inconsiderable task. The flow of information between the optimizer and the multidisciplinary analysis block in this approach is outlined in Figure 9.4. Here, the variables controlled by the optimizer are $\mathbf{x} = \{\mathbf{x}_m, \mathbf{x}_1, \mathbf{x}_2\} \in \mathbb{R}^p$. Given $\mathbf{x}$, multidisciplinary analysis is carried out to compute the objective and constraint functions. Formally, the FIO formulation can be posed as a nonlinear programming problem

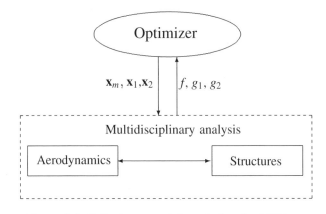

Figure 9.4 Fully integrated formulation for MDO.

of the form

$$\begin{aligned}
\underset{\mathbf{x}}{\textbf{Minimize}}: &\quad f(\mathbf{x}) \\
\textbf{Subject to}: &\quad g_1(\mathbf{x}_m, \mathbf{x}_1, \mathbf{w}_1, \mathbf{y}_{21}) \leq 0 \\
&\quad g_2(\mathbf{x}_m, \mathbf{x}_2, \mathbf{w}_2, \mathbf{y}_{12}) \leq 0.
\end{aligned} \tag{9.11}$$

Here, $f(\mathbf{x})$ is the objective function for the overall design process.[6] In practice, a linear combination of the disciplinary objective functions f_1 and f_2 could be used as the overall objective function. When designing jet-powered aircraft, for example, the Breguet range formula is often used as the objective function to be maximized (Martins et al. 2004), that is,

$$\text{Range} = \frac{V}{c} \frac{C_1}{C_d} \ln \frac{W_i}{W_f}, \tag{9.12}$$

where V is the cruise velocity, c is the thrust-specific fuel consumption of the engine, C_1/C_d is the ratio of lift to drag and W_i/W_f is the ratio of initial and final cruise weights of the aircraft. The Breguet range equation can be interpreted as an aggregate objective function that allows the designer to trade-off between the drag and the empty weight of the aircraft, and hence it is a good indicator of the overall design performance.

Theoretically, the FIO approach can be interpreted as a variable reduction technique since the state variables $\mathbf{w}_1$ and $\mathbf{w}_2$ are eliminated from consideration by solving for them independently via multidisciplinary analysis. Hence, all the iterates generated by an optimization algorithm applied to the FIO formulation satisfy the interdisciplinary compatibility constraints. As a consequence, the attendant computational cost can be significant because of the requirement of iterations between the disciplines to arrive at a multidisciplinary feasible solution at each function evaluation. Additionally, calculation of the sensitivities of the objective and constraint functions is also computationally expensive if gradient-based optimization algorithms are applied to solve (9.11). Note that finite-difference approximations for the sensitivities can be erroneous if the multidisciplinary analysis iterations are not fully converged. In practice, it is preferable to apply the formulations based on global sensitivity equations described previously in Section 4.9.

From a computational viewpoint, the FIO formulation is not straightforward to parallelize unless of course the parallel Jacobi iteration scheme in Algorithm 9.2 is employed to carry out multidisciplinary analysis within each optimization step. An alternative way to achieve parallelism is to employ a population-based evolutionary search technique such as those described in Section 3.2.

In summary, the disadvantages of the fully integrated MDO approach include: (1) the high costs involved in software integration and maintenance of the integrated design system, (2) specialist disciplinary groups do not have any decision making power in the design process, and (3) the increase in computational cost because of the use of a single optimizer to handle all the design variables and the requirement of carrying out multidisciplinary analysis at each function evaluation.

In spite of its obvious disadvantages, the FIO approach cannot be dismissed outright from consideration in design practice. For problems with high coupling bandwidth, this may be the only method applicable from a computational viewpoint. An additional advantage of the FIO approach is that, since multidisciplinary feasibility is established at each function evaluation, even if we prematurely terminate the optimization iterations, the solution obtained will

---

[6]Note that we have suppressed the dependency of the objective function on the state variables since they are computed independently via multidisciplinary analysis.

respect the coupling in the problem physics. As we shall see later, this is a desirable characteristic not found in most other MDO formulations.

9.3 System Decomposition and Optimization

At the beginning of this chapter, we presented a simplified overview of the disciplinary analysis modules that need to be run to calculate the performance of an aircraft. This served to motivate the need for specialized algorithms for MDO. However, in practical design problems, the number of analysis codes as well as the dependencies between them can be significantly more complex.

System decomposition is essentially concerned with dividing the required analyses into subsystems with minimal coupling so as to enable a degree of concurrency in the analysis. In many multidisciplinary design problems, it may be preferable to decompose the system under consideration along the lines of the disciplines involved. However, the idea of system decomposition is very general and can be applied to any problem where it is necessary to run (or iterate between) a sequence of modules to calculate objective and constraint functions. The aim is to reorder the analysis tasks so as to reduce the degree of feedback between them. A number of planning tools (also sometimes referred to as project management tools) exist to assist in the task of system decomposition; see, for example, Browning (2001) or Rogers (1999) and the references therein. These techniques essentially permute the way the analysis modules are ordered to minimize a merit function. For example, it is often desired to decompose the analysis modules into say "m" blocks that have minimal coupling between them. Kroo et al. (1994) describe such a decomposition strategy that aims to reduce the computation time required to carry out gradient-based optimization.

Systems can be classified into different categories depending on the type of coupling between the subsystems. The model problem described earlier in Figure 9.2 is referred to as a *nonhierarchic* system since data is transferred from discipline 1 to 2 and vice versa. The decomposed version of the model problem illustrated in Figure 9.3 is referred to as a *hierarchic* system since here there is no feedback between the disciplines. In more general terms, systems with multidirectional data flow are nonhierarchic, whereas systems with one-way data flow are hierarchic. *Hybrid* systems involve a combination of hierarchic and nonhierarchic subsystems.

System decomposition is useful when deciding on strategies for carrying out distributed analysis. For example, the decomposition in Figure 9.3 was used to formulate a generalized Jacobi iteration scheme. This was a simple exercise in decomposition since we merely divided the analysis blocks along disciplinary lines. However, the idea can be generalized to problems where there is no obvious basis for decomposing the analysis block. System decomposition schemes can also be used to formulate strategies for distributed optimization of complex coupled systems.

As discussed earlier, organizational challenges and computational cost are generally perceived as the major obstacles to the application of MDO methodology in design practice. This motivates the development of distributed architectures for MDO, which address some of these challenges. The idea of distributed optimization is closely coupled to the topic of system decomposition. We next provide an overview of distributed optimization strategies.

The term *multilevel optimization* is commonly used to refer to an architecture where a number of coupled optimization problems are solved independent of each other with data

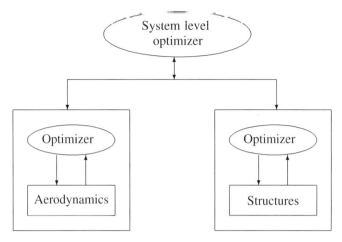

Figure 9.5 Bilevel architecture for MDO.

transferred both in a top-down and bottom-up fashion after each subproblem is solved.[7] For illustration, a bilevel optimization architecture is shown in Figure 9.5. It can be noted that this approach involves the use of a system-level optimizer, which guides a number of disciplinary optimizers to improve the overall system performance, satisfy the disciplinary constraints, and ensure multidisciplinary feasibility of the converged solution. Since multidisciplinary feasibility is not enforced at each iteration, this approach circumvents the problem of "disciplinary sequencing", in which some of the disciplines have to wait for data from the other disciplines before carrying out their design studies. Hence, distributed optimization allows for the possibility of tackling MDO problems in a concurrent fashion (thereby achieving calendar-time compression), and it also enables disciplinary specialists to have a degree of autonomy in the design process. The latter advantage is believed to be an important factor in the acceptance of formal MDO methods by industry. Distributed MDO methods are also well suited to the organizational structure and heterogeneous computing platforms found in many large corporations.

Even though bilevel architectures offer significant advantages in engineering design practice, theoretically they are notoriously difficult to analyze. Specifically, the system-level optimization problem may be highly discontinuous, which may cause difficulties when gradient-based algorithms are used. Further, there are no convergence results for most bilevel architectures and they continue to be the focus of active research. Exceptions are the bilevel optimization formulation of Alexandrov (1993) and the analytical target cascading approach of Michelena et al. (2003).

Sometimes, it is possible to simplify such architectures because of the nature of the couplings involved. Consider the model problem described earlier. Now, suppose that the wing can be assumed to be rigid. In other words, the jig shape of the wing structure can be designed to compensate for deformations induced by aerodynamic loading. This leads to a one-way coupling, that is, $\mathbf{y}_{21} = 0$: this is in fact an example of a hierarchic decomposition since the feedback between the disciplines is eliminated. It is then possible to apply a

---

[7]Not to be confused with multifidelity optimization where we have a range of codes for evaluating a single objective or constraint.

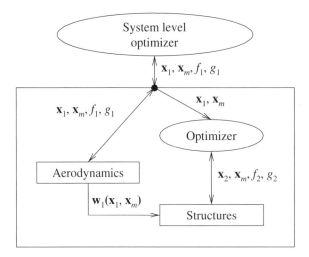

Figure 9.6 A two-level strategy.

two-level strategy where two optimizers are used (Sobieszczanski and Haftka 1997). The top-level optimizer controls the design variables $(\mathbf{x}_m, \mathbf{x}_1)$ to minimize the aerodynamic objective function f_1 subject to the constraints $g_1 \leq 0$. At each function evaluation, given values of $\mathbf{x}_m$ and $\mathbf{x}_1$ and the results of the aerodynamic calculation, a structural optimization is carried out to define the local structural variables $\mathbf{x}_2$ such that f_2 is minimized subject to the constraints $g_2 \leq 0$; see Figure 9.6. Later on in this chapter, we discuss collaborative and current subspace optimization, which are more general bilevel algorithms appropriate for MDO.

9.4 Simultaneous Analysis and Design (SAND)

In the simultaneous analysis and design (SAND) formulation, the state variables are also considered as design variables and the analysis equations are imposed as equality constraints. This formulation is hence the other extreme of the FIO formulation where only $(\mathbf{x}_m, \mathbf{x}_1, \mathbf{x}_2)$ are considered as design variables. The motivation for this formulation arises from the observation that there is no need to impose multidisciplinary feasibility when the current iterate is far from the optima. In order to reduce computational cost, it is preferable to impose multidisciplinary feasibility only in the vicinity of the final optimal solution.

The SAND formulation is also referred to in the literature as the all-at-once formulation or the one-shot method since the processes of obtaining optimal designs and performing analyses are carried out simultaneously by solving one single optimization problem. The optimization problem to be solved can be formally stated as

$$
\begin{aligned}
&\underset{\mathbf{x},\mathbf{w}_1,\mathbf{w}_2}{\text{Minimize}:} \quad f(\mathbf{x}, \mathbf{w}_1, \mathbf{w}_2) \\
&\text{Subject to}: \quad g_1(\mathbf{x}_m, \mathbf{x}_1, \mathbf{w}_1, \mathbf{y}_{21}(\mathbf{w}_2)) \leq 0 \\
&\qquad\qquad\quad g_2(\mathbf{x}_m, \mathbf{x}_2, \mathbf{w}_2, \mathbf{y}_{12}(\mathbf{w}_1)) \leq 0 \\
&\qquad\qquad\quad \mathbf{R}_1(\mathbf{x}_m, \mathbf{x}_1, \mathbf{w}_1, \mathbf{y}_{21}(\mathbf{w}_2)) = 0 \\
&\qquad\qquad\quad \mathbf{R}_2(\mathbf{x}_m, \mathbf{x}_2, \mathbf{w}_2, \mathbf{y}_{12}(\mathbf{w}_1)) = 0.
\end{aligned}
\tag{9.13}
$$

The data flow between the optimizer and the analysis modules in the SAND approach is shown in Figure 9.7. Here, the aerodynamics module is set up to return f_1, g_1 and the

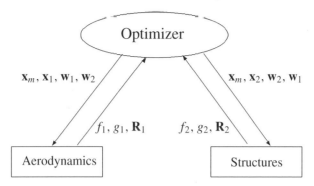

Figure 9.7 Data flow between the optimizer and analysis codes for the simultaneous analysis and design (SAND) formulation.

residual error vector $\mathbf{R}_1$ given $(\mathbf{x}_m, \mathbf{x}_1, \mathbf{w}_1, \mathbf{w}_2)$. Note that complete aerodynamic analysis is avoided and only the residual is computed at each function evaluation.[8] Similarly, the structures module returns f_2, g_2 and the residual error vector $\mathbf{R}_2$, given $(\mathbf{x}_m, \mathbf{x}_2, \mathbf{w}_2, \mathbf{w}_1)$. Note also that the state variables of all disciplines are passed to each disciplinary module in order to calculate the coupling variables $\mathbf{y}_{12}$ and $\mathbf{y}_{21}$.

Clearly, the SAND formulation leads to an increase in the total number of optimization variables and constraints compared to the FIO formulation. Specifically, the latter formulation has only $p = \dim(\mathbf{x}_m, \mathbf{x}_1, \mathbf{x}_2) = \dim(\mathbf{x})$ optimization variables, whereas the SAND formulation leads to a total of $p + \dim(\mathbf{w}_1) + \dim(\mathbf{w}_2)$ variables. Hence, the total number of optimization variables in the SAND formulation can be potentially massive when high-fidelity models with many state variables are used. Fortunately, in some cases, the structure of the problem can be exploited to compute the optimal solution efficiently; see, for example, Orozco and Ghattas (1997) which discusses an efficient reduced SAND formulation.

Note that multidisciplinary feasibility is achieved only when the optimization iterations have converged. Hence, the intermediate iterates do not have any physical meaning. This is a disadvantage compared to the FIO approach. However, from a computational viewpoint, SAND is much cheaper per iteration than FIO since each function evaluation only involves residual calculations – no disciplinary or multidisciplinary analysis is necessary. However, in some cases, it is probable that at least some of the cost savings associated with function evaluation may be offset by the fact that when optimizing over many extra variables it is likely that many more iterations will be needed as well as significantly greater costs if sensitivities must be computed (for example, by finite differencing).

It is also worth noting that the SAND formulation allows for the possibility of achieving massive parallelism at various levels during each function evaluation. This is because the residuals corresponding to each discipline can be computed independent of each other. Further, in some cases, it is also possible to parallelize the process of computing the residual vector for a given discipline. For example, in linear structural analysis, the residual error vector can be calculated using a matrix-vector operation that is easy to parallelize. The SAND

---

[8]To illustrate the difference between analysis and residual calculation, consider a linear system whose governing equations are given by $\mathbf{K}(\mathbf{x})\mathbf{w}_1 = \mathbf{f}$. Here, analysis involves solving for $\mathbf{w}_1$ given $\mathbf{x}$. In contrast, residual calculation involves computing the vector $\mathbf{K}(\mathbf{x})\widehat{\mathbf{w}}_1 - \mathbf{f}$ given $\mathbf{x}$ and an approximation for the state vector $\widehat{\mathbf{w}}_1$. Hence, analysis involves a matrix factorization, whereas the residual can be calculated by one matrix-vector product operation.

formulation allows for disciplinary autonomy in analysis, since, at each function evaluation, the residuals for each discipline can be computed independent of each other. However, since a single optimizer is used to handle all the variables, no disciplinary autonomy is permitted in making decisions on the values that the design variables may take.

In principle, implementation of the SAND approach is straightforward, provided we have a disciplinary analysis code that, given the design variables and an approximation for the state variables, returns the local objective and constraint functions and the residual error vector. However, in practice, many disciplinary analysis codes may not be equipped with the capability of calculating the residual given the approximate state vector. Hence, modifications to the disciplinary analysis code may become necessary. In practice, the solution to the SAND formulation is often computed by solving a sequence of linear or quadratic subproblems. This is achieved by employing either a first- or second-order Taylor series expansion of the objective, constraint and residual error functions in (9.13); see, for example, Gumbert et al. (2001a); Orozco and Ghattas (1997).

We also note that the SAND approach is not limited to multidisciplinary optimization and was originally conceived in the context of efficiently solving structural optimization problems (Schmit and Fox 1965). Numerical studies in the literature suggest that this approach can be significantly more efficient than the standard approach where the optimizer only controls the original design variables. For some problems, it is the case that the optimum solution can be reached at a computational cost comparable to a few full analysis. For the case of multidisciplinary problems, numerical studies by Gumbert et al. (2001a,b) suggest that the computational cost savings may not be that dramatic compared to the FIO formulation, particularly because of the computational effort associated with sensitivity analysis.

9.5   Distributed Analysis Optimization Formulation

The distributed analysis optimization (DAO) formulation lies between the fully integrated and SAND approaches. The basic idea is to define the coupling variables as optimization variables along with $(\mathbf{x}_m, \mathbf{x}_1, \mathbf{x}_2)$. Since the dimensions of these vectors are less than the state variables $\mathbf{w}_1$ and $\mathbf{w}_2$, a reduction in the problem size is achieved compared to the SAND formulation. To illustrate the DAO formulation, let us first rewrite the equations governing the multidisciplinary system given by (9.3) and (9.4) as follows:

$$\mathbf{a}_1 - \mathbf{A}_1(\mathbf{x}_m, \mathbf{x}_1, \mathbf{t}_2) = 0, \quad \mathbf{a}_2 - \mathbf{A}_2(\mathbf{x}_m, \mathbf{x}_2, \mathbf{t}_1) = 0,$$

$$\mathbf{t}_1 - \mathbf{a}_1 = 0 \quad \text{and} \quad \mathbf{t}_2 - \mathbf{a}_2 = 0. \tag{9.14}$$

Here, $\mathbf{t}_1$ and $\mathbf{t}_2$ are auxiliary variables whose dimensions equal those of the coupling variables $\mathbf{a}_1$ and $\mathbf{a}_2$, respectively. The preceding equations lead to the DAO formulation:

$$
\begin{aligned}
&\underset{\mathbf{x},\mathbf{t}_1,\mathbf{t}_2}{\textbf{Minimize}}: \quad && f(\mathbf{x}, \mathbf{t}_1, \mathbf{t}_2) \\
&\textbf{Subject to}: \quad && g_1(\mathbf{x}_m, \mathbf{x}_1, \mathbf{w}_1) \leq 0 \\
& && g_2(\mathbf{x}_m, \mathbf{x}_2, \mathbf{w}_2) \leq 0 \\
& && \mathbf{t}_1 - \mathbf{A}_1(\mathbf{x}_m, \mathbf{x}_1, \mathbf{t}_2) = 0 \\
& && \mathbf{t}_2 - \mathbf{A}_2(\mathbf{x}_m, \mathbf{x}_2, \mathbf{t}_1) = 0.
\end{aligned}
\tag{9.15}
$$

It can be noted that the last two constraints in the DAO formulation impose interdisciplinary compatibility. Hence, the additional degrees of freedom introduced by expanding the

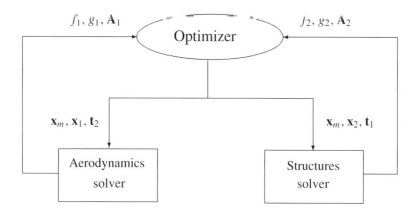

Figure 9.8 Data flow between the optimizer and analysis codes for the distributed analysis optimization (DAO) formulation.

set of optimization variables are removed by these constraints. The total number of design variables in the DAO formulation is $p + \dim(\mathbf{a}_1) + \dim(\mathbf{a}_2)$.

The data transferred between the optimizer and the disciplinary analysis codes is shown in Figure 9.8. Each function evaluation requires one pass through the structural and aerodynamic solvers to calculate the state variables and the local objective and constraint functions. This is in contrast to the SAND formulation where only residuals are calculated. It can be noted that the DAO formulation allows for autonomy in disciplinary analysis, since these can be conducted in parallel and independent of each other. This feature is useful from the viewpoint of reducing organizational and software integration complexities, since each disciplinary group can execute their own analysis codes given the appropriate design variables and $\mathbf{t}_1$ or $\mathbf{t}_2$. However, the DAO formulation only enforces single-discipline feasibility at each function evaluation, that is, only the aerodynamic and structural governing equations are satisfied. Hence, multidisciplinary feasibility is achieved only after the optimizer has converged – intermediate iterates do not necessarily satisfy the coupling compatibility constraints.

Similar to the FIO and SAND formulations, the DAO approach does not impart any decision-making power to the disciplinary groups since a single optimizer is used to control all the design variables. Further, for the case of tightly coupled systems, the number of design variables will grow significantly, which in turn may result in convergence difficulties. It can be shown, however, that the DAO formulation is mathematically equivalent to the FIO formulation, which makes it easier to analyze theoretically. The DAO formulation is also referred to as the individual discipline feasible (IDF) strategy by Cramer et al. (1994). The term IDF was used there to suggest that single-discipline feasibility is ensured at each function evaluation and not that all the disciplinary constraints are satisfied. To avoid the potential confusion associated with this, Alexandrov and Lewis (2000a) suggested the use of the term DAO, which we follow here.

The DAO formulation would be more attractive from a practical viewpoint if all the optimization iterates satisfied the disciplinary constraints. Given an initial guess for $(\mathbf{x}, \mathbf{a}_1, \mathbf{a}_2)$ that satisfies the disciplinary constraints in (9.15), we could then employ an optimization algorithm that maintains feasibility with respect to these constraints at each iteration. However, finding a common set of values for the multidisciplinary variables $\mathbf{x}_m$ such that all the disciplinary constraints are satisfied can be difficult in practice. To circumvent the difficulty associated with finding an initial guess that satisfies the local disciplinary constraints, Alexandrov and Lewis (2000a) proposed the following reformulation of (9.15).

$$
\begin{aligned}
&\underset{\mathbf{x}_m,\sigma_1,\sigma_2,\mathbf{x}_1,\mathbf{x}_2,\mathbf{t}_1,\mathbf{t}_2}{\textbf{Minimize}:} && f(\mathbf{x}_m, \mathbf{x}_1, \mathbf{x}_2, \mathbf{t}_1, \mathbf{t}_2) \\
&\textbf{Subject to}: && g_1(\sigma_1, \mathbf{x}_1, \mathbf{t}_1) \le 0 \\
& && g_2(\sigma_2, \mathbf{x}_2, \mathbf{t}_2) \le 0 \\
& && \mathbf{t}_1 - \mathbf{A}_1(\sigma_1, \mathbf{x}_1, \mathbf{t}_2) = 0 \\
& && \mathbf{t}_2 - \mathbf{A}_2(\sigma_2, \mathbf{x}_2, \mathbf{t}_1) = 0 \\
& && \sigma_1 = \mathbf{x}_m, \ \sigma_2 = \mathbf{x}_m.
\end{aligned}
\tag{9.16}
$$

Here, σ_1 and σ_2 are auxiliary variables with the same dimension as $\mathbf{x}_m$, that is, local copies of the multidisciplinary variables. The values of these variables can be initially set such that the disciplinary constraints are satisfied – since multiple copies of the multidisciplinary variables are used, this step has some degree of flexibility compared to the original DAO formulation. Subsequently, by applying a feasible method, we can ensure that all the iterates satisfy the disciplinary constraints. The last two equality constraints ensure that the copies of the multidisciplinary variables used for evaluating discipline-specific constraints agree at the optimum solution. This further increases the number of variables in use as well as the number of constraints to be satisfied. It is also possible to construct a variant on the DAO formulation where, instead of carrying out merely disciplinary analysis to calculate the residuals, each discipline is also asked to return new values for the local design variables that satisfy the disciplinary constraints (Braun et al. 1996).

9.6 Collaborative Optimization

The collaborative optimization (CO) architecture was originally conceived to meet organizational and computational challenges encountered in the application of MDO to complex coupled systems (Braun 1996; Braun and Kroo 1997; Sobieski 1998). The fundamental idea used is to introduce auxiliary variables to relax the interdisciplinary compatibility requirements. CO belongs to the class of bilevel MDO approaches and hence this architecture involves a system-level optimizer guiding discipline-specific optimizers as sketched in Figure 9.5. As a consequence, the CO architecture allows for disciplinary autonomy in analysis as well as optimization, which is a very attractive property for tackling multidisciplinary problems in an organization with traditional disciplinary groupings. It is worth noting here that the CO approach is rather general in scope and can be applied to solve a wide class of optimization problems, provide they are amenable to decomposition.

A number of variants of the CO architecture have been proposed in the literature and here we focus on two commonly used formulations. An outline of the basic CO architecture is shown in Figure 9.9. As indicated in the figure, the basic idea is to carry out concurrent disciplinary optimization that aims to converge to a design that satisfies the disciplinary constraints while minimizing a measure of discrepancy in the interdisciplinary compatibility. The system-level optimizer coordinates the disciplinary optimizers by setting targets to be met while optimizing the overall design objective. We next discuss the mathematical formulation of the CO architecture and outline some of its theoretical and computational

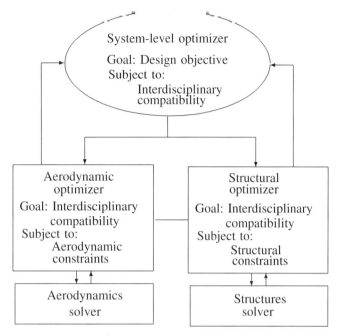

Figure 9.9 The collaborative optimization (CO) architecture.

properties. Our discussion in this section is based on the papers by Alexandrov and Lewis (2000a,b), which focus on the analytical and computational aspects of CO.

The system-level optimization problem in CO can be stated as

$$\begin{aligned}
\underset{\mathbf{x}_m, \mathbf{t}_1, \mathbf{t}_2}{\textbf{Minimize}}: \quad & f(\mathbf{x}_m, \mathbf{t}_1, \mathbf{t}_2) \\
\textbf{Subject to}: \quad & \mathbf{h}(\mathbf{x}_m, \mathbf{t}_1, \mathbf{t}_2) = 0.
\end{aligned} \tag{9.17}$$

The system-level optimizer manipulates the multidisciplinary variables $\mathbf{x}_m$ along with the coupling variables $\mathbf{t}_1$ and $\mathbf{t}_2$ to minimize the overall objective and satisfy the equality constraints $\mathbf{h} = 0$, which ensure that multidisciplinary feasibility is achieved at the optimum solution. In other words, the constraints $\mathbf{h} = 0$ are calculated by solving a sequence of disciplinary optimization problems that we discuss next.

The disciplinary optimization problem for aerodynamics can be stated as

$$\begin{aligned}
\underset{\mathbf{x}_1, \boldsymbol{\sigma}_1}{\textbf{Minimize}}: \quad & \tfrac{1}{2}\left(||\boldsymbol{\sigma}_1 - \mathbf{x}_m||^2 + ||\mathbf{t}_1 - \mathbf{A}_1\left(\boldsymbol{\sigma}_1, \mathbf{x}_1, \mathbf{t}_2\right)||^2 \right) \\
\textbf{Subject to}: \quad & g_1\left(\boldsymbol{\sigma}_1, \mathbf{x}_1, \mathbf{w}_1\left(\boldsymbol{\sigma}_1, \mathbf{x}_1, \mathbf{t}_2\right)\right) \le 0.
\end{aligned} \tag{9.18}$$

Here, $\mathbf{x}_m$ and $\mathbf{t}_1$ are target values for the multidisciplinary design variables and the coupling variables, respectively, that are set by the system-level optimizer. The coupling variable vector $\mathbf{t}_2$ is also passed on since it is required for aerodynamic analysis. $\boldsymbol{\sigma}_1$ is a local copy of the multidisciplinary design variables, which is independently manipulated along with the local disciplinary variables $\mathbf{x}_1$ to satisfy the aerodynamic constraints $g_1 \le 0$. The first term in the objective function essentially enforces the requirement that the local copy of the multidisciplinary design variables $\boldsymbol{\sigma}_1$ matches the system-level target $\mathbf{x}_m$. The second term enforces interdisciplinary compatibility. To calculate the objective and constraint functions in (9.18), we first solve the aerodynamics governing equations $\mathbf{R}_1(\sigma_1, \mathbf{x}_1, \mathbf{w}_1, \mathbf{t}_2) = 0$ for $\mathbf{w}_1$. Subsequently we calculate the coupling variables required for structural analysis, that is, $\mathbf{A}_1(\sigma_1, \mathbf{x}_1, \mathbf{t}_2)$.

On similar lines, the disciplinary optimization problem for structures can be stated as follows:

$$\begin{aligned} \textbf{Minimize}: \quad & \tfrac{1}{2}\left(||\sigma_2 - \mathbf{x}_m||^2 + ||\mathbf{t}_2 - \mathbf{A}_2\,(\sigma_2, \mathbf{x}_2, \mathbf{t}_1)\,||^2\right) \\ \underset{\mathbf{x}_2, \sigma_2}{} \quad & \\ \textbf{Subject to}: \quad & g_2\,(\sigma_2, \mathbf{x}_2, \mathbf{w}_2\,(\sigma_2, \mathbf{x}_2, \mathbf{t}_1)) \leq 0. \end{aligned} \tag{9.19}$$

Here, σ_2 is a local copy of the multidisciplinary design variables $\mathbf{x}_m$ that the structural optimizer is allowed to manipulate independently along with the local design variables $\mathbf{x}_2$ to satisfy the structural constraints. The structural optimizer is also passed values of $\mathbf{t}_1$ and $\mathbf{t}_2$. Here, $\mathbf{t}_1$ are the coupling variables required for structural analysis, whereas $\mathbf{t}_2$ contains targets for the coupling variables computed during structural analysis. Clearly, we need to ensure that the local copies of $\mathbf{x}_m$ match at the optimum solution in addition to ensuring that the interdisciplinary compatibility constraints are satisfied. This is achieved by the two terms in the disciplinary objective function in (9.19).

In the more commonly encountered formulation of CO, which we refer to as CO_1, the constraints in the system-level optimization problem are defined as

$$\begin{aligned} h_1(\mathbf{x}_m, \mathbf{t}_1, \mathbf{t}_2) = \; & \frac{1}{2}\left(||\sigma_1^*(\mathbf{x}_m, \mathbf{t}_1, \mathbf{t}_2) - \mathbf{x}_m||^2 \right. \\ & + \left. ||\mathbf{A}_1\left(\sigma_1^*(\mathbf{x}_m, \mathbf{t}_1, \mathbf{t}_2), \mathbf{x}_1^*(\mathbf{x}_m, \mathbf{t}_1, \mathbf{t}_2), \mathbf{t}_2\right) - \mathbf{t}_1||^2\right) = 0, \end{aligned} \tag{9.20}$$

$$\begin{aligned} h_2(\mathbf{x}_m, \mathbf{t}_1, \mathbf{t}_2) = \; & \frac{1}{2}(||\sigma_2^*(\mathbf{x}_m, \mathbf{t}_1, \mathbf{t}_2) - \mathbf{x}_m||^2 \\ & + ||\mathbf{A}_2\left(\sigma_2^*(\mathbf{x}_m, \mathbf{t}_1, \mathbf{t}_2), \mathbf{x}_2^*(\mathbf{x}_m, \mathbf{t}_1, \mathbf{t}_2), \mathbf{t}_1\right) - \mathbf{t}_2||^2) = 0. \end{aligned} \tag{9.21}$$

We use the superscript $*$ to indicate values that are obtained by solving the disciplinary optimization subproblems for given values of system-level targets. It can be noted from the preceding equations that the two constraints h_1 and h_2 ensure that the copies of the multidisciplinary variables agree and the interdisciplinary compatibility constraints are satisfied.

In the second variant of CO, which we refer to as CO_2, the system-level constraints are written as

$$h_1(\mathbf{x}_m, \mathbf{t}_1, \mathbf{t}_2) = \sigma_1^*(\mathbf{x}_m, \mathbf{t}_1, \mathbf{t}_2) - \mathbf{x}_m = 0 \tag{9.22}$$

$$h_2(\mathbf{x}_m, \mathbf{t}_1, \mathbf{t}_2) = \mathbf{A}_1(\sigma_1^*(\mathbf{x}_m, \mathbf{t}_1, \mathbf{t}_2), \mathbf{x}_1^*(\mathbf{x}_m, \mathbf{t}_1, \mathbf{t}_2), \mathbf{t}_2) - \mathbf{t}_1 = 0 \tag{9.23}$$

$$h_3(\mathbf{x}_m, \mathbf{t}_1, \mathbf{t}_2) = \sigma_2^*(\mathbf{x}_m, \mathbf{t}_1, \mathbf{t}_2) - \mathbf{x}_m = 0 \tag{9.24}$$

$$h_4(\mathbf{x}_m, \mathbf{t}_1, \mathbf{t}_2) = \mathbf{A}_2(\sigma_2^*(\mathbf{x}_m, \mathbf{t}_1, \mathbf{t}_2), \mathbf{x}_2^*(\mathbf{x}_m, \mathbf{t}_1, \mathbf{t}_2), \mathbf{t}_1) - \mathbf{t}_2 = 0 \tag{9.25}$$

It can be seen that the preceding constraints are similar to those of the CO_1 formulation, the only difference being that we separately impose the constraints that local copies of the multidisciplinary design variables match and that interdisciplinary compatibility is achieved.

When a gradient-based optimization algorithm is employed to solve the system-level problem, we need to compute the sensitivities of the constraints $\mathbf{h} = 0$ with respect to the system-level design variables. It can be noted that in both CO_1 and CO_2 formulations these constraints involve terms in $\mathbf{x}_1^*$, σ_1^*, $\mathbf{x}_2^*$ and σ_2^*, that is, the optimum solution of the disciplinary optimization problems. These sensitivity terms can be computed using the approach described earlier in Section 4.8.

9.6.1 Computational Aspects

It can be noted that at each function evaluation of the system-level optimization problem a sequence of disciplinary optimization problems are solved (one per discipline). All the disciplinary design variables and constraints are associated only with a specific discipline and need not be passed to all the disciplinary groups involved. Since the disciplinary optimizations are independent of each other, they can be conducted in parallel. Hence, the CO architecture enables the disciplines to use their own specialized optimization algorithms and, further, the whole design process can be implemented on heterogeneous computing platforms.

In contrast to most of the approaches discussed earlier, the optimizers involved in the CO architecture solve subproblems with fewer design variables. In conjunction with the possibility of parallel implementation, this reduction in problem size may lead to computational cost savings compared to the FIO approach. In general, the CO architecture is expected to be efficient for systems with low coupling bandwidth, since this implies that the dimensions of t_1 and t_2 are small. However, for tightly coupled systems, the size of the system-level optimization problem can be very large because of the increased number of coupling variables. This can increase computational cost as well as create potential convergence problems.

9.6.2 Theoretical Properties

There are two essential steps in theoretical analysis of the convergence of a multilevel MDO algorithm: (1) show that the solution set of the new formulation is equivalent to the FIO formulation and (2) show that both the system-level and subsystem-level optimization algorithms converge. Braun (1996) showed that the solution sets of the CO architecture and the FIO formulation are the same. However, recent studies have revealed that it is not straightforward to prove convergence for the system- and subsystem-level optimization problems in the CO architecture.

More specifically, Alexandrov and Lewis (2000b) showed that for the CO_1 formulation, the system-level constraints can have discontinuous derivatives. They also show that Lagrange multipliers may not exist for the system-level optimization problem in the CO_2 formulation. As a consequence, the system-level optimization problem in CO can be difficult to solve using traditional optimization algorithms. For example, when the Lagrange multipliers do not exist, it is not practical to employ methods such as sequential quadratic programming or feasible directions methods that rely on the Karush–Kuhn–Tucker conditions. DeMiguel and Murray (2000) proposed modifications to the CO formulation that alleviate some of these theoretical difficulties.

9.7 Concurrent Subspace Optimization

Concurrent subspace optimization (CSSO) is an MDO architecture introduced by Sobieszczanski-Sobieski (1989) that employs decomposition strategies and approximations to efficiently implement the FIO approach; see also Renaud and Gabriele (1993) and Wujek et al. (1995). CSSO achieves concurrency in the design process by solving the disciplinary (subspace) optimization problems independent of each other. Here, each subspace/discipline

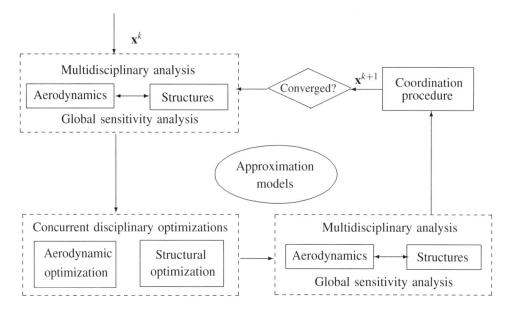

Figure 9.10 Outline of steps in the concurrent subspace optimization (CSSO) approach.

operates on its own unique subset of variables. This is in contrast to the CO architecture where the disciplines are allowed to independently manipulate the multidisciplinary variables. The disciplinary optimizers are brought into agreement by solving a so-called coordination problem, which can be thought of as an approximate system-level optimization problem. An outline of the CSSO approach is shown in the form of a flowchart in Figure 9.10. The steps in the process are:

1. Perform multidisciplinary analysis at current design point $\mathbf{x}^k$.

2. Calculate multidisciplinary sensitivities using the GSE approach described in Section 4.9.

3. Solve disciplinary optimization problems (also referred to as subspace optimization) to update local variables assigned to the disciplines.

4. Apply a coordination procedure to calculate a new design $\mathbf{x}^{k+1}$.

5. If convergence criteria are met stop, else got back to step (1).

Note that in step (3) of the CSSO method, only the local disciplinary variables are manipulated during subspace optimizations – the nonlocal variables are kept fixed. The nonlocal states that are needed for disciplinary analysis are approximated using the global sensitivities computed during step (2). The data generated during subspace optimizations are used to update a database that can be used in later stages to construct approximation models. For example, given sufficient points in the database, a polynomial response surface model can be employed to efficiently solve the coordination problem in step (4). The coordination problem essentially involves the solution of an approximate FIO formulation. The solution of this approximate problem is used to update all the design variables.

It can be noted that in contrast to the CO architecture, which does not involve any multidisciplinary analysis, the CSSO architecture uses multidisciplinary analysis and system-level sensitivities to solve the coordination problem. Since approximate models are used to facilitate the solution of the disciplinary optimization and the coordination problem, appropriate move limit management strategies to be employed to ensure that calculations are made within their ranges of validity. The Bi-Level Integrated System Synthesis (BLISS) approach outlined earlier in Section 2.3 is an MDO architecture that employs ideas similar to CSSO (Sobieszczanski-Sobieski et al. 1998, 2003, 2000).

9.8 Coevolutionary Architectures

We finally describe a coevolutionary architecture for distributed optimization of complex coupled systems proposed by Nair and Keane (2002a). This architecture is inspired by the phenomena of coevolutionary adaptation occurring in ecological systems. In this architecture, the optimization procedure is modeled as the process of coadaptation between sympatric species in an ecosystem. Each species is entrusted with the task of improving discipline-specific objectives and the satisfaction of local constraints. Coupling compatibility constraints are accommodated via implicit generalized Jacobi iteration, which enables the application of the architecture to systems with arbitrary coupling bandwidth between the disciplines, without an increase in the problem size. This architecture is essentially based on the blackboard approach described in Section 2.3.

A brief overview of coevolutionary algorithms is first presented and, subsequently, the issues involved in applying coevolutionary search to distributed design problems are highlighted.

9.8.1 Coevolutionary Genetic Algorithms (CGAs)

A Coevolutionary Genetic Algorithm (CGA) models an ecosystem consisting of two or more sympatric species having an ecological relationship of mutualism. Consider the problem of maximizing $f(\mathbf{x})$, which is a function of p variables. Potter and De Jong (1994) used an approach in which the problem is decomposed into p species – one species for each variable. Hence, each species contains a population of alternative values for each variable. *Collaboration* between the species involves selection of representative values of each variable from all the other species and combining them into a vector, which is then used to evaluate $f(\mathbf{x})$. An individual in a species is rewarded on the basis of how well it maximizes the function within the context of the variable values selected from the other species. This procedure is analogous to the univariate optimization algorithm wherein 1-D search is carried out using only one free variable. Note here that it is also possible to decompose the original problem variables into several blocks, with each species evolving a block rather than a single variable; see, for example, Nair and Keane (2001).

The steps involved in a canonical CGA are summarized in Algorithm 9.3. In general, a CGA approach to function optimization can lead to faster convergence as compared to a conventional GA, for low to moderate levels of variable interdependencies (also referred to as *epistasis*[9] in the GA literature). This can be primarily attributed to the attendant reduction

---

[9]In the context of function optimization, epistasis can be defined as the extent to which the contribution of one variable to the fitness depends on the values of the other variables. From a statistical viewpoint, epistasis can be defined in terms of the magnitude of the interaction effects.

Algorithm 9.3: Outline of a canonical coevolutionary genetic algorithm.

Require: An initial population of individuals for each species which is generated randomly
 or based on some previous design knowledge.

 1: **while** convergence criteria not met **do**
 2:    Evaluate the fitness of the individuals in each species. Fitness evaluation in a CGA
       involves the following steps:
 3:    Choose representatives from all other species.
 4:    **for all** each individual i in the species being evaluated **do**
 5:       Form collaboration between individual i and the (fixed) representatives from other
          species.
 6:       Evaluate the fitness of the collaborative design by applying it to the target problem,
          and assign it to the individual i.
 7:    **end for**
 8:    If termination criteria are not met, then apply a canonical GA involving the operators
       of reproduction and genetic recombination to arrive at a new population for each
       species.
 9: **end while**

---

in the active search space due to coevolution of each variable concurrently, for example,
for the function optimization problem described earlier, the original search space of size
$(2^k)^p$ has been reduced to a series of p constrained search spaces, each of size 2^k, where
k is the number of bits used to represent each variable (assuming a binary encoded GA).
The underlying premise of the CGA approach is that, when the optimization problem under
consideration has a moderate degree of epistasis, faster progress in the search can be made
by decomposing the design space.

In CGA-based search, the term *generation* is used to refer to a cycle involving selection,
recombination and fitness evaluation for a single species, and the term *ecosystem generation*
refers to an evolutionary cycle through all the species being coevolved. The reader is referred
to the dissertation of Potter (1997) for a more detailed description of the theoretical aspects
of CGAs. Further details, including applications of coevolutionary algorithms, can be found
in the literature (Nair and Keane 2001; Potter and De Jong 1994).

9.8.2 Some Issues in Coevolutionary MDO

The issues that need to be addressed in order to apply a coevolutionary search strategy to
optimization of the model problem described earlier are:

> *Problem decomposition*: This issue is mainly concerned with how the coupled system
> should be decomposed into various species. A natural procedure for decomposition
> would be to divide the design variables into groups on disciplinary lines. The chro-
> mosome of each species consists of the variables it is allowed to control. In general,
> it may be preferable to decompose the multidisciplinary design variable vector **x** into
> completely disjoint sets to circumvent the difficulty of arriving at a consensus on the
> independently evolved multidisciplinary variables.

> *Choice of representative individuals*: As mentioned earlier, the species interact with
> each other via representative individuals. The evolution of each species is thus

constantly driven by evolutionary changes in the species it interacts with. In the genetics literature, this is referred to as the Red Queen hypothesis, where each species must constantly adapt just to remain in parity with the others. Hence, a fundamental issue in coevolutionary computation is how to choose the representative individuals from each species.

Bull (1997) proposed two strategies for choosing representative individuals. The first strategy was to select the individual with the highest fitness in a species as its sole representative. In the second strategy, two representative individuals were chosen. The first representative was the individual with highest fitness while the second representative was chosen randomly. Both the representatives are used for fitness evaluations, and the maximum fitness of the two possible collaborations is assigned to the individual under consideration. Notice that this will lead to the requirement of 2^{m-1} fitness evaluations for each individual in a discipline, where m is the total number of species. It was found via numerical experiments that the first strategy, although superior in terms of convergence speed, may not be robust for problems with high epistasis. In contrast, the second strategy is more robust at the expense of slower convergence. Of course, the number of representatives need not be the same for each species nor remain fixed during optimization. However, the computational cost of fitness evaluation grows rapidly with increase in the number of representatives, since a domain-specific analysis must be carried out for each combination of collaborating representatives considered.

Coupling variables: A fundamental issue in distributed optimization of coupled systems is how to ensure multidisciplinary feasibility of the optimal solution. MDO architectures such as CO explicitly treat the coupling variables as auxiliary design variables. The major disadvantage of such a strategy is that the problem size grows significantly with increase in the coupling bandwidth. In the coevolutionary approach, this fundamental difficulty is alleviated by treating the coupling variables implicitly in the optimization process. Details of the strategy used for handling the coupling variables are described later. In contrast to other distributed MDO formulations, a major advantage of this approach is that it can tackle problems with arbitrary coupling bandwidth without an attendant increase in the problem size.

Fitness evaluation: The fitness of the disciplinary species is defined as a function of the respective disciplinary objectives and constraints. One could use here any of the approaches described previously in Chapter 3 for constrained problems.

9.8.3   A Coevolutionary MDO (CMDO) Architecture

Various coevolutionary MDO (CMDO) architectures adopting different considerations for tackling the coupling compatibility constraints and problem decomposition characteristics can be developed within the framework of the coevolutionary genetic adaptation paradigm. For example, a coevolutionary approach can be employed within the framework of the CO architecture described earlier. However, to ensure applicability to systems with broad coupling bandwidth, here the *coupling variables* are not explicitly considered as design variables. The coupling variables are instead initialized to values computed from system analysis of a baseline design. In the subsequent stages of the coadaptation procedure, the disciplinary species exchange values of the coupling variables computed during the disciplinary analysis of their respective *representative individuals*. This implicit iteration scheme reduces the extent of violation of the coupling compatibility constraints as the coevolution reaches stasis.

In the CMDO architecture, the design variables are decomposed into disjoint sets, and hence, no variables are shared between the disciplines. The chromosome of the disciplinary species are therefore composed of independent sets of variables. The chromosome representation and fitness evaluation details for both the species are summarized below. Here, the variables x_m^1 and x_m^2 are the disjoint sets formed from decomposing the vector of multidisciplinary variables x_m, that is, all the elements of x_m are placed in x_m^1 or in x_m^2.

Disciplinary Species 1

Genotype: [x_m^1, x_1], that is, the first part of the multidisciplinary design vector and the variables local to discipline 1.

Fitness evaluation: Computation of the objective and constraint functions for discipline 1 (f_1 and g_1) requires the values of x_m^1, x_1, x_m^2, and y_{21}. The values of x_m^1 and x_1 are readily available since they are directly controlled by species 1. In the first ecosystem generation, the value of y_{21} is initialized either from system analysis of a baseline design or randomly. Similarly, x_m^2 could be initialized randomly in the first ecosystem generation or set to the value corresponding to a baseline design. In the subsequent ecosystem generations, the values of y_{21} and x_m^2 provided by discipline 2 for their representative individual are used.

Disciplinary Species 2

Genotype: [x_m^2, x_2], that is, the second part of the multidisciplinary design vector and the discipline-specific variables.

Fitness evaluation: To compute the objective and constraint functions for discipline 2 (f_2 and g_2), the values of x_m^2, x_2, x_m^1, and y_{12} are required. The values of x_m^2 and x_2 are readily available. The value of y_{12} computed via system analysis of a baseline design (or randomly initialized values) is used in the first ecosystem generation. The subset of multidisciplinary variables controlled by discipline 1, that is, x_m^1, is either initialized randomly or set to the baseline values. In the subsequent ecosystem generations, the values of y_{12} and x_m^1 provided by discipline 1 for their representative individual are used.

It can be noted that the implicit procedure used here to ensure satisfaction of the interdisciplinary coupling compatibility constraints turns out to be similar to the generalized Jacobi scheme discussed earlier for distributed multidisciplinary analysis. It is known that the generalization Jacobi scheme may fail to converge for some problems (Arian 1998). Hence, it becomes important to examine when the implicit procedure used in the CMDO architecture may fail. The approach used here for satisfying the coupling constraints may be interpreted as a *randomly restarted* generalized Jacobi iteration scheme. The term "random" is used here to indicate that the coupling variables y_{12} and y_{21} are reset on the basis of the progress of the coevolution procedure, that is, they may change radically as the representatives change. However, as the coevolutionary process reaches stasis (i.e., as x_m,

x_1 and x_2 converge), the implicit procedure reduces to the conventional generalized Jacobi scheme. Numerical studies conducted in the past suggest that an optimal solution satisfying the interdisciplinary compatibility constraints can often be found (Nair and Keane 2002a). However, for problems where difficulties arise in satisfying the compatibility constraints, alternative iterative schemes (such as the damped Newton method discussed previously in Section 9.1.2) may have to be employed in the later stages of the coevolutionary search.

9.8.4   Data Coordination, Surrogate Modeling, Decomposition, and Other Issues

A common blackboard model for the coupling variables can be employed for transferring data between the disciplines. The disciplinary species post the values of the coupling variables and multidisciplinary design variables corresponding to their representative individuals on this blackboard during the coadaptation procedure. It is to be noted here that, since EAs constitute a global search paradigm, high-quality space-filling computational data can potentially be obtained as a by-product of the coevolution procedure. Such data can be used for constructing surrogate models to accelerate disciplinary optimization and archive the design space – see Section 7.4 for a description of approaches for construction and management of surrogates in evolutionary optimization algorithms.

It is also of interest to examine the possibility of using the information generated during the coevolutionary adaptation procedure to make decisions on how the set of multidisciplinary design variables should be decomposed. This would allow for the possibility of the optimal problem decomposition structure to emerge rather than be fixed *a priori* by the designer. However, this is a topic that remains to be fully investigated. An alternative procedure would be to decompose the multidisciplinary variables into disjoint sets by computing their main effects for the disciplinary objectives via orthogonal array-based experimental designs before starting the optimization.

Since the various species can interact via a single representative individual, the optimization algorithms that may be used within the CMDO architecture are not restricted to evolutionary methods alone. In fact, arbitrary optimization formulations/algorithms may be used within the disciplines.

The CMDO architecture allows two avenues for massive parallelization of the optimization process. First, the evolution of the disciplinary species may be carried out concurrently. Further, there also exists the possibility of computing the fitness of the individuals in each species concurrently.

Having summarized a range of possible ways to tackle MDO problems, we turn finally to a series of case studies. In these, we adopt the FIO approach when multidisciplinary issues arise.

Part IV

Case Studies

The final part of the book is devoted to a series of design search and optimization case studies taken from the work carried out by members of the authors' research group and their industrial collaborators (contributors are listed below). These span problems in structures, control and aerodynamics and in which the analysis is carried out using a variety of university level, aerospace industry written and commercial solvers, including one in which direct computation of derivatives is possible. The approaches taken include the straightforward manipulation of geometry via search engines, the use of bespoke geometry generation codes and also commercial parametric CAD systems. The searches are carried out with gradient descent, pattern and stochastic methods applied directly to the problem in hand, the use of formal design of experiment and updating schemes to construct metamodels, Pareto multiobjective methods, robustness modeling and the fusion of results coming from multiple fidelity codes. The frameworks used include a university research code, a prototype grid-based design system and public domain search tools. These studies therefore span most of the approaches and classes of problems likely to be encountered in aerospace design, from the university research group level through to full-scale industrial design.

Each of the case studies is presented in a stand-alone fashion so that they may be read independently. Inevitably, this involves some repetition of material already covered in earlier parts of the book.

We conclude the book with some guidance on how to tackle a problem in which Design Search and Optimization (DSO) is going to be used. This attempts to steer the user down a sensible path, although, by its very nature, such broad-brush advice must always be treated with caution. Throughout, we maintain a heavy emphasis on choosing an efficient and appropriate parameterization when building models for optimization. We now recommend the use of parametric CAD tools to achieve this as such frameworks provide, in our view, the best place in which to generate such descriptions. We would also emphasize the fact that DSO studies can never do more than help steer the design process – they are no substitute for good engineering judgment. Moreover, though they offer undoubted benefits, automated search schemes place additional burdens on the design team in terms of the increased skill set that must be mastered. As ever, there are no free lunches.

Contributors to the Case Studies

The following researchers helped work on the case studies described in this part of the book and their contributions, efforts and the support of their organizations are gratefully acknowledged.

David K. Anthony – University of Southampton

Steven Bagnall – Rolls-Royce plc.

Atul Bhaskar – University of Southampton

Mike Brennan – University of Southampton

Liming Chen – University of Southampton

Simon Cox – University of Southampton

Steve Elliott – University of Southampton

Hakki Eres – University of Southampton

Alex Forrester – University of Southampton

Shaun Forth – Cranfield University (RMCS Shrivenham)

Carren Holden – BAE Systems plc.

Zhuoan Jiao – University of Southampton

Yash Khandia – Applied Computing and Engineering Ltd.

Steve Leary – BAE Systems plc.

Mohammed Moshrefi-Torbati – University of Southampton

Graeme Pound – University of Southampton

Steve Ralston – Airbus

Gordon Robinson – University of Southampton

Eric Rogers – University of Southampton

Efstratios Saliveros – Airbus

Nigel Shadbolt – University of Southampton

Shahrokh Shahpar – Rolls-Royce plc.

Andras Sóbester – University of Southampton

Wenbin Song – University of Southampton

Mohamed Tadjouddine – Cranfield University (RMCS Shrivenham)

Feng Tao – University of Southampton

Ivan Voutchkov – University of Southampton

Jasmin Wason – University of Southampton

Geoff Williams – Airbus

Janet Worgan – Rolls-Royce plc.

Andy Wright – BAE Systems plc.

Fenglian Xu – University of Southampton

In addition, a number of these individuals have been funded by or worked under the auspices of the BAE Systems/Rolls-Royce University Technology Partnership for Design, which is a collaboration between Rolls-Royce, BAE Systems and the Universities of Cambridge, Sheffield and Southampton. Financial assistance was provided by EADS Astrium for work on the satellite case study. Other supporting companies include Compusys, Condor, Epistemics, Fluent, Intel and Microsoft. The U.K. Engineering and Physical Sciences Research Council also provide support for some of the work described here under grant references GR/R67705/01, GR/L04733, GR/M33624, GR/M53158 and GR/R85358/01. This support is also gratefully acknowledged. We would also like to thank the many students of our research group who have helped throughout in the writing of the book.

Further information on the satellite case study may be found in the following publications:

A. J. Keane and A. P. Bright, "Passive vibration control via unusual geometries: experiments on model aerospace structures," J. Sound Vib. 190(4) pp. 713–719 (1996).

M. Moshrefi-Torbati, A. J. Keane, S. J. Elliott, M. J. Brennan, and E. Rogers, "Passive vibration control of a satellite boom structure by geometric optimization using genetic algorithm", J. Sound Vib. 267(4) pp. 879–892 (2003).

10

A Simple Direct Application of Search to a Problem in Satellite Design

Our first case study concerns the vibration behavior of a satellite boom. Such booms are used to support a range of equipment in space and, when these include the elements of synthetic aperture radars or interferometer-based telescopes, they must hold the instruments with extreme levels of accuracy and at very low vibration levels. Typically, they are made up of multiple beam elements; see, for example, Figure 10.1.

In the design of all engineering structures and machines, it is, of course, desirable to minimize mechanical noise and vibrations – unchecked, they can have catastrophic effects on the performance of a structure. When considering solutions to vibration and noise problems in engineering, at least three possible strategies may be considered. The first, and presently most popular, option is to incorporate some form of vibration-absorbing material into the design of the structure. For example, one might elect to coat elements with a heavy viscoelastic damping material. Similarly, pieces of vibration-isolating material could be placed at mounting points, as is the case with most automobile engines. However, this method has rather significant weight and cost penalties. In the design of aerospace structures, this is a major concern, since any increase in weight results in subsequent increases in the cost of deploying the structure or reductions in payload.

The second method is so-called active vibration control. This employs the use of "antinoise" or "antivibrations" to cancel out unwanted vibrations and hence block noise propagation. However, active vibration control methods are inevitably complex and expensive, but they can yield impressive results.

A third option is passive vibration control. In this method, the design of the structure is modified so that it has intrinsic noise filtration characteristics and modified sensitivities to external disturbances. For example, the work described here shows that the frequency response of a specific structure can be considerably improved by making changes to the structure's geometry. The geometric regularity or irregularity of a structure affects its

Computational Approaches for Aerospace Design: The Pursuit of Excellence.   A. J. Keane and P. B. Nair
© 2005 John Wiley & Sons, Ltd

Figure 10.1 NASA Astronaut, anchored to the foot restraint on the remote manipulator system (RMS) arm, checks joints on the Assembly Concept for Construction of Erectable Space Structures (ACCESS) extending from the payload bay of the Atlantis. Shuttle Mission STS-61B, 27th November 1985, NASA photo STS61B-120-052.

vibrational response. In the case of a satellite boom, a geometrically regular structure is one in which the lengths of beam elements and the angles between them are repeated along the length of the structure. An irregular geometry would be one in which no two beam lengths or angles were equal.

There has been extensive research in this area over the last 20 years and a number of passive and active vibration control (AVC) strategies have been developed (Melody and Briggs 1993; Melody and Neat 1996; Sirlin and Laskin 1990). Here, we consider a purely passive approach and then an active design followed by a combined or hybrid active/passive scheme. We also describe use of both a directly differentiated analysis code and robustness measures that deal with the inevitable variations present in manufactured structures. The design problem tackled is to try and prevent vibrational disturbances from traveling along the structure and this is accomplished by modifying the geometry of the beam (by altering the joint positions) and/or by inserting piezoelectric actuators using a suitable control strategy in up to three of the beam elements. The results achieved are compared to experimental laboratory-based measurements. Figure 10.2 shows a typical structure and Figure 10.3 shows the equivalent structure on test in the lab.

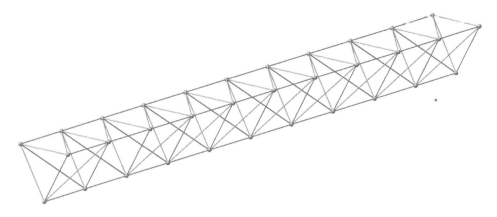

Figure 10.2 Regular three-dimensional boom structure.

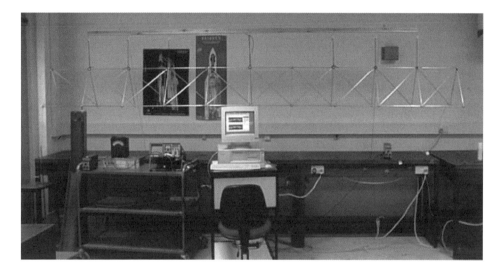

Figure 10.3 Regular three-dimensional structure on test in the lab.

10.1   A Problem in Structural Dynamics

The vibration performance of structures such as those illustrated in Figure 10.1 and 10.2 can be analyzed using a variety of approaches: most commonly, this is carried out via commercial finite element solvers working frequency by frequency or using a modal solution where the eigenvectors and natural frequencies are first obtained and then combined to predict the forced response behavior at frequencies of interest. Here, however, we take an alternative approach that obviates the need for meshing inherent in finite element methods – since the structures of interest are constructed from simple beams, their behavior can be predicted directly by employing classical Euler–Bernoulli beam theory and then constructing the behavior of the whole structure by combining these responses using receptance methods (Shankar and Keane 1995). Using this approach, the authors and coworkers have built and validated a simple Fortran code that reads in the geometry and topology of the structure under consideration, along with the associated material properties and which then computes the displacements and energy levels of the structure in response to random forcing in the frequency domain. Since the source for this code is available, a numerically differentiated version has been produced using the Adifor 3.0 system[1] (Tadjouddine et al. 2005) and this allows direct calculation of gradients, and so on. The receptance code has been validated against the Ansys®[2] FEA package, and we begin by briefly reviewing this process.

10.1.1   The Structure

The initial structure studied (Figure 10.4, which also shows the forcing and response points used) consisted of 90 Euler–Bernoulli beams all having the same properties per unit length. In the design of this structure, 6.35-mm (0.25-in.) diameter aluminum rods were selected with a basic bay length of 0.3428 m (i.e., giving an overall structure length of 3.428 m). Using typical properties for aluminum, this gives individual beam properties of EA equal to 2.25 MN (axial rigidity – Young's modulus times cross-sectional area), EI of 5.67 Nm^2 (flexural rigidity – Young's modulus times second moment of cross-sectional area), and a mass per unit length of 0.087 kg/m.

For the purposes of analysis, it was assumed that the three joints at the left-hand end of the structure (i.e., points (0,0,0), (0,0,0.3428), and (0,−0.29687,0.1714)) were clamped so as to prevent motion in any degree of freedom (N.B., some subsequent studies made use of free–free structures). To simulate the likely excitation experienced in practice, it was further assumed that the structure would be driven by a point transverse force halfway along the length of the bottom left-most beam (i.e., halfway between the points (0,−0.29687,0.1714) and (0.3428,−0.29687,0.1714)). The damping ratio of all beams in the structure was taken to be 0.005 (i.e., frequency proportional).

10.1.2   Finite Element Analysis

Once the basic dimensions of the boom had been chosen, the first step in the validation process was to produce frequency response curves of the base design using the commercial

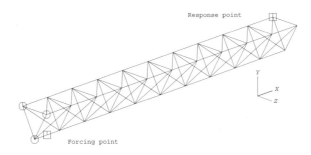

Figure 10.4   Geometry of initial, regular three-dimensional design, also showing forcing and response points.

---

[1]www.adifor.org
[2]www.ansys.com

Finite Element Analysis (FEA) package. The output obtained from the receptance method could then be compared against these finite element calculations to confirm that accurate results were being obtained.

To generate the finite element mesh, the boom was first modeled as a wire frame. Then, to define the mesh, the geometric cross sections of the beams were input along with their properties per unit length. The elements of the mesh were defined as having a maximum size of one-tenth the length of the nondiagonal beams in the structure. This choice of finite element length was validated by showing that, for a single beam, the distance between nodes of the mode shapes for all relevant modes was significantly greater than this element length. Although the frequency range of interest in this case study is 150–250 Hz, frequency response curves covering the range 0–500 Hz were studied to ensure confidence in the results. For a single 0.48-m beam (the longest in the structure), the seventh mode has a frequency of 659 Hz. As can be seen in Figure 10.5, the length of one finite element is significantly less than the distance between the nodes of the mode shape, justifying the mesh adopted.

As a final test of the mesh density, the element length was halved and the analysis for a single beam performed a second time. Whereas modes six and seven had frequencies of 336 and 659 Hz with an element length of 0.03428 m, they have frequencies of 336 and 658 Hz using an element length of 0.01714 m. From these results, it is apparent that the mesh density used during the FEA produces results with accuracy sufficient for the present study. It is, of course, desirable to keep mesh densities low to reduce loads on computer CPU time and memory.

Next, the boundary conditions and excitation force were applied as described earlier to the whole structure, that is, the three joints at the left-hand end were restrained in all degrees of freedom, and excitation applied as a point transverse force in the positive Y direction at a point halfway along the bottom left-hand beam. The frequency response (i.e., displacement versus frequency) was calculated at the right-hand end of the boom at joint 28, having coordinates (3.428,0,0.3428).

The finite element computation was initially solved for the first 100 modes of the structure, in this case up to a natural frequency of 184 Hz. Subsequently, the model was solved to mode 500 (1,090 Hz) and to mode 800 (2,170 Hz). Solving for 800 modes took approximately 12 h of CPU time using a Silicon Graphics R4400 based machine (using an Intel Xeon based machine at 3.0 GHz, this would take around 20 min). In the range of interest (150–250 Hz), no noticeable discrepancy can be seen between the frequency response

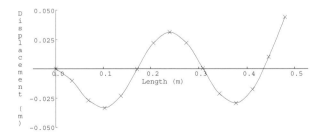

Figure 10.5 Mode shape of a single beam element calculated using FEA, with crosses marking the individual element ends.

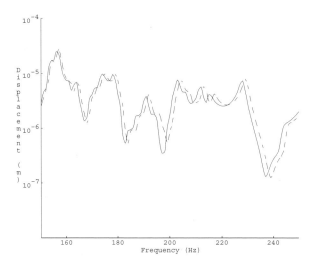

Figure 10.6   Frequency response of the initial three-dimensional design calculated using FEA shown dotted, with response found using receptance theory shown solid.

curves obtained during the 500 mode solution and the 800 mode solution. Thus, confidence can be expressed in the results obtained for this range. Figure 10.6 shows the displacement in the global Y direction of joint 28 as a function of driving frequency in the range 150–250 Hz (along with that from the receptance theory).

10.1.3   Receptance Theory

The next step was to produce a frequency response curve similar to the dotted line in Figure 10.6, but using receptance theory. The program used reads a formatted description of the three-dimensional boom from a data file. The information contained within this file is identical to that input during generation of the finite element model. Essentially, the data file consists of (1) a listing of all beams meeting at each of the 30 joints and (2) a listing of all properties for each of the 90 beams.

To produce a frequency response that can be compared with that produced by the FEA, the program was run for a frequency range of 150–250 Hz using 101 data points. The output was set as the displacement of joint 28 and the number of modes used to describe each individual beam set at 200, which guarantees convergence of the Green functions used. This calculation took about one-third of the time for running the FEA code. The resulting frequency response curve is also shown in Figure 10.6. On comparison with the results of the FEA, the two curves are virtually identical (the slight frequency shift of around 1% being due to the inherent limitations of FEA with finite mesh sizes and solution accuracy). Therefore, confidence may be expressed in the output of the receptance code.

It should also be noted that the time taken to carry out the FEA analysis is dominated by the number of elements used while the receptance method is influenced strongly by the number of frequency points to be studied (it is, of course, also affected by model size). Thus, although the FEA takes around three times as long to carry out as the receptance code when examining 101 frequencies, it takes roughly ten times as long when dealing with the reduced set of frequencies needed for optimization (here a 21 point integration rule is applied to assess band-averaged frequency response). The receptance method's principal attraction over FEA is therefore the speed with which an individual design can be analyzed for its band-averaged performance. Moreover, the receptance code can be interfaced more easily to the optimization software used here and can readily be run in parallel on multiple processors when carrying out design searches without the need for multiple software licenses. Since its source is available, it may also be numerically differentiated.

10.2 Initial Passive Redesign in Three Dimensions

Following a successful study of passive vibration control on a simplified two-dimensional structure (Keane and Bright 1996), a program of geometrical optimization work on three-dimensional booms was initiated. The structure analyzed was that of Figure 10.4, that is, it had a basic bay length of 0.3428 m, with EA of 2.25 MN, EI of 5.67 Nm^2, and a mass per unit length of 0.087 kg/m. It was assumed that the three joints at the left-hand end of the structure were clamped and that the structure would be driven by a point transverse force halfway along the length of the bottom left-most beam. The objective function was taken as the normalized energy in the end horizontal beam averaged over the frequency range 150–250 Hz, searched for using a genetic algorithm (GA) optimizer.

To begin with the number of GA generations was set as 10 and the population size as 100. Using Silicon Graphics R4400 processors, the completion of these 10 generations took approximately 2,000 CPU hours, which, using parallel processing, took around 20 days (at the time – such jobs can now be completed overnight). This extremely large CPU effort explains why so few generations were used and why only 100 members were allowed for each one. Clearly, with $3 \times 27 = 81$ variables, a rather larger generation size would have been desirable, as would more generations. Nonetheless, significant performance improvements were obtained.

The geometry producing the best frequency response curve from the 100 random geometries created during the first generation is displayed in Figure 10.7. As can be seen, the locations of all of the "interior" joints have been significantly changed. To provide a clearer picture of the overall geometry of the new design, the "diagonals" have been omitted from this figure; they were, of course, included during the computations.

To confirm that the optimizer had indeed produced a geometry with an improved frequency response curve, the new structure was again modeled and analyzed using the

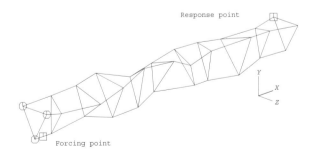

Figure 10.7 Geometry of best three-dimensional design found in first, random generation (diagonal elements omitted for clarity – cf., Figure 10.14).

commercial finite element software. The same boundary conditions were applied as before, that is, the three extreme left-hand joints were restrained in all degrees of freedom. A point transverse force was applied in the positive Y direction at a point 0.1714 m along the length of the bottom left-hand beam (i.e., half the length of the original beam). The response was measured at the same joint as before (which remains at the same coordinates in space), also in the positive Y direction. The resulting frequency response curve is shown in Figure 10.8, along with that produced using the regular geometry (i.e., that of Figure 10.6). It is obvious from this graph that this initial random search has indeed produced a design with an improved frequency response. The magnitude of the displacement at joint 28 has been decreased across most of the frequency range of 150–250 Hz.

The following step in the optimization process was to run the GA for more generations. Figure 10.9 shows the change in the mean, standard deviation, and minimum value of the objective function of the GA over the 10 generations used. As has already been noted, the objective function is the normalized average energy in the end horizontal beam. Therefore, at any given stage, the boom design with the best frequency response curve is represented by the "minimum" curve in Figure 10.9. The initial value of the objective function (i.e., before generation 1) was 0.0365. At the conclusion of the first generation, the value of the objective function for the best design was 0.000650, an improvement of over 5,000% (17.5 dB). This was followed by less dramatic, but nonetheless substantial, improvements, with the final objective function being 0.000173, a reduction of a further 380% (23.2 dB in total). Figure 10.9 also shows that the generational mean has steadily decreased across all generations and that the generational normalized standard deviation showed a generally downward trend. The fact that the deviation was still roughly equal to the mean at the last generation indicates that the populations had by no means become stagnant and that further generations could have been expected to improve the design still further. It is unlikely that any massive additional improvement would be made to the objective function value, but it should be possible to further tune the designs so that the generational mean and minimum start to

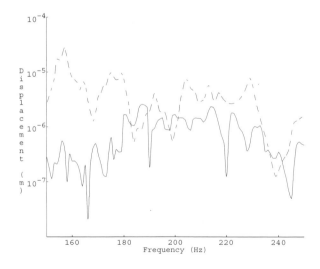

Figure 10.8   Frequency response of best three-dimensional design found in first, random generation, with response for initial design shown dotted.

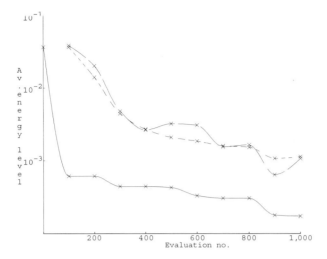

Figure 10.9 Variation of objective function with evaluation number, with generational mean shown dotted and generational normalized standard deviation dashed.

converge and the standard deviation drops to significantly lower levels. It should be noted that with 81 variables the search space being studied is huge and even a GA has only sampled this in a very limited way.

This behavior (a significant initial improvement followed by more modest refinements) reflects what one would expect. It is clear that a geometrically regular design does not produce a particularly favorable frequency response curve. Conversely, an irregular geometry has the potential to produce significant improvements in the frequency response (N.B., many irregular designs are, in fact, worse). This knowledge represents the foundation of the approach. Therefore, it is not surprising that a very large improvement has been achieved by choosing the best out of the 100 irregular geometries in the first generation. During subsequent generations, this irregular geometry is then "fine tuned". With the first generation, a good base design is found. Subsequent generations provide steady, but less dramatic, improvements of a similar magnitude throughout the run. This is a direct consequence of the limited number of trials allowed by the time-consuming analysis underpinning the work. In such cases, it is unrealistic to aim at achieving globally optimal designs; instead, good improved designs have to be sufficient. Even so, steady performance improvements have been obtained; see Figure 10.10, which shows the frequency response for the final design, along with that for the best in the first, random generation. Figure 10.11 shows the final geometry achieved; compare to figures 10.4 and 10.7. Although these modified geometries may appear extreme when compared to the original design, they do provide the required noise-isolation characteristics. Clearly, they would not be simple to build or deploy, but then neither are "active" control systems.

In summary, it may be said that these initial three-dimensional passive redesigns show that significantly improved frequency responses may be obtained for boom structures using GA optimization procedures. However, in order to accurately assess the designs considered by the GA, very significant computations are required, even when using a highly tuned and customized code to carry out the calculations. This leads to long run times, necessitating the

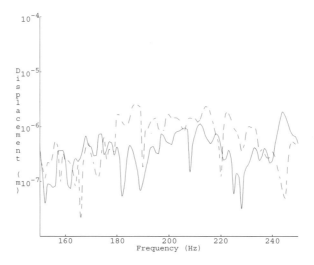

Figure 10.10 Frequency response of final three-dimensional design, with that for the best design found in the first, random generation shown dotted.

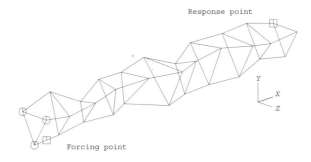

Figure 10.11 Geometry of the final three-dimensional design (diagonal elements omitted for clarity).

use of parallel processing, if realistic studies are to be undertaken. The nature of the changes made suggests that, despite the very significant improvements obtained, the GA is not likely to have found the globally optimum design in such short runs. This would seem to be an inevitable result of dealing with a problem with 81 variables using only 1,000 function evaluations. Moreover, the shapes resulting from this three-dimensional optimization are extremely complex, and so, attention was next turned to how they might be realized in practice so that experimental verification might be made of the improvements predicted.

10.3   A Practical Three-dimensional Design

The key difficulty associated with the three-dimensional designs produced by geometric optimization is that all the beam elements differ in length and, crucially, they all meet at differing angles at the joints. To construct this kind of structure in such a way that the correct geometric relationships are maintained is by no means trivial. A number of schemes were considered, including the use of sophisticated clamping systems during welding and custom-made joints of various kinds. In the end, the simplest workable scheme that could be devised was to use five-axis computer numerically controlled (CNC) machining to manufacture special joints into which the beam elements could be assembled. These joints consist of small aluminum spheres that are predrilled at known angles and to known depths using the CNC machine. Then, if carefully sized rods are used to form the beam elements, the correct structure can be accurately constructed using epoxy resins to glue them into place. This does however introduce lumped masses into the structure at each joint and these must be allowed for in the receptance code (and finite element) analyses.

To test this construction method, a 10-bay geometrical periodic structure was again used as the starting point for redesign, but this time in a free–free configuration. All the 93 Euler–Bernoulli beams in the structure were chosen to have an axial rigidity EA of 2.184 MN, a bending rigidity EI of 5.503 Nm^2 and mass per unit length of 0.085 kg/m. On the basis of initial trial experiments, the value of the proportional viscous damping coefficient was again assumed to be 0.005. The structure was taken to be excited by a point transverse force on the fourth joint (0.45,0.0,0.0)m and, during optimization, the aim was to minimize the total vibrational motions in the right-hand three joints between 150 and 250 Hz.

The beams were all initially either 0.450 m or 0.636 m long and they were joined together by 33 aluminum spheres of 25 mm diameter. The design for connecting the elements of the boom structure is based on the fact that at any stage of the assembly there is only ever one element that may require bending temporarily. More specifically, each sphere contains six holes that accommodate the rods coming from various angles. In constructing each bay, the nondiagonal elements will easily slide into the corresponding holes and another two elements belonging to the following adjacent bay will also fit in with ease. In order to place the sixth (i.e., a diagonal element), a hole is drilled all the way through the sphere. This design proved to be both simple and accurate. Moreover, it is capable of dealing with arbitrary geometries.

10.3.1 The Regular Boom Experiment

Having devised a suitable manufacturing scheme, experiments were first carried out on the purely periodic initial structure. The experimental setup for the boom is in fact that shown in Figure 10.3 and schematically in Figure 10.12. The structure is hung from a beam via several elastic strings and the beam is subsequently suspended from the ceiling by two equal-length wires. Free–free boundary conditions were chosen here as they are simple to achieve in practice. A force transducer (B&K 8200) was used in conjunction with a shaker (LDS V201) that was screwed to the structure. The joint with coordinates (0.45,0.0,0.0) was chosen as the input force location. The response was measured at the location of one of the end joints with coordinates (4.5,0.3897,−0.225), as shown in Figure 10.12. A random input force from 1 to 500 Hz, generated by the analyzer was supplied to the shaker via an amplifier (type TPA100-D). An ammeter was then used in series with the amplifier to avoid damaging the shaker's coil. The force transducer's output was fed back to the analyzer's first acquisition channel via a charge amplifier (B&K 2635). The translational responses of the structure in the X, Y and Z directions at the right-hand end joint (i.e., joint 32) were measured by using a Bruel & Kjaer triaxial accelerometer (type 4326A) via charge amplifiers (B&K 2635) whose gains were also recorded for later use. As the accelerometer

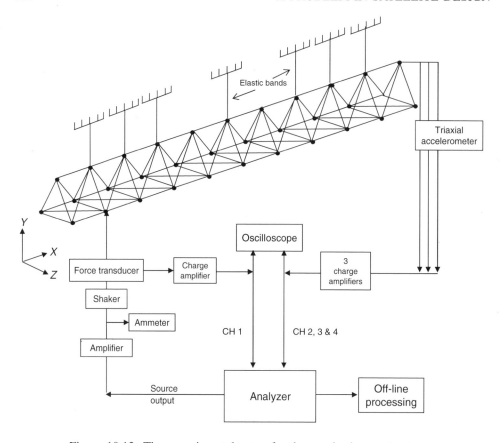

Figure 10.12 The experimental setup for the regular boom structure.

weighs around 13 g, during any joint measurement, the other two end joints had masses of equal weight attached to them. The measured accelerations were integrated to obtain the corresponding velocities. These outputs were fed back to the analyzer.

Figure 10.13 shows the experimental forced response of the structure at one of the joints (joint 32 and in the Y direction) for the frequency band of 1 to 500 Hz compared with the same result obtained using the receptance theory. As can be seen from the results shown in the figure, there is a convincing agreement between the experimental results and the theoretical curve. This agreement implies that the theoretical model is capable of predicting the vibrational behavior of the structure. This gives the confidence to carry out the search for a geometrically optimized structure with reduced vibration transmission levels.

10.3.2 Passive Optimization

In this study, the aim was again to reduce the frequency-averaged response of the boom in the range 150–250 Hz, J, where

$$J = \sum_{150\ \text{Hz}}^{250\ \text{Hz}} \sum_{j=31}^{33} \sqrt{v_{xj}^2 + v_{yj}^2 + v_{zj}^2} \, , \quad j = \text{joint}. \tag{10.1}$$

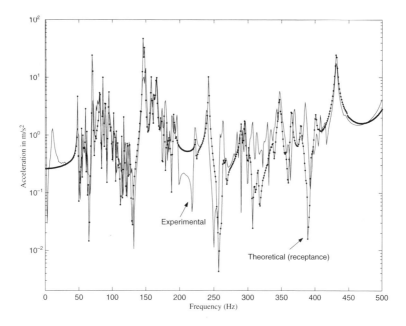

Figure 10.13 Comparison of the experimental and theoretical responses of the boom structure at joint 32 in the *Y* direction.

Once again, to meet other design requirements, the optimization is constrained to keep the
end beams unchanged in length and position with respect to each other. Further, to meet
structural requirements, all of the joints within the structure have been kept within fixed
distances from their original positions. This ensures that no beam is too long or too short
and also restricts the overall envelope of the structure. The free variables in the problem
are thus set as the *X*, *Y* and *Z* coordinates of the 27 midspan joints, that is, 81 variables in
all. To achieve the above objectives, the GA optimizer was used to produce an improved
design with the number of generations now set as 100 and the population size as 300.[3] In
the creation of a new geometry, the coordinates of all joints in the structure were varied
within ±20% bay length maximum deviation from their original positions. The geometry
producing the best frequency response curve is illustrated in Figure 10.14.

As can be seen from the figure, the locations of all the "interior" joints have been
significantly changed. In order to confirm that the optimizer has indeed produced a geometry
with an improved frequency response curve, the response of the new structure may be
compared with the original regular boom structure. This is shown in Figure 10.15, where the
frequency response of the optimized structure is plotted along with that produced using the
regular geometry. From this figure, it is evident that the new design has greatly improved
response within the range of 150–250 Hz. The theoretical attenuation for this optimized
structure within the frequency range of 150–250 Hz was found to be 19.4 dB. To verify
these results experimentally, the optimized design was manufactured and the forced response
tests repeated.

---

[3]The large number of trials used here being made possible by the use of a computing cluster of 500 Pentium
machines.

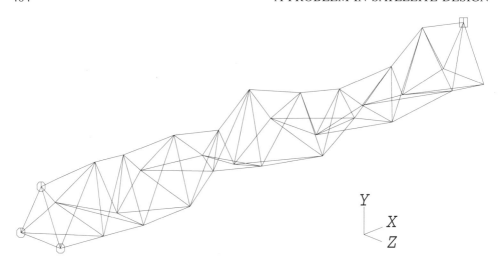

Figure 10.14 Geometry of the best design after 30 generations.

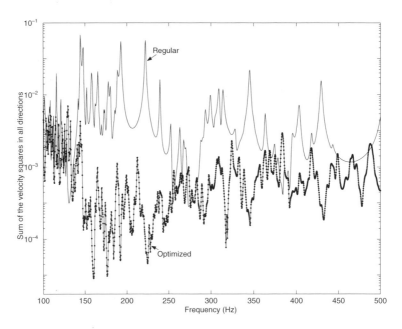

Figure 10.15 Comparison of the theoretical frequency response of the regular boom structure (solid line) with that of the optimized structure (dotted line).

10.3.3 The Optimized Boom Experiment

Figure 10.16 shows the manufactured passively optimized boom structure. The experimental setup for this experiment was exactly as before, when the regular structure was tested. The

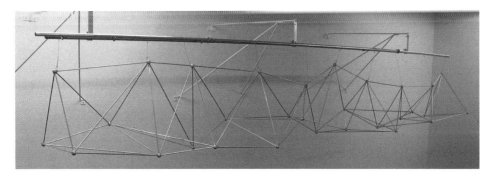

Figure 10.16 The optimized boom structure.

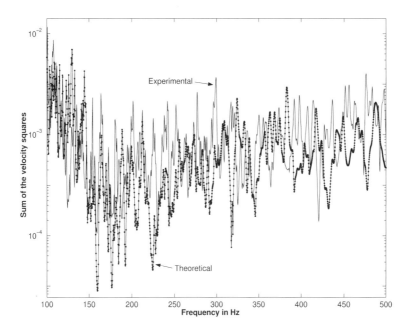

Figure 10.17 Comparison of the experimental and theoretical frequency response curves for the optimized boom structure.

structure was again excited by a unit random input force from 1 to 500 Hz, generated by the analyzer and supplied to the shaker via the amplifier. For the purpose of comparing the objective functions for both boom structures as the criteria for design improvement, the velocities at the three end joints with coordinates (4.5,0.0,0.0), (4.5,0.3897,−0.225) and (4.5,0.0,−0.45) were measured. This experiment was also repeated for the regular boom structure. The squares of all these velocities were then combined and the results compared with the equivalent theoretical curve; see Figure 10.17. Despite the fact that results are drawn from nine different measurements, the differences between the curves are acceptable, particularly for the frequency range of 150–250 Hz, which is of interest here. To demonstrate

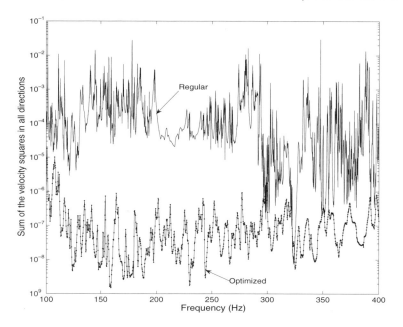

Figure 10.18 Comparison of the experimental frequency response curves for the regular and optimized boom structures.

the improvement achieved in practice by the design, the experimental objective functions for the regular and optimized structures are shown together in Figure 10.18. It can clearly be seen that the optimized design displays much improved vibration characteristics at most frequencies and, in particular, within the band of interest.

10.4   Active Control Measures

Having demonstrated what may be achieved by purely passive means, attention is next turned to completely AVC methods. The approach adopted is via the insertion of up to three piezoelectric axial forcing cells that can be mounted in-line with any of the beam elements in the structure. Studies have been carried out to decide the locations for these actuators. For one or two actuators, the possible number of combinations of positions is not so great as to prevent fully exhaustive searches. However, when positioning three actuators, this is not possible and the genetic algorithm is again employed. Having found suitable actuator positions, experiments were then carried out to verify the quality of the theoretical predictions.

10.4.1   Active Vibration Control (AVC)

The baseline structure used in this work was the same as for the previous section and once more, during optimization, the goal is to minimize the mean vibrational energy level of the three joints at the right-hand end. In order to reduce the vibration of the boom at these three joints, AVC was applied using up to three actuators. Here, the AVC is taken to be a feed-forward implementation with each frequency considered separately (this is most suited to tonal vibrations and their harmonics) and therefore the problem of noncausality is avoided. Feed-forward control requires a coherent reference of the vibration source that is not subject to feedback from the control actuators. In the experiments, the source of vibration disturbance is a force applied to one of the joints near the left-hand end of the structure, while the reference signal is independently available from the source.

In applications of AVC, the most commonly used cost functions are based on the square of a quantity (e.g., acceleration or velocity). The formulation of cost functions for this application is discussed by Anthony and Elliott (2000a). Here, the previously specified function comprising the translational velocity components at the three right-hand end joints was used.

The base vibration (i.e., the primary force) is modeled as a single force of 1N applied in the direction of the Y-axis at the fourth joint. The translational velocity components at joints 31, 32 and 33 of the structure (the right-hand end), at a single frequency, are represented by the velocity vector $\mathbf{v} = [\begin{array}{ccc} \mathbf{v}^{31} & \mathbf{v}^{32} & \mathbf{v}^{33} \end{array}]^T$ which is comprised of three individual joint velocity vectors in the same format. The format of $\mathbf{v}^{31}$ for example comprises the components in the directions X, Y, Z, $\mathbf{v}^{31} = \{\begin{array}{ccc} v_x^{31} & v_y^{31} & v_z^{31} \end{array}\}^T$. With the AVC operational, the net velocities at the three joints, $\mathbf{v}$, are the result of constructive interference between the velocities resulting from the vibration disturbance (the primary velocity, $\mathbf{v}_p$) and the velocities resulting from the AVC, $\mathbf{v}_s$,

$$\mathbf{v} = \mathbf{v}_p + \mathbf{v}_s. \tag{10.2}$$

The control forces $\mathbf{f}_s$ are the quantities controlled by the AVC controller and are applied using double-acting axial actuators in order to reduce the effect of the primary vibration, $\mathbf{v}_p$. The secondary velocity vector, which describes the effect of the AVC, can be derived from the secondary control forces by means of a transformed mobility transfer matrix, $\mathbf{Y}$, so equation (10.2) can be written as,

$$\mathbf{v} = \mathbf{v}_p + \mathbf{Y}\mathbf{f}_s. \tag{10.3}$$

$\mathbf{Y}$ is comprised of two matrices, $\mathbf{U}$, the mobility matrix, and $\mathbf{T}$, the transformation matrix, so (10.2) can be written as (Anthony and Elliott 2000b), $\mathbf{v} = \mathbf{v}_p + \mathbf{UTf}_s$. The formats of the above matrices may be illustrated by means of an example. Equation (10.4) shows equation (10.3) expanded explicitly for the case of two secondary actuators, where f_{s_1} and f_{s_2} are the individual forces of the two actuators,

$$\begin{bmatrix} \mathbf{v}^{31} \\ \mathbf{v}^{32} \\ \mathbf{v}^{33} \end{bmatrix} = \begin{bmatrix} \mathbf{v}_p^{31} \\ \mathbf{v}_p^{32} \\ \mathbf{v}_p^{33} \end{bmatrix} + \begin{bmatrix} \mathbf{U}_{1A}^{31} & -\mathbf{U}_{1B}^{31} & \cdots \\ \mathbf{U}_{1A}^{32} & -\mathbf{U}_{1B}^{32} & \cdots \\ \mathbf{U}_{1A}^{33} & -\mathbf{U}_{1B}^{33} & \cdots \end{bmatrix} \begin{bmatrix} \{1\} & & \\ & \{1\} & \\ & & \cdot \\ & & \cdot \end{bmatrix} \begin{bmatrix} f_{s_1} \\ f_{s_2} \\ \cdot \\ \cdot \end{bmatrix}. \tag{10.4}$$

Here, $\mathbf{U}$ is made up of a number of submatrices, each representing the transfer mobility from the force applied at one end of an actuator to all the velocity components at a single joint. For example, $\mathbf{U}_{1A}^{31}$ is,

$$\mathbf{U}_{1A}^{31} = \mathrm{diag}\left(u_{1A_x}^{31} \quad u_{1A_y}^{31} \quad u_{1A_z}^{31} \right), \tag{10.5}$$

where each component is a single complex transfer mobility. Here, $u_{1A_x}^{31}$, is the transfer mobility from end A of actuator 1 to the velocity in the X direction at joint 31. $\mathbf{U}$ is shown partitioned for each actuator, where each partition contains submatrices representing the transfer mobilities from each end of the actuator to joints 31, 32 and 33. The signs of the submatrices for drive end B are negative in order to represent the double-acting operation of the actuators. The transformation matrix, $\mathbf{T}$, is made up of vectors of ones, which are given by $\{1\} = \{\begin{array}{cccccc} 1 & 1 & 1 & 1 & 1 & 1 \end{array}\}^T$. The transformation matrix is required as each secondary control force affects each of the six joint velocities components. $\mathbf{T}$ maps each force onto the transfer mobilities relating it to these velocities.

In the application of AVC (Fuller et al. 1996; Nelson and Elliott 1992), and as already noted, it is usual to minimize the sum of the square of the responses, that is, a quadratic function in $\mathbf{f}_s$. Here, the resulting cost function, J, is given by $J = \mathbf{v}^H \mathbf{v}$. Substituting for $\mathbf{v}$ from (10.3) gives

$$J = \mathbf{f}_s^H \mathbf{A} \mathbf{f}_s + \mathbf{f}_s^H \mathbf{b} + \mathbf{b}^H \mathbf{f}_s + c \tag{10.6}$$

where $\mathbf{A} = \mathbf{T}^T \mathbf{U}^H \mathbf{U} \mathbf{T}$, $\mathbf{b} = \mathbf{T}^T \mathbf{U}^H \mathbf{v}_p$ and $c = \mathbf{v}_p^H \mathbf{v}_p$. The minimum value of (10.6), $J_{\min}$, is achieved with the optimum vector,

$$\mathbf{f}_s^* = \min(J) = -\mathbf{A}^{-1} \mathbf{b} \tag{10.7}$$

and so, $J_{\min} = c - \mathbf{b}^H \mathbf{A}^{-1} \mathbf{b}$. The nominal (uncontrolled) value of (10.6) without AVC ($\mathbf{f}_s = 0$) is given by $J_{\text{nom}} = c$.

It is common to use AVC to control a number of frequencies within a specified band. In this case, the optimal secondary force vector and the minimized value of the vibration are found on a frequency-by-frequency basis, as shown above, but then the performance is taken as the arithmetic mean of the vibration at each frequency considered. Hence,

$$\overline{J_{\min}} = \frac{1}{n} \sum_{k=1}^{n} J_{\min} \left(\omega_L + (k-1) \Delta \omega \right) \tag{10.8}$$

is the average minimized value of the cost function of the frequency band, starting at frequency ω_L and with frequency increment $\Delta \omega$. $\overline{J_{\text{nom}}}$, the average value of the cost function without AVC, is similarly calculated. The frequency-averaged performance (the attenuation of the mean of the squares of the velocities) is given in decibels by $10 \times \log_{10} \left(\frac{\overline{J_{\text{nom}}}}{\overline{J_{\min}}} \right)$.

10.4.2 AVC Experimental Setup

The experimental setup for the active control is illustrated in Figure 10.19. The actuators used in these experiments are of preloaded open loop Piezoelectric Translator (PZT) type. They are equipped with high reliability multilayer PZT ceramic stacks protected by an internally spring preloaded steel case. The maximum displacement provided by these actuators is 90 μm and they produce pushing and pulling forces of 1000 and 100N, respectively. Without cables, they weight around 62 g and their stiffness is 15N/μm. The input to the actuator is provided by the analyzer through a power amplifier (PI model 790A01). The force applied by the PZT actuator is measured by a force transducer and is taken to the analyzer through a charge amplifier. As already noted, the translational responses of the structure in the X, Y and Z directions at the end three joints (i.e., 31, 32 and 33) were measured by using a Bruel & Kjaer triaxial accelerometer (type 4326 A). The measured accelerations were again integrated in order to obtain the corresponding velocities.

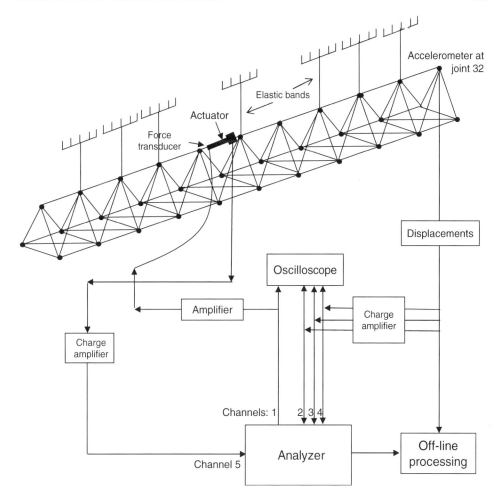

Figure 10.19 The experimental setup for the regular boom structure.

10.4.3 Selection of Optimal Actuator Positions on the Boom Structure

Because of dynamic mechanical coupling between the primary forces and the secondary actuators and their effects at the beam ends, the success of AVC heavily depends on the actuator positions in the structure, and this can be studied using the receptance code. It should be noted that the model used to predict the response of the actively controlled structure takes into account the passive effects of the actuators (overall weight, stiffness, etc.).

Single Secondary Force

The best positions for a single actuator were obtained through an exhaustive search since this is guaranteed to find the best results (being possible here as there are only 93 beams); see Table 10.1. The optimum position is on the middle bay closest to the primary force. This implies that within the bandwidth considered the most effective control with one actuator

Table 10.1   The five best single actuator
position

Actuator Positions	Attenuation (dB)	Control Effort (N^2)
41	15.1	3.01×10^5
50	13.8	4.64×10^5
40	12.9	3.38×10^5
49	12.6	4.31×10^5
31	10.8	4.06×10^5

is obtained by blocking the wave propagation path along the boom structure. For these actuator positions, the best achievable values of attenuation range from 10.8 to 15.1 dB with normalized total control effort ranging from 3.01×10^5 to 4.31×10^5 N^2. It can be seen that the secondary forces are large compared to the primary unit force. This may be explained by the fact that the secondary forces are acting along the structure (i.e., into a high impedance), whereas the primary force acts transverse to the structure (i.e., a low impedance). In this case, the minimum control effort corresponds to the maximum attenuation that is obtained using the best actuator position.

To validate this result, an experiment was carried out with the actuator positioned on beam 41. The input force signal to the shaker and the actuator was a swept-sine wave covering the frequency range of 150–250 Hz. The frequency resolution for all measurements was chosen to be 0.03125 Hz. The response to the secondary input force on element 41 at joints 31, 32 and 33 in all three directions were measured using the triaxial accelerometer. Having obtained the velocities due to the secondary force and recalling the velocities due to the primary input force, it was then possible to calculate the experimental objective function and subsequently compare it with the theoretical predictions. The theoretical results and their experimental counterparts are shown in Figure 10.20. The graphs now show much greater differences between theory and experiment. In particular, the inclusion of the actuator has led to a much "noisier" response and the regions of the frequency response that are free of predicted resonant behavior are not realized in practice. This is perhaps to be expected when moving from an essentially pure metallic structure to one containing complex assembled components such as piezoelectric stacks. Nonetheless, the degrees of attenuation achieved are of similar magnitudes: the experimentally measured value was found to be 16.6 dB, which is slightly higher than the predicted value of 15.1 dB. It is clear that the design approach being followed still allows the controlled reduction of vibration levels in a predictable fashion when band-averaged results are considered.

Two Secondary Forces

In order to find the optimum position for two secondary forces, an exhaustive search was once again carried out. The search indicated that beams number 8 and 45 provided the best positions for two actuators acting together. The amount of attenuation obtained by these two actuators is predicted as 26.1 dB, which is significantly higher than for one actuator. Table 10.2 lists the best computed positions for two actuators. The table shows that the achievable attenuation ranges from 23.0 to 26.1 dB and the total control effort required to achieve these attenuations varies from 10.42×10^5 to 17.24×10^5 N^2. Note that the control

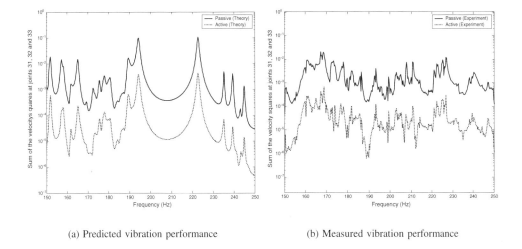

(a) Predicted vibration performance   (b) Measured vibration performance

Figure 10.20 Results for a single actuator positioned on beam 41 of the regular structure.

Table 10.2   The five best locations for two actuators

Actuator Positions	Attenuation (dB)	Control Effort (N²)
8, 45	26.1	10.68×10^5
8, 53	24.8	10.66×10^5
20, 45	24.6	17.24×10^5
11, 45	23.4	10.97×10^5
44, 8	23.0	10.42×10^5

effort of 10.68×10^5 N² corresponding to the maximum attenuation falls in the lower end of the above control effort range.

To carry out an experiment with new actuator positions involves removing the piezo-electric actuators from their current positions and replacing the original rods in their places and then placing the actuators in their new positions. Repeating this procedure can damage both the structure and the actuators. For this reason, it was decided to carry out further experiments only for the three-actuator case.

Three Secondary Forces

Finally, to find the best positions for three secondary forces, the GA was once more employed (an exhaustive search being too time-consuming). As a result of this search, the best positions for the three actuators were found to be on beams 23, 60 and 71. This gave a predicted attenuation of 33.5 dB. Here, the equivalent active control experiment is not carried out in real time. Instead, a simpler experiment is repeated four times and the response of the structure to the primary force and secondary forces are measured separately and the overall J_{min} calculated off-line, using (10.8). Therefore, the experimental setup for three actuators is the same as when one actuator was used. As before, the structure was first excited by an external force at joint 4 with coordinates (0.45, 0.00, 0.00) meters and the response measured using a triaxial accelerometer at the end three joints in the X, Y and Z directions. Then, the external force (i.e., the shaker) was removed and the structure was excited by the three actuators, one at a time. Having measured the responses to the primary and the three secondary forces, J_{min} was calculated. The theoretical values were again found by the use of the receptance code. Figure 10.21 shows the results obtained. The discrepancies that arise are slightly worse than those seen for the one-actuator experiment regarding the detailed variations in the curve, since now the structure contains three highly nonlinear actuators. Nonetheless, the experimental frequency averaged attenuation is still in broad agreement with predictions, being found to be 31.7 dB, that is, slightly less than the predicted value of 33.5 dB.

Measurements versus Predictions

It is clear from the various experimental measurements presented here that the agreement between theory and experiment is not so good as when dealing with the purely passive systems. The inclusion of actuators in the systems introduces a number of complications. When exciting the actuators, one of the main sources of error is the low signal-to-noise ratios used. Increasing the input gain could push the actuators' outputs into their nonlinear regions and so their outputs were necessarily small. Moreover, many of the fine details present in the three actuators (e.g., nuts, screws, force transducers, etc.,) are only approximately accounted for in the theoretical model. Nonetheless, it is clear that significant vibration-isolation performance may be achieved using feed-forward active control designed in this way. The amount of vibration reduction predicted from one end of the boom to the other

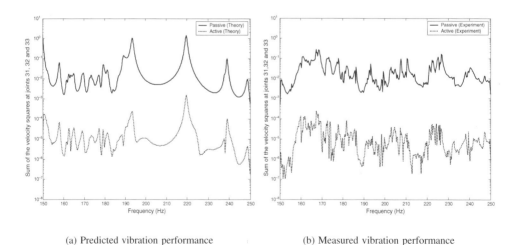

(a) Predicted vibration performance   (b) Measured vibration performance

Figure 10.21 Results for three actuators positioned on beams 23, 60 and 71 of the regular structure.

ranges from 15.1 dB with one, through 26.1 dB with two to 33.5 dB with three secondary forces. That found experimentally varied from 16.6 to 31.7 dB – a broadly similar range.

10.5 Combined Active and Passive Methods

Having considered active and passive approaches separately, attention is next turned to combined active and passive vibration control. Clearly, there are two approaches to designing such a combined system. First, the passive structure can be designed to give optimum, passive-only control and then suitable locations on this structure can be identified for placing any active elements. Alternatively, the combined active/passive system can be designed in one go by using a search engine to look for a geometry that works well in active mode. It would be expected that this second approach should achieve the best performance but it runs the risk that when the active elements are not working the resulting geometry may give rather less isolation than one designed to work in a passive-only way. This is obviously a risk/performance trade-off and a fall-back, passive-only system has many attractions for deployment in space where robustness of energy supplies and maintenance of active systems are major concerns. Both schemes are considered here and backed up by experimental measurements. The results are compared to those presented in the previous two sections.

10.5.1 Optimal Actuator Positions on the Passively Optimized Boom

To test the effect of placing actuators on an already passively optimized structure, the boom of Figure 10.14/10.16 was used as the baseline structure for an extensive GA search. This showed that elements 16, 27 and 48 were the best locations, giving a predicted (passive plus active) reduction of 48.7 dB. Having already built this structure, the actuators were then inserted in their optimal positions and the previous AVC experiments repeated. The response of the structure was measured at the three joints on the right-hand end in all three translational directions giving a total of 36 sets of measurements. These measurements were used off-line to calculate the overall cost function due to active control, as before. Figure 10.22 shows the results achieved and it is clear that the experimental and theoretical predictions show broad agreement in terms of their frequency-averaged behavior. The experimental attenuation was measured to be 40.4 dB which is some 8 dB lower than the predicted value.

10.5.2 Simultaneous Active and Passive Optimization

Lastly, a GA search was used to simultaneously search for a geometry and combination of actuator locations that would achieve good performance with less radical geometric perturbations than considered thus far. To do this, the joint movements were restricted to only 10% of bay length when finding new geometries. Note that this search is a mixed real and integer-valued search, as it is dealing with both joint positions and beam numbers for active element insertion; this is one of the strengths of GAs – when using binary encodings, it is straightforward to deal with such mixed problems. Despite the limited geometric changes, a design was found with three actuators that gave isolation considerably in excess of that possible by carrying out the optimization in two separate stages and with up to 20% deviations. The geometry is shown in Figure 10.23. The best actuator locations were

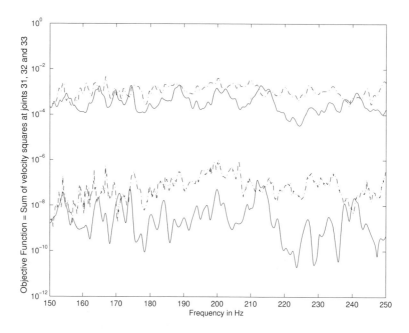

Figure 10.22 Theoretical and experimental passive and active objective functions for actuators placed on beams 16, 27 and 48 of the passively optimized boom structure (solid lines theory, dashed lines experiment, upper curves passive, lower curves active).

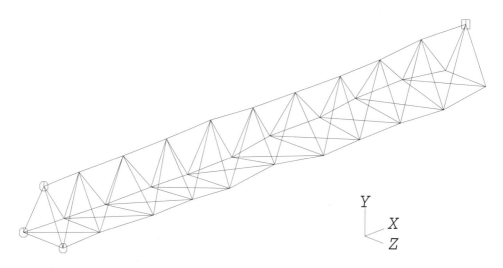

Figure 10.23 Resulting geometry for design optimized for combined passive and active vibration control.

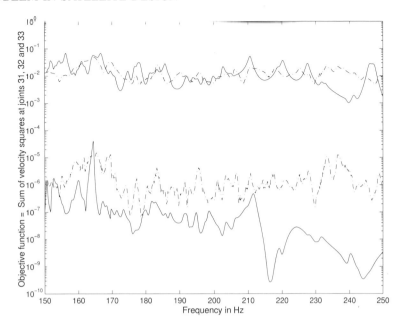

Figure 10.24 Theoretical and experimental passive and active objective functions for actuators placed on beams 6, 7 and 8 of the simultaneously optimized boom structure (solid lines theory, dashed lines experiment, upper curves passive, lower curves active).

found to be on elements 6, 7 and 8. According to the theoretical predictions, the amount of attenuation for this design should be 49.5 dB. The corresponding attenuation value from an experimental study was in fact 38.8 dB, which is around 11.0 dB lower than expected. Even so, this is some 5 dB greater than achieved by simply applying the active elements to a regular structure and nearly 9 dB better than for the passive design with much greater geometric variations. The experimental and theoretical objective functions are plotted in Figure 10.24. Unfortunately, such a design has rather poorer performance when the control system is turned off than a purely passive design. This is a trade-off the designer must consider bearing in mind likely mission profiles and the degree of resilience desired in the vibration suppression system.

10.5.3 Overall Performance Curves

To demonstrate and compare the effects of passive and active control using the various configurations tested, a general performance graph has been produced; see Figure 10.25. This graph serves to summarize all the information on both passive and active configurations and acts as a guide to the designer. By referring to the graph, a designer should be able to decide on the level of passive control (i.e., percentage deviation from regular geometry), active control (i.e., number of actuators) or combination of the two in order to achieve a desired degree of noise reduction. This graph is produced here for the two-stage optimization procedure: if combined geometric and active optimization is carried out, further improvements are possible.

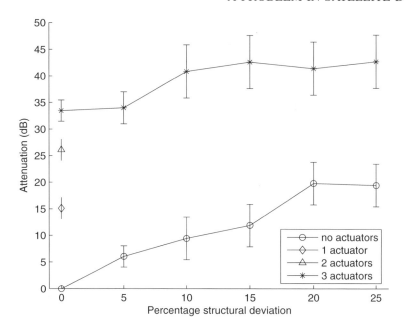

Figure 10.25 Attenuation graph for various passive and active designs.

Before interpreting the figure, it should be noted that although an optimizer is used here to produce the improved designs it is not suggested that the resulting designs are in fact globally optimal. In fact, without carrying out an exhaustive search, it would not be possible to guarantee such an outcome. Given the number of variables involved and the cost of analysis, this would not be feasible within realistic timescales. Such a situation is quite common in practical engineering design problems and the use of optimization in these circumstances is best termed "design search": the goal is merely to improve the design as much as possible within a given computing budget. A key part of the study presented is to demonstrate the magnitudes of the improvements that might be expected from this essentially practical stand point. Design teams need to know if design modifications will produce worthwhile benefits at affordable cost. Nonetheless, from the figure it can be concluded that:

1. Similar amounts of attenuation can be achieved either using one actuator or applying 20% deviation to the joints;

2. The amount of additional attenuation that is achieved on the regular structure (i.e., with 0% deviation) falls off as first two and then three actuators are added;

3. With no actuators, increasing the structural deviation increases the attenuation until 20 % deviation is reached, after which further gains are difficult to achieve;

4. The fall-off in added attenuation with increased structural deviation occurs earlier when three actuators are used.

In addition it was noticed that the location of an actuator within a beam was found to be an important quantity that has not been studied here explicitly – further work should examine whether a judicious choice of location within (along) each beam would further enhance the capabilities of each actuator.

10.5.4 Summary

This case study has demonstrated fairly conclusively that geometric modifications to space structures can be used to achieve significant vibration-isolation capabilities. Such reductions are compatible with active control systems and the complex geometries required can be readily manufactured and assembled using typical numerically controlled machining facilities. Whether or not the passive capability was designed independently of the active system would depend on issues of system redundancy and mission profile. The key impact of the work undertaken during this project is the identification of appropriate design strategies for producing satellite structures with very low vibration signatures. It has been shown, for example, that different combinations of geometric modification and numbers of active elements can be used to select the degree of noise isolation desired, during the design process. Thus, designers may trade the complexity of the active system against the complexity of the geometric redesign when producing new structures.

The structural dynamics of the booms analyzed have been modeled throughout by a Fortran computer code based on receptance theory, whereby the behavior of the global structure is predicted from the Green functions of the individual components, evaluated as summations over their mode shape. Thus far, the minimization of the frequency average response has been accomplished using a GA directly applied to the code and using the nominal response as the objective function. To round off this study, we finally consider alternative metrics that provide more robust results and the use of gradient information – here, derived from an adjoint code produced via numerical differentiation.

10.6 Robustness Measures

It is, of course, a key assumption in any optimization-based design that the code being used to study the problem under consideration accurately reflects the performance of any design variant. In particular, it is critical that the code correctly ranks designs so that sensible improvement choices may be made. In many problems, especially those that are highly nonlinear (as is always the case in structural dynamics), small perturbations to a design can make significant differences to the predicted performance. In such circumstances, heavy focus on optimization of nominal performance can lead to designs that do not realize the predicted abilities in practice.

To some extent, this has been avoided in the experimental studies shown here by the use of a genetic algorithm over only a modest number of generations and applied to a band-averaged objective. Since GAs construct new designs by crossover and mutation, they tend to sample many basins of attraction and these only mildly, especially in the early stage of a search. They are unlikely to land in deep-sided holes in the objective function landscape; rather, they tend to locate regions where less extreme but more robust designs exist. This tends to prevent the process becoming overly fixated on descending a single steeply sided basin. The use of a band-averaged figure of merit also helps preserve robustness: since the optimizer must simultaneously control 21 different frequency points here, which are in turn affected by many modes of the structure, it is much less likely that the search can become bogged down fine tuning the behavior of one or two modes to produce severe antiresonant notches in the frequency response curves that would be unlikely to be realized in practice.

Nonetheless, there are a number of schemes that can be adopted to directly tackle issues of robustness during optimization; see Chapter 8. These generally work in one of two ways. First, it is possible to always oversample the search space when making any calculation, that is, once the nominal performance has been found, a series of small perturbations are made to the current design and an average (or even worst case) result returned to the optimizer in lieu of the nominal performance – though expensive, such an approach is very reliable and should always be adopted if it can be afforded. Furthermore, if it is possible to carry out significant oversampling, this can make use of Design of Experiment (DoE) methods to uniformly study the local objective function landscape and, if a fixed DoE is used, the resulting robustness measures will be repeatable and thus not add noise to the function seen by the optimizer.

A more efficient approach is to sample much more sparingly, either by using only one or two random perturbations to the nominal design or to just add a random perturbation to the nominal design itself each time it is to be evaluated – the so-called Noisy Phenotype (NP) approach (Tsutsui and Ghosh 1997). In this way, the optimizer sees a rather blurred objective function surface and this forces it to take account of any likely lack of robustness. It should be noted that when such a scheme is adopted the introduction of noise requires that the optimizer be capable of working in such an environment – many of the most highly tuned gradient-based methods will not cope well with such formulations. As has been shown earlier, function noise is something that pattern and stochastic-based searches are better able to deal with.

We have carried out some brief experiments using both DoE-based oversampling and the addition of random perturbations to the design for a two-dimensional boom design (Anthony and Keane 2003).

10.6.1   The Optimization Problem

When dealing with the passive optimization of a two-dimensional variant of the boom structure, the optimizer controls the position of 18 joints in the X and Y directions, that is, 36 variables in all. For this study, reductions in the average vibration transmission over the frequency band 175–195 Hz were examined by simple geometric redesign. The structure's performance was again calculated using Euler–Bernoulli beam models and the receptance code. The best geometry found by direct application of the GA to the nominal performance is shown in Figure 10.26, which achieved an average reduction in the energy level of the right-most beam of 47.5 dB over the above frequency range.

10.6.2   Robustness Metrics

Clearly, some structures are more robust than others, and so a structure that is nominally the best is not always the best in practice. Conversely, another which might not have as good a nominal performance may be more robust, and will suffer less deterioration of performance in the face of geometric perturbations. The robustness of any optimized design can be studied in a number of ways. Normally this is achieved by adding sets of uniformly distributed random changes and reevaluating the performance of the design at each point.

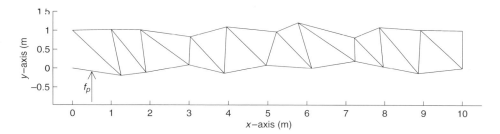

Figure 10.26   Best geometry achieved by optimizing nominal performance to minimize vibration transmission along the structure.

If a large-enough set of perturbations is carried out, the relevant statistics can be found reasonably accurately. One quantitative measure of robustness that can be constructed from such data is the 95% probability limit, f_{95}. This gives the performance that is only expected to be exceeded by 5% of perturbances. Unfortunately, carrying out large Monte-Carlo studies to converge the statistics can rarely be afforded – here, we limit trials to 300 points, leading to a metric that we term $\hat{f}_{95:300}$.

   If we apply perturbations of maximum size ± 0.01 m to the structure of Figure 10.26 and reevaluate the performance 300 times, we find that the robustness deteriorates by 5.0 dB, that is, the 95% probability limit lies 5.0 dB above the nominal performance of 47.5 dB at 42.5 dB. Since carrying out 300 additional receptance analyses is not possible within a design optimization process of reasonable length, we have examined five other cheaper perturbation schemes. These methods fall into two categories: first, the use of the noisy phenotype and second, DoE methods.

Noisy Phenotype One – Tsutsui and Ghosh's Method (NP)

In this method, noise is added to the phenotype (i.e., the optimization variables) during the evaluation of the performance (objective function) by the GA. The noise does not change the genetic information manipulated by the GA. If, because of the addition of noise the candidate solution is evaluated as having better than its natural performance, it will have a higher probability of succeeding (as a whole or in part) to the next generation. However, in the subsequent generations, a *different* value of noise will be added, which is unlikely to produce the same artificially high performance, and so this solution is unlikely to compete well against more robust solutions. Similarly, if the effect of the noise is to artificially deteriorate the true performance, the solution will immediately be treated as if it were less fit. In comparison, a robust solution is one that is unaffected by noise. Thus, the GA will tend to produce both *good and robust* solutions. This method is distinct from those that simply add noise to the final evaluated value. The reader is referred to Tsutsui and Ghosh (1997) for a full description of the method. The 95% probability limit achieved using the NP approach is denoted here as the estimate $\hat{f}_{95:NP}$.

Noisy Phenotype Two – Modified Tsutsui and Ghosh Method (NP2)

The NP method strives to produce good and robust solutions with no additional overhead as each solution is evaluated only once. In the modified scheme, the true performance and the noisy performance are both evaluated, the objective function value being set as the worst of these two measures. Functionally, this is only different from the NP method when an unrobust solution is evaluated with artificially high noise-induced performance. This will limit the short-term higher fitness of these solutions, which would subsequently be evaluated with poor fitness due to their lack of robustness. This modification may help to increase the speed of convergence of the genetic algorithm. This method is subsequently referred to as the NP2 method. The 95% probability limit achieved using NP2 is denoted here as $\hat{f}_{95:NP2}$.

Design of Experiment One – One-at-a-time Experiments (OAT)

So-called *one-at-a-time* experiments are extensively used in product development. They provide the smallest experiment size by varying all the individual factors (variables) separately. The main disadvantage is that they do not consider any possible interactions between factors. If interactions exist (unless of small significance), this strategy is expected to produce poor results. Despite this, the approach is commonplace in engineering, usually without due concern for potentially misleading results. The 95% probability limit achieved using the One-at-a-time Experiments (OAT) strategy is here termed $\hat{f}_{95:OAT}$.

Design of Experiments Two and Three – Orthogonal Arrays (L64 & L81)

A better strategy than the OAT method is to conduct a full-factorial experiment, in which the response to all the combinations of factor levels is determined. However, for more than a small number of factors, the number of experiments required often becomes prohibitively large. Fractional factorial design can be used if it is a reasonable assumption that higher-order interactions are small and this requires significantly less number of experiments. Taguchi published a number of fractional factorial designs (Taguchi 1987), and two of these are used here: (1) the L64 array, which is a 64-experiment design for 63 two-level factors, and (2) the L81 array, which is an 81-experiment design for 40 three-level factors. Only 36 factors are required for our problem and so the remaining design columns were not used. In order to maintain the balanced properties of the table, all of the experiments must still be run, however. The L64 and L81 arrays are defined using factor values $(1, 2)$ and $(1, 2, 3)$ respectively. These are mapped on to perturbation values of $(0, +v)$ and $(0, +v, -v)$, where v is the maximum perturbation. The first line of both arrays contain all 1's and therefore the unperturbed geometry is inherently included in the experiment. The 95% probability limits achieved using the L64 and L81 experimental designs are denoted as $\hat{f}_{95:L64}$ and $\hat{f}_{95:L81}$, respectively.

Comparison of Metrics

To enable a comparison of the DoE metrics, Figure 10.27 shows the results of applying them to 10 optimal designs, each of which was produced using the GA, but with differing random number seeds. The average deterioration here (as measured by $\hat{f}_{95:300}$) is 5.2 dB, although for one structure this figure is as high as 10.5 dB. Although $\hat{f}_{95:300}$ is still an estimate, it is the highest quality estimator and is used here as a reference. $\hat{f}_{95:L64}$ appears to give the closest agreement to $\hat{f}_{95:300}$ in this trial. The modified phenotype approaches do not, of course, produce sufficient information to warrant including in this figure – rather, they are aimed at guiding the GA search process into robust regions of the design space.

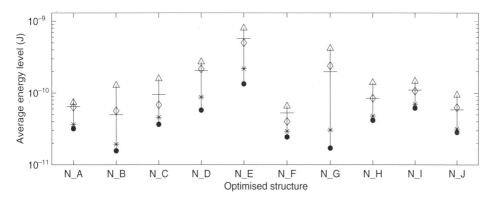

Figure 10.27 A comparison of the 95% probability limits evaluated for 10 structures optimized (minimized) on unperturbed performance. Limits evaluated using $\hat{f}_{95:300}$ denoted by the horizontal intersecting line, $\hat{f}_{95:OAT}$ by $*$, $\hat{f}_{95:L64}$ by $\diamond$ and $\hat{f}_{95:L81}$ by $\triangle$. The nominal performance of the structure is shown by $\bullet$.

10.6.3 Results of Robust Optimization

Next, 10 optimized structures were generated with the genetic algorithm using the above methods of incorporating robustness into the objective function evaluation. The approach followed was as for the structure shown in Figure 10.26: the GA used five generations, each of population size 200, $p_{\text{crossover}} = 0.8$, $p_{\text{mut}} = 0.005$, with an elitist survival strategy and, in each case, differing random number sequences were used. The maximum value of the perturbation size in each case was this time set as 5 mm. Of course, for four of the methods, the time to produce each optimized structure was increased over the nominal-only case by the amount of additional evaluations required. Thus, for the L81 method, each structure took 81 times as long to optimize. Finally, after each optimization, the robustness of the design solutions were reevaluated using the higher quality measure $\hat{f}_{95:300}$. The results are shown in Figure 10.28, which, for each method, shows the change in the values of average nominal and perturbed performance, and also the robustness, compared with the average performance of the 10 structures previously optimized using a nominal performance-only measure. Robustness, r, is a measure of the variability of the performance in the face of such perturbances, and is defined here as $r = |f_{95} - f(\mathbf{x}^*)|$, where $\mathbf{x}^*$ is the optimum design variable vector, defining the optimum geometry resulting from the optimization. Depending on the specific objective of the optimization, either the worst level of absolute performance (perturbed performance) or the variability (robustness) may be of primary concern.

Interestingly, optimizing for both robust and optimal structures has not compromised the average nominal structure performance, except for the L81 method, where a small increase is seen. Indeed, when using the OAT method, an average reduction in the nominal performance of about 2 dB results, although as discussed above, this has dubious practical significance. On average, all methods produce improvements in the 95% probability limit and the robustness – most notably, the OAT method, by about 3 and 4 dB, and then the NP2 method, by about 2 and 3 dB, respectively.

Because of the wide variation in the additional computational overhead for each robust optimization method, a figure of merit for the methods may be calculated by normalizing

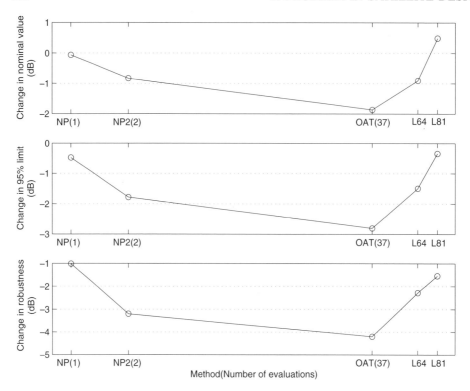

Figure 10.28 The average improvement (negative change) in nominal performance, 95% probability limit and the robustness for the optimized structures achieved using the GA, against structures produced using a nominal-performance objective function.

the results against the additional computational cost, and this is shown in Figure 10.29. Now, it is seen that the success of the OAT method is achieved at the cost of a large increase in computational effort; however, the NP2 method appears to provide worthwhile improvements for little additional overhead.

It is interesting to note that, for the DoE methods used, the best estimate of the 95% probability limit, $\hat{f}_{95:L64}$, when used as the value of the objective function did not yield the best results. Also given that the NP and NP2 methods require either no, or only a doubling of computational effort, for GA optimization, the explicit consideration of robustness need not be expensive. Comparing the convergence between the NP and NP2 genetic algorithm runs, only the latter showed consistent average improvements at each generation of the optimization.

10.7 Adjoint-based Approaches

Generally, GAs do not require the gradient of the cost function. However, the application of GAs in large-scale applications is limited because of the large number of evaluations of the objective function often needed. Even today, work on the structures studied here has been limited to 100 generations using 500 processor clusters. The resulting designs are therefore

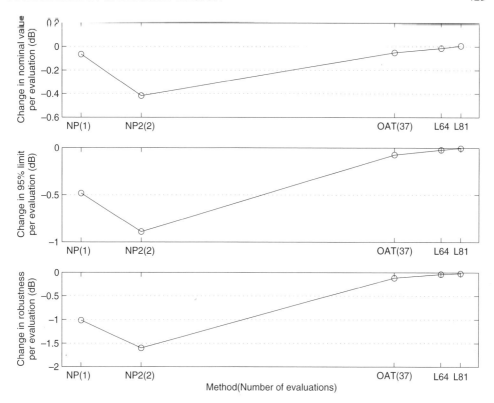

Figure 10.29 The results given in Figure 10.28 when normalized against the additional computational expense required.

unlikely to be fully converged, although when using nominal performance, this is perhaps a good idea. Having set out means to deal with this issue, however, it would be desirable to be able to refine any promising designs to get the best from them, while preserving robustness. One way of doing this would be via gradient descent type methods. If, however, gradient-based calculations are performed using finite-differencing (FD), then each set of gradients would involve up to 90 additional function evaluations – a prohibitive expense for routine use. Work has therefore been carried out on producing an adjoint code which produces such information at more reasonable cost (see Chapter 4 for a more complete treatment of this topic).

In due course, the use of the adjoint code will allow the use of both direct gradient-based searches and also hybrid genetic algorithm–local search approaches that combine Darwinian and Lamarckian evolution models. GAs use Darwinian evolution, which is based on "natural selection" and is set up so that the most fit individuals are likely to survive. Lamarckian evolution takes the view that individuals may improve within their environment as they become adapted to it and that the resulting changes are inheritable. Consequently, the genotype of an improved individual is forced to reflect the result of such an improvement. Typically, Lamarckian improvement is achieved using gradient descent methods in frameworks like the Glossy template described in Chapter 3.

In our application, any local search would be carried out using a gradient-based method, which requires the calculation of the gradients of the code outputs $\mathbf{F} : \Re^{90} \rightarrow \Re$ (i.e., up to 90 independents representing the coordinates of the joint points in the design geometry). Automatic differentiation (AD) software tools enable differentiation of the computer program defining $\mathbf{F}$ to produce a new code that will calculate $\mathbf{F}$'s gradient (Griewank 2000). In our application, the computation of $\mathbf{F}(\mathbf{x})$ involves complex variable calculations, which are handled by Adifor 3.0 (Carle and Fagan 2000) since the dependents and independents are real values (Pusch et al. 1995). In theory, the gradient $\nabla\mathbf{F}$ of such a function is cheaply calculated using the reverse or adjoint mode of AD, since the cost of the gradient is independent of the number of variables and is bounded above by a small factor of the cost of evaluating the function. In practice, large memory requirement may prohibit use of the adjoint code (reverse AD generated code). In this section, we outline the differentiation of the receptance code by the Adifor 3.0 AD tool. Adifor 3.0 employed in reverse mode produces an adjoint code that, after being tuned manually for performance enhancement, calculates the function and its gradient in 7.4 times the CPU time required for its function evaluation. Moreover, the adjoint code runs 12.6 times faster than one-sided FD.

As before, the initial boom structure studied is three dimensional and composed of 90 Euler–Bernoulli beams each having the same properties per unit length. The three joints at the left-hand end of the structure are here assumed fixed. Once again, the structure is taken to be excited by a point transverse force applied to the left-hand end of the structure. The vibrational energy level is calculated at the right-hand end using the receptance code. The optimization aims at minimizing vibrations by minimizing the frequency-averaged response in the range 150–250 Hz. To achieve this, the optimizer is allowed to modify the geometry of the boom by changing the coordinates of the free joints in the structure.

10.7.1 Differentiation of the Receptance Code

The initial receptance code, as sketched in Figure 10.30, starts by reading in data from files representing the boom geometry (listing all beams and their connections) and the properties of each beam. Given extra information such as the range of frequencies over which to solve, the number of data points within the specified frequency range, the joint numbers at which to calculate the energies and the number of axial, torsional and transverse modes in the modal summations for the Green functions, the program builds up a linear complex system $\mathbf{A}\mathbf{f} = \mathbf{b}$ for nodal forces $\mathbf{f}$ and solves it for each frequency. An averaged energy function is calculated at the specified end beam(s). As described earlier, the boom structure is composed of 90 beams connected by three fixed and 30 free joints. We aim to calculate the sensitivities of the energy function with respect to the coordinates of the free joints. This represents a gradient calculation with 90 independent variables. The computer code is differentiated using FD and AD via Adifor 3.0.

We first computed a single directional derivative $\dot{y} = \nabla\mathbf{F}(\mathbf{x})\dot{\mathbf{x}}$ for a random direction $\dot{\mathbf{x}}$, by using one-sided FD, AD in forward mode and a single adjoint $\bar{x} = \nabla\mathbf{F}(\mathbf{x})^{T}\bar{\mathbf{y}}$ for $\bar{\mathbf{y}} = 1$ via reverse mode AD. By definition of the adjoint operator, we have the following equality: $\langle\bar{\mathbf{x}}, \dot{\mathbf{x}}\rangle = \langle\bar{\mathbf{y}}, \dot{y}\rangle$. This allows us to validate the results of the differentiation. The initial Adifor 3.0 generated codes gave inconsistent results with those from FD. This is caused partly by nondifferentiable statements in the code for the function $\mathbf{F}$ that were replaced by equivalent but differentiable coding.

Read in:
  No. of beams, beam properties, and list of connections
  frequency range $[\omega_{min}, \omega_{max}, N]$ ($N \equiv$ No. of frequencies)
  co-ordinates of beam ends $\mathbf{x}_{n,j}$
calculate some geometric information
Initialize $I = 0$ (integral of power)
$\Delta = (\omega_{min} - \omega_{max})/(N - 1)$
For $k = 1, N$
  $\omega = \omega_{min} + (k - 1) \times \Delta\omega$
  Assemble:
    Green Function Matrix $\mathbf{A}(\mathbf{X}, i\omega)$
    r.h.s. forcing $\mathbf{b}(\mathbf{X}, i\omega)$
    Solve $\mathbf{A}(\mathbf{X}, i\omega)\mathbf{f} = \mathbf{b}(\mathbf{X}, i\omega)$ (using lapack, which takes 50% of the CPU time)
    Obtain displacement $\mathbf{D}_{n,j}$ at ends of beam n
    Obtain power $P = Re(\mathbf{f}_j.\mathbf{D}_{n,j})/2$
    Update integral $I = I + P^2 \times \Delta\omega$
End For
Set objective function $\mathbf{F} = I$

Figure 10.30 Schematic of the receptance code.

In addition, the complex linear solver $\mathbf{A}\mathbf{f} = \mathbf{b}$ employs the lapack routine zgesv (Anderson et al. 1995). Differentiating the lapack source code routines for zgesv using an AD tool without taking account insights into the nature of the linear solver would give inefficient code. Mechanical generation of the zgesv derivative by Adifor 3.0 gave not only inefficient code but also inconsistent results with FD. Instead of using Adifor 3.0 to differentiate zgesv, we instead use hand-coding for both forward and reverse mode. So, for example, differentiating $\mathbf{A}\mathbf{f} = \mathbf{b}$ using the matrix-equivalent of the product rule, we obtain $\dot{\mathbf{A}}\mathbf{f} + \mathbf{A}\dot{\mathbf{f}} = \dot{\mathbf{b}}$, and so the derivatives $\dot{\mathbf{f}}$ are given by the solution of $\mathbf{A}\dot{\mathbf{f}} = \dot{\mathbf{b}} - \dot{\mathbf{A}}\mathbf{f}$ and in the forward mode we may reuse the LU-decomposition of $\mathbf{A}$ to efficiently solve for the derivatives $\dot{\mathbf{f}}$. The following procedure is used:

- Perform an LU-decomposition of the matrix $\mathbf{A}$

- Solve $\mathbf{A}\mathbf{f} = \mathbf{b}$

- Form $\mathbf{b}_{new} = \dot{\mathbf{b}} - \dot{\mathbf{A}}\mathbf{f}$

- Reuse LU-decomposition to solve $\mathbf{A}\dot{\mathbf{f}} = \mathbf{b}_{new}$

For the reverse mode, deriving the adjoint update is more tedious but a similar scheme may be followed. Using this approach, we obtain the following procedure for the adjoint of the linear solve:

1. In the forward sweep,

    - Perform an LU-decomposition of the matrix $\mathbf{A}$
    - Store $\mathbf{L}$ and $\mathbf{U}$ and the pivot sequence IPIV
    - Solve $\mathbf{A}\mathbf{f} = \mathbf{b}$

2. In the reverse sweep,

- Load $\mathbf{L}$ and $\mathbf{U}$ and the pivot sequence IPIV
- Solve $\mathbf{A}^T \bar{\mathbf{b}} = \bar{\mathbf{f}}$
- Update $\bar{\mathbf{A}} = \bar{\mathbf{A}} - \bar{\mathbf{b}}\mathbf{f}^T$

Here, the memory storage is dramatically reduced compared with black-box application of Adifor 3.0. If $\mathbf{A}$ is an $N \times N$ matrix, we store only N^2 complex coefficients instead of $\mathcal{O}(N^3)$ when Adifor 3.0 tapes all variables on the left of assignment statements in the LU-decomposition.

10.7.2 Initial Results and Validation

After implementing these procedures described on the Adifor 3.0 generated code, we obtained tangent and adjoint derivative codes that calculate directional derivatives consistent with FD. The obtained codes were compiled with maximum compile optimizations and run on a Sun Blade 1000 machine. Table 10.3 shows the results and timings of forward mode AD, reverse mode AD, and one-sided FD for that calculation. These results showed that forward and reverse AD gave the same directional derivative value within round off while the maximum difference with the FD result is around 10^{-6}. This difference is of the order of the square root of the machine-relative precision. This validates the AD results as being in agreement with the FD result.

From Table 10.3, we can also deduce that while the AD reverse mode calculates the gradient in around 5 min, one-sided FD and forward AD will require 91 function evaluations and 90 directional derivatives, respectively, and consequently run times of over 35 min. We see that using reverse mode AD can speed up the gradient calculation by a factor of around seven over FD while giving accurate derivatives. However, the core of the calculation (building up the linear system, solving it and calculating the local energy contribution for each frequency) of the receptance code is an independent loop and therefore can be differentiated in parallel.

10.7.3 Performance Issues

Usually, after checking that the AD in forward and reverse modes agrees with the finite differences, we seek to improve efficiency of the automatically generated code. As shown by the results of Table 10.3, the reverse mode is superior to the finite differences and forward mode but we found it requires a very large amount of memory to run. This memory is due to the size of the tape, which required 12 GB. By hand-coding the adjoint of the linear solver,

Table 10.3   Results for a single directional derivative, timings are in CPU seconds

Method	$\langle \bar{\mathbf{x}}, \dot{\mathbf{x}} \rangle$	$\langle \bar{\mathbf{y}}, \dot{\mathbf{y}} \rangle$	CPU($\nabla \mathbf{F}(\mathbf{x})\dot{\mathbf{x}}$)
FD (1-sided)		0.124578003587	48.7
Adifor 3.0(fwd)		0.124571139127	54.0
Adifor 3.0(rev)	0.124571139130		311.5

Table 10.4 CPU Timings (in seconds) on a SUN Blade 1000, UltraSparcIII

Method	CPU(∇F)	CPU(∇F)/CPU(F)
Adifor 3.0(rev.)	311.5	13.3
Adifor 3.0(rev.,par.)	174.7	7.4
FD (1-sided)	2215.9	93.1

we reduced the size of the tape to around 6 GB. Furthermore, the core of the calculation of the receptance code is carried out by a parallel loop. Because the iterations of such a loop are independent, we can run the loop body taping all the required information for just one iteration, and then immediately adjoin the body of the loop (Hascoet et al. 2001). This reduced the tape size of the adjoint code down to around 0.3 GB. The second row of Table 10.4 shows that after this optimization the ratio between the gradient calculation and the function is 7.4. It also shows a speed up factor of 12.6 over the popular one-sided FD method.

In summary, Adifor 3.0 allows us to build up an adjoint for a code that makes extensive use of complex variable arithmetic (in its original form) to accurately calculate the gradient of the objective function. The adjoint code requires only 7.4 times the CPU time of the original function code and the memory requirement for taping is a modest 0.3 GB. It also runs 12.6 times faster than the gradient calculated using one-sided FD when dealing with 90 variables.

11

Airfoil Section Design

It will be clear from the previous case study that a great deal can be accomplished with a problem-specific analysis code directly coupled to a search engine. It will also be evident that routine use of evolutionary search can lead to designs that have enhanced and robust performance in practice. It is, however, the case that many design studies must deal with problems where "black-box" solvers have to be used, typically CFD or FEA codes. Often, these have very extended run times that must also be allowed for. We therefore turn next to problems in aircraft aerodynamics using commercially available CFD codes. Such problems require more complex geometric models and so we begin by examining airfoil representation schemes.

Clearly, the combination of optimization algorithms and CFD solvers offers promise for the development of improved aerodynamic designs. Such optimization strategies have a common requirement for representation of geometry by a number of design parameters. For wing design, the parameterization is generally separable into a representation of the planform and the airfoil sections at a number of span-wise positions. Here, a series of NASA SC(2) transonic two-dimensional airfoil sections has been used as a set of base functions that can then be used to build new, but similar, section designs. It essentially consists of finding a set of orthogonal shape functions that, when combined together in appropriate ratios, can be used to construct those foils in the original NASA set and, by using other mixes, other related section shapes. Although these functions allow less design variation than B-spline or Bézier representations, they are ideal for conceptual design where the aim is selection of sections appropriate for local flow conditions. The work described here allows for up to six such base shapes to be used in this process but demonstrates that the NASA foils can be recovered to reasonable precision with just two functions, which essentially reflect thickness-to-chord ratio and camber.

In this case study, the design of airfoil sections is then considered from a multiobjective standpoint whereby section drag is traded against robustness to deviations in section shape, that is, a good design is considered to be one that exhibits low drag over a range of speeds and whose drag does not vary much as small changes are made to the section shape. We consider overall wing planform design in the next case study.

Computational Approaches for Aerospace Design: The Pursuit of Excellence.   A. J. Keane and P. B. Nair
© 2005 John Wiley & Sons, Ltd

11.1   Analysis Methods

When carrying out aerodynamic design using computational methods, calculations can be made using a variety of approaches with varying levels of fidelity. We begin be summarizing the choices normally available during concept design.

11.1.1   Panel Methods

Panel or boundary element methods solve the linearized potential equation for inviscid, irrotational, incompressible flow. With the addition of a compressibility correction in the free-stream flow direction (e.g., Prandtl–Glauert, Karman–Tsien), panel methods can also be applied to a limited extent to compressible flows around slim bodies. However, this does not extend to the transonic cruise design point. In general, even with a compressibility correction, panel methods are limited to free-stream Mach numbers of less than 0.7.

Within their domain of validity, however, panel methods have two substantial advantages. First, as a boundary element method, only a surface discretization is necessary, there being no requirement for grid generation in the flow field. Secondly, solutions are very rapid. This combination of low computational cost and ease of automatic discretization makes panel methods attractive candidates for the low level analysis level in a multifidelity optimization framework. However, the problem of their fundamentally subsonic nature remains. Being unable to faithfully model weak shocks, panel methods cannot directly predict wave drag. In addition, the inability to model shocks may lead to incorrect estimation of skin friction drag, as the effect of the shock pressure gradient on boundary layer development is also absent. Furthermore, the absence of a compression shock in the panel solutions leads to overestimation of the coefficient of lift. This is likely to result in severe difficulties in transonic optimization unless additional constraints are imposed. For example, as a result of the absence of wave drag, optimization will be driven away from supercritical airfoil shapes.

11.1.2   Full-potential Methods

The full-potential equation describes inviscid, irrotational, compressible flow. As all compressibility terms are included, full-potential solutions are valid for high Mach number flows. However, the irrotational condition, necessary for a potential to be definable, imposes a condition of constant entropy throughout the flow field. This results in the relations across the shocks being isentropic rather than obeying the correct Rankine–Hugoniot relations. Full-potential solutions therefore strictly do not apply to flows with strong shocks. At the transonic cruise design point of a well-designed supercritical airfoil, the shocks are at most weak and the isentropic relations provide acceptable accuracy. However, without additional constraints, optimizations are likely to pass through regions of the design space associated with strong shocks.

The application of the constant entropy condition across strong shocks results in the overestimation of wave drag. However, even in the presence of strong shocks, the errors are moderate. For example, for a normal shock with an upstream Mach number of 1.25, the error in downstream Mach number is only 0.03 (Hirsch 1988). Perhaps of greater consequence is the tendency for the predicted location of strong shocks to be too far aft, leading to overestimation of lift. Full-potential solutions of flows with strong shocks therefore overestimate both drag and lift. The effect of this on optimization depends on whether the overprediction of drag is proportionately greater than the overprediction of lift. If so, full-potential optimizations will be driven away from regions with strong shocks. This is presumed to be the case as the full-potential method has been successfully used in several transonic optimizations (e.g., Obayashi and Yamaguchi (1997)). An alternative approach is the use of shock-fitting full-potential methods in which the numerical relations across the shock are empirically modified in order to approximate the Rankine–Hugoniot relations. Such methods (e.g., Klopfer and Nixon (1983)) have been shown to be capable of producing excellent agreement with experiment.

11.1.3   Field Panel Methods

Although the basic panel method is incapable of representing compressible flows, various attempts have been made to extend the method to solution of the full-potential equations by the addition of sources in the flow field. These methods require specification of a volume grid in addition to the surface panels. However, the volume grid is not required to be body fitted and, in general, penetrates the body. Generation of the volume grid is therefore much more straightforward than is the case with mesh generation for full potential and Euler codes. The solution requires the integration of the influence of the volume sources on each other. For a full volume grid extending to the far field, these integrations are very expensive. Most codes therefore eliminate sources below a specific threshold strength from the solution. This results in field sources only appearing in regions of high Mach number, such as the supercritical region of transonic wings. Workers at Boeing have also reported reducing the computational cost by selecting flow grid divisions such that the integrals can be carried out by fast Fourier integration. These methods allow solutions to be obtained with computational costs and accuracy comparable to full-potential methods. Maskew (1987) has reported a modified version of VSaero implementing a field panel method and has shown good agreement with experiment for a NACA 0012 airfoil. Sinclair (1986) has also reported a field panel method used within BAE Systems.

11.1.4   Euler Methods

The Euler equations represent inviscid, rotational, compressible flow. The admissibility of rotational flow means that the flow field may no longer be represented by a potential distribution. However, as noted above, the solutions are no longer isentropic and therefore the correct Rankine–Hugoniot relations across shocks apply. The shocks therefore appear at the correct chord-wise position and are of the correct strength. The computational cost of Euler methods are somewhat greater than those for full-potential methods. In this case study, Euler methods are the most accurate considered: despite the increasing take-up of RANS codes in aircraft design, they are not routinely used in concept work at the time of writing.

11.2   Drag-estimation Methods

The drag on transonic transport aircraft wings can be considered to comprise three components:

1. wave drag due to the presence of shocks;

2. viscous wake or profile drag due to the boundary layer;

3. vortex or induced drag due to the tip vortex of the 3-d wing.

Recovering accurate drag values from CFD calculations for these components has proved to be considerably more difficult than the accurate calculation of pressure distributions. The state of the art in drag recovery was reviewed in the AGARD report AR-256 (AGARD 1989). The document drew several conclusions including the following:

"Accurate and consistent computation through CFD of (absolute) drag levels for complex configurations is, not surprisingly, beyond reach for a considerable time to come. Pacing items are basically the same as those of CFD in general (grid generation, turbulence modeling, grid resolution, speed and economics of computation). However, for drag prediction purposes the importance of some factors, such as grid resolution and speed/economics of computation, is amplified by one or several orders of magnitude."

"For attached flow around simple configurations ... CFD prediction has met with some, though limited, success. It appears that for two-dimensional airfoils most but not all codes can now predict drag with an accuracy of 5%. For three-dimensional wings this figure appears to be of the order of 10%; possibly somewhat less for transport aircraft wings, ..."

"Navier-Stokes code typically do not (yet) involve drag prediction except for two-dimensional airfoil flows. Even then they do not do a better job than zonal methods involving potential flow or Euler schemes coupled to boundary layers."

"The application for drag prediction purposes of the current generation of Euler codes, in particular in three-dimensions, is hampered by (over) sensitivity to grid density and quality through spurious (artificial) dissipation. For 3d wings and wing-bodies with attached flow only full potential methods with or without boundary layers appear to have met with some success."

"Most authors seem to agree that a "far-field" type of drag assessment based on the application of the momentum theorem is to be preferred over a "near-field" type of procedure..."

Although the document was written in 1989, the calculation of accurate drag values remains a difficult and technically intricate task. The choice of drag calculation method continues to have a significant effect on the accuracy of the CFD code.

11.2.1 Surface Pressure Integration

The simplest algorithm for recovering drag from a CFD flow field is to integrate the pressure coefficients over the surface. This near-field drag recovery method is commonly referred to as surface pressure integration (SPI). Surface integration generally results in reasonably accurate lift coefficients. However, the net drag value results from the near cancellation of relatively much greater thrust and drag components. This integral is therefore sensitive to small errors in the pressure integration and, in particular, to errors in the position of the forward stagnation point. In the case of potential methods, the forward stagnation point is generally predicted reasonably well. However, spurious entropy in Euler methods often leads to errors in the prediction of the stagnation point. This is especially the case for three-dimensional analyses for which the surface pressure integration method is particularly inappropriate.

For use in aerodynamic design, SPI also has the disadvantage of returning total drag rather than separate induced and wave drag components. However, for optimization, there is less of a requirement for the separation of drag into its components.

11.2.2 Far-field Integration of Induced Drag

An alternative to surface panel integration is to consider momentum balance over a control volume a large distance from the body. In principle, wave drag can be calculated by such far-field integration; however, in CFD drag recovery, far-field methods have been mainly used for the estimation of induced drag. The calculation of induced drag is carried out over the so-called Treffitz plane, a plane normal to the flow direction, a large (in theory infinite) distance downstream of the body. Integration of the cross flow velocity (vorticity in the case of integration over a wake) gives the induced drag.

Many Euler codes implement a "far-field" drag calculation by integration of the vorticity distribution over a plane. However, as the numerical dissipation of the Euler solution attenuates vorticity with distance, the integration of vorticity should be carried out over a scan plane immediately downstream of the wing trailing edge.

11.2.3 Calculation of Wave Drag by Integration over the Shock

Wave drag can be calculated by integration over the shock. One approach is to integrate the flow conditions both at entry and exit from the shock. Such an algorithm has been reported by van Dam and Nikfetrat (1992). The wave drag is calculated from the entropy difference between two scan planes positioned immediately upstream and downstream of the shock location. A simpler approach is to integrate only across the entry face of the shock, using some approximation of the shock relations to calculate the exit conditions. This approach has been adopted in ARA's MACHCONT procedure.

The above methods, although providing reasonable accuracy, require the shock position throughout the flow volume to be located. The MACHCONT algorithm performs a search along coordinate lines circumscribing the airfoil. However for the three-dimensional case, finding the shock position in the flow field is nontrivial. Lock (1986) has proposed a simplification that requires only the flow parameters at the wing surface to be considered. The method is based on the approximation that the shock lies along a normal to the local surface. Although, in practice shocks are curved, most wave drag is generated close to the surface, where the normal assumption is reasonable. With this assumption, wave drag can be calculated from the shock entry conditions at the surface and the local surface curvature. The method has been shown to produce satisfactory wave drag estimates on realistic configurations.

11.3 Calculation Methods Adopted

The CFD methods used in this case study are a full-potential code (VGK) and an Euler code (MG2D – the two-dimensional version of MGaero). Optimization is carried out using VGK while MG2D is used to provide validation. The drag recovery methods used are:

1. *Viscous drag:* based on far-field momentum thickness applying the Cooke (1973) implementation of the Squire and Young (1937) approximation.

2. *Wave drag:* Lock (1986) second-order method (also known as the ESDU (1987) method). This method requires the estimation of shock position, so an algorithm for searching the results for compression shocks on the upper surface of the airfoil has been developed. The wave drag estimation then uses relations based on the Mach number immediately upstream of the shock together with the mean airfoil surface curvature over the region upstream of the shock.

The DRA Farnborough written code VGK (ESDU 1996) is distributed as part of the ESDU Transonic Aerodynamics pack. VGK is a two-dimensional viscous coupled finite difference code that solves the full-potential equations, modified so as to approximate the Rankine–Hugoniot relations across shocks. The code can solve for either a specified angle of attack or for a target lift. For typical airfoils, a target lift, viscous coupled solution takes a few seconds.

To validate the results coming from VGK, the viscous coupled multigrid Cartesian Euler code MGaero is used in its two-dimensional mode – MG2D. The code can analyze finite trailing edge airfoils: the blunt bases of such airfoils contribute to the wake thickness and therefore to drag. The approach used here is to estimate this "base drag" and to increment the boundary layer thicknesses by the base thickness. The Squire and Young estimate of wake development is then based on a modified momentum thickness and shape factor, the resulting viscous drag coefficient including the base drag.

11.3.1   Comparison between VGK and MG2D on RAE2822 Airfoil

The drag recovery methods used here have been validated by comparison with experimental data. In order to assess the drag estimation of VGK and MG2D, the RAE2822 test case was compared with experimental results (test case 6 of AGARD AR138, M = 0.73, Re = 6.5e6, transition fixed on upper and lower surface at 3% chord; see Redeker (1994)). The resulting pressure distribution is presented in Figure 11.1.

The Squire & Young derived viscous drag coefficients for both codes are shown in Figure 11.2(a). The viscous drag values at low Mach number are comparable; however, it is apparent that while MG2D predicts only a moderate dependence of viscous drag coefficient on Mach number, VGK shows a dramatic increase with Mach number above $M = 0.75$.

The Lock method–estimated wave drag coefficients are presented in Figure 11.2(b). The agreement between VGK and MG2D is again reasonable for the lower Mach numbers. However, the reduction of wave drag with Mach number predicted by VGK for Mach numbers above 0.75 is nonphysical. The VGK results demonstrate the limitations of the full-potential method in the presence of strong shocks, whereas the MG2D results appear to behave as expected.

The total drag coefficients ($C_d = C_{dv} + C_{dw}$) estimated by both methods are compared with the experimental results in Figure 11.2(c). It is apparent that the VGK results agree with the experiment values slightly better than those of MG2D. This is despite VGK's obvious difficulties with strong shocks: the errors in viscous and wave drag appear to cancel. Also presented on the same figure are the drag values calculated by MGaero using SPI. It can be seen that the drag recovery algorithms used provide a massive improvement upon the very poor drag estimates obtained by SPI.

11.4   Airfoil Parameterization

Having set out an analysis process for dealing with two-dimensional airfoil design, we turn next to the issue of geometric parameterization. A representation has been sought here that both minimizes the number of design parameters and has a global influence on geometry. As has been discussed throughout this book, parameterization is a key issue when setting up an optimization process. A further, and often neglected, requirement is that the design

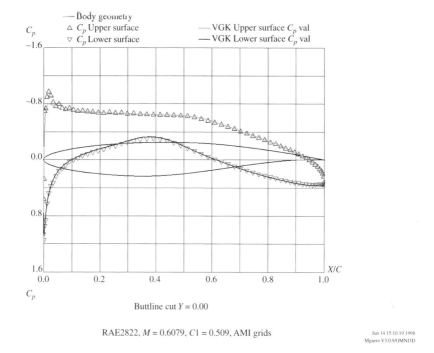

Figure 11.1 C_p distribution for RAE2822 using MG2D (shown as triangles) and VGK.

functions should be as geometrically orthogonal as possible. Lack of orthogonality implies a nonunique mapping of the parameter values to the geometry. The resulting spurious multimodality of the objective function can significantly degrade the search process. We begin by noting a number of the methods discussed in the literature.

11.4.1 Previous Nonorthogonal Representations

The direct use of airfoil coordinates is ruled out from most optimization studies by the number of variables required and the potential for strong coupling between the geometric specification and discretization errors. Perhaps, the closest nonorthogonal approach to the direct use of coordinates is via collections of B-spline or Bézier curves to represent all or part of an airfoil surface (e.g., Cosentino and Holst (1986); Jameson (1988)). This allows innovative design at the cost of a large number of parameters. However, as much of the design space consists of non "airfoil-like" shapes, this representation is better suited to airfoil rather than wing design studies. Another popular approach, which guarantees airfoil-like shapes, is linear combination of existing airfoil sections (e.g., Obayashi and Tsukahara (1997)). Ramamoorthy and Padmavathi (1977) have represented airfoils by approximations of their surface slope by superposition of Wagner functions. Perhaps the most commonly adopted parameterization is modification of an existing base airfoil with a set of functions such as the widely used Hicks–Henne functions (Hicks and Henne (1978); Lores et al. (1980)). None of the modifying functions mentioned so far have been geometrically orthogonal. However, Aidala et al. (1983) and Desterac and Reneaux (1990) have used inverse methods to derive

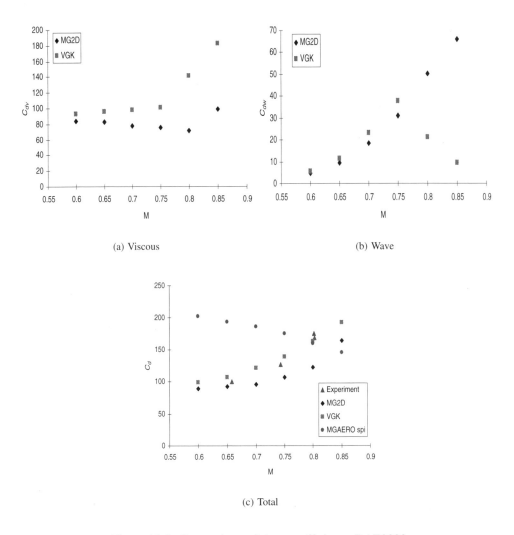

(a) Viscous   (b) Wave

(c) Total

Figure 11.2 Comparison of drag coefficients: RAE2822.

modifying "aerofunctions" whose effects on the base airfoil are approximately orthogonal in an aerodynamic sense.

11.4.2 Previous Orthogonal Representations

Linear combination of existing airfoils may have the disadvantage of leading to spurious modality. This arises where similar airfoil geometries can be created by different linear combinations of the base functions, that is the base functions are not orthogonal. When using airfoils directly as base functions, this will often be the case. Obayashi and Tsuka-hara (1997) have shown that when several airfoils are combined extremely high modality can result. A superior approach is therefore likely to be use of a set of orthogonal basis functions derived from existing airfoils. Kuruvila et al. (1995) used a set of orthogonal functions derived from the polynomial representation of NACA 4 digit airfoils to recover the geometry of the NACA 0012. Chang et al. (1995) used a similar approach, deriving 20 orthogonal shape functions from the polynomial terms used in specification of the NACA 6 digit airfoils. Although derived from subsonic airfoils, they also showed the functions to be capable of faithfully representing a supercritical airfoil (Korn airfoil). However, the use of 20 coefficients for defining an airfoil seems somewhat excessive, probably indicating that the functions are also capable of representing non airfoil-like shapes.

High numbers of orthogonal functions can result from inclusion of all the polynomial terms without consideration of their significance in airfoil representation. An alternative approach is to derive basis functions from a set of existing airfoils. This can result in a set of basis functions, able to represent the variation of shape among the airfoils from which they are derived, without being so generic as to represent many non airfoil-like geometries. Here, work has been carried out to derive such a set of functions from the NASA SC(2) family, using the approach outlined below.

11.4.3   Choice of Source Airfoil Sections

The objective here was derivation of basis functions suitable for conceptual optimization of supercritical wings by selection of appropriate rather than innovative sections. The NASA SC(2) family of supercritical airfoils presented in NASA TP-2969 (Harris 1990) was chosen as the starting point. This family of supercritical airfoils was designed using a hodograph method during the late 1970s. The airfoils may therefore be somewhat less efficient than more recent designs; however, their performance was considered competitive enough for NASA to delay publication of the report until 1990. The family of airfoils cover the design space from 2 to 14% thickness/chord and from symmetrical to $C_L = 1.0$. The 22 airfoils are clustered in areas of design space associated with transport aircraft, business jet and fighter applications. Each airfoil was designed using a hodograph method for a Reynolds number of 30 million with transition on upper and lower surfaces fixed at 3%, the design Mach number being allowed to float in the design process. The designation of each airfoil reflects the thickness to chord and the design point (e.g., SC(2)0612 is 12% thick with a design C_L of 0.6). In the work described here, nine airfoils (SC(2)0406, SC(2)0606, SC(2)0706, SC(2)0410, SC(2)0610, SC(2)0710, SC(2)0412, SC(2)0612 and SC(2)0712) were selected as the most appropriate for mid and outer sections of transport aircraft wings.

Preliminary analyses of the raw coordinates given in the NASA report using the full-potential code VGK (ESDU 1996) gave unexpectedly large drag values. As the hodograph method used in the derivation of the airfoils is closely related to the VGK method used to calculate these drags, good agreement had been anticipated. Examination of the coordinates revealed the second derivatives to be discontinuous. It was suspected that the discontinuity arose from the rounding applied to the figures in the tables used to overcome "limitations on space". The coordinates were therefore passed through a third-order Savitzky–Golay smoothing algorithm (Press et al. 1992) and the smoothed airfoils analyzed with VGK. The drag of the smoothed airfoils was much reduced and in good agreement with the values in

NASA TP-2969. These smoothed airfoils were used in the derivation of the basis functions, and are referred to here with the suffix 's'.

11.4.4   Derivation of Basis Functions

The aim here was derivation of a set of orthogonal basis functions that approximated the nine smoothed airfoils. Rather than using analytically designed functions, each base function is defined numerically in a similar manner to the original airfoils. Each of the airfoils is defined as sets of lower and upper coordinates at 103 chord-wise positions. For each airfoil, there are therefore 206 (x_i, z_i) pairs. It is then assumed that a set of nonorthogonal basis functions exists, similarly defined as coordinate pairs. The z coordinates of the jth airfoil can therefore be approximated as linear combinations of the z coordinates of a smaller number of, say n, basis functions:

$$z_i^j = \sum_{k=1}^{n} b_k^j z_i^k.$$

A nonlinear least-square fit was used to determine the single base function that best fitted the nine smoothed airfoils. The results of the fit are the basis function coordinates z_i^l and the weighting coefficients b_l^j. The fitted z values were subtracted from the airfoil coordinates and a similar fit carried out to obtain the second base function coordinates and weights. The process of successive fitting was repeated to derive the remaining base functions. As a result of numerical error, the functions obtained were not exactly orthogonal. The functions were therefore orthogonalized by Gram–Schmidt normalization (Press et al. 1992). Table 11.1 shows the residuals from fitting various numbers of basis functions to the nine airfoils. The first five functions are illustrated in Figures 11.3 and 11.4.

The residual values suggest that to achieve a geometric accuracy comparable to that of the specification of the original airfoils ($\sim$1e-4) requires five basis functions, the inclusion of the sixth function having little effect on the quality of the fit. However, it is possible that a different number of functions are required to reproduce the aerodynamic behavior of the sections.

To determine the aerodynamic matching of the sections, VGK was used to calculate the drag of airfoils reconstructed using different numbers of base functions. The actual design Mach numbers of the airfoils are not tabulated in TP-2969, however, Figure 29 of the report allows the design Mach number to be estimated. Table 11.2 shows the estimated design Mach numbers used in the calculations.

Table 11.1   Residuals from fitting base functions to the original nine airfoils

No. of Basis Functions	Maximum Residual	σ Residual
1	7.5e-3	2.9e-3
2	3.0e-3	5.1e-4
3	1.1e-3	2.7e-4
4	5.2e-4	1.5e-4
5	3.0e-4	4.7e-5
6	3.0e-4	3.9e-5

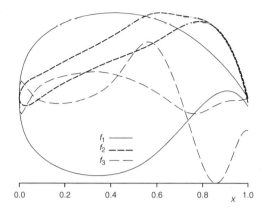

Figure 11.3 First three orthogonal base functions.

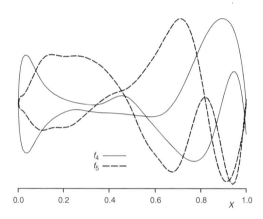

Figure 11.4 Fourth and fifth orthogonal base functions.

Table 11.2   Estimated design points for nine original airfoils

	0406	0606	0706	0410	0610	0710	0412	0612	0712
C_L	0.4	0.6	0.7	0.4	0.6	0.7	0.4	0.6	0.7
t/c	0.06	0.06	0.06	0.10	0.10	0.10	0.12	0.12	0.12
M	0.84	0.82	0.80	0.79	0.78	0.77	0.78	0.76	0.75

The maximum and standard deviation of the differences between the drag coefficients of the original and reconstructed airfoils are presented in Table 11.3 and are summarized in Table 11.4. It is apparent that fitting only the first three base functions to each airfoil approximates the aerodynamic properties sufficiently closely for the drag to be within two counts. Whereas including the fourth and fifth functions lead to some improvement in fitting to the geometry, this is not reflected in any correspondingly significant change in the drag values. The ability to represent the variation in aerodynamic characteristics of nine

Table 11.3   Drag values calculated by VGK for original and reconstructed airfoils

No of Basis Functions	0406	0606	0706	0410	0610	0710	0412	0612	0712
Original Airfoil	**74.0**	**73.8**	**69.7**	**71.7**	**83.4**	**87.1**	**84.1**	**89.7**	**90.0**
1	65.9	96.0	121.5	74.7	86.2	100.0	92.8	90.0	93.1
2	68.7	73.8	70.7	71.8	85.4	89.6	85.3	89.8	91.2
3	72.7	76.1	68.2	71.1	83.2	86.7	85.1	89.8	91.3
4	74.9	76.8	69.1	70.9	84.2	87.1	85.0	89.5	89.7
5	75.3	76.8	70.6	70.9	83.0	87.0	85.2	89.5	89.7
6	75.4	76.4	70.6	71.4	83.5	86.3	84.6	89.6	90.4

Table 11.4   Drag values calculated by VGK for original and reconstructed airfoils

No. of Basis Functions	Maximum Residual	σ Residual
1	51.8	19.7
2	5.3	2.2
3	2.3	1.2
4	3.0	1.2
5	3.0	1.2
6	2.6	1.1

airfoils by just three parameters illustrates the potential for high degrees of redundancy in representations using linear combinations of airfoils.

In order to verify the VGK drag predictions, the airfoils reconstructed from the three base functions were also analyzed with MG2D. The results are presented in Table 11.5 along with the VGK values and the wave drag coefficients calculated by MG2D. The drag values from the Euler and full-potential analyses agree reasonably well. As might be expected, given the limitations of the full-potential method, the agreement is worst for the airfoils with higher wave drag.

Table 11.6 shows results of a further VGK study using the first two base functions only. The first function can be described as representing the mean airfoil while the second has the character of a camber adding function. In this case, rather than fitting geometrically the base functions to the airfoils, the results presented are after optimization of β (the ratio between the camber adding function and the mean airfoil), so as to minimize drag at the design point. While the results do not infer other aerodynamic properties, they indicate that with suitable choice of β, the drag performance of the original airfoils can be matched to within a few counts using airfoils constructed from only two base functions (the drags are never more than two counts worse and are sometimes significantly better). Figure 11.5 illustrates the section geometries resulting from this process for two of these cases (0406s and 0610s). The figure demonstrates that when the optimized airfoil has similar drag properties to those of the original section the geometric differences are slight. In the case of the 0406s airfoil where the optimized design has around 8% less drag, the differences are more marked. Investigations

Table 11.5   Drag values calculated by VGK and MGaero for the airfoils approximated using three base functions

	0406	0606	0706	0410	0610	0710	0412	0612	0712
VGK	72.7	76.1	68.2	71.1	83.2	86.7	85.1	89.8	91.3
MG2D	68.9	75.0	66.0	74.0	81.6	96.0	88.6	97.2	108.2
C_{dw}(MG2D)	8.7	9.0	8.2	6.1	13.7	22.7	8.8	11.8	18.9

Table 11.6   Comparison of original and "two base function" airfoils

SC(2)				Drag (counts)	
Airfoil	C_L	t/c	M	Original	$f_1 + \beta_{opt} f_2$
0406s	0.4	0.06	0.84	74	66
0606s	0.6	0.06	0.82	74	72
0706s	0.7	0.06	0.80	70	65
0410s	0.4	0.10	0.79	72	70
0610s	0.6	0.10	0.78	83	85
0710s	0.7	0.10	0.77	87	89
0412s	0.4	0.12	0.78	84	85
0612s	0.6	0.12	0.76	90	88
0712s	0.7	0.12	0.75	90	90

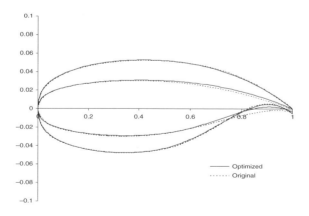

Figure 11.5   Original and optimized 0406s and 0610s airfoil geometries.

at design points other than those of the original airfoils suggest that low drag airfoils can be obtained throughout the design space; $0.4 < C_L < 1.0$, $0.06 < t/c < 0.12$. This suggests that for the purposes of conceptual design supercritical airfoils could be represented by just two parameters: t/c and β. The selection of airfoils in wing conceptual design could then be reduced to optimization of the span-wise distribution of t/c and β. Optimizations of wing geometries using such an approach are described in the next case study.

11.4.5   Mapping the Influence of the First Three Base Functions on Drag

The ability to represent the aerodynamic performance of the airfoil by as few as two or three parameters offers the possibility of visualizing the effect of each parameter on drag. Runs of VGK have therefore been carried out to generate maps of the variation of drag with the base function weighting parameters. Figure 11.6 shows a plot at the SC(2)0610 design point with the weights of the second and third base functions (labeled a2 and a3 in the figure) varied over a range covering the nine reconstructed airfoil values. In each case, the value of the first weight has been set to give a t/c of 10%.

Although derived to be geometrically orthogonal, the three functions also appear to be approximately orthogonal in an aerodynamic sense, the main features of the drag landscapes being oriented with the axes. It is thought that the subsidiary valleys present in the landscape result from differences in convergence. Figure 11.7 shows the drag landscape obtained from a lesser number of runs of MG2D.

As the overall shape of the drag landscapes calculated with VGK and MG2D are similar, the use of these airfoil parameterizations for developing and testing multifidelity optimization algorithms seems reasonable.

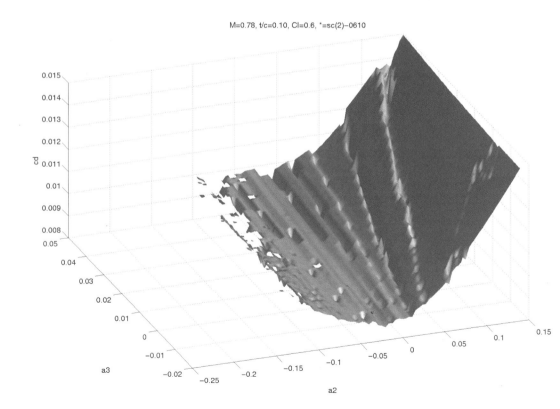

Figure 11.6   VGK drag landscape for three base functions at SC(2)0610 design point.

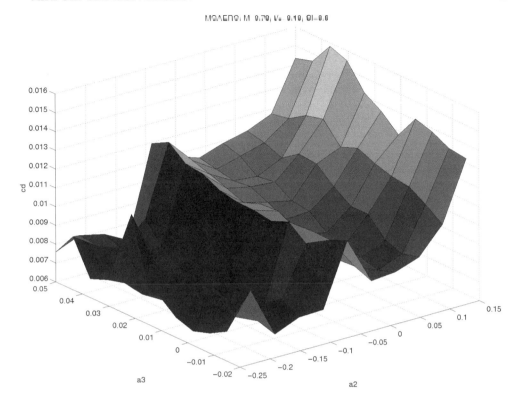

Figure 11.7 MG2D drag landscape for three base functions at SC(2)0610 design point.

11.4.6   Summary of Airfoil Analysis

To summarize, it is clear that using the approach described here, orthogonal basis functions can be found that are capable of faithfully representing the NASA SC(2) airfoil family with between three and five parameters. The results indicate that only three base functions are required to fit the original smoothed airfoils so as to effectively reproduce their aerodynamic performance. Moreover, good section designs can be produced using just two functions. Although not reported here, similar results have been obtained from orthogonalization of other more-modern supercritical airfoil families. This compares with the 20 parameters required to represent the korn airfoil using the basis functions derived by Chang et al. (1995) from polynomial representations.

11.5   Multiobjective Optimization

Having established a concise way of representing supercritical airfoils using an orthogonal parameterization, attention is next turned to the multiobjective design problem of producing low drag designs that are also robust in the face of minor variations in geometry and changes in operating speed. As will be clear from the tables presented in the previous section, the aerodynamic performance of the sections produced using the orthogonal basis functions developed here can give significant variation in performance. In this section, we set up a multiobjective optimization problem aimed at producing low drag designs capable of operating at a range of speeds and which have reduced sensitivity to small variations in shape (see Chapter 8 for a more complete discussion of uncertainty in design).

11.5.1   Robustness at Fixed Mach Number

To begin with, we examine sections operating at fixed lift and Mach number. We set the two goal functions of low drag and minimum variation in drag for small random perturbations in the geometric definition. The free variables in the problem are the weights of the basis functions and the overall thickness-to-chord ratio – six variables in all if we normalize weights two to six against weight one and then scale the resulting shape to the desired thickness-to-chord ratio. We evaluate the robustness of any design by making a series of twenty random perturbations to the weighting values of $\pm 2\%$ of the range of weight (or t/c) being considered. The resulting 20 drag values are then used with the nominal drag to calculate the standard deviation in C_d, which is then minimized when seeking robust designs. To prevent the sections becoming overly thin, we add constraints for minimum section depth at 25% and 65% chord of 0.095 and 0.075, respectively, which are meant to represent the positions of typical main spars.

Figure 11.8 shows the results of two Pareto front searches on this problem at Mach 0.78 and Mach 0.8, in each case using 10,000 steps of an evolution strategy search using the logic of Figure 3.30. It is clear that, as might be expected, the lowest drag designs are not the most robust with respect to small parameter variations. Although most designs have standard deviations in drag values of between two and four drag counts (0.0002 to 0.0004), there are a number of designs that show only around one drag count standard deviation. At Mach 0.78, of the very low drag designs, there are some with deviations of only just more than one drag count and the lowest drag design of all has a standard deviation of just over two counts, that is, below the average. The lowest drag design at Mach 0.8, however, has

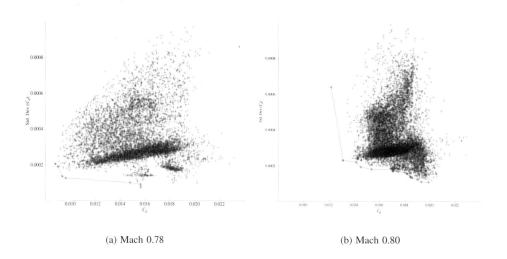

(a) Mach 0.78   (b) Mach 0.80

Figure 11.8 Pareto fronts for nominal C_d versus standard deviation in C_d due to geometric changes ($\pm 2\%$).

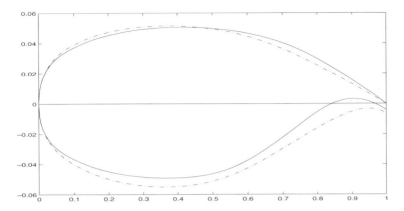

Figure 11.9 Variation in section shape for two Pareto optimal designs at Mach 0.80 (solid line is for lowest drag design, dotted line for most robust).

a deviation of over six counts and is worse than most of its peers in this regard. At this speed, a drag rise of some 25 counts must be tolerated to gain a design with less than two drag counts standard deviation.

Figure 11.9 shows the variation in shape of the two designs lying at the extreme ends of the Pareto front for Mach 0.80. The low drag design has a C_d of 121 counts and a standard deviation of six while the robust design has a C_d of 199 counts and a standard deviation of one count. Here, the low drag design is noticeably thinner (it is hard against both the minimum depth constraints at 25% and at 65% of chord, while the robust design is not constrained by either restriction) and has significantly more curvature at the rear on the upper surface. Further checks on this design show that it is close to producing significant shocks, although not so close that any of the random perturbations shows dramatic drag rises – this being the whole point of the multiobjective search. No doubt, a sensible choice would be to accept a slightly higher nominal drag of around, say, 135 drag counts where good all-round designs can be found. Given the impact of speed rise on drag variation, attention is next turned to designs that are robust with respect to both shape and speed variation.

11.5.2 Robustness against Varying Geometry and Mach Number

Since no aircraft flies at fixed speed, it is always desirable that a design be capable of operating at a variety of speeds. We therefore next add the additional requirement that drag rise beyond the design speed also be minimized. To do this, we study sections with a design Mach number of 0.78, but additionally integrate the drag values over the range Mach = 0.78 to 0.81 in 15 equal steps. This then leads to a multiobjective problem with three goals: minimum drag at Mach = 0.78, minimum standard deviation in drag at this speed due to ±2% variations in the weights of the basis functions and minimum average drag over the range Mach = 0.78 to 0.81. This is a demanding set of goals and so an extended VGK Pareto front search has been carried out over 50,000 evaluations. Figure 11.10 shows one of the best-balanced airfoil sections from the Pareto front that results from this search. Also shown on the plot are the designs with C_d of 121 and 199 counts at Mach 0.8 from

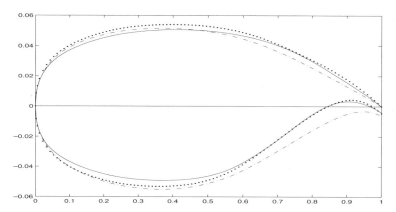

Figure 11.10 Variation in section shape for design with good performance in all three goal areas (shown dotted) together with the two designs from Figure 11.9.

Figure 11.9. Plate 1 illustrates the three-dimensional front itself, along with many of the design points studied.

It is clear from the Pareto front that nominal drag can be traded against both robustness and drag rise due to speed increase. Notice that there is a very severe drag cliff at higher speeds if the design is pushed too far at the nominal speed, as might be expected. There are, however, designs that are comfortably away from this cliff that have good nominal performance and reasonable robustness to geometric variation. These have standard deviations in drag of around one and a half counts, a nominal C_d of some 110 counts and an average of 145 counts. The airfoil section shown in Figure 11.10 has a nominal C_d of 106 counts, a standard deviation of 1.5 counts and a mean C_d of 142 counts (the drag range is from 106 counts at Mach 0.78 to 181 at Mach 0.81). Interestingly, this design is not bound by the section thickness constraints. Its lower rear shape is very similar to that of the previous 121 drag count design, but it has more gentle upper section shape. It is clear that such a design would offer good all-round performance and would be a better candidate for further research than one merely optimized for nominal drag performance.

12

Aircraft Wing Design – Data Fusion between Codes

The design of the wings for a transonic civil transport aircraft is an extremely complex task. It is normally undertaken over an extended time period and at a variety of levels of complexity. Typically, simple empirical models are used at the earliest stages of concept design, followed by ever more complex methods as the design process proceeds toward the final detailed stages. Moreover, this process is increasingly dominated by computational methods for both analysis (such as CFD) and synthesis (such as optimization). In this concept design study, the Southampton multilevel wing design environment (Keane and Petruzelli 2000) is used to examine the merits of data fusion when applied to three-dimensional CFD solvers over a transonic wing system. This system has been developed to support fundamental studies in aircraft design. Its central thrust has been to address what Malone et al. (1999) have called the "zoom axis" in design. To do this it allows the designer to rapidly and easily change the sophistication of the analysis codes being used to study a particular design feature at a given stage in the design process. Here, the wing drags predicted by an Euler code are supplemented by a linearized potential method and an empirical code to provide a multifidelity capability. Both CFD codes are available as commercial packages and have been integrated into the design system used in such a way as to require no additional input data from the designer, over and above that used by the empirical method.

The relative accuracy of the three aerodynamic codes used, however, is not easily quantifiable. Nonetheless, for the purposes of demonstrating multifidelity algorithms, the predictions of the Euler code are assumed to be accurate. At the same time, the empirical code is only well calibrated for conventional wing designs. For unconventional designs, the validity of the assumptions implicit in its models may be limited, resulting in significant differences from the Euler predictions. Although the potential code is capable of reasonable drag predictions at low Mach numbers, the use of an essentially subsonic boundary element code for drag prediction at transonic Mach numbers would not normally be contemplated. To reduce any errors as far as possible, the two CFD codes have been carefully integrated using similar drag estimation routines with particular care being placed on gaining acceptable estimates of wave drag from the panel code by integrating two-dimensional Euler data into this routine.

Computational Approaches for Aerospace Design: The Pursuit of Excellence.   A. J. Keane and P. B. Nair
© 2005 John Wiley & Sons, Ltd

It is of course the case that wing concept design must also consider a number of other aspects such as strength, fuel-carrying capacity, undercarriage housing, and so on, while overall aircraft design must focus on payload, range, operating costs, and so on. Here, however, the more limited task of low wing drag is set, but paying some regard to overall wing design by placing limits on certain key elements of wing geometry and adding a few basic layout and structural constraints. Thus, constraints are placed on fuel-tank volume, pitch-up margin, and root triangle volume, all of which are estimated empirically. At the same time, a relatively sophisticated wing weight model is used to tension the searches carried out, so that overly slender or swept geometries are suitably penalized by the addition of added structural weight, which is then subject to a hard constraint.

The resulting system allows the concept design team direct access to more sophisticated drag prediction methods than they would habitually use. Of course, these teams do already use results from such methods. However, they normally achieve this by turning over their putative wing planform and other basic data to specialist aerodynamicists. One of the thrusts of the work described here is to demonstrate that this process can, at least to some extent, be automated.

The use of CFD codes in concept design does, however, introduce run-time issues. Here, the empirical drag prediction model plus design of experiment (DoE), response surface and data-fusion methods are brought together with the empirical structural modeling and CFD to provide a reasonably fast optimization system. It allows high-quality designs to be found using full three-dimensional CFD codes without the expense of always using direct searches. The fused response surface built is accurate over wider ranges than the initial empirical drag model or simple response surfaces based on the CFD data alone. Data fusion is achieved by building response surface Krigs of the differences between drag prediction tools, which are working at varying levels of fidelity. The Krigs are then used with the empirical tool to predict the drags coming from the CFD code. This process is much quicker to use than direct searches of the CFD, although it is still more expensive than simply searching the empirical model alone.

12.1 Introduction

The initial specification of the overall wing planform parameters is a crucial stage in the development of any aircraft, since changes to planform further down the design process are often extremely difficult to cope with and usually represent an unacceptable delay or cost to the process. One of the main reasons for researching concept design systems is the fact that such fundamental decisions are made at this point. Moreover, it is quite common for concept design teams to lack access to the sophisticated analysis codes routinely used further on in the design process. Rather, they typically use empirical codes based on previous designs to estimate a good set of planform parameters. These make no attempt to solve the flow conditions over the wings being studied but can be accurately calibrated to deal with familiar geometries. They experience difficulties, however, whenever extreme or even moderately novel configurations are considered. Conversely, they are very easy to use and are capable of giving rapid estimates of likely drag levels (Cousin and Metcalfe 1990).

It would, of course, be useful if full three-dimensional CFD solvers could be used in lieu of the empirical codes contained in normal concept tools: currently, this is not standard practice because of two major difficulties. First, empirical concept tools do not require complete geometric details of the wing being studied, since they typically work with gross wing parameters such as aspect ratio, span, mean camber, and so on, while CFD solvers of course require a complete surface and mesh description. Such descriptions can be quite time consuming to develop to sufficient levels of realism to be of use in practical applications. Secondly, even Euler-based CFD codes typically have quite significant run times when considering three-dimensional geometries.

Enabling the "zoom axis" at concept design therefore involves allowing the design team to employ the full range of drag analysis methods available to the company, along with techniques to mitigate any issues stemming from the increased run times of CFD solvers. This should enhance the capability of the concept team to investigate more radical solutions rather than to stick to just the tried and tested designs spanned by the empirical tool (noting, of course, that radical here might mean a change in sweep of less than 5°). To provide this zoom facility requires that the design system be able to mimic, with reasonable fidelity, the follow-on processes used to carry out full CFD of the wings, that is, as full CFD requires a complete geometric description of the wings and the concept team have only the overall planform geometry, the system must supply the missing data in a realistic fashion. Clearly, if this is not possible, any use of more complex methods of analysis may well be of dubious value. At the same time, issues of run time must be addressed so that rapid searches over widely varying geometries can be carried out in reasonable timescales. The Southampton multilevel wing design environment addresses these concerns by allowing the user to employ differing levels of sophistication when studying the aerodynamics of a wing at the planform layout stage, combined with various structural analysis and weight models. It also permits use of DoE, response surface and data-fusion methods to help mitigate the increased run times that are incurred when using full CFD solvers.

In this case study, an empirical drag estimation tool, Tadpole (Cousin and Metcalfe 1990), and two CFD codes are used to carry out the desired aerodynamic analysis, along with two empirical structural weight estimation tools. The CFD codes employed are a linearized potential method, VSaero (Maskew 1982) and an Euler method, MGaero (Epstein et al. 1989), both of which are available as commercial codes and both of which allow for viscous coupled three-dimensional solutions. These codes can be used in place of the empirical method whenever the designer wishes, and without the designer supplying any additional input data. This provides the concept design team with the desired multifidelity or "zoom" capability that allows them to control the amount of computational effort used in aerodynamic analysis. The system used prepares all the additional data and mesh details such codes need, using an approach that mimics some of the activities of an aerodynamics team. It makes various judgments on lift distributions and the appropriate sections to be used along the wing, based on two-dimensional section data, so as to return a wing that exhibits good practice in this area. The resulting wing can then be analyzed by the fully three-dimensional CFD codes to return data equivalent to that provided by the empirical method that would otherwise be used, ensuring that the empirical results are supported by additional calculations, even when significant design changes are proposed. These codes are connected to the Options optimization system,[1] which allows designs to be produced with minimum drag while meeting specific constraints using a variety of simple, multilevel and multifidelity optimization strategies. Moreover, design spaces can be searched using any combination of the empirical or CFD methods and then validated using full Euler analysis, all within the one system. The Options system also provides all the tools necessary to construct and use response surfaces and to provide fusion between these at varying levels of fidelity.

---

[1]http://www.soton.ac.uk/~ajk/options/welcome.html

As first built, the Southampton wing design system did not offer much to mitigate problems of run time. All that was proposed in the original tool was that designers should start work with the concept tools usually available and then switch to the full CFD solvers whenever sufficient progress had been made. The ability to make this switch without supplying extra geometric data was its main advance. So, for example, when carrying out an optimization study, initial searches would be made using the empirical concept tool and these designs could then be seamlessly passed to the Euler solver for further, less wide ranging and necessarily shorter searches. Although this approach was a step forward, it is not of great benefit during design synthesis, since most of the design trade-offs considered were still made with the empirical code. Rather, it was most useful to rapidly check that the empirical methods of the original concept tool were still valid at any more radical design points of interest.

To overcome this limitation, data-fusion approaches have been adopted that draw information from the empirical and CFD-based drag routines using Design of Experiment (DoE) techniques (McKay et al. 1979; Mead 1988) and Kriging (Jones et al. 1998) to build Response Surface Models (RSMs) (Myers and Montgomery 1995). Variants on these methods have been used in aerospace design for some time. However, they have mostly been used to accelerate direct optimization approaches using expensive codes. It is only relatively recently that it has been proposed that they might be helpful in multifidelity or zoom analysis (Hutchinson et al. 1994; Liu et al. 1999; Malone et al. 1999; Vitali et al. 1999). The idea in multifidelity analysis is to use the DoE and Krig to produce a RSM that models corrections to the low cost, empirical analysis so that the correction model, together with the drag model of the original concept tool, may be used in lieu of the full CFD code. This provides results that are both well calibrated and capable of being used outside of the scope of the original concept tool in a seamless fashion. The basic approach for this kind of multifidelity modeling is described in more detail in Chapter 6; see also Leary et al. (2003a); here, the aim is to demonstrate its application to a modern wing design problem to illustrate its potential benefits and shortcomings in this relatively realistic setting.

To begin with, we use the various individual drag estimation routines, both singly and in concert, in relatively straightforward direct searches for improved wing planform geometries. We then speed up the process by constructing RSMs of both empirical and CFD data to model the variations in drag. Finally, we combine data from the empirical and CFD codes in a fusion process where the RSM models differences between the codes, enabling both speed improvements and a multifidelity approach.

12.2 Overall Wing Design

The parameters used to describe the wing design problem considered in this work consist of the planform shape together with several span-wise geometry variations plus data at individual sections; see Table 12.1 (N.B., span-wise variations in thickness and twist are assumed linear between root, kink and tip). These quantities are normally fixed by the concept design team on the basis of their experience of similar designs or by the use of optimization search tools. This is the approach adopted here.

12.2.1 Section Geometry

The detailed shapes of the individual sections of the wing are the critical elements that define its performance. In many concept design tools, this task is ignored entirely and an assumption

Table 12.1   Wing design variables

Symbol	Definition
SG	Wing area
AR	Aspect ratio
$SWPI$	Leading edge sweep
λ_1	Inner panel taper ratio
λ_2	Outer panel taper ratio
h_k	Trailing edge kink position
t/c_r	Root thickness/chord
t/c_k	Kink thickness/chord
t/c_t	Tip thickness/chord
w	Washout at tip
k	Fraction of tip washout at kink

made about likely section performance based on previous data and other empirical methods. Such an approach has the advantage of not needing to call upon sophisticated CFD codes but runs the risk of limiting improvements in design, even if only the overall planform and other major parameters are being investigated. Here, a compromise approach is adopted to section modeling.

For transonic designs perhaps the most important aspect of aerodynamic performance considered is the position and nature of the compression shock on the upper surface and how this can be controlled. Unfortunately, airfoil sections are some of the most subtle and complex shapes encountered in engineering and their design is never simple. Describing these sections in an efficient way is a critical part of any optimal design process. If too much freedom is allowed in the section shapes, the space that must be investigated in the search for new designs becomes vast and there is little hope of finding really good, new designs. Conversely, if too tight a description is used, although the space to search may be small, the chances of finding interesting new designs is again radically reduced. Moreover, when providing a concept design aid, it is important that designs show consistency from project to project and designer to designer, so that trade-off studies can be carried out with confidence (i.e., a house style should exist).

Here, the input geometries to the CFD solver are created using the set of orthogonal functions derived from NACA transonic foils described in the previous case study (Harris 1990; Robinson and Keane 2001). Although these functions allow less design variation than B-spline or Bezier representations, they are ideal for conceptual design where the aim is selection of sections appropriate for local conditions. As already noted, this approach allows for up to six such base shapes to be used, but that reasonable airfoils can be recovered with just two functions, which essentially reflect thickness-to-chord ratio and camber. This is an important feature of the method since any additional data needed for defining wing shapes would require significant effort so as to ensure consistent and good-quality designs.

Given a prescribed thickness-to-chord ratio and twist distribution along the wing, which are part of the input data here (i.e., they are either specified by the designer or searched for by the optimizer), this reduces the geometric design problem to one of finding the correct span-wise distribution of camber. These span-wise variations of the orthogonal function weights are here represented by five internal parameters that are set so as to give a reasonable span-wise pressure loading on the wing. Following this approach, once the span-wise distribution of camber along the wing is known, the full geometric description of the wing may be constructed for subsequent analysis by the CFD codes (or any other form of analysis). For the system used here, dedicated software expands from the section shapes to generate the appropriate input files automatically. In the case of the boundary element code, these specify surface and wake panel geometries. For the Euler solver, the files specify the surface discretization and the Cartesian grids.

12.2.2   Lift and Drag Recovery

Having specified the overall envelope, the aerodynamicist's task is to calculate the performance of the wing, usually as total lift and drag coefficients. Here, the lift coefficient is derived from the aircraft weight with angle of attack being a resulting output, that is, the CFD codes are used to find an angle of attack that yields the desired lift. As has already been mentioned, the drag is found either by the use of empirical data, as housed in the Tadpole code, or one of the two CFD codes. All three codes have been implemented to allow some form of drag recovery for a specified lift, but given their underlying differences, these also differ in how the various elements of drag are calculated internally. All three methods return the drag coefficient in three components, however: wave drag, due to the presence of shocks; viscous wake or profile drag, due to the boundary layer; vortex or induced drag, due to the tip vortex of the three-dimensional wing. Here, to make comparisons between the methods more direct and also to allow interworking between the methods, a common approach to drag recovery has been used where possible.

Tadpole is the quickest method, taking some 50 ms, and simply returns drags based on curve fits to previously analyzed wings making assumptions about the kinds of rooftop pressure profiles now commonly achieved in transonic wing design.

The next method in terms of computational cost is the viscous coupled boundary element code, VSaero, requiring approximately 5 min of compute time per evaluation (using a 500 MHz Pentium III processor). The code uses a Morino formulation with constant singularity panels. Prandtl–Glauert and Karman–Tsien compressibility corrections can be specified. In addition, a viscous coupled boundary layer model can be coupled with the inviscid solution via surface transpiration. Using this code, the induced drag is calculated from Treffitz plane integration, and the profile drag using the Cooke (1973) implementation of Squire and Young's approximation (Squire and Young 1937). For each streamline, the momentum thickness, shape factor and separation point are calculated. The Mach number at the separation point is determined from the output of panel centroid values. These parameters are the inputs to Cooke's method to determine the momentum thickness of the wake at downstream infinity. This process is repeated for each streamline on both upper and lower surfaces and the resulting momentum thicknesses integrated across the span. The viscous drag coefficient is twice the integral divided by the reference wing area. The linearized potential equations used by the method of course preclude the possibility of calculating wave drag directly. However, a routine (Petruzzelli and Keane 2001) applying simple sweep theory and data based on the off-line analysis of all the possible two-dimensional sections in use, by the two-dimensional version of the Euler code MGAero (MG2D) and Lock's 2nd order method ((Lock 1986), also known as the ESDU method (ESDU 1987)), provides a wave drag estimate from the calculated span-wise loading. This process involves four-dimensional interpolation in two look up tables. When combined with an internal search to optimize the camber distribution profile, the total analysis of the wings using this method takes approximately 7 min.

The most expensive method used here is the viscous coupled multigrid Cartesian Euler code MGaero. The code is unusual in using Cartesian grids rather than the more common body-fitted structured or unstructured grids. Like VSaero, the Euler inviscid analysis can be coupled with a boundary layer model via surface transpiration. Therefore, when using this code a similar drag recover process is used to that for VSaero, that is, the viscous drag from Cooke's implementation of the Squire and Young method and the wave drag using Lock's second order method, this time directly applied to the flow solution. The induced drag is obtained from the integration of vorticity immediately downstream of the trailing edge as dissipation due to the presence of numerical viscosity precludes integration in the Treffitz plane. Of course, such drag values are subject to the usual inaccuracies associated with drag recovery from Euler codes (van Dam and Nikfetrat 1992). Nonetheless, the results returned by this code are here taken to be accurate for the purposes of wing optimization. The main aim is to provide an aid for concept designers, not an improved drag estimation facility for aerodynamics specialists. The total run time of this method when used with the internal camber optimization procedure is around 1 h.

12.2.3   Wing Envelope Design

To begin the design process, as has already been noted, the user specifies the overall geometry of the wing and its thickness to chord and twist distributions, while the desired lift coefficient is derived from aircraft weight. If a Tadpole calculation is being performed this is sufficient to allow the return of a drag estimate from the empirical data held within the code. However, if a full CFD method is to be used, a complete wing envelope must be designed following good aerodynamic practice. Here, to start with, the assumption is made that a good span-wise distribution of lift is an elliptic one (since this minimizes induced drag) and also that the wing section around midspan works hardest in terms of wave drag. The section at 52% span is thus used as the starting point for shaping the wing envelope.

Off-line two-dimensional Euler computations for the sections in use here have been performed over a wide range of camber, t/c and Cl values and it is clear from these that low values of wave drag occur for section β values between -0.5 and 0.2 (the weighting parameter for the second base function that fixes camber), the lowest values being located in the neighborhood of β equal to -0.2. The initial value of β of the critical section is therefore set equal to -0.45. This choice of camber weighting at a shock sensitive section is a good starting point for design but is refined as the process goes forward. The aim is to ensure that all other sections are slightly less cambered and thus less likely to suffer shocks than this critical section. By placing its camber weighting to one side of the range where such shocks do not occur, maximum flexibility is available for setting the cambers of the remaining sections. Given this starting point, the local section lift coefficients and angles of attack for the rest of the wing can be found (twist is already specified bilinearly from root to kink and kink to tip and lift is assumed directly proportional to angle of attack). This allows the local camber values to be designed, based on two-dimensional Euler data. Thus, the whole wing shape can be defined assuming that the local camber weighting β, varies linearly from root to kink and quadratically from kink to tip: the detailed distribution being chosen to give the desired elliptic loading and total lift with the chosen twist.

Having set up the wing in this way using a lift distribution chosen purely to minimize down-wash, the next step is to investigate the trade-off of induced drag with wave drag. Given that the two-dimensional performance of each section is known the wave drag can be estimated for the whole wing on the basis of the two-dimensional Euler data. Also, the induced drag can be found by simply integrating the lift, based on camber data. At this stage, it is then possible to carry out a small suboptimization search to find a lift distribution that minimizes the sum of wave and induced drag without calling either of the CFD codes. Here, a straightforward Simplex algorithm is used, with multiple restarts to ensure a good result. Using a relatively highly cambered section near midspan results in all the other sections having camber weights in the desired range. Although no calls to CFD codes are required for this step, it is still quite time consuming since it involves a significant number of calls to the wave drag and lift look-up tables holding the two-dimensional Euler data.

Next, if VSaero is being used, a call is made to the code to get an improved estimate of the span-wise distribution of lift for the current geometry and an improved estimate of the angle of attack as well. Then, the Simplex search can be repeated using this data in the wave and induced drag estimation. A sequence of VSaero calls and suboptimizations of camber distribution can then be carried out until the process converges, while at the same time iterating the overall angle of attack to generate the desired lift coefficient. Then, the viscous coupled feature of VSaero is invoked and another series of calls made until the boundary layer has converged, but holding the wing shape fixed. Finally, the drag recovery methods are applied to the viscous solution to extract the required data. This completes the panel code–based analysis.

If, in contrast, MGaero is to be used to recover the drag, after the initial suboptimization from the elliptic distribution of lift, MGaero can be called directly and its internal angle of attack control feature used to provide the correct total lift. More normally, however, the use of MGaero is preceded with the inviscid VSaero section redesign process, since although this process is costly compared to the use of Tadpole, it still only adds marginally to the total run time of the Euler code. It also ensures full comparability with the potential code (i.e., then both CFD codes use identical wing geometries when run).

Various experiments with these methods have demonstrated that the wave drag recovered following a design using just VSaero and one that then goes on to run the full three-dimensional compressible calculation using MGaero are sufficiently close for the methods to be used in harness together over a range of designs (the agreement begins to break down for highly swept designs). Given such a method, it is then possible to use a three-dimensional panel code such as VSaero to carry out initial camber design prior to calls to a compressible code such as MGaero. It also allows results from VSaero calls to be used to generate drag data in multifidelity design strategies where the whole wing geometry is being optimized and reduced use of the Euler code is desirable. Nonetheless, it is important to always bear in mind that, even when augmented with Euler-based section data, a linearized panel code like VSaero should not be trusted too far when analyzing transonic flows.

12.3 An Example and Some Basic Searches

To begin our design studies, we first illustrate the use of the wing design environment for direct searches, using either the empirical or CFD solvers. Here, the response being studied is the drag of a transonic civil transport wing. A simple test problem has been constructed with the aim of optimizing the wing for operation at Mach 0.785 and a Reynolds number of 7.3 million The objective is minimization of wing D/q as calculated by the MGaero CFD solver with target lift, wing volume, pitch-up margin and root triangle layout chosen to be representative of a 220-seat wide-body airliner. The initial planform and thickness distribution are based on the DLR F4 wing (Redeker 1994). Limits are placed on the design variables that are typical of work in this area (although they still admit designs that would be considered radical in practice – it is not common to use sweep angles as high as 45 (in a civil aircraft for example).

In addition to simple geometrical constraints, an empirical wing weight model is incorporated in the system that reflects the impact of geometrical change on structural weight. The resulting overall wing weight is then subject to a hard constraint. This is a critical aspect of any wing design process since, without such models, there is a natural tendency for optimizers to consider wings with high aerodynamic efficiency without regard to any structural penalties. It is well known that long slender wings tend to be aerodynamically desirable but structurally inefficient, for example.

12.3.1 Drag Results

To begin with, the drag is computed using the Tadpole tool or either of the two CFD codes working in isolation. Typical results from these systems are detailed in Table 12.2 while Figure 12.1 illustrates the equivalent geometry. Notice that in addition to the 11 parameters defining the wing constraints are placed on the wing volume, undercarriage bay length, pitch-up margin and weight. At all times, the angle of attack is set to generate the required lift and

Table 12.2 Initial design parameters, constraint values and objective function values (Tadpole weight calculation)

Lower Limit	Value	Upper Limit	Quantity (units)
100	168	250	Wing area (m²)
6	9.07	12	Aspect ratio
0.2	0.313	0.45	Kink position
25	27.1	45	Sweep angle (degrees)
0.4	0.598	0.7	Inboard taper ratio
0.2	0.506	0.6	Outboard taper ratio
0.1	0.150	0.18	Root t/c
0.06	0.122	0.14	Kink t/c
0.06	0.122	0.14	Tip t/c
4.0	4.5	5.0	Tip washout (degrees)
0.65	0.75	0.84	Kink washout fraction
	13,050	13,766	Wing weight (kg)
40.0	41.73		Wing volume (m³)
	4.179	5.4	Pitch-up margin
2.5	2.693		Undercarriage bay length (m)
	3.145		D/q (m²) – Tadpole
	2.970		D/q (m²) – VSaero
	2.922		D/q (m²) – MGaero

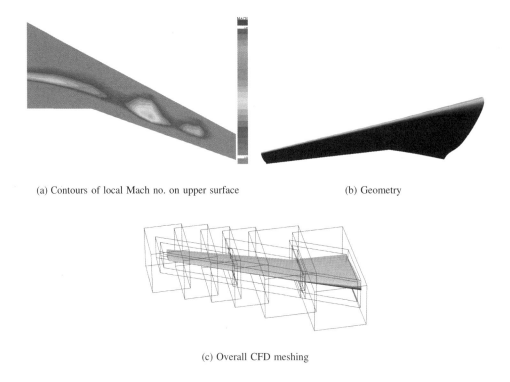

(a) Contours of local Mach no. on upper surface   (b) Geometry

(c) Overall CFD meshing

Figure 12.1 Initial wing geometry (Tadpole weight calculation).

the wing weight changes in a realistic fashion allowing for necessary structural modifications as its dimensions alter. Here, the methods yield drag estimates that differ by some 8% despite the careful validation of the Tadpole code and considerable effort in attempting to get the drag recovery from the Panel and Euler codes to work in a directly compatible fashion. This is partially due to the public domain wing airfoil sections used to generate the CFD geometry, which differ from the commercial sections for which Tadpole is calibrated.

12.3.2 Wing Weight Prediction

The most common way to calculate the weight of the load carrying part of the wing is to split the structure into the parts that resist bending and those that resist shear. For the aircraft category under consideration (subsonic transport and executive aircraft), the wing is generally built up from the subparts depicted in Figure 12.2.

Here, the wing weight is estimated using an analytical-empirical approach that is a development of Torenbeek's method (Torenbeek 1984, 1992) developed by Airbus, and also embedded in the Tadpole tool (Cousin and Metcalfe 1990). In such methods, the wing weight is computed as the sum of several functional components, each of which is estimated via a rational and/or statistical approach. The methods are especially suitable for the preliminary design stage, when sensitivity studies are required on the effects of geometric and other variations on the design characteristics.

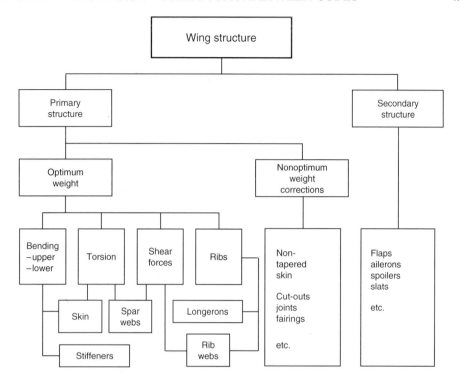

Figure 12.2 Subdivision of wing and weight distributions.

The structural principles assumed are as follows:

1. The structure may be considered as a statically determined equivalent system:

- Bending is absorbed by the stiffened skin panels and the spar flanges, while the stiffeners are "buttered" with the cover plates to an equivalent plate thickness. Simple beam theory is used for the analysis.

- The shear force loads are transmitted by the spar webs, again using equivalent thicknesses.

- Torsional loads will not be taken into account explicitly, but the wing box will be checked in a simplified fashion for sufficient torsional stiffness and, if necessary, locally strengthened, mainly in the outer wing.

- The ribs support the coverplates, transmit shear loads and introduce concentrated forces into the wing box.

2. Only maneuvering and gust loads are considered as the determining cases.

3. Stress levels will be obtained from the most important failure criteria and types of construction, taking into account material characteristics.

4. The "idealized" theoretical (minimum required) weight will be corrected with so-called nonoptimum corrections, which are intended to account for practical deviations from theory, for example, due to nontapered skins, joints, splices, openings, access panels, and so on.

5. Secondary structure weight is estimated for each category separately, using an area-based method and functional and statistical parameters as input.

The main differences between the Torenbeek and Tadpole methods lies in the use of a different level of detail for the description of the wing structure and in a different level of accuracy for the calculation of the aerodynamic loads. In Torenbeek's method, wing plan-form parameters and t/c values at key sections (centerline, root, kink and tip) are the only geometry data required as input. Moreover, assumptions are made for the mass (structure, fuel, and external loads) and loading distributions based on existing configurations. In the Tadpole approach, a more detailed description of the wing box, leading and trailing edge structures, including high lift devices and control surfaces, is required. Fuel mass distri-butions and external load weights and positions are also needed as input data. The basic aerodynamic loads are evaluated using the ESDU 83040 method (ESDU 1983), which gives a rapid and accurate estimation of the span-wise lift distribution of wings with camber and twist in subsonic attached flow. The method also allows for the use of the more accurate lift distribution recovered from the flow analyses. Moreover, "adjustment" factors allow for the calibration of the predicted weight terms against actual values of existing configurations. In both methods, several of the input variables that specify the properties of the materials to be used in the construction of the wing box can be modified to cater to different technology standards.

Of the modeled wing structure, primary structure (i.e., structural items between and including the front and rear spars) is sized directly by configuration or structural design variables, whereas secondary structure (i.e., structural items in front of the front spar and behind the rear spar, plus any miscellaneous item) changes size and weight only as needed to remain consistent (through linked design variables) with the primary structure. As the calculation of the weight relief due to the wing structural weight effect on the bending depends on the wing weight itself, which is a result of the present methods, an iterative procedure is used to find the primary structure size.

Once the wing weight has been evaluated, this has to be included in the aircraft weight buildup. The design aircraft weight can be broken into crew weight, payload weight, fuel weight and operational empty weight (structure, engines, landing gear, fixed equipment, avionics and anything else not considered part of crew, payload or fuel). The crew and payload weights are both known since they are given in the design requirement. The only unknowns are the fuel weight and the operational empty weight, both of which are dependent on the total aircraft weight. Thus, an iterative process must also be used for aircraft sizing.

For the purposes of this study, it has been assumed that part of the operational empty weight (i.e., excluding the weight of the wing structure) is fixed (in the range of our search) and that the maximum fuel weight is dependent on the size of the wing box (which houses the fuel tanks). The maximum take off weight is then given by:

$$W_0 = W_{\text{wingless}} + \rho_{\text{fuel}} V_{\text{fuel}} + W_{\text{wing}}, \tag{12.1}$$

where ρ_{fuel} and V_{fuel} are, respectively, the fuel density and the maximum fuel capacity, and

$$W_{\text{wingless}} = W_{\text{crew}} + W_{\text{payload}} + W_{\text{empty}} - W_{\text{wing}}. \tag{12.2}$$

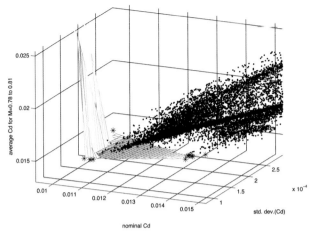

(a)   nominal C_d versus standard deviation in C_d versus mean C_d

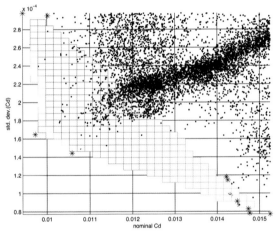

(b)   nominal C_d versus std. deviation in C_d

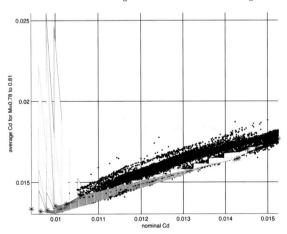

(c)   nominal C_d versus mean C_d

Plate 1   Pareto front for nominal C_d versus std. deviation in C_d due to geometric changes ($\pm 2\%$) and mean C_d due to variation in Mach number (0.78 to 0.81).

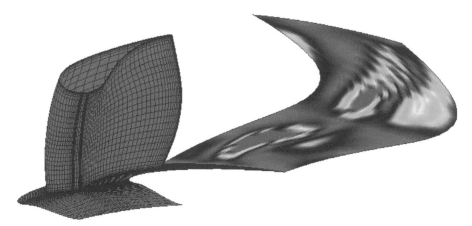

Plate 2 Initial guide vane geometry and map of secondary kinetic energy in the slot behind the vane.

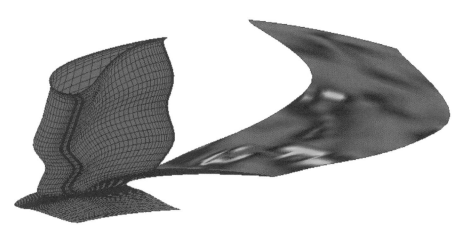

Plate 3 Best guide vane geometry found after 100 point DoE and three best SKE updates together with map of the secondary kinetic energy in the slot behind the vane.

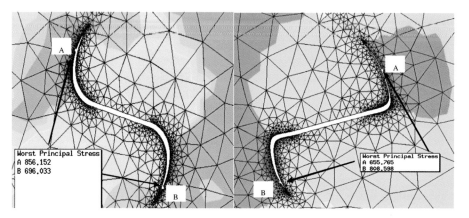

Plate 4 Stress contour maps for the notch region before and after optimization.

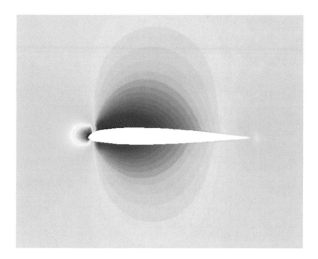

Plate 5 Visualizing results in the Matlab environment.

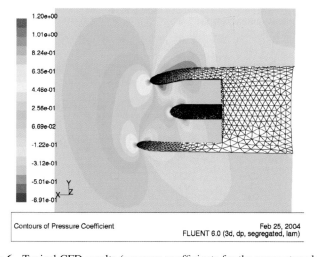

Plate 6 Typical CFD results (pressure coefficients for the symmetry plane).

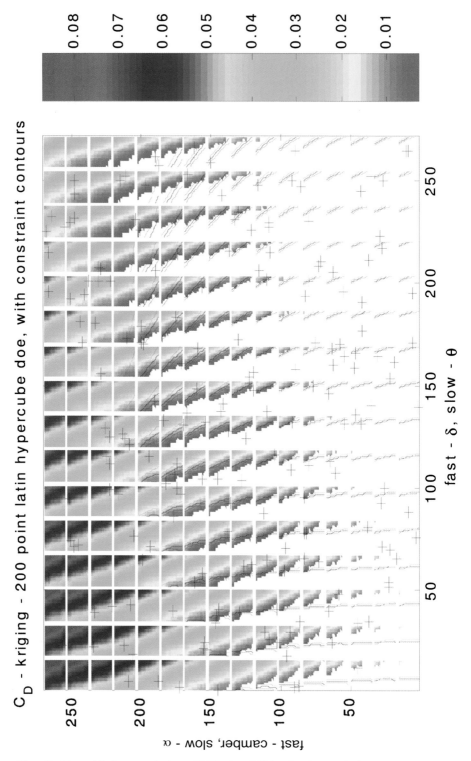

Plate 7 Hierarchical axes technique (HAT) plot of 200 point data set in four dimensions (also showing constraints and sample points) – courtesy of C. M. E. Holden.

The only unknown in (12.1) is now the weight of the wing structure, W_{wing}, the maximum fuel capacity V_{fuel} being derived from merely geometric considerations. The effects of improved lift/drag ratio and aircraft weight changes due to wing weight variations on the range can then be reflected in the fraction of the maximum allowable fuel used when calculating cruise weight, *fuel_frac*. Alternatively, and to simplify the process, the fuel fraction can be left fixed so that the range of the aircraft varies slightly as the geometry changes. In either case, wing optimization is then carried out considering the aircraft in steady straight and level flight.

12.3.3 Weight Sensitivities

The Tadpole method is thought to be more accurate than the original Torenbeek approach as it has been calibrated to suit recent wing weights of twin-engined wide-body aircrafts of medium capacity, which represent the range in which our design search is performed. The differences between the two methods can be seen by considering the weights generated for the geometry given in Table 12.2; see Tables 12.3 and 12.4.

In both the methods, weight items represented by components such as trailing and leading edges, undercarriage pickup structure, special purpose ribs, openings, and so on, are computed using semiempirical formulae, based on the information that are available according to the sophistication of the model. Where similar formulas are used in the Torenbeek method, the same adjustment factors are used as in the Tadpole methods. This is the case for the calculation of the weight of the leading edge, trailing edge and undercarriage pickup structure. For the remaining components, in order to prevent inconsistencies between the methods due to the use of different empirical formulas, it is assumed that their overall weight amounts to 10% of the wing weight.

The results shown in the tables indicate an underestimation of the wing weight prediction of Torenbeek's method with respect to Tadpole of about 2%. It should be noted, however,

Table 12.3 Wing structure parameters

Quantity (units)	Value
Wing LE as a fraction of local chord	0.153
Wing TE as a fraction of local chord	0.284
Wing LE as a fraction of centerline chord	0.153
Wing TE as a fraction of centerline chord	0.350
Number of wing mounted engines	2
Position of the engine	0.598
Mass of 1 wing mounted engine (kg)	9,000
Inner tank boundary limit	0.0
Inner/outer tank boundary position	0.58
Outer tank boundary limit	0.82
Chord-wise fraction of TE that the flap occupies	0.7555
Chord-wise fraction of TE that the slat occupies	0.9364
Span-wise extent of outboard flap	0.7714
Rib pitch	0.75

Table 12.4   Aircraft weight build up

Item	Weight (kg) Tadpole	Weight (kg) Torenbeek
Lower panel	3,325.7	
Upper panel	2,621.1	6,985.2
Spars	1,445.0	
Light rib	322.9	413.7
U/C pickup structure	625.4	674.5
Trailing edge	2,350.2	2,350.2
Leading edge	1,055.0	1,055.0
Miscellaneous	1,305.0	1,275.3
Total wing structure	13,050.3	12,753.9
Wingless weight	78,515.3	78,515.3
Maximum Fuel mass	33,507.5	33,507.5
Maximum Takeoff weight	125,073.1	124,776.7
Cruise Fuel mass (*fuel_frac* = 0.37)	12,397.8	12,397.8
Cruise weight	103,963.4	103,667.0

that this relatively small difference arises as a result of a cancellation of errors: an overestimate of the U/C pickup structure and rib weights compensates for an underestimate of the spar and upper and lower panel weights. For the configurations considered, this appears to occur throughout.

A very brief sensitivity study of the wing weight with respect to four independent parameters including planform and wing section geometry inputs has been performed for the reference aircraft. The independent variables considered are the aspect ratio, AR, the gross wing area, SG, the thickness-to-chord ratio at the root, t/c_r, and the inboard leading edge sweep angle, $SWPI$. Figure 12.3 shows the wing weight variations with respect to each of these parameters in the chosen range. Notice that the wing weight sensitivities are similarly predicted for AR, SG, and $SWPI$, in that an increase of each of these variables leads to an increase of weight. On the other hand, a different trend is predicted for t/c at the root, where the wing weight appears to be relatively insensitive in Tadpole but highly sensitive in Torenbeek's method. Figure 12.4 shows cross sections of the D/q objective function as computed by the empirical Tadpole drag recovery routine, with respect to these parameters in the corresponding ranges. The plots indicate that despite the differences the impact of wing weight prediction on the objective function is not severe. It is, nonetheless, sufficient to affect the outcome of optimization searches.

12.3.4   Direct Optimization

Having set up this simplified design problem, it may then be very rapidly optimized if the empirical code is used to estimate the drag. Here, a 25-generation Genetic Algorithm (GA) search with a population size of 200 members has been used followed by a gradient descent search to fine-tune the final optimum. As has already been noted, GAs are not deterministic in nature and rely on the random number sequences being used. When testing or developing such stochastic search engines, it is therefore normal practice to average results over a number of statistically independent trials – when using them in design, this is a luxury that

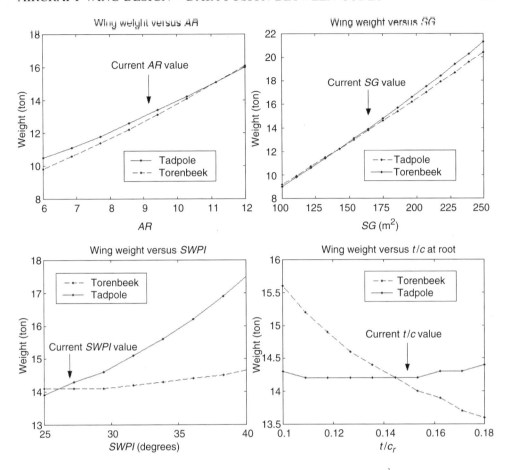

Figure 12.3 Wing weight computation – Sensitivity analysis.

rarely can be afforded. Here, each GA search is carried out for just one set of random numbers. It is therefore the case that some caution must be exercised when considering the results, particularly those that depend on single, relatively short GA searches. The results are more certain where several such searches are combined in producing designs (such as when using multiple RSM updates) or when rather longer GA searches with larger populations can be afforded (as here for the Tadpole search). Like all evolutionary methods, the GA is rather slow and inaccurate for problems with few variables but comes into its own as the number of variables grows. It is also not suitable for problems without bounds on all the variables. Currently, GAs seem to the best of the commonly used stochastic methods. Because of these limitations, where possible, the GA is followed up by a traditional downhill search, normally the well-known Simplex method (Nelder and Mead 1965), but if this stalls, which it sometimes does, by using Rosenbrock's rotating coordinate search (Rosenbrock 1960).

Application of this hybrid GA/downhill optimization strategy to the Tadpole code drives the wing volume constraint down to its limit and also increases the sweep angle considerably, although the total wing area is little changed; see Table 12.5. This reduces the drag by over

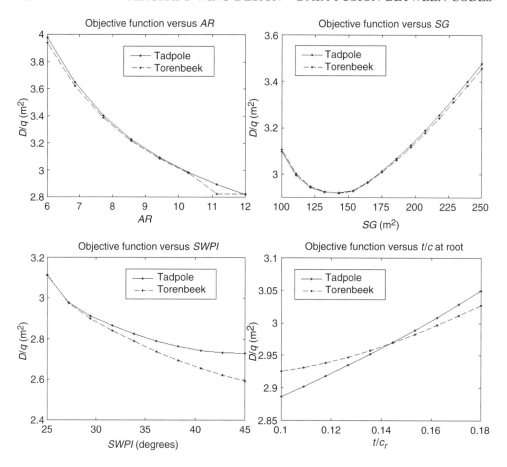

Figure 12.4   Wing weight computation – Effect on the objective function evaluation.

9% (as predicted by Tadpole). Such a search process represents the current everyday activity of a concept design team. Having carried out this study, the Southampton system then allows the drag to be checked by invoking the MGaero CFD solver – this result is also recorded in the·table, and it is seen that again the predictions still differ, now by 12%. The CFD predicted drag has, however, been decreased by nearly 13%.

Given the difference in drag between the two predictions, it is interesting to check whether a direct search applied to the CFD code would have produced a similar design geometry. Table 12.5 also gives the results of such a study, although now the GA optimization has been reduced to 15 generations and a population size of only 100 and the final hill-climbing search has been omitted, all to save time. Even so, this search represents some 150 *days* of computing effort; here carried out on a cluster of PCs running in parallel over two weeks (the Tadpole search took ten minutes!). The extreme cost of such searches makes them infeasible for everyday use – but they do provide benchmarks against which to compare other results. Notice that in this case the drag is reduced by some 14% (as predicted by the Euler CFD code) and that the two codes still do not agree on the resulting drag, now differing by 19%. Comparing the results in Table 12.5, it is apparent that the two methods

Table 12.5 Final design parameters, constraint values and objective function values for the best designs produced by the direct Tadpole and MGaero searches (Tadpole weight calculations)

Lower Limit	Tadpole	MGaero	Upper Limit	Quantity (units)
100	168.5	177.5	250	Wing area (m^2)
6	9.32	9.30	12	Aspect ratio
0.2	0.244	0.406	0.45	Kink position
25	31.8	25.2	45	Sweep angle (degrees)
0.4	0.516	0.683	0.7	Inboard taper ratio
0.2	0.227	0.259	0.6	Outboard taper ratio
0.1	0.104	0.143	0.18	Root t/c
0.06	0.115	0.096	0.14	Kink t/c
0.06	0.063	0.069	0.14	Tip t/c
4.0	4.7	4.5	5.0	Tip washout (degrees)
0.65	0.68	0.67	0.84	Kink washout fraction
	13,653	13,273	13,766	Wing weight (kg)
40.0	40.0	41.6		Wing volume (m^3)
	5.04	3.67	5.4	Pitch-up margin
2.5	3.51	2.56		Undercarriage bay length (m)
	2.853	2.998		D/q (m^2) – Tadpole
	2.555	2.524		D/q (m^2) – MGaero

converge to somewhat different optima for this design study – the Tadpole predicted drag for the design coming from the MGaero search being nearly 5% higher than for the design resulting from searching tadpole directly, while at the same time the CFD-predicted drag is slightly more than 1% lower. Moreover, the CFD-based wing has a significantly larger area.

These results demonstrate the power of this wing design system to explore interesting geometries. It is, however, noticeable that the difference in drag estimation between the two approaches has risen significantly after the searches and that they converge toward different geometries. Presumably, the differences arise because the empirical code is working rather far away from the calibrated zones for which it has been extensively tested.

12.4 Direct Multifidelity Searches

Direct multifidelity (or zoom) analysis assumes that the designer has at least two different ways of computing results of interest for the design under consideration. These are then combined during the search to speed up and hopefully improve the process. We therefore next consider a number of such direct multifidelity searches, now using the Torenbeek weight model and an increased fuel fraction of 0.55 (with suitably increased upper wing weight constraint), but again starting from the DLR F4 wing; see Table 12.6. Figure 12.5 shows the geometry, wing sections and pressure profile for this initial configuration. A number of direct multifidelity optimization runs have been carried out on this geometry, limited to at most 1,000 MGaero evaluations.

Table 12.6   Initial design parameters and constraints values (Torenbeek weights)

Lower Limit	Value	Upper Limit	Quantity (units)
100	168	250	Wing area (m^2)
6	9.07	12	Aspect ratio
0.2	0.313	0.45	Kink position
25	27.1	45	Sweep angle (degrees)
0.4	0.6	0.7	Inboard taper ratio
0.2	0.51	0.6	Outboard taper ratio
0.1	0.15	0.18	Root t/c
0.06	0.122	0.14	Kink t/c
0.06	0.122	0.14	Tip t/c
4.0	4.5	5.0	Tip washout (degrees)
0.65	0.75	0.84	Kink washout fraction
	19,568	20,394	Wing weight (kg)
40.0	42.35		Wing volume (m^3)
	4.179	5.4	Pitch-up margin
2.5	2.693		Undercarriage bay length
	3.12		D/q (m^2) – Tadpole
	3.13		D/q (m^2) – VSaero
	2.58		D/q (m^2) – MGaero

To provide a basis for comparison, two single-level searches have been carried out, as before. In the first case, a search with a population size of 250 was run for 40 generations using Tadpole; see line 1 of Table 12.7. In the second, the population was restricted to 100 and only 20 generations used, but the panel code VSaero was employed; see line 2. It is clear from these two lines that, although the initial design has very similar drag values as predicted by the two methods, the optimized designs do not. In fact, although optimizing the wing using these two codes has reduced the drag predicted by either method for both final designs, the MGaero results for the VSaero-optimized design have deteriorated (2.58 to 2.95) while those for the Tadpole-optimized design have improved (2.58 to 2.38). These difficulties can be attributed to the geometry produced by the VSaero search: it has a sweep of 31°, which is rather high for this kind of operational speed. At such sweeps, the conditions are highly three dimensional and the drag estimates of the panel-based method, with its two-dimensional wave drag corrections, are beginning to be misleading. Moreover, the development of the boundary layer, and therefore the viscous drag estimations, are also not well modeled in these conditions, i.e., although the panel code is helpful in the area where the various drag recovery routine described earlier work well, the code is increasingly more deceptive for more radical designs involving strongly three-dimensional flows.

Next, the final population of the Tadpole run was used to form the initial, partially converged generation for eight generations of a VSaero GA search with a population of 250; see line 3 of the Table. This results in an even lower VSaero drag but unfortunately the MGaero drag has worsened even further. So, although this two-stage search has helped achieve a better outcome from VSaero, this is only by adopting even more extreme geometries: the

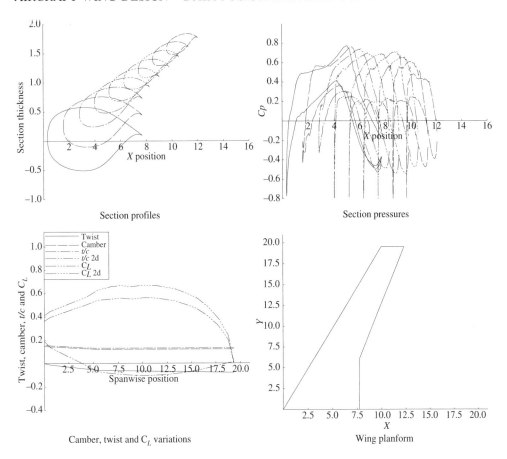

Figure 12.5 Geometry, section and pressure plots for the initial configuration (Torenbeek weight calculation).

Table 12.7   Results of direct multifidelity wing optimization (Torenbeek weights)

Search and Analysis Method	Tadpole D/q (m²)	VSaero D/q (m²)	MGaero D/q (m²)
1 GA search, 10,000 Tadpole evaluations	2.75	2.83	2.38
2 GA search, 2,000 VSaero evaluations	3.06	2.59	2.95
3 GA search, 10,000 Tadpole evaluations followed by 2,000 VSaero evaluations	2.94	2.47	3.06
4 GA search, 10,000 Tadpole evaluations followed by 2,000 VSaero evaluations followed by 500 MGaero evaluations	2.81	2.77	2.43
5 GA search, 10,000 Tadpole evaluations followed by 1,000 MGaero evaluations	2.75	2.87	2.37
6 GA search, 1,000 MGaero evaluations	3.02	3.28	2.54

sweep is now at $32°$. This demonstrates that the multifidelity strategy allows VSaero to be explored more efficiently, but also that this is likely to drive the code toward unrealistic designs due to the deficiencies of the drag calculations made with VSaero.

To follow this approach to its conclusion, the output of the VSaero search has been used to seed a genetic algorithm search using MGaero with five generations each of 100 members. To do this, the best 100 members of the final population of the VSaero search were used to form the initial population of the MGaero search; see line 4. Now, of course, the MGaero drag has improved, since this is being acted on directly for the first time. Even so, the result is still worse than the result of using just the Tadpole search, although it is 15% better than the initial design. It is clear that the VSaero part of this multifidelity search has not helped the process of producing low MGaero drags, although, as will be seen shortly, the final result is still better than can be obtained by using MGaero alone.

Given that the last stage of the previous three-stage search began with a worse MGaero drag than the initial design, a two-stage Tadpole/MGaero approach is considered next. In this case, the best 100 members of the final population of the 10,000-step Tadpole search were used to form the initial population of the MGaero search, which consisted of 10 generations of 100 members each; see line 5 of Table 12.7. This approach gives the best design overall, although the improvements over just using the Tadpole search are slight. Nonetheless, the results are considerably better than those obtained from using the same effort on the MGaero search alone (recall that 10,000 Tadpole evaluations cost less than

Table 12.8   Final design parameters, constraint values and objective function values for the best designs produced by the combined three-stage Tadpole/VSaero/MGaero search, Tadpole/MGaero two-stage search and MGaero-only search (Torenbeek weight calculations)

Lower Limit	3 Stage	2 Stage	1 Stage	Upper Limit	Quantity (units)
100	172.9	165.8	168.0	250	Wing area (m^2)
6	8.97	9.85	9.07	12	Aspect ratio
0.2	0.281	0.217	0.314	0.45	Kink position
25	33.7	31.6	27.1	45	Sweep angle (degrees)
0.4	0.469	0.459	0.495	0.7	Inboard taper ratio
0.2	0.217	0.202	0.273	0.6	Outboard taper ratio
0.1	0.155	0.174	0.177	0.18	Root t/c
0.06	0.070	0.068	0.077	0.14	Kink t/c
0.06	0.060	0.078	0.125	0.14	Tip t/c
4.0	4.0	4.5	4.8	5.0	Tip washout (degrees)
0.65	0.80	0.66	0.75	0.84	Kink washout fraction
	20,070	19,433	19,140	20,394	Wing weight (kg)
40.0	42.7	40.8	45.5		Wing volume (m^3)
	5.23	5.31	3.94	5.4	Pitch-up margin
2.5	3.73	3.84	3.34		Undercarriage bay length
	2.81	2.75	3.02		D/q (m^2) – Tadpole
	2.77	2.87	3.28		D/q (m^2) – VSaero
	2.43	2.37	2.54		D/q (m^2) – MGaero

a single MGaero evaluation); see line 6 of the table, although this final approach also significantly improves on the initial design. Overall, the design from the Tadpole/MGaero two-stage search generates a 17% reduction in drag as predicted by MGaero (12% as predicted by Tadpole but only 8% by VSaero).

In all it is clear that searches using more approximate methods should be carried out before transferring effort to the more complex methods and that multifidelity strategies are capable of delivering good results provided the methods are in reasonable agreement with each other as regards the locations of good designs. It should also be recalled that, despite the fact that the VSaero searches do not produce helpful designs for seeding searches by MGaero, the code is also being used within the MGaero process to help decide local camber variation, where the induced drag results from VSaero are utilized alongside two-dimensional wave drag data from MGaero. These are clearly of value in this role given the good correlation between Tadpole and MGaero they afford and also the low drags of the final designs.

To complete this example, the parameters for the design produced by the three-stage combined Tadpole/VSaero/MGaero search are detailed in Table 12.8, along with those for the best Tadpole/MGaero and direct MGaero searches. Figures 12.6–12.8 show the equivalent

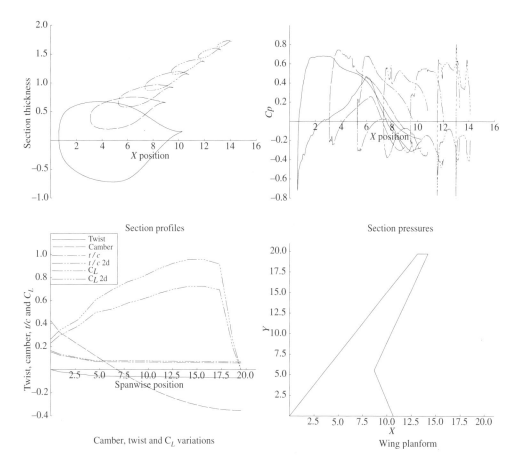

Figure 12.6   Geometry, section and pressure plots for the best design produced by the combined three-stage Tadpole/VSaero/MGaero search (Torenbeek weight calculation).

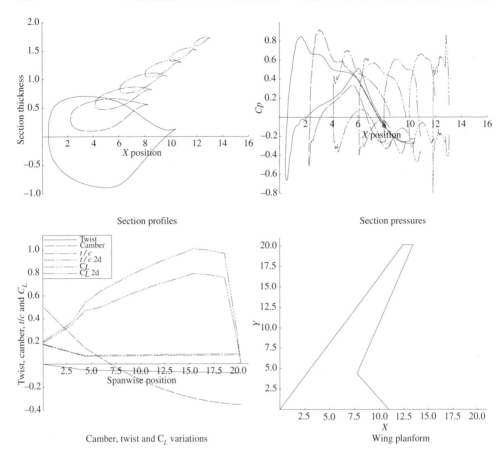

Figure 12.7   Geometry, section and pressure plots for the best design produced by the Tadpole/MGaero two-stage search (Torenbeek weight calculation).

geometries, wing sections and pressure profiles for these three results. It may be seen from the table that significant improvements in the drags may be achieved without changes to the wing area, aspect ratio, kink position or sweep, even when using Tadpole drag calculations, which do not use detailed section geometry. It is also clear from the Table that the best design is limited by the wing volume constraint (in fact, during the MGaero searches, a number of designs with even lower drags were seen that violated this or other members of the constraint list). The optimized designs also show some common features:

1. the chord inboard of the kink is significantly increased while that at the tip is reduced, that is, the wings are much more highly tapered;

2. the inboard sections are de-cambered to prevent shocks occurring on these enlarged sections;

3. the pressure plots show small shocks outboard combined with increased leading edge suction profiles inboard;

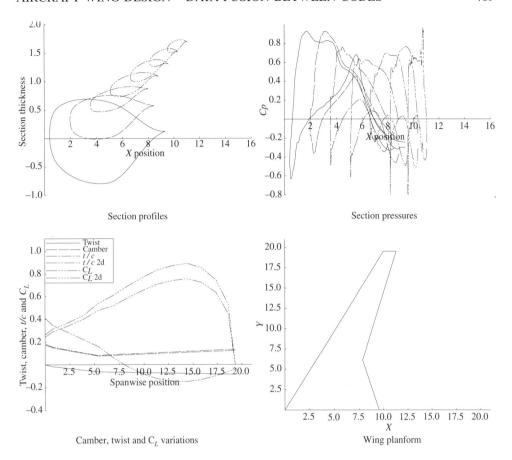

Figure 12.8 Geometry, section and pressure plots for the best design produced by the MGaero only search (Torenbeek weight calculation).

4. the wings have much increased section lift coefficients outboard; and

5. the wings tend to have increased sweep angles.

In doing this, the direct multifidelity search process has reduced the drag by some 17%. It has improved on the use of a simple search applied to MGaero alone by some 7%, because, despite the differences between the codes used, sufficient improvement was achieved by the empirical method that the CFD-based search was able to capitalize on the better starting position. It should also be noted, however, that the constraints and design objective of the problem presented here are much simplified and that more realistic problems, including fuselage modeling and finite element structural models, would need to be tackled before confidence could be expressed in these results.

It should also be noted that, despite the improvement over a direct search, two fundamental problems remain with this multistage approach: (1) unless the methods agree very well, there is a danger that the results coming from one may mislead the next – such differences are apparent in all the results in Tables 12.2, 12.5 and 12.8, and (2) the final direct search of the Euler CFD code is still too computationally expensive for routine use. Clearly what would be preferable is some way to speed up the searches and also a more sophisticated approach to integrating these sources of design information.

12.5   Response Surface Modeling

The searches reported in the previous sections simply involved applying optimization methods directly to the analysis codes, using a Genetic Algorithm for wide-ranging searches and then downhill methods for local improvement (if they can be afforded – recall that downhill methods cannot normally make great use of parallel computing environments), or by using a sequence of searches to the different fidelity codes, passing results between them. In all cases the final searches involved many calls to the Euler code. Even with parallel computing, searches on this code are still very expensive to carry out. Consequently, many workers in this field advocate the use of Response Surface Models (RSMs) where surrogate metamodels are produced by curve fitting techniques to samples of the expensive data (Jones et al. 1998; Myers and Montgomery 1995).

As has already been set out in Chapters 3, 5 and 7, the basic RSM process involves selecting a limited number of points at which the expensive code will be run, normally using formal DoE (DoE) methods (McKay et al. 1979; Mead 1988). Then, when these designs have been analyzed, usually in parallel, a response surface (curve fit) is constructed through or near the data. Design optimization is then carried out on this surface to locate new and interesting combinations of the design variables, which may then, in turn, be fed back into the full code. This data can then be used to update the model and the whole process repeated until the user either runs out of effort, some form of convergence is achieved or sufficiently improved designs are reached. This process has already been illustrated in Figure 3.21.

12.5.1   Design of Experiment Methods and Kriging

It is no surprise that there are a number of variations and refinements that may be applied to the basic RSM approach – the literature offers many possible alternatives. Here, by way of example, an $LP\tau$ DoE sequence (Statnikov and Matusov 1995) is used to generate the initial set of points and a Kriging or Gaussian Process model applied to build the RSM.

Most DoE methods seek to efficiently sample the entire design space by building an array of possible designs with relatively even but not constant spacing between the points. Notice that this is in contrast to a pure random spacing, which would result in some groups of points occurring in clumps while there were other regions with relatively sparse data – this might be desirable if there were no correlation between the responses at points, however close they were to each other (i.e., the process resembled white noise) but this is highly unlikely in engineering design problems. A particular advantage of the $LP\tau$ approach is that not only does it give good coverage for engineering purposes, but that it also allows additional points to be added to a design without the need to reposition existing points – this can be useful if the designer is unsure how many points will be needed before commencing work. Then, if the initial build of the RSM is found to be inadequate, a new set of points can be inserted without invalidating the statistical character of the experiment (similarly, if for some reason the original experiment cannot be completed, the sequence available at any intermediate stage will still have useful coverage).

Having built up an array of data points from which a surface can be constructed, the user's next major decision is whether or not to regress (as opposed to interpolate) the data. Regression of course allows for noise in the data (which may occur because of convergence errors in CFD, for example) but also allows the response surface to model rapidly changing data without excessive curvature. Consider trying to fit a polynomial to a function like a square wave – it is well known that such curve fits typically have difficulty modeling the steps in the function without overshoot if they are forced to interpolate the data points – in such circumstances, a degree of regression may give a more usable model, particularly when optimization searches are then applied to the problem, as functions using regression usually have fewer basins of attraction and are thus easier to search.

The most obvious forms of regression are those using least squares (often quadratic) polynomials. These are commonly used in the statistics community and much of the RSM literature is based on them. They are not good, however, at modeling complex surfaces that have many local basins and bulges in them. Here, a Kriging approach is used instead (Jones et al. 1998), since this allows the user to control the amount of regression as well as providing a theoretically sound basis for judging the degree of curvature needed to adequately model the user's data. Additionally, Kriging provides measures of probable errors in the model being built that can be used when assessing where to place any further design points. It also allows for the relative importance of variables to be judged as the hyperparameters produced may be used to rank the variables if the inputs are normalized to a unit range before the Krig is tuned.

This basic approach can be used to model any response quantity, including constraints. Here, since the constraints may be rapidly computed, there is no need to apply the RSM process to them at all (recall that empirical structural and weight models are used for the calculations underlying the constraints). Thus, a Krig is built just for the predicted drag.

Kriging is not a panacea for all evils, however. It is commonly found that it is difficult to set up such models for more than 10–20 variables and also that the approach is numerically expensive if there are more than a few hundred data points, since the setup (hyperparameter tuning) process requires the repetitive LU decomposition of the correlation matrix, which has the same dimensions as the number of points used. Moreover, the number of such LU steps is strongly dependent on the number of variables, since the likelihood has to be searched over the hyperparameters linked to each variable and it is also commonly highly multimodal. The authors have found that it is difficult to deal with Krigs involving more than 20 variables and 1000 points.

12.5.2 Application of DoE and Kriging

To demonstrate basic RSM production, 250 points of an LPτ array have been applied to the example problem of Table 12.2, which has 11 variables, using the inexpensive Tadpole code and a Krig built using a Genetic Algorithm and gradient descent two-stage search of the concentrated likelihood function to tune the hyperparameters (see Table 12.9 – note that $Log_{10}(\theta)$ hyperparameter values less than -2 indicate variables with relatively little impact on the Krig – here, the dominant variables are therefore wing area, aspect ratio, sweep angle and inboard taper ratio). To demonstrate the accuracy of this model, 390 further random design points were also computed with Tadpole and then the results at these further points were predicted using the Krig. Figure 12.9 shows the correlation plot for this test data, and it may be seen that while some differences occur, the overall correlation coefficient is 0.991.

Table 12.9   Krig hyperparameters for Tadpole model with 11 variables produced from 250 LPτ data points (regularization constant $= 10^{-18}$)

Log$_{10}(\theta)$	p_h	Quantity
−0.593	1.573	Wing area
−1.428	1.826	Aspect ratio
−2.294	1.778	Kink position
−0.910	1.443	Sweep angle
−1.574	1.569	Inboard taper ratio
−2.006	1.596	Outboard taper ratio
−2.177	1.824	Root t/c
−2.320	2.000	Kink t/c
−2.472	1.352	Tip t/c
−8.615	1.886	Tip washout
−2.761	1.468	Kink washout fraction

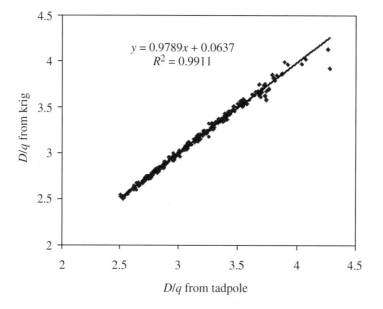

Figure 12.9 Correlation between 390 random Tadpole D/q calculations and those predicted by the Krig trained on a separate set of 250 LPτ calculations (Tadpole weight calculation).

This relatively good predictive capability is also indicated by a standardized cross-validated (SCV) residual test on the original data, where the mean SCV residual turns out to be 0.541 with just two of the 390 residuals being greater than three (values of less than one represent a good model, while those over three indicate poor correlations, i.e., outliers). Moreover, negligible regularization (regression) is needed to model the data.

Table 12.10   Krig hyperparameters for MGaero model with 11 variables produced from 250 LPτ data points (regularization constant $= 1.22 \times 10^{-2}$)

$\mathrm{Log}_{10}(\theta)$	p_h	Quantity
−0.866	1.718	Wing area
−1.698	1.983	Aspect ratio
−8.228	1.513	Kink position
−2.108	1.001	Sweep angle
−0.282	1.004	Inboard taper ratio
−3.782	2.000	Outboard taper ratio
−1.848	2.000	Root t/c
−0.967	1.830	Kink t/c
−2.854	1.012	Tip t/c
−3.293	1.749	Tip washout
−1.668	2.000	Kink washout fraction

These results show that it is possible to build Krigs successfully with this many dimensions using 250 data points. This is hardly necessary for Tadpole given its run time, however, which is barely more than that for using the Krig itself. Moreover, tuning the hyperparameters for this model takes much longer than the generation of the original Tadpole data. The real use of the approach arises when attempting to model expensive data coming from the CFD code itself. This process is not so successful since the CFD data is intrinsically much less smooth and contains significant noise. Table 12.10 and Figure 12.10 show an equivalent set of results for a Krig built on CFD data, which yields a correlation coefficient of only 0.4903 – it is clear from the figure that there is much more scatter in these results, fortunately, mostly for the higher drag data. With this model the mean SCV is 0.929 and now 10 residuals are greater than three, again indicating that this data is harder to model with many more outliers. Significant regularization is also required. Note further that the relative significances of the variables have changed between the two tables. In Table 12.10, the sweep angle is seen to be much less important than in Table 12.9 while two of the thickness-to-chord ratios and the kink washout fraction are more important. In both tables, the wing area, aspect ratio and inboard taper ratio remain significant.

Of course, the real test for the Krig of the MGaero data is whether it can be successfully used to optimize the wing design as predicted by the CFD code. So, next a two-stage GA and gradient descent search has been carried out on the Krig RSM and the resulting design evaluated with the CFD code (and Tadpole for comparison); see column two of Table 12.11. This design has significantly worse drag than either of those in Table 12.5 using either Tadpole or MGaero to predict the drag, despite the fact that the Krig predicts much lower drags for this design – clearly, the Krig is not modeling the data well in this location. This is quite normal when dealing with problems in high dimensions as the initial DoE cannot be expected to give sufficient coverage in all possible areas of interest. This is why it is almost always necessary to update the data set used to build the Krig, by adding new points in areas where good designs are being predicted (there are other schemes for positioning these added points such as the use of expected improvement criteria, but these are not discussed in this case study; see instead the study on turbine guide vane design and also Section 7.3).

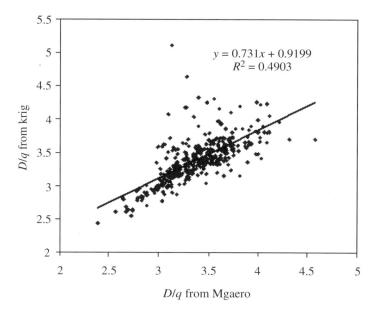

Figure 12.10 Correlation between 390 random MGaero D/q calculations and those predicted by the Krig trained on a separate set of 250 LPτ calculations (Tadpole weight calculation).

Table 12.11 Parameters and objective function values for the best design produced by the search on the initial MGaero Krig and after 10 updates (Tadpole weights)

Lower Limit	Value	After Updates	Upper Limit	Quantity (units)
100	169.7	158.7	250	Wing area (m^2)
6	8.135	9.781	12	Aspect ratio
0.2	0.411	0.367	0.45	Kink position
25	30.74	30.18	45	Sweep angle (degrees)
0.4	0.471	0.467	0.7	Inboard taper ratio
0.2	0.309	0.300	0.6	Outboard taper ratio
0.1	0.117	0.133	0.18	Root t/c
0.06	0.0732	0.106	0.14	Kink t/c
0.06	0.103	0.0627	0.14	Tip t/c
4.0	4.005	4.838	5.0	Tip washout (degrees)
0.65	0.849	0.679	0.84	Kink washout fraction
	13,758	13,764	13,766	Wing weight (kg)
40.0	40.0	40.0		Wing volume (m^3)
	4.14	4.99	5.4	Pitch-up margin
2.5	3.30	3.05		Undercarriage bay length (m)
	2.429	2.316		D/q (m^2) – Krig
	3.046	2.879		D/q (m^2) – Tadpole
	2.777	2.543		D/q (m^2) – MGaero

Following the update strategy of Figure 3.21, each new design point is then added to the set used to produce the Krig, and the hyperparameters retuned before it is again used to try and find an improved design. This process can be repeated as many times as the designer wants or until some form of convergence is achieved. Here, 10 such iterations are carried out to yield the result of column three in Table 12.11. This final design, although better than the initial design, fails to give D/q values as good as those achieved either by the direct search on the empirical Tadpole code, or on the Euler-based MGaero CFD code. Its performance is 0.7% worse than the best design achieved by Tadpole optimization (and using Tadpole predictions for comparison) and 0.6% worse than that from the direct CFD optimization. This result indicates that although the RSM approach commonly yields improved results these may well not be as good as direct searches on the underlying codes. This can occur even when suitable steps are taken to update the surface as part of the process and represents a fundamental limitation of metamodeling. The approach is, however, much faster than the direct CFD search since it requires nearly six times fewer CFD evaluations.

Having shown what may be achieved with simple optimization, direct multifidelity searches and the direct response surface approach, attention is lastly turned to fusion of the information coming from Tadpole and MGaero using response surfaces.

12.6 Data Fusion

Direct fusion of the data coming from the various aerodynamic analysis codes may be achieved if, instead of using the RSM to model an expensive CFD code directly, it is used to capture the differences between this and a cheaper (typically empirical) alternative. The RSM then serves as an on-line correction service to the empirical code so that, when designs are studied where it is less accurate, the corrections derived from full three-dimensional CFD are automatically included (see also Section 6.1). To begin this process, data coming from the DoE run on MGaero is taken and an equivalent set of drag results computed for each point using Tadpole. The differences between the two are then used to form the Krig. Then, when searches are carried out and new predictions are needed, these are calculated by calling *both* Tadpole and the Krig and summing their contributions; see Figure 12.11.

Again the Krig is built using a Genetic Algorithm and gradient descent two stage search of the concentrated likelihood function to tune the hyperparameters (see Table 12.12 – note that now the $Log_{10}(\theta)$ hyperparameters indicate the dominant *differences* in the effects of variables and the sweep angle is the critical one here, that is, it is in the impact of sweep that the two codes differ most). Note that the simple and explicit fusion model used here allows such interpretations to be made very easily.

Again, this model may be tested by its ability to predict unseen data. Figure 12.12 shows such a plot where the same 390 results used earlier are compared with the drag values coming from direct calls to MGaero, and it may be seen that while significant differences do still occur the overall correlation coefficient is now 0.7319 as compared to 0.4903 for the Krig based solely on the MGaero data. This improved predictive capability arises despite the mean SCV residual of the Krig being 1.224 with 22 of the 390 residuals being greater than three. This is because the Krig is now not used alone, but as a corrector to an already well set up empirical method, that is, a combination of black-box and physics-based estimators is being used and so deficiencies in the Krig are compensated for by Tadpole and vice versa. The correlation coefficient measures the effectiveness of this combined process.

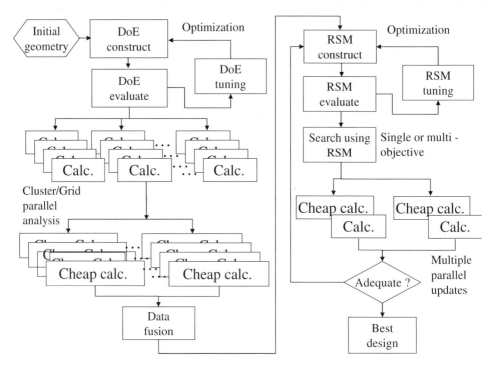

Figure 12.11 Improved response surface method calculation, allowing for data fusion with parallel calculations, DoE and RSM tuning plus updates.

Table 12.12 Krig hyperparameters for fused Tadpole/MGaero model with 11 variables produced from 250 LPτ data points (regularization constant $= 2 \times 10^{-2}$)

Log$_{10}(\theta)$	p_h	Quantity
−1.875	2.000	Wing area
−3.795	1.999	Aspect ratio
−8.414	1.858	Kink position
2.844	1.807	Sweep angle
−1.381	1.003	Inboard taper ratio
−7.707	1.058	Outboard taper ratio
−9.603	1.033	Root t/c
−2.811	1.726	Kink t/c
−9.746	1.064	Tip t/c
−7.744	1.019	Tip washout
−2.388	1.997	Kink washout fraction

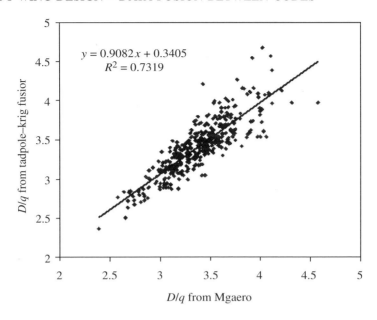

Figure 12.12 Correlation between 390 random MGaero D/q calculations and those pre-dicted by the Tadpole–Krig fusion trained on a separate set of 250 LPτ calculations (Tadpole weight calculation).

It is, of course, possible to build more sophisticated data-fusion models in this way. Here, the Krig is just modeling the differences between the two analysis methods. This will work well if there is good correlation between the models and, in particular, if changes of a given magnitude in any design variable tend to lead to similar shifts in the behaviors within the two analysis models. Sometimes, it can be more effective to use the response surface to model the *ratio* between the two codes rather than their differences. More complex relationships can be posed and again used with the Krig: in the end, that used will be a matter of personal preference, experience and the degree to which the underlying functions are understood. This is also an area of ongoing research and a number of workers have proposed schemes whereby the relationship between the analysis codes considered is modeled at the same time as the correction surface is built. One way of doing this is to treat the outputs of the cheap code as an additional input to a Krig of the expensive code and then to rely on the resulting metamodel to relate correctly all the terms – in such an approach, we should expect the "variable" that is in fact the output of the simpler model to strongly correlate to the expensive code and thus to take a dominant role in the resulting metamodel; see, for example, the work of El-Beltagy (2004).

12.6.1 Optimization Using the Fusion Model

Having produced a fusion model, it can then be used to try and optimize the design being studied. Table 12.13 shows the results from using the model without any further updates and if 10 updates are added following the strategy already outlined.

Now, the improvement in MGaero drag before updates is almost as good as that from the direct search on the code while after updates it is 0.3% better. Moreover, after the updates,

Table 12.13   Parameters and objective function values for the designs produced from the initial MGaero/Tadpole difference Krig and after 10 updates (Tadpole weights)

Lower Limit	Value	After Updates	Upper Limit	Quantity (units)
100	153.3	156.6	250	Wing area (m^2)
6	10.56	10.25	12	Aspect ratio
0.2	0.297	0.436	0.45	Kink position
25	27.04	31.2	45	Sweep angle (degrees)
0.4	0.424	0.438	0.7	Inboard taper ratio
0.2	0.201	0.200	0.6	Outboard taper ratio
0.1	0.129	0.118	0.18	Root t/c
0.06	0.106	0.124	0.14	Kink t/c
0.06	0.0820	0.0667	0.14	Tip t/c
4.0	4.054	4.601	5.0	Tip washout (degrees)
0.65	0.655	0.717	0.84	Kink washout fraction
	13,633	13,142	13,766	Wing weight (kg)
40.0	40.1	40.0		Wing volume (m^3)
	4.61	5.4	5.4	Pitch-up margin
2.5	3.42	2.96		Undercarriage bay length (m)
	2.262	2.012		D/q (m^2) – Krig
	2.875	2.817		D/q (m^2) – Tadpole
	2.531	2.515		D/q (m^2) – MGaero

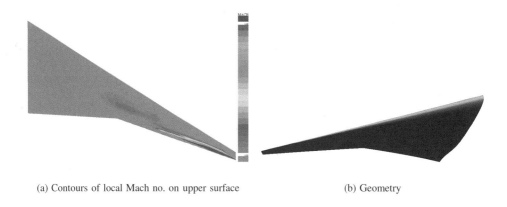

(a) Contours of local Mach no. on upper surface   (b) Geometry

Figure 12.13   Geometry of final design produced by the search on the refined MGaero/Tadpole difference Krig with 10 updates (plan view shows upper surface supersonic Mach contours, Tadpole weight calculation).

the Tadpole drag is over 1% better than for the direct search on the Tadpole code at the same time. The final design is illustrated in Figure 12.13. This optimization process again uses around one-sixth of the computing effort of the direct search on the CFD code, while achieving better results.

Table 12.14 Summary of D/q results from previous tables

	Tadpole	% Table 12.2	MGaero	% Table 12.2	Krig	Notes
Table 12.2	3.145	100.0	2.922	100.0		Initial design
Table 12.5	2.853	90.7	2.555	87.4		Tadpole search
Table 12.5	2.998	95.3	2.524	86.4		MGaero search
Table 12.11	3.046	96.9	2.777	95.0	2.429	Krig search
Table 12.11	2.879	91.5	2.543	87.0	2.316	Krig + updates
Table 12.13	2.875	91.4	2.531	86.6	2.262	Fusion Krig search
Table 12.13	2.817	89.6	2.515	86.1	2.012	Fusion Krig + updates
Table 12.15	2.831	90.0	2.471	84.6	2.260	Best design ever

It is interesting to compare the initial and final designs: the initial design has slightly greater area, less sweep and a much higher percentage of its lift generated by the outboard sections. Notice also that in Figure 12.1 there is a weakly supersonic flow over a large region of the upper surface (although there is virtually no wave drag on this design), while in Figure 12.13, there is a much more localized but stronger shock near the wing tip. The impact of this shock on the overall drag is more than compensated for, however, by the large, low-drag inboard section of wing – despite the shock, only 1.1% of the total drag on the design is wave drag. Note also that the optimized design is at the limits of two of the constraints: wing volume and pitch-up margin. This is a common feature in optimization processes and means that great care must be placed in setting the values of such hard constraints. The various drag values from all the searches based on the Tadpole weight model are summarized in Table 12.14 for ease of comparison.

Finally, it should be remembered that none of the designs produced using direct, simple RSM or fusion-based RSM searches can be guaranteed to be at the optimum of the problem being tackled: if different search strategies or starting points are employed, different optima can be located. Table 12.15 shows one such design that is better than all those detailed so far from the MGaero point of view. This design was in fact produced without using the Tadpole-based fusion strategy although locating it was something of a fluke: it was found when testing the system described with all 640 MGaero evaluations to build RSMs. Using bigger data sets is clearly an advantage when carrying out data modeling. It turns out to be 1.5% better than the design given by the fusion-based approach in terms of MGaero predictions (but note that it is worse in terms of Tadpole drag calculations, as might be expected).

12.7 Conclusions

In this case study, the central difficulties posed by the inevitably limited computing resources deployable on any design problem have been discussed. It is clear that the belief that faster computers will make this problem go away is a mirage: as faster computers emerge, designers inevitably adopt more sophisticated models and see no reason why search engines in optimization systems should use outmoded design methods when looking for better results. This requirement for the most sophisticated design methods to be applied by evaluation-hungry search engines leads naturally to the multifidelity or "zoom" optimization problem

Table 12.15   Final design parameters, constraint values and objective function values for the lowest MGaero drag design ever found for the problem being searched (Tadpole weight calculation)

Lower Limit	Value	Upper Limit	Quantity (units)
100	147.6	250	Wing area (m^2)
6	11.20	12	Aspect ratio
0.2	0.436	0.45	Kink position
25	26.3	45	Sweep angle (degrees)
0.4	0.491	0.7	Inboard taper ratio
0.2	0.215	0.6	Outboard taper ratio
0.1	0.176	0.18	Root t/c
0.06	0.0941	0.14	Kink t/c
0.06	0.0605	0.14	Tip t/c
4.0	4.992	5.0	Tip washout (degrees)
0.65	0.849	0.84	Kink washout fraction
	13,438	13,766	Wing weight (kg)
40.0	40.1		Wing volume (m^3)
	4.75	5.4	Pitch-up margin
2.5	2.58		Undercarriage bay length (m)
	2.260		D/q (m^2) – Krig
	2.831		D/q (m^2) – Tadpole
	2.471		D/q (m^2) – MGaero

where few expensive design calculations are augmented by many cheaper, but less accurate, methods.

Here, a wing design problem has been studied making use of three levels of aerodynamic analysis. It is shown that even though these methods differ in computational cost by four orders of magnitude and have quite deceptive relationships to each other it is possible to make effective use of a multifidelity strategy to achieve results that cannot be accomplished by direct searches at any one level in reasonable time.

A number of methods for carrying out searches are described: direct optimization of the user's analysis codes, multifidelity combinations of these codes, searches of response surfaces derived from the codes and search of a response surface derived from two related but different fidelity codes. The latter multifidelity or "fusion-based" response surface approach seeks to combine the speed of fast empirical codes with the precision of full three-dimensional CFD solvers. This requires an integrated system of analysis that allows multiple codes to be used alongside each other: this is provided by the Southampton wing design system which allows use of various levels of aerodynamic analysis and also a sophisticated weight estimation model. The fusion-based approach outperforms direct search of the Euler CFD code at considerably reduced cost while also being more accurate than a simple response surface method using only data from the CFD codes.

13

Turbine Blade Design (I) – Guide-vane SKE Control

This case study describes the application of Design of Experiment (DoE) techniques combined with Response Surface Models (RSMs) and optimizers from the Options[1] toolkit to help design a turbine guide vane using the Rolls-Royce *FAITH* design system (Shahpar 2001) and *sz02* CFD code (Moore and Moore 1987). These codes allow design geometries to be modified and full three-dimensional Navier–Stokes solutions with pressure correction plus upwinding for high-speed transonic flows to be computed. They have been applied to a geometry supplied by Rolls-Royce and using Rolls-Royce provided meshes and parameters.

In the study carried out here, a single turbine guide vane has been analyzed using a Linux variant of the codes. Using the 2 GHz processors available, such calculations typically take 2 h for each design geometry considered. All the work carried out addresses the Secondary Kinetic Energy (SKE) in the flows as the objective and deviations of the flow capacity (*Capac*) as a possible constraint. The variables considered are the circumferential lean (Y_{cen}), axial lean (X_{cen}) and skew (*Skew*) of the vane at each of seven radial positions, giving, at most, a 21-variable problem.

13.1 Design of Experiment Techniques, Response Surface Models and Model Refinement

As has been set out in Chapters 3, 5 and 7 and the previous case study, response surface methods may be used to provide curve fits to data coming from computer codes that can be used in lieu of these codes when carrying out design tasks. Usually, these methods involve first the production of a set of data points spread through the space of interest (often using a DoE formalism) to build an initial metamodel, followed by a series of refinement steps

---

[1] http://www.soton.ac.uk/~ajk/options/welcome.html

Computational Approaches for Aerospace Design: The Pursuit of Excellence. A. J. Keane and P. B. Nair
© 2005 John Wiley & Sons, Ltd

to improve the quality of the model locally in regions that show promise, as discussed previously; see Figure 3.21.

13.1.1 DoE Techniques

DoE techniques are used when deciding where to place trial calculations in the absence of prior knowledge, that is, they allow a systematic placing of trials so as to maximize the information returned from the tests. There are various methods that can be used in designing such experimental sequences: those commonly used are random sequences, Latin hypercube sequences, LPτ sequences and geometric sequences (such as face-centered cubic designs). Throughout the work reported here, LPτ sequences have been used as experience has shown that they work well when the quantities of data available are extremely limited (Statnikov and Matusov 1995). Figure 13.1 illustrates (for an unconnected case – the beam design problem used in Chapter 1) a typical random sequence, a Latin hypercube sequence and a LPτ sequence in two dimensions, in each case showing 100 points. Note that the random sequence tends to include regions with few points and also clumps of points while the Latin hypercube sequence and LPτ sequence give increasingly more even coverage. The smoother the problem being considered, the more appropriate even coverage becomes – random designs are only efficient when the function being modeled may contain discontinuities and features of widely varying length scales. A further consideration when using DoE methods is the choice of the number of points to include in any experiment design. Here, two problems are considered, one of seven variables and one of 21. In both cases, experiments of 25, 50 and 100 points have been used when constructing the initial models (recall that each design requires 2 h to analyze, although these runs can be carried out in parallel – 10 processors were used here).

13.1.2 Response Surface Models

As has already been set out, RSMs are used to construct metamodels (curve fits) given previously computed data. Again, there are many ways of carrying out this process, stretching over many years of research (Franke and Neilson 1980; Shepard 1968), including polynomial methods, splines, radial basis functions (RBFs) of various types, Kriging, Stochastic Process Models (SPMs), Neural Networks and Support Vector Machines (SVMs), again of various types, see Chapter 5. In this case study, RBFs using cubic splines and Kriging have been compared.

It should be noted that the construction of the Kriging models requires the selection of hyperparameters to tune the model – see Section 5.6. For the Krigs used here, two are required for each variable plus an additional one to deal with any noise in the function via a regression mechanism (i.e., for the 21-variable problem, 43 are needed). These have been found via a five-stage hybrid optimization search across the hyperparameter space, as these can be highly multimodal. The first stage consisted of a traditional Simplex search to gain initial values for the parameters. This was then followed by a Genetic Algorithm (GA) search using a population of 200 designs over 10 generations. The best resulting design was then refined using the Simplex method again. The final design from this process was then used to initiate a second GA search, again with populations of 200 over 10 generations followed finally by a third Simplex search. Such searches for the 43-parameter model using 100 data points take roughly 30 min, that is, a significant fraction of the effort needed to analyze any given design so that five new designs can be added to the RBF model in the

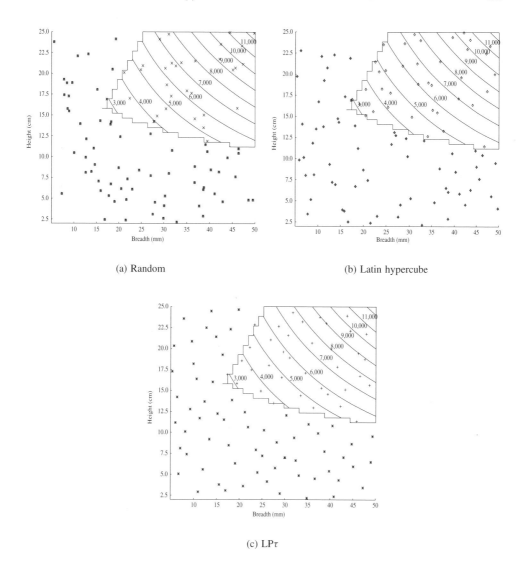

(a) Random

(b) Latin hypercube

(c) LPτ

Figure 13.1 Various DoE designs in two dimensions (for the example beam problem of Chapter 1).

same time that four are added to the Kriging model. This time is very strongly linked to the number of data points in use as well as to the number of parameters. An equivalent search on the seven-variable problem with 25 data points takes only a few tens of seconds. Even so, experience has shown that this very forceful search over the hyperparameters is warranted. If poor values are used, the resulting Krig can be quite misleading and this can waste all the effort placed in computing added data points. As a general rule, we believe that hyperparameter searches taking up to 30% of the cost of each of the points being modeled can usually be justified.

13.1.3   Model Refinement

Having carried out an initial series of experiments and built a model of the data, it may be that further refinement of the model is desired. Typically, this occurs because optimization searches on the model suggest new configurations that show promise but these are in regions where confidence in the model is not high because the designs of interest lie some way from the data points used to construct the initial model. Various strategies can be used to update models but all tend to be based on two, often competing, aspects. First, there is a natural desire to add information to the model in regions where the best designs appear to be (exploitation of the metamodel), but secondly, there is a desire to improve the accuracy of the model in regions where it is thought the current accuracy is sufficiently low that estimates made there may be falsely obscuring promising designs (exploration of the design space). Given estimates of model uncertainty, it is possible to balance these two needs by constructing a combined measure of quality and uncertainty. This measure is normally termed the expected improvement operator $I(\mathbf{x})$, see section 7.3.1. In the work reported here, the expected improvement operator is applied to the Krigs as such models naturally provide the required uncertainties. Adding points at the most promising location is also tested here, for both the RBF and Kriging models.

It should be noted that the location of any model refinement point requires an optimization search across the model constructed. Such models can be highly multimodal, as will be illustrated later. Here, these searches have been conducted using a four-stage hybrid search. The first stage consisted of a GA search using a population of 200 designs over ten generations. The best resulting design was then refined using a traditional Simplex method. The final design from this process was then used to initiate a second GA search, again with populations of 200 over 10 generations followed finally by a second Simplex search. An extensive search is warranted because the RSM is cheap to evaluate and one should therefore not stint on trying to gain the best information from it – this becomes particularly true as the dimensions of the model and size of the dataset rise – model construction is then so expensive that only forceful searches of the RSM should be contemplated.

13.2   Initial Design

The initial design considered here is illustrated in Plate 2, which shows the geometry considered and the SKE in the exit plane of the slot behind the vane. As can be seen from the figure, a considerable degree of swirl occurs in the flow – the total SKE is here 0.007376. Another important aspect of the design process used by Rolls-Royce is the desire to decouple the effects of changes to one stage in the turbine from those in subsequent stages. To do this, attempts are made to hold the capacity of each stage fixed once sufficient design effort has been expended on the overall system. Thereafter, work to improve individual stages seeks to work at fixed flow capacity. Here, this leads to an equality constraint on the capacity at its initial value of 75.75, which while not completely rigid should be held to as far as possible.

13.3   Seven-variable Trials without Capacity Constraint

For the seven-variable trials, only the circumferential lean of the vane (Y_{cen}) was varied within limits of $\pm 10°$ while the capacity of the design (*Capac*) was not constrained but merely recorded for interest (as already noted, the current design process would need to be significantly changed if vane capacities were free to alter in this way). After some initial studies, three sets of trials were conducted. These were:

1. LPτ experiments followed by cubic spline RBF model construction followed by five model updates each located at the most optimal design point found in the model constructed;

2. LPτ experiments followed by Kriging model construction followed by five model updates each located at the most optimal design point found in the model constructed; and

3. LPτ experiments followed by Kriging model construction followed by five model updates each located at the design point indicated by the $I(\mathbf{x})$ operator when applied to the model constructed.

In each case, this process was carried out for an initial experiment with 25 designs; one with 50 designs; and one with 100 designs. The results from these trials are reported in Tables 13.1–3.

The results in the tables indicate that the Kriging model gives the best-quality results and, moreover, it is more consistent during the update process. The RBF model only helps improve the design when at least 50 points are available to construct the initial model and then only after five updates. The $I(\mathbf{x})$ update strategy with the Kriging model seems,

Table 13.1 Seven-variable trial results; SKE, no constraints (results in bold are the best for a given DoE size, that underlined is the best overall, fail indicates the *sz02* run failed to converge) – initial SKE = 0.007376

Model Type	Update Strategy	Parameter	No. of Updates	25 Points	50 Points	100 Points
RBF	Best SKE	SKE	0	0.003614	0.003614	0.003138
			1	0.005524	Fail	0.004366
			2	0.003582	0.003737	0.003635
			3	0.008577	0.005518	0.003214
			4	0.005277	0.004299	0.004182
			5	0.004929	0.003499	0.003298
Kriging	Best SKE	SKE	0	0.003614	0.003614	0.003138
			1	0.004654	0.003466	0.003497
			2	0.003863	0.003704	0.003493
			3	**0.003038**	0.004312	0.003228
			4	0.003830	0.005083	0.003339
			5	0.003069	0.003769	0.003546
Kriging	$I(\mathbf{x})$	SKE	0	0.003614	0.003614	0.003138
			1	0.004003	**0.003237**	0.003415
			2	0.003466	Fail	**0.003044**
			3	0.003752	0.004124	0.003307
			4	0.003301	0.007694	Fail
			5	0.004744	0.003446	0.00529

Table 13.2 Seven-variable trial results; Capacity, no constraints (results in bold are the best for a given DoE size, that underlined is the best overall, fail indicates the *sz02* run failed to converge) – initial *Capac* = 75.75

Model Type	Update Strategy	Parameter	No. of Updates	25 Points	50 Points	100 Points
RBF	Best SKE	*Capac*	0	76.3262	76.3262	76.1176
			1	76.643471	Fail	76.295685
			2	77.004295	76.026489	76.887245
			3	75.48291	75.904442	76.080421
			4	76.300644	76.062164	76.978493
			5	77.299232	76.444	76.630554
Kriging	Best SKE	*Capac*	0	76.3262	76.3262	76.1176
			1	75.8863	76.3580	76.270905
			2	76.6154	76.0889	76.989792
			3	**77.1084**	76.7170	76.319572
			4	77.1456	75.9280	76.256088
			5	76.9157	76.6727	76.495995
Kriging	*I*(**x**)	*Capac*	0	76.3262	76.3262	76.1176
			1	75.982887	**76.297646**	76.712936
			2	76.066071	Fail	**76.21106**
			3	75.988388	76.157265	77.305862
			4	76.004761	77.032295	Fail
			5	75.809647	75.519928	77.00325

Table 13.3 Seven-variable trial results; variations of quantities found in the initial 100-point experiment

Quantity	SKE	*Capac*
Mean	0.006215	76.1362
Minimum	0.003138	71.4044
Maximum	0.026167	77.5396
Standard deviation	0.002683	0.6713

however, to need at least 50 data points to give greater improvements during update than the straightforward approach of just adding points at the best found value of predicted SKE (i.e., at the best-predicted objective function point). Notice also that the initial 100 designs considered have an average SKE below that of the base design although with a lowered capacity and also the variations in these designs are quite wide, as might be expected given the significant changes to circumferential lean being explored.

Table 13.4 gives more detailed results of the best designs found from the 25-point trial with simple SKE-based updating. This compares the initial base design, that found after the initial test of 25 designs and the best found during model update. It can be seen that

Table 13.4 Detailed parameter variations (in degrees) and resulting response quantities for the designs found during the 25-trial experiment, Kriging model construction and updating using SKE values, for the seven-parameter problem

Quantity	Initial Design	Best after 25 DoE Points	Best after Five Added Points using Kriging
$Y_{cen}(1)$	0.0	6.2500	−1.67399
$Y_{cen}(2)$	0.0	3.7500	−5.89621
$Y_{cen}(3)$	0.0	−1.2500	−1.18315
$Y_{cen}(4)$	0.0	−8.7500	−8.54335
$Y_{cen}(5)$	0.0	8.7500	9.18559
$Y_{cen}(6)$	0.0	−3.7500	2.77289
$Y_{cen}(7)$	0.0	3.7500	3.83272
SKE	0.007376	0.003614	0.003038
Capac	75.75	76.33	77.12

the SKE value has been reduced from 0.007376, through 0.003614 to 0.003038 during this process, a reduction of some 59% while at the same time the capacity has increased from 75.75 to 77.12. The resulting designs have a significant 'S' shape due to the circumferential lean shifts at the center of the vane being in opposite directions.

A further output that can be gained from the Kriging model is the hyperparameter set found to best model the data. Of these, the first set of variable hyperparameters allows the importance of each quantity to be ranked, the second set of variable parameters allows the smoothness of the model in each direction to be estimated and the last the degree of regression needed to maximize the quality (likelihood) of the model. These 15 parameters are listed in Table 13.5 for the best design found noted in Table 13.4. They show that Y_{cen} at stations four and five are easily the most critical variables in the design, that these variables have smooth modeling parameters and that while there is noise in the problem this is not causing significant regression to be needed. Conversely, Y_{cen} values at stations 7, 2 and, in particular, 1, that is, near the tip and root, have almost no effect on the SKE values.

Lastly, to better understand the models constructed using the three approaches, contour plots of the SKE model predictions for variations in the four most significant design variables have been produced; see Figures 13.2–13.4. In each case, $Y_{cen}(4)$ has been plotted against $Y_{cen}(5)$ and $Y_{cen}(3)$ against $Y_{cen}(6)$, that is, the most significant pair and the next most significant pair of variables. These six graphs further indicate the inadequacy of the RBF model when constructed with so few data points, and also the relatively poorer model produced using the $I(\mathbf{x})$ update strategy in this case (the $Y_{cen}(3)$ versus $Y_{cen}(6)$ variation is rather poorly captured). The RBF model lacks the flexibility of the Krig and so does not adequately reflect the subtle changes in the data, while the Krigs built on the $I(\mathbf{x})$ data are less well conditioned and so have extreme curvature. Figures 13.3(a) and 13.4(a) do agree on the desirability of setting $Y_{cen}(4)$ low and $Y_{cen}(5)$ high to reduce the SKE, however. The small square marker on each plot indicates the location of the last update point.

Table 13.5   Hyperparameters for the final Kriging model found during the 25-trial experiment and updating using SKE values, for the seven-parameter problem (smoothness hyperparameters vary from 1–2 with 2 indicating maximum smoothness; values of Log(Regression) greater than −2.0 indicate significant noise in the data)

Hyperparameter	Symbol	Value
Log(Y_{cen}(1) sensitivity)	θ_1	−8.29705
Log(Y_{cen}(2) sensitivity)	θ_2	−6.01469
Log(Y_{cen}(3) sensitivity)	θ_3	−0.381093
Log(Y_{cen}(4) sensitivity)	θ_4	−0.631272E-02
Log(Y_{cen}(5) sensitivity)	θ_5	0.220536
Log(Y_{cen}(6) sensitivity)	θ_6	−0.496623
Log(Y_{cen}(7) sensitivity)	θ_7	−3.70932
Y_{cen}(1) smoothness	p_1	1.62156
Y_{cen}(2) smoothness	p_2	1.10752
Y_{cen}(3) smoothness	p_3	1.28887
Y_{cen}(4) smoothness	p_4	1.99966
Y_{cen}(5) smoothness	p_5	1.99965
Y_{cen}(6) smoothness	p_6	1.00325
Y_{cen}(7) smoothness	p_7	1.35997
Log(Regression)	Λ	−4.33116

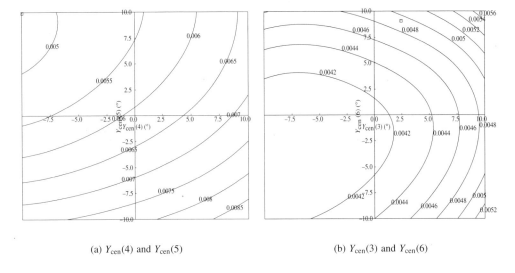

(a) Y_{cen}(4) and Y_{cen}(5)   (b) Y_{cen}(3) and Y_{cen}(6)

Figure 13.2   Variation of predicted SKE: cubic RBF model after 25 initial points and five updates (all other variables at the predicted optimum).

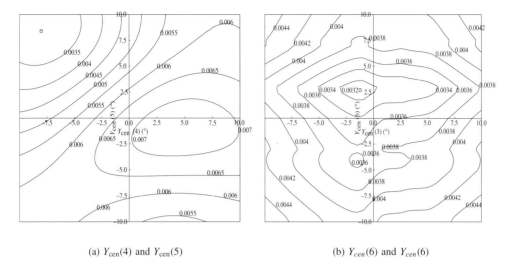

(a) $Y_{cen}(4)$ and $Y_{cen}(5)$ (b) $Y_{cen}(6)$ and $Y_{cen}(6)$

Figure 13.3   Variation of predicted SKE: Kriging model after 25 initial points and five updates (all other variables at the predicted optimum).

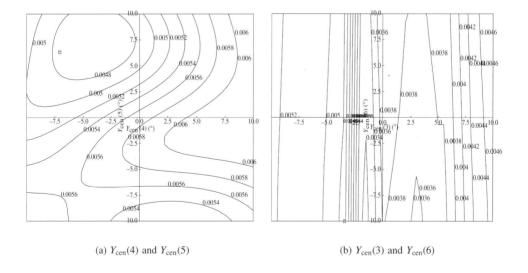

(a) $Y_{cen}(4)$ and $Y_{cen}(5)$ (b) $Y_{cen}(3)$ and $Y_{cen}(6)$

Figure 13.4   Variation of predicted SKE: Kriging model after 25 initial points and five $I(\mathbf{x})$ updates (all other variables at the predicted optimum).

13.4   Twenty-one-variable Trial with Capacity Constraint

The next set of trials considered the problem of modifying all 21 variables and, in addition, addressed the requirement of preventing significant changes in flow capacity. For these trials, a requirement was imposed that the capacity be with ± 0.5 of the original value. While a tolerance of this magnitude on capacity might well not be acceptable in practice, it is sufficient to ensure that final designs could be fine-tuned back to the original capacity using a modest line search with little difficulty (although at the cost of further computing effort). This level of capacity change is adopted to allow the searches used to find points at which to carry out updates – as will be seen later, the actual changes occurring in some of the resulting optimized designs are negligible.

An interesting aspect of dealing with a constraint like this, as well as the principal design objective, is the need to produce a separate response surface model to deal with the constraint data (since it is just as expensive to compute as the SKE). Here, for simplicity, a cubic RBF constraint model has been used throughout and this is found to be sufficient to achieve the desired goals provided suitable steps are taken in the way that this is invoked.

A similar set of nine test trials has been carried out for the 21-variable problem, but now with changes of only $\pm 5°$ at each station rather than the $\pm 10°$ used for the 7-variable model. The results are reported in Tables 13.6–8.

The results in these tables again indicate that the Kriging model gives the best-quality results and it remains more consistent during the update process. The RBF model now

Table 13.6   Twenty-one variable trial results; SKE, capacity constraint (results in italics invalidate the constraint requirement on capacity of ± 0.5 change, bold underlined result is the best overall design) – initial SKE $= 0.007376$

Model Type	Update Strategy	Parameter	No. of Updates	25 Points	50 Points	100 Points
RBF	Best SKE	SKE	0	0.005701	0.004877	0.004877
			1	*0.004504*	*0.006062*	*0.003428*
			2	*0.005174*	*0.003927*	*0.003759*
			3	*0.004703*	*0.005644*	0.003853
			4	*0.007236*	0.004135	0.002529
			5	*0.004837*	0.007384	0.002397
Kriging	Best SKE	SKE	0	0.005701	0.004877	0.004877
			1	*0.006375*	*0.003733*	0.003563
			2	*0.003635*	*0.005598*	*0.003497*
			3	*0.005572*	*0.005491*	0.002392
			4	*0.004869*	*0.005848*	**<u>0.002326</u>**
			5	*0.004145*	0.006053	0.00261
Kriging	$I(\mathbf{x})$	SKE	0	0.005701	0.004877	0.004877
			1	*0.006328*	*0.003732*	*0.002681*
			2	*0.004577*	*0.006884*	*0.003394*
			3	*0.004949*	*0.006384*	0.002529
			4	*0.003454*	*0.004005*	0.002491
			5	0.008378	*0.002664*	*0.002858*

Table 13.7 Twenty-one variable trial results, Capacity, capacity constraint (results in italics invalidate the constraint requirement on capacity of ± 0.5 change, bold underlined result is the best overall design) – initial $Capac = 75.75$

Model Type	Update Strategy	Parameter	No. of Updates	25 Points	50 Points	100 Points
RBF	Best SKE	*Capac*	0	75.8683	75.9588	75.9588
			1	*76.566109*	*69.607201*	*74.972198*
			2	*73.457344*	*72.222183*	*75.125381*
			3	*79.164261*	*71.440376*	76.020523
			4	*78.676636*	76.072243	75.432663
			5	*72.817093*	75.921814	75.538437
Kriging	Best SKE	*Capac*	0	75.8683	75.9588	75.9588
			1	*73.90773*	*73.547256*	75.735619
			2	*71.994026*	*73.743164*	74.749214
			3	*72.884346*	*74.061867*	75.704109
			4	*74.561142*	*74.160316*	**<u>75.360435</u>**
			5	*71.902092*	75.429993	75.335785
Kriging	$I(\mathbf{x})$	*Capac*	0	75.8683	75.9588	75.9588
			1	*72.980759*	*72.47361*	*75.192192*
			2	*73.995018*	*73.057594*	*75.019981*
			3	*74.188087*	75.208855	75.518761
			4	*73.696198*	*71.250114*	75.499619
			5	76.043182	*72.560226*	*75.185783*

Table 13.8 Twenty-one variable trial results; variations of quantities found in the initial 100 point DoE

Quantity	SKE	*Capac*
Mean	0.005661	75.4224
Minimum	0.002720	66.9723
Maximum	0.009384	84.5676
Standard deviation	0.001374	3.6064

really only helps improve the design when at least 100 points are available to construct the initial model. The $I(\mathbf{x})$ update strategy with the Kriging model also seems to need at least 100 data points to give improvements during update that are as good as the straightforward approach of just adding points at the best-predicted value of SKE, although now both strategies need 100 points to give consistent improvements. All strategies find it difficult to meet the constraint requirement, probably indicating that the RBF constraint model is not as precise as could be desired. Note also that the range of SKEs seen in the initial 100-point experiment is now less than that for the seven-variable trial. Conversely, that for capacity is much greater, despite the fact that changes of only $\pm 5°$ are being allowed in the leans and skew (in fact, only 56 of these designs lie within the initial limit of ± 0.5 change in capacity).

It should also be noted that if tighter limits are placed on the capacity changes then update 3 using the best SKE point of the Kriging model with an initial experiment of 100 points achieves an SKE value of 0.002392 and a capacity of 75.70, that is, a change of less than 0.1 in capacity with an SKE value within 0.00007 of the best found during this work. That at update 1 achieves an even smaller change in capacity of around 0.01 while still reducing SKE from its initial value of 0.007376 to a level of 0.003563. that is, a 52% improvement.

Table 13.9 gives more detailed results of the best designs found. This compares the initial base design, that found after the initial test of 100 designs, the best found during model update and the design with minimal shift in capacity. Plate 3 shows the geometry for this last design and the equivalent map of SKE in the outlet slot. These results come from the Kriging model being updated at the best-predicted SKE locations. It can be seen that the SKE value has been reduced from 0.007376, through 0.004877 to 0.002610 during this process, a reduction of some 65% while at the same time the capacity has reduced slightly from 75.75

Table 13.9 Detailed parameter variations (in degrees) and resulting response quantities for the designs found during the 100-trial experiment, Kriging model construction and updating using SKE values, for the 21-parameter problem

Quantity	Initial Design	Best Feasible after 100 DoE Points	Best Feasible after Five Added Points Using Kriging	Best Design with Minimal Capacity Change
$Y_{cen}(1)$	0.0	−0.0769058	1.68877	−4.82686
$Y_{cen}(2)$	0.0	3.66638	−2.06105	2.54933
$Y_{cen}(3)$	0.0	0.115324	3.68669	4.91514
$Y_{cen}(4)$	0.0	−2.84086	−4.80000	−4.98718
$Y_{cen}(5)$	0.0	2.24659	3.98071	3.51843
$Y_{cen}(6)$	0.0	−4.46719	2.19878	−4.69451
$Y_{cen}(7)$	0.0	3.72969	2.20122	−4.52516
$X_{cen}(1)$	0.0	4.14198	0.813918	−1.81099
$X_{cen}(2)$	0.0	−1.62100	3.18193	2.72442
$X_{cen}(3)$	0.0	4.56274	−2.36472	4.50769
$X_{cen}(4)$	0.0	2.50482	2.48803	4.94933
$X_{cen}(5)$	0.0	1.27997	−4.92528	−4.97009
$X_{cen}(6)$	0.0	−3.66580	1.13602	4.94933
$X_{cen}(7)$	0.0	−1.15237	−0.878757	2.72002
$Skew(1)$	0.0	−2.23247	−1.07424	−4.98962
$Skew(2)$	0.0	−4.05839	2.88815	4.40745
$Skew(3)$	0.0	−1.31992	3.36897	−4.77473
$Skew(4)$	0.0	−3.24696	−4.62223	−2.79732
$Skew(5)$	0.0	3.98602	0.899634	4.63272
$Skew(6)$	0.0	4.66425	−0.799024	1.90769
$Skew(7)$	0.0	−4.20391	3.17228	4.06642
SKE	0.007376	0.004877	0.002326	0.003563
Capac	75.75	75.96	75.36	75.74

to 75.34. The minimal capacity change design yields an SKE improvement of 52% while holding capacity essentially fixed. The resulting designs again have a significant "S" shape because of the circumferential lean shifts at the center being in opposite directions although the overall shape changes are somewhat complex given the 21 variations involved. It should be noted, of course, that such a design is rather impractical since the need to provide internal cooling flows would make such a shape very difficult indeed to manufacture. Nonetheless, it demonstrates what may be achieved using this RSM-based approach.

The hyperparameter set from the Kriging model again allows the importance of each quantity to be ranked, the smoothness of the model in each direction to be found and the degree of regression needed. These 43 parameters are listed in Table 13.10 for the best design found noted in Table 13.9. They show that all the variables at stations 4 and 5 impact strongly on the design and that while there is slightly more noise in the problem this is still not causing a great deal of regression to be needed. In addition, variable values at stations 7, 2 and, in particular, 1, again have almost no effect on the SKE values. Interestingly, for the 21-variable model, Y_{cen} at station 6 is now also having little impact, which may stem from the more limited range of variable changes used in this part of the study.

The variations of the key design parameters are further illustrated in Figure 13.5, which show the effects on SKE of changes to the circumferential lean, axial lean and skew at radial stations 4 and 5, as predicted by the final Kriging model updated at the best-predicted

Table 13.10 Hyperparameters for the final Kriging model found during the 100-trial experiment and updating using SKE values, for the 21-parameter problem (smoothness hyperparameters vary from 1 to 2 with 2 indicating maximum smoothness; values of Log(Regression) greater than −2.0 indicate significant noise in the data)

Hyper-parameter	Value	Hyper-parameter	Value	Hyper-parameter	Value
$Log(Y_{cen}(1)$ sensitivity)	−7.38805	$Log(X_{cen}(1)$ sensitivity)	−7.55349	$Log(Skew(1)$ sensitivity)	−5.07127
(2)	−7.02317	(2)	−7.38714	(2)	−2.86383
(3)	−2.72264	(3)	−2.20653	(3)	−3.35046
(4)	−2.58333	(4)	−2.51984	(4)	−1.96843
(5)	−1.50397	(5)	−1.85635	(5)	−1.70397
(6)	−8.33445	(6)	−2.34879	(6)	−2.70397
(7)	−7.07171	(7)	−6.82540	(7)	−4.06851
$Y_{cen}(1)$ smoothness	1.49924	$X_{cen}(1)$ smoothness	1.64565	$Skew(1)$ smoothness	1.16372
(2)	1.96957	(2)	1.50495	(2)	1.50128
(3)	1.99911	(3)	1.17089	(3)	1.99936
(4)	1.60618	(4)	1.99993	(4)	1.18131
(5)	1.78450	(5)	1.99960	(5)	1.18114
(6)	1.52281	(6)	2.00000	(6)	1.99982
(7)	1.63418	(7)	1.48982	(7)	1.99963
Log(Regression)	−2.08485				

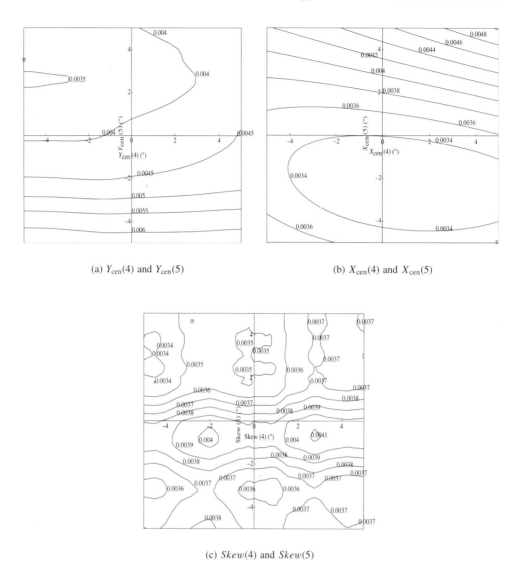

(a) $Y_{cen}(4)$ and $Y_{cen}(5)$ (b) $X_{cen}(4)$ and $X_{cen}(5)$

(c) $Skew(4)$ and $Skew(5)$

Figure 13.5   Variation of predicted SKE: Kriging model after 100 initial points and five updates (all other variables at the minimum capacity change position).

objective points (and for all other parameters set as per the minimum capacity change design from Table 13.9). Notice that the plots are quite smooth for the lean parameters but much more complex when skew is altered. Also, the absolute effects of the circumferential lean changes are about twice those of axial lean, which are, in turn, about 50% greater than those of skew. Examination of the error predictions provided by the Kriging model indicate that the standard errors across these plots are all similar and typically range from 0.0002 to 0.001 at worst.

13.5 Conclusions

This brief study has examined optimization of a Rolls-Royce turbine guide vane aerodynamic design. It has compared two response surface modeling methods when applied to data generated from LPτ experiment design techniques applied to the *FAITH* design system working in harness with the *sz02* Navier–Stokes CFD solver. The work demonstrates that significant reductions of the order of 50% can be achieved in secondary kinetic energy while maintaining capacity unchanged, for the design studied. The resulting geometries would be extremely difficult to manufacture but illustrate what might be achieved if only aerodynamic aspects are considered.

Two sizes of problem have been considered: a 7-variable problem where only circumferential lean was varied and a 21-variable model where circumferential and axial lean together with skew were all varied. The study shows that useful improvements can be obtained in the 7-variable problem with only 30 calls to the CFD solver while the 21-variable problem requires over 100 calls.

Of the two RSMs examined, it is clear that a Kriging model is significantly better than a radial basis function model using cubic splines. Interestingly, the expected improvement update strategy was seen not to be particularly helpful in this context when compared to the simpler idea of adding new points at the estimated best objective function locations.

A useful additional output of the Kriging model is the hyperparameter set that allows the importance of the variables considered to be ranked. This clearly shows that geometric changes near the hub or tip have little effect and could be safely excluded from design optimization considerations of the type studied here, thus significantly reducing the number of parameters to be considered. Moreover, the variables at midspan dominate this design optimization process and significant reductions in SKE could be achieved by considering only these quantities. Circumferential lean is seen to have more impact on SKE than axial lean, which in turn has more effect than skew.

A simple study like this could be enhanced by a number of items of further work:

1. Alternative DoE methods could be examined, although, given the very small number of data points available, this is unlikely to be of great significance.

2. Other RSM methods could be investigated:

   (a) Kriging could be applied to the constraint model as well as to the objective function model;

   (b) other RBFs could be investigated, including those using hyperparameters for model tuning; and

   (c) polynomial regression models could be used.

3. Tighter constraint limits could be investigated to see if the capacity of the design could be held under even tighter control during search – alternatively, line searches from final designs to bring the capacity back to its initial value could be investigated.

4. Geometric constraints could be imposed to ensure that designs were readily manufacturable.

5. Very large numbers of *sz02* runs could be carried out to plot the actual variation of SKE near a predicted optimum to see if the RSM surfaces are accurate (this would of course require access to a rather large computing cluster).

14

Turbine Blade Design (II) – Fir-tree Root Geometry

This case study describes the use of a modern parametric CAD system in the shape optimization of a turbine blade structure. It involves using a combined feature-based and B-spline-based free-form geometry modeling approach. The geometry is defined by a number of features for the global shape and by NonUniform Rational B-splines (NURBS), using control points for the local shape. The coordinates of the control points along with the overall dimensions are chosen as design variables. A knot insertion/removal process is introduced to increase flexibility and to obtain a compact definition of the final optimized shape. Here, the geometry is modeled using the ICAD®[1] parametric CAD system, which enables complete automation of the design-to-analysis process. Stresses are obtained from finite element analysis. A two-stage hybrid strategy is used to solve the resulting optimization problem. Its application to the design of a turbine disk fir-tree root is demonstrated. The approach is compared to a more traditional analysis using geometry models based on straight lines and circular arcs. Improvements are achieved in the design performance and the flexibility of the geometry definition.

14.1 Introduction

As has been noted in earlier chapters, the use of computer-aided design tools and computational analysis techniques such as finite element analysis has become common practice in aerospace structural design. In addition, optimization techniques are increasingly used where weight reduction is a goal, typically subject to various stress constraints at points within the structure. A comprehensive survey on structural optimization is provided by Vanderplaats (1999). Structural optimization problems commonly involve sizing, topology or shape optimization. Sizing optimization is used to define size-related variables, such as the cross-sectional areas of bars and trusses. Topology optimization mainly involves the layout of materials, while shape optimization considers the optimum shape of the component

[1]http://www.ds-kti.com

Computational Approaches for Aerospace Design: The Pursuit of Excellence.   A. J. Keane and P. B. Nair
© 2005 John Wiley & Sons, Ltd

boundaries. In general, shape optimization problems can be further classified into two types. The first type involves the determination of the dimensions of predefined shapes (geometric features) such as beam lengths or the radii of circular holes, and so on. This type of design method can be referred to as feature-based design. The satellite structure studied in the first case study works with predefined features in just this way. The second type involves the determination of the shape of an arbitrary open or closed boundary. The definition of airfoil sections falls into this class of model.

As has been noted in Chapter 2, there is a large variety of approaches that can be used when parameterizing open boundaries, for example, in a study carried out by Braibant and Fleury (1984), three shape optimization problems were examined. These were the determination of the optimal shape of a beam in bending, the optimal shape of a hole and the optimal shape of a fillet. They were solved using the concept of design elements that were defined using B-splines composed of a number of elements. In the work of Chang and Choi (1992), the shape was parameterized using a parametric cubic representation of primitives supported in Patran.[2] Equidistant mapping in parametric space of the curve was used to create the finite element discretization. This is also referred to as an iso-parametric mapping in which the curve or surface is discretized evenly in the parametric space. Using these methods, attention must be paid to the creation of internal nodes during FE analysis to avoid the generation of distorted elements. A geometry-based approach for coupling CAD with finite element methods was presented by Schramm and Pilkey (1993), where NURBS were used to model the shape of a cross section in a torsion problem. More recent work is reported by Waldman et al. (2001) in which an optimum free-form shape for a shoulder fillet was obtained in tabular form. The use of NURBS can also be seen in CFD-based shape optimization, for example, in the work carried out by Lepine et al. (2001), where it is mainly used to reduce the number of design variables. One of the major drawbacks in the above applications is that the optimized shape is often not CAD-ready and additional effort is required to utilize results in the design process.

The development of parametric feature-based solid modeling capabilities in CAD systems has enabled the implementation of more complex shape optimization approaches to be developed that output the optimized geometry in a CAD-ready form. Such capabilities have also enabled designers to focus on design innovation by allowing them to quickly develop design concepts utilizing the available geometric features. The limitation of this approach is that it is difficult to generate shapes that do not exist in the library of predefined features provided by the system. Although free-form geometric modeling can alleviate this problem to some extent, it is still difficult to design a product purely using free-form geometry, because of the prohibitively large number of control points required. The effects caused by movement of the control points then become very difficult to control. This is particularly important when the design is embedded in an optimization loop. Here, to demonstrate the ability to embed design search methods directly in tools used by designers, a CAD-based approach is adopted. The parametric modeling is performed using the ICAD design automation tool.

In this study, a combined approach using feature-based design and free-form geometry is adopted in the design optimization of a turbine blade fir-tree root. The notch stresses of a fir-tree root are known to be the critical factors affecting the final design of the root, but traditional modeling methods using arcs and straight lines provide little flexibility to further optimize many designs. NURBS are therefore introduced to define the two-dimensional local tooth profile of the root. A general approach for defining the position of control points

---

[2]www.mscsoftware.com.au/products/software/msc/patran

is described using nondimensional quantities that are dependent on the geometric features. One of the advantages of this approach is that flexibility can be achieved without loss of the geometric significance of the design variables used to formulate the optimization problem. It also provides good control over the range of design variables and the resultant geometry.

The use of ICAD to construct the geometry makes this method applicable to a broad range of problems. A direct link has been established between ICAD and the finite element package to enable the modified geometry to be passed directly to the analysis code along with the associated boundary conditions and mesh control properties. Each time the geometry is modified, a new mesh is generated on the geometry using automatic meshing tools. This is preferable to utilizing the internal parametric description of the geometry within the analysis code, allowing the method to tackle more complex problems than those dealt with in most previous work (Braibant and Fleury 1984; Chang and Choi 1992; Schramm and Pilkey 1993).

14.2 Modeling and Optimization of Traditional Fir-tree Root Shapes

A multilobe fir-tree root is commonly used in turbine structures to attach blades to the rotating disk. This feature is usually identified as a critical component in an aero-engine and is subject to high mechanical loads. The loading on the root is mainly due to centrifugal forces, which are dependent on the mass of the whole blade. Today, the design of such a structure relies on extensive use of finite element methods and judgments based on both analysis results and industrial experience gained in the design and operation of such structures.

The geometry of turbine disk fir-trees can be described using a number of dimension-based features and modeled using feature-based CAD systems. The inclusion of a CAD system in the optimization cycle offers a number of benefits, the most obvious of which is that it is consistent with the overall engineering design process and results can be directly used by the design team.

The two-dimensional shape of this type of structure can be described using a set of straight lines and circular arcs. This is referred to as a traditional design in this study. A typical fir-tree root shape consisting of three teeth is illustrated in Figure 14.1 (left). Other forms of notch representations, such as elliptical fillets and compound arcs, could also be used in the notch shape modeling, as noted in the literature (Saywer 1975). An example using elliptical fillets was reported in an effort to reduce the notch peak stress by Lee and Loh (1988).

Although a two-dimensional model of a fir-tree cannot reveal the significance of the effect of skew angle and stress variation along the axial direction, it is a useful simplification that can be used in the preliminary design stage (an estimate of the effect of the skew angle can be included as a scalar factor, called the skew factor, in the stress calculation). Three-dimensional detailed analysis can then be carried out at the component design stage to verify the choice of design candidates.

The tooth profile used in the traditional fir-tree root geometry is shown in Figure 14.1 (right). It is defined using a set of geometric dimensions including lengths, angles and arc radii. There are 10 parameters used here in defining the basic tooth geometry: root-wedge-angle, tooth-pitch, top-flank-angle, under-flank-angle, next-top-flank-angle, blade-crest-radius, blade-trough-radius, disk-crest-radius, disk-trough-radius, and non-contact-face-clearance. When used in

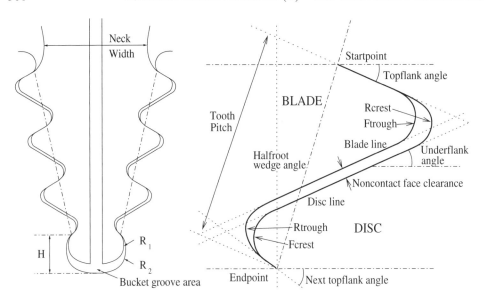

Figure 14.1   Typical geometry definition of fir-tree root (left) and single-arc fillet tooth (right).

the optimization process, these geometric features are treated as design variables and are varied by the optimizers in an effort to minimize the rim load subject to geometric and mechanical constraints drawn from both the structural analysis domain and industrial experience. The area outside of the last continuous radius of the turbine disk, referred to as the fir-tree frontal area, is chosen as the objective since it is proportional to the rim load by virtue of a constant axial depth, assuming that the blade load remains constant.

This geometric definition method has been used by Song et al. (2002b) in an effort to improve the design by first automating the repetitive design modification process and then applying various search techniques. However, it suffers from two major drawbacks: first, it becomes more difficult to characterize the effect of different parameters as the total number of parameters increases. Although some global search techniques such as GAs can be used to explore large design spaces, the computational cost related to such methods can prove impractical given the requirements for reduced design times. Also, understanding the characteristics of a detail shape design problem like this can be best achieved if attention is focused on local changes. In practice, almost all new designs can be traced back to earlier models and often a large amount of experience exists for the heuristic determination of some parameters. This can be used to reduce the number of design variables and makes it possible to focus on local areas where improvements can be critical to the quality of the whole product. At the same time, the designer may have more confidence in applying such local changes to the actual product. Further, although minimizing rim load will reduce overall weight, the life of the blade and disk is highly dependent on the notch stresses, and therefore, the notch stress should ultimately be minimized for the required life target. An example of minimizing notch peak stress using elliptical fillets can be found in the study by Lee and Loh (1988).

In the previous study by Song et al. (2002b), this traditional geometry definition was optimized using various optimization approaches including gradient-based methods, population-based genetic algorithms, and a two-stage hybrid method. It was observed that the notch stresses remain the critical factor affecting the final design. In addition, the notch stress minimization studied in the previous work further revealed that any reduction of notch stress almost inevitably resulted in an increase in the fir-tree frontal area and an increase in the crushing stresses because of a reduction in the bedding length resulting from an increase in the notch radii. It is this observation that motivated the use of a free-form shape in the definition of the tooth profile in an effort to further minimize the maximum notch stress within the notch region without increasing the fir-tree frontal area or the magnitude of the crushing stresses.

14.3 Local Shape Parameterization using NURBS

Geometry modeling using polynomial and spline representations has been incorporated into most CAD packages, and these methods provide a universal mathematical approach to represent and exchange geometry in engineering applications.

The curves and surfaces represented in the ICAD system are, at their lowest level, represented by piecewise parametric polynomial functions, including NURBS. There are a number of properties that have resulted in NURBS being considered desirable mathematical representations in CAD systems, so that most systems now use NURBS as their internal shape representation. In terms of shape optimization, perhaps the two most useful characteristics are their strong convex hull property and their local control property. The strong convex hull property means that the curve is contained in the convex hull of its control points; in fact, the control polygon constitutes its approximation; therefore, the choice of the control points as design variables has clear geometric meaning. The local control property means that moving the control points (P_i) only changes the curve in a limited interval; in contrast, the change of the position of one control point affects the whole of a Bezier curve. NURBS thus look set to be the future geometry standard for free-form curve and surface representations within CAD/CAM (and scientific visualization, solid modeling, numerically controlled machining and contouring). By way of example, a cubic NURBS curve defined using four control points is illustrated in Figure 14.2. The control of the curve, achieved by changing the control point positions and weights (w_i), are of interest for structural shape optimization. The effect of modifying the weight of point P_2 is also shown in Figure 14.2.

The motivation behind the introduction of NURBS into the definition of the tooth profile comes from the idea that it may permit the reduction of the stress concentration around the notch regions of the fir-tree root while maintaining a constant contact length between the blade and disk. The first model considered is simply an extension of an existing traditional single-arc model, referred to as the conic fillet model. It is derived from a NURBS representation of the single-arc fillet, using the simplest NURBS arc of degree two. A second model, using a NURBS curve of degree three, is derived from a double-arc fillet model; this model is referred to as a cubic fillet model.

14.3.1 NURBS Fillet of Degree Two – Conic Fillet

The simplest NURBS arc of degree two can be defined using three control points, as shown in Figure 14.3 (defined by points P_0^0, P_1^0, P_2^0). The complete single tooth profile defined

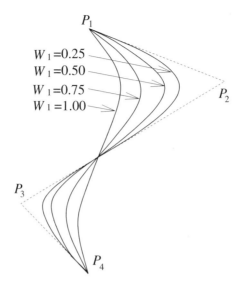

Figure 14.2   A NURBS curve and effect of changing the weight w_2 of point P_2 on the shape.

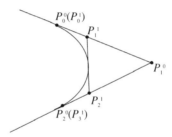

Figure 14.3   NURBS representation of single-arc fillet using three or four control points.

using this 3-point NURBS is shown in Figure 14.4. Instead of using absolute coordinates, the nondimensional coordinates are used here to define the position of the control points. This definition of the tooth profile is used as a starting point. Now, moving the middle control points will generate a series of cubic curves, while modifying the weight associated with the middle control point will generate a different series of curves. Initial comparison results showed that the benefit in terms of stress reduction from this model is not as good as a double-radii fillet, which is the normal alternative to the single-arc fillet. Therefore, instead of using the simplest NURBS arc, a more general form of degree three involving four control points was adopted in this study; see again Figure 14.3 (defined by P_0^1, P_1^1, P_2^1, P_3^1).

14.3.2   NURBS Fillet of Degree Three – Cubic Fillet

A double-arc fillet can be described using seven control points if the component arc is defined by a cubic NURBS, and the points are properly positioned, as illustrated in Figure 14.5, Describing the double-arc using NURBS instead of two radii gives much greater flexibility.

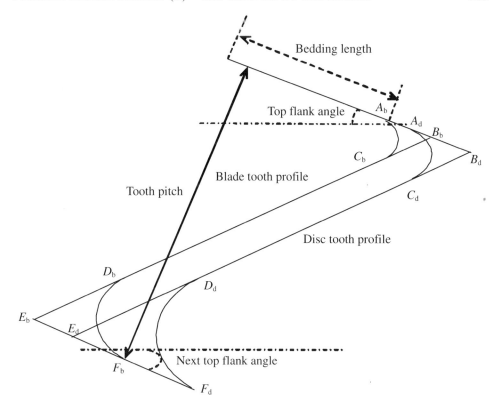

Figure 14.4 Notch conic fillet design using three control points.

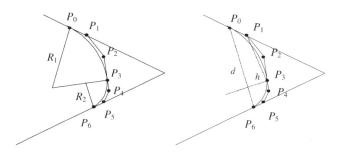

Figure 14.5 NURBS representation of double-arc fillet using seven control points and its defining coordinates.

The choice of cubic curve is also consistent with the fact the most CAD systems use cubic curves to represent geometry.

Describing the positions of the control points becomes difficult if the actual coordinates of the points are used, especially when each tooth profile is varied independently. Even for a uniform tooth profile, defining one tooth will involve 14 coordinates for these seven points, and it is very difficult to manipulate these coordinates to ensure that an acceptable geometry is produced. Considering the convex property of NURBS curves and the fact that the tooth profile should always lie within the polygon prescribed by the points P_0^0, P_1^0, P_2^0 shown in Figure 14.3, a number of nondimensional coordinates are used to define the position of the control points, as described next.

The point P_0 will be defined by the bedding length (i.e., the end of the straight line segment) as illustrated in Figure 14.4, which is the main quantity affecting the crushing stress. The point P_6 is the endpoint of the curve, essentially the point C_d in Figure 14.4 in the case of the disk profile. The point P_3 is defined by two nondimensional quantities h and d as shown in Figure 14.5 (b) using points P_1 and P_5 as reference points. The points P_2 and P_4 are defined in a similar manner using P_1, P_3 and P_3, P_5 as reference points. This approach of defining control points provides improved control as well as sufficient flexibility. Continuity requirements can then be built in to reduce the number of independent coordinates; in this case, the three points P_2, P_3, P_4 should form a straight line to maintain at least G^2 continuity of the curve at point P_3.

To define a tooth profile using these seven control points thus requires nine independent variables. The total number to be optimized will be several times more if each tooth has a different profile, of course. However, from initial analyses, the stress distributions along each tooth notch follow a similar pattern and, therefore, the same nondimensional values are used for all the teeth. Although this could be further relaxed to consider different tooth profiles, as a first step it is useful to focus on a constant profile to obtain the optimum shape in terms of a smooth distribution of stresses.

14.4   Finite Element Analysis of the Fir-tree Root

The Rolls-Royce in-house finite element analysis program *sc03* is used here to perform the stress analyses. An automatic and smooth coupling between the ICAD system and the finite element analysis package *sc03* has been previously established for use in optimization. The geometry and related information such as boundary conditions and loads are defined in the CAD system and transferred automatically into *sc03*. For further details about how these two systems are coupled and information exchanged, see Song et al. (2002b).

As the fir-tree geometry is constant along the root center line, it is possible to think of the stresses as two dimensional. However, the loading applied along the root center line is not uniform, so, strictly speaking, the distribution of stresses will be three dimensional. Nonetheless, it is still possible to assume that each section behaves essentially as a two-dimensional problem with different loadings applied to it. The difference of loading on each section is affected by the existence of skew angle, which will increase the peak stresses in the obtuse corners of the blade root and increase the stresses in the acute corners of the disk head. From previous root analysis research, it has been found convenient to use a factor to estimate the peak stresses at each notch of the blade and disk, and this factor takes different values for different teeth.

It is generally assumed that there are two forms of loading that act on the blade, the primary radial centrifugal tensile load resulting from the rotation of the disk, and bending of the blade as a cantilever that is produced by the action of the gas pressure on the airfoil and forces due to tilting of the airfoil. The resulting stress distribution in the root attachment area is a function of geometry, material and loading conditions (which are of course related to the speed of rotation).

Different types of mechanical constraints are involved in the design of such a structure, based on industrial experience. Finite element analysis is utilized to obtain these stresses:

- *Crushing stress:* This describes the direct tensile stress on the teeth: bedding width is the factor affecting the stress.

- *Disk neck creep:* the disk posts are subject to direct tensile stress that causes material creep. Too much creep, combined with low cycle fatigue, can dramatically reduce the component life.

- *Peak stresses:* peak stresses occur at the inner fillet radii of both the blade and the disk. If the fillet radii are too small and produce unacceptable peak stresses, some bedding width has to be sacrificed to make the radii bigger.

A single sector model is analyzed to obtain the required stress data. Moreover, as finite element analysis is computationally expensive, a compromise between accuracy, numerical noise and computation cost is required to enable its practical implementation within an optimization run. This compromise is made here by an appropriate choice of mesh density parameters. The aim is to find a compromise between the high costs incurred with very fine meshes and the accuracy needed to capture stresses in the notch area. Following setup, a series of systematic evaluations has been carried out to establish appropriate mesh density parameters. These were performed by varying the local edge node spacing. Reducing the notch edge node spacing increases the mesh density and therefore reduces errors in the computed maximum notch stress. These studies resulted in a spacing of 0.001 mm being used for the notch edge nodes. Figure 14.6 shows the results of finite element analysis for two differing fir-tree root designs.

It has been identified in previous work that the notch stress constraints are the limiting factors in the optimization and remain active in the final results. It is possible to reduce

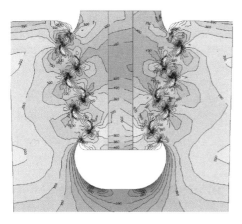

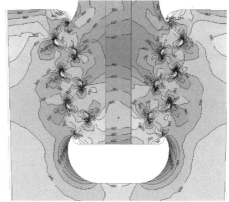

(a) Initial design – peak contact stress 1,482 MPa   (b) Reduced radius design – peak contact stress 1,107 MPa

Figure 14.6 Effect of varying tooth radius on fir-tree root structure.

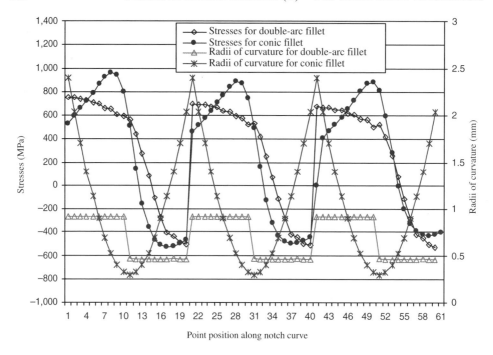

Figure 14.7 Variation of notch stresses and curvature along the notch curve.

the notch stress by increasing the notch radius; however, in a traditional design, such an increase will inevitably result in a reduction in the bedding length, which then will lead to an increase in crushing stress. This is because the traditional shape parameterization does not provide sufficient local geometric flexibility in the notch region to reduce both the crushing stress and notch stress to the required levels.

The aim of introducing NURBS in defining the notch fillet is to balance these two conflicting factors in the design. Achieving the minimum notch stress is essentially equivalent to achieving the most uniform possible stress distribution along the notch region. The stress distribution along the notch length may be plotted along with the radius of curvature to compare the effects of different shapes on the stress distribution and the radius of curvature. Such plots for a conic notch and a double-arc notch are given in Figure 14.7. It can be seen that the peak stress along the conic fillet is higher than for a double-arc notch. This is why a cubic fillet derived from a double-arc fillet was chosen as a basis for notch shape parameterization in this optimization study.

14.5 Formulation of the Optimization Problem and Two-stage Search Strategy

The formulation of the shape optimization problem involves the choice of a number of important geometric features as design variables and a set of geometric and mechanical constraints based on industrial experience and previous analysis results. It is known that a number of critical geometric features, such as flank angle, fillet radii and skew angle, have a significant effect on the stress distribution. The effect of these quantities has been explored previously (Song et al. 2002b). In this study, the notch region is the main area of interest. The parameters used to define the notch local profile were designated as design variables while the parameters describing the global shape were held fixed, based on previous results. Three types of mechanical constraints are used, as described earlier.

The optimization is performed here using the Options search package[3] which provides designers with a flexible structure for incorporating problem-specific code as well as access to more than 30 optimization algorithms. The critical parameters to be optimized, known as the design variables, are stored in a design database, which also includes the objective, constraints and limits. The design variables are transferred to ICAD by means of a property list file, which contains a series of pairs of variable names and values. This file is updated during the process of optimization and reflects the current configuration. The geometry file produced by ICAD is then passed to the FE code *sc03*, which is controlled by a command file. The analysis results are written out to another file, which is read in by the optimization code. The design variables are then modified according to the optimization strategy in use until convergence or a specified number of loops has been executed.

A simple two-stage search strategy combining a Genetic Algorithm (GA) with a gradient search was used in this problem. The GA is first employed in an attempt to give a fairly even coverage on the search space, and then gradient-based search methods are applied on promising individuals. The main consideration behind this strategy is that as the GA proceeds, the population tends to saturate with designs close to all the likely optima including suboptimal and globally optimal designs, while gradient-based methods are efficient at locating the exact position of individual optima. Moreover, the GA is capable of dealing with discrete design variables such as the number of teeth (held fixed in this study but of interest in the overall design of the fir-tree system). Hence, the GA results are used to provide good starting points for the gradient search methods.

Several gradient-based search methods have been applied to the problem after the GA search; these include the Hooke and Jeeves direct search method (see Section 3.2.4) plus various other methods discussed by Schwefel (1995).

14.6   Optimum Notch Shape and Stress Distribution

Three different tooth profiles have been optimized using the two-stage optimization strategy described in the previous section. Comparisons are made here between the optimized single-arc fillet and conic fillet; the optimized conic fillet and double-arc fillet and the optimized double-arc fillet and cubic fillet.

14.6.1   Comparison between Conic Fillet and Single-arc Fillet

Replacing the single-arc fillet with a conic fillet brings explicit control over the bedding length, and therefore on the crushing stress. It therefore becomes possible to decrease the crushing stress below the material criteria while at least maintaining the notch stress level at those seen in single-arc fillet designs, as shown in Figure 14.8. However, it is difficult to reduce the notch stress as the conic fillet offers little more flexibility than the single-arc fillet.

---

[3]http://www.soton.ac.uk/~ajk/options/welcome.html

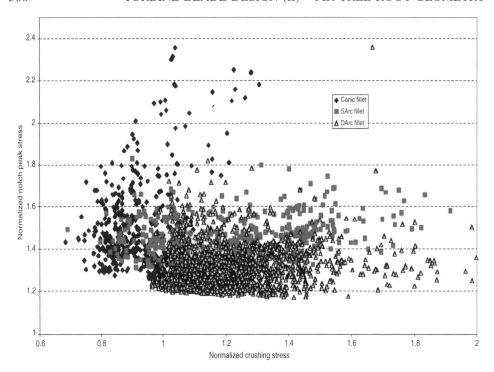

Figure 14.8 Normalized maximum notch stress against maximum crushing stress (normalization is defined as stress/material yield stress).

14.6.2   Comparison between Double-arc Fillet and Conic Fillet

Before moving on to the more flexible cubic fillet, a double-arc fillet is next used to provide a basis for a general notch profile. The use of a double-arc fillet can further reduce the peak notch stress, and this can also be seen from Figure 14.8, where the objective is to minimize the notch stress. It may be seen that both the crushing stress and notch stress are reduced compared to single-arc fillet, which is very difficult to achieve in this example if the notch shape is restricted to a traditional single-arc design. The comparison between conic and single-arc fillet shows that the use of NURBS can bring benefits and that a double-arc fillet provides a good basis for a NURBS model of the notch profile.

14.6.3   Comparison between Cubic Fillet and Double-arc Fillet

The introduction of the cubic fillet allows further reductions to the disk notch peak stress, while all the other stresses can be maintained below the material allowable levels. Therefore, an increase of fatigue life can be expected using a cubic fillet. Convergence curves are shown in Figure 14.9 for the different types of notch fillet, from which the benefits of introducing NURBS fillets can be clearly seen. A stress contour map for the notch region is shown in Plate 4, in which peak stresses before and after optimization are noted, a 6% reduction being achieved. However, whether the introduction of the more flexible cubic notch fillet will cause difficulties in manufacture or incur significant cost increases remains to be further investigated.

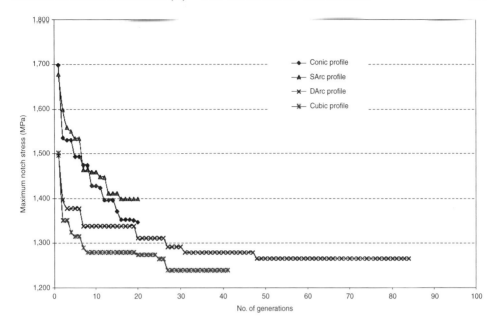

Figure 14.9 Convergence curves for different types of notch fillet using the GA.

14.7 Summary

In this study a combined geometric parameterization method using both basic geometric features and free-form shapes is applied to the local shape optimization of a turbine blade fir-tree root problem. This is used to improve the fir-tree design compared with traditional tooth profiles consisting of straight lines and circular arcs only. NURBS are used to model the tooth profile. The optimization problem is defined in terms of various geometric properties and stresses. The geometry is constructed from a number of geometric features and localized free-form shape control points. The control points and related weights of the NURBS curves along with several geometric feature dimensions are then chosen as design variables. The geometry is modeled using the parametric CAD system ICAD. Stresses are obtained from finite element analysis. A fully integrated system of CAD modeling and finite element analysis is incorporated into the optimization loop. A two-stage hybrid search strategy is used to find improved designs. The resulting designs have stresses some 6% lower than those obtained using traditional manual experience-based methods.

15

Aero-engine Nacelle Design Using the Geodise Toolkit

In this final case study, we examine a problem that is defined in the commercial CAD system ProEngineer[®][1] and carry out a design search using the Geodise[2] grid-based design optimization toolkit. The problem tackled is the noise radiated from a gas-turbine engine and how this might be controlled via alterations to the engine nacelle geometry. The calculations underpinning the design are carried out using the commercial Navier–Stokes CFD solver Fluent[®].[3] Since the run times of this code are quite long for such problems, the calculations carried out are placed on machines connected to the UK's National Grid Service (NGS). This is a collection of widely dispersed university-based computing clusters.

The Geodise system is the latest tool set to emerge from the authors' research group and thus encapsulates much of the thinking lying behind this book. It builds on the authors' experience of a number of commercial search tools. These frameworks have been largely based on fixed intranet infrastructures and can lack the scalability and flexibility required for dynamic composition of large scale, possibly long-lived application models from different software packages. Our aim is to develop tools that are able to deal with complicated optimization tasks such as those seen in the domain of multidisciplinary design across the boundaries of multiple establishments. Another issue for the approach adopted in many commercial packages is the limited use of open standards, protocols, and so on, which restricts generic plug-and-play capabilities. To overcome these difficulties, a number of commercial and research systems such as ModelCenter[®],[4] Fiper[®],[5] Nimrod/G,[6] Triana[7] and Dakota[8] are adopting service-oriented architectures. All aim to enable users to rapidly integrate simulation-based design processes, drawing on commercial CAD/CAE software,

---

[1]`http://www.ptc.com/`
[2]`http://www.geodise.org`
[3]`http://www.fluent.com`
[4]`http://www.phoenix-int.com/`
[5]`http://www.engineous.com/`
[6]`http://www.csse.monash.edu.au/~davida/nimrod/`
[7]`http://www.triana.co.uk/`
[8]`http://endo.sandia.gov/DAKOTA/index.html`

Computational Approaches for Aerospace Design: The Pursuit of Excellence. A. J. Keane and P. B. Nair
© 2005 John Wiley & Sons, Ltd

various classes of solvers including commercial and in-house tools, spreadsheets, and so on, across widely distributed and varied computing infrastructure. They thus represent the future of integrated optimization-based design tools.

When constructing such tools, typically two sorts of interfaces are required. First, there is the need for simple to use and easily understood graphical programming environments that allow users to rapidly set up and solve straightforward problems. Secondly, as users grow in confidence and sophistication, there is the need to be able to set out workflows that are more readily described in programming scripts. It is also important to be able to transition between these two worlds in a simple way. With these requirements in mind, the widely used scripting environment Matlab®[9] was chosen as the hosting environment for building the Geodise system. Matlab provides a fourth-generation language for numerical computation, built-in math and graphics functions and numerous specialized toolboxes for advanced mathematics, signal processing and control design. Matlab is widely used in academia and industry to prototype algorithms, and to visualize and analyze data. The NetSolve system (Casanova and Dongarra 1997), which uses a client–server architecture to expose platform-dependent libraries, has also successfully adopted Matlab as a user interface.

To provide the capabilities needed to support our vision within Matlab, various sets of low-level workflow composition, computation, optimization and database toolkits have been built. These make use of the emerging computing grid infrastructure[10] and compute cycle farming technologies.[11] From version 6.5, support for Java[12] within Matlab provides the just-in-time acceleration necessary to take advantage of the Java CoG tools[13] to leverage the current development of grid middle-ware (Foster et al. 2002, 2001). The use of the Java platform also allows the toolkits to be easily migrated to other scripting environments supporting Java such as Jython and integrated into various GUI tools. This has allowed a drag and drop graphical interface to be developed that allows high-level composition of workflows by selecting elements from the available toolkits. Although this does not provide access to all the capabilities available within Matlab, it nonetheless allows useful size problems to be tackled without recourse to the development of scripts by hand.

Engineering design search is also data intensive, with data being generated at different locations with varying characteristics. It is often necessary for an engineer to access a collection of data produced by design and optimization processes to make design decisions, perform further analysis and carry out postprocessing. It is common practice to use the file system as a medium for transferring such data between these various activities, which works fine on a local system, but falls short in a grid environment where processes can be dispatched to run on different machines with varied hardware and software configurations: in such an environment secure, reliable, transacted messages coupled to databases offer a better way to provide resilience and quality of service. Databases also allow the state of the design process to be exposed to context-sensitive design advisers, enabling them to provide dynamic advice to the user. They therefore play an essential role in our architecture, where it is important to capture processes in addition to storing any results obtained.

---

[9]http://www.mathworks.com
[10]http://www.globus.org
[11]http://www.cs.wisc.edu/condor
[12]http://java.sun.com/
[13]http://www.globus.org/cog

Three main elements are normally required in CFD-based shape optimization problems: these are geometry design, mesh generation and solution. These elements can be loosely coupled within the Geodise environment. In this case study, the STEP data exchange standard is used to exchange geometry data between the CAD and mesh generation packages. A parametric geometry model suitable for CFD analysis is first generated using ProEngineer. The mesh is then generated with minimum reference to the geometry information, as only the top-level entities in the CAD model are referenced in the meshing script executed by Gambit (the Fluent mesh generation tool). This is also consistent with the fact that a top-down approach is often adopted in the geometry modeling. Lastly, the CFD solver can then be run using a predefined journal file on any appropriate computing platform.

Having set up the problem in the Geodise environment, grid infrastructure is then used to support a response surface modeling approach to design search. This consists of farming out a series of design of experiment chosen jobs using grid middle-ware. These jobs are subsequently harvested into a dedicated database system where they can be monitored before finally being modeled and exploited using Krig tools within the optimization toolkit.

15.1 The Geodise System

The primary aim of the Geodise project is to provide a powerful design process system for engineers that brings grid technologies into an environment and mode of working that they find familiar. To do this, it makes available a powerful workflow construction engine and incorporates optimization services, compute facilities and database tools, while being easily deployable alongside a range of commercial and in-house analysis codes. This capability is provided via a series of Matlab toolboxes that allow designers to use their favorite editor to write a script, incorporating calls to the toolkits, and then to run it from the Matlab command prompt. Geodise also provides a workflow construction tool to assist users with the task of developing and executing their design process without recourse to script editing. Both approaches emphasize the routine use of database-driven repositories that allow data and workflows to be passed from process to process on the grid, and retrieved and exploited by other tools or packages. Security issues are addressed in the Geodise toolboxes using the Grid Security Infrastructure (GSI) implementation in the Globus toolkit.

15.1.1 Architecture

The Geodise architecture is illustrated in Figure 15.1. As can be seen from the figure, it consists of the following elements:

- a graphical workflow construction environment;

- a knowledge repository;

- a database service; and

- five Matlab toolboxes that deal with knowledge, databases, optimization, compute resources and XML-enabled data transfers, respectively.

These are coupled to the user's computing environment, which is presumed to consist of local machines, condor pools and/or a computing grid accessible via Globus middle-ware,

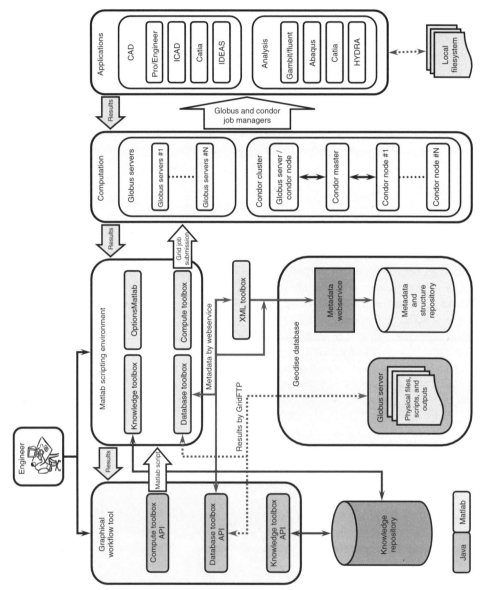

Figure 15.1 The Geodise grid-based optimization system architecture – http://www.geodise.org.

web services, and so on. Mounted on these facilities are the user's CAD and analysis tools that talk to their respective local file stores and which run in some form of batch mode using input and output files. In common with all such process integration systems, the user must have a design parameterization and well-automated analysis codes available before commencing to use the system.

Within Geodise, the database service is built on an Oracle 9i®[14] relational database that hold all the users' metadata. This system can keep track of all ongoing work via the database toolbox. The knowledge repository, which is held in a Description Logic (DL) based Instance Store (iS),[15] holds semantically enriched information stored by the team who have built the Geodise toolkit. This includes semantic descriptions of the toolkits and typical resources related to the design search processes that can be used via the knowledge toolbox. This aids designers in setting up and solving design problems. The graphical workflow construction environment makes direct use of all of the Geodise toolboxes and the knowledge and data services to provide an easy-to-use and intuitive front end to the system. In addition, since all of the toolboxes are available as Matlab functions, more-experienced users can access this functionality directly when this is appropriate. In our experience, the most-sophisticated users of designs systems often prefer to work in this way, even if they initially began by using a graphical programming approach.

15.1.2   Compute Toolbox

The user of the Geodise compute toolbox acts as a client to remote compute resources that are exposed as grid services. Users must be authenticated, and then authorized to access resources to which they have rights. They can submit their own code to compute resources, or run software packages that are available as services. The user can discover the available resources, decide where to run a job and monitor its status. It is obviously essential that the user be able to easily retrieve the results of a simulation. Additionally, the requirements of design search and optimization mean that compute resources must be available programmatically to algorithms that may initiate a large number of computationally intensive jobs, serially or in parallel.

In this case study, two different mechanisms are used to submit computational jobs to the computing servers. The first uses a web service interface to Condor to submit ProEngineer jobs to a cluster of Windows machines to generate geometry. The ClassAd mechanism provided in Condor is used to match computational requests to available Windows resources with ProEngineer installed. The second mechanism is implemented via the use of Globus and the Java CoG tools, which enables computational jobs to be submitted to computing servers using a proxy certificate. The first mechanism does not require that users know in advance which resources have the desired software installed, but the second does.

The Globus toolkit provides middle-ware that allow the composition of computational grids through the agglomeration of resources that are exposed as grid services. This middle-ware provides much of the functionality required by Geodise including authentication and authorization (GSI), job submission (GRAM), data transfer (GridFTP) and resource monitoring and discovery (MDS). Client software to Globus Grid services exists natively on a number of platforms and also via a number of Commodity Grid (CoG) kits that expose grid services to "commodity technologies", including Java, Python, CORBA and Perl. By using

---

[14]http://otn.oracle.com/products/oracle9i
[15]http://instancestore.man.ac.uk/

Table 15.1 Principal compute commands

Function Name	Description
gd_createproxy	Creates a Globus proxy certificate from the user's credentials.
gd_certinfo	Returns information on the user's certificate credentials.
gd_proxyinfo	Returns information on the user's current proxy certificate.
gd_jobsubmit	Submits a GRAM job, specified by a RSL string, to a Globus server. Returns a job handle to the user.
gd_jobstatus	Returns the status of the GRAM job specified by a job handle.
gd_jobpoll	Polls the status of the GRAM job specified by a job handle.
gd_jobkill	Terminates the GRAM job specified by a job handle.
gd_listjobs	Returns job handles for all GRAM jobs associated with the users credentials registered on a MDS server.
gd_makedir	Makes a directory on the remote host.
gd_listdir	Lists the contents of a directory on the remote host.
gd_getfile	Retrieves a file from a remote host using GridFTP.
gd_putfile	Transfers a file to a remote host using GridFTP.
gd_fileexists	Checks for the existence of a file on the remote host.
gd_rmdir	Removes a directory on the remote host.
gd_rmfile	Removes a file on the remote host.

client software to grid services written for these commodity technologies, the developer of a problem solving environment (PSE) is able to remain independent of the platform and operating system.

Table 15.1 details the compute functions available in the Geodise system. This set of functions describes the minimum functionality required to allow the user to run jobs on remote compute resources. The functions may be loosely categorized into those concerned with the user's credentials, job submission to the Globus Resource Allocation Manager (GRAM), and file transfer.

The GSI used by the Globus toolkit is based upon the Public Key Infrastructure (PKI). Under the PKI, an individual's identity is asserted by a certificate that is digitally signed by a Certificate Authority within a hierarchy of trust. In an extension to this standard, the GSI allows a user to delegate their identity to remote processes using a temporally limited proxy certificate signed by the user's certificate. The toolkit command gd_createproxy allows a user to create a Globus proxy certificate within the Matlab environment, essentially creating a point of single sign-on to the grid resources that the user is entitled to use.

The gd_jobsubmit command allows users to submit compute jobs to a GRAM job manager described by a Resource Specification Language (RSL) string. The gd_jobsubmit command returns a unique job handle that identifies the job. The job handle may be used to query or terminate the user's job. In addition, the gd_listjobs command may be used to query a Monitoring and Discovery Service (MDS) to return all the job handles associated with the user's certificate.

Two file-transfer commands are provided to allow users to transfer files to and from grid-enabled compute resources to which they have access. These commands support the high performance file-transfer protocol GridFTP, which defines a number of extensions to the FTP protocol to enable transfer of high volumes of data.

15.1.3   Database Toolbox

Design search is a data-intensive process, where significant amounts of unstructured, related data are produced. These include plain text files, graphics items and proprietary binary data. Providing a way to find files of interest based on their characteristics rather than their location in a file system gives users a more effective way to manage their own data, provides a means for reuse and collaboration, and can act as a source for advice based on similar examples of a given problem. Moreover, typical searches may have significant run times and rather than repeat parts of the process it is advantageous to store files for future reuse. For performance reasons, it is desirable to store such files close to where they are most likely to be used, usually the site where they were produced. However, accessibility by users at other sites should also be considered if collaboration and sharing is to be encouraged. Managing these unstructured data is a challenging task, and within the Geodise system, the database toolbox makes use of the concept of metadata: files and structured data can be augmented with metadata, which describes the information stored in a way that is both human and machine friendly.

The metadata used in this process need to be generated automatically whenever possible and only be requested from the user when necessary. Users must also be able to specify who else can discover and retrieve these files. A query mechanism should facilitate interactive and noninteractive use, so that files can be located programmatically in scripts. Engineers should be able to specify what kind of file they are looking for in an intuitive manner without necessarily needing knowledge of a database query language (e.g., SQL). Ideally, the user should not need to know the name of the database or machine their data is stored on, or its underlying storage mechanism. It should also be possible to locally record a unique identification number for accessing the file once it has been archived and to be able to use that handle at a later date to retrieve the file.

If appropriately defined, metadata enables data to be more effectively used and managed. Metadata becomes even more powerful in knowledge reuse when described in a semantically consistent way to allow key concepts and relationships to be defined and shared using onto-logical language. Using the database toolbox, related data, such as that referring to a whole design job, may be logically grouped together as a data-group and metadata can be added so that the entire collection can be described. Data may belong to multiple data-groups, which may also contain sub-data-groups, so users can describe and exploit relationships or hierarchies. For example, in a CAD-based shape optimization problem, the CAD model and related meshing and solver journal files are grouped and archived together; see Figure 15.2. This helps improve efficiency and reliability in the use of this information when the number of models grows. Another use of grid-enabled databases is the remote monitoring of an optimization process given the job handle. As results are delivered into the database during optimization, the user can monitor the progress of the search from arbitrary locations, even on a 3G mobile phone.

Within Geodise, metadata are also archived in an Oracle 9i relational database system. All database access is supplied through secure web services for storage and query of meta-data, file location and authorization, which can be invoked with platform-independent client code anywhere on the grid. Web service communication is secure, allowing certificate-based authentication and authorization using the same delegated credentials that provide a point of single sign-on for computational jobs. Secure requests are signed with the user's proxy certificate, which the web service verifies and uses to authorize database access, record data ownership and allow users to grant access permissions to others. The architecture permits

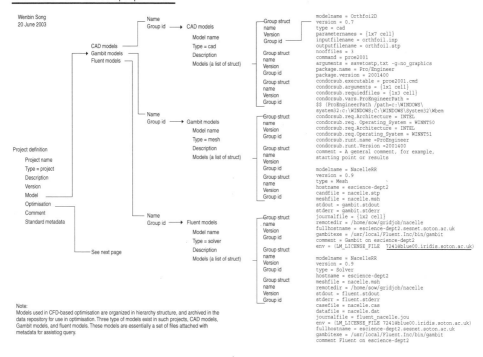

Figure 15.2   Hierarchy of data models for CFD-based shape optimization within the data repository.

Table 15.2   Principal database commands

Function Name	Description
gd_archive	Stores a file in a repository with associated metadata.
gd_retrieve	Retrieves a file from the repository to the local machine.
gd_query	Retrieves metadata about a file based on certain restrictions.
gd_datagroup	Creates a new data-group, used to group together archived files and variables.
gd_datagroupadd	Adds an archived file, variable or subdatagroup to a data-group.

data archiving, querying and retrieval from any location with web access, Java support, the required APIs and a valid proxy certificate. In addition to its use from scripts written directly in Matlab, the database toolkit has also been integrated into the Geodise workflow construction environment (WCE). Table 15.2 details the main Geodise database functions. These functions provide users with the ability to store files or variables in a repository with associated metadata, query the metadata and retrieve the data, provided they have the correct access rights.

15.1.4 Optimization Toolbox

The Geodise optimization toolbox is built on top of the Options[16] design exploration system. It provides a powerful set of tools within Matlab via an extremely simple interface function called OptionsMatlab. In addition to over 30 direct optimization routines, OptionsMatlab also supports a number of Design of Experiment (DoE) and Response Surface Modeling (RSM) algorithms that allow optimization to be carried out cheaply using approximations to the values of the objective function and/or constraints.

OptionsMatlab can be called in one of two basic modes. First, it can be used to initiate direct searches using one of the wide range of built-in optimizers on user-supplied problems, with computations carried out using the grid resources made available via the compute toolbox. Where appropriate, these searches can be carried out in parallel. The user's problem-specific codes are linked to the toolkit using simple Matlab functions that invoke the user's code and parse it for results, which are then returned to the optimizer. As in all such toolkits, it is the user's responsibility to set up this interfacing code. Within Geodise, this process is supported by the availability of the other toolkits (and, in particular, the compute and database tools).

Secondly, OptionsMatlab can be called with results coming from prior computations (typically search runs or calls to DoE-based surveys of the fitness landscape). It can then build one of a number of RSMs on the supplied data and search this, again using any of the available optimization routines. When used in this way, multiple results may be returned to allow parallel update schemes to be adopted to refine the RSM through further calls to the user's code.

In either mode, the user specifies the names of the routines that link the problem-specific codes, the number of design variables, their initial values and any bounds on them, the number and type of constraints with their limits and whether any additional parameters are to be passed to the models in use via OptionsMatlab. The user also sets out the desired quantity of monitoring information to be displayed during a search, the maximum number of parallel process that may be used, the maximum number of search steps desired, the type of search to be deployed and, if appropriate, the structure containing results from which the RSM is to be built, along with the type of RSM desired. Table 15.3 lists the available search methods (including DoE methods), while Table 15.4 lists the available RSM types.

In addition to the set of mandatory inputs, there are a very large number of control parameters ranging from population sizes to convergence tolerances that can be optionally specified to tune the behavior of searches and the construction and tuning of the response surfaces. One key capability is the recursive ability to apply the search engines within OptionsMatlab to tune the hyperparameters of its own RSMs. The system also makes available trace information of its progress that can be used to monitor and steer the behavior of searches. Moreover, since this capability is all defined within Matlab, it is then simple to build customized metasearches by sequenced calls to OptionsMatlab that suit the problem in hand. A number of popular approaches are supplied with Geodise in this way, including trust-region methods and multiple parallel update RSM based methods.

In addition to the direct availability of Options within Matlab via the OptionsMatlab function, the toolkit can also be accessed via the Java Native Interface (JNI), which exposes the API directly and thus provides more flexibility for implementing complicated search strategies such as complex hybrids of Generic Algorithms and local gradient descent methods

---

[16]http://www.soton.ac.uk/~ajk/options/welcome.html

Table 15.3   Search methods available in OptionsMatlab

Code	Search Method
0	just evaluate the user's problem code at the point specified
1.1	OPTIVAR routine ADRANS
1.2	OPTIVAR routine DAVID
1.3	OPTIVAR routine FLETCH
1.4	OPTIVAR routine JO
1.5	OPTIVAR routine PDS
1.6	OPTIVAR routine SEEK
1.7	OPTIVAR routine SIMPLX
1.8	OPTIVAR routine APPROX
1.9	OPTIVAR routine RANDOM
2.1	user specified routine OPTUM1
2.2	user specified routine OPTUM2
2.3	NAG routine E04UCF
2.4	bit climbing
2.5	dynamic hill climbing
2.6	population-based incremental learning
2.7	numerical recipes routines
2.8	design of experiment based routines:
	• random numbers
	• LPτ sequences
	• central composite design
	• face centered cubic full factorial design
	• Latin hypercubes
	• user supplied list of candidate points
3.11	Schwefel library Fibonacci search
3.12	Schwefel library Golden section search
3.13	Schwefel library Lagrange interval search
3.2	Schwefel library Hooke and Jeeves search
3.3	Schwefel library Rosenbrock search
3.41	Schwefel library DSCG search
3.42	Schwefel library DSCP search
3.5	Schwefel library Powell search
3.6	Schwefel library DFPS search
3.7	Schwefel library Simplex search
3.8	Schwefel library Complex search
3.91	Schwefel library twomembered evolution strategy
3.92	Schwefel library multimembered evolution strategy
4	genetic algorithm search
5	simulated annealing
6	evolutionary programming
7	evolution strategy

Table 15.4   Response surface model types available in OptionsMatlab

Code	Response Surface Model Type
0	if the underlying user supplied function is to be called directly
1.0	a Shepard response surface model
2.1	linear Radial Basis Function
2.2	thin plate Radial Basis Function
2.3	cubic spline Radial Basis Function
2.4	cubic spline Radial Basis Function with regression via reduced bases
3.1	mean polynomial regression model
3.2	first-order polynomial regression model
3.3	first-order polynomial regression model plus squares
3.4	first-order polynomial regression model plus products (crossterms)
3.5	second order polynomial regression model
3.6	second order polynomial regression model plus cubes
4.1	a Stochastic Process Model (Krig)
4.2	the root mean square error of the Stochastic Process Model (Krig)
4.3	the expected improvement of the Stochastic Process Model (Krig)

with user steering, and so on. For example, it is possible to set up optimizers as step-by-step services, which are then repeatedly invoked by the user whenever needed. This gives total flexibility over the search process without the need for callbacks from the search engine to the user's code, making steering simpler to implement, without loosing the power of the available search tools.

15.1.5   Knowledge Toolbox

The knowledge toolbox within Geodise is a set of auxiliary tools that assist better understanding and more effective operation of the system by allowing a knowledge advisor to be called where appropriate. This reuses preacquired knowledge stored in a distributed knowledge repository. It consists of a set of Matlab functions that invoke remote knowledge advisor services via their web service interfaces (Table 15.5). The toolkit is used to serialize workspace entities in XML and pass them to the knowledge advisor services that then process instances in the knowledge repository for relevant semantic content. This provides intelligent advice such as suggestions on workflow composition and function configuration.

15.1.6   XML Toolbox

The XML Toolbox allows the conversion of variables and structures between the internal Matlab format and human-readable XML format. This format is required to be able to store parameter structures, variables and results from applications in nonproprietary files or XML-capable databases and can also be used for the transfer of data across the grid. The toolbox contains bidirectional conversion routines implemented as four small intuitive and easy-to-use Matlab functions that can be called directly within the environment. As an additional feature, the toolbox allows the comparison of internal Matlab structures through their XML string representations, which was not previously possible. The main use of the toolbox within Geodise is to convert users' Matlab data structures to XML, which

Table 15.5 Principal knowledge service commands

Function Name	Description
gdk_functions	query functions in the instance store that match the wild-card, only return functions that have function signatures
gdk_next	get function/function signatures that can be deployed in the next step in a workflow composition
gdk_before	get function/function signatures that can be deployed in the previous step
gdk_semantics	get XML/Matlab structure representation of the semantics for any concrete instance in the instance store
gdk_options	advice on OptionsMatlab control parameters and their appropriate values
gdk_signatures	query function signatures that match a given wild-card
gdk_function	get XML/Matlab structure representation of function semantics in the instance store, including one-shot contextual advice for function assembly
gdk_matlab	get the Matlab structure representation of a function semantic in the instance store
gdk_functiontree	return all functions and subfunctions for a specified function node in the instance store
gdk_edit	launch the Domain Script Editor (DSE), which helps in editing Geodise Matlab scripts

are then transferred across the grid and stored in the XML-enabled Oracle database. This comprises metadata and standard user-defined data. Queries to the database return results in XML, which are subsequently converted at client level to Matlab structures, again using the toolbox.

15.1.7   Graphical Workflow Construction Environment (WCE)

Since many users require a simple to use and intuitive interface to design exploration tools, the Geodise system provides a graphical workflow construction environment (WCE). Workflows are used within Geodise to set out the order in which calculations are carried out, the data-flows between them, archiving strategies, search strategies, and so on. Figure 15.3 shows the main roles of the WCE in assisting engineers to build and reuse workflows, verify resources, execute workflow scripts, and monitor or control the workflow as it runs locally or on the grid.

Workflow Construction

A user's first task in the WCE is to load-up or visually construct a series of connected tasks (the workflow); see for example, Figure 15.4. The component view on the left pane of the GUI contains a hierarchy of tasks and related computational components (e.g. Matlab functions). The hierarchy is created from an XML document describing the available named components, their inputs and outputs. A component can be dragged onto the main workflow view and may be linked to other components. The default initial values of function inputs

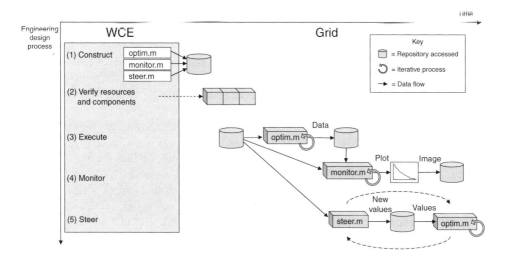

Figure 15.3 Roles of Workflow Construction Environment (WCE) in workflow construction, resource verification, execution, monitoring and steering.

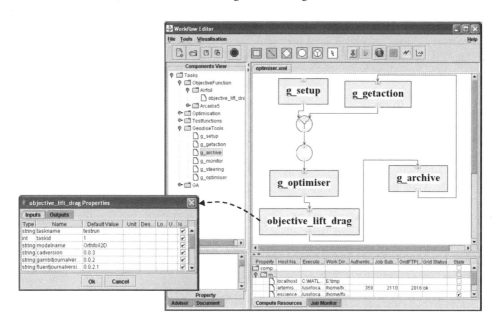

Figure 15.4 The workflow construction environment allows users to select and configure workflows, run them locally or on the grid and monitor their progress.

and outputs can be altered in the component property window. Data-flow is configured by making associations between the output and input parameters of different components. The complete workflow description is saved as XML for reuse in the graphical editor, but more importantly as a Matlab script (e.g., optim.m in Figure 15.3) which can be stored in the Geodise repository, reused and edited outside the WCE without engineers learning new workflow formats, and run on compute resources with Matlab support. By using data-groups, it is also possible to link the script to any data generated during its execution. The available compute resources are predescribed in a user-editable configuration file and also displayed by the WCE, which is able to verify the resources are accessible to the user and configured properly before execution. The WCE also assists in constructing scripts to monitor and steer workflow execution (e.g., monitor.m and steer.m), which may be stored in the repository for later use.

In order to provide additional information on component selection, the lower left panel of the GUI is used to show the properties of selected components. Moreover, to assist the user to compose a complicated workflow, likely next and previous candidates for any selected component can be retrieved using knowledge-based support.

It should also be noted that workflows can be constructed that cannot actually be executed. Therefore, the environment allows for a check and debugging stage to be carried out before job submission. The validation checks the availability of task nodes against the selected Matlab server. After validation, visual feedback is presented to the user by the tool: valid nodes are colored green while invalid ones are shown red.

Monitoring

Optimizations based on high-fidelity analysis are typically time consuming; therefore, it may be desirable to monitor the progress of the search as it is running and make changes if necessary. Most integrated packages implement monitoring on the local machine controlling the optimization. However, it is often difficult, if not impossible, to monitor the process from a remote machine in such systems. By exploiting the central data repository within Geodise, it is possible to retrieve and postprocess monitoring data independently and asynchronously from the running script. Furthermore, the postprocessed information (e.g., a plot or image) may itself be returned to the database, from where it can be accessed and retrieved from any location.

While technologies exist that allow monitoring of grids for scheduling, performance analysis and fault detection, for example, the NetLogger Toolkit (Gunter et al. 2002) and Gridscape (Gibbins and Buyya 2003), Geodise focuses on monitoring the progress of the user's application. This progress is described by the ongoing deposition of variables and data into the repository as transactions within and between grid processes as they occur. The user can supply a script that periodically performs a query to retrieve the data, render it, and then return the image to a predefined location, for example, the database. Data deposited for monitoring purposes are aggregated together into a data-group assigned with an identifier (a UUID) and additional metadata (e.g., job index, job name) to help the monitor script, or a user operating away from the WCE query and retrieve the data. Along with bespoke visualizations a user might develop for their data, a number of standard plots are available to monitor the progress of the search, such as viewing the objective function at the sample design points explored, or its convergence, or RSMs; see Figure 15.5.

Steering

Since the Geodise repository provides a convenient location to deposit and then monitor variables as the search progresses, steering can be achieved by updating values in the database, which are then retrieved by the running script to control the design search. A number of uses for steering have been identified, for example, to modify design variables when the objective functions show no further improvements early in the search. In practical implementations,

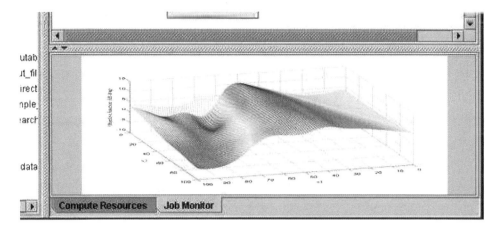

Figure 15.5   Screen shot of Response Surface Modeling monitoring.

two important issues arise: (1) side effects from arbitrary variable updates must be carefully understood and controlled; (2) information in the database must be "write-once", to prevent data being modified or overwritten; instead, scripts use queries to locate the latest version of the variable. Thus, as variables are updated, a complete log of all intermediate values is recorded, which is important for establishing provenance and repeatability.

Figure 15.6 illustrates how the system works using a typical test function with multiple local optima. To achieve steering, workflows are augmented with a set of grid-enabled query and archive functionalities that users can plug into their existing scripts at various points;see Figure 15.6(a). During each iteration, the optimization process retrieves the current user action request from the action history in the database and takes the appropriate action to achieve steering. The user can then monitor progress and interactively steer the process. The landscape of the test objective function is shown in Figure 15.6(b). The search progress is monitored from another Matlab session based on the data-group ID that provides the entry point to the database. The search history is shown in Figure 15.6(c) along with the effect of user steering.

15.1.8   Knowledge Services

To get the best out of a complex tool set like Geodise requires some form of information-rich support service. This needs to support the engineer throughout the design process by providing advice and guidance on what to do at any stage. Geodise has adopted the concept of a knowledge service to provide such advice in a robust and easily managed fashion that is both semantically rich and machine interpretable. This involves providing ontology services, annotation services and context-sensitive advice based on the states of the computation.

In order to achieve this vision, Geodise has adopted semantic web-based knowledge management. To capture the knowledge needed to describe functions and workflows so that they may be best reused by less-experienced engineers, this has involved the development of a layered semantic infrastructure that provides:

- a generic knowledge development and management environment to develop the ontology that underpins the more domain specific components (OntoView[17]);

---

[17]http://www.geodise.org/publications/papers.htm

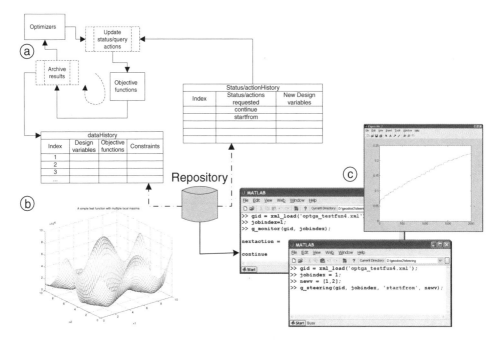

Figure 15.6   Illustration of optimization monitoring and steering using a grid-enabled database.

- a Matlab function annotation mechanism to associate concepts with functions;

- a mechanism that allows engineers to locate functions based on ontology-driven search criteria;

- a GUI-based component integrated with the Geodise workflow construction environment and a domain script editor that uses function semantic information to recommend candidates for function assembly and workflow composition; and

- an ontology developed by domain experts in collaboration with knowledge workers using the RDF-based DAML + OIL language.[18]

Capturing Knowledge

In order to form a conceptualization of the engineering design search and optimization domain, various experts were interviewed, information taken from application manuals and source code studied for key concepts and relationships. This information has been used to build an ontology that is represented in the DAML + OIL machine readable language. This language is based on DLs, contains formal semantics and supports reasoning. Ontologies built in DAML + OIL can be elaborate and expressive, but the temptation is to overcomplicate the interface to them, rendering them daunting and incomprehensible to the user. In Geodise, a simplified presentation interface has been adopted that loses little of the expressivity of the language but hides complexity from the user. In Geodise, this is called

---

[18]http://www.daml.org

OntoView – it provides a "domain expert sympathetic" view over the ontology, which is configurable by knowledge engineers in collaboration with domain specialists.

OntoView also includes a Java API that allows client code to perform ontology reasoning and instance manipulation, as well as query formation and execution. The domain-specific components of the Geodise knowledge services – the function annotator, the function query GUI and the knowledge advisor – are all built on this API.

In addition to capturing knowledge from domain experts, it is important to be able to capture and reuse knowledge from ordinary Geodise users. To support this, Geodise provides a function annotator that allows users to semantically annotate their own Matlab functions, while incorporating them into the Geodise environment. This makes them available for use in building workflows and available for reuse by others. To help support this process, information about input and output parameters and location details is parsed directly from the Matlab code. Other types of semantic information can be entered manually.

Reusing Knowledge

The fundamental purpose of knowledge capture is, of course, its reuse. Within Geodise, tools are available to support function queries and advice to assist in function discovery and configuration, script assembly and workflow composition. Each function can be viewed as a domain-specific service that must be configured correctly and assembled with other services to form a problem-solving workflow. The granularity of the services varies from low-level atomic functions (usually generic) to high-level workflow building blocks (often more problem specific), which are themselves made up of low-level functions.

Three specific types of advice can be provided within the Geodise system. First, function queries can be supported where users input keywords to match against the criteria defined for each function within the Geodise toolkits. For example, it is possible to ask for all functions written by a particular author or for those that invoke a particular computational facility. Secondly, it is possible to ask for function configuration advice so that a user can find out what the requirements of the elements in the toolkit are. Since such elements can be higher-level compound functions, this process is recursive and the user can probe down into functions to increase their understanding of how to configure and use such tools. Finally, it is possible to ask for function assembly advice. This advice details the compatibility between functions that are to be connected together in a workflow and can also provide information on likely successor or precursor functions that the user might wish to consider when assembling a workflow. In this way, engineers can focus on compatible elements that will be of use to further assemble the workflow without tediously investigating the semantic interfaces of all irrelevant functions.

In each case, the advice process is underpinned by the semantically enriched knowledge structure held in the DAML + OIL ontology. Moreover, Description Logic–based languages such as DAML + OIL support queries that are not possible using standard database queries. OntoView not only provides a means by which complex DL semantics can be represented within the ontology and instance descriptions, but it also takes advantage of these semantics to allow the formation of complex DL queries, which allow end-users, who are totally unaware of the existence of DLs, to formulate queries employing DL-based constructs:

In summary, the key features of the Geodise knowledge service are:

- the provision of advice on component configuration – exposing the semantic interface, tool-tipping annotations, auto-completions, and so on;

- the provision of advice on component assembly – interface matching and reasoning for contextual component recommendation;

- the knowledge capture process is decentralized – semantic instances are collected separately from their use;

- it is also generic – the tools can be used to advise on different domain scripts when loaded with corresponding semantic annotations, for example, on solver scripts, Geodise functions and problem-specific function scripts in Matlab, and so on, and finally

- it is component based – services can be delivered as java swing GUI components that can be used in any java application (e.g., in the GUI-based workflow construction environment as an alternative view of the workflow).

15.1.9 Simple Test Case

To demonstrate the capabilities of Geodise, we begin by examining the two-dimensional flow over a NACA four digit airfoil, using the direct scripting approach. A sketch of the problem is given in Figure 15.7. At the velocity inlet, the assumed free-stream velocity profile is constant while the angle of attack is measured in the counterclockwise direction to the horizontal. The upper and lower boundaries are periodic and there is a pressure outlet on the right-hand side of the computational domain. Gambit and Fluent are used for the mesh generation and solution processes, respectively, deployed via grid middle-ware.

In general, the process of obtaining a numerical solution on a remote Globus server involves at least the following steps (see Figure 15.8):

1. The user generates a grid proxy by entering their password.

2. The user generates data and input files required for the mesh generation and analysis software and transfers them to the remote Globus server.

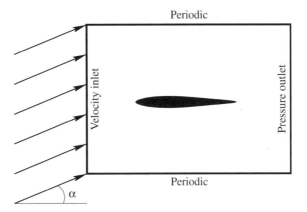

Figure 15.7 NACA four digit airfoil problem with boundary conditions. Here, α is the angle of attack.

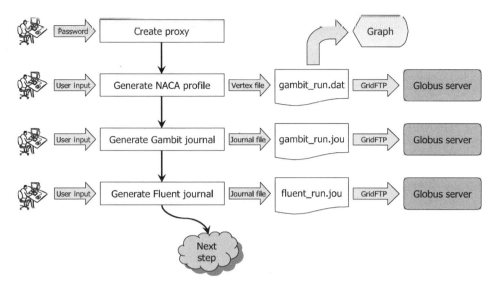

Figure 15.8   Initial data-flow during the generation of the grid proxy and preparation of data and journal files.

3. The mesh generation and analysis tools are run on the Globus server and intermediate files are queried to retrieve information regarding the mesh quality and the convergence of the solution.

4. If everything seems satisfactory, the user transfers the simulation results to the local file system or grid-based database, checks the objective function values and possibly visualizes the simulation results in Matlab, or a third-party product/plug-in.

Since a grid proxy is required to use the grid-enabled resources, the user creates their grid proxy certificate by using the gd_createproxy command. This command invokes the Java CoG, which in turn pops a window where the user can enter their password and create a proxy.

   Next, the vertex file describing the airfoil under consideration is transferred to the remote Globus server, where it will be used by the mesh generation tool Gambit. Meshing and solution is controlled by a pair of user-prepared journal files that define the work to be carried out by Gambit and Fluent. These files are tailored according to various input parameters that are entered by the user. The Gambit journal file informs Gambit to use the given vertex file as input, to mesh the domain using given mesh size parameters, and to export a Fluent compatible mesh file as output. Similarly, the Fluent journal file instructs that program to use the mesh file as input, to use inlet velocity and angle of attack parameters in the numerical solution, and to export a data file after the solution converges. When the journal files are ready, the user transfers them to the remote Globus server by using the gd_putfile(⟨FQHN⟩, ⟨Local File⟩, ⟨Remote File⟩) Matlab command. Here, ⟨FQHN⟩ is the fully qualified host name of the remote Globus server.

   After the data is prepared, the user submits jobs to the remote Globus server using the gd_jobsubmit(⟨RSL⟩, ⟨FQHN⟩) command. Here, the ⟨RSL⟩ describes the name of the executable on the Globus server, the executable's command line arguments, the names of standard input, output and error files, and so on. After job submission, the gd_jobsubmit

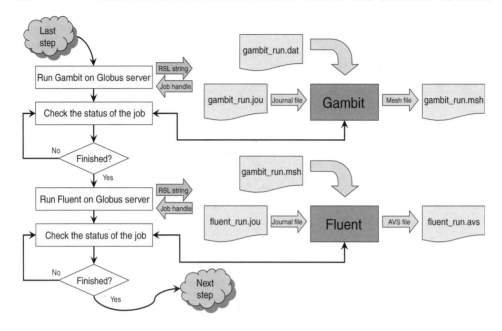

Figure 15.9 Process flow diagram for mesh generation and analysis tools.

command returns a job handle back to the Matlab environment, which is later used to check the status of the submitted job.

The mesh generation and analysis steps are summarized in Figure 15.9. Note that a correctly generated mesh file is required before analysis can start. At least to begin with, the user should wait until Gambit finishes meshing (i.e., the Globus server changes its status from "ACTIVE" to "DONE") and check that the quality of the generated mesh is acceptable for analysis. Therefore, before running the analysis tool, the standard error file from the Gambit run is returned to the local file system and parsed for mesh quality information.

If everything is satisfactory, the user can submit the analysis job on the remote Globus server, get back a job handle, check the status of the job, and retrieve convergence information and objective function values by using a very similar process. Finally, the solution can be transferred back to the local file system and visualized in Matlab; see Plate 5. Throughout this process, intermediate and solution files can be archived in the grid-enabled Geodise repository, with the gd_archive command. By associating metadata with the files, the design archive may be accessed interactively or programmatically when required using gd_query. Note that when using this tool there is no requirement for metadata and run data to be stored in the same place – often, it is more convenient to leave run data on the disk stores attached to the relevant compute servers and record these locations in the metadata repository.

Having illustrated the use of the Geodise toolkits on a simple test case they are next used to study a complex multi-dimensional problem in gas turbine noise control.

15.2   Gas-turbine Noise Control

Control of the noise generated by modern gas-turbine engines is now a major feature of the design of these systems. The need to take off earlier in the morning and later at night

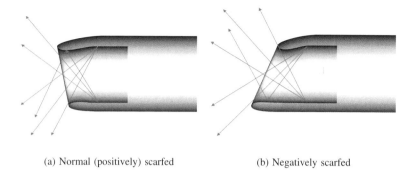

(a) Normal (positively) scarfed   (b) Negatively scarfed

Figure 15.10 Scarfed aero-engine inlets.

from runways lying in densely populated urban centers has driven recent legislation on aircraft noise levels, particularly at takeoff. A number of schemes have been adopted to reduce aero-engine noise levels, including fan redesign, insertion of nacelle linings and nacelle shape changes. In this final study, we concentrate on nacelle shape. In particular, we consider so-called negatively scarfed inlets where the extended lower lip of the nacelle is used to reduce the perceived noise levels (PNLs) measured on the ground during takeoff; see Figure 15.10. Clearly, extending the lower lip of the nacelle inlet tends to mask noise being radiated downward. Such changes do, however, have significant impact on the engine inflow during flight, since it is normal practice to use positively scarfed inlets to account for the slightly nose-upward attitude of most aircraft in flight and, in particular, during takeoff.

In this case study, we define improved noise performance by the degree of negative scarfing achieved with increasingly negatively scarfed inlets being considered increasingly better. We then constrain this aim by the need to ensure good pressure recovery at the fan face during cruise. We thus have two goal functions, scarf angle (which can be set directly, as it is an input to the CAD system that generates the geometry used in flow analysis) and fan face pressure recovery (which requires an expensive Navier–Stokes solve). Since these two goals are in direct opposition, we set this problem up as one requiring multiobjective optimization via the generation of Pareto fronts.

15.2.1   CAD Model

The geometry used here is defined parametrically using the ProEngineer CAD system with 33 input variables. Figure 15.11 illustrates the geometry created using this system, while Table 15.6 lists the parameter set used in the model. Parametric geometries have been generated here using feature-based solid modeling (FBSM). The results of this modeling are then exported via the use of the STEP standard data exchange format for meshing and analysis.

The variables are divided into three groups. The first group contains 13 variables from Scarf_angle to Silater; the second group contains 12 variables from Droop to Var_rmb; the third group contains eight variables from Var_d90 to Var_d236. Although all the geometry parameters are interlinked with each other in controlling the geometry, the first group is more directly involved in defining the shape of scarf and base lip shape; the second group

Table 15.6   Design parameters in the ProEngineer model

Variable Names	Lower Limits	Initial Values	Upper Limits	Description (units)
Scarf_angle	−10	−5	25	Negative scarf angle (deg.)
Teaxis	5	10	20	Axial coord. of top ext. profile (mm)
Telater	5	10	20	Radial coord. of top ext. profile (mm)
Beaxia	10	18.5	20	Axial coord. of bottom ext. profile (mm)
Belater	10	19	20	Radial coord. of bottom ext. profile (mm)
Seaxis	5	10	15	Axial coord. of side ext. profile (mm)
Selater	10	15	20	Radial coord. of side ext. profile (mm)
Tiaxis	1.5	2	2.5	Ratio of top inner profile coord. in axial direction against radial direction
Tilater	1	1.34	1.6	Coefficient used to determine top inner profile coord. in lateral direction
Biaxis	2.5	2.8	3.0	Ratio of bottom inner profile coord. in axial direction against radial direction
Bilater	1.0	1.27	1.5	Coefficient used to determine bottom inner profile coord. in lateral direction
Siaxis	2	2.45	3	Ratio of side inner profile coord. in axial direction against radial direction
Silater	1	1.34	1.6	Side inner profile coord. in lateral direction (mm)
Droop	−5	0	5	Radial offset of h.l. plane center (mm)
Var_d115	1,800	1,825	1,900	Axial length of h.l. plane center (mm)
Var_d119	1,800	1,850	1,900	Top lip axial length (mm)
Var_d122	1,700	1,750	1,800	Bottom lip axial length (mm)
Var_d124	3,300	3,340	3,400	Nozzle axial length (mm)
Var_d126	−5	0	5	Nozzle end point offset (mm)
Var_d138	85	90	95	Nozzle exit angle (mm)
Var_d220	1,100	1243	1,300	Top lip radius (mm)
Var_d221	1,300	1,360	1,400	Bottom lip radius (mm)
Var_rhs	1,400	1,430	1,500	Side lip radius (mm)
Var_rmt	1,700	1,710	1,800	Top lip end point radius (mm)
Var_rmb	1,600	1,625	1,700	Bottom lip end point radius (mm)
Var_d90	40	50	60	Forward length of inner barrel (mm)
Var_d92	600	620	650	Half length of inner barrel (mm)
Var_d225	22	25.4	28	Radial control of top lip profile (mm)
Var_d226	22	25.4	28	Axial control of top lip profile (mm)
Var_d230	22	25.4	28	Radial control of side lip profile (mm)
Var_d231	22	25.4	28	Axial control of side lip profile (mm)
Var_d235	22	25.4	28	Radial control of bottom lip profile (mm)
Var_d236	22	25.4	28	Axial control of bottom lip profile (mm)

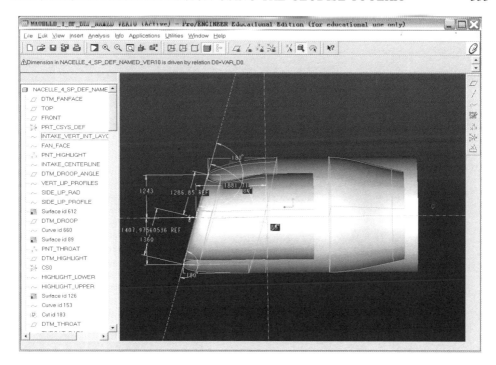

Figure 15.11 ProEngineer CAD model of intake geometry.

controls the general directions of the lip and nacelle axis; the third group controls the shape of inner barrel, the nozzle and lip profile.

15.2.2   Mesh Generation and CFD Analysis

In this problem, inlet flow conditions are studied for an inflow cruise velocity of Mach 0.85 with 4° upward inclination of the nacelle. Fluent is used to solve the problem adopting a subsonic laminar model. The nacelle is placed in a pressure far-field modeled using a rectangular box. Left/right symmetry is exploited to reduce the problem size and hence run times. The CFD goal function is the average pressure just behind the fan face exit. An important part of the analysis is accurate capture of the strongly three-dimensional boundary layers around the lip of the nacelle.

When carrying out CFD analysis as part of an optimization study, automatic mesh generation is essential, and this should rely on the highest-level entities in the model. Here, we decide node spacing based on the relative length of the edges in critical regions of the model and use size functions to control the local mesh pattern. It is crucial for a CFD-ready mesh that nodes on disjoint edges in close proximity have matching spacing-scales, and also that triangles on adjacent surfaces have matching length-scales. In most systems, it remains the user's role to identify such situations and enforce length-scale compatibility between nearby entities, although a limited number of automated facilities can often be provided to support such checks. For example, by partly recovering the topological information pertinent to the original geometry data, consistency conflicts on the boundary between entities can be avoided.

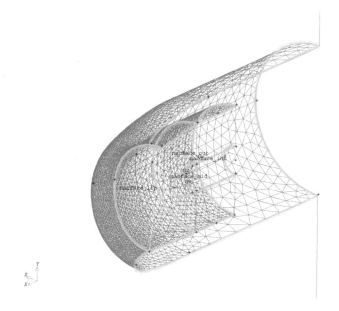

Figure 15.12 Unstructured surface mesh for engine nacelle.

Here, the edge mesh is generated by specifying an appropriate node spacing based on both the relative length of the edge being meshed and the relative lengths of adjacent edges. This process is fairly robust for a wide range of geometries with scarf angles varying from −10 to +25 degrees. A typical unstructured surface mesh is shown in Figure 15.12, while typical CFD results are shown in Plate 6.

15.2.3 Optimization Strategy

The aim of the optimization process here is to return a Pareto front of scarf angle versus average fan face total pressure recovery, as computed via Gambit/Fluent for the 33 variables in the ProEngineer model. Since each calculation takes 8 h (serial processing on a 3 GHz processor), a hybrid approach is adopted: first, we carry out a DoE-based survey using a grouping strategy. This allows us to rank the variables according to their impact on the pressure recovery – here, the values of the hyperparameters in a Krig model are used, although various other methods could be adopted. One by-product of using Krigs is that the metamodels can be inspected graphically, which can aid the designer in understanding the problem being studied. Having ranked the variables in the three separate groups, the most dominant ones can then be used in a second DoE study: here, we reduce the number of variables being considered to eleven. Finally, we search the reduced problem using a fourth Krig metamodel and updates together with a multiobjective Pareto strategy. These steps may be summarized as:

- Split the 33 design parameters into three groups (of 13, 12 and 8 – based on their geometric relationships).

- Run LPτ DoEs on each group separately using Gambit/Fluent. Continue until the number of successful results returned is around 8 to 10 times as many as there are variables in the group and then Krig the average pressure recovery versus the parameters (with careful hyperparameter tuning and regression using a hybrid GA/DHC search). Note that many combinations of parameters do not lead to feasible geometries, some of these will not mesh with the journal file used and a very few will not solve – typically around a half of those designs sent for analysis yield final results during the initial DoE runs.

- Use the hyperparameters to rank the design variables and take the most important from each model (here, five from group one, four from group two and two from group three).

- Run a fourth LPτ DoE on the resulting 11-dimensional space, again versus average pressure recovery from Gambit/Fluent until around 100 results are available.

- Take the resulting points to build a Krig and use it alongside scarf angle in a multi-objective search to define a Pareto front of scarf angle versus Krig estimated values of pressure recovery. The front is identified using a nondominated sorting GA in a search to generate around 20 evenly spaced Pareto optimal points.

- Take the points from the surrogate-based Pareto front and run full CFD calculations to update the Krig. Design points far from the Pareto front are rejected during updating to maintain the size of the dataset used to form the Krig – this improves tuning speeds and prediction accuracy as the global model slowly localizes in the region of the front.

- Repeat the Pareto search and update process until we have assessed some 200 points.

- Return the best Pareto set from all 200 CFD-based points (as used to in the fourth Krig) as the resulting *real* Pareto surface and thus the solution to the initial problem.

To do this efficiently requires that we can run multiple copies of the CFD code in parallel. Assuming that 25 licenses were available on 25 separate processors and we have to run full calculations on 4 × 4 batches of 25 for the initial DoEs plus 6 batches of 15 updates, this leads to 22 batches in all. Given a typical solution time of 8 h per batch, this leads to a run time of around one week.

The somewhat complicated workflow needed to carry out this process can be readily set up using the Geodise toolkits. This can take advantage of the heterogeneous computing resources required for the different tasks such as CAD, mesh generation and flow solution. The aim is to provide an automatic process with the greatest possible robustness. It is quite common to have failed CAD geometry regeneration in this process because of the nature of the design space being studied. Moreover, the mesh generation can also fail because of poor geometry features and weakness in the mesh generation process arising from inadequate geometry information being passed into the meshing tool through the geometry data exchange standard. The DoE-based process outlined here can readily tolerate failed or missing runs, however – these runs are simply omitted from the Krig construction and, provided a reasonable coverage of the design space is achieved, the overall process can still be followed. Note that using an LPτ DoE allows run sizes to be adapted to the success rates being achieved.

Table 15.7 Parameters and resulting hyperparameters for the initial groups, along with design values for a good compromise design taken from the final Pareto front (ranking is within groups)

Variable Names	Lower Limits	Initial Values	Upper Limits	Hyper-parameters	Rank	Final Design
Scarf_angle	−10	−5	25	−1.7452	6	9.250
Teaxis	5	10	20	−0.6119	3	5.012
Telater	5	10	20	−8.9167	11	10
Beaxia	10	18.5	20	−8.2143	9	18.5
Belater	10	19	20	−1.542	5	19
Seaxis	5	10	15	−9.1343	12	10
Selater	10	15	20	−2.2151	7	15
Tiaxis	1.5	2	2.5	−0.5357	2	1.599
Tilater	1	1.34	1.6	−1.4785	4	1.100
Biaxis	2.5	2.8	3.0	−0.2945	1	2.790
Bilater	1.0	1.27	1.5	−9.2881	13	1.27
Siaxis	2	2.45	3	−6.523	8	2.45
Silater	1	1.34	1.6	−8.8341	10	1.34
Droop	−5	0	5	−7.7871	6	0
Var_d115	1,800	1,825	1,900	−8.3255	9	1825
Var_d119	1,800	1,850	1,900	−8.1841	8	1850
Var_d122	1,700	1,750	1,800	−2.0404	4	1779.5
Var_d124	3,300	3,340	3,400	0.0418	1	3331.7
Var_d126	−5	0	5	−6.1258	5	0
Var_d138	85	90	95	−0.0226	2	93.55
Var_d220	1,100	1243	1,300	−8.4515	10	1,243
Var_d221	1,300	1,360	1,400	−1.7391	3	1,399.9
Var_rhs	1,400	1,430	1,500	−9.2943	12	1,430
Var_rmt	1,700	1,710	1,800	−7.905	7	1,710
Var_rmb	1,600	1,625	1,700	−8.8214	11	1,625
Var_d90	40	50	60	1.2238	1	44.04
Var_d92	600	620	650	0.3443	2	638.9
Var_d225	22	25.4	28	−7.9516	7	25.4
Var_d226	22	25.4	28	−6.6808	6	25.4
Var_d230	22	25.4	28	−1.3783	4	25.4
Var_d231	22	25.4	28	−1.4108	5	25.4
Var_d235	22	25.4	28	−8.5261	8	25.4
Var_d236	22	25.4	28	−0.018	3	25.4

   This flexibility is particularly important when remote computing and database resources are used, since, although in a perfect world these resources would be available around the clock, in practice it is impossible to avoid unscheduled downtimes. The use of the Geodise repository and client-side toolkit allows jobs to be restarted without reevaluating existing

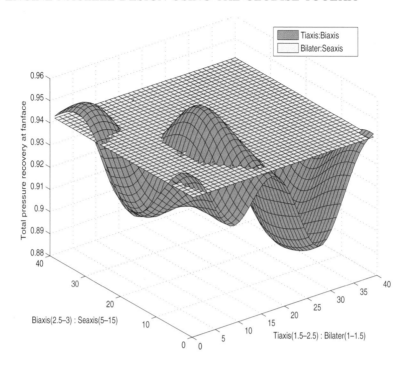

Figure 15.13 Illustration of the effects of four different design variables on the total pressure recovery for the first group (Biaxis v's Tiaxis and Bilater v's Seaxis).

points. In addition, the restart can be driven from different client machines pointing to the same data repository.

Design variables, datum values and associated hyperparameters obtained for the Kriging models are listed in Table 15.7. To illustrate the effects of the different design variables, contour maps can be produced for the most important and least important parameters from each set, while fixing the remaining parameters at datum values; see, for example, Figure 15.13. In total, 93 DoE points returned values for the first group, 76 for the second and 131 for the third. The choice of design variables to take forward to the combined group is based on a combination of their hyperparameters, contour plots like those in Figure 15.13 and engineering judgment. The process of initial group selection and subsequent combination making use of such information is discussed more extensively by Dupplaw et al. (2004).

Figure 15.14 illustrates the effects of the two most important variables on the Krig for the combined problem (besides scarf angle) after the update strategy has been completed, while Figure 15.15 shows the final Pareto front. Table 15.8 details the resulting hyperparameters and variable ranking. Notice that the rank ordering has changed even for variables within the same initial grouping. It may be seen from Figure 15.15 that there is a trade-off possible between scarf angle and fan face total pressure recovery – as the angle increases, the pressure recovery slowly drops. Figure 15.16 shows a typical design taken from the Pareto front that has useful negative scarf and good pressure recovery. The variables for this design are also detailed in Table 15.7.

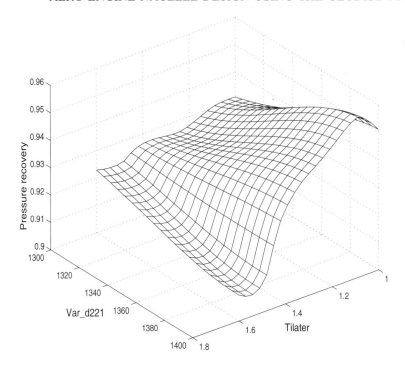

Figure 15.14   Effects of variables from the combined group on the pressure recovery (Var_d221 v's Tilater).

Table 15.8   List of parameters and hyperparameters (after updating) for the combined group of most influential parameters

Variable Names	Lower Limits	Initial Values	Upper Limits	Hyper-parameters	Rank
Scarf_angle	−10	−5	25	−1.9202	3
Biaxis	2.5	2.8	3.0	−5.3170	6
Tiaxis	1.5	2	2.5	−4.9529	4
Teaxis	5	10	20	−7.1523	8
Tilater	1	1.34	1.6	−0.5073	1
Var_d124	3,300	3,340	3,400	−11.4759	10
Var_d138	85	90	95	−6.7069	7
Var_d221	1,300	1,360	1,400	−1.8832	2
Var_d122	1,700	1,750	1,800	−10.3956	9
Var_d90	40	50	60	−17.1866	11
Var_d92	600	620	650	−5.2611	5

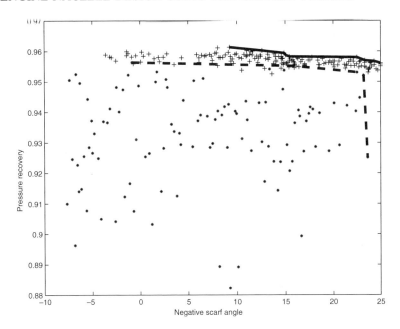

Figure 15.15 Evolution of Pareto front for total pressure recovery versus scarf angle based on the final Krig (dotted line shows front after initial DoE, solid line shows final front, stars show DoE CFD results and crosses show update points).

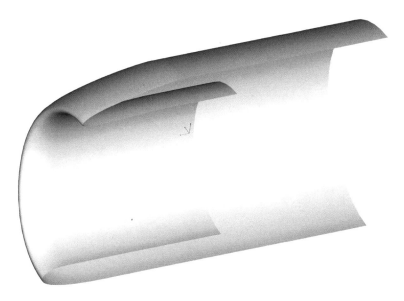

Figure 15.16 Design geometry from the Pareto front for the column marked "Final Design" in Table 15.7.

15.3 Conclusions

The choice of the Matlab environment to host the Geodise toolkit was a pragmatic decision. Matlab provides a flexible and robust user interface for grid computing from a familiar and powerful scripting tool. The Java CoG provides a valuable platform-independent client-side API from which to access grid-enabled resources. Database technologies have an important role to play in engineering design and optimization and are useful for managing contextual and technical metadata associated with design processes. They also facilitate data sharing among engineers. The use of web services for API access to the databases is beneficial for creating usable client-side functions and facilitating communication between different languages and platforms.

By exposing compute and data functionality as toolkit components, it is possible to construct high-level functions that utilize grid resources for analysis and design search tasks. Given commands in a high-level interpretive language, it is then straightforward for the engineer to exploit the available grid-enabled resources to tackle computationally complex and data-intensive design tasks.

In this example, an expensive CFD search has been undertaken in a multiobjective setting, making use of group screening and response surface modeling. The Geodise toolkit permits all of these activities to be undertaken from a single environment using state-of-the-art tools across large-scale distributed computing facilities in a straightforward fashion. It represents current best practice in the authors' research group and combines many of the ideas set out in this book.

16

Getting the Optimization Process Started

When first facing a design problem where formal optimization methods are to be applied, the engineer is confronted with choosing between a great many possible approaches – this book aims to discuss most of these and to try and set out what is currently possible. In this final chapter, we attempt to condense some of this information into a concise guide on how to begin such studies. It should be stressed, however, that such a guide can never be completely foolproof and so the engineer is, as ever, required to make professional judgments when carrying out his or her work. This guide is mainly intended to deal with single-domain problems or those where multiple domains have been closely coupled together – it does not address cases where there are competing loosely coupled domains that must be balanced against each other. Moreover, it only provides a point of departure – once work has begun, the results produced must inevitably influence the course of the design process and there are too many possible directions things can take to attempt to summarize them here (there is also a growing and ever changing array of web sites on the internet that offer advice in these areas, see for example `http://plato.la.asu.edu/guide.html`).

Of course, the very first thing that must be achieved if optimization methods are to be deployed is the creation of some form of design automation system that allows alternative designs to be created, analyzed and then ranked. Such automation must define inputs and outputs. It must allow the inputs to be simply specified and the outputs to be processed into the quantities of interest in a robust and efficient form. This is not to say that this process cannot involve manual intervention but that an automated scheme is normally the aim. If any manual input is required, this will necessarily limit the kinds of search that can be carried out and generally slow things down. In some circumstances, this is unavoidable but this fact should not preclude the use of relatively formalized searches. The discipline of clearly setting out goals and constraints, listing those variables open to manipulation by the designer and those that are not, before attempting to rank competing alternatives can, by itself, lead to better design.

Throughout this book, it has been noted that a parsimonious approach to inputs is desirable – the fewer quantities the designer must consider, the better. To this end, it is again stressed that giving significant thought and effort to design parameterization is a key

Computational Approaches for Aerospace Design: The Pursuit of Excellence.   A. J. Keane and P. B. Nair
© 2005 John Wiley & Sons, Ltd

activity that is likely to lead to successful searches and thus better designs. In far too many cases, design optimization is rejected because of bad experiences that have arisen from the adoption of poor or clumsy parameter choices. Designers are often much better at choosing appropriate measures of merit and constraints than on setting out the way the problem is to be described and manipulated – this is almost a cultural phenomenon that stems from the training most engineers receive – nonetheless, it is one that must be overcome if the best designs are to be obtained.

Having set up and linked together the appropriate analysis modules (probably at the cost of significant effort) and tested them on some preliminary design calculations, work can then begin on assessing how to set up a search process.

16.1   Problem Classification

Almost by virtue of necessity, once automation of the analysis process is complete, the designer will have at least one set of inputs for which the process can and will have been run. Very often, this will represent some previous design or initial study from which the search for better things can begin. A good deal of information can be extracted from such a test case and this forms the starting point in attempting to classify the problem at hand so that a sensible strategy can be decided upon.

16.1.1   Run Time

The first thing that the designer will know about this test case is how long it takes to run on the available computing facilities. This will typically vary from fractions of a second to weeks. The designer is also likely to know how far in the future the search for improved designs is likely to have to stop – he/she thus knows, to at least an order of magnitude, how many design analyses can be afforded at most. This quantity forms the first item of information needed in planning a search strategy (see Figure 16.1). This run-time calculation must allow for any multidisciplinary factors in the problem and whether concurrent calculations are possible with, say, aerodynamic and structures analyses taking place alongside each other, or if a strictly serial approach must be taken. Broadly, searches may be classified directly by the number of calculations that can be afforded as follows:

- Very-expensive analysis – those where fewer than 10 runs are possible in the available time (typically, analyses that make use of unsteady CFD or impact-based CSM codes).

- Expensive analysis – those where between 10 and 1,000 runs are possible (steady CFD using RANS or Euler codes or simpler nonlinear CSM).

- Cheap analysis – those for which thousands or even tens of thousands or runs are possible (panel-based CFD codes or linear finite element analysis).

- Very cheap analysis – those where hundreds of thousands of calls or more can be made (empirical methods or those where high-fidelity PDE solvers are not needed).

The reason for this initial focus on run time is that in all practical design processes the available computing budget will be circumscribed by cost and also by the need to conclude the search process within some reasonably (often unreasonably) short product definition period.

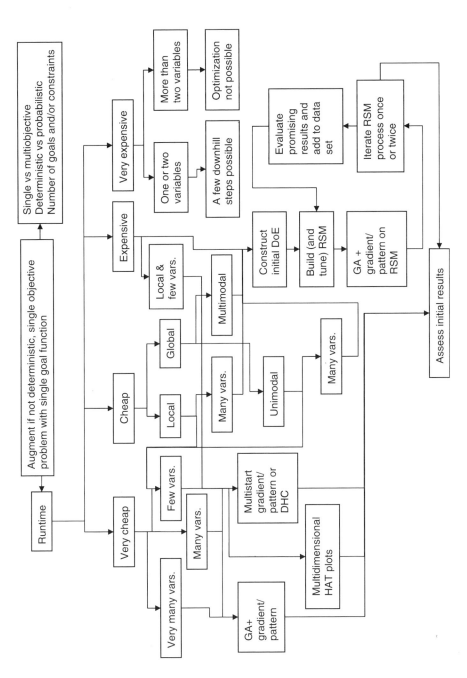

Figure 16.1 Flow chart for initial search method selection – this follows the discussion in the text by first considering run time and the problem scale, so as to lead to a suggested initial search method.

16.1.2 Deterministic versus Probabilistic Analyzes

Following thoughts about run time, the designer must next decide whether the analyses being carried out are deterministic or not, and if not, how design goals and constraints are to be described – will the designer treat them as essentially deterministic anyway and apply some safety factors after search is complete or will a more formal treatment of uncertainty be used. Sometimes, it is possible to arrange for the analysis codes in use to yield statistical measures directly, while in other cases, the aim is to try and explore any uncertainty as part of the search process by multiple calls to the available codes. If direct calculation of probabilistic measures is available, then no further actions will be necessary except in deciding how to treat the resulting statistical descriptions in terms of goals and constraints. If not, a trade-off between run time and the quality of any statistics that can be found has to be made. It is clear that it is always possible to perturb any analysis being made in a Monte-Carlo sense and thus to glean information on mean values, standard deviations and so on. Unfortunately, gaining accurate measures in this way typically involves hundreds of runs and thus moves all but the simplest problems into the area of expensive or very-expensive analysis. If this cannot be afforded, then less-accurate approaches must be adopted.

16.1.3 Number of Variables to be Explored

Another key feature of any search process is the number of designer-chosen variables that are to be explored (the design vector). This will vary widely between problems and also between parameterizations of a given problem (as illustrated in the case study on airfoil-section design). It is also likely to be strongly linked to the expense of the analysis being considered, simply because when a problem permits only a very few runs to be undertaken designers will simply not even begin to think of search processes if they know that hundreds or more variables must be considered. Here, we distinguish between problems with few variables (less than five or so), those with many (tens) and those with very many (hundreds and more). Distinctly different approaches will be taken in dealing with problems at the extremes of these ranges.

16.1.4 Goals and Constraints

Finally, in completing an initial classification of our problem, we must note how many goals we have (one or more than one) and whether there are any constraints that must be met. As has already been pointed out, if there is more than one goal (or if some of the constraints are to be treated as "soft" targets), then a multiobjective approach will be needed unless the designer is happy to combine all these quantities into a single measure of merit using some of the schemes outlined in Chapter 3. If formal multiobjective techniques are to be used, this will also have a knock-on effect on run times, since the construction of sets of possible solutions (such as Pareto sets) will usually add significantly to run time.

16.2 Initial Search Process Choice

Given a problem classification, the designer can start to map out an initial search process, following the logic of Figure 16.1. We begin by observing that design is almost never a one-shot activity, even if a fully automated search process is available. Usually, the classification of the problem will have some lack of precision; commonly, the designer is not certain about the need to manipulate some variables, the importance of some goals is unknown and some of the constraints may be relaxed/removed if need be. The designer will thus approach the problem in an iterative spirit and treat the search process as a design guidance system that must be steered by hand between searches. Therefore, it is wise to split up the likely available computing budget into several chunks. We would advise committing no more than 15–25% of the available budget to initial search runs.

16.2.1 Problems where Function Cost is not an Issue

If the analysis process is "very cheap", then almost any directly coupled optimization approach can be taken and the designer's aim will be to try and ensure that a globally optimum solution (or set of solutions) is obtained, with each point being sampled multiple times if probabilistic measures are to be computed on the fly. In these (rather rare) circumstances, the best approach will probably be to use a population-based search followed by local improvement applied to promising solutions. It will also be important (probably implicit) that any data-flows and transfers be accomplished very efficiently – for example, transferring data around a series of analysis codes by the use of files will probably not be sensible – rather, all codes should be linked directly together or in core, memory-based data transfers used. This will also mean that all function evaluations will most probably happen on a single processor (or perhaps a multiprocessor-shared memory machine). The precise choice of search method used will depend on the number of variables being investigated. If there are "many" or "very many", then tools such as the genetic algorithm will be most appropriate, followed by a gradient descent method if there is little numerical noise or a pattern search if there is. When there are "few" variables, then the use of a sequence of gradient descent or pattern searches working from multiple (DoE selected) start points can be more effective. Alternatively, extensive use of dynamic hill climbing (Yuret and de la Maza 1993) combines these in some sense and can prove simple to manage. Unfortunately, most design problems do not allow such a profligate use of function calls and a trade-off must be made between depth and breadth of search.

16.2.2 Moderately Difficult Problems

Where the problem being considered is "cheap" and allows some thousands of function calls, it will still be plausible to directly couple the search engine to the analysis codes but now more care must be used in the choice of search process. In particular, the designer will need to decide whether a global search is desired or alternatively if simple local improvements are all that is needed. If a global search is envisaged, it will be important to decide if the function space being searched is multimodal or not. If it is multimodal, then it will be difficult to search all the most promising basins of attraction with a few thousand analyses, especially if there are many variables to be dealt with. In consequence, it may be wise to consider a response surface–based approach.

If the problem is thought to contain only a few basins of attraction, then a global search and a local one differ little and the best approach will be multiple starts of an appropriate down-hill search method (gradient based if the function contains little noise or pattern based if it does). Such an approach will work provided there are not too many variables so that generating gradients or sensitivities is not too expensive by finite differencing (this can, of course, be mitigated by an adjoint code or other means of direct gradient evaluation).

If there are many variables or a global search is needed over many basins of attraction, then some kind of scheme will be needed to limit run times. The most obvious methods involve parallelized Design of Experiment (DoE) searches to build a database of results followed by some kind of Response Surface Modeling (RSM). The choice of DoE and RSM will depend on the nature of the problem being tackled, as will any update strategy envisaged (it is always wise to reserve at least some of the computational budget for model refinement when constructing RSMs). When there are many or very many variables, it will probably be necessary to limit any RSMs to local coverage so as to increase the density of sample points.

16.2.3 Expensive Problems

If the problem being dealt with is sufficiently expensive that only tens or hundreds of runs are possible, direct searches will only be possible if there are few variables, unless gradients can be calculated cheaply. Even then, only local searches will normally be possible. More commonly, it will be more appropriate to switch to DoE and RSM approaches so that a metamodel can be searched in lieu of the full problem. Ideally, this should allow around ten evaluations per variable being examined if a global RSM is to be constructed. If only a local model is being built, many fewer calls will generate a useful RSM, at least for initial evaluation purposes.

16.2.4 Very Expensive Problems

If the problem being dealt with permits less than ten runs, it will be impossible to carry out other than the most simple of limited surveys over the design space. When there are more than two variables to deal with, the designer will be forced to reject some variables from the search completely – this will preferably be based on experience but may have to be arbitrary. Having reduced the problem to the two most promising variables, the designer should then try and assess the most likely direction for improvement and place a single design in that direction, varying both variables by a modest step. This alone will have probably expended the exploratory phase of any search.

16.3 Assessment of Initial Results

Once the initial calculations have been performed, the design team should stand back and assess what to do next. It is never wise to plow on consuming the computational budget until one is sure that the effort can be justified. At this stage, the following information should be available:

1. For a very cheap problem, a significant number of minima should have been located, most likely to a good level of precision – if only a single solution has emerged then, in all probability, the function being dealt with is unimodal – even so, no guarantees can be made that the design basins found contain the global optimum; if no improvement over the initial design has been found, it is most likely that there is some problem with the analysis being used or the severity of any hard constraints – it is extremely rare to have a well-posed problem that is to be optimized and where tens of thousands of design shots produce no improvement at all.

2. For cheap problems where only a local search is required over a few variables, again one or more (probably local) optima should have been located; otherwise, some kind of response surface model will have been constructed and searched to reveal promising locations that have been tested at least once; a failure to find improvements is now more likely though still rare, especially where direct search has been carried out; a failure to make improvements via a RSM may be due to a poor choice of RSM or insufficient data being available to construct the model.

3. For expensive problems, a response surface model will have been built and tested, although probably the RSM will be somewhat limited in accuracy or scope (it might, for example, be a localized RSM) – it will be clear from any updates whether the RSM is accurately predicting good new designs.

4. For very-expensive problems, the initial testing phase will have focused on locating a single design change in probably two variables to run – hopefully, this will have improved the design – if not, at least the sensitivity of the design goal to changes in the variables will be clearer.

The next stage of the search process involves assessing this information and committing the bulk of the available computing effort, although perhaps 10–15% of the budget should still be retained for final refinement of the design.

16.3.1 Multiple Solutions found for Very Cheap Problems

When multiple good solutions are being found for very cheap problems, then, in all likelihood, the search is working well. Little needs to be done except to ensure that some of the best designs are fully converged and that the designs being produced are sufficiently spaced out across the range of interest. If all the good designs being found are too bunched together, it may be wise to simply use bound constraints to remove such regions from the search before proceeding – this should force other designs to come forward or to make the search system butt up against such bounds. If, on the other hand, the designs being produced are not fully converged or are adjacent to bound constraints, then it may be simpler to search a reduced and bounded region or set of regions using the current techniques or closely related ones (sometimes, a change of gradient or pattern search may be sufficient to improve matters). If no improved designs are being returned at all, something is most likely wrong with the problem specification – if there are no analysis problems, it may be that constraints are preventing any design improvements. In any case, an investigation should be made to see why improvements cannot be found before proceeding.

16.3.2 Single Solution found for Cheap or Very Cheap Problem

If only a single solution was found despite a reasonable number of searches of the problem, then it is probable that the function being dealt with is unimodal – a study of the results coming from the evaluations carried out thus far should confirm if this is true. Detailed convergence calculations will then be warranted, along with some wider-ranging investigations to see if the current search boundaries or any constraints are restricting progress. Again if no improvements are being found, the problem setup should be examined and explored as before.

16.3.3   Good Predictive RSM Built

When a good RSM has been built, the problem being dealt with is essentially transformed into one in which very large numbers of function evaluations can be afforded. The problem then becomes one of maintaining the accuracy of the RSM so that it can be exploited. When there are few variables being considered, this is normally straightforward, either using local or global models, though, if global models are being used, care must be taken that they offer sufficient flexibility as the search converges toward any optimum (for example, a regression-based model will converge only very slowly if many outlier points are included in the final stages – this is why trust-region methods strictly control the region in which search is carried out).

When there are many or very many variables, it will be harder to ensure that the RSM remains effective and so it will be necessary to continue to use trust-region approaches to limit the range over which the RSM is valid to maintain the density of samples being used, that is, to move the validity of the RSM slowly over the search landscape rejecting old data points as new values representing better-quality results are found and used.

16.3.4   Graphical Presentation of Results

When assessing initial results from searches, it is often helpful to use pictorial methods of assessing the available data. Two particular forms of graphical representation can be very helpful: first, there is the plot of objective function and constraint values against evaluation count – the search history over elapsed time. Such plots show whether things are getting better and also how quickly. They will reveal if any constraints are being explored and whether the search is working continuously or in stages. The second plot of interest displays the goals and constraints as functions of the design variables. For a single variable, this is obviously a simple graph, while for two variables, a contour plot is used. For more than two variables, a series of contour plots is of value – up to four can readily be shown on a single page if a hierarchical axes technique is used; see Plate 7. Plots such as this commonly require a sophisticated data interpolation/regression strategy to produce, since it is rare to have the tens of thousands of data points needed as a result of direct function calls – this is where a RSM can again be of value, this time based on the results of the search points made to date, whether or not they have been chosen using DoE methods. Plate 7 shows a remarkable quantity of information: the variation of the primary goal function via the colored contour map, the position of the data points used to construct the map via the crosses, the presence of an inequality constraint boundary via the uncolored region and of an equality constraint by the sets of solid lines (which show the equality and $\pm 5\%$ variations either side of it). This particular plot has been produced using a Krig to generate the data needed for the figure from the search runs. Armed with such information, the designer is better positioned to carry on the pursuit of excellence that is the raison d'etre of this book.

Bibliography

I.H. Abbot and A.E. Von Doenhoff. *Theory of Wing Sections*. Dover Publications, 1959.

H.M. Adelman and R.T. Haftka. Sensitivity analysis of discrete structural systems. *AIAA Journal*, 24: 823–832, 1986.

S. Adhikari. Reliability analysis using parabolic failure surface approximation. *ASCE Journal of Engineering Mechanics*, 130(12): 1407–1427, 2004.

S. Adhikari and M.I. Friswell. Eigenderivative analysis of asymmetric non-conservative systems. *International Journal for Numerical Methods in Engineering*, 51: 709–733, 2001.

AGARD. Technical status review on drag prediction and analysis from computational fluid dynamics: state of the art. Technical Report AR-256, AGARD, 1989.

H. Agarwal, J.E. Renaud, E.L. Preston, and D. Padmanabhan. Uncertainty quantification using evidence theory in multidisciplinary design optimization. *Reliability Engineering and System Safety*, 85: 281–294, 2004.

S. Ahmed, L.T.M. Blessing, and K.M. Wallace. The relationships between data, information and knowledge based on a preliminary study of engineering designers. *Proceedings of ASME Design Theory and Methodology*, Las Vegas, NV, 12–15, September, 1999, Paper number: DETC99/DTM-8574.

S. Ahmed and C.T. Hansen. An analysis of design decision-making in industrial practice. *Design 2002 – 7th International Design Conference*, Dubrovnik, Croatia, 1: 145–150, 2002.

P.V. Aidala, W.H. Davis, and W.H. Mason. Smart aerodynamic optimization. AIAA Paper 83-1863, 1983.

M. Akgün. New family of modal methods for computing eigenvector derivatives. *AIAA Journal*, 32: 379–386, 1994.

M.A. Akgün, J.H. Garcelon, and R.T. Haftka. Fast exact linear and non-linear structural reanalysis and the Sherman-Morrison-Woodbury formulas. *International Journal for Numerical Methods In Engineering*, 50: 1587–1606, 2001a.

M.A. Akgün, R.T. Haftka, K.C. Wu, J.L. Walsh, and G.H. Garcelon. Efficient structural optimization for multiple load cases using adjoint sensitivities. *AIAA Journal*, 39: 511–516, 2001b.

N.M. Alexandrov. *Multilevel Algorithms for Nonlinear Equations and Equality Constrained Optimization*. PhD thesis, Rice University, 1993.

N.M. Alexandrov, S.J. Alter, H.L. Atkins, K.S. Bey, K.L. Bibb, R.T. Biedron, M.H. Carpenter, F.M. Cheatwood, P.J. Drummond, P.A. Gnoffo, W.T. Jones, W.L. Kleb, E.M. Lee-Rausch, N.R. Merski, R.E. Mineck, E.J. Nielsen, M.A. Park, S.Z. Pirzadeh, T.W. Roberts, J.A. Samareh, R.C. Swanson, J.L. Thomas, V.N. Vasta, K.J. Weilmuenster, J.A. White, W.A. Wood, and L.P. Yip. Opportunities for breakthroughs in large-scale computational simulation and design. NASA/TM-2002-211747, June 2002.

N.M. Alexandrov, J.E. Dennis Jr., R.M. Lewis, and V. Torczon. A trust region framework for managing the use of approximation models in optimization. *Structural Optimization*, 15: 16–23, 1998b.

N.M. Alexandrov, R.M. Lewis, C.R. Gumbert, L.L. Green, and P.A. Newman. Optimization with variable-fidelity models applied to wing design. AIAA Paper 2000-0841, 2000.

N.M. Alexandrov and R.M. Lewis. Algorithmic perspectives on problem formulations in MDO. AIAA 2000-4719, 2000a.

N.M. Alexandrov and R.M. Lewis. Analytical and computational aspects of collaborative optimization. NASA/TM-2000-210104, 2000b.

M. Allen and K. Maute. Reliability-based design optimization of aerelastic structures. *Structural and Multidisciplinary Optimization*, 27: 228–242, 2004.

J.J. Alonso, J.R.R.A. Martins, J.J. Reuther, R. Haimes, and C.A. Crawford. High-fidelity aero-structural design using a parametric CAD-based model. AIAA Paper 2003-3429, 2003.

N.R. Aluru and J. White. A multilevel Newton method for mixed-energy domain simulation of MEMS. *Journal of Microelctromechanical Systems*, 8: 299–308, 1999.

E. Anderson, Z. Bai, C. Bischof, J. Demmel, J. Dongarra, J. Du Croz, A. Greenbaum, S. Hammarling, and A. McKenney. *LAPACK User's Guide*. SIAM, Philadelphia, second edition, 1995.

G.L. Ang, A.H-S.Ang, and W.H. Wang. Optimal importance-sampling density function. *ASCE Journal of Engineering Mechanics*, 118: 1146–1163, 1992.

D.K. Anthony and S.J. Elliott. Comparison of the effectiveness of minimizing cost function parameters for active control of vibrational energy transmission in a lightly damped structure. *Journal of Sound and Vibration*, 237: 223–244, 2000a.

D.K. Anthony and S.J. Elliott. Robustness of optimal design solutions to reduce vibration transmission in a lightweight 2-d structure, part II: application of active vibration control techniques. *Journal of Sound and Vibration*, 229: 520–548, 2000b.

D.K. Anthony, S.J. Elliott, and A.J. Keane. Robustness of optimal design solutions to reduce vibration transmission in a lightweight 2-d structure, part I: geometric design. *Journal of Sound and Vibration*, 229: 505–528, 2000.

D.K. Anthony and A.J. Keane. Robust-optimal design of a lightweight space structure using a genetic algorithm. *AIAA Journal*, 41: 1601–1604, 2003.

E. Arian. Convergence estimates for multidisciplinary analysis and optimization. ICASE Report No. 97-57, NASA Langley Research Center, 1998.

E. Arian and M.D. Salas. Admitting the inadmissible: adjoint formulation for incomplete cost functionals in aerodynamic design. *AIAA Journal*, 37: 37–44, 1999.

D.V. Arnold and H.G. Beyer. Local performance of the (1+1)-ES in a noisy environment. *IEEE Transactions on Evolutionary Computation*, 6: 30–41, 2002.

P. Audze and V. Eglais. Chapter New Approach to Planning out of Experiments. *Problems of Dynamics and Strength*, Volume 35, pages 104–107. Zinatne Publishing House, Riga, 1977 (in Russian).

M. Avriel. *Nonlinear Programming: Analysis and Methods*. Prentice Hall, Englewood Cliffs, 1976.

E.H. Baalbergen and H. van der Ven. Spineware: a framework for user-oriented and tailorable meta-computers. NLR-TP-98463, 1998.

E.H. Baalbergen and H. van der Ven. Spineware – a framework for user-oriented and tailorable meta-computers. *Future Generation Computing Systems*, 15: 549–558, 1999.

T. Back, F. Hoffmeister, and H.-P. Schwefel. A survey of evolution strategies. In R.K. Belew and L.B. Booker, editors, *Proceedings of the 4th International Conference on Genetic Algorithms (ICGA IV)*, pages 2–9. Morgan Kaufmann Publishers, 1991.

M.H. Bakr, J.W. Bandler, R.M. Biernacki, S.H. Chen, and K. Madsen. A trust region aggressive space mapping algorithm for EM optimization. *IEEE Transactions on Microwave Theory and Techniques*, 46: 2412–2425, 1998.

M. Bakr, J. Bandler, K. Madsen, and J. Sondergaard. Review of the space mapping approach to engineering optimization and modeling. *Optimization and Engineering*, 1: 241–276, 2000.

V.O. Balabanov. *Development of Approximations for HSCT Wing Bending Material Weight Using Response Surface Methodology*. PhD thesis, Virginia Polytechnic Institute and State University, 1997.

V.O. Balabanov, A.A. Giunta, O. Golovidov, B. Grossman, W.H. Mason, L.T. Watson, and R.T. Haftka. Reasonable design space approach to response surface approximation. *Journal of Aircraft*, 36: 308–315, 1999.

M.J. Balas. Trends in large space structure control theory: fondest hopes, wildest dreams. *IEEE Transactions on Automatic Control*, AC-27: 522–535, 1982.

J.M. Baldwin. A new factor in evolution. *American Naturalist*, 30: 441–451, 1896.

E. Balmes. Parametric families of reduced finite element models. Theory and applications. *Mechanical Systems and Signal Processing*, 10: 381–394, 1996.

J.W. Bandler, R.M. Biernacki, S.H. Chen, P.A. Grobelny, and R.H. Hemmers. Space mapping technique for electromagnetic optimization. *IEEE Transactions on Microwave Theory and Techniques*, 42: 2536–2544, 1994.

J.W. Bandler, R.M. Biernacki, S.H. Chen, R.H. Hemmers, and K. Madsen. Electromagnetic optimization exploiting aggressive space mapping. *IEEE Transactions on Microwave Theory and Techniques*, 43: 2874–2882, 1995.

J.W. Bandler, M.A. Ismail, J.E. Rayas-Sánchez, and Q.J. Zhang. Neuromodeling of microwave circuits exploiting space mapping technology. *IEEE Transactions on Microwave Theory and Techniques*, 47: 2417–2427, 1999.

K.J. Bathe. *Finite Element Procedures*. Prentice Hall, 1996.

L.M. Beda. Programs for automatic differentiation for the machine BESM. Technical report, Institute for Precise Mechanics and Computation Techniques, Academy of Science, Moscow, 1959.

R.K. Belew and L.B. Booker, editors. *Proceedings of the 4th International Conference on Genetic Algorithms*. Morgan Kaufmann, San Diego, CA, July 1991.

T. Belytschko, Y. Krongauz, D. Organ, M. Fleming, and P. Krysl. Meshless methods: an overview and recent developments. *Computer Methods in Applied Mechanics and Engineering*, 139: 3–47, 1996.

Y. Ben-Haim. *Information Gap Decision Theory: Decisions Under Severe Uncertainty*. Academic Press, 2001.

R. Benayoun, J. de Montgolfier, J. Tergny, and O. Laritchev. Linear programming with multiple objective functions: Step method (stem). *Mathematical Programming*, 1: 366–375, 1971.

Y. Ben-Haim and I. Elishakoff. *Convex Models of Uncertainty in Applied Mechanics*. Elsevier Science, 1990.

M.P. Bendsoe. Optimal shape design as a material distribution problem. *Structural Optimization*, 1: 193–202, 1989.

G. Berkooz, P. Holmes, and J.L. Lumley. The proper orthogonal decomposition in the analysis of turbulent flows. *Annual Review of Fluid Mechanics*, 25: 539–575, 1993.

F. Berman, G. Fox, and A.J.G. Hey. *Grid Computing: Making the Global Infrastructure a Reality*. John Wiley & Sons, Chichester, 2003.

C. Bishop. *Neural Networks for Pattern Recognition*. Oxford University Press, 1995.

M.I.G. Bloor and M.J. Wilson. The efficient parametrization of generic aircraft geometry. *Journal of Aircraft*, 32: 1269–1275, 1995.

P.T. Boggs and J.W. Tolle. A strategy for global convergence in a sequential quadratic programming algorithm. *SIAM Journal of Numerical Analysis*, 26: 600–623, 1989.

P.T. Boggs and J.W. Tolle. Sequential quadratic programming. *Acta Numerica*, 4: 1–51, 1995.

A.J. Booker, J.E. Dennis, P.D. Frank, D.B. Serafini, V. Torczon, and M.W. Trosset. A rigorous framework for optimization of expensive functions by surrogates. *Structural Optimization*, 17: 1–13, 1998.

D. Borglund and I.M. Kroo. Aeroelastic design optimization of a two-spar flexible wind-tunnel model. *Journal of Aircraft*, 39: 1074–1075, 2002.

A. Bouazzouni, G. Lallement, and S. Cogan. Selecting a Ritz basis for the reanalysis of the frequency response of modified structures. *Journal of Sound and Vibration*, 199: 309–322, 1997.

G.E.P. Box and N.R. Draper. *Empirical Model Building and Response Surfaces*. John Wiley & Sons, 1987.

W.J. Boyne. *Beyond the Horizons: The Lockheed Story*. St. Martins Press, 1999.

R.H. Bracewell. Synthesis based on function-means trees: Schemebuilder. In *Engineering Design Synthesis Issues, Understanding and Methods*, ed. A. Chakrabarti, 199–212, Springer-Verlag, 2002.

K.R. Bradley. A sizing methodology for the conceptual design of blended-wing-body transports. NASA/CR-2004-213016, 2004.

V. Braibant and C. Fleury. Shape optimal design using B-splines. *Computer Methods in Applied Mechanics and Engineering*, 44: 247–267, 1984.

R.D. Braun. *Collaborative Optimization: An Architecture for Large-Scale Distributed Design*. PhD thesis, University of Stanford, 1996.

R. Braun, P. Gage, I. Kroo, and I. Sobieski. Implementation and performance issues in collaborative optimization. AIAA 1996-4017, 1996.

R.D. Braun and I.M. Kroo. Development and application of the collaborative optimization architecture in a multidisciplinary design environment. In N.M. Alexandrov and M.Y. Hussaini, editors, *Multidisciplinary Design Optimization: State of the Art*, pages 98–116. SIAM, 1997.

R.P. Brent. *Algorithms for Minimization Without Derivatives*. Prentice Hall, 1973.

P. Brezillon, J.F. Stauf, A.M. Peraultstaub, and G. Milhaud. Numerical estimation of the 1st order derivative – approximate evaluation of an optimal step. *Computers and Mathematics With Applications*, 7: 333–347, 1981.

T.R. Browning, "Applying the Design Structure Matrix to System Decomposition and Integration Problems: A Review and New Directions", *IEEE Transactions on Engineering Management*, 48(3): 292–306, August 2001.

C.G. Broyden. A class of methods for solving nonlinear simultaneous equations. *Mathematics of Computation*, 19: 577–593, 1965.

R.J. Buck and H.P. Wynn. Optimization strategies in robust engineering design and computer-aided design. *Quality and Reliability Engineering International*, 9: 39–48, 1993.

L. Bull. Evolutionary computing in multi-agent environments: partners. *Proceedings of the Seventh International Conference on Genetic Algorithms*, pages 370–377. Morgan Kaufmann, 1997.

D.G. Cacuci. Sensitivity theory for non-linear systems 1. Non-linear functional-analysis approach. *Journal of Mathematical Physics*, 22: 2794–2802, 1981a.

D.G. Cacuci. Sensitivity theory for non-linear systems 2. Extensions to additional classes of responses. *Journal of Mathematical Physics*, 22: 2803–2812, 1981b.

R.H. Cameron and W.T. Martin. The orthogonal development of nonlinear functionals in series of Fourier-Hermite functions. *Annals of Mathematics*, 48: 385–392, 1947.

A.S. Campobasso, M.C. Duta, and M.B. Giles. Adjoint calculation of sensitivities of turbomachinery objective functions. *Journal of Propulsion And Power*, 19: 693–703, 2003.

R.A. Canfield. High quality approximation of eigenvalues in structural optimization. *AIAA Journal*, 28: 1116–1122, 1990.

Y. Cao, M.Y. Hussaini, and T.A. Zang. On the exploitation of sensitivity derivatives for improving sampling methods. AIAA Paper 2003-1656, 2003.

C.M.M. Carey, G.H. Golub, and K.H. Law. A Lanczos-based method for structural dynamic reanalysis problems. *International Journal for Numerical Methods in Engineering*, 37: 2857–2883, 1994.

A. Carle and M. Fagan. ADIFOR 3.0 overview. Technical Report CAAM-TR-00-02, Rice University, 2000.

J.C. Carr, W.R. Fright, and R.K. Beatson. Surface interpolation with radial basis functions for medical imaging. *IEEE Transactions on Medical Imaging*, 16: 96–107, 1997.

H. Casanova and J. Dongarra. NetSolve: a network enabled server, examples and users. In *Proceedings of the Heterogeneous Computing Workshop*, Orlando, FL, 19–28, 1998.

H. Casanova and J. Dongarra. Netsolve: a network server for solving computational science problems. *International Journal of High Performance Computing Applications*, 11: 212–223, 1997.

K.-H. Chang and K.K. Choi. A geometry-based parameterisation method for shape design of elastic solids. *Mechanics of Structures and Machines*, 20: 215–252, 1992.

C. Chang, F.J. Torres, and C. Tung. Geometric analysis of wing sections. NASA TM 110346, 1995.

O. Chapelle, V. Vapnik, and J. Weston. Transductive inference for estimating values of functions. In *Advances in Neural Information Processing Systems* 12: 421–427, 1999.

C. Chatfield. Model uncertainty, data mining and statistical inference. *Journal of the Royal Statistical Society, Series B*, 158: 419–466, 1995.

W. Chen, J.K. Allen, K.-L. Tsui, and F. Mistree. A procedure for robust design. *ASME Journal of Mechanical Design*, 118: 478–485, 1996.

Y.-M. Chen, A. Bhaskar, and A.J. Keane. A parallel nodal-based evolutionarystructural optimization algorithm. *Structural and Multidisciplinary Optimization*, 23: 241–251, 2002.

W. Chen, R. Jin, and A. Sudjianto. Analytical uncertainty propagation in simulation-based design under uncertainty. AIAA Paper 2004-4356, 2004.

W. Chen, A. Sahai, A. Messac, and G.J. Sundararaj. Exploration of the effectiveness of physical programming in robust design. *ASME Journal of Mechanical Design*, 122: 155–163, 2000.

S.H. Chen and Z.J. Yang. A universal method for structural static reanalysis of topological modifications. *International Journal for Numerical Methods in Engineering*, 61: 673–686, 2004.

V. Cherkassky and F. Mulier. *Learning from Data*. Wiley, 1998.

J.B. Cherrie, R.K. Beatson, and G.N. Newsam. Fast evaluation of radial basis functions: methods for generalized multiquadrics in $\mathbb{R}^n$. *SIAM Journal of Scientific Computing*, 23: 1549–1571, 2002.

K.K. Choi and W. Duan. Design sensitivity analysis and shape optimization of structural components with hyperelastic material. *Computer Methods in Applied Mechanics and Engineering*, 187: 289–306, 2000.

K.K. Choi and N.H. Kim. *Structural Sensitivity Analysis and Optimization*. Vol. 1. Linear Systems. Springer-Verlag, 2005.

K.K. Choi and N.H. Kim. *Structural Sensitivity Analysis and Optimization*. Vol. 2. Nonlinear Systems and Applications. Springer-Verlag, 2005.

A. Choudhury, P.B. Nair, and A.J. Keane. A data parallel approach for large-scale Gaussian process modeling. In *Proceedings of the Second SIAM International Conference on Data Mining*, Arlington, VA, 2002.

H.-S. Chung and J.J. Alonso. Using gradients to construct cokriging approximation models for high-dimensional design optimization problems. AIAA Paper 2002-0317, 2002.

S.M. Clarke, J.H. Griebsch, and T.W. Simpson. Analysis of support vector regression for approximation of complex engineering analyses. *ASME Journal of Mechanical Design*, to appear.

C.W. Clegg, C.M. Axtell, L. Damodaran, B. Farbey, R. Hull, R. Lloyd-Jones, J. Nicholls, R. Sell, and C. Tomlinson. Information technology: a study of performance and the role of human and organizational factors. *Ergonomics*, 40: 851–871, 1997.

W.J. Cody, W.B. Rouse, and K.R. Boff. Designers' associates: intelligent support for information access and utilization in design. *Human/Technology Interaction in Complex Systems*, 7: 173–260, 1995.

R. Collobert, S. Bengio, and Y. Bengio. A parallel mixture of SVMs for very large scale problems. *Neural Computation*, 14(5), 1105–1114, 2002.

A.R. Conn, N.I.M. Gould, and P.L. Toint. *Trust-region Methods*. SIAM, 2000.

T.A. Cooke. Measurements of the boundary layer and wake of two aerofoil sections at high reynolds numbers and subsonic mach numbers. Technical Report R&M 3722, Aircraft Research Centre, 1973.

G. Corliss, C. Faure, A. Griewank, L. Hascoët, and U. Naumann, editors. *Automatic Differentiation of Algorithms: From Simulation to Optimization*. Computer and Information Science Series. Springer, 2001.

R. Cornish, J.T. Mills, J.P. Curtis, and D. Finch. Degradation mechanisms in shaped charge jet penetration. *International Journal of Impact Engineering*, 26: 105–114, 2001.

G.B. Cosentino and T.L. Holst. Numerical optimization of advanced transonic wing configurations. *Journal of Aircraft*, 23: 129–199, 1986.

S.F. Cotter, J. Adler, B.D. Rao, and K. Kreutz-Delgado. Forward sequential algorithms for best basis selection. *IEE Proceedings on Vision, Image, and Signal Processing*, 146: 235–244, 1999.

J. Cousin and M. Metcalfe. The BAE Ltd transport aircraft synthesis and optimization program. AIAA Paper 90-3295, 1990.

E.J. Cramer, J.E. Dennis Jr., P.D. Frank, R.M. Lewis, and G.R. Shubin. Problem formulation for multidisciplinary optimization. *SIAM Journal on Optimization*, 4: 754–776, 1994.

N.A.C. Cressie. *Statistics for Spatial Data*. John Wiley & Sons, 1993.

N. Cristianini and J. Shawe-Taylor. *An Introduction to Support Vector Machines*. Cambridge University Press, 2000.

L. Csató and M. Opper. Sparse representation for Gaussian process models. *Advances in Neural Information Processing Systems*, 13: 444–450, 2001.

C. Currin, T. Mitchell, M. Morris, and D. Ylvisaker. Bayesian prediction of deterministic functions, with applications to the design and analysis of computer experiments. *Journal of the American Statistical Association*, 86: 953–963, 1991.

P. Cusdin and J.-D. Müller. Deriving linear and adjoint codes for CFD using automatic differentiation. Technical Report QUB-SAE-03-06, The Queen's University of Belfast, 2003.

R.L. Dailey. Eigenvector derivatives with repeated eigenvalues. *AIAA Journal*, 27: 468–491, 1989.

J.W. Daniel, W.B. Bragg, L. Kaufmann, and G.W. Stewart. Reorthogonalization and stable algorithms for updating the Gram Schmidt QR factorization. *Mathematics of Computation*, 30: 772–795, 1976.

D. DeLaurentis and D. Mavris. Uncertainty modeling and management in multidisciplinary analysis and synthesis. AIAA Paper 2000-0422, 2000.

A.-V. DeMiguel and W. Murray. An analysis of collaborative optimization methods. AIAA Paper 2000-4720, 2000.

J.E. Dennis and R.B. Schnabel. *Numerical Methods for Unconstrained Optimization and Nonlinear Equations*. Prentice-Hall, 1983.

D. Desterac and J. Reneaux. Transport aircraft aerodynamic improvement by numerical optimization. In *Proceedings of the 17th ICAS Congress*, Stockhol, pages 1427–1438, 1990.

G.T.S. Done. Past and future progress in fixed and rotary wing aeroelasticity. *Aeronautical Journal*, 100: 269–279, 1996.

X. Du and W. Chen. Towards a better understanding of modeling feasibility robustness in engineering design. *ASME Journal of Mechanical Design*, 122: 385–394, 2000.

X. Du and W. Chen. Efficient uncertainty analysis methods for multidisciplinary robust design. *AIAA Journal*, 40: 545–552, 2002.

Q. Du, V. Faber, and M. Gunzburger. Centroidal Voronoi tessellations: applications and algorithms. *SIAM Review*, 41: 637–676, 1999.

Q. Du and M. Gunzburger. Model reduction by proper orthogonal decomposition coupled with centroidal Voronoi tessellation. *Proceedings of the Fluids Engineering Division Summer Meeting*, Paper number FEDS2002–31051. ASME, 2002.

D.P. Dupplaw, D. Brunson, A.-J.E. Vine, C.P.P. Please, S.M. Lewis, A.M. Dean, A.J. Keane, and M.J. Tindall. A web-based knowledge elicitation system (gisel) for planning and assessing group screening experiments for product development. *ASME Transactions - Journal of Computing and Information Science in Engineering*, 4: 218–225, 2004.

A.R. Dusto. A method for predicting the stability characteristics of an elastic airplane, FLEXSTAB theoretical description. NASA CR-114-712, 1974.

B. Efron. *The Jackknife, the Bootstrap, and Other Resampling Plans*, CBMS Monograph 38. SIAM, 1981.

M.A. El-Beltagy. A risk adjusting fusion framework for optimizing models with variable fidelity. AIAA Paper 2004-4577, 2004.

M.A. El-Beltagy, P.B. Nair, and A.J. Keane. Metamodelling techniques for evolutionary optimization of computationally expensive problems: promises and limitations. *Proceedings of the Genetic and Evolutionary Computation Conference*, pages 196–203. Morgan Kaufmann, 1999.

M.S. Eldred, A.A. Giunta, and S.S. Collis. Second-order corrections for surrogate-based optimization with model hierarchies. AIAA Paper 2004-4457, 2004.

M.S. Eldred, A.A. Giunta, S.F. Wojtkiewicz, and T.G. Trucano. Formulations for surrogate-assisted optimization under uncertainty. AIAA Paper 2002-5585, 2002.

M.S. Eldred, P.B. Lerner, and W.J. Anderson. Higher order eigenpair perturbations. *AIAA Journal*, 30: 1870–1876, 1992.

M. Emmerich and B. Naujoks. Metamodel-assisted multiobjective optimization and their application in airfoil design. *Proceedings of the Fifth International Conference on Adaptive Design and Manufacture*, pages 249–260. Springer, 2004.

B. Epstein, A. Luntz, and A. Nachson. Multigrid Euler solver about aircraft configurations, with cartesian grids and local refinement. AIAA Paper 89-1960, 1989.

ESDU. Method for the rapid estimation of spanwise loading of wings with camber and twist in subsonic attached flow. Technical Report 83040, ESDU, 1983.

ESDU. A method for determining the wave drag and its spanwise distribution on a finite wing in transonic flow. Technical Report 87003, ESDU, 1987.

ESDU. VGK method for two-dimensional aerofoil sections. Technical Report 96028, ESDU, 1996.

M. Evans and T. Swartz. Methods for approximating integrals in statistics with special emphasis on Bayesian integration problems. *Statistical Science*, 10: 254–272, 1995.

M. Evans and T. Swartz. *Approximating Integrals Via Monte Carlo and Deterministic Methods*. Oxford University Press, 2000.

T. Evgeniou, M. Pontil, and T. Poggio. Regularization networks and support vector machines. *Advances in Computational Mathematics*, 13: 1–50, 2000.

G.M. Fadel, M.F. Riley, and J.F.M. Barthelemy. Two-point exponential approximation method for structural optimization. *Structural Optimization*, 2: 117–124, 1990.

K.-T. Fang, D.K.J. Lin, P. Winker, and Y. Zhang. Uniform design: theory and application. *Technometrics*, 42: 237–248, 2000.

C. Farhat, M. Lesoinne, and P. LeTallec. Load and motion transfer algorithms for fluid/structure interaction problems with non-matching discrete interfaces: Momentum and energy conservation, optimal discretization and application to aeroelasticity. *Computer Methods in Applied Mechanics and Engineering*, 157: 95–114, 1998.

G.E. Fasshauer. Solving differential equations with radial basis functions: multilevel methods and smoothing. *Advances in Computational Mathematics*, 11: 139–159, 1999.

J.H. Ferziger and M. Peric. *Computational Methods for Fluid Dynamics*. Springer-Verlag, Berlin, 1996.

G.B.R. Fielden and W. Hawthorne. Sir Frank Whittle, O.M., K.B.E. *Biographical Memoirs of Fellows of the Royal Society of London*, 44: 433–452, 1988.

R.A. Fisher. *The Design of Experiments*. Oliver and Boyd, Edinburgh, 1935.

D.B. Fogel. Applying evolutionary programming to selected travelling salesman problems. *Cybernetics and Systems*, 24: 27–36, 1993.

D.R. Fokkema, L.G. Sleijpen, and H.A. Van der Vorst. Accelerated inexact Newton schemes for large systems of nonlinear equations. *SIAM Journal of Scientific Computing*, 19: 657–674, 1998.

C.M. Fonseca and P.J. Fleming. An overview of evolutionary algorithms in multiobjective optimization. *Evolutionary Computing*, 3: 1–16, 1995.

S. Forrest, editor. *Proceedings of the 5th International Conference on Genetic Algorithms*. Morgan Kaufmann, Urbana-Champaign, IL, June 1993.

A. Forrester. *Efficient Global Aerodynamic Optimisation Using Expensive Computational Fluid Dynamics Simulations*. PhD thesis, University of Southampton, 2004.

I. Foster and C. Kesselman. *The Grid: Blueprint for a New Computer Infrastructure*. Morgan Kaufmann, 1999.

I. Foster, C. Kesselman, J.M. Nick, and S. Tuecke. The physiology of the grid: an open grid services architecture for distributed systems integration. http://www.globus.org/research/papers/ogsa.pdf, 2002.

I. Foster, C. Kesselman, and S. Tuecke. The anatomy of the grid: Enabling scalable virtual organizations. *International Journal of High Performance Computing Applications*, 15: 200–222, 2001.

R.L. Fox and M.P. Kapoor. Rates of change of eigenvalues and eigenvectors. *AIAA Journal*, 6: 2426–2429, 1968.

R.L. Fox and H. Miura. An approximate analysis technique for design calculations. *AIAA Journal*, 9: 177–179, 1971.

R. Franke. Scattered data interpolation: tests of some methods. *Mathematics of Computation*, 98: 181–200, 1982.

R. Franke and G. Neilson. Smooth interpolation of large sets of scattered data. *International Journal for Numerical Methods in Engineering*, 15: 1691–1704, 1980.

J.H. Friedman. Greedy function approximation: a gradient boosting machine. *Annals of Statistics*, 29 (5): 1189–1232, 2001.

C.R. Fuller, S.J. Elliott, and P.A. Nelson. *Active Control of Vibration*. Academic Press, London, 1996.

S. Gallopoulos, E. Houstis, and J. Rice. Future research directions in problem solving environments for computational science. Technical Report CSD-TR-92-032, Purdue University, 1992.

J.H. Garcelon, R.T. Haftka, and S.J. Scotti. Approximations in optimization of damage tolerant structures. *AIAA Journal*, 38: 517–524, 2000.

J.S. Gero, J.L. Sushil, and S. Kundu. Evolutionary learning of novel grammars for design improvement. *Artificial Intelligence for Engineering Design, Analysis and Manufacturing*, 8: 83–94, 1994.

R. Ghanem and P. Spanos. *Stochastic Finite Elements: A Spectral Approach*. Springer-Verlag, 1991.

H. Gibbins and R. Buyya. Gridscape: a tool for the creation of interactive and dynamic grid testbed web portals. In *Proceedings of the 5th IWDC International Workshop*, pages 131–142, India, 2003.

M. Gibbs and D.J.C. MacKay. Efficient implementation of Gaussian processes. Technical report, University of Cambridge, Cavendish Labs, 1997.

J.P. Giesing and J.-F. Barthelemy. A summary of industry MDO applications and needs. AIAA Paper 98-4737, 1998.

J.C. Gilbert and J. Nocedal. Global convergence properties of conjugate gradient methods. *SIAM Journal on Optimization*, 2: 21–42, 1992.

M.B. Giles, M.C. Duta, J.D. Müller, and N.A. Pierce. Algorithm developments for discrete adjoint methods. *AIAA Journal*, 41: 198–205, 2003.

M.B. Giles and N.A. Pierce. An introduction to the adjoint approach to design. *Flow Turbulence And Combustion*, 65: 393–415, 2000.

M.B. Giles and N.A. Pierce. Adjoint error correction for integral outputs in error estimation and adaptive discretization methods in computational fluid dynamics. In T. Barth and H. Deconinck, editors, *Lecture Notes in Computational Science and Engineering*, pages 47–96. Springer-Verlag, 2002.

P.E. Gill, W. Murray, and M.H. Wright. *Practical Optimization*. Academic Press, 1981.

A.A. Giunta. *Aircraft Multidisciplinary Design Optimization Using Design of Experiments Theory and Response Surface Modeling Methods*. PhD thesis, Virginia Polytechnic Institute and State University, 1997.

A.A. Giunta. Sensitivity analysis for coupled aero-structural systems. NASA TM-1999-209367, 2000.

A.A. Giunta and M.S. Eldred. Implementation of a trust region model management strategy in the DAKOTA optimization toolkit. AIAA Paper 2000-4935, 2000.

A.A. Giunta, S.F. Wojtkiewicz Jr., and M.S. Eldred. Overview of modern design of experiments methods for computational simulations. AIAA Paper 2003-0649, 2003.

R.H. Goddard. A method of reaching extreme altitudes. *Smithsonian Institution Miscellaneous Collections*, 71(2), 1919.

D.E. Goldberg. *Genetic Algorithms in Search, Optimization and Machine Learning*. Addison-Wesley, 1989.

G.H. Golub and C.F. Van Loan. *Matrix Computations*. The John Hopkins University Press, 1996.

S.J. Gould. *The Panda's Thumb*. Norton & Co., New York, 1980.

D.J. Graham. The development of cambered airfoil sections having favorable lift characteristics at supercritical mach numbers (8-series). NACA Report 947, NACA, 1949.

W.H. Greene and R.T. Haftka. Computational aspects of sensitivity calculations in transient structural analysis. *Computers and Structures*, 32: 433–443, 1989.

J.J. Grefenstette, editor. *Proceedings of the 1st International Conference on Genetic Algorithms*. Lawrence Erlbaum Associates, Pittsburgh, PA, July 1985.

J.J. Grefenstette, editor. *Proceedings of the 2nd International Conference on Genetic Algorithms*. Lawrence Erlbaum Associates, Cambridge, MA, July 1987.

A. Griewank. *Mathematical Programming: Recent Developments and Applications*, chapter On Automatic Differentiation, pages 83–108. Kluwer Academic Publishers, 1989.

A. Griewank. *Evaluating Derivatives: Principles and Techniques of Algorithmic Differentiation*, Number 19 in Frontiers in Applied Mathematics. SIAM, Philadelphia, PA, 2000.

M.J. Grote and T. Huckle. Parallel preconditioning with sparse approximate inverses. *SIAM Journal on Scientific Computing*, 18: 838–853, 1997.

C.R. Gumbert, G.J.-W. Hou, and P.A. Newman. Reliability assessment of a robust design under uncertainty for a 3-d flexible wing. AIAA Paper 2003-4094, 2003.

C.R. Gumbert, G.J.-W. Hou, and P.A. Newmann. Simultaneous aerodynamic analysis and design optimization (saado) for 3-d flexible wing. AIAA Paper 2001-1107, 2001a.

C.R. Gumbert, G.J.-W. Hou, and P.A. Newmann. Simultaneous aerodynamic analysis and design optimization (saado) for 3-d wing. AIAA Paper 2001-2527, 2001b.

D. Gunter, B. Tierney, K. Jackson, J. Lee, and M. Stoufer. Dynamic monitoring of high-performance distributed applications. In *Proceedings of IEEE HPDC-11*, pages 163–170, Edinburgh, 2002.

X. Guo, K. Yamazaki, and G.D. Cheng. A new three-point approximation approach for design optimization problems. *International Journal for Numerical Methods in Engineering*, 50: 869–884, 2001.

W.G. Habashi, J. Dompierre, Y. Bourgault, D. Ait-Ali-Yahia, M.Fortin, and M.-G. Vallet. Anisotropic mesh adaptation: towards user-independent, mesh-independent and solver-independent CFD. Part I: general principles. *International Journal for Numerical Methods in Engineering*, 32: 725–744, 2000.

A. Habbal, M. Thellner, and J. Petersson. A Nash game approach for multidisciplinary topology design. In *Fifth World Congress on Structural and Multidisciplinary Optimization*, Venice, 2003.

R.T. Haftka. Sensitivity calculations for iteratively solved problems. *International Journal for Numerical Methods In Engineering*, 21: 1535–1546, 1985.

R.T. Haftka. Combining global and local approximations. *AIAA Journal*, 29: 1523–1525, 1991.

R.T. Haftka and Z. Gurdal. *Elements of Structural Optimization*. Kluwer Academic Publishers, third edition, 1992.

A. Haldar and S. Mahadevan. *Reliability Assessment Using Stochastic Finite Element Analysis*. John Wiley & Sons, 2000.

M.A. Hale and D.N. Mavris. Enabling advanced design methods in an internet-capable framework. In *World Aviation Congress and Exposition*. San Francisco, CA, October 19–21, SAE/AIAA 199901 -5578, 1999.

J.M. Hammersley and D.C. Handscomb. *Monte Carlo Methods*. Methuen and Co., London, 1964.

M.H. Hansen and B. Yu. Model selection and the principle of minimum description length. *Journal of the American Statistical Association*, 96: 746–774, 2001.

C.D. Harris. NASA supercritical airfoils: a matrix of family-related airfoils. NASA TP 2969, 1990.

L. Hascoet, S. Fidanova, and C. Held. Adjoining independent computations. In G. Corliss, C. Faure, A. Griewank, L. Hascoet, and U. Naumann, editors, *Automatic Differentiation: From Simulation to Optimization, Computer and Information Science*, pages 285–290. Springer, New York, 2001.

T. Hastie, R. Tibshirani, and J. Friedman. *The Elements of Statistical Learning: Data Mining, Inference, and Prediction*. Springer-Verlag, 2001.

E.B. Haugen. *Probabilistic Approaches to Design*. John Wiley & Sons, London, 1968.

A.S. Hedayat, N.J.A. Sloane, and J. Stufken. *Orthogonal Arrays: Theory and Applications*. Springer, 1999.

J.C. Helton, J.D. Johnson, and W.L. Oberkampf. An exploration of alternative approaches to the representation of uncertainty in model predictions. *Reliability Engineering and System Safety*, 85: 39–71, 2004.

F.M. Hemez and Y. Ben-Haim. Info-gap robustness for the correlation of tests and simulations of a non-linear transient. *Mechanical Systems and Signal Processing*, 18: 1443–1467, 2004.

R.M. Hicks and P.A. Henne. Wing design by numerical optimization. *Journal of Aircraft*, 15: 407–412, 1978.

G.E. Hinton and S.J. Nowlan. How learning can guide evolution. *Complex Systems*, 1: 495–502, 1987.

C. Hirsch. *Numerical Computation of Internal and External Flows*. John Wiley, 1988.

C.M.E. Holden. *Aeroelastic Optimization using the Collocation Method*. PhD thesis, University of Stanford, 1999.

C.M.E. Holden. *Visualization Methodologies in Aircraft Design Optimization*. PhD thesis, University of Southampton, 2004.

R. Hooke and T.A. Jeeves. *Direct search solution of numerical and statistical problems*. Westinghouse Research Labs. Scientific Paper, 1960.

J.R.M. Hosking, E.P.D. Pednault, and M. Sudan. A statistical perspective on data mining. *Future Generation Computing Systems*, 13: 117–134, 1997.

P.J. Huber. *Robust Statistics*. John-Wiley & Sons, 1981.

T.J.R. Hughes. *The Finite Element Method*. Prentice Hall, 1987.

K.F. Hulme, C.L. Bloebaum, and Y. Nozaki. A performance-based investigation of parallel and serial approaches to multidisciplinary analysis convergence. AIAA Paper 2000-4812, 2000.

M.G. Hutchinson, E.R. Unger, W.H. Mason, and R.T. Haftka. Variable complexity aerodynamic optimization of a high speed civil transport wing. *Journal of Aircraft*, 31: 110–116, 1994.

L. Huyse. Free-form airfoil shape optimization under uncertainty using maximum expected value and second-order second-moment strategies. NASA/CR-2001-211020, 2001.

L. Huyse and R.M. Lewis. Aerodynamic shape optimization of two-dimensional airfoils under uncertain conditions. NASA/CR-2001-210648, 2001.

ICAD. ICAD user guide, http://www.ktiworld.com/our_products/icad.shtml. Knowledge Technologies International, 2002.

IGES. *Initial Graphics Exchange Specification (IGES) (fips pub 177-1)*. National Technical Information Service, U.S. Department of Commerce and National Computer Graphics Association, http://www.itl.nist.gov/fipspubs/fip177-1.htm, 1996.

I.C.F. Ipsen and C.D. Meyer. The idea behind Krylov methods. *American Mathematical Monthly*, 105: 889–899, 1998.

K. Ito and S.S. Ravindran. A reduced-order method for simulation and control of fluid flows. *Journal of Computational Physics*, 143: 403–425, 1998.

A.J. Izenman. Recent developments in nonparametric density estimation. *Journal of the American Statistical Association*, 86: 205–224, 1991.

R.H.F. Jackson and G.P. McCormick. Second order sensitivity analysis in factorable programming. *Mathematical Programming*, 41: 1–28, 1988.

A. Jameson. Aerodynamic design via control theory. *Journal of Scientific Computing*, 3: 233–260, 1988.

A. Jameson. A perspective on computational algorithms for aerodynamic analysis and design. *Progress in Aerospace Sciences*, 37: 197–243, 2001.

A. Jameson and S. Kim. Reduction of the adjoint gradient formula for aerodynamic shape optimization problems. *AIAA Journal*, 41: 2114–2129, 2003.

A. Jameson and J. Vassberg. Computational fluid dynamics (CFD) for aerodynamic design: its current and future impact. AIAA Paper 2001-0538, 2001.

Y. Jin, M. Olhofer, and B. Sendhoff. A framework for evolutionary optimization with approximate fitness functions. *IEEE Transactions on Evolutionary Computation*, 6: 481–494, 2003.

T. Joachims. Making large scale support vector machine learning possible. In B. Schölkopf, C. Burges, and A. Smola, editors, *Advances in Kernel Methods*. MIT Press, 1999.

E.H. Johnson. Adjoint sensitivity analysis in msc/nastran. In *MSC 1997 Aerospace Users Conference Proceedings*, Newport Beach, CA, 1997.

M. Johnson, L. Moore, and D. Ylvisaker. Minimax and maximin distance designs. *Journal of Statistical Planning and Inference*, 26: 131–148, 1990.

D.R. Jones, M. Schonlau, and W.J. Welch. Efficient global optimization of expensive black-box functions. *Journal of Global Optimization*, 13 (4): 455–492, 1998.

G.K. Kanji. *100 Statistical Tests*. Sage Publications, 1999.

M.K. Karakasis, A.P. Giotis, and K.C. Giannakoglou. Inexact information aided, low-cost, distributed genetic algorithms for aerodynamic shape optimization. *International Journal for Numerical Methods in Fluids*, 43: 1149–1166, 2003.

M. Karpel, B. Moulin, and M. Idan. Robust aeroservoelastic design with structural variations and modeling uncertainties. *Journal of Aircraft*, 40: 946–954, 2003.

A.J. Keane. Wing optimization using design of experiment, response surface, and data fusion methods. *Journal of Aircraft*, 40: 741–750, 2003.

A.J. Keane. Design search and optimisation using radial basis functions with regression capabilities. In I.C. Parmee, editor, *Proceedings of the Conference on Adaptive Computing in Design and Manufacture VI*, pages 39–49. Springer-Verlag, Bristol, 2004.

A.J. Keane and A.P. Bright. Passive vibration control via unusual geometries: experiments on model aerospace structures. *Journal of Sound and Vibration*, 190: 713–719, 1996.

A.J. Keane and N. Petruzzelli. Aircraft wing design using GA-based multi-level strategies. AIAA Paper 2000-4937, 2000.

A.J. Keane, P. Temarel, X.-J. Wu, and Y. Wu. Hydroelasticity of non-beamlike ships in waves. *Philosophical Transactions of the Royal Society of London A*, 334: 339–355, 1991.

M. Kennedy and A. O'Hagan. Predicting the output from a complex computer code when fast approximations are available. *Biometrika*, 87: 1–13, 2000.

J.Y. Kim, N.R. Aluru, and D.A. Tortorelli. Improved multi-level Newton solvers for fully coupled multi-physics problems. *International Journal for Numerical Methods in Engineering*, 58: 563–480, 2003.

S. Kirkpatrick, C.D. Gelatt Jr., and M.P. Vecchi. Optimization by simulated annealing. *Science*, 220: 671–680, 1983.

U. Kirsch. Reduced basis approximation of structural displacements for optimal design. *AIAA Journal*, 29: 1751–1758, 1991.

U. Kirsch. Efficient sensitivity analysis for structural optimization. *Computer Methods in Applied Mechanics and Engineering*, 117: 143–156, 1994.

U. Kirsch. Effective move limits for approximate structural optimization. *ASCE Journal of Structural Engineering*, 123: 210–217, 1997.

U. Kirsch. A unified reanalysis approach for structural analysis, design, and optimization. *Structural and Multidisciplinary Optimization*, 25: 67–85, 2003.

U. Kirsch and M. Bogomolni. Procedures for approximate eigenproblem reanalysis of structures. *International Journal for Numerical Methods in Engineering*, 60: 1969–1986, 2004.

U. Kirsch and S. Liu. Structural reanalysis for general layout modifications. *AIAA Journal*, 35: 382–388, 1997.

U. Kirsch and P.Y. Papalambros. Structural reanalysis for topological modifications – a unified approach. *Structural and Multidisciplinary Optimization*, 21: 333–344, 2001.

G.H. Klopfer and D. Nixon. Non isentropic potential formulation for transonic flows. AIAA Paper 83-0375, 1983.

J.D. Knowles and D.W. Corne. The pareto archived evolution strategy: A new baseline algorithm for pareto multiobjective optimization. *Proceedings of the 1999 Congress on Evolutionary Computation, CEC'99*. IEEE Press, Piscataway, NJ, 1999.

P.N. Koch, R.-J. Yang, and L. Gu. Design for six sigma through robust optimization. *Structural and Multidisciplinary Optimization*, 26: 235–248, 2004.

J.R. Koehler and A.B. Owen. Computer experiments. In S. Ghosh and C.R. Rao, editors, *Handbook of Statistics*, pages 261–308. Elsevier, 1996.

T.G. Kolda, R.M. Lewis, and V. Torczon. Optimization by direct search: new perspectives on some classical and modern methods. *SIAM Review*, 45: 385–482, 2003.

I. Kroo, S. Altus, R. Braun, P. Gage, and I. Sobieski. Multidisciplinary optimization methods for aircraft preliminary design. AIAA Paper 94-4325-CP, 1994.

G. Kuruvila, S. Ta'asan, and M. D. Salas. Airfoil design and optimization by the one shot method. AIAA Paper 95-0478, 1995.

C.L. Ladson and C.W. Brooks Jr. Development of a computer program to obtain ordinates for the NACA 6-and 6a-series airfoils. NASA TM X-3069, 1974.

C.L. Ladson and C.W. Brooks Jr. Development of a computer program to obtain ordinates for NACA 4-digit, 4-digit modified, 5-digit, and 16-series airfoils. NASA TM X-3284, 1975.

J.C. Lagarias, J.A. Reeds, M.H. Wright, and P.E. Wright. Convergence properties of the Nelder-Mead simplex method in low dimensions. *SIAM Journal on Optimization*, 9: 112–147, 1998.

R.S. Langley. Unified approach to probabilistic and possibilistic analysis of uncertain systems. *ASCE Journal of Engineering Mechanics*, 126: 1163–1172, 1999.

E. Larsson and B. Fornberg. A numerical study of some radial basis function based solution methods for elliptic PDEs. *Computers and Mathematics with Applications*, 46: 891–902, 2003.

C.T. Lawrence and A.L. Tits. A computationally efficient feasible sequential quadratic programming algorithm. *SIAM Journal on Optimization*, 11: 1092–1118, 2001.

A. Le Moigne and N. Qin. Variable-fidelity aerodynamic optimization for turbulent flows using a discrete adjoint formulation. *AIAA Journal*, 42: 1281–1292, 2004.

S.J. Leary, A. Bhaskar, and A.J. Keane. Screening and approximation methods for efficient structural optimization. AIAA Paper 2002-5540, 2002.

S.J. Leary, A. Bhaskar, and A.J. Keane. A knowledge-based approach to response surface modelling in multifidelity optimization, *Journal of Global Optimization*, 26: 297–319, 2003.

S.J. Leary, A. Bhaskar, and A.J. Keane. Surrogate modeling for optimization adjoint CFD codes. *AIAA Journal*, 42: 631–641, 2004b.

R.L. Lee and D.L. Loh. Structural optimization of turbine blade firtrees. AIAA Paper 88-2995, 1988.

P.A. LeGresley and J.J. Alonso. Airfoil design optimization using reduced order models based on proper orthogonal decomposition. AIAA Paper 2000-2545, 2000.

P.A. LeGresley and J.J. Alonso. Dynamic domain decomposition and error correction for reduced order models. AIAA Paper 2003-0250, 2003.

J.P. Leiva and B.C. Watson. Automatic generation of basis vectors for shape optimization in the genesis programme. AIAA Paper 1998-4852, 1998.

J. Lepine, F. Guibault, J.-Y. Trepanier, and F. Pepin. Optimized nonuniform rational B-spline geometrical representation for aerodynamic design of wings. *AIAA Journal*, 39: 2033–2041, 2001.

D. Levin. The approximation power of moving least-squares. *Mathematics of Computation*, 67: 1517–1532, 1998.

R.M. Lewis. Numerical computation of sensitivities and the adjoint approach. NASA/CR-97-206247, 1997.

S. Li and W.K. Liu. Meshfree and particle methods and their applications. *Applied Mechanics Review*, 5: 1–34, 2002.

K. H. Liang, X. Yao, and C. Newton. Evolutionary search of approximated n-dimensional landscapes. *International Journal of Knowledge-Based Intelligent Engineering Systems*, 4: 172–183, 2000.

Y. Liao, S.C. Fang, and H.L. Nuttle. Relaxed conditions for radial basis networks to be universal approximators. *Neural Networks*, 16: 1019–1028, 2003.

J.L. Lions. Optimal Control of Systems Governed by Partial Differential Equations. (translated by S.K. Mitter) Springer-Verlag, 1971.

T.J. Liszka, C.A.M. Duarte, and W.W. Tworzydlo. Hp-meshless cloud method. *Computer Methods in Applied Mechanics and Engineering*, 139: 263–288, 1996.

B. Liu, R.T. Haftka, and M.A. Akgün. Two-level composite wing structural optimization using response surfaces. *Proceedings of the First ASMO UK / ISSMO conference on Engineering Design Optimization*, Ilkley, pages 19–27, 1999.

R.C. Lock. Prediction of the drag of aerofoils and wings at high subsonic speeds. *Aeronautical Journal*, 90: 207, 1986.

M. Loève. *Probability Theory*. Springer, fourth edition, 1977.

M.E. Lores, P.R. Smith, and R.A. Large. Numerical optimization – an assessment of its role in transport aircraft aerodynamic design through a case study. In *12th ICAS Congress*, Munich, pages 41–42, 1980.

D.J. Lucia. *Reduced Order Modeling for High Speed Flows with Moving Shocks*. PhD thesis, Air Force Institute of Technology, Wright-Patterson Air Force Base, Ohio, 2001.

J.N. Lyness and G. Trapp. Numerical differentiation of analytic functions. *SIAM Journal of Numerical Analysis*, 4: 202–210, 1967.

D.J.C. MacKay. *Bayesian Methods for Adaptive Models*. PhD thesis, California Institute of Technology, 1992.

D.J.C. MacKay. *Information Theory, Inference, and Learning Algorithms*. Cambridge University Press, 2003.

H.O. Madsen, S. Krenk, and N.C. Lind. *Methods of Structural Safety*. Prentice Hall, 1986.

P.G. Maghami, S.M. Joshi, and E.S. Armstrong. An optimisation-based integration controls-structure design methodology for flexible space structures. NASA-TP-3283, 1993.

S. Mallat and Z. Zhang. Matching pursuit in a time-frequency dictionary. *IEEE Transactions on Signal Processing*, 41: 3397–3415, 1993.

J.B. Malone, J.M. Housner, and J.K. Lytle. The design of future airbreathing engine systems within an intelligent synthesis environments. In *Proceedings of the 14th International Symposium on Air Breathing Engines (XIV ISABE)*, ISABE Paper 99-7173, ed. P.J. Waltrup, Florence, page 87, 1999.

B. Malone and M. Papay. *ModelCenter®; An Integration Environment for Simulation Based Design*. Phoenix Integration, Inc., Blacksburg, VA, 2000.

A. Marshall. The use of multi-stage sampling schemes in Monte Carlo computations. In M. Meyer, editor, *Symposium on Monte Carlo Methods*, pages 123–140. Wiley, 1956.

J.G. Marshall and M. Imregun. A review of aeroelasticity methods with emphasis on turbomachinery applications. *Journal of Fluids and Structures*, 10: 237–267, 1996.

S. Martello and P. Toth. *Knapsack Problems: Algorithms and Computer Implentations*. Wiley, Chichester, 1990.

J.D. Martin and T.W. Simpson. On the use of Kriging models to approximate deterministic computer models. In *Proceedings of DETC, DETC2004/DAC-57300*, Salt Lake City, UT, 2004.

J.R.R.A. Martins. *A Coupled-Adjoint Method for High-Fidelity Aero-Structural Optimization*. PhD thesis, Stanford University, 2002.

J.R.R.A. Martins, J.J. Alonso, and J.J. Reuther. High-fidelity aerostructural design optimization of a supersonic business jet. *Journal of Aircraft*, 41: 523–530, 2004.

J.R.R.A. Martins, J.J. Alonso, and J.J. Reuther. A coupled-adjoint sensitivity analysis method for high-fidelity aero-structural design. *Optimization and Engineering*, 6: 33–62, 2005.

J.R.R.A. Martins, P. Sturdza, and J.J. Alonso. The complex-step derivative approximation. *ACM Transactions on Mathematical Software*, 29: 245–262, 2003.

B. Maskew. Prediction of subsonic aerodynamic characteristics: a case for low-order panel methods. *Journal of Aircraft*, 19: 157, 1982.

B. Maskew. Calculation of flow characteristics for general configurations using panel methods. In *Proceedings of the Third GAMM-Seminar, Panel Methods in Fluid Mechanics with Emphasis on Aerodynamics*. Friedr. Vieweg & Sohn, 1987.

G. Matheron. Principles of geostatistics. *Economic Geology*, 58: 1246–1266, 1963.

K. Maute, M. Nikbay, and C. Farhat. Coupled analytical sensitivity analysis and optimization of three-dimensional nonlinear aeroelastic systems. *AIAA Journal*, 39: 2051–2061, 2001.

L.A. McCullers. Aircraft configuration optimization using optimized flight profiles. In J. Sobieszczanski-Sobieski, editor, *Proceedings of the Symposium on Recent Experiences in Multidisciplinary Analysis and Optimization*. NASA CP-2327, pages 395–412, 1984.

M.D. McKay, W.J. Conover, and R.J. Beckman. A comparison of three methods for selecting values of input variables in the analysis of output from a computer code. *Technometrics*, 21: 239–245, 1979.

R. Mead. *The Design of Experiments*. Cambridge University Press, 1988.

M. Meckesheimer, R.R. Barton, T.W. Simpson, F. Limayem, and B. Yannou. Metamodeling of combined discrete/continuous responses. *AIAA Journal*, 39: 1950–1959, 2001.

M. Meckesheimer, A.J. Booker, R.R. Barton, and T.W. Simpson. Computationally inexpensive metamodel assessment strategies. *AIAA Journal*, 40: 2053–2060, 2002.

R.E. Melchers. *Structural Reliability Analysis and Prediction*. John Wiley & Sons, 1999.

J.W. Melody and H.C. Briggs. Analysis of structural and optical interactions of the precision optical interferometer. *Proceedings of SPIE*, 1947: 44–57, 1993.

J.W. Melody and G.W. Neat. Integrated modeling methodology validation using the micro-precision interferometer testbed. *35th IEEE Conference on Decision and Control*, Kobe, 1996.

A. Messac. Physical programming: effective optimization for computational design. *AIAA Journal*, 34: 149–158, 1996.

N. Michelena, H. Park, and P.Y. Papalambros. Convergence properties of analytical target cascading. *AIAA Journal*, 41: 897–905, 2003.

C.A. Michelli. Interpolation of scattered data: distance matrices and conditionally positive definite functions. *Constructive Approximation*, 2: 11–22, 1986.

A.J. Miller. Contribution to the discussion of "regression, prediction and shrinkage" by j. b. copas. *Journal of the Royal Statistical Society, Series B*, 45: 346–347, 1983.

W.C. Mills-Curran, R.V. Lust, and L.A. Schmit. Approximation methods for space frame synthesis. *AIAA Journal*, 21: 1571–1580, 1983.

N. Milton, N. Shadbolt, H. Cottam, and M. Hammersley. Towards a knowledge technology for knowledge management. *International Journal of Human-Computer Studies*, 53(3): 615–664, 1999.

G. Mitchell. *R.J. Mitchell, World Famous Aircraft Designer – Schooldays to Spitfire*. Nelson and Saunders Publishers, 1986.

T.J. Mitchell and M.D. Morris. Bayesian design and analysis of computer experiments: two examples. *Statistica Sinica*, 2: 359–379, 1992.

T.J. Mitchell, M.D. Morris, and D. Ylvisaker. Existence of smoothened stationary processes on an interval. *Stochastic Processes and Their Applications*, 35: 109–119, 1990.

B. Mohammadi and O. Pironneau. *Applied Shape Optimization for Fluids*. Oxford University Press, 2001.

D.C. Montgomery. *Design and Analysis of Experiments*. John Wiley & Sons, fourth edition, 1997.

J.J Mor'e and D. Thuente. Line search algorithms with guaranteed sufficient decrease. *ACM Transactions on Mathematical Software*, 20: 286–307, 1994.

J. Moore, and J.G. Moore. Performance evaluation of linear turbine cascades using 3-dimensional viscous-flow calculations. *ASME Journal of Engineering for Gas Turbines and Power*, 107(4): 969–975, 1985.

M.D. Morris. Factorial sampling plans for preliminary computational experiments. *Technometrics*, 33: 161–174, 1991.

M.D. Morris, T.J. Mitchell, and D. Ylvisaker. Bayesian design and analysis of computer experiments: use of derivatives in surface prediction. *Technometrics*, 35: 243–255, 1993.

M. Moshrefi-Torbati, A.J. Keane, S.J. Elliott, M.J. Brennan, and E. Rogers. The integration of advanced active and passive structural vibration control. *Advances in Vibration Engineering*, 2: 1–12, 2003.

R.L. Muhanna and R.L. Mullen. Uncertainty in mechanics problems – interval-based approach. *ASCE Journal of Engineering Mechanics*, 127: 557–566, 2001.

D.V. Murthy and R.T. Haftka. Approximations to eigenvalues of modified general matrices. *Computers and Structures*, 29: 903–917, 1988.

R.H. Myers and D.C. Montgomery. *Response Surface Methodology: Process and Product Optimization Using Designed Experiments*. John Wiley & Sons, 1995.

D. Nagy. Modal representation of geometrically nonlinear behavior by the finite element method. *Computers and Structures*, 10: 683, 1979.

V.N. Nair. Taguchi's parameter design: a panel discussion. *Technometrics*, 34: 127–161, 1992.

P.B. Nair. On the theoretical foundations of stochastic reduced basis methods. AIAA Paper 2001-1677, 2001.

P.B. Nair. Equivalence between the combined approximations technique and Krylov subspace methods. *AIAA Journal*, 40: 1021–1023, 2002a.

P.B. Nair. Physics-based surrogate modeling of parameterized PDEs for optimization and uncertainty analysis. AIAA Paper 2002-1586, 2002b.

P.B. Nair. Projection schemes in stochastic finite element analysis. In E. Nikolaidis, D.M. Ghiocel, and S. Singhal, editors, *CRC Engineering Design Reliability Handbook*. CRC Press, 2004.

P.B. Nair, A. Choudhury, and A.J. Keane. A Bayesian framework for uncertainty analysis using deterministic black-box simulation codes. AIAA Paper 2001-1676, 2001.

P.B. Nair, A. Choudhury, and A.J. Keane. Some greedy algorithms for sparse regression and classification with Mercer kernels. *Journal of Machine Learning Research*, 3: 781–801, 2003.

P.B. Nair and A.J. Keane. Passive vibration suppression of flexible space structures via optimal geometric redesign. *AIAA Journal*, 39: 1338–1346, 2001.

P.B. Nair and A.J. Keane. A coevolutionary architecture for distributed optimization of complex coupled systems. *AIAA Journal*, 40: 1434–1443, 2002a.

P.B. Nair and A.J. Keane. Stochastic reduced basis methods. *AIAA Journal*, 40: 1653–1664, 2002b.

P.B. Nair, A.J. Keane, and R.S. Langley. Improved first-order approximation of eigenvalues and eigenvectors. *AIAA Journal*, 36: 1721–1727, 1998a.

P.B. Nair, A.J. Keane, and R.P. Shimpi. Combining approximation concepts with genetic algorithm-based structural optimization procedures. AIAA Paper 98-1912, 1998b.

B.K. Natarajan. Sparse approximate solutions to linear systems. *SIAM Journal of Computing*, 25: 227–234, 1995.

B.K. Natarajan. On learning functions from noise-free and noisy examples via Occam's razor. *SIAM Journal of Computing*, 29: 712–727, 1999.

R.M. Neal. *Bayesian Learning for Neural Networks*, Lecture Notes in Statistics. Springer, 1996.

R.M. Neal. Regression and classification using Gaussian process priors (with discussion). In J.M. Bernardo, editor, *Bayesian Statistics 6*, pages 475–501. Oxford University Press, 1998.

J.A. Nelder and R. Mead. A simplex method for function minimization. *Computer Journal*, 7: 308–313, 1965.

P.A. Nelson and S.J. Elliott. *Active Control of Sound*. Academic Press, London, 1992.

A. Neumaier. *Interval Methods for Systems of Equations*. Cambridge University Press, 1990.

H. Niederreiter. *Random Number Generation and Quasi-Monte Carlo Methods*. SIAM, 1992.

E.J. Nielsen. *Aerodynamic Design Sensitivities on an Unstructured Mesh Using the Navier-Stokes Equations and a Discrete Adjoint Formulation*. PhD thesis, Virginia Polytechnic Institute and State University, 1998.

E. Nikolaidis, S. Chen, H. Cudney, R.T. Haftka, and R. Roca. Comparison of probability and possibility for design against catastrophic failure under uncertainty. *Journal of Mechanical Design*, 126: 386–394, 2004a.

E. Nikolaidis, D.M. Ghiocel, and S. Singhal, editors. *Engineering Design Reliability Handbook*. CRC Press, 2004b.

A.K. Noor. Recent advances in reduction methods for nonlinear problems. *Computers and Structures*, 13: 31, 1981.

J.K. Northrop. The development of the all-wing aircraft. *35th Wilbur Wright Memorial Lecture, Royal Aeronautical Society*, 51: 481–510, 1947.

H. Nowacki. Modelling of design decisions for CAD. In G. Goos and J. Hartmanis, editors, *Lecture Notes in Computer Science No. 89: CAD Modelling, Systems Engineering, CAD-systems*. Springer-Verlag, Berlin, 1980.

S. Obayashi and T. Tsukahara. Comparison of optimization algorithms for aerodynamic shape design. *AIAA Journal*, 35: 1413–1415, 1997.

S. Obayashi and Y. Yamaguchi. Multiobjective genetic algorithm for mutidisciplinary design of transonic wing planform. *Journal of Aircraft*, 34: 690–692, 1997.

W.L. Oberkampf, J.C. Helton, C.A. Joslyn, S.F. Wojtkiewicz, and S. Ferson. Challenge problems – uncertainty in system response given uncertain parameters. *Reliability Engineering and System Safety*, 85: 11–19, 2004.

J.T. Oden, I. Babuska, F. Nobile, Y. Feng, and R. Tempone. Theory and methodology for estimation and control of errors due to modeling, approximation, and uncertainty. *Computer Methods in Applied Mechanics and Engineering*, 194: 195–204, 2005.

A. O'Hagan. Monte Carlo is fundamentally unsound. *The Statistician*, 36: 247–249, 1987.

A. O'Hagan, M.C. Kennedy, and J.E. Oakley. Uncertainty analysis and other inference tools for complex computer codes (with discussion). In *Bayesian Statistics 6*, pages 503–524. Oxford University Press, 1999.

Y.S. Ong and A.J. Keane. Meta-lamarckian learning in memetic algorithms. *IEEE Transactions on Evolutionary Computation*, 8: 99–109, 2004.

Y.S. Ong, P.B. Nair, and A.J. Keane. Evolutionary optimization of computationally expensive problems via surrogate modeling. *AIAA Journal*, 40: 687–696, 2003.

Y.S. Ong, P.B. Nair, A.J. Keane, and Z.Z. Zhou. Surrogate-assisted evolutionary optimization frameworks for high-fidelity engineering design problems. In Y. Jin, editor, *Knowledge Incorporation in Evolutionary Computation*, Studies in Fuzziness and Soft Computing. Springer-Verlag, 2004.

C.E. Orozco and O.N. Ghattas. A reduced SAND method for optimal design of non-linear structures. *International Journal for Numerical Methods in Engineering*, 40: 2759–2774, 1997.

J.M. Ortega and W.C. Rheinboldt. *Iterative Solution of Nonlinear Equations in Several Variables*. Academic Press, 1970.

E. Panier and A.L. Tits. On combining feasibility, descent and superlinear convergence in inequality constrained optimization. *Mathematical Programming*, 59: 261–276, 1993.

M. Papila and R.T. Haftka. Response surface approximations: Noise, error repair, and modeling errors. *AIAA Journal*, 38: 2336–2343, 2000.

J.M. Parks. On stochastic optimization: Taguchi methods demystified its limitations and fallacy clarified. *Probabilistic Engineering Mechanics*, 16: 87–101, 2001.

I.C. Parmee. The maintenance of search diversity for effective design space decomposition using cluster-oriented genetic algorithms (cogas) and multi-agent strategies (gaant). In I.C. Parmee, editor, *Adaptive Computing in Engineering Design and Control – '96*. University of Plymouth, 1996.

I.C. Parmee. Improving problem definition through interactive evolutionary computation. *AI EDAM-Artificial Intelligence for Engineering Design Analysis and Manufacturing*, 16: 185–202, 2002.

S.V. Patankar. *Numerical Heat Transfer and Fluid Flow*. McGraw-Hill, New York, 1980.

M.J. Patil, D.H. Hodges, and C.E.S. Cesnik. Nonlinear aeroelastic analysis of complete aircraft in subsonic flow. *Journal of Aircraft*, 37: 753–760, 2000.

K. Pepper, P. Waterson, and C. Clegg. The role of human and organisational factors in the implementation of new it systems to support aerospace design. *Proceedings of the 13th International Conference on Engineering Design - ICED 01*, Glasgow, 2001.

J.S. Peterson. The reduced basis method for incompressible viscous calculations. *SIAM Journal of Scientific and Statistical Computing*, 10: 777, 1989.

N. Petruzzelli and A.J. Keane. Wave drag estimation for use with panel codes. *Journal of Aircraft*, 38: 778–782, 2001.

M.S. Phadke. *Quality Engineering Using Robust Design*. Prentice Hall, Englewood Cliffs, 1989.

Phoenix Integration Inc. Process integration using ModelCenter®. Technical White Paper, Phoenix Integration Inc., Blacksburg, VA, 2000.

O. Pironneau. On optimum design in fluid mechanics. *Journal of Fluid Mechanics*, 64: 97–110, 1974.

S. Plimpton, S. Attaway, B. Hendrickson, J. Swegle, C. Vaughan, and D. Gardner. Transient dynamics simulations: parallel algorithms for contact detection and smoothed particle hydrodynamics. *Proceedings Supercomputing*, 1996.

T. Poggio and F. Girosi. Networks for approximation and learning. *Proceedings of the IEEE*, 78: 1481–1497, 1990.

T. Poggio, R. Rifkin, S. Mukherjee, and P. Niyogi. General conditions for predictivity in learning theory. *Nature*, 428: 419–422, 2004.

M.A. Potter. *The Design and Analysis of a Computational Model of Cooperative Coevolution*. PhD thesis, George Mason University, 1997.

M.A. Potter and K.A. De Jong. A cooperative coevolutionary approach to function optimization. *Proceedings of the Third Conference on Parallel Problem Solving from Nature*, pages 249–257. Springer-Verlag, 1994.

M.J.D. Powell. Radial basis functions for multivariable interpolation: a review. In J.C. Mason and M.G. Cox, editors, *Algorithms for Approximation*. Oxford University Press, 1987.

J.W. Pratt, H. Raiffa, and R. Schlaifer. *Introduction to Statistical Decision Theory*. MIT Press, 1996.

W.H. Press, B.P. Flannery, S.A. Teukolsky, and W.T. Vetterling. *Numerical Recipes: The Art of Scientific Computing*. Cambridge University Press, 1986.

W.H. Press, S.A. Teukolsky, W.T. Vetterling, and B.P. Flannery. *Numerical recipes in Fortran: the art of scientific computing*. Cambridge University Press, second edition, 1992.

C. Prud'homme, D.V. Rovas, and K. Veroy. Reliable real-time solution of parametrized partial differential equations: reduced-basis output bound methods. *ASME Journal of Fluids Engineering*, 124: 70–80, 2002.

P. Pugh. *The Magic of a Name: The Rolls-Royce Story, Part Two: The Power Behind the Jets*. Icon Books, 2001.

G.D. Pusch, C. Bischof, and C. Carle. On automatic differentiation of codes with complex arithmetic with respect to real variables. Technical Report ANL/MCS›TM›188, Argonne National Laboratory, 1995.

T. Pyzdek. *The Six Sigma Handbook*. McGraw-Hill Trade, 2000.

R. Rackwitz. Reliability analysis–a review and some perspectives. *Structural Safety*, 23: 365–395, 2001.

A. Ralston and P. Rabinowitz. *A First Course in Numerical Analysis*. Dover Publications, second edition, 2001.

S.S. Rao. *Engineering Optimization: Theory and Practice*. John Wiley & Sons, 1996.

S.S. Rao and L.T. Cao. Optimum design of mechanical systems involving uncertain parameters. *Journal of Mechanical Design*, 124: 465–472, 2002.

S.S. Rao and L. Chen. Numerical solution of fuzzy linear equations in engineering analysis. *International Journal for Numerical Methods in Engineering*, 43: 391–408, 1998.

C.E. Rasmussen. *Evaluation of Gaussian Processes and Other Methods for Nonlinear Regression*. PhD thesis, University of Toronto, 1996.

C.E. Rasmussen and Z. Ghahramani. Bayesian Monte Carlo. Advances in Neural Information Processing Systems 15, pages 489–496, MIT Press, 2003.

A. Ratle. Kriging as a surrogate fitness landscape in evolutionary optimization. *Artificial Intelligence for Engineering Design Analysis and Manufacturing*, 15: 37–49, 2001.

P. Ramamoorthy and K. Padmavathi. Airfoil design by numerical optimization. *Journal of Aircraft*, 14: 219–221, 1977.

A.B. Ravi, P.B. Nair, M. Tan, and W.G. Price. Optimal feedforward and feedback control of vortex shedding using trust-region algorithms. AIAA Paper 2002-3076, 2002.

S.S. Ravindran. Proper orthogonal decomposition in optimal control of fluids. NASA TM-1999-209113, 1999.

M.T. Reagan, H.N. Najm, R.G. Ghanem, and O.M. Knio. Uncertainty quantification in reacting-flow simulations through non-intrusive spectral projection. *Combustion and Flame*, 132: 545–555, 2003.

J.N. Reddy. *An Introduction to the Finite Element Method*. McGraw-Hill, New York, second edition, 1993.

G. Redeker. DLR-F4 wing body combination, a selection of experimental test cases for the validation of CFD codes. Technical Report AR-303, AGARD, 1994.

L. Reichel and W.B. Gragg. Algorithm 686: Fortran subroutines for updating the QR decomposition. *ACM Transactions on Mathematical Software*, 16: 369–377, 1990.

J. Renaud and G. Gabriele. Improved coordination in non-hierarchic system optimization. *AIAA Journal*, 31: 2367–2373, 1993.

J.J. Reuther, A. Jameson, J.J. Alonso, M.J. Rimlinger, and D. Saunders. Constrained multipoint aerodynamic shape optimization using an adjoint formulation and parallel computers, parts 1&2. *Journal of Aircraft*, 36: 51–74, 1999a.

J.J. Reuther, A. Jameson, J.J. Alonso, M.J. Rimlinger, and D. Saunders. Constrained multipoint aerodynamic shape optimization using an adjoint formulation and parallel computers, part 2. *Journal of Aircraft*, 36: 61–74, 1999b.

G.M. Robinson and A.J. Keane. Concise orthogonal representation of supercritical airfoils. *Journal of Aircraft*, 38: 580–583, 2001.

J.F. Rodriguez, J.E. Renaud, and L.T. Watson. Convergence of trust region augmented Lagrangian methods using variable-fidelity approximation data. *Structural Optimization*, 15: 1–7, 1998.

J.L. Rogers. Tools and techniques for managing complex design projects. *Journal of Aircraft*, 36: 266–274, 1999.

D.L. Rondeau, E. Peck, S.M. Tapfield, A.F. Williams, and S.E. Allwright. Generative design and optimisation of the primary structure for a commercial transport wing. AIAA Paper 1996-4135, 1996.

D.L. Rondeau and K. Soumil. The primary structure of commercial transport aircraft wings: rapid generation of finite element models using knowledge-based methods. *Proceedings MSC Aerospace Users' Conference, http://www.mscsoftware.com/support/library/conf/auc99/p02999.pdf*, 1999.

M. Rosenblatt. Remarks on a multivariate transformation. *The Annals of Mathematical Statistics*, 23: 470–472, 1952.

H.H. Rosenbrock. An automatic method for finding the greatest or least value of a function. *Computer Journal*, 3: 175–184, 1960.

T.J. Ross. *Fuzzy Logic with Engineering Applications*. John Wiley, second edition, 2004.

W.B. Rouse. A note on the nature of creativity in engineering: Implications for supporting system design. *Information Processing & Management*, 22: 279–285, 1986a.

W.B. Rouse. On the value of information in system design: A framework for understanding and aiding designers. *Information Processing & Management*, 22: 217–228, 1986b.

G.I.N. Rozvany. Aims, scope, methods, history and unified terminology of computer-aided topology optimization in structural mechanics. *Structural and Multidisciplinary Optimization*, 21: 90–108, 2001.

P. Ruffles. Improving the new product introduction process in manufacturing companies. *International Journal of Manufacturing Technology and Management*, 1: 1–19, 2000.

Y. Saad. *Iterative Methods for Sparse Linear Systems*. PWS, 1996.

Y. Saad and H.A. Van der Vorst. Iterative solution of linear systems in the 20th century. *Journal of Computational and Applied Mathematics*, 123: 1–33, 2000.

T.L. Saaty. The analytic hierarchy process. Technical report, University of Pittsburgh, 1988.

J. Sacks, W.J. Welch, T.J. Mitchell, and H.P. Wynn. Design and analysis of computer experiments (with discussion). *Statistical Science*, 4: 409–435, 1989.

I. Sadrehaghighi, R.E. Smith, and S.N. Tiwari. Grid sensitivity and aerodynamic optimization of generic airfoils. *Journal of Aircraft*, 32: 1234–1239, 1995.

J.A. Samareh. Status and future of geometry modeling and grid generation for design and optimization. *Journal of Aircraft*, 36: 97–104, 1999.

J.A. Samareh. Novel multidisciplinary shape parameterization approach. *Journal of Aircraft*, 38: 1015–1024, 2001a.

J.A. Samareh. Survey of shape parameterization techniques for high-fidelity multidisciplinary shape optimization. *AIAA Journal*, 39: 877–884, 2001b.

J.A. Samareh and K.G. Bhatia. A unified approach to modelling multidisciplinary interactions. AIAA Paper 2000-4704, 2000.

S.M. Sanchez. A robust design tutorial. In J.D. Tew, S. Manivannan, D.A. Sadowski, and A.F. Seila, editors, *Proceedings of the 1994 Winter Simulation Conference*, IEEE Press, 1994.

T.J. Santner, B.J. Williams, and W.I. Notz. *The Design and Analysis of Computer Experiments*. Springer, 2003.

M. Sasena, P. Papalambros, and P. Goovaerts. Global optimization of problems with disconnected feasible regions via surrogate modeling. AIAA Paper 2002-25573, 2002.

J.W. Saywer. Design of the gas turbine. In *Gas Turbine Engineering Handbook*, volume 1, Gas Turbine Publications Inc., 1975.

R. Schaback and H. Wendland. Adaptive greedy techniques for approximate solution of large RBF systems. *Numerical Algorithms*, 24: 239–254, 2000.

J.D. Schaffer, editor. *Proceedings of the 3rd International Conference on Genetic Algorithms*, George Mason University, Morgan Kaufmann, Fairfax, VA, June 1989.

L.A. Schmit. Structural design by systematic synthesis. *Proceedings of Second Conference on Electronic Computation*, New York, American Society of Civil Engineers, 105–122, 1960.

L.A. Schmit and B. Farshi. Some approximation concepts for structural synthesis. *AIAA Journal*, 12: 692–699, 1974.

L.A. Schmit and H. Miura. Approximation concepts for efficient structural synthesis. NASA CR-2552, 1976.

L.A. Schmit and R.L. Fox. An integrated approach to structural synthesis and analysis. *AIAA Journal*, 3: 1104–1112, 1965.

B. Schölkopf and A.J. Smola. *Learning with Kernels: Support Vector Machines, Regularization, Optimization, and Beyond*. MIT Press, 2001.

M. Schonlau. *Computer Experiments and Global Optimization*. PhD thesis, University of Waterloo, 1997.

U. Schramm and W.D. Pilkey. The coupling of geometric description and finite elements using nurbs – a study in shape optimisation. *Finite Elements in Analysis and Design*, 15: 11–34, 1993.

H.-P. Schwefel. *Evolution and Optimum Seeking*. John Wiley & Sons, 1995.

T.W. Sederberg and S.R. Parry. Free-form deformation of solid geometric models. *Computer Graphics*, 20: 151–160, 1986.

M. Sefrioui and J. Periaux. Nash genetic algorithms: examples and applications. In *Proceedings of the Congress on Evolutionary Computing (CEC00)*, La Jolla, pages 509–516, 2000.

P. Sen and J.-B. Yang. *Multiple Criteria Decision Support in Engineering Design*. Springer-Verlag, 1998.

S. Shahpar. Three-dimensional design and optimisation of turbomachinery blades using the Navier-Stokes equations. In *Proceedings of the 15th International Symposium on Air Breathing Engines (XV ISABE)*, ISABE Paper 2001-1053, Bangalore, 2001.

K. Shankar and A.J. Keane. Energy flow predictions in a structure of rigidly joined beams using receptance theory. *Journal of Sound Vibration*, 185: 867–890, 1995.

D. Shepard. A two dimensional interpolation function for irregularly spaced data. In *Proceedings of the 23rd National Conference*, pages 517–523. ACM, New York, 1968.

J.N. Siddall. *Optimal Engineering Design: Principles and Applications*. Marcel Dekker, New York, 1982.

T.W. Simpson, A.J. Booker, D. Ghosh, A.A. Giunta, P.N. Koch, and R.-J. Yang. Approximation methods for multidisciplinary analysis and optimization: a panel discussion. *Journal of Structural and Multidisciplinary Optimization*, 27: 302–313, 2004.

T.W. Simpson, D.K.J. Lin, and W. Chen. Sampling strategies for computer experiments: design and analysis. *International Journal of Reliability and Applications*, 2(3): 209–240, 2001.

T.W. Simpson, J.D. Peplinksi, P.N. Koch, and J.K. Allen. On the use of statistics in design and the implications for deterministic computer experiments. In *Proceedings of the ASME Design Engineering Technical Conference*, Paper number DETC97/DTM3881, Sacramento, CA, 1997.

P.M. Sinclair. An exact integral (field panel) method for the calculation of two-dimensional transonic potential flow around complex configurations. *Aeronautical Journal*, 90(896): 227–236, 1986.

S.W. Sirlin and R.A Laskin. Sizing of active piezoelectric struts for vibration suppression on a space-based interferometer. *Proceedings of the 1st US/Japan Conference on Adaptive Structures*, pages 47–63, Maui, Hawaii, 1990.

L. Sirovich. Turbulence and the dynamics of coherent structures. Parts I-III. *Quarterly of Applied Mathematics*, XLV(3): 561–590, 1987.

M.J. Smith, D.H. Hodges, and C.E.S. Cesnik. An evaluation of computational algorithms to interface between CFD and CSD methodologies. Technical Report WL-TR-96-3055, Flight Dynamics Directorate, Wright Laboratory, Wright-Patterson Air Force Base, OH, 1995.

E. Snelson, C.E. Rasmussen, and Z. Ghahramani. Warped Gaussian processes. In *Advances in Neural Information Processing Systems*, 16: 337–344, MIT Press, 2004.

A. Sóbester. *Enhancements to Global Optimisation*. PhD thesis, University of Southampton, 2004.

A. Sóbester and A.J. Keane. Empirical comparison of gradient-based methods on an engine-inlet shape optimization problem. AIAA Paper 2002-5507, 2002.

A. Sóbester, S.J. Leary, and A.J. Keane. On the design of optimization strategies based on global response surface approximation models. *Journal of of Global Optimization*, 2005.

A. Sóbester, P.B. Nair, and A.J. Keane. Evolving intervening variables for response surface approximations. AIAA Paper 2004-4379, 2004.

I.P. Sobieski. *Collaborative Optimization Using Response Surface Estimation*. PhD thesis, University of Stanford, 1998.

J. Sobieszczanski-Sobieski. Optimization by decomposition: a step from hierarchic to non-hierarchic systems. NASA TM-101494, 1989.

J. Sobieszczanski-Sobieski. Sensitivity analysis and multidisciplinary optimization for aircraft design: Recent advances and results. *Journal of Aircraft*, 27: 993–1001, 1990a.

J. Sobieszczanski-Sobieski. Sensitivity of complex, internally coupled systems. *AIAA Journal*, 28: 153–160, 1990b.

J. Sobieszczanski-Sobieski, J.S. Agte, and R.R. Sandusky. Bi-level integrated system synthesis (BLISS). NASA TM-1998-208715, 1998.

J. Sobieszczanski-Sobieski, T.D. Altus, and R.R. Sandusky. Bilevel integrated system synthesis for concurrent and distributed processing. *AIAA Journal*, 41: 1996–2003, 2003.

J. Sobieszczanksi-Sobieski, J.F. Barthelemy, and K.M. Ripley. Sensitivity of optimum solution to problem parameters. *AIAA Journal*, 20: 1291–1299, 1982.

J. Sobieszczanski and R.T. Haftka. Multidisciplinary aerospace design optimization: survey of recent developments. *Structural Optimization*, 14: 1–23, 1997.

J. Sobieszczanski-Sobieski, M.S. Emiley, J.S. Agte, and R.R. Sandusky. Advancement of bi-level integrated system synthesis. AIAA Paper 2000-0421, 2000.

I.M. Sobol. *A Primer for the Monte Carlo Method*. CRC Press, 1994.

W. Song, A.J. Keane, J. Rees, A. Bhaskar, and S. Bagnall. Local shape optimization of turbine disc firtrees using NURBS. AIAA Paper 2002-5486, 2002a.

W. Song, A.J. Keane, J. Rees, A. Bhaskar, and S. Bagnall. Turbine blade fir-tree root design optimisation using intelligent CAD and finite element analysis. *Computers and Structures*, 80: 1853–1867, 2002b.

P. Spellucci. Domin, a subroutine for BFGS minimization. Technical report, Department of Mathematics, Technical University of Darmstadt, Germany, 1996.

W. Squire and G. Trapp. Using complex variables to estimate derivatives of real functions. *SIAM Review*, 40: 110–112, 1998.

H.B. Squire and A.D. Young. The calculation of profile drag of aerofoils. Technical Report R&M 1838, ARC, 1937.

N. Srinivas and K. Deb. Multiobjective optimization using nondominated sorting in genetic algorithms. *Evolutionary Computing*, 2: 221–248, 1995.

R.B. Statnikov and J.B. Matusov. *Multicriteria Optimization and Engineering*. Chapman & Hall, 1995.

STEP. STEP, ISO 10303, industrial automation systems and integration - product data representation and exchange. International Standards Organisation TC 184 / SC4, 1994.

G.W. Stewart and J. Sun. *Matrix Perturbation Theory*. Academic Press, 1990.

D. Stinton. *The Design of the Aeroplane*. Blackwell Sciences, 1983.

G. Stiny. *Pictorial and Formal Aspects of Shape and Shape Grammars*. Birkhauser Verlag, Basel, 1975.

O.O. Storaasli and J. Sobieszczanski-Sobieski. On the accuracy of the Taylor approximation for structural resizing. *AIAA Journal*, 12: 231–233, 1974.

N. Sugiura. Further analysis of the data by Akaike's information criterion and the finite corrections. *Communications in Statistics*, A7: 13–26, 1978.

M. Tadjouddine, S.A. Forth, and A.J. Keane. Adjoint differentiation of a structural dynamics solver. In H.M. Bücker, G. Corliss, P. Hovland, U. Naumann, and B. Norris, editors, *Automatic Differentiation: Applications, Theory, and Tools*, Lecture Notes in Computational Science and Engineering. Springer, 2005.

G. Taguchi. *System of Experimental Design*. UNIPUB/Kraus International Publications, New York, 1987.

G. Taguchi and Y. Wu. Introduction to off-line quality control. Technical report, Central Japan Quality Control Association, Nagoya, 1980.

K. Takeda, O.R. Tutty, and A.D. Fitt. A comparison of four viscous models for the discrete vortex method. AIAA Paper 97-1977, 1997.

L. Tang, R.E. Bartels, P.C. Chen, and D.D. Liu. Numerical investigation of transonic limit cycle oscillations of a two-dimensional supercritical wing. *Journal of Fluids and Structures*, 17: 29–41, 2003.

B.H. Thacker, D.S. Riha, H.R. Millwater, and M.P. Enbright. Errors and uncertainties in probabilistic engineering analysis. AIAA Paper 2001-1239, 2001.

C.-A. Thole, S. Mayer, and A. Supalov. Fast solution of MSC/Nastran sparse matrix problems using a multi-level approach. In *1997 Copper Mountain Conference on Multigrid Methods, Electronic Transactions on Numerical Analysis*, 6: 246–254, Colorado, 1997.

D.M. Tidd, D.J. Strash, B. Epstein, A. Luntz, A. Nachson, and T. Rubin. Application of an efficient 3-d multigrid Euler method (MGAERO) to complete aircraft configurations. AIAA Paper 91-3236, 1991.

V. Torczon and M.W. Trosset. Using approximations to accelerate engineering design optimization. AIAA Paper 98-4800, 1998.

E. Torenbeek. *Synthesis of Subsonic Airplane Design*. Delft University Press and Kluwer Academic Publishers, 1984.

E. Torenbeek. Development and application of a comprehensive, design sensitive weight prediction method for wing structures of transport category aircraft. Report LR 693, Delft University, 1992.

V.V. Toropov and L.F. Alvarez. Development of MARS – multipoint approximation method based on the response surface fitting. AIAA Paper 98-4769, 1998.

V.V. Toropov, A.A. Filatov, and A.A. Polynkin. Multiparameter structural optimization using FEM and multipoint explicit approximations. *Structural Optimization*, 6: 1517–1532, 1993.

V.V. Toropov, F. van Keulen, V.L. Markine, and L.F. Alvarez. Multipoint approximations based on response surface fitting: a summary of recent developments. In V. V. Toropov, editor, *Proceedings of the 1st ASMO UK/ISSMO Conference on Engineering Design Optimization*, pages 371–380. MCB University Press, 1999.

V. Tresp. A Bayesian committee machine. *Neural Computation*, 12: 2719–2741, 2000.

L. Trocine and L. Malone. Finding important independent variables through screening designs. In *Proceedings of the 2000 Winter Simulation Conference*, Orlando, FL, 2000.

M.W. Trosset. Taguchi and robust optimization. Technical report, Department of Computational and Applied Mathematics, Rice University, 1996.

M.W. Trosset, N.M. Alexandrov, and L.T. Watson. New methods for robust design using computer simulations. *Proceedings of the Section on Physical and Engineering Sciences*. American Statistical Association, 2003.

S. Tsutsui and A. Ghosh. Genetic algorithms with a robust solution searching scheme. *IEEE Transactions on Evolutionary Computation*, 1: 201–208, 1997.

S. Tsutsui and A. Ghosh. A comparative study on the effects of adding perturbations to phenotypic parameters in genetic algorithm with a robust solution searching scheme. In *Proceedings of the IEEE Systems, Man, and Cybernetics Conference*, pages 585–591, Tokyo, Japan, 1999.

B.G. Van Bloemen Waanders, M.S. Eldered, A.A. Guinta, G.M. Reese, M.K. Bhardwaj, and C.W. Fulcher. Multilevel parallel optimization using massively parallel structural dynamics. AIAA Paper 2001-1625, 2001.

C.P. van Dam and K. Nikfetrat. Accurate prediction of drag using Euler methods. *Journal of Aircraft*, 29: 516–519, 1992.

G. Vanderplaats. Automated optimization techniques for aircraft synthesis. AIAA Paper 76-909, 1976.

G.N. Vanderplaats. *Numerical Optimization Techniques for Engineering Design*. McGraw-Hill, 1984.

G.N. Vanderplaats. Structural design optimisation status and direction. *Journal of Aircraft*, 36: 11–20, 1999.

F. Van Keulen, R.T. Haftka, and N.H. Kim. Review of options for structural design sensitivity analysis, part 1: linear systems. *Computer Methods in Applied Mechanics and Engineering*, in press.

E. VanMarcke. *Random Fields: Analysis and Synthesis*. The MIT Press, 1983.

V. Vapnik. *Statistical Learning Theory*. John Wiley & Sons, 1998.

V. Vapnik and A. Chervonenkis. On the uniform convergence of relative frequencies of events to their probabilities. *Theory of Probability and its Applications*, 16: 264–280, 1971.

S. Venkataraman and R.T. Haftka. Structural optimization: What has Moore's law done for us? AIAA Paper 2002-1342, 2002.

G. Venter. *Non-Dimensional Response Surfaces for Structural Optimization with Uncertainty*. PhD thesis, University of Florida, Gainsville, 1998.

G. Venter and R.T. Haftka. Response surface approximations in fuzzy set based design optimization. *Structural Optimization*, 18(4): 218–227, 1999.

G. Venter, R.T. Haftka, and J.H. Starnes. Construction of response surface approximations for design optimization. *AIAA Journal*, 36: 2242–2249, 1998.

P. Vincent and Y. Bengio. Kernel matching pursuit. *Machine Learning*, 48: 165–187, 2001.

R. Vitali, R.T. Haftka, and B.V. Sankar. Multifidelity design of stiffened composite panel with a crack. *4th World Congress of Structural and Multidisciplinary Optimization*, Paper 51-AAM2-1, Buffalo, 1999.

T. von Kármán. Turbulence. *Journal of the Royal Aeronautical Society*, 41: 1108–1141, 1937.

T. von Kármán. Solved and unsolved problems of high-speed aerodynamics. In *Conference on High Speed Aerodynamics*, Brooklyn, pages 11–39, 1955.

G. Wahba. *Spline Models for Observational Data*. SIAM, 1990.

W. Waldman, M. Heller, and G.X. Chen. Optimal free-form shapes for shoulder fillets in flat plates under tension and bending. *International Journal of Fatigue*, 23: 509–523, 2001.

K.M. Wallace, C. Clegg, and A.J. Keane. Visions for engineering design – a multi-disciplinary perspective. *Proceedings of the 13th International Conference on Engineering Design – ICED 01*, pages 107–114, Glasgow, 2001.

J.L. Walsh, J.C. Townsend, A.O. Salas, J.A. Samareh, V. Mukhopadhyay, and J.-F. Barthelemy. Multidisciplinary high-fidelity analysis and optimization of aerospace vehicles, part 1: formulation. AIAA Paper 2000-0418, 2000.

B.P. Wang. Parameter optimization in multiquadric response surface approximations. *Structural and Multidisciplinary Optimization*, 26: 219–223, 2004.

L.P. Wang and R.V. Grandhi. Improved two-point function approximation for design optimization. *AIAA Journal*, 33: 1720–1727, 1995.

A. Watt and M. Watt. *Advanced Animation and Rendering Techniques*. Addison-Wesley Publishing, New York, 1992.

N.P. Weatherill, K. Morgan, and O. Hassan. *An Introduction to Mesh Generation*. John Wiley & Sons, 2005.

W.J. Welch, R.J. Buck, J. Sacks, H.P. Wynn, T.J. Mitchell, and M.D. Morris. Screening, predicting and computer experiments. *Technometrics*, 34: 15–25, 1992.

W.J. Welch and J. Sacks. A system for quality improvement via computer experiments. *Communications in Statistics – Theory and Methods*, 20: 477–495, 1991.

W.J. Welch, T.K. Yu, S.M. Kang, and J. Sacks. Computer experiments for quality control by parameter design. *Journal of Quality Technology*, 22: 15–22, 1990.

H. Wendland. Numerical solution of variational problems by radial basis functions. In C.K. Chui and L.L. Schumaker, editors, *Approximation Theory IX*, Volume II: Computational Aspects, pages 361–368. Vanderbilt University Press, Nashville, 1999.

R.E. Wengert. A simple automatic derivative evaluation program. *Communications of ACM*, 7: 463–464, 1964.

Sir F. Whittle. *Gas Turbine Aero-Thermodynamics, with Special Reference to Aircraft Propulsion*. Pergamon Press, 1981.

N. Wiener. The homogeneous chaos. *American Journal of Mathematics*, 60: 897–936, 1938.

S.S. Wiseall and J.C. Kelly. Take-up and adoption of optimisation technology for engineering design. *Optimization in Industry*, 2002.

D.H. Wolpert and W.G. Macready. No free lunch theorems for optimization. *IEEE Transactions on Evolutionary Computation*, 1: 67–82, 1997.

W. Wright. Some aeronautical experiments. *Presented to the Western Society of Engineers*, September 18, 1901, Journal of the Western Society of Engineers, 1901, reprinted Smithsonian Institution Annual Report, 1901.

O. Wright and W. Wright. The Wright brothers aeroplane. *Century Magazine*, 1908.

Y.-T. Wu. Computational methods for efficient structural reliability and reliability sensitivity analysis. *AIAA Journal*, 32: 1717–1723, 1994.

B.A. Wujek, J.E. Renaud, S.M. Batill, and J.B. Brockman. Concurrent subspace optimization using design variable sharing in a distributed computing environment. *Proceedings of the 1995 Design Engineering Technical Conferences, Advances in Design Automation*, pages 181–188. ASME DE-Vol. 82, 1995.

Y.M. Xie and G.P. Steven. *Evolutionary Structural Optimisation*. Springer, London, 1997.

D. Xiu and G.E. Karniadakis. The Wiener-Askey polynomial chaos for stochastic differential equations. *SIAM Journal of Scientific Computing*, 24: 619–644, 2002.

S. Xu and R.V. Grandhi. Effective two-point function approximation for design optimization. *AIAA Journal*, 36: 2269–2275, 1998.

F. Yates. The design and analysis of factorial experiments. Technical Report 35, Imperial Bureau of Soil Science, Harpenden, England, 1937.

B.D. Youn and K.K. Choi. Selecting probabilistic approaches for reliability-based design optimization. *AIAA Journal*, 42: 124–131, 2004.

D. Yuret and M. de la Maza. Dynamic hill climbing: Overcoming the limitations of optimization techniques. In *Proceedings of the 2nd Turkish Symposium on AI and ANN*, pages 254–260, Istanbul, Turkey, 1993.

L.A. Zadeh. Fuzzy sets. *Information and Control*, 8: 338–353, 1965.

J. Zentner, P.W.G. De Baets, and D.N. Marvis. *Formulation of an integrating framework for conceptual object-oriented systems design*. SAE Paper 2002-01-2955, 2002.

G. Zha, M. Smith, K. Schwabacher, A. Rasheed, D. Gelsey, and M. Knight. High performance supersonic missile inlet design using automated optimization. *Journal of Aircraft*, 34: 697, 1997.

O. Zhang and A. Zerva. Accelerated iterative procedure for calculating eigenvector derivatives. *AIAA Journal*, 35: 340–348, 1997.

T. Zhang, G.H. Golub, and K.H. Law. Eigenvalue perturbation and generalized Krylov subspace method. *Applied Numerical Mathematics*, 27(2): 185–202, 1998.

T.A. Zang, M.J. Hemsch, M.W. Hilburger, S.P. Kenny, J.M. Luckring, P. Maghami, S.L. Padula, and W.J. Stroud. Needs and opportunities for uncertainty-based multidisciplinary design methods for aerospace vehicles. NASA/TM-2002-211462, 2002.

W. Zhongmin. Hermite-Birkhoff interpolation of scattered data by radial basis functions. *Approximation Theory and Applications*, 8: 1–10, 1992.

Z.Z. Zhou, Y.S. Ong, and P.B. Nair. Hierarchical surrogate-assisted evolutionary optimization framework. In *Proceedings of the IEEE Congress on Evolutionary Computation*, 1586–1593, Portland, OR, 2004.

O.C. Zienkiewicz. *The Finite Element Method*. McGraw-Hill, London, third edition, 1977.

G. Zoutendijk. *Methods of Feasible Directions*. Elsevier, Amsterdam, 1960.

Index